AAPG TREATISE OF PETROLEUM GEOLOGY

REPRINT SERIES

The American Association of Petroleum Geologists
gratefully acknowledges and appreciates the leadership and support
of the AAPG Foundation in the development of the
Treatise of Petroleum Geology.

AMERICAN ASSOCIATION OF PETROLEUM GEOLOGISTS FOUNDATION

TREATISE OF PETROLEUM GEOLOGY FUND*

Major Corporate Contributors
($25,000 or more)

Mobil Oil Corporation
Pennzoil Exploration and Production Company

Other Corporate Contributors
($5,000 to $25,000)

The McGee Foundation, Incorporated
Union Pacific Foundation

Major Individual Contributors
($1,000 or more)

C. Hayden Atchison
Richard R. Bloomer
A. S. Bonner, Jr.
David G. Campbell
Herbert G. Davis
Lewis G. Fearing
James A. Gibbs
William E. Gipson
Robert D. Gunn
Cecil V. Hagen
William A. Heck
Harrison C. Jamison
Thomas N. Jordan, Jr.
Hugh M. Looney
John W. Mason
George B. McBride
Dean A. McGee
John R. McMillan
Charles Weiner
Harry Westmoreland

The Foundation also gratefully acknowledges the many who have supported this endeavor with smaller contributions, which now total $11,815.50.

*Contributions received as of September 30, 1988.

STRUCTURAL CONCEPTS AND TECHNIQUES I

BASIC CONCEPTS, FOLDING, AND STRUCTURAL TECHNIQUES

COMPILED BY
NORMAN H. FOSTER
AND
EDWARD A. BEAUMONT

TREATISE OF PETROLEUM GEOLOGY
REPRINT SERIES, NO. 9

PUBLISHED BY
THE AMERICAN ASSOCIATION OF PETROLEUM GEOLOGISTS
TULSA, OKLAHOMA 74101, U.S.A.

ISBN: 0-89181-408-6

INTRODUCTION

This reprint volume belongs to a series that is part of the *Treatise of Petroleum Geology*. The *Treatise of Petroleum Geology* was conceived during a discussion we had at the 1984 AAPG Annual Meeting in San Antonio, Texas. When our discussion ended, we had decided to write a state-of-the-art textbook in petroleum geology, directed not at the student, but at the practicing petroleum geologist. The project to put together one textbook gradually evolved into a series of three different publications: the Reprint Series, the Atlas of Oil and Gas Fields, and the Handbook of Petroleum Geology; collectively these publications are known as the *Treatise of Petroleum Geology*. With the help of the Treatise of Petroleum Geology Advisory Board, we designed this set of publications to represent the cutting edge in petroleum exploration knowledge and application. The Reprint Series provides previously published landmark literature; the Atlas collects detailed field studies to illustrate the various ways oil and gas are trapped; and the Handbook is a professional explorationist's guide to the latest knowledge in the various areas of petroleum geology and related fields.

The papers in the various volumes of the Reprint Series complement the different chapters of the Handbook. Papers were selected on the basis of their usefulness today in petroleum exploration and development. Many "classic papers" that led to our present state of knowledge have not been included because of space limitations. In some cases, it was difficult to decide in which Reprint volume a particular paper should be published because that paper covers several topics. We suggest, therefore, that interested readers become familiar with all the Reprint volumes if they are looking for a particular paper.

We have divided the topic of structural concepts and techniques into three volumes. The first volume contains papers that discuss Basic Concepts, Folding, and Structural Techniques. The first paper in Basic Concepts is the classic 1979 paper by Harding and Lowell, "Structural styles, their plate-tectonic habitats, and hydrocarbon traps in petroleum provinces." We have used Harding and Lowell's classification of structural styles to group papers in volumes II and III. Basic Concepts also includes papers on stress analysis and pore-pressure effects. Folding includes papers describing folding processes and geometries. Structural Techniques is a collection of practical papers included to help the petroleum geologists solve structural problems encountered in exploration and development. These papers relate ways to visualize and predict the three-dimensional arrangement and location of strata in the subsurface.

Volume II contains papers related to Basement-Involved Deformation. These papers are subdivided into Extensional, Compressional, and Strike-Slip Deformation. Extensional Deformation includes papers discussing crustal rifting and normal faulting. Compressional Deformation includes papers discussing foreland deformation. In most cases, these papers use the Rocky Mountains example. Strike-Slip Deformation includes papers discussing strike-slip or wrench fault deformational processes and the consequent effects these processes have on folding, faulting, basin formation, and sedimentation.

Volume III, Detached Deformation, is subdivided into three sections: Extensional Deformation, Compressional Deformation, and Salt Tectonics. Extensional Deformation contains papers that discuss listric normal faulting and growth faulting (a species of listric normal faulting). Compressional Deformation contains papers on the processes and mechanics of low-angle thrust faulting. Salt Tectonics is a collection of papers on salt movement and salt dissolution, and their subsequent effect on the structural geometry of associated strata.

Edward A. Beaumont
Tulsa, Oklahoma

Norman H. Foster
Denver, Colorado

Treatise of Petroleum Geology
Advisory Board

*Deceased

TABLE OF CONTENTS

STRUCTURAL CONCEPTS AND TECHNIQUES I

BASIC CONCEPTS

FOLDING

STRUCTURAL TECHNIQUES

TABLE OF CONTENTS

STRUCTURAL CONCEPTS AND TECHNIQUES II

BASEMENT-INVOLVED DEFORMATION

EXTENSIONAL DEFORMATION

COMPRESSIONAL DEFORMATION

STRIKE-SLIP DEFORMATION

TABLE OF CONTENTS

STRUCTURAL CONCEPTS AND TECHNIQUES III

DETACHED DEFORMATION

The upper detachment in concentric folding.
Clinton D. A. Dahlstrom

Passive-roof duplex geometry of the Kirthar and Sulaiman mountain belts, Pakistan.
C. J. Banks and J. Warburton

Three-dimensional geometry and kinematics of experimental piggy-back thrusting.
G. Mulugeta and H. Kogi

The energy balance and deformation mechanisms of thrust sheets.
D. Elliott

SALT TECTONICS

Mechanism of salt migration in northern Germany.
F. Trusheim

External shapes, strain rates, and dynamics of salt structures.
M. P. A. Jackson and C. J. Talbot

Evolution of salt structures, east Texas diapir province: Part 1: Sedimentary record of halokinesis.
S. J. Seni and M. P. A. Jackson

Evolution of salt structures, east Texas diapir province: Part 2: Patterns and rates halokinesis.
S. J. Seni and M. P. A. Jackson

Salt movement on continental slope, northern Gulf of Mexico.
C. C. Humphris, Jr.

Salt tectonics as related to several Smackover fields.
D. J. Hughes

Age, budget, and dynamics of an active salt extension in Iran.
C. J. Talbot and R. J. Jarvis

Oil and geology in Cuanza basin of Angola.
G. P. Brognon and G. R. Verrier

Structural history of Peplardome and the dissolution of Charles Formation salt, Roosevelt County, Montana.
D. M. Orchard

BASIC CONCEPTS

The American Association of Petroleum Geologists Bulletin
V. 63, No. 7 (July 1979), P. 1016-1058, 29 Figs., 1 Table

Structural Styles, Their Plate-Tectonic Habitats, and Hydrocarbon Traps in Petroleum Provinces[1]

T. P. HARDING[2] and JAMES D. LOWELL[3]

Abstract Broadly interrelated assemblages of geologic structures constitute the fundamental structural styles of petroleum provinces. These assemblages generally are repeated in regions of similar deformation, and their associated hydrocarbon traps can be anticipated prior to exploration. Styles are differentiated on the basis of basement involvement or detachment of sedimentary cover. Basement-involved styles include wrench-fault structural assemblages, compressive fault blocks and basement thrusts, extensional fault blocks, and warps. Detached styles are decollement thrust-fold assemblages, detached normal faults ("growth faults" and others), salt structures, and shale structures.

These basic styles are related to the larger kinematics of plate tectonics and, in some situations, to particular depositional histories. Most styles have preferred plate-tectonic habitats: (1) wrench faults at transform and convergent plate boundaries; (2) compressive fault blocks and basement thrusts at convergent boundaries, particularly in forelands and orogenic belts; (3) extensional fault blocks at divergent boundaries in all stages of completion and certain parts of convergent boundaries; (4) basement warps in a variety of plate-interior and boundary settings; (5) decollement thrust-fold belts in trench inner walls and foreland zones of convergent boundaries; (6) detached normal faults, usually in unstable, thick clastic wedges (mostly deltas); (7) salt structures primarily in interior grabens that may evolve to completed divergent boundaries; and (8) shale structures in regions with thick overpressured shale sequences.

Important differences in trend arrangements and structural morphologies provide criteria for differentiation of styles. These differences also result in different kinds of hydrocarbon traps. Wrench-related structural assemblages are concentrated along throughgoing zones and many have en echelon arrangements. The basic hydrocarbon trap is the en echelon anticline, in places assisted by closure directly against the wrench fault itself. Compressive and extensional fault styles typically have multiple, repeated trends, which combine to form zigzag, dogleg, or other grid patterns. Their main trap types are fault closures and drape folds above the block boundaries. Basement warps (domes, arches, etc) are mostly solitary features and commonly provide long-lived positive areas for hydrocarbon concentration in broadly flexed closures.

Most decollement thrust-fold structures are arranged in long, sinuous belts and are repeated in closely spaced, wavelike bands. Effective closures include slightly to moderately disrupted compressive anticlines and lead edges of thrust sheets. Most detached normal faults are listric faults that occur in coalescing, cuspate bands parallel with the strike of contemporaneous sedimentation. Their basic hydrocarbon traps are associated rollover anticlines which are uniquely concentrated along the downthrown sides of major faults. Salt and shale structures are present both as buoyantly rising pillows, domes, ridges, etc, and as highly complex injected features caused by tectonic forces. Stratigraphic factors, such as truncation, wedging, onlap, and unconformity, add to the variety of traps in all styles.

In many places the structures of a petroleum province are either, or both, a gradation between the described fundamental styles and a mix of several styles. These structures can be further complicated by superimposition of fundamentally different tectonic environments. Additional modification of structures can result from still other factors inherent in the deformed region or in the particular tectonic event.

INTRODUCTION

The intent of this paper is to provide an initial framework for better understanding and predictability of structural trends and types of traps in petroleum provinces. Almost all geologic structures, if viewed in enough detail, have unique geometries and histories. On a more regional scale, however, certain general characteristics define broad categories of structures. These characteristics are the substance of this report.

Our structural classification employs the concepts of structural styles and comparative tectonics. The style of a particular region describes its dominant structural geometry. Basic styles are

[1]Manuscript received, March 8, 1978; accepted, February 26, 1979.

[2]Exxon Production Research Co., Houston, Texas 77001.

[3]Consulting geologist, Denver, Colorado 80202.

We are indebted to our colleagues past and present at Exxon Production Research Co. Included are W. H. Bucher and E. Cloos (deceased consultants), J. C. Crowell (former consultant), and R. S. Bishop, A. R. Green, F. A. Johnson, Jr., K. H. Hadley, H. R. Hopkins, T. H. Nelson, D. H. Roeder, D. R. Seely, R. C. Shumaker, P. G. Temple, P. R. Vail, R. E. Wilcox, D. A. White, and J. Zimmerman. Many instructive discussions with them greatly increased our exposure to varied structural settings and contributed to the development of concepts. Kaspar Arbenz, A. W. Bally, J. C. Crowell, W. F. Roux, Jr., J. S. Shelton, and G. R. Stude have critically reviewed the manuscript and provided very helpful suggestions. Grateful appreciation is extended to Exxon Production Research Co. for whom this investigation originated, and to Exxon Co., U.S.A., for their permission to publish.

Article Identification Number
0149-1423/79/B007-0001$03.00/0

1016

defined by an assemblage of tectonically related elements and their spatial arrangement. Although styles are often modified by local rock differences, such as ductility contrasts and preexisting fabric, and tectonic events, including intensity, duration, and timing, the principal style or styles are usually discernible. For basic style definitions, morphologic types, essential repeatable characteristics, and general trend arrangements are emphasized rather than number of structures present in a specific region, their geographic localities, or particular histories. The concept of comparative tectonics involves the use of basic styles, well documented in one area, as guides for structural interpretation of a lesser known but similarly deformed area.

Several structural classifications have been made (Badgley, 1965, p. 50-97). None, however, have dealt directly with subsurface structures and associated hydrocarbon traps in sedimentary basins. Moreover, recent data in support of large-scale lithospheric plate movement show some earlier classifications that featured only a limited number of deformational mechanisms (e.g., the "verticalist" approach of Beloussov, 1959) are incomplete. Plate tectonics has also provided a unified concept of deformational processes and tectonic habitats to which structures can be related. Finally, deep-penetration reflection seismic surveys processed with modern techniques have revealed subsurface structures, such as detached normal faults ("growth faults"), that were poorly known at the time of past classifications. Modern seismic data have also resolved critical older tectonic controversies, demonstrating conclusively, for example, the validity of regional detachment or decollement in thrust-fold belts (Bally et al, 1966).

In this paper we categorize structural styles in petroleum provinces and discuss key characteristics for their differentiation. Each style is treated as to common plate-tectonic settings, typical structural patterns and morphologies, critical differences from other styles, and associated hydrocarbon traps. Last, factors influencing variations in style expression and occurrence are discussed.

STRUCTURAL STYLES IN PETROLEUM PROVINCES

Classification of Structural Styles

Our classification of structural styles is based primarily on the involvement or noninvolvement of basement in the observed structures. Additional criteria are inferred deformational force and mode of tectonic transport inferred from strain features of the structures (Table 1). Although

basement can exhibit widely different competence and mechanical behavior, in petroleum provinces it is usually rigid crystalline igneous or metamorphic rock. The degree of basement involvement is critical to petroleum exploration, for it indicates not only how structures propagated, but, qualitatively, how much of the sedimentary section is in a trap configuration (Fig. 1).

Structural styles are initially treated separately as discrete "end members" to facilitate description and to emphasize their distinctive aspects. Styles apply in scale to trends of tectonically interrelated, contemporaneous structures, but consideration is also given to the components common to single features. Each style has provided hydrocarbon traps (Fig. 1).

Identification of Styles

Identifying structural styles is one of the critical tasks in petroleum exploration and is often difficult because a single characteristic seldom is unique to one structural assemblage, and identification must be made early with few data. Identification depends mostly on (1) recognition of key structural elements such as en echelon folds and faults, trap-door blocks, rollover anticlines, etc; (2) critical differences in local trend arrangements; and (3) gross regional patterns of structures. Following is a résumé of the most common distinguishing characteristics:

Drag fold—Fold formed by drag of sedimentary rocks along a fault.

Drape (forced) fold—Fold formed in sedimentary layers by a forcing member from below, usually a basement fault block.

En echelon—Consistently overlapped structures aligned parallel with one another but oblique to the zone of deformation in which they occur.

Intersecting or *grid*—Structures repeated in multiple alignments crisscrossing over broad areas (Fig. 12); these can combine to form *zigzag* or *dogleg* (Fig. 11, index map) and clustered features.

Irregularly clustered—Concentrations of structures that lack consistent spatial arrangements.

Parallel—Similar elements in parallel alignment. In some places parallel trends may be repeated in closely spaced, wavelike bands, constituting *belts* whose sinuosity delimits *salients* and *reentrants* (Fig. 3).

Relay—Inconsistently overlapped elements aligned parallel with one another and with the zone of deformation in which they occur.

Solitary—Isolated, singular features generally not aligned with other structures of similar characteristics.

Trap-door blocks—Blocks formed by intersection of two faults with maximum relative uplift at or near the point of intersection.

Zonal—Structures in a discrete, elongate, linear trend. Local trends can be longitudinal (parallel), oblique, or transverse to the larger or dominant structural features of an area.

BASEMENT-INVOLVED STRUCTURES

Wrench-Fault Assemblages

Transform plate boundaries are the primary environment of wrench-fault structural assemblages, but divergent and convergent boundaries are also very important habitats (Table 1; Fig. 2). On transform boundaries the movement between lithospheric plates results in a side-by-side motion in concert with the strike-slip kinematics of the style itself. Strike slip in this example is commonly distributed throughout a set of parallel faults (Fig. 2A), or it can be concentrated on a single master fault (Fig. 2B). Parallel transform faults are also present at divergent boundaries, where they offset spreading axes. Where transforms intersect divergent continental margins, they can delimit subbasins with different depositional histories, as along the West African and East Greenland shelves (Lehner and DeRuiter, 1977; Surlyk, 1977).

Master wrench faults also develop subparallel with convergent plate boundaries (Fig. 2C) and are attributed to obliquity of plate encroachment (Fitch, 1972). These are longitudinal wrench faults (a term suggested by D. R. Seely, personal commun., 1971) and typically occupy axial positions in orogenic belts or magmatic arcs (e.g., Zagros Main fault, Figs. 3, 4).

Strike-slip faults trending at an angle to the alignment of the convergent plate boundary are termed "oblique wrench faults." They are present in orogenic belts (Fig. 2D) and forelands, and many seem to have patterns and senses of displacement that conform to conjugate shear sets. Foreland wrench faults generally have smaller displacements and less widely distributed associated structures than those of either the transform or longitudinal type. The region north of the Himalayas may contain notable exceptions (Tapponnier and Molnar, 1976).

Continental lithospheric segments of divergent margins not near oceanic crustal joins and intraplate graben systems incorporate wrench faulting less commonly. Intraplate regions appear to be the least common habitat. In North America, the known midplate wrench faults mostly have small displacements, plus limited associated structures which commonly have a fault (rather than fold) response to the wrench tectonics. The Cottage

Table 1. Structural Styles and Habitats*

Structural Style	Dominant Deformational Force	Typical Transport Mode	Plate-Tectonic Habitats	
			Primary	Secondary
BASEMENT INVOLVED				
Wrench-fault assemblages	Couple	Strike slip of subregional to regional plates	Transform boundaries	Convergent boundaries: 1. Foreland basins 2. Orogenic belts 3. Arc massif Divergent boundaries: 1. Offset spreading centers
Compressive fault blocks and basement thrusts	Compression	High to low-angle convergent dip slip of blocks, slabs, and sheets	Convergent boundaries: 1. Foreland basins 2. Orogenic belt cores 3. Trench inner slopes and outer highs	Transform boundaries (with component of convergence)

Structural style	Cause	Deformation	Plate-tectonic settings	
Extensional fault blocks	Extension	High to low-angle divergent dip slip of blocks and slabs	Divergent boundaries: 1. Completed rifts 2. Aborted rifts; aulacogens Intraplate rifts	Convergent boundaries: 1. Trench outer slope 2. Arc massif 3. Stable flank of foreland and fore-arc basins 4. Back-arc marginal seas (with spreading) Transform boundaries: 1. With component of divergence 2. Stable flank of wrench basins
Basement warps: arches, domes, sags	Multiple deep-seated processes (thermal events, flowage, isostasy, etc)	Subvertical uplift and subsidence of solitary undulations	Plate interiors	Divergent, convergent, and transform boundaries Passive boundaries

DETACHED

Structural style	Cause	Deformation	Plate-tectonic settings	
Decollement thrust-fold assemblages	Compression	Subhorizontal to high-angle convergent dip slip of sedimentary cover in sheets and slabs	Convergent boundaries: 1. Mobile flank (orogenic belt) of forelands 2. Trench inner slopes and outer highs	Transform boundaries (with component of convergence)
Detached normal-fault assemblages ("growth faults" and others)	Extension	Subhorizontal to high-angle divergent dip slip of sedimentary cover in sheets, wedges, and lobes	Passive boundaries (deltas)	
Salt structures	Density contrast Differential loading	Vertical and horizontal flow of mobile evaporites with arching and/or piercement of sedimentary cover	Divergent boundaries: 1. Completed rifts and their passive margin sags 2. Aborted rifts; aulacogens	Regions of intense deformation containing mobile evaporite sequence
Shale structures	Density contrast Differential loading	Dominantly vertical flow of mobile shales with arching and/or piercement of sedimentary cover	Passive boundaries (deltas)	Regions of intense deformation containing mobile shale sequence

*Styles in regions underlain by continental crust are emphasized; see Dickinson (1974) for discussion of most kinds of settings. Nontectonic types of drape folds, such as supratenuous or compaction folds, and structures resulting from igneous or metamorphic activity, are excluded.

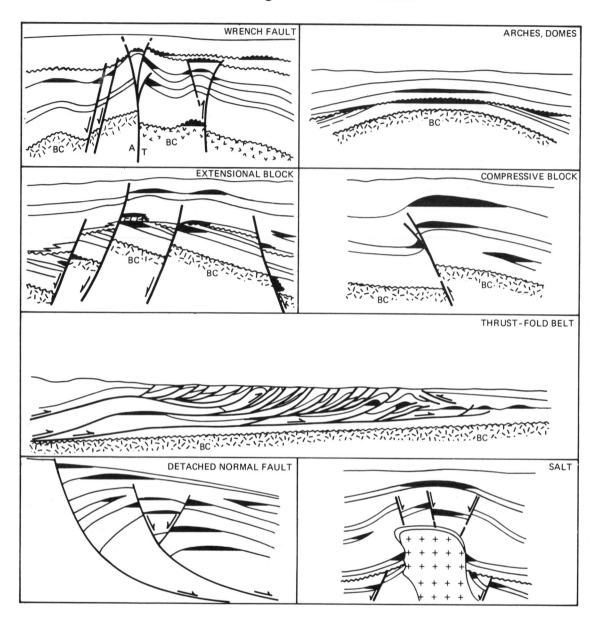

FIG. 1—Schematic diagrams of hydrocarbon traps (black areas) most commonly associated with structural styles of sedimentary basins. Purely stratigraphic type traps and traps associated with basement thrusts are omitted. Salt-related closures modified after Halbouty (1967, Fig. 6). *BC*, basement complex; *T*, displacement toward viewer; *A*, away from viewer.

Grove fault of the southern Illinois basin appears to be a characteristic example (Wilcox et al, 1973). Not all midplate wrench faults developed this way, however. The Rough Creek fault zone, also of the Illinois basin area, has associated en echelon folds and other compressive features suggesting a moderately large right-lateral displacement.

The structures associated with wrench faults (Fig. 5) are more diverse than those of any other style and include most elements that are fundamental to other styles. Wrench assemblages have both compressional and extensional features (Harding, 1973), or are dominated by compressional (Harland, 1971; Lowell, 1972; Harding, 1974), or extensional structures (Harland, 1971; Harding, 1974). These three substyles, side-by-side, convergent, and divergent wrenching, result from the configuration of the laterally moving blocks, or from the orientations of their boundaries relative to regional plate motion, or both (Wilcox et al, 1973).

The structural patterns of a single wrench zone show much repetition. Tectonic relations between several wrench faults, however, are much less consistent. Wrenches develop as singular, major displacement zones more commonly than other kinds of faults, such as thrusts, block faults, or detached normal faults, which tend to occur in sets. The diversity of several documented wrench zones (Fig. 2) demonstrates that their occurrence cannot be adequately explained by any single, unifying scheme. We thus differ, for example, with Moody and Hill's (1956) and Moody's

(1973) universal conjugate wrench sets. They considered such sets to be the result of pervasive meridional and equatorial compression, with faults formed at 30° angles to maximum principal compressive stress. Their approach disregards the kinematic and structural differences between different types of plate boundaries and tectonic settings. Their views are inconsistent with relations demonstrated by earthquake focal mechanisms; moreover, plate rotations would disrupt the orientations of any such global fault system.

The wide variety of features included in wrench-fault assemblages has resulted in confusion with other styles. Where direct offset data

A. GULF OF CALIFORNIA PARALLEL TRANSFORM SET

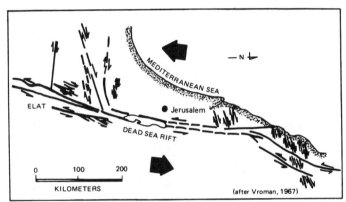

B. DEAD SEA ASYMMETRIC MASTER TRANSFORM SYSTEM

Established or Inferred Plate Motion

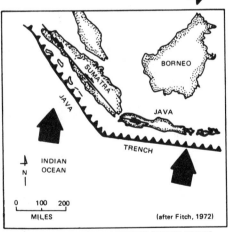

C. WESTERN INDONESIA LONGITUDINAL WRENCH FAULTS

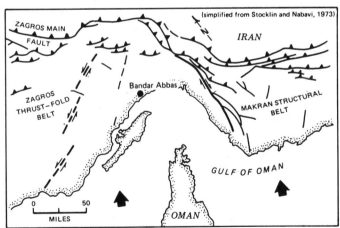

D. OBLIQUE WRENCH FAULTS, SOUTHEASTERN IRAN

FIG. 2—Wrench "set" examples. **A,** from transform plate boundary; strike slip occurs on parallel fault set. **B,** example from transform plate boundary; short, right-lateral antithetic strike-slip faults intersect single, main displacement zone at various oblique angles. **C,** example from convergent plate boundary; aligned left- and right-lateral faults presumably owe their opposing displacements to differences in sense of obliquity of plate encroachment that results from bend of plate boundary. **D,** example from convergent plate boundary; strike-slip displacements may have developed analogous to conjugate shears. Adjoins right margin of Figure 3.

are lacking, two characteristics provide a basis for initial distinction of wrench faults. The oblique resolution of stresses along a finite, throughgoing tectonic boundary (Fig. 5, center) commonly causes structures (1) to be arranged en echelon and (2) to be confined to a relatively narrow, persistent, linear zone.

In profile view, some wrench faults have a characteristic seismic signature termed a "flower structure" (R. F. Gregory, personal commun., 1970). It is expressed as an upward-spreading fault zone, whose elements usually have reverse separations (Fig. 6); the spreading fault system need not be symmetrical; half-flowers are also known (Lowell, 1972). Development of flower structures is enhanced where strike slip is accompanied by components of convergence (Lowell, 1972) and where the rocks are highly mobile. Syl-

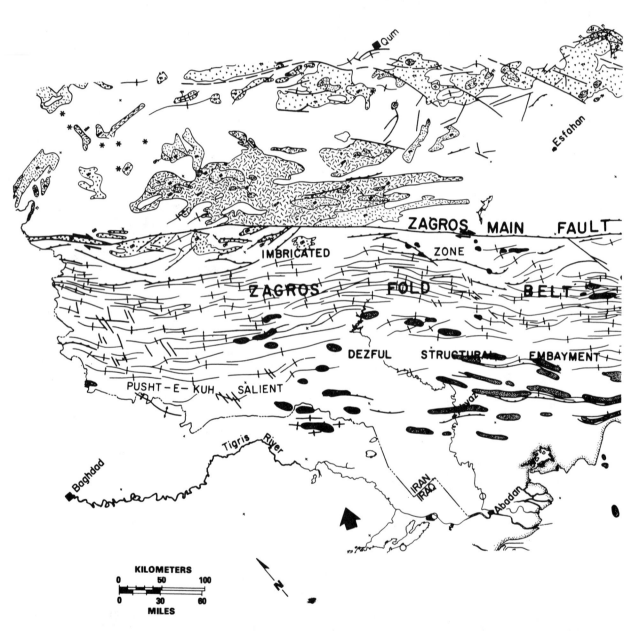

FIG. 3—Zagros orogenic belt, Iran. Frontal or external folds are considered to be detached and probably thrust in part. Basement-involved thrusts are present adjacent to Zagros Main fault, which has a late-stage component of right-lateral strike slip. Buoyantly and tectonically injected salt structures also contribute to mix of structural styles. Generalized from Stöcklin and Nabavi (1973).

vester and Smith (1976) have documented surface examples along the San Andreas fault in Mecca Hills, California, and have described their genesis.

Flower-structure reversals can be differentiated from those in other structural styles by recognizing fault displacements directly below the reversal that indicate a high-angle fault stem involving basement (Fig. 6). "Negative" flower structures

have also been observed and consist of shallow sags overlying upward-spreading strike-slip faults with normal separations.

A great variety of hydrocarbon traps (Fig. 1) occurs with the wrench-fault substyles. By far the most prolific have been those associated with en echelon folds. En echelon normal fault blocks, subthrust bed terminations, and flower structures have also been effective traps. Closure types and

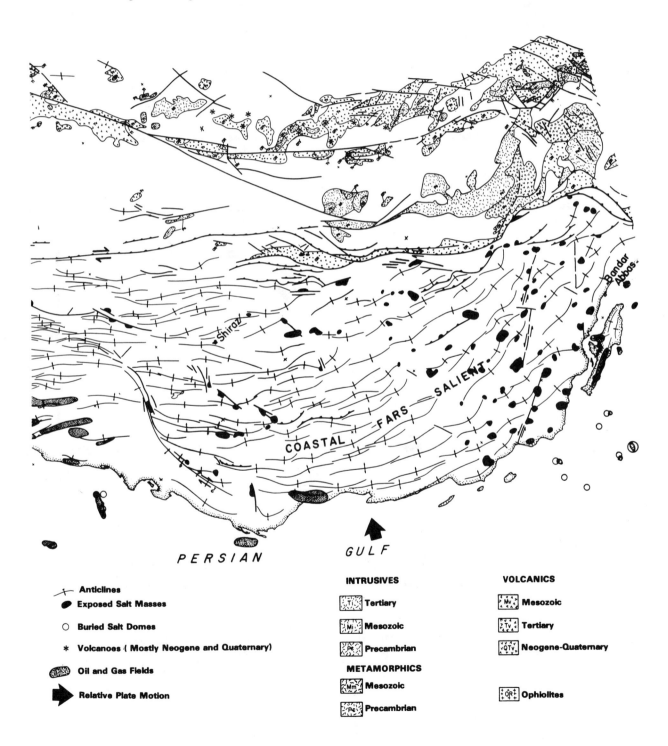

Legend	
Anticlines	
Exposed Salt Masses	
Buried Salt Domes	
Volcanoes (Mostly Neogene and Quaternary)	
Oil and Gas Fields	
Relative Plate Motion	

INTRUSIVES
Ti — Tertiary
Mi — Mesozoic
P€ — Precambrian

METAMORPHICS
Mm — Mesozoic
P€ — Precambrian

VOLCANICS
Mv — Mesozoic
Tv — Tertiary
QTv — Neogene-Quaternary
OR — Ophiolites

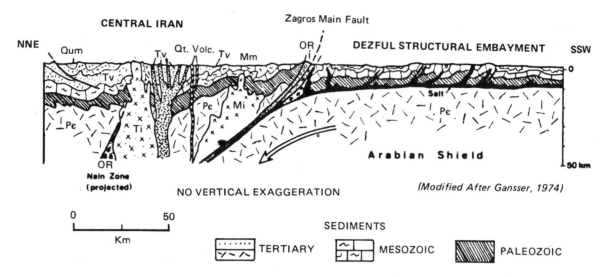

FIG. 4—Generalized cross section of Zagros orogenic belt; extends southwest from city of Qum (see Fig. 3). See Figure 3 for explanation of symbols.

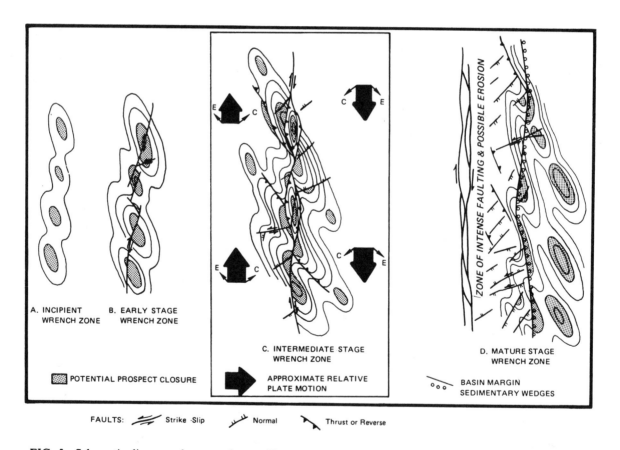

FIG. 5—Schematic diagram of structural assemblage associated with major wrench fault and stages in evolution that cause changes in hydrocarbon trapping; modeled after California examples (discussed in Harding, 1974). Arrows *C, E* represent compressive and extensional vectors, respectively, that arise from coupling motion between adjacent plates.

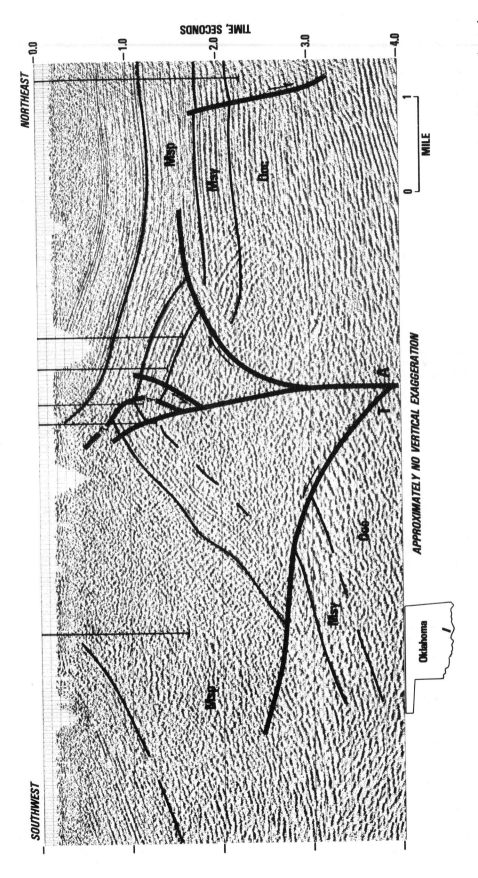

FIG. 6—Migrated seismic profile across wrench-fault zone in Ardmore basin of Oklahoma demonstrating flower-structure geometry (adapted from unmigrated interpretation by R. F. Gregory and E. C. Lookabaugh, 1973). *Msp*, Mississippian Springer; *Msy*, Mississippian Sycamore, and *Ooc*, Ordovician Oil Creek reflectors. *T*, displacement toward viewer, *A* away from viewer.

changes in the prospective fairway that develop with increased strike slip have been described previously (Fig. 5; Harding, 1974, 1976).

Compressive Fault Blocks and Basement Thrusts

Compressive fault blocks (Fig. 7) and basement thrusts preferentially occur on convergent-plate boundaries. The former are typically more areally restricted, being confined mostly to forelands, whereas the latter are found in forelands, orogenic belts, and landward walls of oceanic trenches (Table 1). Compressive blocks are discussed first, then basement thrusts with which compressive blocks seem to be transitional in foreland settings, and finally, basement thrusts in settings other than forelands.

Forelands are developed in two convergent-margin positions, back-arc and peripheral (adapt-ed from Dickinson, 1974). Back-arc forelands lie between the magmatic-volcanic arc and the craton and commonly have thrust-fold belts directed toward the plate interior or craton ("Andean" or "Cordilleran" type, Dewey and Bird, 1970). Peripheral forelands develop with continental collision and have thrust-fold belts lying between the magmatic-volcanic arc and the former trench ("collision" or "Himalayan" type, Figs. 3, 4). Folds and thrusts in this example are directed toward the plate boundary or position of the earlier trench.

Compressive blocks are best known from the Laramide back-arc foreland of Wyoming. In Wyoming and elsewhere, although orogenic belts are laterally extensive, adjacent forelands with significant compressive block faulting are seemingly rare. Lowell (1974) and Dickinson and Snyder

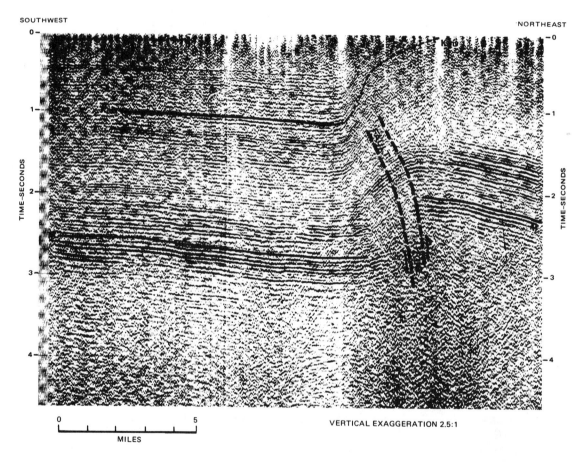

FIG. 7—Southwest-northeast seismic profile across Rangely oil field structure, northwestern Colorado. Inclination of fault is approximated by syncline's axial plane which dips steeply under upthrown block. Flexure is caused dominantly by fault drag and drape over edge of tilted, compressive fault block. Anticlines formed in this way parallel uplifted block edges; their traces reflect pattern of primary block-bounding faults. Compressive effects, down to deepest limit of this control (3 or 4 sec), are concentrated in fault zone (with no apparent deformation in reflectors on either side). *M*, Mississippian; *Kmv*, Cretaceous Mesaverde reflectors.

(1978) have postulated that some of the reverse displacement on fault blocks in Wyoming is caused, respectively, by either buoyancy from or physical contact with an underlying subducted lithospheric slab. The limits of the deep-seated slab would have controlled the regional distribution of foreland structures. Burchfiel and Davis (1975) have attributed structures of the Wyoming province to a thermally weakened, brittle crust.

Prucha et al (1965) used the term "upthrust" for reverse faults that bound the Wyoming uplifts and considered that the uplifts were caused by differential vertical movement. The upthrusts characteristically flatten upward from nearly vertical fault surfaces through reverse faults into less steeply dipping thrusts (Fig. 7). Upthrust or reverse faults, however, can also be important components of the convergent wrench style (Lowell, 1972), both as wrench faults themselves (Fig. 6; Sylvester and Smith, 1976) and as integral elements of the associated fold set that is away from the wrench fault.

We do not use the term "upthrust" for compressive blocks because the proposed fault profile may not be consistent with deeper intrabasement relations revealed by recent seismic data (Smithson et al, 1978) and because the genetic implications seem incompatible with coexisting basement thrusts and wrench faults that require at least some basement compression and lateral movement. Furthermore, Reches (1978a) has recently demonstrated layer-parallel shortening within the sedimentary cover in foreland monoclinal structures in the Colorado Plateau.

In our treatment of compressive blocks we are concerned with the structural geometry of the uppermost basement and sedimentary cover where trap closures develop. At these levels compressive blocks can have bounding fault surfaces that range from subvertical, with reverse (Fig. 7) or, rarely, normal segments (e.g., Rattlesnake Mountain anticline of Bighorn basin, Wyoming; Pierce and Nelson, 1968) to lower dipping thrusts, particularly on zones with greater structural relief. Block faults can also have subordinate amounts of strike slip.

In the Bighorn basin of the Wyoming foreland the variable trend of block-bounding faults and their overlying and paralleling drape folds or monoclines is the most distinctive style characteristic. Trends are dominantly northwest-southeast longitudinal structures and subordinately north-south and east-west oblique structures (Fig. 8). The variously aligned elements can combine to form rhombic or cross-trended patterns, or outline individual discrete blocks. Blocks include large rectilinear features (e.g., Beartooth uplift),

uplifts with dogleg boundaries (northwest terminus of Bighorn uplift, Fig. 8), clusters of trap-door blocks (Pryor Mountains, Fig. 8), and solitary trap doors or drape anticlines of several repeated alignments. These distinctive block features are present within or disrupt a more general relay pattern formed by the northwest-southeast folds (Fig. 8, southwest quadrant).

In profile view (Fig. 7), a threefold vertical zonation is common in higher structural levels and further demonstrates the fault-block geometry of the style. The near top of basement level is a tilted fault block. A steep drag fold develops at intermediate levels, and a gentle drape fold, or tilted monocline, lies at shallow levels where preserved from erosion. Blocks appear as rotated slabs in cross sections, and associated flexures typically are markedly asymmetric (Fig. 7). Symmetric, curvilinear flexures are rare.

Internally, individual structures range from simple to complex; large normal faults that parallel the structural axis but lie well downdip from the backlimb (i.e., flank of flexure away from the block boundary) have been observed (Howard, 1966). Secondary crestal normal faults, both longitudinal and transverse, are common on some drape flexures (Wise, 1963). Some transverse faults have strike-slip components and offset flexure axes, and terminate mostly at high angles at the block boundary. They can also act as basement-involved tear faults that displace block edges (Foose et al, 1961).

Compressive blocks appear to grade into low-angle, large-displacement basement thrusts on some foreland structures. Deep reflection profiling across the Wind River Mountains (see Fig. 20, right side) has shown a thrust surface that can be traced to a depth of at least 15 mi (24 km) at an average dip of 30 to 35°. A minimum horizontal displacement of 8 to 13 mi (13 to 21 km) exceeds the minimum vertical displacement of 8 mi (Smithson et al, 1978).

The multidirectional trends of foreland block structures present problems in interpretation that we will see are similar to those of extensional block faulting. General contemporaneity of the block-associated flexures is demonstrated at block corners along the west flank of the Bighorn uplift, where flanking monoclines have equal relief on either edge of the corner and are not offset at the point of junction (Fig. 8). Similar relations are apparent at the Pryor Mountains trap-door faults and at the corners of the Beartooth uplift. Elsewhere we have not examined in detail the relative timing of the differently aligned structures; tectonic overprint with different orientations remains a possibility.

Buried, older zones of basement weakness control development of several structures (Foose et al, 1961; Reches, 1978a) but we do not know whether this is a universal prerequisite for the generation of grid patterns (Hoppin and Palmquist, 1965). Marked similarities in trend patterns with extensional block-faulted terranes (e.g., clustered trap doors, platforms with dogleg boundaries) suggest that some structures may result from compression and inversion of a preexisting normal-fault fabric imparted to the Wyoming province during earlier phases of rifting (Stewart, 1972). The deep listric basement-involved normal faults could provide listric surfaces for the Laramide thrust displacements and at the same time would impart the fault-block patterns observed at the surface (see Lowell, 1974, Fig. 2).

Because of the diversity of styles and the lack

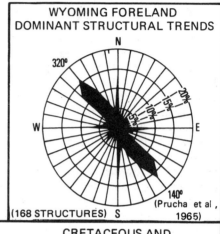

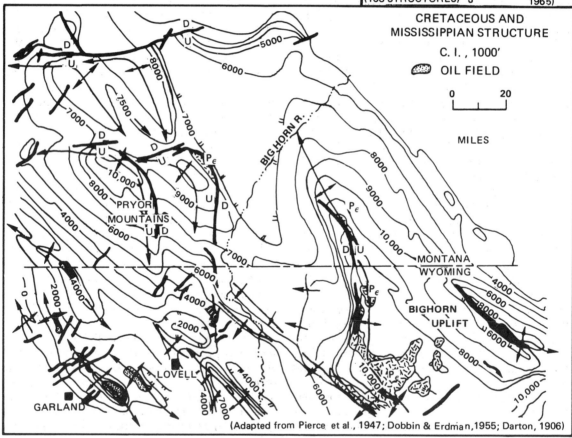

FIG. 8—Structures at northeast corner of Bighorn basin, Wyoming, demonstrate three dominant trend orientations of general Wyoming foreland province (insert plot includes orientations of major faults, folds, uplifts, and lineaments). Fold axes parallel dominant block-bounding faults where latter are exposed at surface or are demonstrated by seismic (not shown). Similar faults may be inferred in structures lacking this control on basis of consistent, fold-fault relations observed elsewhere.

of control for fault attitudes within the basement, Wyoming uplifts have been given many different interpretations. Stone (1970) thought that wrenching controlled the basic tectonics. He explained fault blocks, such as the Pryor Mountains, as bounded by a thrust on one edge and a strike-slip tear fault on the other edge. Structural relations demonstrate that both boundaries have identical drape-flexure style and lack evidence of strike slip. Foreland structures have also been interpreted as due to differential vertical uplift (Stearns, 1975), a mechanism that fails to explain the large thrust overlap of the Wind River Mountains, and the co-occurrence in some areas of wrench faults. Deep, intrabasement thrusting (Bally, 1975) has received increased attention as the primary tectonic control. We believe that the Wyoming foreland contains a mix of Laramide structural styles, compressive fault blocks, basement thrusts, wrench faults, and basement warps, and should not be attributed to a single type of deformation.

Differentiation of block faulting from detached thrusting or convergent wrenching is critical in regions of compressive deformation. With sparse data (e.g., a single seismic profile) differentiation can be difficult, especially where upthrust or reverse faults are a potential element of more than one style. Regionally, the grid pattern of block structures differs distinctly from the wavelike salient and reentrant patterns of conventional fold-thrust belts (cf. Figs. 8 and 3). The grid pattern is also distinctive from the en echelon structures and throughgoing, straight master fault of many wrench zones (cf. Figs. 8 and 5). Trap-door clusters and doglegs are thought to be especially diagnostic of block faulting. Upthrust faults at transform plate boundaries can be distinguished from shallow, steep fault profiles of the compressive block style by the former's more nearly unidirectional orientation and association with en echelon folds.

In cross section, the curvilinear fold profiles of wrench assemblages and thrust-fold belts are commonly quite different from the rotated slab and monoclinal-step appearance of many block structures. The latter are markedly asymmetric, whereas wrench-associated folds are often symmetric. Both wrench flower structures and compressive fault blocks, however, can have similar-appearing reverse dislocations. Identification of a relatively shallow rollover and branching faults that dip inward toward a narrow, common stem is important in recognizing the wrench-associated structure (cf. Figs. 6 and 7); half-flower structures, however, are harder to distinguish.

The two block-fault styles, compressive and extensional, are differentiated by the character of the block-bounding faults and by the steeper flexures and compressive features obviously associated with the former style (cf. Figs. 7 and 11).

Compressive block faulting has created prolific traps (Fig. 1) in the Rocky Mountain foreland, the Permian basin of west Texas (Figs. 9, 10; Elam, 1969), and, perhaps, the Oriente province of Ecuador. Specific closures include culminations on drape anticlines (Fig. 8, southwest quadrant), trap doors (Fig. 10), cross-faulted noses, and backlimb subsidiary flexures and faults. Additional production has come from subthrust warps closed against block-bounding faults and from various associated stratigraphic traps.

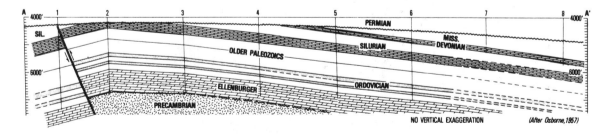

FIG. 9—Cross-section AA′ across Keystone field, Permian basin, Texas. Steep dip of fault at top of basement and in sedimentary cover is indicated by well control. Seismic and subsurface control demonstrate similar steep faults at these levels elsewhere in basin (Elam, 1969). See Figure 10 for location of section.

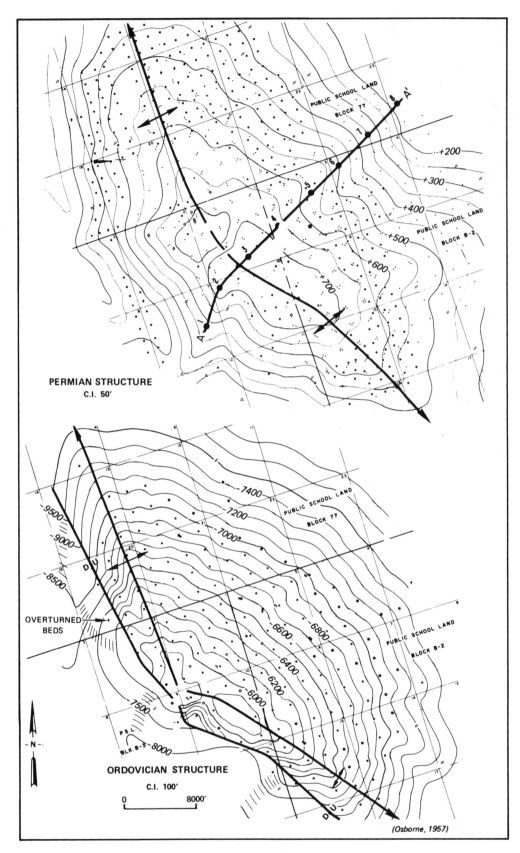

PERMIAN STRUCTURE
C.I. 50'

ORDOVICIAN STRUCTURE

C.I. 100'

0   8000'

OVERTURNED BEDS

(Osborne, 1957)

FIG. 10—Structure at Keystone field, west edge of Central basin platform, Permian basin, Texas (from Osborne, 1957). Shallow flexure (Permian, upper diagram) reflects drape over buried Ordovician trap-door block (lower) and has characteristic culmination opposite obtuse-angle junction of bounding faults. Other angles of fault junctions in area range from obtuse to acute. Trap-door flexures associated with latter appear as triangular faulted domes.

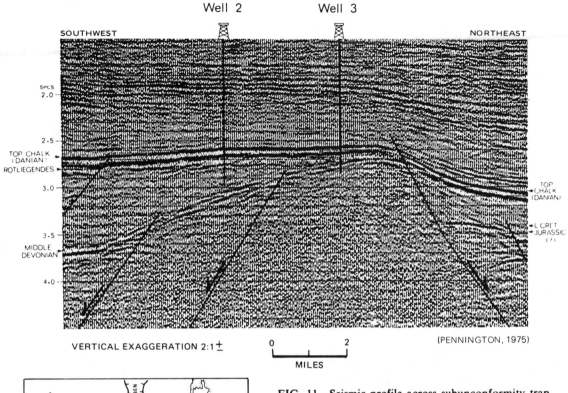

VERTICAL EXAGGERATION 2:1 ±

MILES

(PENNINGTON, 1975)

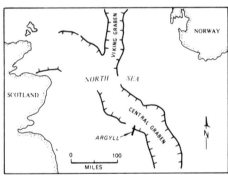

FIG. 11—Seismic profile across subunconformity trap (Rotliegendes sandstone) at Argyll field, western margin of Central graben, North Sea. Three interrelated structural levels are present; slablike, unflexed, rotated normal-fault block at depth (Middle Devonian), fault block with dip-slip fault drag flexure at intermediate level (Danian chalk), and shallow, intact drape flexure (above 2.4 sec). In index map, note dogleg configuration of Central graben (i.e., proceeding from far north end, graben trend is first south, then southeast, and then back to original south trend) and bifurcation at junction with Viking graben.

Basement thrusts in several other convergent-plate settings have not been found prospective for petroleum because of their extremely complex deformation under conditions of ultra-high pressure or temperature, or both. Such basement thrusts include orogenic belts that flank back-arc and peripheral forelands (Fig. 3) and faults beneath the outer high in the landward slope of oceanic trenches. At the latter, the downgoing or underthrust oceanic lithosphere is involved in deep thrust slices together with the detached thrusts (Seely et al, 1974).

Extensional Fault Blocks

Normal faults occur subordinately in all style assemblages, but certain suites of structures (Fig. 11) and tectonic settings (Table 1) are dominated by regional, deep-seated normal faults which constitute a discrete fault-block style. Such normal faulting is perhaps the most widespread of all styles. It dominates divergent margins in early stages of development, the oceanic crust formed at spreading centers, and some intraplate regions. In cross section, normal faulting is one of the least complex styles. In plan view, its patterns are highly variable and difficult to predict.

The Gulf of Suez (Fig. 12) typifies the basic style observed at divergent margins in various stages of development and in intraplate grabens. In the Suez graben, regional faults are distinctly multidirectional and create a grid or intersecting system (Robson, 1971). Blocks with oblique, zig-

17

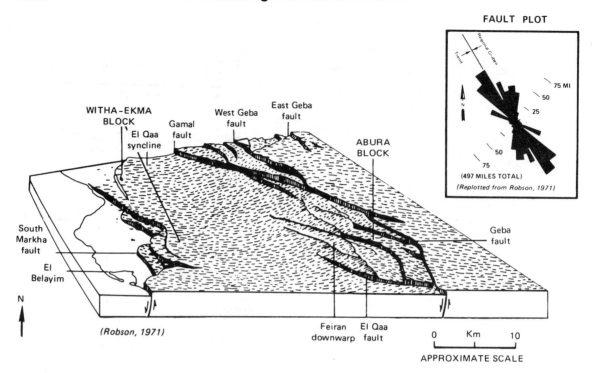

FIG. 12—Perspective block diagram of normal faults at east-central border of Gulf of Suez graben, Egypt. Fault frequency (insert) is plotted by cumulative lengths for each 5° quadrant of fault strike. Structure is controlled by extensive surface exposures (see Robson, 1971, Pl. I).

zag edges and longitudinal fault blocks with relay patterns are both present; either can occur alone or in series. Fault plots (Fig. 12) show a preferred longitudinal orientation that parallels the overall graben trend. Faults oblique in mostly two directions about this regional trend are also important and about equally developed. Transverse regional faults are rare. Alignment plots are similar for both synthetic (i.e., downthrown toward the basin axis) and antithetic (i.e., upthrown toward the basin) faults with either major or minor displacements (Robson, 1971). Overprints of several different episodes of normal faulting result in still more complex patterns.

Individual fault blocks are internally complex (Figs. 13, 14). Second-order faults can repeat the regional fault trends or can have inconsistent transverse orientations, but they generally terminate at block boundaries.

On a regional scale, the normal faults form intraplate rifts that range in complexity from simple fault troughs with persistent straightaways to grabens with many junctions, multiple bifurcations, and doglegs (Fig. 15; index map, Fig. 11). Divergent plate boundaries inherit their outline mainly from the trends of earlier intraplate rifts.

The high frequency of longtitudinal faults ap-

pears to satisfy the common concept that normal faults trend perpendicular to regional extension (Anderson, 1942). The oblique faults do not. They interact at block corners, uplift intersecting sides equally, and appear to be unified with the relay elements into one system to achieve a well-integrated graben subsidence. The various fault directions thus appear to be contemporaneous in some places, which precludes explanation of the pattern by multiple periods of extension.

The oblique faults, because of their roughly 60° intersections in map view, have been considered to be conjugate strike-slip faults. Surface mapping in the Gulf of Suez area (Robson, 1971; El-Tarabili and Adawy, 1972) and elsewhere (King, 1965) and subsurface control demonstrate, however, that the faults are notable for their absence of significant strike slip. Other investigators (El-Tarabili and Adawy, 1972) have assumed that preexisting zones of crustal weakness influenced the later position of oblique normal faults, and this has occurred in some areas (King, 1965). In others, the typical normal-fault system continues to develop without the aid of older zones of basement weakness (Illies, 1970). The persistent recurrence of such faults in many different areas and tectonic settings, and at many scales, suggests that multidirectional fault sets may be the fundamental result of extensional rupture. Reches (1978b) has reached a similar conclusion on theoretical grounds and the multidirectional dip-slip pattern has been reproduced with single-stage tectonic models (Freund and Merzer, 1976).

Asymmetric blocks with steep sides bounded by listric normal faults and gentle sides comprised of constant dips characterize the style in cross section (Fig. 11; Lowell and Genik, 1972). Some graben profiles show rotation and down-stepping of blocks consistently toward the trough axis (Fig. 14, right side) or toward one side of a half-graben. Others show rotation on both synthetic and antithetic faults which causes the basement surface to be at various inclinations and depths. Some graben segments can inherit a consistent regional slope from prior structural events or can have it imposed during a rift-related regional arching phase. Along strike the position of the graben axis relative to these earlier or imposed slopes can change, causing abrupt variation

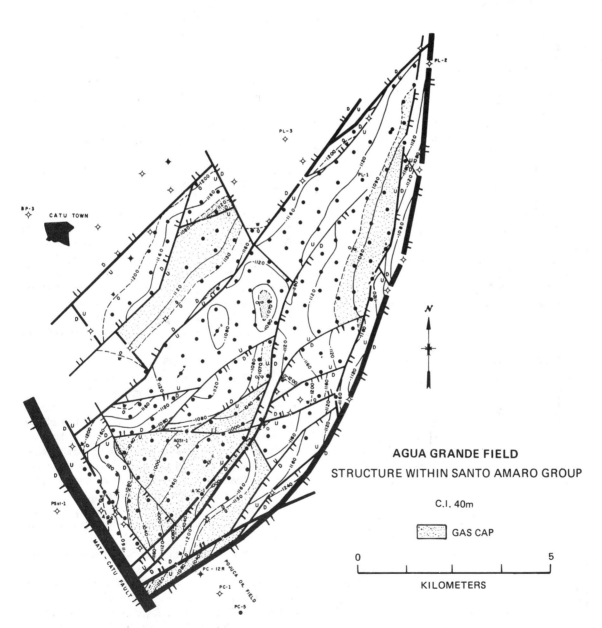

AGUA GRANDE FIELD

STRUCTURE WITHIN SANTO AMARO GROUP

C.I. 40m

GAS CAP

0   5
KILOMETERS

FIG. 13—Trap-door structure at Agua Grande oil field, Recôncavo basin, Brazil. Longitudinal faults trend northeast, parallel with graben axis, and oblique faults trend generally east-northeast and nearly north-south to north-northwest.

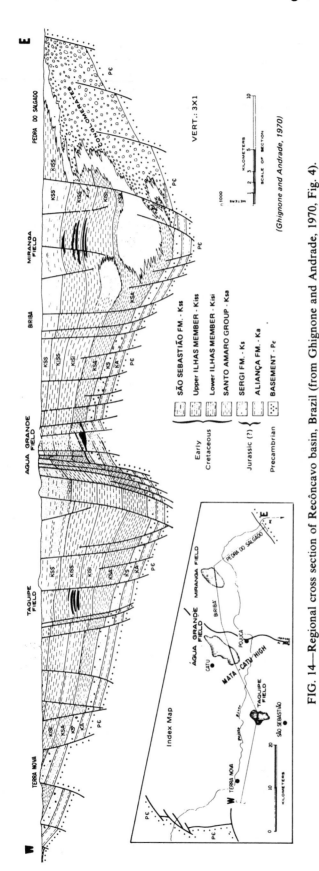

FIG. 14—Regional cross section of Recôncavo basin, Brazil (from Ghignone and Andrade, 1970, Fig. 4).

in profile appearance.

Listric normal faults cause individual block rotations that can result in reverse drag on the downthrown side of faults. When reverse drag is opposite regional tilt a low-side rollover can occur (e.g., apparently at Taquipe field; Fig. 14) comparable to those that form on some detached normal faults. Extreme block rotations result from multiple stages of listric faulting (Proffett, 1977). Rotational effects attributed to listric faults are not consistently apparent, however, even on adjacent blocks. Presumably this is due to differences in amount of displacement and to different levels at which faults flatten, the smaller the displacement and the deeper the level of flattening, the less obvious the rotation of dip into the downthrown side of the fault.

Cross sections are further complicated by the complex patterns of oblique and longitudinal faults that can dip either toward or away from the graben axis. Intersection of different fault sets brings different block rotations into the line of section and causes a loss of lateral structural continuity. Also, block rotation can be modified by postfaulting, en masse subsidence of the rift system.

In many places the sedimentary cover is draped over block edges (Fig. 11). Where faults are downthrown in a direction opposite the direction of block tilt, drape forms reversals on the upthrown sides of and parallel with the faults. The pattern of drape is multidirectional, reflecting the grid patterns of the controlling faults. Where low-side reverse drag is also present graben structures can include a mixture of low- and high-side dip reversals.

Convergent plate boundaries are also an important normal-fault habitat (Table 1). Convergent settings that can be dominated by this style, progressing from plate boundary to plate interior, are: (1) the outer trench slope of downgoing plates in front (seaward) of the trench and arc complex, (2) the inner or arc-massif flank of fore-arc basins, (3) intra-arc basins within the magmatic-volcanic arc proper, (4) back-arc marginal seas produced by back-arc spreading, and (5) the stable or craton flank of back-arc and peripheral foreland basins (see Dickinson, 1974, for elaboration of settings).

On some transform margins, basins oriented parallel with a bounding major strike-slip zone have had significant normal faulting on their opposite, stable flanks. Examples include the Greater Oficina–Temblador region of eastern Venezuela opposite the El Pilar fault, the Bakersfield arch of the San Joaquin Valley basin opposite the San Andreas fault (Fig. 16), and the Magallanes basin of the Andean Scotia arc.

In the San Joaquin Valley, normal faulting was coeval with the period of greatest displacement and deformation along the San Andreas fault (Harding, 1976). At that time thick basin fill was deposited in sags flanking the developing Bakersfield arch (Fig. 16, bottom and left margins). Here and in Greater Oficina, the family of nor-

mal faults is similar to orientation plots for the Suez graben (Fig. 12). Fault-trend maxima are parallel with both regional strike of the shelf and the bounding strike-slip zone on the opposite basin flank. Drape flexing of beds is rare and hydrocarbon traps owe their closure principally to the bends, splays, and updip intersections of the oblique and longitudinal faults (Fig. 16). Similar fault closures have been found on the stable flank of some foreland basins (e.g., German Molasse basin).

The plan-view pattern permits differentiation of this style from others, except for compressive fault blocks (see previous section). In profile, however, compressive effects are notably absent; flexing, if present, is more limited than with compressive blocks. The steps and rotated-slab appearance of normal fault blocks contrast with the curvilinear fold profiles common to most other styles.

Rift systems are a prolific hydrocarbon setting. Here oil and gas traps depend mostly on the multidirectional faults and their associated drape flexures (Figs. 1, 13). Important trap-door closures occur at fault junctions, where oblique faults cause rollovers to intersect, and regional dip or block rotation maximizes structural relief at the point of fault intersection. Horsts and, less commonly, the downthrown sides of faults provide additional structural closures. The high edges of blocks are ideally situated to localize development of reefs and other reservoir facies. Subunconformity truncation traps in rotated blocks (Fig. 11) and other kinds of buried topographic closures are common. Their morphologies also reflect an ultimate control by the multidirectional fault pattern.

Basement Warps—Arches, Domes, and Sags

Basement warps are generally large simple flexures with gently dipping broad flanks. Regional positive elements include arches (Fig. 16), domes, massifs, anteclises, upwarps, swells, platforms, broad noses, and structural terraces. Negative features are downwarps, depressions, syneclises, sags, and troughs. Basement warps develop at both local and regional scales and typically are solitary. More intensely deformed basement folds also occur as elements of well-defined structural trends. These have been treated previously under other styles, but distinctions from the basement-warp category are not always obvious.

Basement warps are present in all habitats and are the dominant style of many plate-interior or craton regions (Table 1). They are present on the stable flanks of some convergent margin basins (e.g., Thornton arch of Sacramento Valley fore-

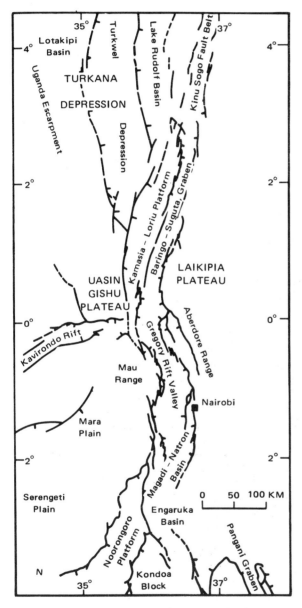

FIG. 15—Distinct pattern of three-armed graben junction at intersection of Kavirondo and Gregory rifts, Kenya. Northern and southern arms splay or spread out from junction, and each arm has multidirectional sets of normal faults. Main elements of southern arm between Gregory rift valley and eastern border of Kondoa block delineate dogleg. From Baker et al (1972).

arc basin), stable flanks of transform margin basins (e.g., Bakersfield arch of San Joaquin Valley, Fig. 16) and in divergent margin basins (e.g., Cape Fear arch of Atlantic seaboard).

The Williston basin and adjoining shelf on the west (Fig. 17) demonstrate the style of basement warps and other structures common to intracratonic settings. Structures are of two general types: warps of various shapes and sizes, and narrow linear-faulted or fault-related features. Structures of the first type include the Miles City arch; Porcupine dome (Fig. 17, no. 7); Nesson anticline (Fig. 17, no. 1), a broad southward-plunging nose or arch with several culminations; the Bowdoin

(Fig. 17, no. 2) and Poplar domes (Fig. 17, no. 3) west of the basin proper; and the central-basin deep (Fig. 17, no. 4), which is an irregularly shaped sag or downwarp. The second type of structure, linear fault-associated features, is exemplified by the Cedar Creek anticline (Fig. 17, no. 5) which is essentially a monocline facing up the basin flank. Most of these Williston basin structures are solitary and do not share common geographic alignments or obvious trend associations; instead, their orientations include all quadrants of the compass.

Additional styles occurring in intracratonic regions are normal faults, either randomly dis-

FIG. 16—Bakersfield arch, on stable east flank of San Joaquin Valley basin, California. Individual hydrocarbon traps are provided by secondary normal faulting, apparently concentrated by transverse arching, and by small basement domes at lower part of arch. Interference from en echelon fold set associated with San Andreas fault (west of map) is evident at southwest plunge of arch. Contours are of top of Tertiary. C.I. = 500 ft.

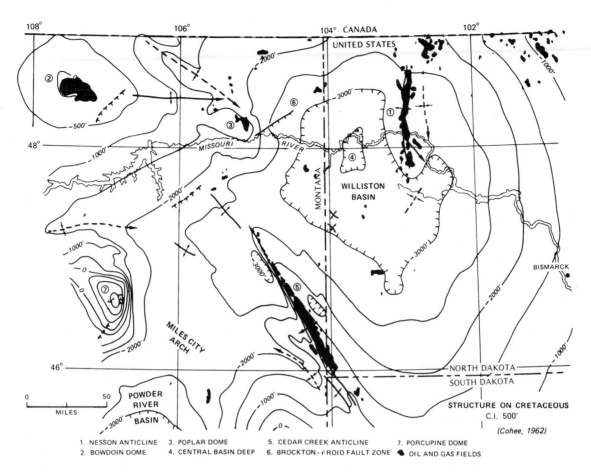

FIG. 17—Regional structure and main elements of Williston basin area. Trend orientations are diverse and continuity from one structure to another is absent (except in Bowdoin and Poplar domes).

persed or in well-defined zones and a few solitary wrench faults.

The structural genesis of basement warps is not well understood and they probably have multiple origins. Warps lack diagnostic features, such as drag and drape folds or trend patterns, to indicate directions of tectonic transport. Also lacking is an obvious plate-boundary association that could suggest a particular type of deformation. Differential regional subsidence or uplift has intuitive appeal as a cause of many intraplate warps. For example, the Miles City arch appears to be a residual high separating the Williston and younger Powder River basins (Fig. 17). The main Williston basin structures had their early growth in Ordovician to Mississippian time (Hansen, 1972), synchronous with the first major phase of basin subsidence (Mallory, 1972). Differential subsidence or uplift could have occurred because of (1) irregularities in fundamental subcrustal processes and (2) large inhomogeneities within the crust.

Green (1977), in a review of crustal processes,

stated that the following have all been considered mechanisms of continental lithospheric thinning and subsidence: differential lithospheric cooling, phase changes, ductile flowage, subcrustal erosion, injection and stoping of dense material and related magma-chamber collapse, and surface erosion following thermal uplift. In addition, deep-seated listric normal faulting can also cause lithospheric thinning and associated subsidence.

Crustal inhomogeneities can modify structural responses to those deep-seated processes. Steep linear zones of weakness such as preexisting wrench faults may be reactivated in a dip-slip mode. The reactivation could develop basement steps or linear monoclines such as the Paleozoic precursor of the Cedar Creek anticline (Fig. 17, no. 5). Some basement warps may be influenced by earlier buried fault blocks to the extent that they merge with the shallow expressions of fault-block styles (perhaps Porcupine dome, Fig. 17, no. 7).

The origin of domal features in contrast to linear or blocklike structures is much more speculative. Irregularly shaped intrabasement masses with differing physical properties perhaps could subtly and locally retard or enhance regional subsidence.

Tectonic loading and sediment loading of foredeeps have been considered mechanisms of warping in foreland settings. Crustal compression may be an additional agent for basement warping in such regions. The Williston basin structures were rejuvenated during the Laramide (Fig. 17) when compression was prevalent elsewhere in the Rocky Mountains province. Using earthquake focal plane solutions, Sykes and Sbar (1974) have demonstrated present compression within the North American plate.

The style and occurrence of basement warps pose problems for the simplistic plate-tectonic approach that restricts structural development to the edges of so-called rigid plates and that implies mainly horizontal plate movement. Many basement warps in intraplate settings suggest first-order vertical movement in their development. Other features suggest midplate compression. All argue against simple plate tectonics as the exclusive cause of deformation. Clearly, vertical movements must occur within a larger scope of horizontal plate translations.

Basement warps can be differentiated from other basement-involved folds by their solitary occurrence, inconsistent orientations, lack of distinct trend patterns, and general lack of dependency on faulting for their development. Basement folds associated with block faulting reflect block-fault patterns as a genetic control. Basement folds associated with thrust or wrench faults demonstrate the characteristic trend patterns of these styles.

Positive basement warps, particularly where they intervene in areas of thick sedimentary cover, are the focal points for oil migration and accumulation (Fig. 16). They commonly persist over long periods of time and are particularly effective in localizing truncation, unconformity, convergence, and onlap in the sedimentary section. All of these factors can enhance entrapment of oil and gas (Fig. 1). Associated closures commonly are considerably larger and less segmented than the more complicated structures of other styles.

DETACHED STRUCTURES

Decollement Thrust-Fold Assemblages

Thrust-fold assemblages are an essential element of many convergent plate boundaries (Table 1). They may be present along the mobile flanks of back-arc and peripheral forelands, and

at trench inner slopes and outer highs (Dickinson, 1974). They occur as wide zones of deformed sedimentary cover on the external side, or outer fringe, of many orogenic belts (Fig. 3). Gradations to basement thrusting are common toward the internal, or core, regions (Fig. 4). The detached structures regionally are festooned into externally gently convex salients and more sharply concave reentrants (Fig. 3). Both salients and reentrants are comprised of parallel bands of thrusts and associated folds having relay trend patterns. Locally, thrusts overlap in cuspate arrangement convex in the direction of tectonic transport. Anticlines, where present, lie in the hanging walls of the thrust faults and have axes parallel with the trends of the thrusts.

In profile, complex listric thrust faults that often merge with bedding at depth are characteristic (Fig. 19). Lithostratigraphic changes and attendant ductility contrasts are responsible for localization of the fault surfaces so that thrusts are generally parallel or subparallel with bedding in incompetent rocks and oblique to bedding in competent rocks. Thus, the thrusts alternately follow bedding, then ramp upward as step faults (Fig. 20). The faults almost invariably cut upsection in the direction of relative tectonic transport of the hanging wall.

Where massive carbonate beds are a significant part of the deformed section, thrust sheets typically consist of repeated slabs. Where more duc-

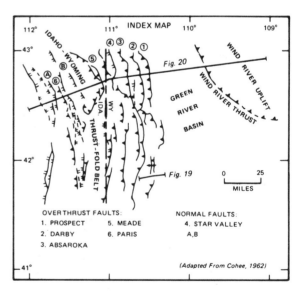

FIG. 18—Index map across Idaho-Wyoming thrust-fold belt. Seismic section and cross sections are shown in Figures 19 and 20.

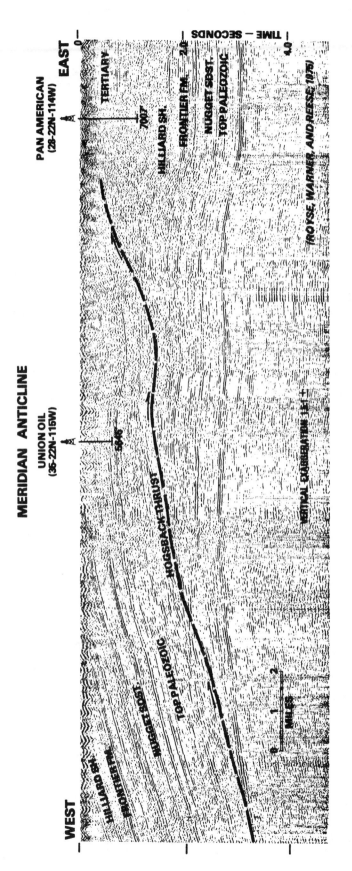

FIG. 19—Seismic line across Meridian anticline and Hogsback thrust on east side of Idaho-Wyoming thrust-fold belt near latitude of Kemmerer, Wyoming (see Fig. 18). Fault alternately parallels bedding and steps upsection. Velocity pull-up on time section lies beneath leading edge of thrust, which has introduced relatively high-velocity carbonate rocks. Paleozoic rocks in hanging wall in part of structure near label "Hogsback thrust" are repeated by imbricate faults that splay from main detachment.

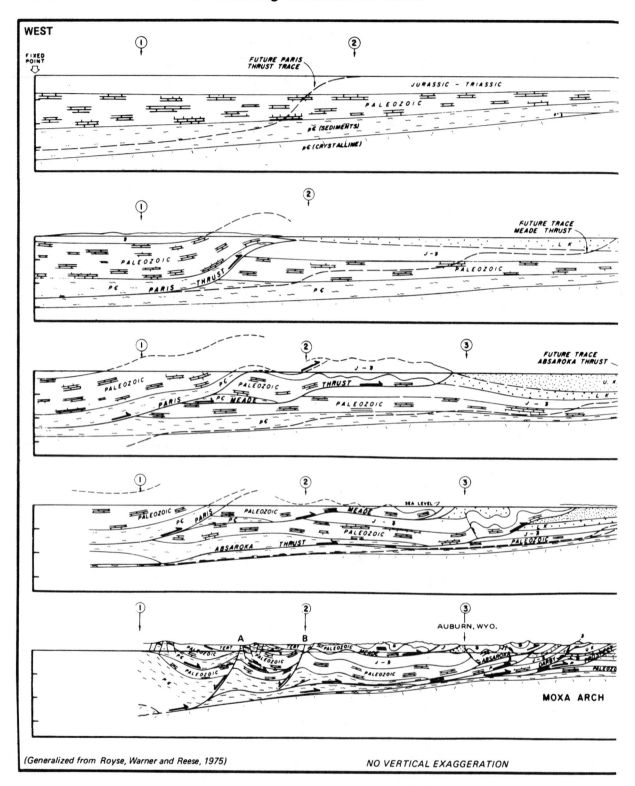

NO VERTICAL EXAGGERATION

FIG. 20—Sequential evolution of Idaho-Wyoming thrust-fold belt at latitude of Auburn and Pinedale, Wyoming (see Fig. 18). Detachments form progressively from west to east, as rigid basement slab is presumably underthrust from east to west (Lowell, 1977). Wedging action of underthrusting lifts earlier formed western thrust sheets (e.g., Paris thrust) so that older rocks are exposed. Reference numbers 1 through 7, which ultimately become present-day 30' longitude positions, demonstrate cumulation of shortening in sedimentary cover. A, B in bottom cross section are later listric normal faults. Generalized from Royse et al (1975).

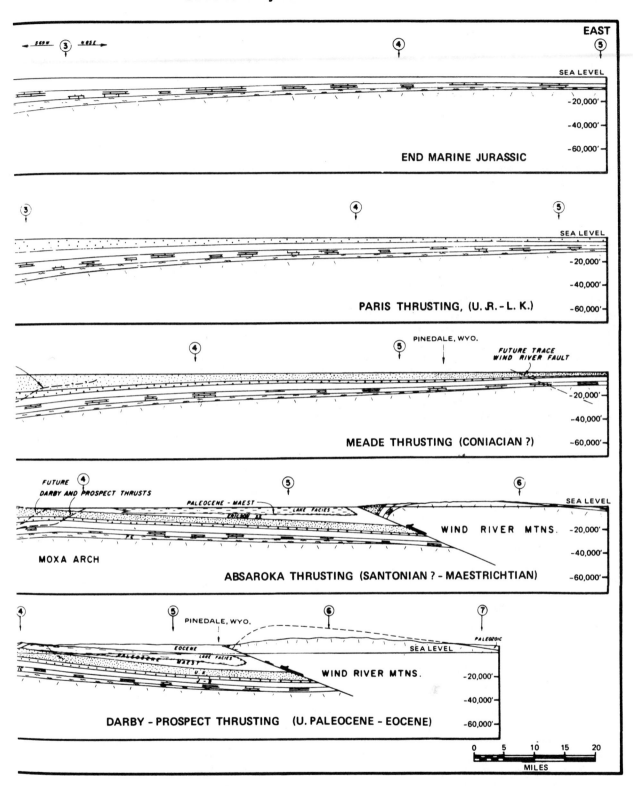

END MARINE JURASSIC

PARIS THRUSTING, (U. JR. - L. K.)

PINEDALE, WYO.

FUTURE TRACE WIND RIVER FAULT

MEADE THRUSTING (CONIACIAN ?)

FUTURE DARBY AND PROSPECT THRUSTS

PALEOCENE - MAEST

LAKE FACIES

WIND RIVER MTNS.

MOXA ARCH

ABSAROKA THRUSTING (SANTONIAN ? - MAESTRICHTIAN)

PINEDALE, WYO.

PALEOZOIC

EOCENE

LAKE FACIES

WIND RIVER MTNS.

DARBY - PROSPECT THRUSTING (U. PALEOCENE - EOCENE)

0 5 10 15 20

MILES

tile lithologies prevail, hanging-wall folds dominate. Such folds are characteristically asymmetric in the direction of relative transport of the upper plate. Structures generally are younger from the internal to the external side of the deformed belt (Fig. 20).

Anticlines with imbricate thrusts and broad folded thrust-fault structures (Fig. 19) occur across the breadth of a thrust-fold belt. The latter seem to be more common on the internal side and are usually located where faults step up in the sedimentary cover. Some thrust faults, backthrusted or directed against the direction of relative transport of the upper plate, bound belts at their fronts and are termed "delta structures."

The process of thrusting areally extensive but thin sheets of sedimentary rocks for long distances has always been an enigma of structural geology. Overthrusting by compression (or "pushing" at the rear of thrust sheets) does not resolve yield-strength and stress-transmission problems in moving relatively weak materials. Gravity sliding, which invokes body forces to overcome these problems, also is not a viable universal mechanism of thrust emplacement because: (1) necessary regions of tectonic denudation ·are rarely observed, (2) penetrative deformation of sedimentary cover in the internal parts of orogenic belts is not explained, (3) basement slopes required for sliding are usually in the wrong direction, (4) the chronologic sequence of thrusting is reversed, and (5) the laterally extensive nature of thrust-fold belts that follow plate boundaries for thousands of miles is not compatible with the local aprons that would be expected from simple sliding of sedimentary cover off of uplifted regions. Although gravity sliding is not a suitable mechanism for emplacement of thrust sheets on a regional scale, it has created smaller deformed belts where a slope on the base of the detached sedimentary section exists over a limited area, such as the Bearpaw Mountains of north-central Montana.

As an alternative to regional gravity sliding, a model of gravity spreading has been proposed (Bucher, 1956; Price, 1969, 1973; Price and Mountjoy, 1970; Elliott, 1976) in which only a slope on top of the sedimentary prism, and not on the actual thrust surfaces, is required to facilitate thrust-sheet emplacement. The structural results of gravity spreading can also be accommodated by the model of underthrusting described in the next paragraph and the two need not be mutually exclusive. However, at least in the Idaho-Wyoming thrust belt, thrusting began before any significant surface slope was attained (Fig. 20).

Large-scale plate movement provides a ratio-

nale for underthrusting between oceanic and continental lithospheric plates and within continental lithosphere. The model of stripping and stacking of sedimentary cover by crustal underthrusting in the landward slopes of oceanic trenches (Seely et al, 1974) appears to satisfy many of the critical relations in thrust-fold belts generally (Lowell, 1977). Underthrusting obviates problems of rock strength and stress transmission in that there need be no deformation of layered sedimentary rocks lying on rigid sialic basement until these rocks reach a zone of uncoupling (Fig. 20). The wedging action provided by the insertion of successive thrust sheets from below is also a mechanism for rotation and uplift of earlier formed thrusts and thereby accounts for the exposure of older rocks in the more internal parts of most orogenic belts. Finally, the age progression of detached thrusting from older to younger toward the underthrust plate is compatible with and even predicted by underthrusting (Fig. 20).

Thrust structures can be identified on seismic profiles by two main criteria: (1) shallow compressive folding above essentially undeformed reflectors and (2) stretches of backlimb dips that ultimately converge with underlying reflectors, generally in a direction opposite to tectonic transport (Fig. 19, left margin). In the latter way, thrusts do not demonstrate the persistent regional uplift present in many compressive fault blocks. The typical convergence of beds on one flank only reflects the asymmetry of the detachment. The bed geometry contrasts with the more symmetric convergence of reflectors apparent down both flanks of other detachments, such as flower structures (cf. Figs. 6 and 19) and salt pillows.

The decollement thrust-fold assemblage is further distinguishable from other thrust styles on the basis of distribution patterns. On a very large scale, many thrust-fold assemblages have great lateral persistence. Wrench-fault-associated assemblages have more finite, limited distributions. The distribution of compressive fault blocks is more random.

The sinuosity of most thrust-fold belts is also distinctive. On an intermediate scale, individual relay trends curve with this sinuosity, thus conforming to the outline of salients or reentrants. Thrusts and folds associated with wrench faults ideally trend obliquely to the outline of the deformed zone (cf. Figs. 3 and 5). On a more local scale, the relay patterns of thrust belts differ from both the en echelon trends of wrench-associated structures and the rhombic or cross-trended patterns of fault blocks (cf. Figs. 3 and 8).

Most petroleum in thrust-fold belts has been trapped in asymmetric, hanging-wall folds and the lead edges of thrust sheets (Fig. 1). Effective closures are commonly located at a relatively external part of a belt and at generally shallow to moderate (10,000 ft; 3,000 m) structural levels. Closures at these positions have been only slightly to moderately disrupted by faulting allowing reservoir continuity and migration pathways to remain essentially intact. Structures in this setting along the Zagros foldbelt of Iran account for more than three-fourths of the production from the world's thrust-fold belts. Enormous volumes of hydrocarbons are contained in fractured carbonate rocks on very high-amplitude, structurally simple, detached anticlines (Fig. 4). Significant production occurs also in the Canadian Rockies where traps are mainly at the lead edges of thrust sheets. Recently, important discoveries have been made in the Idaho-Wyoming thrust-fold belt; all traps found so far are hanging-wall folds.

Detached Normal-Fault Assemblages

Normal faults detached from basement occur in a wide variety of tectonic terranes but many are secondary elements of other structural styles. Most normal faults on the crests of folds and above diapirs (Fig. 1) are secondary faults. In other places, for example the Gulf Coast basin of Texas and Louisiana, detached normal faults occur in regional zones and are a dominant mode of deformation. Such zones commonly have unique structural assemblages and comprise a distinct structural style (Fig. 21).

The most common types of primary detached normal faults are down-to-basin faults that develop in ductile lithologies and have a close relation between displacement history and sedimentation. Interval thicknesses increase abruptly along the downthrown sides of these faults and there is a

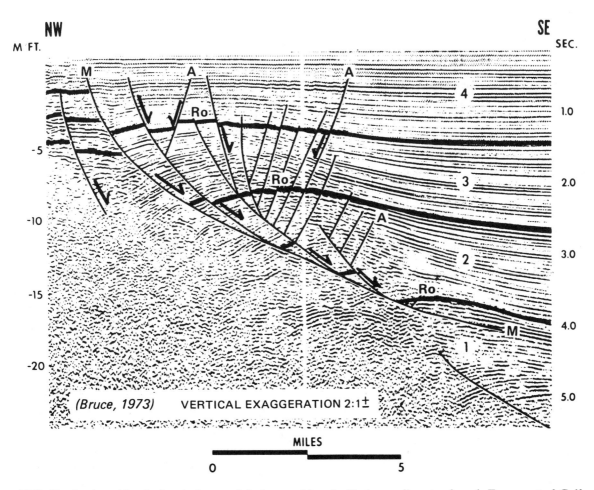

FIG. 21—Syndepositional, detached normal fault assemblage in Tertiary sediments of south Texas part of Gulf Coast basin. *M*, master down-to-basin synthetic fault. *A*, antithetic faults. *Ro*, rollover anticline with basinward migration of crest at depth. Northwest-dipping events beneath *M* fault would migrate to right to form part of rollover anticline. Some assemblages have only synthetic faults.

generally progressive downward increase in throw through much of the fault profile. These syndepositional faults also have been termed "regional faults," "growth faults" (Ocamb, 1961), "contemporaneous faults" (Hardin and Hardin, 1961), and "down-to-basin faults" (Cloos, 1968). These attributes, however, are obviously not restricted to just one type of fault and such terminology does not describe a discrete style.

Postdepositional detached normal faults, whose structural geometries are similar to the syndepositional types, have developed where a gradient has been present for sliding, for example, the topographically highest end of the Heart Mountain fault (Pierce, 1957). Detached normal faults, some of which flatten on the ramped portion of earlier thrust faults, are also present in thrust-fold belts (Fig. 20, faults A, B). Such faulting occurred in the Idaho-Wyoming thrust belt during a later episode of regional extension that was unrelated to thrusting but did affect deposition and preservation of section on the downthrown side of faults. Others apparently developed at the conclusion of thrusting and do not show contemporaneous expansion of sedimentary section. All of these postdepositional detached normal faults developed in relatively competent rocks. Their importance in petroleum provinces is minor compared to the Gulf Coast variety.

Syndepositional detached normal faults develop a dominant style in thick, generally regressive clastic sequences that have been built out into unconfined depositional sites, such as large deltas and the edges of continental shelves on passive continental margins (Todd and Mitchum, 1977, Fig. 9). There, rapidly deposited, semiconsolidated sediment is free to move basinward. In addition to the process of creep, faulting is often closely associated with the growth of salt (Lehner, 1969; Spindler, 1977) and shale structures (Bruce, 1973; Dailly, 1976). The largest displacement faults, termed "synthetic," are downthrown toward the basin. Some secondary faults on individual structures have up-to-basin displacements ("antithetic") and generally terminate against synthetic faults (Fig. 21). Other antithetic faults are primary, and have major displacements and structural characteristics similar to the regional synthetic faults (Spindler, 1977).

In profile view, many syndepositional faults dip subparallel with the flank of an underlying shale ridge or salt diapir, and may use the contacts with the overlying sediments for their detachment zones (Fig. 22). The faults are distinctively listric and can flatten downdip to merge parallel with bedding (Fig. 21) or dissipate within a chaotic slide-flow zone (J. S. Shelton, personal commun., 1978; Rider, 1978). In some places the depositional top of salt is a detachment surface (W. F. Bishop, 1973). Syndepositional faults can also terminate in an adjacent shale ridge (Fig. 22) or die out in the bottom of "depopods" between shale or salt diapirs (McGookey, 1975, Fig. 13). Compressional features have been reported at the toes or downdip ends which are sites of crumpling and telescoping of mobile section within the detached masses (Amery, 1969; Humphris, 1976; Dailly, 1976). Upward the displacement can decrease to nil owing to greater sediment filling of the downthrown side of the fault. Continued basin subsidence and late rejuvenation of some buried faults can reextend them to the surface, which has produced faults with both earlier growth and later postdepositional characteristics. The shallower, rejuvenated segments have displacement magnitudes that are constant throughout their profile, and formation thicknesses are the same in upthrown and downthrown blocks.

Detached normal-fault assemblages are complex (Fig. 21). The listric nature of many detached faults requires a backward rotation of beds into the steeper part of the fault (Cloos, 1968). In some places where rotation reverses basinward dip, a detached "rollover" anticline is formed on the downthrown side of and parallel with the master fault (top of Fig. 23). Downthrown beds are thickest adjacent to the fault and generally thin away from it in this example. Structural warping and consequent closure parallel with fault strike are thought to be the result of fault curvature and displacement variations along the trend of the fault that also cause changes in stratigraphic thickness within the downthrown block (Roux, 1977). The broad shallow rollover is commonly complicated at depth by increased faulting which segments the reverse dip into separated rotated fault blocks and increases the breadth of the counter-regional dip (Fig. 23, lower map). The overall rollover structure shifts basinward with depth but dip reversal can occur within different fault blocks at different locations and depths, and some blocks have no internal flexing (Fig. 24, at level of Kings Bayou sand).

The extensional tectonics in the Gulf Coast basin are further complicated by the presence of other kinds of normal faults. These include numerous local keystone faults above salt or shale domes and turtle structures, and more areally extensive detached "collapse faults" (Seglund, 1974) which circumscribe salt or shale withdrawal basins. Older basement-involved normal faults, thought to have facilitated opening of the Gulf of Mexico (Freeland and Dietz, 1971), may have been rejuvenated during prolonged subsidence of

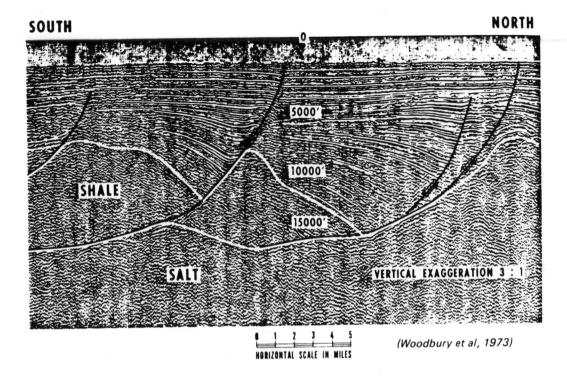

FIG. 22—Syndepositional detached normal faults in outer continental shelf of Gulf Coast basin, Louisiana and Texas. Faults sole out at salt-overburden contact identified by strong reflector and associated diffractions below 15,000 ft (4,500 m) in center of cross section. Presence of diapiric shale core (center) or salt diapir (right) below upthrown side of large growth fault is reported to be very common in this province (Woodbury et al, 1973).

the basin. If so, they could cut upward into the regional detached structures and could even trigger the shallower detached faulting.

Regional elements of the detached normal fault style in the United States Gulf Coast (Cohee, 1962) and in the Niger delta (Fig. 25) are present in wide, arcuate, parallel bands that trend subparallel with basin outline and/or depositional strike. Individual faults in map view are usually cuspate, concave in the direction of tectonic transport, and link to form an internally complex, anastomosing pattern. Fault zones become progressively younger in the direction of sedimentary progradation. Within a particular zone, however, the local timing for individual faults may reverse this sequence.

The detached normal-fault style in the Gulf Coast basin has been attributed to several causes. Large basement subsidence is required to provide a site for relatively rapid, thick deposition and to maintain a regional slope for basinward transport. Positive uplifts, other than those caused by diapirism, are lacking. Yorston and Weisser (1976) and W. F. Bishop (1973) have emphasized gravity sliding in several areas. Lehner (1969),

W. F. Bishop (1973), and Spindler (1977) have cited salt withdrawal in areas with significant salt. Differential sedimentary loading and compaction of shales have been emphasized by Bruce (1973); Roux (1977) associated the style with differential subsidence and compaction. Cloos (1968) has reproduced the basinwide fault pattern and characteristics of individual structural assemblages with regional gravity creep and local sliding models, respectively. Many authors (e.g., Bruce, 1973) have noted the importance of overpressured shales in the faulting process.

Yorston and Weisser (1976) suggested that two different detached normal-fault assemblages can result from downslope gravity sliding: peripheral breakaway faults, and local slope transport faults. The first occurs near the updip edge of the regional fault system. In the northeastern Gulf Coast, they utilize salt for their detachment. Movement in this region apparently was facilitated by sediment loading of downdropped blocks and consequent flowage of salt from beneath the loaded blocks (W. F. Bishop, 1973). The development of local slope transport faults is thought to involve successive detachment and basinward creep of

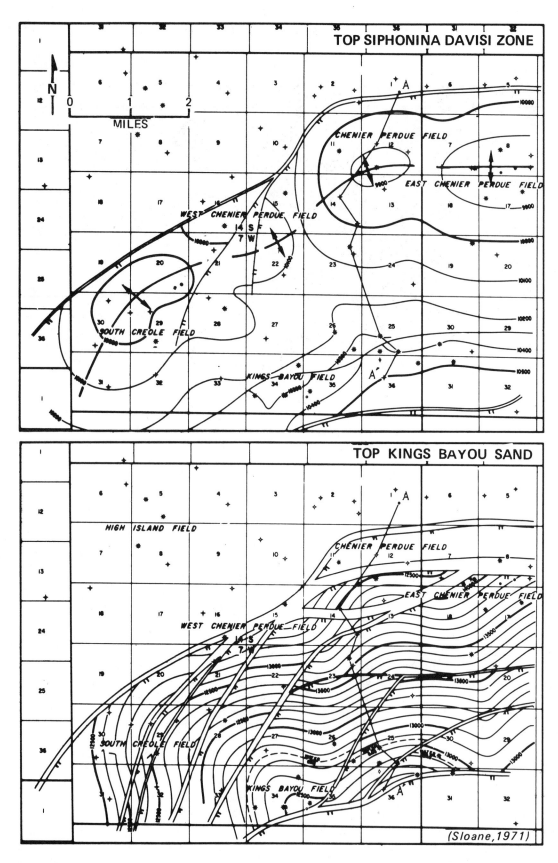

FIG. 23—Detail of detached normal fault pattern, southern Louisiana Gulf Coast. Parallelism of axis of rollover anticline on downthrown side of associated down-to-basin fault is apparent in top map. Rotation of regional dip into deeper faults increases landward-dipping segments that form individual fault traps in bottom map.

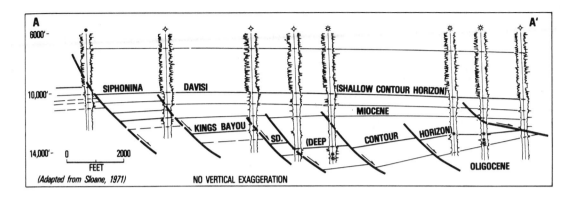

FIG. 24—North-south cross-section AA', across East Cameron area, United States Gulf Coast, showing normal fault pattern. Rollover has been removed by rotation of regional dip (Fig. 23). Location of cross section is shown on Figure 23.

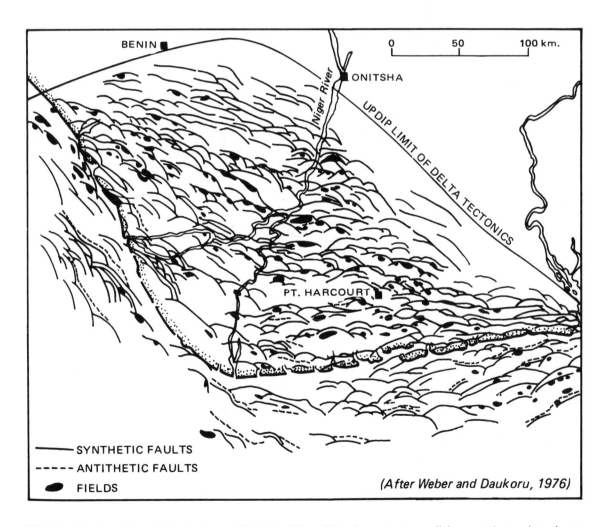

FIG. 25—Syndepositional detached normal faults in Niger delta demonstrate parallel crescentic trends and complex linkage of individual faults that are concave toward direction of transport.

deltaic lobes deposited on depositional slopes closer to the delta front.

In south Louisiana and the adjacent Gulf of Mexico, where salt is thick and extensive, salt mobility is an important factor in regional detached normal faulting. McGookey (1975, Figs. 12, 13) has documented syndepositional detached normal faults that occur as boundaries between rising salt diapirs and depopods that subside deeply into underlying salt. Regional detached normal-fault systems develop primarily on the seaward side of salt ridges or stocks. Synchroneity of salt withdrawal and displacements on a major up-to-basin antithetic fault has been demonstrated in this region (Spindler, 1977).

Salt is not usually present in south Texas, where Bruce (1973, Fig. 2) has attributed the regional detached faulting mostly to local differential loading of basin fill by sand depocenters. Faults have formed at the landward flanks of these depocenters and flatten with depth. The result is a basinward-dipping normal fault that utilizes the flank of a residual shale mass as its detachment zone. Basinward-dipping faults on the landward side of the shale ridges in south Texas are usually postdepositional and have relatively small displacements and no associated rollover anticline (Bruce, 1973, Figs. 3, 9). These faults mostly dip into the adjacent shale masses at relatively steep angles and do not have an associated glide plane.

Roux (1977) differed from other authors and proposed that the listric fault profile is in part at least a result of postfaulting compaction of shales and, to a lesser degree, sands; the fault profile is flattened appreciably as the original bed thicknesses are compacted during subsidence of the fault system. Roux (1977) stated that because fault segments exhibiting growth must have formed at surfaces of sedimentation these faults initially had steep dips throughout large segments of their profiles similar to faults present today at modern depositional surfaces. Rotation of beds into the listric faults during continued displacement may also rotate the adjoining fault segments and further decrease basinward fault dip.

The localization of anticlinal rollovers consistently within the downthrown block differs from the typical expression of all other structural styles in sedimentary basins. This characteristic and the direct evidence of detachment accompanying listric extensional faulting are key recognition criteria. Style similarities with basement-involved extensional block faulting do occur, particularly where low-side rotation of beds into the fault accompanies the block faulting, but this is far from universal in the latter style (Fig. 14). Further-

more, the rollover anticline has a flexed curvilinear profile that is different from the slablike profiles of most extensional fault blocks (cf. Figs. 11 and 21). In map view, syndepositional detached normal faults lack the grid aspects of fault blocks, and trap-door structures are absent. Tectonic settings and correlations across the fault are helpful in establishing the downthrown block for differentiation from overthrust anticlines, which also lie above listric fault surfaces and have similarly asymmetric profiles on seismic lines (cf. Figs. 19 and 21).

Syndepositional detached normal-fault assemblages of the type present in the Gulf Coast basin and Niger delta are by far the most prolific hydrocarbon-producing features of this style. Closures are mainly rollover anticlines and, less commonly, the upthrown sides of faults (Figs. 1, 24). Exploitation can be complex because basinward shift of structure and increased trap segmentation by faulting both occur with depth. A vertical stacking of quite different closures (fold versus fault) can result (Fig. 24).

Salt Structures

Salt has a capacity for mobile rise that is independent of tectonic forces and it is in this context that salt structures are considered a separate style (Fig. 26). Salt also can be mobilized by tectonic forces to form stocks, sills, dikelike masses, and other irregularly shaped bodies which are commonly injected along faults, folds, and fault-fold intersections. The Romanian Carpathians provide examples of salt features that have been clearly modified and reshaped by tectonic forces that formed the thrust-fold belt (Fig. 27). Differentiation between buoyant and tectonic salt structures is sometimes difficult. Distinctions can be arbitrary, for virtually every transition exists between the two types (O'Brien, 1968). Those demonstrably of tectonic origin are here considered as modifiers of the particular structural assemblage with which they have developed.

The preferred habitats of buoyant salt structures are divergent continental margins, their superimposed passive margin sags, and aborted rift systems (Table 1). The narrow restricted troughs, which form at the initiation of rifting, provide ideal settings for thick evaporite accumulations, particularly in hot, arid, low latitudes. Tectonic salt structures occur mostly where original salt basins (typically grabens) have been transported by large-scale plate movement into other, commonly compressive, deformational environments (H. R. Hopkins, personal commun., 1977).

Basin history and configuration control original limits, thicknesses, and composition of salt and overburden. In this way, they influence the general distribution (Fig. 28) and types of buoyant salt features. Basinward increase in original salt thickness is ideally expressed by the following progression of structures (Trusheim, 1960; Hughes, 1968): (1) downbends in sedimentary rocks above areas of salt withdrawal along the original depositional edge of the salt; (2) pillows of salt that can coalesce along strike to form low ridges; (3) anticlines and low domes of salt that can increase in relief to piercement features; and (4) walls of piercement salt.

In the coastal plain of the Gulf Coast and southward to the lower continental shelf the progression of structure is from (1) small, elongate, isolated spines to (2) progressively closer spaced, larger spines, and (3) almost continuous ridges and irregularly shaped large massifs (Stude, 1978). Growth sequences of offshore Gulf Coast structures are from initial salt swells to narrower ridges or stocks and finally to narrow salt chimneys or spines 2 to 4 n. mi in diameter (Lehner, 1969). Both the form of the salt features themselves and the effect on the adjacent sedimentary section of simultaneous lateral and vertical salt movement are important in understanding and exploring salt basins.

Distribution of salt domes in the absence of tectonic influence is typically in irregular clusters (Fig. 28, upper). In some places salt ridges or walls tend to be subparallel (Fig. 28, middle) and may be aligned with basin margins. Extensive massifs and ridges are closely spaced in a large area of the lower continental shelf of the Gulf Coast basin, but mostly lack discernible trend patterns (Martin, 1976).

A widely accepted mechanism for diapirism has been the movement of salt by buoyancy (Nettleton, 1934). Alternatives, which explain certain aspects of diapirism not resolved by the buoyancy model, have been proposed by Barton (1933), Bornhauser (1969), and R. S. Bishop (1978). According to R. S. Bishop (1978) initial salt movement occurs during progradation of sedimentary overburden across essentially exposed salt deposits on the basin floor (McGookey, 1975). The mobile evaporites flow both horizontally and vertically as waves or pillows in front of the advancing overburden. Flowage can occur before overburden is sufficiently thick to cause a density inversion between salt and sedimentary cover. The sedimentation rate, progradation rate, regional dip, sediment density, and thickness of both the mobile substrate and overburden all affect the size and shape of incipient salt structures. Slow, even accumulation of overburden results in salt structures with random distribution. Focused

loading concentrates structures in specific sites and can align them parallel with thick areas of overburden (R. S. Bishop, 1978, Fig. 11).

As overburden buries the initial salt structure, horizontal flow is restricted but salt buoyancy may develop and vertical growth may continue as an unrestrained extrusion to the seafloor. Extrusion alleviates the space problems that arise where the rising diapir is required to force aside its overburden.

If the density inversion (buoyancy) disappears, extrusion will cease and diapirism continues only as a buried intrusion. The cause of diapirism at this stage may be selective dissolution and removal of salt by migrating subsurface fluids (upsetting mass equilibrium) or differential loading (e.g., salt flowing out from under a preferentially loaded rim syncline). According to R. S. Bishop (1978) growth of many Gulf Coast diapirs alternated between extrusion and intrusion.

Continuous vertical diapirism can be restrained by significant increases in overburden thickness or strength. Growth then becomes sporadic and occurs only by fracturing and forceful injection of salt into the overburden, which occur only when the pressure within the diapir, arising from sediment loading, becomes sufficient to rupture the overburden (R. S. Bishop, 1978). Deeply buried diapirs may expand laterally prior to vertically fracturing their overburden. When fracturing does not occur, the diapir ceases to grow, even if evaporite supply is sufficient.

The vertical rise of salt cores causes localized stretching and detached normal keystone faulting of the sedimentary overburden. Faulting can be highly complex and patterns are mostly radial and transverse. Stude (1978) has documented the faulting sequence at the South Timbalier Block 54 salt dome in offshore Louisiana, "The older graben faults moved farther from the crest of the salt and became inactive (as the dome continued to pierce through the overburden and widen). These faults were replaced by younger faults over the salt crest." Currie (1956) has modeled salt-dome growth and demonstrated similar timing but other sequences are also possible (Stude, 1978).

Distinctive profile characteristics facilitate identification of salt structures on seismic lines. On Figure 26, residual salt highs are considered remnants of the original salt bed left by incomplete evacuation of salt. Turtle structures are inverted sediment-filled lows that formed above areas of early salt withdrawal. Rim synclines reflect the growth history of the associated salt dome.

The most prolific salt-related traps are those that are not pierced by evaporites at reservoir levels (Fig. 1). These include reservoirs concordant

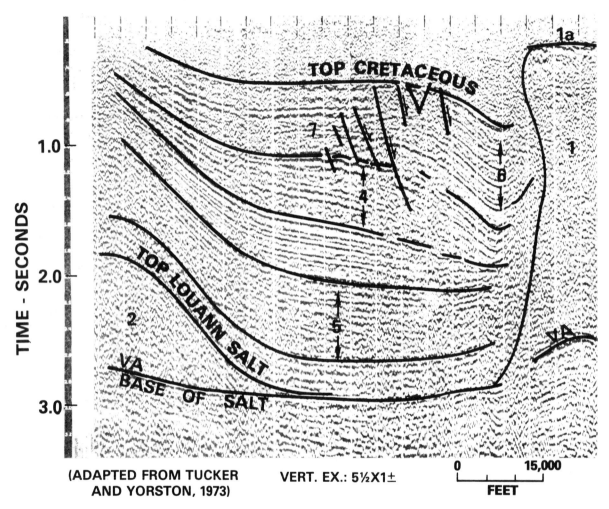

FIG. 26—Seismic section across Hainesville dome (center), east Texas, with structures as indicated. Higher velocity of salt relative to surrounding rocks at velocity anomalies at left and center results in "pull-up" of time section. Velocity anomaly at right shows as depression, because salt here has lower velocity than laterally adjacent rocks.

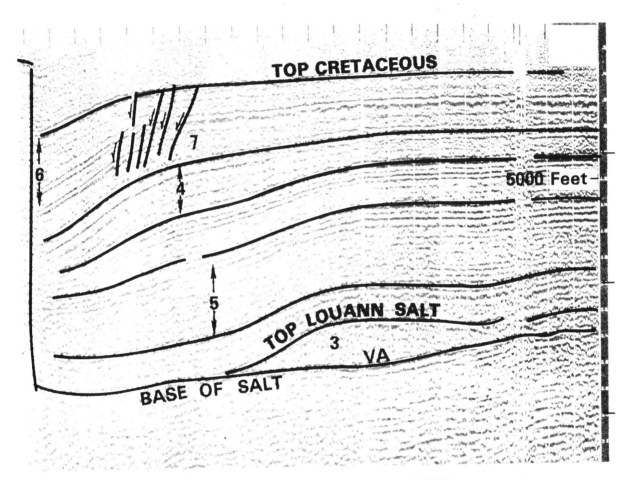

STRUCTURES: SALT RELATED

1. PIERCEMENT SALT DOME
   (1a. CAP ROCK)
2. NON - PIERCEMENT SALT ANTICLINE
3. RESIDUAL SALT HIGH
4. TILTED TURTLE STRUCTURE

5. PRIMARY SALT WITHDRAWAL
   SYNCLINE
6. RIM SYNCLINE
7. NORMAL FAULTS
VA - VELOCITY ANOMALY

above salt pillows and anticlines. Much less common are turtle structures and passive drape closures over residual highs. Other significant traps are formed by reservoirs turned up against and sealed by salt, in places beneath salt overhangs. The two basic trap types can be further modified by stratigraphic pinchouts and unconformities. Moreover, the crestal normal faulting of the sedimentary section in response to arching above salt features can disrupt and segment closures (Gussow, 1968; Lafayette Geol. Soc., 1970).

Shale Structures

Shale structures (Fig. 29) range from nondiapiric residual shale masses such as those that underlie and can help motivate detached normal faults (Fig. 22), to highly mobile, diapiric mud

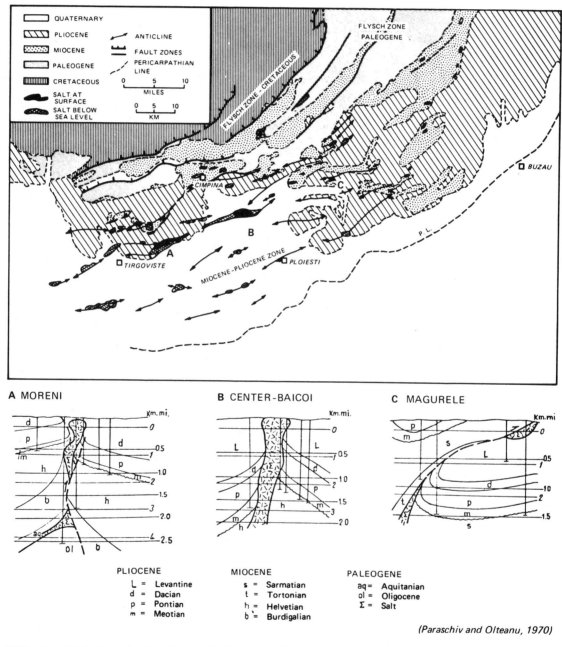

PLIOCENE
L = Levantine
d = Dacian
p = Pontian
m = Meotian

MIOCENE
s = Sarmatian
t = Tortonian
h = Helvetian
b = Burdigalian

PALEOGENE
aq = Aquitanian
ol = Oligocene
Σ = Salt

(Paraschiv and Olteanu, 1970)

FIG. 27—Salt features in Ploesti area of Romanian Carpathian thrust-fold belt are influenced by tectonic forces, as attested by diapir, **B** and by diapirs localized along fault surface, **A, C.**

38

lumps. They are not limited to any particular structural or plate-tectonic environment, as attested by their widespread occurrence. Shale structures, however, are perhaps best known in thick, modern clastic deltas (Table 1; R. S. Bishop, personal commun., 1977) where high depositional rates create thick sequences that retard water expulsion from fine mud fractions. Muds are consequently undercompacted, have abnormally high pore-fluid pressures, and thus can react

in a highly mobile buoyant manner.

In south Texas, shale ridges develop by differential compaction resulting from differential loading and shale diagenesis. Ridges here are aligned subparallel with regional strike, and are commonly tens of miles long, up to 25 mi (40 km) wide, and 10,000 ft (3,000 m) high (Bruce, 1973). In Trinidad, shale diapirism appears to be tectonically motivated, with diapirs present along major strike-slip faults. Shale structures can also

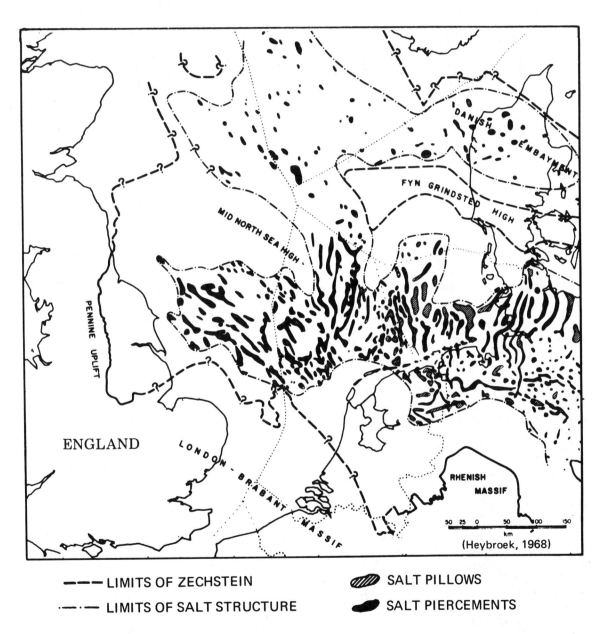

LIMITS OF ZECHSTEIN

LIMITS OF SALT STRUCTURE

SALT PILLOWS

SALT PIERCEMENTS

FIG. 28—Distribution of primarily buoyant salt features in southern North Sea and adjacent parts of Netherlands, northwest Germany, and northern Denmark as known in 1968. Randomly trended, clustered structures prevail in north part; trends are more linear and locally parallel in south.

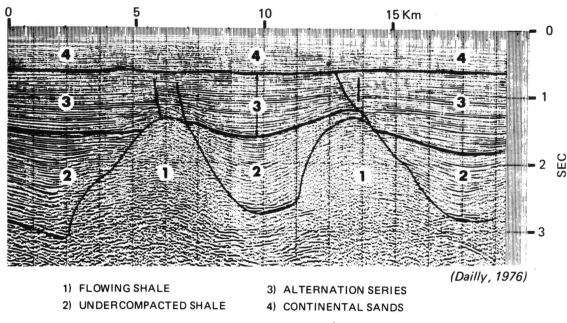

(Dailly, 1976)

1) FLOWING SHALE   3) ALTERNATION SERIES

2) UNDERCOMPACTED SHALE   4) CONTINENTAL SANDS

VERTICAL EXAGGERATION 2:1±

FIG. 29—Seismic expression of shale diapirs within deltaic sequence.

result from crowding and shortening of the sedimentary cover during downslope creep as at the toes of syndepositional detached normal-fault assemblages.

Morphologic characteristics of salt structures are also shared by most abnormally pressured shale structures formed in passive tectonic environments. Indeed, the two are known to occur together to form compound domal masses (Atwater and Forman, 1959; Musgrave and Hicks, 1968). On reflection seismic sections, both salt and shale flowage usually are manifested by reflection cutouts (Musgrave and Hicks, 1968).

Differentiation of salt and shale is possible when acoustic velocity information is available inasmuch as shale flowage masses typically have only about half the velocity of salt. Moreover, shale features seem to lack the well-defined rim synclines frequently associated with salt bodies, probably because salt flowage can occur at substantial distances from a developing dome, whereas shale flowage involves less lateral movement. Slower flow rates over longer periods in salt-structure growth may also be important factors.

Greater shale compactibility may be a main cause of differences in the development of shale and salt structures according to R. S. Bishop (1978). Salt diapirs can pierce thicker overburden and rise to greater heights. Bishop (1978) wrote:

During burial, the ratio of salt to the overburden density increases, and the salt diapir may stay near the surface and remain vertically unconstrained throughout sedimentation. In contrast, the ratio of shale density to its overburden may decrease during burial (due to shale compaction), thereby causing intrusion at a slower rate than sedimentation. This results in deeper burial of the shale diapirs; the consequence is vertical constraint and perhaps cessation of the diapirism.

Closures related to shale masses have not trapped nearly so much petroleum as that attributable to most other structural styles. O'Brien (1968), however, has cited numerous shale domes and features associated with mud volcanoes as significant sources of hydrocarbons.

CONCLUSIONS AND APPLICATIONS

In general terms, the common geologic structures of sedimentary basins are distributed in definable, interrelated suites that constitute styles. Their preferred tectonic habitats result broadly from lithospheric plate movements but are also controlled by other factors. The type of deformation and tectonic history of specific regions can only be determined by correct differentiation of styles. The style approach can then provide a first approximation of the kinds of hydrocarbon traps and the general distribution patterns to be expected during exploration.

Variations in Structural Style

Many factors contribute to the development of individual closures and can influence the expression of structural style or control its areal limits. Of prime importance are the physical properties of the deformed region, such as thickness and lithology of the sedimentary cover, and mobility and preexisting structural fabric of the basement. Mobile lithologies in the basin fill, such as evaporites and undercompacted shales, directly control the occurrence of salt and shale structures and are critical in the development of syndepositional detached normal faults. Greater basement ductility can enhance or even control the presence of folds in basement-involved styles (e.g., west side of San Joaquin Valley, Harding, 1976) although a lack of ductility may result in a brittle structural response (e.g., Wyoming foreland). Older basement fabric can localize fault-block closures.

Style expression is also influenced by tectonic variables such as intensity, rate, duration, and stage of deformation (Fig. 5), and differences of applied stress. In addition, sedimentation contemporaneous with deformation and the structural level of available control can determine the structural geometry observed. Gentle rejuvenation of dip-slip faulting concurrent with sedimentation, for example, results in drape flexures of greater amplitude that become the shallow expression of block faulting. In each of the described situations, however, the basic style category is still discernible.

Diversity of Styles Occurrence

For convenience, we have described each style individually, but actual occurrence of structures frequently indicates transition between styles. Detached thrust systems, for example, can ultimately root downdip in basement thrusts and in places merge along strike with blocklike compressive features. In other settings, syndepositional detached normal faults change downdip into detached thrust features. Transitions in the wrench style from structures that are dominantly compressive to those that are extensional can occur along a single zone.

The structures of a sedimentary basin commonly include several styles and transitions between styles and these can have either separate or overlapped distributions. In the San Joaquin Valley (Harding, 1976, Fig. 1) a wrench-fault assemblage dominates the mobile southwest flank, a salient of locally detached thrust faults predominates at the south end where right- and left-lateral

strike-slip faults converge, and basement warps with secondary normal faults dominate the stable northeast flank.

Some extensional settings can combine basement fault blocks with salt and/or shale structures and syndepositional detached normal faults, often superimposed at different levels (Fig. 22, in part). Compressive settings can mix detached thrusts and folds, basement thrusts or compressive block faults, wrench-related features, and, with the presence of mobile substrata, tectonic salt and/or shale structures (Fig. 3). Each basin has a unique history and styles can be further complicated by the overprints of fundamentally different stress systems through time.

No style is universally present within its preferred habitat, nor is a style always consistently developed where it is present. Foreland settings present the greatest extremes. In contrast with the Wyoming foreland, the Alberta basement foreland of the Canadian Rockies on strike has very few structures (King, 1969). Late Paleozoic structures along the foreland of the approximately coeval Ouachita-Llanoria fold-thrust belt are widespread and highly varied (King, 1969). Included are (1) normal block faults on the stable north flank of the Arkoma basin, (2) wrench faults along the trend of the Ardmore-Anadarko basin, (3) broad basement arch at the Llano uplift, and (4) dominantly compressive fault blocks in the Permian basin. Obviously, forces of plate convergence can be manifested in different ways. Earlier structures apparently controlled the style and/or distribution of some of this foreland deformation, but still other factors may have been important.

Identification of Structural Style

Two major factors—variations in style expression and complications in style occurrence—can make differentiation of styles exceedingly difficult; this may be especially true of convergent margin settings and their orogenic belts.

Many structural patterns can be developed in several ways, and "false" patterns arising from unique circumstances must be ruled out. For example, an overprint of faults trending obliquely to an earlier thrust-fold belt could result in a misleading en echelon appearance. Linear strike-slip fault sets superimposed on each other could result in a fault-block appearance. A pattern usually typical of a particular tectonic regime can be simply the result of unique physical properties in the deformed terrane such as a strong basement grain or anisotropy within the sedimentary cover. Determination of several recognition criteria is often essential. Consideration of alternatives is always necessary and requires equal familiarization with all styles.

REFERENCES CITED

Amery, G. B., 1969, Structure of Sigsbee Scarp, Gulf of Mexico: AAPG Bull., v. 58, p. 2480-2482.

Anderson, E. M., 1942, The dynamics of faulting and dyke formation with application to Britain: Edinburgh and London, Oliver Boyd, 206 p. (rev. ed. 1951).

Atwater, G. I., and M. J. Forman, 1959, Nature of growth of southern Louisiana salt domes and its effect on petroleum accumulation: AAPG Bull., v. 43, p. 2592-2622.

Badgley, P. C., 1965, Structural and tectonic principles: New York, Harper and Row, 521 p.

Baker, B. H., P. A. Mohr, and L. A. J. Williams, 1972, Geology of the Eastern rift system of Africa: Geol. Soc. America Spec. Paper 136, 67 p.

Bally, A. W., 1975, A geodynamic scenario for hydrocarbon occurrences: 9th World Petroleum Cong. Proc., v. 2, p. 33-44.

——— D. L. Gordy, and G. A. Stewart, 1966, Structure, seismic data, and orogenic evolution of southern Canadian Rocky Mountains: Bull. Canadian Petroleum Geology, v. 14, p. 337-381.

Barton, D. C., 1933, Mechanics of formation of salt domes, with special reference to Gulf Coast salt domes of Texas and Louisiana: AAPG Bull., v. 17, p. 1025-1083.

Beloussov, V. V., 1959, Types of folding and their formation: Internat. Geol. Rev., v. 1, p. 1-21.

Bishop, R. S., 1978, Mechanism for emplacement of piercement diapirs: AAPG Bull., v. 62, p. 1561-1583.

Bishop, W. F., 1973, Late Jurassic contemporaneous faults in north Louisiana and south Arkansas: AAPG Bull., v. 57, p. 858-877.

Bornhauser, M., 1969, Geology of Day dome (Madison County, Texas)—a study of salt emplacement: AAPG Bull., v. 53, p. 1411-1420.

Bruce, C. H., 1973, Pressure shale and related sediment deformation—mechanisms for development of regional contemporaneous faults: AAPG Bull., v. 57, p. 878-886.

Bucher, W. H., 1956, Role of gravity in orogenesis: Geol. Soc. America Bull., v. 67, p. 1295-1318.

Burchfiel, B. C., and G. A. Davis, 1975, Nature and controls of Cordilleran orogenesis, western United States—extensions of an earlier synthesis: Am. Jour. Sci., v. 275-A, p. 363-396.

Cloos, E., 1968, Experimental analysis of Gulf Coast fracture patterns: AAPG Bull., v. 52, p. 420-444.

Cohee, G. V., 1962, Tectonic map of the United States: U.S. Geol. Survey and AAPG, scale, 1:2,500,000.

Currie, J. B., 1956, Role of concurrent deposition and deformation of sediments in development of salt-dome graben structures: AAPG Bull., v. 40, p. 1-16.

Dailly, G. C., 1976, A possible mechanism relating progradation, growth faulting, clay diapirism and overthrusting in a regressive sequence of sediments: Bull. Canadian Petroleum Geology, v. 24, p. 92-116.

Darton, N. H., 1906, Cloud Peak-Fort McKinney folio,

Wyoming: U.S. Geol. Survey Geol. Atlas, Folio 142, 16 p.

Dewey, J. F., and J. M. Bird, 1970, Mountain belts and the new global tectonics: Jour. Geophys. Research, v. 75, p. 2625-2647.

Dickinson, W. R., 1974, Plate tectonics and sedimentation, in Tectonics and sedimentation: SEPM Spec. Pub. 22, p. 1-27.

——— and W. S. Snyder, 1978, Plate tectonics of the Laramide orogeny, in Laramide folding associated with basement block faulting in western United States: Geol. Soc. America Mem. 151, p. 355-366.

Dobbin, C. E., and C. E. Erdman, 1955, Structure contour map of the Montana plains: U.S. Geol. Survey Oil and Gas Inv. Map OM 178A, scale, 1:500,000.

Elam, J. G., 1969, The tectonic style in the Permian basin and its relationship to cyclicity, in Cyclic sedimentation in the Permian basin: West Texas Geol. Soc. Pub. 69-56, p. 55-79.

Elliott, D., 1976, The motion of thrust sheets: Jour. Geophys. Research, v. 81, p. 949-963.

El-Tarabili, E., and N. Adawy, 1972, Geologic history of Nukhul-Baba area, Gulf of Suez, Sinai, Egypt: AAPG Bull., v. 56, p. 882-902.

Fitch, T. J., 1972, Plate convergence, transcurrent faults, and internal deformation adjacent to southeast Asia and the western Pacific: Jour. Geophys. Research, v. 77, p. 4432-4460.

Foose, R. M., D. U. Wise, and G. S. Garbarini, 1961, Structural geology of the Beartooth Mountains, Montana and Wyoming: Geol. Soc. America Bull., v. 72, p. 1143-1172.

Freeland, G. L., and R. S. Dietz, 1971, Plate tectonic evolution of Caribbean-Gulf of Mexico region: Nature, v. 232, p. 20-23.

Freund, R., and A. M. Merzer, 1976, The formation of rift valleys and their zigzag fault patterns: Geol. Mag., v. 113, p. 561-568.

Gansser, A., 1974, Ophiolitic melange, a world-wide problem on Tethyan examples: Eclogae Geol. Helvetiae, v. 67, p. 479-507.

Ghignone, J. I., and G. de Andrade, 1970, General geology and major oil fields of Recôncavo basin, Brazil, in Geology of giant petroleum fields: AAPG Mem. 14, p. 337-358.

Green, A. R., 1977, The evolution of the earth's crust, in The earth's crust—its nature and physical properties: Am. Geophys. Union Geophys. Mon. 20, p. 1-17.

Gussow, W. C., 1968, Salt diapirism: importance of temperature, and energy source of emplacement, in Diapirism and diapirs: AAPG Mem. 8, p. 16-52.

Halbouty, M. T., 1967, Salt domes—Gulf region, United States and Mexico: Houston, Gulf Pub. Co., 425 p.

Hansen, A. R., 1972, The Williston basin, in Geologic atlas of the Rocky Mountain region, United States of America: Rocky Mtn. Assoc. Geologists, p. 265-269.

Hardin, F. R., and G. C. Hardin, Jr., 1961, Contemporaneous normal faults of Gulf Coast and their relation to flexures: AAPG Bull., v. 45, p. 238-248.

Harding, T. P., 1973, Newport-Inglewood trend, California—an example of wrenching style of deformation: AAPG Bull., v. 57, p. 97-116.

—— 1974, Petroleum traps associated with wrench faults: AAPG Bull., v. 58, p. 1290-1304.

—— 1976, Tectonic significance and hydrocarbon trapping consequences of sequential folding synchronous with San Andreas faulting, San Joaquin Valley, California: AAPG Bull., v. 60, p. 356-378.

Harland, W. B., 1971, Tectonic transpression in Caledonian Spitzbergen: Geol. Mag., v. 108, p. 27-42.

Heybroek, P., 1968, Geologische Waarnemingen op de Noordzee: Geologie en Mijnbouw, v. 47, no. 3, p. 209-210 (illustrations with no. 4).

Hoppin, R. A., and J. C. Palmquist, 1965, Basement influence on later deformations, the problem, techniques of investigation, and examples from Bighorn Mountains, Wyoming: AAPG Bull., v. 49, p. 993-1003.

Howard, J. H., 1966, Structural development of the Williams Range thrust, Colorado: Geol. Soc. America Bull., v. 77, p. 1247-1264.

Hughes, D. J., 1968, Salt tectonics as related to several Smackover fields along the northeast rim of the Gulf of Mexico basin: Gulf Coast Assoc. Geol. Socs. Trans., v. 18, p. 320-330.

Humphris, C. C., Jr., 1976, Salt movement on continental slope, northern Gulf of Mexico (abs.): AAPG Bull., v. 60, p. 683.

Illies, J. H., 1970, Graben tectonics as related to crust-mantle interaction, in J. H. Illies, and St. Muellar, eds., Graben problems: Stuttgart, Schweizerbart, p. 4-27.

King, P. B., 1965, Geology of the Sierra Diablo region, Texas: U.S. Geol. Survey Prof. Paper 480, 185 p.

—— 1969, Tectonic map of North America, scale, 1:5,000,000: U.S. Geol. Survey.

Lafayette Geological Society, 1970, Typical oil and gas fields of southwest Louisiana, v. 2, variously paged.

Lehner, P., 1969, Salt tectonics and Pleistocene stratigraphy on continental slope of northern Gulf of Mexico: AAPG Bull., v. 53, p. 2431-2479.

—— and P. A. C. DeRuiter, 1977, Structural history of Atlantic margin of Africa: AAPG Bull., v. 61, p. 961-981.

Lowell, J. D., 1972, Spitsbergen Tertiary orogenic belt and the Spitsbergen fracture zone: Geol. Soc. America Bull., v. 83, p. 3091-3102.

—— 1974, Plate tectonics and foreland basement deformation: Geology, v. 2, p. 275-278.

—— 1977, Underthrusting origin for thrust-fold belts with applications to the Idaho-Wyoming belt: Wyoming Geol. Assoc. Guidebook, 29th Ann. Field Conf., p. 449-455.

—— and G. J. Genik, 1972, Sea-floor spreading and structural evolution of southern Red Sea: AAPG Bull., v. 56, p. 247-259.

—— et al, 1975, Petroleum and plate tectonics of the southern Red Sea, in A. G. Fisher and S. Judson, eds., Petroleum and global tectonics: Princeton Univ. Press, p. 129-153.

Mallory, W. W., ed., 1972, Geologic atlas of the Rocky Mountain region, United States of America: Rocky Mtn. Assoc. Geologists, 331 p.

Martin, R. G., 1976, Geological framework of northern and eastern continental margins, Gulf of Mexico, in Beyond the shelf break: AAPG Marine Geology Comm. Short Course, v. 2, p. A-1-A-28; also in AAPG Studies in Geology, no. 7, 1978, p. 21-42.

McGookey, D. P., 1975, Gulf Coast Cenozoic sediments and structure, an excellent example of extra-continental sedimentation: Gulf Coast Assoc. Geol. Socs. Trans., v. 25, p. 104-120.

Moody, J. D., 1973, Petroleum exploration aspects of wrench-fault tectonics: AAPG Bull., v. 57, p. 449-476.

—— and M. J. Hill, 1956, Wrench-fault tectonics: Geol. Soc. America Bull., v. 67, p. 1207-1246.

Moore, D. G., and E. C. Buffington, 1968, Transform faulting and growth of the Gulf of California since the late Pliocene: Science, v. 161, p. 1238-1241.

Musgrave, A. W., and W. G. Hicks, 1968, Outlining shale masses by geophysical methods, in Diapirism and diapirs: AAPG Mem. 8, p. 122-136.

Nettleton, L. L., 1934, Fluid mechanics of salt domes: AAPG Bull., v. 18, p. 1125-1204.

O'Brien, G. D., 1968, Survey of diapirs and diapirism, in Diapirism and diapirs: AAPG Mem. 8, p. 1-9.

Ocamb, R. D., 1961, Growth faults of south Louisiana: Gulf Coast Assoc. Geol. Socs. Trans., v. 11, p. 139-173.

Osborne, W. C., 1957, Keystone field, in Occurrence of oil and gas in west Texas: Texas Univ. Pub. 5716, p. 156-171.

Paraschiv, D., and Gh. Olteanu, 1970, Oil fields in Mio-Pliocene zone of eastern Carpathians (District of Ploiesti), in Geology of giant petroleum fields: AAPG Mem. 14, p. 399-427.

Pennington, J. J., 1975, Geology of Argyll field, in A. W. Woodland, ed., Petroleum and the continental shelf of northwest Europe: New York, John Wiley & Sons, p. 285-297.

Pierce, W. G., 1957, Heart Mountain and South Fork detachment thrusts of Wyoming: AAPG Bull., v. 41, p. 591-626.

—— and W. H. Nelson, 1968, Geologic map of the Pat O'Hara Mountain quadrangle, Park County, Wyoming: U.S. Geol. Survey Quad. Map GQ-755, scale, 1:62,500.

—— D. A. Andrews, and J. K. Keroher, 1947, Structure contour map of the Big Horn basin, Wyoming and Montana: U.S. Geol. Survey Oil and Gas Inv. Prelim. Map 74, scale approx. 1 in. = 3 mi.

Ponte, F. C., J. D. R. Fonseca, and R. G. Morales, 1977, Petroleum geology of eastern Brazil continental margin: AAPG Bull., v. 61, p. 1470-1482.

Price, R. A., 1969, The southern Canadian Rockies and the role of gravity in low-angle thrusting, foreland folding, and the evolution of migrating foredeeps (abs.): Geol. Soc. America Abs. with Programs, pt. 7, p. 284-286.

—— 1973, Large-scale gravitational flow of supracrustal rocks, southern Canadian Rockies, in K. A. deJong and R. Scholten, eds., Gravity and tectonics: New York, Wiley-Interscience Pub., p. 491-502.

—— and E. W. Mountjoy, 1970, Geologic structure of the Canadian Rocky Mountains between Bow and Athabasca Rivers—a progress report: Geol. Assoc. Canada Spec. Paper 6, 25 p.

Profett, J. M., Jr., 1977, Cenozoic geology of the

Yerington district, Nevada, and implications for the nature and origin of Basin and Range faulting: Geol. Soc. America Bull., v. 88, p. 247-266.

Prucha, J. J., J. A. Graham, and R. P. Nicholson, 1965, Basement controlled deformation in Wyoming province of Rocky Mountains foreland: AAPG Bull., v. 49, p. 966-992.

Reches, Z., 1978a, Development of monoclines, pt. 1, Structure of the Palisades Creek branch of the East Kaibab monocline, Grand Canyon, Arizona, in Laramide folding associated with basement block faulting in western United States: Geol. Soc. America Mem. 151, p. 235-271.

———— 1978b, Analysis of faulting in three-dimensional strain field: Tectonophysics, v. 47, p. 109-129.

Rider, M. H., 1978, Growth faults in western Ireland: AAPG Bull., v. 62, p. 2191-2213.

Robson, D. A., 1971, The structure of the Gulf of Suez (clysmic) rift, with special reference to the eastern side: Jour. Geol. Soc., v. 127, p. 247-276.

Roux, W. F., Jr., 1977, The development of growth fault structures: AAPG Structural Geology School Course Notes, 33 p.

Royse, F., Jr., M. A. Warner, and D. C. Reese, 1975, Thrust belt structural geometry and related stratigraphic problems, Wyoming-Idaho-northern Utah, in Symposium on deep drilling frontiers in the central Rocky Mountains: Rocky Mt. Assoc. Geologists, p. 41-54.

Seely, D. R., P. R. Vail, and G. G. Walton, 1974, Trench slope model, in C. A. Burk and C. C. Drake, eds., The geology of continental margins: New York, Springer-Verlag, p. 249-260.

Seglund, J. A., 1974, Collapse-fault systems of Louisiana Gulf Coast: AAPG Bull., v. 58, p. 2389-2397.

Sloane, B. J., 1971, Recent developments in the Miocene *Planulina* gas trend of south Louisiana: Gulf Coast Assoc. Geol. Socs. Trans., v. 21, p. 199-210.

Smith, R. B., D. R. Mabey, and G. P. Eaton, 1976, Penrose Conference report—Regional geophysics and tectonics of the intermountain west: Geology, v. 4, p. 437-438.

Smithson, S. B., et al, 1978, Nature of the Wind River thrust, Wyoming, from COCORP deep-reflection data and from gravity data: Geology, v. 6, p. 648-652.

Spindler, W. M., 1977, Structure and stratigraphy of a small Plio-Pleistocene depocenter, Louisiana continental shelf: Gulf Coast Assoc Geol. Socs. Trans., v. 27, p. 180-196.

Stearns, D. W., 1975, Laramide basement deformation in the Bighorn basin—the controlling factor for structures in the layered rocks: Wyoming Geol. Assoc. 27th Ann. Field Conf., p. 149-158.

Stewart, J. H., 1972, Initial deposits in the Cordilleran geosyncline: evidence of a Late Precambrian (>850 m.y.) continental separation: Geol. Soc. America

Bull., v. 83, p. 1345-1360.

Stöcklin, J., and M. H. Nabavi, 1973, Tectonic map of Iran: Iran Geol. Survey, scale 1:2,500,000.

Stone, D. S., 1970, Wrench faulting and Rocky Mountains tectonics: The Mountain Geologist, v. 6, no. 2, p. 67-69.

Stude, G. R., 1978, Depositional environments of the Gulf of Mexico South Timbalier Block 54—salt dome and salt dome growth models: Gulf Coast Assoc. Geol. Socs. Trans., v. 28, p. 627-646.

Surlyk, F., 1977, Stratigraphy, tectonics, and palaeogeography of the Jurassic sediments of the areas north of Kong Oscars Fjord, East Greenland: Grønlands Geol. Undersøgelse Bull. 123, 56 p.

Sykes, L. R., and M. L. Sbar, 1974, Focal mechanism solutions of intraplate earthquakes and stresses in the lithosphere, in L. Kristjansson, ed., Geodynamics of Iceland and the North Atlantic area: Dordrecht, Holland, D. Reidel Pub. Co., p. 207-224.

Sylvester, A. G., and R. R. Smith, 1976, Tectonic transpression and basement-controlled deformation in San Andreas fault zone, Salton trough, California: AAPG Bull., v. 60, p. 2081-2102.

Tapponnier, P., and P. Molnar, 1976, Slip-line theory and large-scale plate tectonics: Nature, v. 264, p. 319-324.

Todd, R. G., and R. M. Mitchum, Jr., 1977, Seismic stratigraphy and global changes of sea level, part 8: identification of Upper Triassic and Lower Cretaceous seismic sequences in Gulf of Mexico and offshore West Africa, in Seismic stratigraphy—applications to hydrocarbon exploration: AAPG Mem. 26, p. 145-163.

Trusheim, F., 1960, Mechanism of salt migration in northern Germany: AAPG Bull., v. 44, p. 1519-1540.

Tucker, P. M., and H. J. Yorston, 1973, Pitfalls in seismic interpretation: Soc. Explor. Geophysicists Mon. Ser. 2, 50 p.

Vroman, A. J., 1967, On the fold pattern of Israel and the Levant: Israel Geol. Survey Bull., v. 43, p. 23-32.

Weber, K. J., and E. Daukoru, 1976, Petroleum geology of the Niger delta: 9th World Petroleum Cong. Proc., v. 2, p. 209-221.

Wilcox, R. E., T. P. Harding, and D. R. Seely, 1973, Basic wrench tectonics: AAPG Bull., v. 57, p. 74-96.

Wise, D. U., 1963, Keystone faulting and gravity sliding driven by basement uplift of Owl Creek Mountains, Wyoming: AAPG Bull., v. 47, p. 586-598.

Woodbury, H. O., et al, 1973, Pliocene and Pleistocene depocenters, outer continental shelf, Louisiana and Texas: AAPG Bull., v. 51, p. 2428-2439.

Yorston, H. J., and G. H. Weisser, 1976, The seismic analysis of glide plane fault structures; styles, mechanisms, and relationship to sedimentation (abs.): Soc. Explor. Geophysicists, 46th Annual Internat. Mtg. Abs., p. 85.

Reprinted from *Bulletin of Canadian Petroleum Geology*, v. 12, no. 2 (1964), p. 263-278. Copyright Canadian Society of Petroleum Geologists.

BULLETIN OF CANADIAN PETROLEUM GEOLOGY

VOL. 12, NO. 2, PP. 263-278

VARIATION IN TECTONIC STYLE[1]

L. U. DE SITTER

Geological Institute

University of Leiden, Netherlands

INTRODUCTION

Every field geologist knows from his own experience how greatly the style of tectonic deformation varies. In one mountain chain he passes from the centre with smooth flow folds in highly metamorphic rocks to large flat lying nappes in an external zone and to gentle folding in the marginal trough and finally to fault tectonics in the outer border. In another mountain chain he meets with vertical almost isoclinal cleavage folds in the centre, finds isoclinal recumbent folds in a border zone and passes through steep upthrusts into small gliding nappes. In great contrast to his experience in such mobile belts is the picture offered to him by the cratonic regions where often the only tectonical disturbances are faults.

These observations teach him that the deformation of the earth's crust is a complex phenomenon, in which many factors play a role and which no simple and single theory can explain.

In the following pages an attempt is made in classifying these strongly variable tectonical disturbances of the visible part of the earth's crust and their relationship. We will follow the outline presented in Table I, below:

TABLE I.

Synsedimentary and other non-tectonical deformations
Cratonic style of deformation
    faults
    monoclines
paratectonic style of deformation
    concentric folds
    flat overthrusts and gliding nappes
Orthotectonic style of deformation
    axial plane cleavage
    schistosity flow folding

Many transitional types occur between the above mentioned styles of rock deformation and it is these which often give us the means of understanding the relations between the pure types. They will be mentioned when we proceed with our descriptions. Obviously only some aspects of the problems we are faced with can be mentioned in this short review.

SYNSEDIMENTARY OR PENECONTEMPORANEOUS DEFORMATIONS

Synsedimentary or penecontemporaneous deformations will only be mentioned here with the object of pointing out their similarity in shape with

---

[1]Manuscript received October 5, 1963. This paper was delivered by Dr. de Sitter as the Seventh Annual Honorary Address to the Alberta Society of Petroleum Geologists, May, 1963.

tectonic deformation of solid rock. Fig. 1 gives an example of rather strong folding of synsedimentary origin in Pleistocene gypsiferous marls (Lissan Marl) of the Dead Sea basin. The undisturbed position of the underlying and overlying sediments prove the non-tectonical origin of these folds but their shape, the detachment plane at the bottom, the complications in the anticlinal and synclinal cores are typical of concentric folds. Figure 2 gives an example taken from Carboniferous shales from Clare Country on the west coast of Ireland; purely recumbent almost isoclinal folds originated on the border of a channel slump. No doubt the flattening of the folds must be partly due to compaction of the mud to shale after their folding, but their present configuration shows no distinction from a recumbent cleavage fold, common in orthotectonic regions.

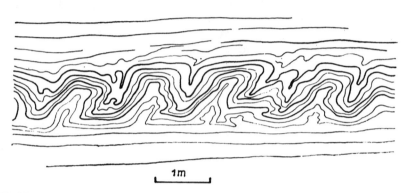

1m

Fig. 1.—Folds due to slumping in the gypsiferous Lissan Marl (Pleistocene) of the Dead Sea, outcropping west of S'dom salt horst.

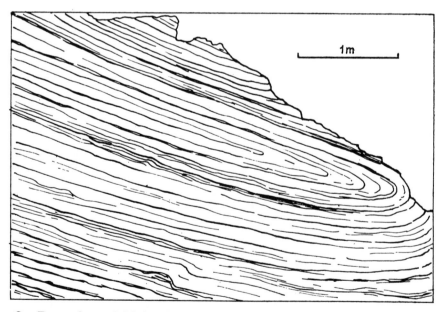

1m

Fig. 2.—Recumbent fold in slumps in Carboniferous shales, Clare County, Ireland.

The same lack of distinctive features of penecontemporaneous disturbances is offered by the chaos- or melange-structures of olistostromes (Wildflysch) due to the tumbling down of small or large blocks of adjacent rocks in the mud deposits of troughs. They are hard to distinguish from breccias due to tectonic disturbance, or from tillites.

There exist many other kinds of transitional types of deformation between purely synsedimentary and gravitational gliding structures which we will not mention here.

These few examples must suffice to illustrate the fact that apparently the shape and other distinctive features of rock disturbance is only dependent on relative factors, the relation between the stress fields or the rate of strain and the strength of the material for instance, and not on absolute values.

THE CRATONIC STYLE OF DEFORMATION

Cratonic type of deformation is characterized by the restriction of deformation to narrow fault zones between large blocks of undeformed rocks. It is often amazing how rare even these faults can be over large distances as for instance in the Paleozoic sedimentary cover of the Canadian Shield in northern North America. Whether the faults have a strike-slip or dip-slip character is often difficult to decide. A particularly interesting type of deformation is offered by the sedimentary cover of faulted basement rocks, especially when the faults are thrust faults. In such cases the sedimentary cover is warped in the form of a flexure over the nearly vertical fault offset of the basement. The Arabian Shield in Israel shows many of these monoclines, which are characterized by a very flat and broad flank on one side and a narrow steep flank on the other (de Sitter, 1962, Fog. 3A). A surprising feature is the fact that in these simple structures the sharp syncline next to the steep flank is often not overthrusted, although crumpling is often observed. Faults do occur in the anticlinal region, where the erosion brings us nearer to the basement fault.

Such structures are common also in the Laramide region of the Rocky Mountains, of which Berg (1962) has given many examples (Figs. 3B and 3C). The boundaries of the Beartooth Mountains, described by Foose *et al.* (1961) are of the same type. There can be no doubt that the steeply dipping thrustfault in the basement flattens upwards causing the flexure in the sedimentary cover to become recumbent. This relation between recumbent folds and vertical movement plays often an important role in the outer border zones of orogenes. Figure 3D gives an example from the Bergamasc Alps (de Sitter, 1949) where a combination of an overturned syncline below the overthrust and a gliding nappe coming from the uplifted block and deposited on the lower block, demonstrate how the surficial structure can mask the original and fundamental vertical block movement. Beloussov (1962) has called particular attention to these structures, which abound in the Caucasus. This example of cratonic structures on the border of a real orogene demonstrates also that the boundary between cratonic and orogenic deformation types is not sharp but transitional. A similar transition is also very apparent in the Irsaelean structures when one passes from the simple monocline to a simple paratectonic anticline when the sedimentary cover becomes thicker.

We will omit the big graben structures accompanied by tilting of the blocks of typical cratonic character, which are seldom accompanied by folds because a tensional stress field can only break the rocks. The relation of wrench faults to folds, both originating by a horizontal stress field, has often been demonstrated (de Sitter, 1956, pp. 166-168). These smaller wrench faults belong more to the paratectonic type. Large wrench faults like the San

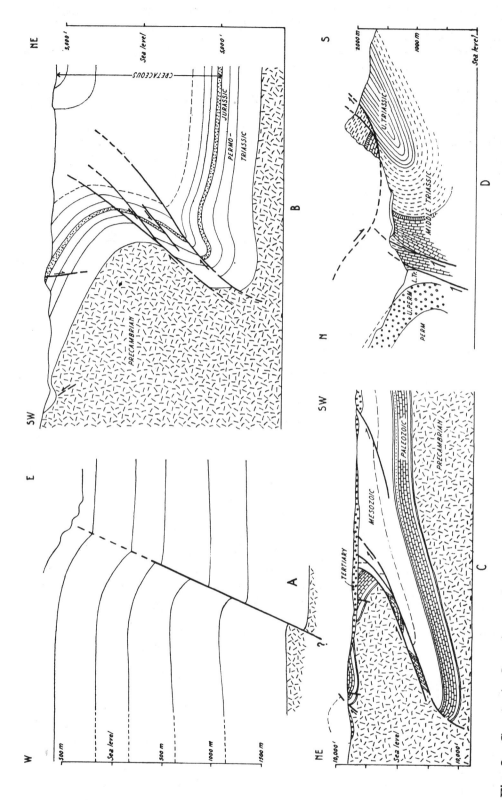

Fig. 3.—Crantonic flexuring and monoclines: A. Type of monocline structure in the Negev, Israel; B. Golden thrust, Jefferson County, Colorado; C. EA thrust near DuBois, Fremont County, Wyoming (b & c after Berg, 1962); D. A thrust in the Bergamasc Alps.

Andreas fault in California have a long and variable history resulting in very complicated structures.

A very typical example of another kind of folding related to wrench faulting has been mapped on the southern border of the Cantabric Mountains in northern Spain (Fig. 4). A wrench fault of Tertiary age cut through a massive carboniferous limestone, strongly folded in the Hercynian period and the same offset is repeated in the post-Hercynian Upper-Carboniferous and Cretaceous by a kind of flexure with two almost vertical fold axes.

PARATECTONIC STYLE OF DEFORMATION

The relatively simple concentric or parallel folding type with its anticline-syncline alternation grading from the gentle to the steep type with dips ranging from a few to 90 degrees is familiar to all of us. The accompanying faults are closely related to the folding itself and the more or less constant folding depth in a certain region, related to the thickness of the sedimentary series involved in the folding process, proves that the stress field is compressive and tangential to the surface. Its relation to the cratonic style when deep seated basement faults are involved is clear and often it can be noticed that the paratectonic style replaces the cratonic deformation when the sedimentary cover gradually thickens from the higher regions towards a basin. On the other hand the transition towards the orthotectonic style is also evident, not only in the horizontal sense when we approach the mountain chain coming from the marginal zone, but also in the vertical sense going downwards inside the mountain chain from the most recent sedimentary layers deposited just before the major phase of deformation to deeper levels of older rocks.

It is certainly true that in paratectonic folding the difference in reaction to the stress field of competent and incompetent rocks is particularly well developed. A very spectacular instance of this difference in reaction is the composite wedge formed by a thick limestone compressed near the surface which

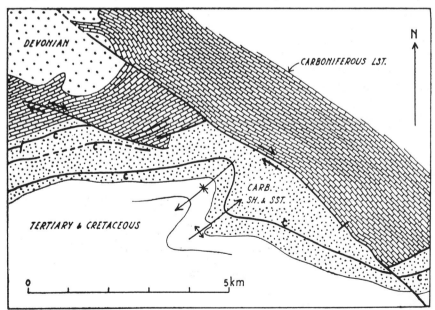

Fig. 4.—Wrench faulting accompanied by a vertical flexure fold in the southern border of the Cantabric Mountains, west of Cervera de Pisuerga, Palencia, northern Spain.

broke up in segments thrusted over one another in order to follow the anticlinal bending of the more incompetent rocks below. Minor folds of competent rocks in a shale surrounding are also common.

The purely concentric folding of these competent members in contrast to
the oblique shear deformation resulting in pencil shales is another example of
difference in reaction to the same stress by rocks of different competency.

Of particular interest is the phenomenon of cross-folding in concentric folds.
Their analysis is more difficult than that of the same feature in cleavage folding because the available data are only related to the bedding. In consequence in pure concentric cross-folding we often have no conclusive evidence
which is the earlier and which the later stress field. Crossfolding of this type
can show up in two different ways:

      a. On the map we see that anticlinal and synclinal axis are bent,
      showing two different strikes in adjoining regions (*e.g.*, Ouachita Moun-
      tains of the southern United States).

      b. The same beds have been folded twice along different strikes, *e.g.*
      an anticline superimposed on a syncline.

In the first case the region where the two directions meet is often full of
faults, which are less prominent where the folding is unidirectional.

If in the second case the first of the folding systems is accompanied by
thrusting then the succession can be established because a thrust plane is originally essentially a flat plane, and has been folded later by the second phase.

A good example of refolded thrust sheet is the Esla nappe of some 10 km
width. The thrusting from south to north is proved to be pre-Upper Carboniferous (pre-Stephanian) by unconformable Stephanian conglomerates trun-
cating the thrust-plane. The Stephanian itself is strongly folded however,
and the thrust plane became involved in this younger folding process (Fig. 5).

The same phenomenon of refolding of thrust sheets is typical of the Canadian Foothills and Frontal Range thrusts. These thrust structures of the
Cantabric Mountains and the Canadian Rockies are very similar in many
respects, although the imbricated structures of the Rockies are absent in
Cantabria.

Concentric folds often develop into thrusted structures, in particular when
they were asymmetric from the beginning. Further compression makes flat
overthrusts of these original steep thrusts and finally a real flat nappe structure can be formed. This kind of nappe structure has the typical property
that an overturned flank is completely missing. Its development from a
thrusted anticline is very well demonstrated by the thrusts in the Bernesga
zone of the Cantabric Mountains in northern Spain, where west of this river
we find faulted anticlines and to the east the overthrusts develop in real nappes

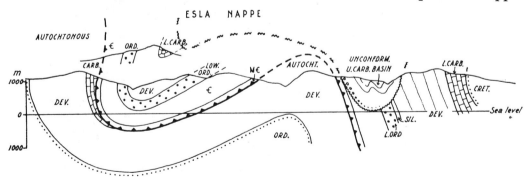

Fig. 5.—The folded thrustplane of the Rio Esla nappe, southern Cantabric Mountains, northern Spain.

of 5 to 10 km thrust movement, which still further to the east are replaced by one 15-20 km thrusted nappe, the Esla nappe (de Sitter, 1963, Fig. 6).

These structures are difficult to distinguish from gravitational gliding nappes, when the roots are not exposed and often the two kinds are closely related when the original flat thrust starts gliding due to an uplift of its root region as happened for instance with the Helvetian nappes.

Flat gravitational gliding nappes can originate both as a result of late orogenic and early orogenic uplifts. The early or pre-orogenic type is due to steepening of the slopes of the sedimentary basin at a late stage of its development as for instance in the French circum-alpine Stephanian basins (Pruvost, 1963, Fig. 7). The late orogenic type is typical of the border zones of the orogene, where the necessary slope is created by the uplift of the folded axial one, as on the southern slope of Pyrenees and High Atlas or in the southern Alaps. An extreme case of sliding of the orogenic type has been mapped in the Tell geosyncline of the northern Algeria (Kieken, 1962) where distances of at least 100 km gliding are involved.

Such gravitational gliding structures are more of an epeirogenic character and the smaller they become the more they approach the synsedimentary structures of slumping and convolute bedding. This continuous transition from synsedimentary to epeirogenic and to paratectonic structures demonstrates very clearly that no sharp boundaries can be drawn between these deformation styles.

Nappe structures of the Helvetian type have often been regarded as the most typical "Alpine" structure, and consequently of the orogenic style of deformation. The short review we gave here shows that they can also occur in extreme cratonic structures and as such form a direct transition from cratonic to orogenic types of deformation.

A particular kind of a similar folding but still of the concentric type is the accordion style of folding (often called chevron folding). It is restricted to thin and well bedded sedimentary series. The folds are relatively small (200 m wave length and smaller) and form a transitional type to axial plane cleavage folding. It is concentric because most of the movement takes place along the bedding planes in the flanks, only in the anticlinal and synclinal

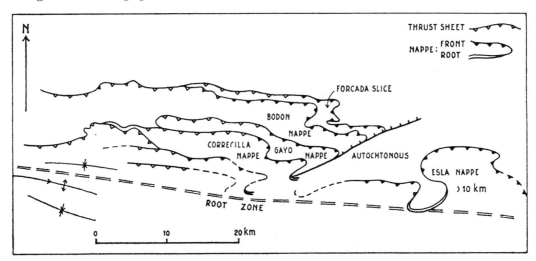

Fig. 6.—The thrusted anticlines of the Bernesga region developing eastward into nappes and further east into one nappe, the Esla nappe, Cantabric Mountains, northern Spain.

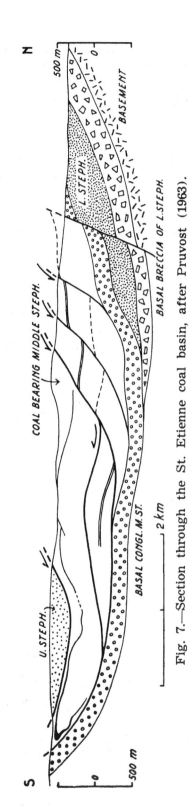

Fig. 7.—Section through the St. Etienne coal basin, after Pruvost (1963).

axes a movement parallel to the axial plane is apparent often resulting in cleavage development here.

The dip of the flank is seldom steeper than 60°-70° because beyond this limit the amount of slip along the bedding plane by further steepening would become prohibitive. Further compression results in shear faults often parallel to the bedding in one flank and cutting hrough the bedding in the other flank (Fig. 8).

Its transitional character between concentric and cleavage folding is demonstrated by its occurrence between concentric folding in the top beds and cleavage folding in the deep units, as for instance in Devonshire, in the Ouachita Mountains and in the west German coal basin.

For some reason or other a thin and well bedded series gets knicked instead of bent, already in the initial stage of folding. The size of the folds seems to be related to the thickness of the original layers. In some cases one can see a transition of accordion folding in the core of a minor fold to concentric folding in the periphery of the fold. The knicking type of deformation is common also on a much smaller scale as a late deformation of cleavage beds, which have acquired the property of good foliation by the cleavage development. This knicking can occur even on a microscopic scale as typical accordion folding (Fig. 9A) or as minor structures of 10 cm size in which only one flank has been submitted to deformation (Fig. 9B). They have been imitated experimentally by Paterson and Weiss (1962) by compression of well foliated phyllite with a stress acting parallel to the foliation. If the knicking deformation affects only one flank it can often be observed that the cohesion of the rock in this disturbed strip has been loosened and that recrystallization has taken place only in this strip. When this phenomenon is well developed a tectonic banding has been formed which obliterates the original schistosity

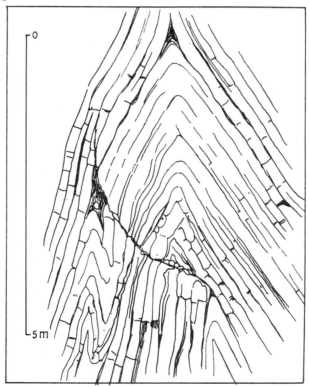

Fig. 8.—Accordion fold with later overthrust and small extra fold in one flank.

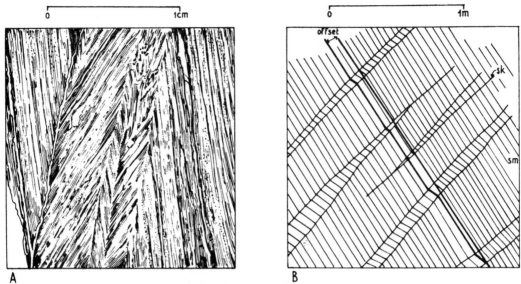

Fig. 9.—A. Accordion folding on a microscopic shale; B. Schematic drawing of knicking, both in Ordovician slates of the Pyrenees.

and bedding. Even the initial stage of granitization takes advantage of such knick zones in gneisses.

The accordion folding ranging from 100 m to 1 mm size is obviously always conditioned by the well foliated or bedded condition of the rock. Whether the relation stress field/yield value of the rock is an important factor remains doubtful.

ORTHOTECTONIC STYLE OF DEFORMATION

The orthotectonic style of deformation is in principal restricted to orogenic belts and to their infrastructure (de Sitter and Zwart, 1960).

Epizonal Cleavage Folding

Up till now the deformation styles which we reviewed all give witness of the strong dependance of the shape of folds from the original bedding of the layers. Thrust sheets glide along an incompetent layer, and the internal movement of concentric folds is in principal parallel to the bedding. In similar cleavage folding the deformation is in principal independent of the bedding. The pure type of cleavage folding is presented to those folds where the cleavage plane is also the axial plane of the fold. This independency of the fold shape is particularly well illustrated in transitional types of cleavage folding where the shales show the development of cleavage, but the competent sandstone or limestone layers still behave as they do in concentric folds. The flattening of the shale layers is proved by the thinning in the flanks and thickening on the hinges, which does not occur in the concentrically folded competent beds. When cleavage folding is complete the competent layers show the same phenomena of flattening and thickening as the shale layers.

That the similar cleavage folds are really due to a flattening process, with the principal stress perpendicular to the cleavage and the axial plane is proven also by the flattening and extension of pebbles and fossils (e.g. oolites, E. Cloos, 1947; fossils, Breddin, 1955, Kurtman, 1960; pebbles, Flinn, 1956).

The different behaviour of the incompetent and competent layers in the same fold is also well demonstrated by the parasitic folds (often called erroneously "drag folds") and boudins of the competent members in a shale surrounding (de Sitter, 1958).

If the parallelogram of Fig. 10a is compressed to the shape of Fig. 10b, its diagonal d is shortened until the angle α reaches 45°, beyond that it gets lengthened (Fig. 10c) because its length

$$d = \frac{y}{\sin \alpha \cos \alpha} = \frac{y}{\tfrac{1}{2} \sin 2\alpha}$$

in which formula y is the constant surface of the parallelogram. From this it follows that if the diagonal d represents a competent layer in an incompetent shale surrounding which becomes flattened it is crumpled when its original position was flat, e.g. parasitic folds are formed, and it becomes stretched when its original position was steeper than 45°, e.g. boudins, shown in Fig. 11, are formed (de Sitter, 1958). Mullions are a form of compression of a thicker competent layer, they are related to the parasitic folds. That such folds can not be due to "drag" and therefore ought not to be called "drag folds" is also proven by the fact that the crumpling continues over the hinge of the fold where all drag due to concentric folding is absent anyhow.

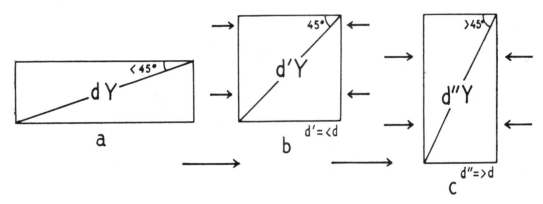

Fig. 10.—Flattening of a parallelogram by a horizontal stress.

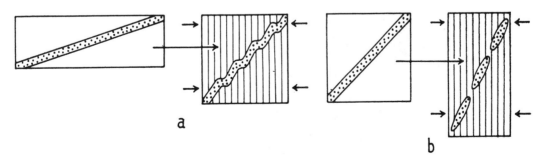

Fig. 11.—Parasitic folds (a) and boudins (b).

Often one can see how during the flattening process first the competent layers were crumpled into parasitic folds, which later became flattened together with surrounding shales (Fig. 12A). Fig. 12B shows an example of an overthrusted parasitic fold witnessing of the shortening it was subjected to.

The explanation of the origin of parasitic folds offered here presumes of course that a certain inclination of the beds probably due to folding, existed already before the flattening process and the formation of cleavage sets in. This presumption is also necessary for the formation of axial plane cleavage folds, and we have to assume that every cleavage fold was preceded by a concentric fold, otherwise the flattening process would only produce a thickening of the layers and no folding. The thickening and thinning of the layers can then give us some information about the total flattening that took place, which only in extreme cases exceeds 70 percent.

Because the yield value of incompetent shale is lower than that of competent rocks the stress has obviously been higher in axial plane stress cleavage folds than in concentric folds, the larger stress interval in the case of the incompetent rocks leading to cleavage flattening, the smaller interval of the competent member leading to bending of the concentric type in the same fold. A higher stress field apparently does not lead to quicker folding but to another type of deformation. In other words, when the rate of increase of the stress is high, more energy becomes available for deformation and flattening is preferred to bending.

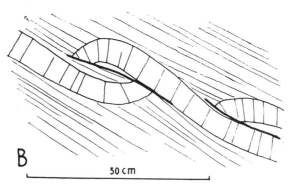

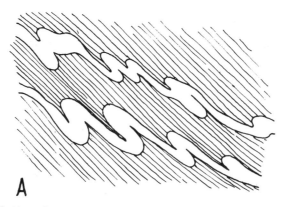

Fig. 12.—Flattened parasitic folds; B. Overthrusted parasitic fold.

This conclusion is also confirmed by the precedence of concentric folding to cleavage folding, *e.g.* the rising stress field first caused concentric folding, and only when the rate of increase is high, enough energy becomes available to replace this kind of folding by flattening. Then cleavage is superimposed on the concentric fold, all in the same process. It is quite possible of course that during this process the original concentric folds became tilted by some larger scale movement of the basement, for instance, and then the dip and even the strike of the superimposed cleavage need no longer be parallel to the axial plane of the original concentric fold. Such tilting can also happen at a later stage and then we will find two superimposed cleavages, the second one often showing up as a fracture cleavage cutting through the original slaty cleavage (Hoeppener, 1957).

Meso- and Catazonal Schistosity Folding

Meso- and catazonal schistosity folding is typical of the infrastructure of a fully developed orogene. There is no doubt that a temperature rise during the deformational process has here been an active factor, because recrystallization of new mineral plays an important role in the metamorphic zoning. In principal the schistosity plane is horizontal in direct contrast to the vertical cleavage of the epizonal cleavage folding.

Also in contrast with the axial plane cleavage, here the "S" shaped structure of the relict schistosity in the porphyroblasts proves that much of the movement is shearing along the schistosity plane (Zwart, 1960). The original bedding has largely disappeared and has often been replaced by tectonic banding due to differential recrystallization of dark (biotite) and white (quartz and feldspar) minerals in alternating bends. In incipient stages of such migmatization this process can be followed accurately. When the temperature rise is intensive enough real granitization takes place accompanied by randomly oriented flow structures. It is probable that the shear-flow folding of these metamorphic rocks is stimulated by the large amount of water that is expelled from the minerals in this process.

The experiments by Winkler (1961) demonstrating the partial melting of sedimentary rocks have given us an excellent theoretical and experimental background for understanding these flow-structure phenomena of migmatites, and their relation to both the stress field and the temperature rise.

Detailed work in the Scottish Highlands, in the Pyrenees, in Scandinavia and in many other regions has shown that in these infrastructural regions superimposed cleavages and schistosities are common. Both the strike and dip of the succeeding phases varies largely and the question arises whether the superimposed succeeding phases belong together to one and the same major stress field with variable yielding of the rock, or whether the succeeding phases can be compared to those we observe in non-metamorphic rocks where the main stress fields are variable in direction and stratigraphically defined as belonging to different phases.

I think the answer to this question is that in metamorphic rocks many of these succeeding phases actually belong to one and the same major stress field, but some exceptions can be expected.

In general the succeeding phases have superimposed small deformations on the main phase cleavage or schistosity, but this relation varies in different orogenes. In Caledonian and Archean orogenes for instance the later phases are often of a very large size. It is also true that granitization is in principle directly related to one or more sets of the later phases.

When the different stages have been followed upwards from the mesozonal area into the epizonal area of axial cleavage deformation, similar sets of cross-folding can be discerned, often less strongly developed, and the higher up we come the less the cross-folding is evident. And, on the other hand sometimes an early concentric folding superimposed by a cleavage development of quite another strike of the suprastructure can not be found back in the metamorphic zone. It must be admitted, however, that until now there have been few efforts in correlating stratigraphically defined cross-folding with that of the infrastructure. The opportunities for studying this problem are rare, because it is only seldom that the metamorphic infrastructure and the non-metamorphic suprastructure belonging to the same orogenesis are both exposed.

Nevertheless shear schistosity, axial plane cleavage and concentric folding can occur simultaneously in the same structural unit distributed from deep to shallow levels where it is highly probable that the same ultimate stress field causes different styles of deformation at different levels, partly due to the increase of temperature and liberated water, partly to the increase of hydrostatic pressure in depth. In some of the metamorphic domes of the Pyrenees it can easily be proved that the rise of temperature towards depth is not due to a sinking of that part of the mountain chain but to an active elevation of a high temperature front in a limited area, causing catazonal recrystallization at a depth of burial of 2 to 4 km only, which result is confirmed by the fact that this mesozonal facies is characterized by the absence of kyanite in the Al rich pelites. This feature proves that the horizontal schistosity development, is somehow related to the rise of temperature, presumably to a level near 700° because granitization is common. The schistosity shear folding, which occasionally becomes of the non-oriented flowage type is apparently due to a de-decrease of the yield level of the rock due to high temperature mostly accompanied by liberateion of water.

As we have no means to determine the time needed for the complete process of the succeeding phases in the infrastructure, and the process is there much more continuous than nearer the surface, it is still quite possible that stratigraphically defined successive phases of the suprastructure together belonging to one orogenic period, are simply a reflection of a slow continuous process at depth. The fact that intrusive plutons of orogenic type are generally late in relation to the orogenic phases of the suprastructure and the origin of palingenetic granitic melts, certainly related to the granitization process of the infrastructure, is also a late orogenic phenomenon points to this conception.

CONCLUSIONS

In the course of this very much abbreviated survey of deformation styles we have met with different factors which influence the style development.

We have seen that both the alternation of competent and incompetent rocks and the character of the bedding itself influence the style. Moreover we have found that transitions of style exist everywhere and in every sense. The lithology of the rocks and the temperature under which they are deformed both influence their yield value to the stress. Flow folds in salt domes are very much like flow folds in migmatites. A more fundamental factor is therefore the relation of the yield value of the rock to the stress field, and even this factor is only relative.

More important is the rate of increase of the stress. If this rate is small the energy available for the rock deformation is small and absorbed by concentric folds. If this rate is larger, more energy is available and cleavage folding sets in, and if it is still larger flow folding with shear schistosity or shear movement as in salt can develop.

In Fig. 13a the stress/strain diagram of one rock type with a growing stress field is presented. If the stress stays between the d_1 and d_2 levels that kind of rock will fold concentrically, if the stress can rise to a level between d_2 and d_3, flattening with axial plane cleavage will develop and when it can rise still higher (or the resistance of the rocks decreases by a considerable rise in temperature) flow folding sets in. When the same stress fields acts on three different kinds of rocks at the same time it rises to a level at which the competent rock (quartzite) is folded concentrically, the incompetent rock (slate) develops cleavage and the soft rock (salt) starts to flow (Fig. 13b). It is the interval between the stress level and the strength of the rock which determines the folding style, and this interval is dependent of the rate of increase of stress.

If we apply this simple rule to an orogene, it becomes clear that as soon as the yield value of the more competent rock or that of a whole series of rocks, has been exceeded by the growing stress, the first thing that happens will be concentric folding in the supracrust. Rise of temperature will then weaken the resistance of the deeper levels and will allow the development of clavage folding in the higher levels. If this temperature rise does not happen, the concentric folding dies out with the decrease of the stress and we find only paratectonic folding. When the temperature rise does take place we get the development of a real orogene.

Therefore, there are really two factors which determine the folding style: 1) the rate of increase of the stress field and 2) the temperature increase. Why, in an orogene, both are restricted to a narrow zone of the earth's crust we do not know, but they must have a common cause. Moreover, we have learned that vertical uplift is an important factor, but it can occur in a cratonic region or can be related to an orogene.

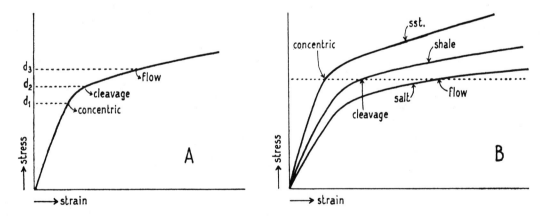

Fig. 13.—Stress-strain diagrams and the reaction of rocks of variable competency to the stress.

REFERENCES

Beloussov, V. V., 1962, "Basic Problems in Geotectonics," McGraw-Hill, New York.

Berg, R. R., 1962, "Mountain Flank Thrusting in Rocky Mountain Foreland, Wyoming and Colorado," Bull. Amer. Assoc. Petrol. Geol., Vol. 46, pp. 2019-2032.

Breddin, H., 1955, "Tektonische Gesteinsdeformation im Karbongürtel Westdeutschlands und Süd-Limburg," Z. D. Geol. Ges. 107, 231-260.

Cloos, E., 1947, "Oolite Deformation in the South Mountain Fold, Maryland," Bull. Geol. Soc. Amer., Vol. 58, pp. 843-918.

Flinn, D., 1956, "On the Deformation of the Funzie Conglomerate, Fetlar, Shetland," Jour. Geol., Vol. 64, pp. 480-505.

Foose, R. M., Wise, D. U., and Barbarini, G. S., 1961, "Structural Geology of the Beartooth Mountains, Montana and Wyoming," Bull. Geol. Soc. Amer., Vol. 72, pp. 1143-1172.

Hoeppener, R., 1957, "Zur Tektonik des SW-Abschnittes der Moselmulde," Geol. Rundschau, Vol. 46, No. 2, pp. 318-348.

Kieken, M., 1962, "Les Traits Essentiels de la Géologie Algérienne," Livre Mém. Paul Fallot, I, pp. 545-614.

Kurtman, F., 1960, Fossildeformation und Tektonik im Nördlichen Rheinischen Schiefergebirge," Geol. Rundschau, Vol. 49, No. 2, pp. 439-459.

Paterson, M. S., and Weiss, L. E., 1962, "Experimental Folding in Rocks," Nature 195, pp. 1046-1048.

Pruvost, P., 1963, "Les Jeux propres du Socle, révélés par l'Histoire de certains Bassins Houillers à la Périphérie du Domaine Alpin," Livre Mém. Paul Fallot, II, pp. 11-18.

de Sitter, L. U., 1956, "Structural Geology," McGraw-Hill, New York.

—————, 1958, "Boudins and Parasitic Folds in Relation to Cleavage and Folding," Geol. and Mijnbouw, Vol. 20, pp. 277-286.

—————, 1962, "Structural Development of the Arabian Shield in Palestine," Geol. and Mijnbouw, Vol. 41, pp. 116-124.

—————, 1963, "The Structure of the Southern Slope of the Cantabrian Mountains," Leidse Geol. Med., Vol. 26, pp. 255-264.

—————, and de Sitter-Koomans, C. M., 1949, "Geology of the Bergamasc Alps," Leidse Geol. Med., Vol. 14B, pp. 1-257.

—————, and Zwart, H. J., 1960, "Tectonic Development in Supra- and Infrastructures of a Mountain Chain," International Geol. Congress, Copenhagen, Pt. 18, pp. 248-256.

Winkler, H. G. F., 1961, "Genesen von Graniten und Migmatiten auf Grund neuer Experimente," Geol. Rundschau, Vol. 51, pp. 347-364.

Zwart, H. J., 1960, "The Chronological Succession of Folding and Metamorphism in the Central Pyrenees," Geol. Rundschau, Vol. 50, pp. 203-218.

Reprinted by permission of the Geological Society of America
from M. King Hubbert, *Geological Society of America Bulletin*, v. 62, no. 4 (1951), p. 355-372.

BULLETIN OF THE GEOLOGICAL SOCIETY OF AMERICA
VOL. 62, PP. 355-372, 16 FIGS., 2 PLS.   APRIL, 1951

MECHANICAL BASIS FOR CERTAIN FAMILIAR GEOLOGIC STRUCTURES

By M. King Hubbert

Abstract

A simple experiment with loose sand shows that this material exhibits faulting under deformational stresses in a manner remarkably similar to rocks. Moreover, the sand experiment is amenable to theoretical analysis with good agreement between predicted and observed behavior. The same theoretical treatment, with slight modification, is alsoa pplicable to the behavior of rocks, and appears to afford a basis of understanding for a variety of empirically well-known geologic structures.

CONTENTS

ILLUSTRATIONS

Introduction

During the last century and a half, a large volume of empirical data on the mechanical behavior of rocks in response to tectonic stresses has been accumulated in the course of geologic mapping over all parts of the earth. In addition geologists have given considerable attention to the relation between various well-defined classes of geologic structures and the associated forces or stresses. This has included both small-scale structures such as slaty and fracture cleavage and joints, and large-scale structures such as faulting and folding. During the same period in the fields of physics and engineering, a great deal of exact knowledge has been acquired concerning the properties of stress in extended media and the response of matter to

355

various stress situations, but only a beginning has been made toward an amalgamation of this knowledge with geologically well-known situations.

The present paper will examine some of the more common types of large-scale geologic structures in the light of physical relations which have to be satisfied, to see whether a better understanding of these structures may thus be acquired. Historically, the paper is based upon a series of lectures in structural geology delivered at Columbia University during the period 1931–1940. For his original interest in the geological aspects of the problem, the author is indebted to Professors J Harlen Bretz and Rollin T. Chamberlin; for the experiment to be described and the applicable theory, he is principally indebted to the writings of, and personal conferences with, Dr. A. Nádai (1928, 1931a, 1931b).

The essential theory appears to have first been established by W. J. M. Rankine in 1857 in a study of the conditions of stability of loose earth. This was elaborated in a series of papers by Otto Mohr (1871; 1872; 1882; 1900) who, in addition to developing simple graphical methods of stress analysis, also enquired into the conditions of failure. Within recent decades, these principles have been extensively employed in the newly developed subject of soil mechanics. In geological literature it was noted by Chamberlin and Shepard (1923, p. 511–512), in some model experiments on geologic structures, that loose sand exhibited reverse faulting, with dip angles ranging from 20° to 35°. W. J. Mead (1925) reported further upon this property as related to dilatancy, but without analysis otherwise. Recently E. M. Anderson (1942) has reverted to the original work and methods of Rankine and has reached by a somewhat different route conclusions in essential agreement with those presented here.

EXPERIMENTAL FAULTING IN SAND

Experimental Apparatus

Plate 1, Figure 1 shows a box with a plate-glass front which is divided into two compart-ments by a rigid partition. The partition is attached to a longitudinal screw which can be turned by a crank. When the screw is rotated, the partition is moved by translation to the right or the left, parallel to the long axis of the box.

The box is filled to a depth of about $2\frac{1}{2}$ inches with a fine-grained, loose, dry sand obtained from the beach at Galveston. Embedded in the sand next to the plate glass are three horizontal white lines composed of dry powdered plaster of Paris placed on the sand when the box was being filled; except as visible markers, they are without significance.

The sand in the box will stand at a maximum angle of repose of about 30° on a free slope.

When the partition is moved from left to right the left-hand compartment is lengthened, the right-hand compartment is shortened by the same amount and since the behavior of the sand in one is independent of that in the other, two separate experiments are performed simultaneously.

The first thing that happens (Pl. 1, fig. 2) is the development in the left-hand compartment of a distinct normal fault whose measured angle of dip is 63°. In the meantime, nothing other than a slight bulging is observable in the right-hand compartment. As the partition is cranked farther to the right, reverse faults (Pl. 2 figs. 1, 2) with a measured angle of dip of about 28° develop. These extend from the bottom of the box near the foot of the partition to the surface of the sand where clearly visible escarpments are produced. This experiment has been repeated a number of times and the results are presented in Tables 1 and 2.

Average Dip of Normal Faults

In the case of the normal faults, the frictional drag of the sand on the glass plate invariably produced a steeper angle at the plate than in the interior of the box. Since the fault plane extends from the scarp on the surface to the foot of the partition its angle in the interior can be determined approximately by the equation,

---

PLATE 1. EXPERIMENTAL FAULTING IN SAND

FIGURE 1. EXPERIMENTAL APPARATUS
Box is filled with loose, dry sand. Markers are powdered plaster of Paris.
FIGURE 2. OBLIQUE VIEW OF NORMAL FAULT SHOWING ESCARPMENT

FIGURE 1

FIGURE 2

EXPERIMENTAL FAULTING IN SAND

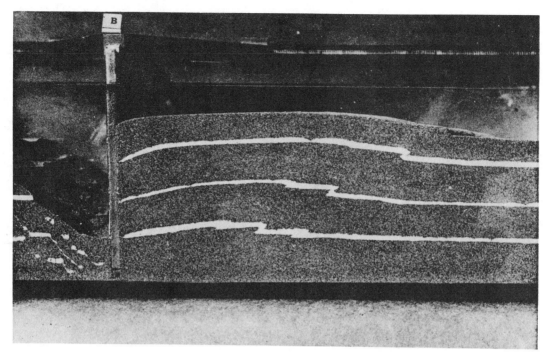

FIGURE 1

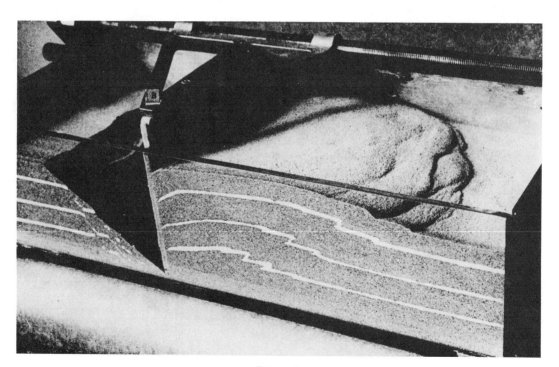

FIGURE 2

EXPERIMENTAL FAULTING IN SAND—REVERSE FAULTS

$$\alpha = \tan^{-1} z/x \qquad (1)$$

where x is the distance of the escarpment from the partition in the interior of the box and z the depth of the sand.

TABLE 1.—ANGLES OF DIP OF NORMAL FAULTS

Experiment No.	Depth z of sand (inches)	Distance z of fault scrap from partition (inches)	Angle of Dip	
			Interior of sand	At glass surface
1	2.56	1.4	61.3°	64°
2	2.7	1.4	62.6	63
3	2.65	1.3	63.9	67
4	2.8	1.5	61.8	63
5	2.65	1.6	58.9	60
6	2.65	1.6	58.9	60
7	2.6	1.4	61.7	62.5
8	2.75	1.4	63.1	64.5
9	2.65	1.35	63.0	62.5
10	2.7	1.45	61.8	62
11	2.6	1.5	60.0	60
12	2.7	1.7	57.8	59
13	2.5	1.4	60.8	64
Average			61.2°	63.2°

Four of the measurements for normal faults were made with loosely packed sand; the remainder were made after the sand had been shaken to a closer packing. The angles for loosely packed sand were somewhat less steep than those for the closely packed sand, and the fault planes were less sharply defined.

Average Dip of Reverse Faults

It was not possible to compute the angle of dip of the reverse faults in the interior, because the fault planes did not originate at the bottom edge of the partition, but at a variable distance in front. Angles of dip of the reverse faults as observed against the glass plate appear to be somewhat steeper than the angles in the interior of the box. This is indicated by the arcuate shape of the fault scarp which is convex in the direction of the thrusting. The difference, however, is probably not more than a degree or two.

COMPARISON OF RESULTS WITH GEOLOGICAL FAULTS

No field geologist can observe this experiment without being impressed by the remarkable

TABLE 2.—ANGLES OF DIP OF REVERSE FAULTS IN SAND

Experiment No.	Angle of Dip
1	23°
2	22
3	28
4	23
5	25
6	28
7	27
8	26.5
9	26.5
10	29.5
11	24.5
12	22
13	26
14	21.5 / 27.5
15	24.5 / 20.5
16	27
Average..................	25.2°

similarity between the faults produced in the sand and those which occur in rocks in the field. Although statistical evidence on the observed angles of dip of faults is sparse, it is a matter of common geological observation that in areas of medium topographical relief and mild tectonic deformation (where the deformation is insufficient to seriously change the initial angle of dip of faults) normal faults occur with dips consistently greater than 45°—rarely less—and reverse faults have dips consistently less than 45°.

This generalization is confirmed strikingly by a recent study by Sax (1946, p. 42–46) of faulting in the coal measures in the Netherlands. In this study, measurements were made underground of the angles of dip of 2102 separate faults. Of these, 1651 (79 per cent) were

PLATE 2. EXPERIMENTAL FAULTING IN SAND—REVERSE FAULTS

FIGURE 1. DEVELOPMENT OF REVERSE FAULTS
FIGURE 2. OBLIQUE VIEW OF REVERSE FAULTS

normal and 451 (21 per cent) were reverse faults (thrusts and overthrusts). Frequency curves in 5°—interval groups were also plotted. The curve for each class of fault showed a well-defined peak, that for normal faults at 63° and that for reverse faults at 22°.

In 1913, C. K. Leith (p. 55) reported on the same object as follows:

"The dip of a thrust fault is usually low, that of a normal fault, high. An average from the United States Geological Survey Folios gives a dip of 36° for reverse fault planes and 78° for normal fault planes."

These figures are of only qualitative significance since without doubt many faults where placed on the cross sections for which accurate dip measurements were unobtainable.

Many normal faults are known from subsurface work in the Texas and Louisiana Gulf Coast region. In most of these, dips steeper than 45° are reported, though some have dips as low as 35°. In this region, the faulting in most cases has occurred in incompletely compacted sediments which subsequently have become compacted by amounts up to as much as 50 per cent of their initial thickness. Hence, an initial dip of 60° could be reduced to as little as 41°, or an initial dip of 50° to 31°.

Billings (1946, p. 206) shows a line drawing of a graben structure made by Hans Cloos with soft mud as his experimental material. The principal normal fault even in this "incompetent" material has a dip of 50°.

Thus, not only is there a qualitative resemblance between the faults in the sand box and those in rocks, but the angles of dip in the two cases are in reasonable agreement. This strongly suggests that the mechanical basis for the phenomena in both cases is essentially the same.

MECHANICAL BASIS FOR THE FAULTING

Nature of Stresses Produced

Consider the state of stress at a point in the interior of the sand in the right-hand compartment of the box (Fig. 1). In the undisturbed state, the vertical and horizontal normal stresses will be approximately equal, and the stress state will be hydrostatic with the normal stress σ on any surface equal approximately to the

pressure of the overburden, or

$$\sigma = p = \rho g z \qquad (2)$$

where ρ is the bulk density of the sand, g the acceleration of gravity, and z the depth beneath the surface of the sand.

As the partition is advanced to the right

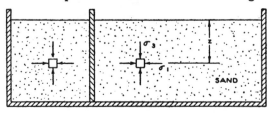

$$\sigma_1 = \sigma_3 = \rho g z$$

FIGURE 1.—INITIAL STATES OF STRESS

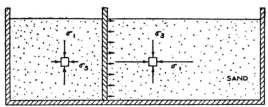

FIGURE 2.—STATES OF STRESS WHEN PARTITION IS MOVED TO THE RIGHT

(Fig. 2), the vertical stress σ_z remains constant but the horizontal stress σ_x is gradually increased. From the geometry of the deformation, these two stresses will also be, respectively, approximately the minimum and the maximum stresses acting, so we may regard them as the minimum and maximum principal stresses σ_3 (minimum) and σ_1 (maximum). An intermediate principal stress, σ_2, will exist at right angles to these two, but in a two-dimensional problem it does not require consideration.

Starting with the maximum and minimum principal stresses σ_1 and σ_3, let us first enquire as to the nature and the magnitude of the stresses acting across a plane which is obliquely oriented with respect to the axes of greatest and least principal stresses. Let such a plane be parallel to the axis of the intermediate principal stress, and let it make an angle α with the axis of the least principal stress σ_3.

Stress on Arbitrary Plane

Figure 3 is a section in the plane of the greatest and least principal stresses. Let this section be of unit thickness and let the line MN be the trace of the plane across which the stresses are to be determined. Let ABC be a small right-triangular prism whose sides are respectively perpendicular to the axes of the stresses σ_1 and σ_3. Let dS be the area of its hypotenuse, and let σ and τ be respectively the normal and shear stresses which act upon this surface.

We take the prism small enough that its weight is negligible as compared with the surface forces which act upon it. It will be accordingly in equilibrium, under the influence of the surface forces due to the stresses σ_1, σ_3, σ, and τ. Resolving these forces into horizontal and vertical components, the sum of the vertical components must be zero, and also the sum of the horizontal components.

The forces acting upon the prism are the products of the stresses and the areas acted upon. The areas of three sides are:

$$\left.\begin{array}{l} \text{side } AB = dS, \\ \text{side } BC = dS \sin \alpha, \\ \text{side } CA = dS \cos \alpha. \end{array}\right\} \tag{3}$$

The sums of the horizontal and vertical components are respectively:

$$\left.\begin{array}{l} \sigma_1 \, dS \cos \alpha - \sigma \cos \alpha \, dS - \tau \sin \alpha \, dS = 0; \\ \sigma_3 \, dS \sin \alpha - \sigma \sin \alpha \, dS + \tau \cos \alpha \, dS = 0. \end{array}\right\} \tag{4}$$

If we eliminate dS and solve equations (4) for the two unknowns τ and σ, we obtain

$$\left.\begin{array}{l} \sigma = \sigma_1 \cos^2 \alpha + \sigma_3 \sin^2 \alpha, \\ \tau = (\sigma_1 - \sigma_3) \sin \alpha \cos \alpha. \end{array}\right\} \tag{5}$$

From trigonometry:

$$\cos^2 \alpha = \frac{1 + \cos 2\alpha}{2},$$

$$\sin^2 \alpha = \frac{1 - \cos 2\alpha}{2},$$

$$\sin \alpha \cos \alpha = \frac{\sin 2\alpha}{2}.$$

When these values are substituted into equations (5), those reduce to the simpler forms:

$$\left.\begin{array}{l} \sigma = \dfrac{\sigma_1 + \sigma_3}{2} + \dfrac{\sigma_1 - \sigma_3}{2} \cos 2\alpha, \\[2ex] \tau = \dfrac{\sigma_1 - \sigma_3}{2} \sin 2\alpha, \end{array}\right\} \tag{6}$$

wherein τ and σ are expressed as simple functions of σ_1 and σ_3 and the angle 2α.

Mohr's Stress Circle

These equations are capable of a very simple geometrical representation if we choose coordinates with τ as the axis of ordinates and σ as the axis of abscissas. In such a system, σ_1

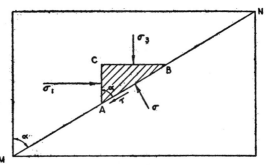

FIGURE 3.—STRESSES ACROSS A SURFACE MAKING ANGLE α WITH AXIS OF LEAST PRINCIPAL STRESS

and σ_3 (Fig. 4) will be points on the σ-axis. The quantity $(\sigma_1 + \sigma_3)/2$ will be a point midway between σ_1 and σ_3, and $(\sigma_1 - \sigma_3)/2$ will be a length equal to half the distance between σ_1 and σ_3.

Now if we let the point $(\sigma_1 + \sigma_3)/2$ be an origin for a radius vector which makes an angle 2α with the positive σ-axis, and if we let $(\sigma_1 - \sigma_3)/2$ be the length of this radius vector, we shall find that for every value of the angle α the coordinates of the terminus of this vector satisfy equations (6) and hence give the values of σ and τ for the plane which makes an angle α with the axis of least stress. Also, since $(\sigma_1 - \sigma_3)/2$ is constant, as α is changed the radius vector describes a circle which is the locus of all of the values σ and τ for all possible orientations of the reference plane.

This construction (Fig. 4), originally described by the German engineer Otto Mohr (1882) and known as Mohr's circle, enables one to determine by inspection the magnitudes of both the normal and the shear stresses upon a plane of arbitrary orientation α through a given point when the principal stresses σ_1 and σ_3 at that point are known. For example, when α is zero, $\sigma = \sigma_1$, and $\tau = 0$; upon a surface perpendicular to the greatest principal stress, the normal stress is equal to the greatest principal stress and the shear stress is zero. Likewise, when the surface is perpendicular to the least principal stress ($\alpha = 90°$, $2\alpha = 180°$) $\sigma = \sigma_3$, $\tau = 0$. We can also see that the shear stress is a maximum at $2\alpha = \pm 90°$, or $\alpha = \pm 45°$, and that magnitude of the maximum shear stress is

$$\tau_{max} = \frac{\sigma_1 - \sigma_3}{2}. \qquad (7)$$

Another interesting observation pertains to the shear stresses upon any two mutually perpendicular surfaces α and $\alpha + 90°$. The points on Mohr's circle for these two planes will be at 2α and $2\alpha + 180°$ and so will be

a given surface slippage will occur. We can get this information by a separate investigation. If we place the sand in a box (Fig. 5) whose upper half can be skidded with respect to the lower half, thereby producing slippage in the sand along the plane of shear, we can vary the normal stress σ by increasing the load N pressing upon the sand. For a given value of σ, we can increase the shear stress τ by increasing the force T until slippage occurs.

This experiment upon sand has been performed many times by the students of soil mechanics. It is found, to a first approximation, that failure occurs when

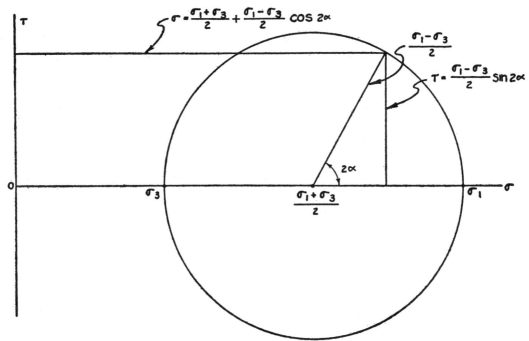

FIGURE 4.—MOHR'S STRESS CIRCLE

diametrically opposite one another Consequently

$$\tau_\alpha = -\tau_{90+\alpha}, \qquad (8)$$

which indicates that the two shear stresses will be of equal magnitudes but of opposite sense.

Properties of the Material

We now require some additional knowledge of the physical properties of the material. We have observed that the material (sand) fails by slippage along discrete surfaces; it will be of interest, therefore, to consider under what conditions of tangential and normal stress upon

$$\frac{T}{N} \text{ or } \frac{\tau}{\sigma} \cong \text{const.} \qquad (9)$$

This experiment is closely analogous to that for determining the friction between two solids. In the latter case a block of one solid is rested upon the horizontal surface of another and loaded by various amounts. For each load, the force required to slide the block is determined. In this case, too, it has been found that for given materials

$$\frac{T}{N} \text{ or } \frac{\tau}{\sigma} \cong \text{const.}$$

The forces T and N or the stresses τ and σ

at failure can be thought of as being the two components of a force or a stress, respectively, acting across the surface of contact at an angle ϕ to the surface normal which is a constant of the material. Hence

$$\frac{T}{N} \text{ or } \frac{\tau}{\sigma} = \tan \phi. \tag{10}$$

In ordinary friction, ϕ is said to be the angle of friction between the given materials; in continuous materials, such as the sand here discussed, ϕ is said to be the *angle of internal friction* of the material. Extensive tests (Krynine, 1941, p. 179) have shown that for sand the angle ϕ is not strictly constant but ranges

about 30°, it follows that this is the steepest angle at which the slope is stable (Fig. 6).

Fracture Lines on Mohr Diagram

Let us now combine our knowledge of the property of the material as expressed by equation (10) with the general properties of stress as expressed by Mohr's diagram. Upon the $\tau\sigma$-plane of Mohr's diagram (Fig. 7), equation (10) represents a straight line of slope ϕ passing through the origin. From symmetry there will be two such lines, one with a slope of $+ \phi$ and the other with a slope of $- \phi$. These lines mark the boundaries between regions of stable

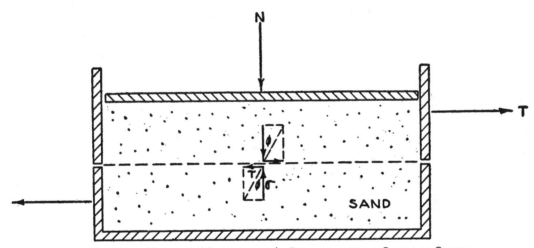

FIGURE 5.—TEST BOX FOR MEASURING τ/σ RATIO AT WHICH SLIPPAGE OCCURS

from about 30° to 35°. The angle of 30° is characteristic of loose sand, whereas the higher angles occur when the sand is in a state of close packing.

The lower angle of 30° is also the approximate angle of repose of loose sand. That this should be so can be seen by noting that upon a slope of inclination β, the normal and tangential stresses across the bottom of a surface lamina of sand will be respectively

$$\sigma = w \cos \beta, \tag{11}$$
$$\tau = w \sin \beta,$$

or

$$\frac{\tau}{\sigma} = \tan \beta. \tag{12}$$

Hence, when the slope is tilted until $\beta = \phi$, slippage will occur. Since in this case ϕ is

and unstable states of stress for the sand. The state of stress corresponding to any point in the region between these lines and the σ-axis has a value

$$\frac{\tau}{\sigma} < \tan \phi,$$

and hence is a condition of stability; for any point outside the lines

$$\frac{\tau}{\sigma} > \tan \phi,$$

so that fracture will occur before this state can be reached. These lines may also be regarded as *fracture lines*, since when the stresses upon any surface reach the values represented by points on either of the two lines, fracture will occur.

Analysis of Reverse Faulting

Combining now the fracture lines of Figure 7 with Mohr's diagram, we can anticipate what

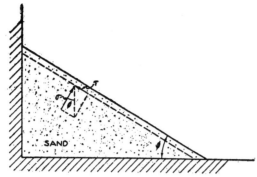

FIGURE 6.—NORMAL AND SHEAR STRESSES ACTING UPON SURFACE LAYER AT ANGLE OF REPOSE

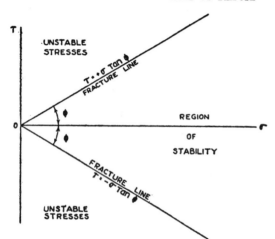

FIGURE 7.—REGIONS OF STABLE AND OF UNSTABLE STRESSES ON THE $\sigma\tau$ PLANE

should happen in the sandbox experiment (Fig. 8). In the initial stage

$$\sigma_1 \cong \sigma_3 = \rho g h.$$

As the partition is pushed forward, the vertical stress σ_3 remains constant while the horizontal stress σ_1 is gradually increased. For each value of σ_1 a particular Mohr's circle will correspond, of which σ_3 and σ_1 will lie at opposite ends of the horizontal diameter. As σ_1 is increased, the Mohr's circle will get larger until finally it will become tangential to the two lines of fracture. At this point, fracture will occur upon one or more of those planes in the body whose trangential and normal stresses have reached the critical ratio. These will be the planes whose angles α correspond to the points

of tangency of the Mohr's circle with lines of fracture. From the diagram the points of tangency occur at

$$2\alpha = \pm(90° + \phi)$$

or

$$\alpha = \pm\left(45° + \frac{\phi}{2}\right). \tag{13}$$

Since, as we have noted, ϕ for sand ranges from 30° to 35°, fracture should occur along one or both planes, making an angle of about 60° to 62.5° to the least principal stress, which is vertical, or 27.5° to 30° to the horizontal. This is in fair agreement with the dip of the reverse faults shown in the photographs. It is a few degrees higher than the average dips observed for the reverse faults—a discrepancy to which we shall return later.

Analysis of Normal Faulting

Let us now consider what happens in the compartment where the normal faults occur. Here, as in the reverse-fault compartment, the initial state of stress at a point inside the sand is given approximately by

$$\sigma_1 = \sigma_3 = \rho g z.$$

In this case, however, as the partition begins to recede, the horizontal stress, which now becomes the least stress σ_3, begins to diminish, while the vertical stress, which is now the greatest stress σ_1, remains stationary, as before.

On Mohr's diagram (Fig. 9) we have, in this case, a fixed maximum stress and a diminishing minimum stress, with again a Mohr's circle of increasing radius. Ultimately, as σ_3 is further decreased, the circle will again become tangential to the fracture lines $\tau = \sigma \tan \phi$, and fracture will occur. As before, the angles α of the planes along which fracture will occur will be given by

$$2\alpha = \pm(90° + \phi),$$

or

$$\alpha = \pm\left(45° + \frac{\phi}{2}\right) \tag{14}$$

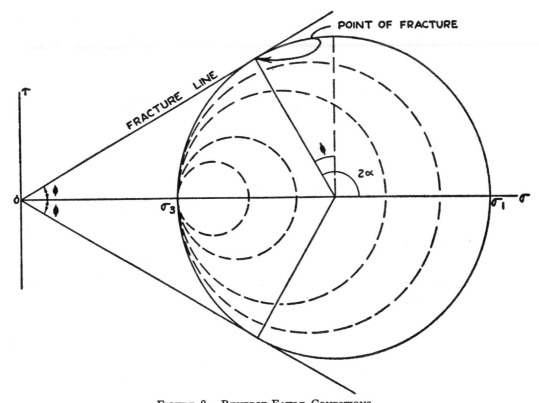

FIGURE 8.—REVERSE-FAULT CONDITIONS

Stress σ_1 increases until Mohr's circle reaches fracture lines. Fracture occurs at angle $\alpha = 45° + \dfrac{\phi}{2}$.

so that in this case also the fracture should be at an angle of about 60° to 62.5° to the least principal stress. In this instance, the least principal stress is horizontal, so the fracture plane should dip about 60° to 62.5°.

The average dip for normal faults in Table 1 was 61.2°, which is a very satisfactory agreement.

The normal-fault case is instructive in still another respect. Geometrically, normal faults correspond to a horizontal elongation of the section in which the faulting occurs. These faults accordingly have commonly been attributed to tensile stress and are often referred to as "tension faults." Loose, dry sand has zero tensile strength and is incapable of sustaining a tensile stress. We observe, however, that normal faults occur in it as readily as in solid rocks, and entirely as the result of compressive stresses. In fact, in both cases the fracture occurs along planes making an angle of 45°—$\phi/2$ to the greatest principal stress. In the reverse-fault case this stress is horizontal; in the normal-fault case it is vertical.

APPLICATION TO COHESIVE MATERIALS AND TO GEOLOGIC STRUCTURES

Coulomb Equation

So far our theory has been restricted to a noncohesive material; rocks, however, are cohesive materials, so it remains to be seen how we may modify the theory to include these also. Since Mohr's circle represents a general property of stress and needs no modification, we need only modify equation (10) which describes the behavior of the material. This modification consists only in adding a cohesive shear stress τ_0 to the stress τ of equation (10) to overcome the cohesion or initial shear strength of the material. The equation then reads

$$\left.\begin{array}{c} \dfrac{\tau - \tau_0}{\sigma} = \tan \phi, \\[2mm] \text{or} \\[2mm] \tau = \tau_0 + \sigma \tan \phi. \end{array}\right\} \quad (15)$$

This equation appears to have been first proposed by the French physicist Coulomb (1776), and, while it may not be rigorously correct, it serves as an excellent first approximation to the behavior of cohesive solids,

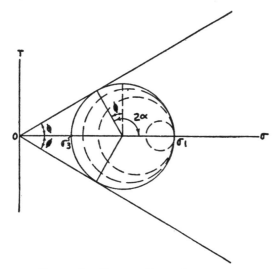

FIGURE 9.—NORMAL-FAULT CONDITIONS
Minimum stress σ_3 decreases until Mohr's circle reaches fracture lines.

including rocks. When plotted upon the $\sigma\tau$-diagram (Fig. 10), equation (15) represents a pair of straight lines with slopes $+\phi$ and $-\phi$ intersecting on the σ-axis at a point $\sigma = \tau_0/\tan\phi$.

The geometrical relations (Fig. 11) of the stresses at the time of fracture in this case are the same as in the noncohesive case. Fracture occurs when the Mohr's circle becomes tangential to the lines of fracture on the $\sigma\tau$-diagram, and, as before, the planes of fracture make an angle of $45° + \phi/2$ to the axis of least stress, or of $45° - \phi/2$ to the axis of greatest stress.

Data upon Angle ϕ for Rocks

Two methods are available for determining the angle of internal friction, ϕ, for rocks. One consists in computing ϕ from the observed angle between the surfaces of slip and the axes of principal stress when the latter are known; the second is by means of tests of rock specimens under known conditions of triaxial stress.

Prandtl and Rinne (Nádai, 1931a, p. 109–

110) carried out a series of tests on polished cylindrical specimens of rock by compressing them axially to incipient failure. On the cylindrical surfaces, two families of helical lines of fracture were observed intersecting at an angle 2θ bisected by the axis of greatest stress. The angle θ is then complementary to the angle α, and from the Mohr diagram (Fig. 11)

$$\phi = 2\alpha - 90 = 90 - 2\theta.$$

On two photographs of marble specimens (shown by Nádai) the angle 2θ was 41° and 45°, respectively. The corresponding values of ϕ would therefore be 49° and 45°.

The same principal may be used in connection with the dip angles of simple normal and reverse faults. When faulting occurs at or near the surface of the ground in nearly level topography, the principal stresses must be respectively nearly vertical and horizontal. Here also

$$\phi = 90 - 2\theta$$

where θ is the angle of dip of reverse faults, or the angle of hade of normal faults.

Thus from fault data, with the angles of dip of simple reverse faults, or the hades of normal faults lying between the limits 20° and 40°, the angle of internal friction would lie between 10° and 50°. For the peak frequency in the angle of hade for normal faults of 27°, as reported by Sax, the peak frequency for the angle ϕ would be 36°.

The second method, that of testing specimens under known conditions of triaxial stress, consists in testing a cylinder under simultaneous radial and axial stress. The radial stress is supplied by a fluid pressure and the axial stress by a piston. The fluid is set at a chosen pressure $p = \sigma_2 = \sigma_3$ (or $p = \sigma_1 = \sigma_2$) and the axial stress, σ_1, is increased (or decreased if the axial stress is σ_3) until failure occurs. The Mohr's circle for the values of σ_1 and σ_3 at which failure occurs is constructed. This procedure is repeated with additional specimens of the *same* material until a family of over-lapping circles of increasing radii is obtained. The envelopes of these circles obviously represent the empirically determined loci of the fracture curves of the diagram. The angle of slope of the curves with respect to the σ-axis is the angle of internal friction for the given material.

Tests of this kind appear first to have been made on rocks by von Kármán (1912); he

few results on rock and concrete have now been published (Jones, 1946; Balmer, 1946; McHenry, 1948). On rock cores from Boulder Dam (rock type not specified), the results of

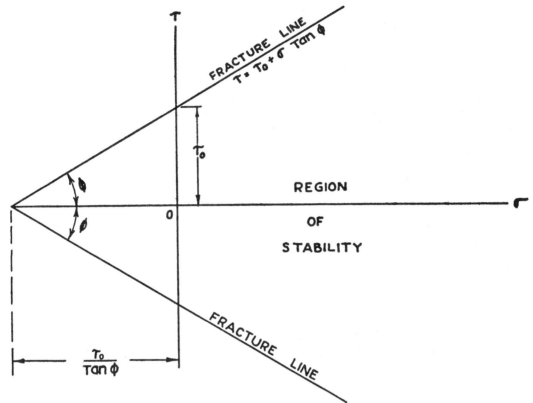

FIGURE 10.—POSITION OF THE FRACTURE LINES ON $\sigma\tau$ DIAGRAM FOR COHESIVE MATERIALS

ran a series of tests on marble which was found to behave as a brittle material at lower radial pressures but to flow plastically at higher pressures. The envelopes of the Mohr's circles thus obtained were convex outward with ϕ varying from an initial value of about 40° (corresponding to $\theta = 25°$) to a final value of near zero. Also conjugate slip surfaces on the slightly deformed specimens show an acute angle $2\ \theta \cong 50°$.

Triaxial tests on concrete were reported by Ros and Eichinger (1928). The angle θ as observed directly varied from 22° to 33°, which would correspond to values from 24° to 46° for the angle ϕ.

The U. S. Bureau of Reclamation at Denver has recently installed a large triaxial testing machine for testing cores of rock and concrete in connection with the construction of large dams (Blanks and McHenry, 1945), and a

16 tests gave for ϕ the value of 35°, which would correspond to $\theta = 27.5°$. Tests on concrete by both Jones and McHenry have given values of $\phi \cong 45°$, or of $\theta \cong 22.5°$.

It is interesting to compare these results with tests on asphalt aggregates for road building recently reported by Nijboer (1942). Triaxial tests were made upon dozens of different mixes under a wide range of conditions. In most cases the envelopes of the Mohr circles were straight lines though occasionally one with a small outward convexity was observed.

The angles ϕ found by these tests ranged between the extremes of 19° and 38°. The most frequently observed value, however, was 28°, and, for more than 95 per cent of all the measurements made, the observed values lay within the range 28° $\pm$ 4°. For rocks, this would correspond to normal faults with hades, or reverse faults with dips, of 31° $\pm$ 2°. Kry-

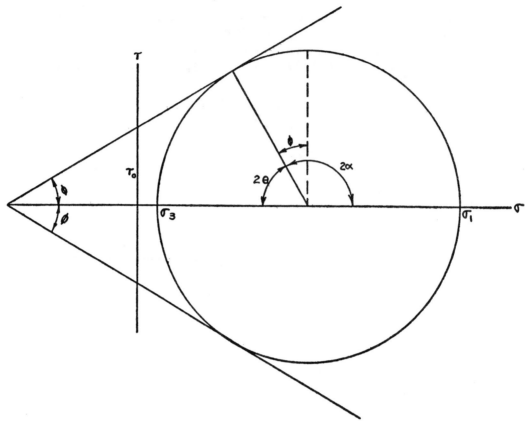

FIGURE 11.—STRESS CONDITIONS FOR FAULTING IN COHESIVE MATERIALS

Faults occur at angle $\alpha = 45° + \frac{\phi}{2}$ to the axis of least stress.

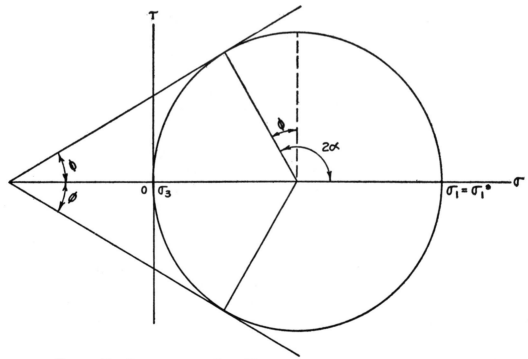

FIGURE 12.—CONDITIONS FOR ZERO HORIZONTAL STRESS IN NORMAL FAULTING

nine (1941, p. 172) reports that, for wet clay, ϕ ranges from a few degrees to 15° or 20°. This would correspond to normal-fault dips between 45° and 55°.

Thus, although data upon the angle of internal friction for rocks are as yet sparse, the computed angles of faulting are in good agreement with field observations.

Tension in Normal Faulting

We have already seen that, in the case of cohesionless materials, normal faults are produced only under compression. For cohesive materials (Fig. 12), it will be seen that, for any given material, there must exist at the time of faulting a certain critical value of σ_1, say σ_1^* for which σ_3 will be zero. Then, for all values of $\sigma_1 > \sigma_1^*$, the horizontal stress σ_3 in normal faulting will be compressive, and for all values of $\sigma_1 < \sigma_1^*$ the horizontal stress σ_3 will be one of tension.

Since, in the case of normal faulting, the vertical stress at a given depth is equal to the pressure of the overburden, there will be some critical depth for a given material below which the horizontal stress producing normal faults will be compressive and the existence of a tensile stress will be impossible. At less than this depth, the horizontal stress will be one of tension.

The approximate depth at which this transition will occur can be obtained from crushing-strength data of rocks. The conditions

$$\sigma_3 = 0,$$

$$\sigma_1 = \sigma_1^*,$$

are those which are satisfied approximately in the crushing-strength tests for rocks at atmospheric pressure. Consequently, for a given rock the critical stress σ_1^* is equal approximately to its crushing strength. For dry rocks the depth at which this vertical stress occurs would be

$$z^* = \sigma_1^*/\rho g$$

where ρ is the bulk density of the material. In the earth, however, the rocks ordinarily are saturated with ground water, which exerts an upward force of buoyancy. In this case, the effective stress is that transmitted by the solid system. Hence this effective stress $\bar{\sigma}_1$ at a given depth is

$$\bar{\sigma}_1 = (\rho_s - \rho_w) \, g \, (1 - \epsilon) \, z,$$

or for the critical depth

$$z^* = \bar{\sigma}_1^*/[(\rho_s - \rho_w) \, g \, (1 - \epsilon)], \qquad (16)$$

where ρ_s is the grain or mineral density ρ_w the density of water, and ϵ the porosity of the rock.

The crushing strengths of rocks range from values near zero, for recent sediments, to as high as 3×10^9 dynes/cm² for granites. Such tests, however, are made upon nearly flawless specimens, so that, for large volumes of rock broken by numerous joints, the effective values will no doubt be considerably smaller. The Beaumont clay underlying Houston for example has a bearing strength of only about 10^6 dynes/cm². Recorded crushing strengths of sedimentary rocks are commonly of the order of $5-15 \times 10^8$ dynes/cm².

Introducing these values of σ_1^* into equation (16) and solving for z^*, we find the maximum depths at which tension can exist in rocks of various types. In the Gulf Coast sediments, for example, this depth is found to be of the order of 10^3 cm., or 10 meters (about 30 feet). In the more consolidated sediments, it may increase to a few hundred meters; and, in the strongest crystalline rocks, it may reach values of several kilometers.

Yet the Gulf Coast sediments are cut by numerous normal faults extending from the surface to depths of thousands of meters. It is clear, therefore, that aside from a surface veneer, these sediments have been broken by a series of normal faults in response exclusively to compressive stresses.

PROBLEM OF ASYMMETRICALLY FOLDED AND FAULTED BELTS

One-sided Thrust Hypothesis

With this background, let us consider an equally familiar but more complex problem. One of the commoner types of orogenic structure is the belt of asymmetrically folded and faulted sediments as exemplified in North America by the Appalachian Mountains, the Ouachitas, the foothills of the Rocky Mountains, and some of the West Coast ranges, and in Europe by the Alps, the Juras, and the Scottish Highlands. In these cases, the folds are unidirectionally overturned and accompanied by reverse faulting, with the direction of thrusting —that is, the direction of the relative dis-

Application of the Newtonian Laws of Motion

That this picture is somehow deficient becomes evident when considered in the light of the Newtonian laws of motion, according to which the sum of all the forces acting upon the

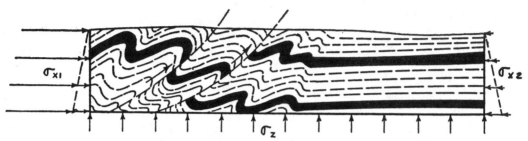

FIGURE 13.—COMPRESSIVE STRESSES ON ENDS OF SECTION OF ASYMETRICALLY FOLDED AND FAULTED SEDIMENTS

placement of the upper block with respect to the lower—in the same sense as the overturning of the folds. Moreover, where data are available, as in the foothills region of Alberta where seismic surveys have been made and numerous oil wells drilled, the fault surfaces have been found to be concave upward, the dip decreasing with depth.

Considerable attention has been given to the causes of these structures, and the hypothesis most often favored has been that formulated by J. D. Dana (1847a; 1847b)—that such structures are formed by a one-sided "active thrust" from the direction opposite to that of the overturning. Thus, the Appalachian Mountains would have been formed by an active thrust from the southeast, which presumably died out with distance toward the northwest. An auxiliary question often considered is how far rocks of a given strength are capable of transmitting such a thrust.

The state of stress thus postulated and its associated orogeny is illustrated in Figure 13, where a large "active thrust" σ_{x_1} acts on the left-hand end of the section and dies out to the much smaller stress σ_{x_2} on the right-hand end. Not only does this appeal to the intuition, but it is supported by the deformational evidence as well. Since the rocks on the left are intensely deformed whereas those on the right are undistorted, the inference is justified that σ_{x_1} is very much greater than σ_{x_2}.

block in a given direction must equal the product of the mass of the block and the component of its acceleration in that direction. Taking the x-axis as horizontal, this requires that in the x-direction

$$\Sigma F_x = \int_0^{z_1} (\sigma_{x_1} - \sigma_{x_2})dz = ma_x \quad (17)$$

As an order of magnitude, $(\sigma_{x_1} - \sigma_{x_2})$ may be taken to be about 10^{+9} dynes/cm². Then if the block of Figure 13 is assumed to be 100 kilometers wide, 10 kilometers deep, and 1 centimeter thick, ΣF_x will be about 10^{18} dynes.

For the ma-term, the mass of the block will be about 3×10^{13} grams. Velocities in known orogenies are rarely more than a few centimeters per year, so that secular accelerations much greater than a centimeter per year per year or about 10^{-15} cm/sec² are unlikely. Hence, the ma-term would have a magnitude only of the order of 10^{-2} gm cm/sec², which is infinitesimal as compared with the forces acting, and the sum of all forces must effectively be zero throughout the orogenic process.

Since, with the stress distribution postulated in Figure 13, the sum of the horizontal forces acting upon the block is manifestly not zero, some essential element must be lacking. Since body forces are ineffective horizontally, the only remaining possibility is shear stresses upon the vertical faces parallel to the plane of the drawing and along the bottom of the block. In arcuate structures, shear stresses on the vertical faces normal to the strike exist, but they vanish as the curvature approaches zero and the structure becomes rectilinear.

Assuming the structure to be rectilinear,

of opposite rotational senses. Hence there must also be a downward-directed shear stress on the left-hand vertical face of the block equal in magnitude at the bottom corner to the stress on the bottom, and approaching zero toward the top. A similar shear stress, but of

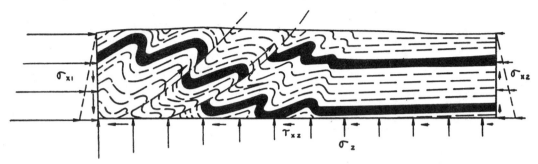

FIGURE 14.—COMPLETE SYSTEM OF TWO-DIMENSIONAL STRESSES ACTING UPON BOUNDARIES OF BLOCK

we are then left only with the bottom shear stress τ_{zs}, whose magnitude at any given distance x along the bottom must just be sufficient to balance the force due to the stress difference $d\sigma_x$ in the horizontal distance dx. Thus, at distance x

$$\tau_{zs} dx = + \int_0^{z_1} \left[\left(\sigma_x + \frac{\partial \sigma_z}{\partial x} dx \right) - \sigma_s \right] dz$$

or

$$\tau_{zs} = + \int_0^{z_1} \frac{\partial \sigma_x}{\partial x} dz ; \qquad (18)$$

that is the shear stress on the bottom of the block at a given point is proportional to the gradient of the normal stress in that direction. The gradient of the normal stress would be approximately the same as the gradient of the intensity of the deformation. Hence, in Figure 13 the shear stress would have a maximum value near the lower left-hand corner of the block and would decline to near zero at the lower right-hand corner. The complete equation of forces in the x-direction must therefore be

$$\Sigma F_x = \int_0^{z_1} (\sigma_{x1} - \sigma_{x2}) dz - \int_{x_1}^{x_2} \tau_{zs} dx = 0. \quad (19)$$

As noted, the components of shear stress parallel to a plane normal to both are the same on any two mutually perpendicular planes, but

magnitude near zero and directed upward, must exist on the right-hand end.

In addition, all surface forces and body forces must be so related that all turning moments acting on the block are zero. When these several conditions are satisfied the approximate set of boundary stresses shown in Figure 14 is obtained as being necessary to produce the generalized type of orogeny shown.

Interior Stress Distribution

From the boundary stresses shown in Figure 14, the approximate pattern of the stress distribution in the interior of the block can be inferred.[1] Trajectories of principal stress are curves which at every point are tangent to a given principal stress. Since the principal stresses are mutually perpendicular, then three perpendicular stress trajectories, one each for σ_1, σ_2, and σ_3, must pass through each point. In a two-dimensional stress system on a plane parallel to σ_1 and σ_3, two families of orthogonal stress trajectories, one everywhere parallel to σ_1 and the other to σ_2, will exist.

Recalling that, when the principal stresses are unequal, shear stresses exist upon all sur-

[1] The reasoning here is intentionally only qualitative, the objective being to convey an intuitive sense of the principles involved. A more formal quantitative analysis is given in a companion paper by the author's colleague, W. Hafner (1951).

faces except those perpendicular to the stress trajectories, the approximate pattern of the stress trajectories in the interior of the block may be determined from the combination of normal and shear stresses on its boundaries.

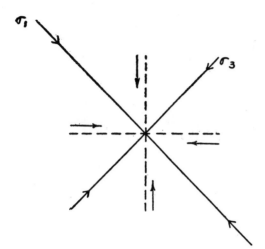

FIGURE 15.—DIRECTIONS OF SHEAR STRESSES IN VARIOUS QUADRANTS DEFINED BY PRINCIPAL STRESSES

can be definitely known. Across any plane on which a compressive normal stress σ and shear stress τ exists, the axis of greatest principal stress, σ_1, will lie in the quadrant between the plane normal and the direction *from* which the shear stress occurs (Fig. 15).

The boundary stresses of Figure 14 will produce the approximate pattern of principal-stress trajectories in the interior shown by the solid curves of Figure 16. The family of greatest-stress trajectories, σ_1, is tangential to the upper surface of the block, plunging downward and divergent to the right, and convex upward. The family of least-stress trajectories, σ_3, is orthogonal with this.

Surfaces of Potential Faulting

If we now assume that faulting is most likely to occur along surfaces tangent to the intermediate stress, σ_2, and at an angle of about $45° - \dfrac{\phi}{2}$ to the greatest stress, where ϕ

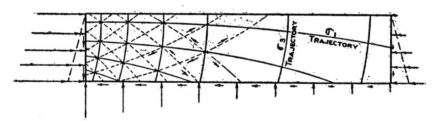

FIGURE 16.—TRAJECTORIES OF PRINCIPAL STRESSES (SOLID LINES), AND OF POTENTIAL REVERSE-FAULT SURFACES (BROKEN LINES) COMPATIBLE WITH THE BOUNDARY STRESSES OF FIGURE 14

Arrows indicate directions of potential slip

On the upper surface of the block, the normal stress is the atmospheric pressure and the shear stresses are zero. Therefore, at any point upon this surface, one of the stress trajectories must terminate perpendicularly, and the other must be tangential to the surface.

Across the left-hand end of the block, and the bottom, the stress trajectories must cross obliquely, with the angle of obliquity approaching zero near the upper surface, and increasing downward. Without knowing the actual magnitudes of both the normal and shear-stress components, the angles of incidence of the stress trajectories can be determined only qualitatively, but the quadrant in which each falls

may be taken to be about 30°, we are able to sketch in on Figure 16 the traces of the surfaces of potential faults, and indicate their directions of relative displacement. These comprise two families of curves (shown by dashed lines) which everywhere intersect the σ_1-trajectories at an angle of about 30°. In virtue of the divergence and curvature of the stress trajectories, these surfaces are also curved, one set convex and the other concave upward.

At any given point, the stresses on each of these conjugate surfaces are the same; so that if they are at the critical value for fracturing on one surface, the same is true for the other. However, slippage along a finite surface involves an integral of conditions along the entire surface. Whatever this integral should be, in homogeneous materials with symmetrical stress distributions, its value over each of a pair of conjugate surfaces of potential slippage will be the same, so that equal slip should occur on each family of surfaces.

In asymmetrical cases, the integral of any stress quantity over one surface or family of surfaces will in general be different from that over the conjugate surface or family. Because of this inequality, slippage should occur on one set before on the other, or on only one of the two sets. In the asymmetrical geological system here considered, faulting develops almost exclusively on the concave-upward set of surfaces. The same was true for the reverse faulting in the sandbox experiment.

In experiments with symmetrical systems, slippage on both sets of surfaces is observed.

Returning to the sandbox experiment, it will be recalled that the average angle of hade for the normal faults was about 29°, and of dip for the reverse faults was 25°, whereas our theory indicated that the two should be the same. Perhaps we can now explain this discrepancy. We assumed the principal-stress trajectories to be strictly horizontal and vertical. However, in the reverse-fault case, the sand was pushed along the bottom which exerted a frictional reaction in the form of a bottom shear stress. The effect of this would be to deflect the horizontal stress trajectories downward and thus to lower the dip angle of the reverse faults.

Other shear stresses along the vertical faces of the partition must also have existed which would likewise tend to produce some deviation in observed angles from the idealized conditions assumed.

CONCLUSION

It thus appears that some of the more common large-scale geological structures are in satisfactory agreement with theoretical deductions based upon the stress patterns inferred from the observed deformation and Newton's laws of motion, and the empirically determined properties of rocks under known stress conditions.

REFERENCES CITED

Anderson, E. M. (1942) *Dynamics of faulting and dyke formation*, Oliver and Boyd, Edinburgh and London, p. 191.

Balmer, Glenn G. (1946) *A revised method of interpretation of triaxial compression tests for the determination of shearing strength*, U. S. Bur. Reclamation Basic Structural Lab. Rep. No. SP-9.

Billings, Marland P. (1946) *Structural geology*, Prentice-Hall, Inc., New York, p. 206.

Blanks, R. F., and McHenry, Douglas (1945) *Large triaxial testing machine built by Bureau of Reclamation*, Eng. News-Record, vol. 135, p. 171–173.

Chamberlin, R. T., and Shepard, F. P. (1923) *Some experiments in folding*, Jour. Geol., vol. 31, p. 490–512.

Coulomb, C. A. (1776) *Essai sur une application des regles des maximis et minimis à quelques problèmes de statique relatifs a l'architecture*, Acad. Sci., Paris, Mem. pres. divers savants, vol. 7.

Dana, J. D. (1847a) *Geological results of the earth's contraction*, Am. Jour. Sci. (2), vol. 3, p. 176–188; vol. 4, p. 88–92.

—— (1847b) *On the origin of continents*, Am. Jour. Sci. (2), vol. 3, p. 381–398.

Hafner, W. (1951) *Stress distributions and faulting*, Geol. Soc. Am., Bull., vol. 62, p. 373-398.

Jones, Valens (1946) *Tensile and triaxial compression tests of rock cores from the passageway to penstock tunnel N-4 at Boulder Dam*, U. S. Bur. Reclamation, Basic Structural Res. Rept. No. SP-6.

von Kármán, Th. (1912) *Festigkeits Versuche unter allseitigen Druck*, Mitteilungen über Forschungzarbeiten auf dem Gebiete des Ingenieurswesens, H. 118, V.d.I.

Krynine, D. P. (1941) *Soil mechanics*, McGraw-Hill, Inc., New York, p. 451.

Leith, C. K. (1913) *Structural geology*, 1st Ed., Henry Holt & Co., p. 55.

McHenry, Douglas (1948) *The effect of uplift pressure on the shearing strength of concrete*, International Congress on Large Dams, June 1948.

Mead, W. J. (1925) *Geologic role of dilatancy*, Jour. Geol., vol. 33, p. 685–698.

Mohr, Otto (1871; 1872) *Beiträge zur Theorie des Erddrucks*, Zeitschr. Architekten und Ingenieur—Ver. Hannover, vol. 17, p. 344; vol. 18, p. 67 and 245.

—— (1882) *Über die Darstellung des Spannungzustandes eines Körpelementes*, Zivil Ingenieure, p. 113.

—— (1900) *Welche Umstände bedingen die Elastizitätsgrenze und den Bruch eines Materials*, Zeitschr. Vereins deutsches Ing., p. 1524.

Nádai, A. (1928) *Plasticität und Erddruck*, Handbuch der Physik, VI, J. Springer, Berlin, p. 428–500.

—— (1931a) *Plasticity*, McGraw-Hill, Inc., New York, p. 349.

—— (1931b) *Phenomenon of slip in plastic materials*, Edgar Marburg Lecture, Am. Soc. Test. Mat., Pr., vol 31, pt. II, p. 11–46.

Nijboer, L. W. (1942) *Onderzoek naar den weerstand*

van bitumen-mineraalaggregaat mengsels tegen plastische deformatie, N. V. Noord-Hollandsche Uitgevers Maatschappij, Amsterdam, p. 232.

Rankine, W. J. M. (1857) *On the stability of loose earth*, Royal Soc. London, Philos. Tr., vol. 147.

Sax, H. G. J. (1946) *De tectoniek van het Carboon in het Zuid-Limburgsche mijngebied*, Mededeelingen van de Geologische Stichting, Ser. C-I-I no. 3, p. 1–77.

Ros, H. C. M., and Eichinger, A. (1928) *Experimental attempt to solve the problem of failure in materials—nonmetallic materials*, Federal Mat. Testing Lab., E. T. H., Zürich, Rept. No. 28. Translation by F. Stenger, U. S. Bur. Reclamation, Denver.

SHELL OIL CO., 3737 BELLAIRE BLVD. HOUSTON 5, TEXAS.
MANUSCRIPT RECEIVED BY THE SECRETARY OF THE SOCIETY OF AMERICA, SEPTEMBER 23, 1949.

The American Association of Petroleum Geologists Bulletin, v. 47, no. 5 (May 1963), p. 717-755.

BULLETIN

of the

AMERICAN ASSOCIATION OF
PETROLEUM GEOLOGISTS

MAY, 1963

EXPERIMENTAL DEFORMATION OF SEDIMENTARY ROCKS UNDER CONFINING PRESSURE: PORE PRESSURE TESTS[1]

JOHN HANDIN,[2] REX V. HAGER, JR.,[2] MELVIN FRIEDMAN,[2] AND JAMES N. FEATHER[2]

Houston, Texas

ABSTRACT

Berea sandstone, Marianna limestone, Hasmark dolomite, Repetto siltstone, and Muddy shale have been subjected to triaxial compression tests in which the external confining pressures and internal pore pressures (to 2 kilobars) are applied and measured independently. The interstitial water pressure is maintained constant throughout the test, and porosity changes are determined as functions of permanent shortening.

The ultimate strength and ductility of porous rocks are found to depend on effective confining pressure—the difference between external and internal pressures when the pore fluid is chemically inert, the permeability is sufficient to insure pervasion and uniform pressure distribution, and the configuration of pore space is such that the interstitial hydrostatic (neutral) pressure is transmitted fully throughout the solid framework.

At high effective pressures (about 1 kilobar) porosity decreases with progressive permanent strain. At intermediate pressures (about 500 bars) the porosity remains essentially constant for compressions as great as 20 per cent, and at low pressures (about 200 bars and less) the rocks are dilatant. Microscopic examination of the sandstone reveals that grain breakage becomes progressively less important as pore pressure is increased until the deformation becomes entirely intergranular. Since the aggregate is initially closely packed, the shortening leads inevitably to increased void volume.

An explanation of the pore pressure effects on the basis of Coulomb friction is consistent with the empirical data. High pore pressure reduces internal friction (but does not modify the coefficient). The rock is weak and relatively brittle; faulting is favored. These facts support the Hubbert-Rubey theory of large-scale overthrust faulting and enhance our understanding of diapir structures in shales, of sandstone dikes, of localized faulting in zones of "abnormal" formation pressures, and of high interstitial pressures in young intensely folded rocks of low permeability.

INTRODUCTION

In two previous papers the writers emphasized the importance for both structural geology and engineering of the dynamics of rock deformation, which is best learned through controlled laboratory experiments realistically simulating the four significant natural environmental factors—confining pressure, interstitial fluid (formation)pressure, temperature (geothermal), and time of loading. Because of the interdependence of these factors, the heterogeneity of crustal materials, and the lack of generally applicable theories, an adequate empirical understanding of the mechanical properties of rocks can be acquired only by investigating the effects of the variables separately.

In the first paper on the experimental deformation of sedimentary rocks (Handin and Hager, 1957) are recorded the results of room-temperature tests on 23 dry rocks in which only the confining pressure was varied in the range of 0 to 2 kilobars.[3] Several-fold increases of ultimate

---

[1] Manuscript received, July 2, 1962.

[2] Shell Development Company (A Division of Shell Oil Company), Exploration and Production Research Research Division (Publication No. 294).

[3] The *bar*, the fundamental unit of pressure, is 10^6 dynes/cm.^{2}; 1 bar is approximately 14.5 psi.; 1 kilobar $= 10^3$ bars $= 10^9$ dynes/cm.$^2 \approx 14.5 \times 10^3$ psi.

717

strength are found in all the rocks studied. In anhydrite, some dolomites, limestone, shale, and siltstone, ductility is enhanced by confining pressure, but silica-cemented sandstone, quartzite, slate, and some dolomites remain brittle.[4]

The data from high-temperature experiments are reported in the second paper (Handin and Hager, 1958). Eleven rocks were tested in the pressure range of 0 to 2 kilobars at temperatures from 24° to 300°C. to simulate depths of burial to 30,000 feet. Heating tends to enhance ductility and to reduce yield stress. The increase of temperature at a given confining pressure results in the reduction of ultimate strength along with yield stress, except rarely in work-hardening rocks, where greater ductility permits the attainment of larger permanent strains and hence larger stress differences. Except for halite, the ultimate strengths of all materials tested at any simulated depth to 30,000 feet exceed the crushing strengths at atmospheric conditions.

This paper deals with the third variable—the pressure of interstitial fluids in combination with confining pressure and temperature. Because there are now three interdependent variables, their effects can be adequately evaluated only from many experiments on a single sample. This severely restricts the number of different rocks that can be tested in one laboratory in a reasonable length of time. Included here are the results for only five rocks—Hasmark dolomite, Marianna limestone, Berea sandstone, Muddy shale, and Repetto siltstone. The writers believe that the significant principles have nonetheless been delineated for the following experimental conditions: confining pressures (kerosene) of 0 to 2 kilobars, pore pressure (water) of 0 to 2 kilobars, temperatures of 24° to 300°C., and constant strain rate of 1 per cent per minute. Samples are loaded perpendicular to bedding, fissility, or foliation.

In attempting to understand the influence of pore pressure, the writers have (1) determined stress-stain relations from triaxial compression tests in which confining and pore pressures were applied and measured independently, (2) studied the effects of deformation on porosity and permeability, and (3) made microscopic examinations of experimentally deformed sandstone.

[4] For definitions of the technical terms used to describe rock deformations, see Handin and Hager (1957, p. 3–5).

Since both confining and pore pressures must remain constant throughout an experiment, a specimen must be permeable enough to allow pervasion by the interstitial fluid so that initial saturation will be assured and the pore pressure will attain equilibrium throughout. The fluid must be free to flow in or out of the specimen during the deformation so that any void volume change can be precisely determined and the pore pressure will remain uniformly distributed and truly measurable externally. This requirement is probably met by all the rocks except the shale, which can not be saturated within 24 hours. The effects of interstitial fluids on the deformational behavior of rocks can be both physical and chemical. Both are interesting and are doubtlessly important in nature. However, chemical processes may be exceedingly complicated and difficult to understand. In the first attempt to comprehend the influence of pore pressure, it is wise to exclude chemical effects by using relatively inert distilled water in the experiments. (In this regard, an inert gas would be even better but would make volume measurements impractical.) Even pure water can alter the properties of the swelling clays, but except for the siltstone, the rocks chosen for study contain little or none of this material. Except in the siltstone, there is evidence that the pore pressure effects of these experiments are purely physical; that is, the constituents of the rocks are unaffected by the introduction of water and are not attacked during the deformation.

One purpose of this study is to evaluate the concept of *effective stress*, which has already been found useful in soil mechanics (Terzaghi, 1943, p. 51).

Consider a dry, homogeneous, jacketed cylindrical test specimen which is initially under an external hydrostatic *confining pressure* p_c. Let S_1, S_2, and S_3 designate the maximum, intermediate, and minimum principal total stresses, respectively; regard compressive stresses as positive. Then

$$S_1 = S_2 = S_3 = p_c.$$

Let the axial pressure be increased or decreased by an amount $\Delta S = S_1 - S_3$, the *differential pressure*, while the radial stresses all remain equal to the confining pressure. Then

$$S_1 > S_2 = S_3 = p_c, \qquad S_1 = S_3 + \Delta S = p_c + \Delta S$$

in a *compression* test, and

$$S_3 < S_2 = S_1 = p_c, \qquad S_3 = S_1 - \Delta S = p_c - \Delta S$$

in an *extension* test (Fig. 1).

Now suppose that an internal hydrostatic *pore pressure* p_p is applied. The concept of effective stress can be expressed analytically as

$$\sigma_1 = S_1 - p_p, \quad \sigma_2 = S_2 - p_p, \quad \sigma_3 = S_3 - p_p,$$

where σ_1, σ_2, and σ_3 are the maximum, intermediate, and minimum principal *effective* stresses, respectively.

In the compression test

$$\sigma_2 = \sigma_3 = (S_2 - p_p) = (S_3 - p_p) = (p_c - p_p),$$

$$\sigma_1 = (S_1 - p_p) = (S_3 + \Delta S - p_p) = \Delta S + (p_c - p_p),$$

$$(1a)$$

and in the extension test

$$\sigma_1 = \sigma_2 = (S_1 - p_p) = (S_2 - p_p) = (p_c - p_p),$$

$$\sigma_3 = (S_3 - p_p) = (p_c - p_p) - \Delta S.$$

$$(1b)$$

The quantity $(p_c - p_p)$ is here defined as the *effective confining pressure.*

All normal pressures in the solid framework are merely reduced by an amount equal to the "neutral" pressure p_p; therefore, from the data of triaxial tests the effective principal stresses are calculated by subtracting the pore pressure from the external pressures (Fig. 1).

Another purpose is to test the applicability of the theory of *internal friction* proposed by Coulomb (1776) and later generalized by Mohr (1900). The total shearing resistance offered by an isotropic material to failure (shear fracturing, faulting, or distributed shear) is supposed to be the sum of a *cohesive strength* τ_0 (independent of direction) and a term reflecting frictional resistance to slip along the potential failure plane. This term is the product of the effective normal stress σ_n across that plane and a *coefficient of internal friction* $n = \tan \phi$, where ϕ is an angle analogous with that of ordinary sliding friction. Thus

$$\tau = \tau_0 + \sigma_n \tan \phi. \qquad (2)$$

The normal stress S_n and shear stress T on any plane parallel to the direction of intermediate principal stress and inclined at an angle θ with respect to the direction of maximum principal stress are given by

$$S_n = \frac{S_1 + S_3}{2} - \frac{S_1 - S_3}{2} \cos 2\theta,$$

$$T = \frac{S_1 - S_3}{2} \sin 2\theta. \qquad (3)$$

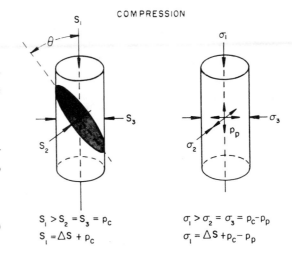

COMPRESSION

$$S_1 > S_2 = S_3 = p_c$$
$$S_1 = \Delta S + p_c$$

$$\sigma_1 > \sigma_2 = \sigma_3 = p_c - p_p$$
$$\sigma_1 = \Delta S + p_c - p_p$$

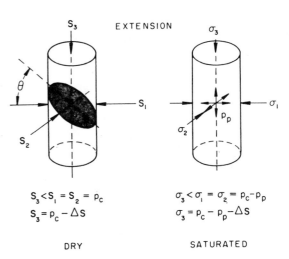

EXTENSION

$$S_3 < S_1 = S_2 = p_c$$
$$S_3 = p_c - \Delta S$$

$$\sigma_3 < \sigma_1 = \sigma_2 = p_c - p_p$$
$$\sigma_3 = p_c - p_p - \Delta S$$

DRY   SATURATED

FIG. 1.—States of stress developed in dry (left) and fluid-saturated (right) homogeneous triaxial test specimens. Symbols S and σ denote total and effective stresses, respectively. p_c and p_p denote confining and pore pressures, respectively. ΔS is axial differential stress. Angles θ measure inclinations of faults relative to maximum principal pressure.

When the pore pressure is subtracted from the principal total stresses, the effective stresses are

$$\sigma_n = S_n - p_p = \frac{\sigma_1 + \sigma_3}{2} - \frac{\sigma_1 - \sigma_3}{2} \cos 2\theta,$$

$$\tau = T = \frac{\sigma_1 - \sigma_3}{2} \sin 2\theta, \qquad (4)$$

and equation 2 becomes

$$\tau = \tau_0 + (S_n - p_p) \tan \phi. \qquad (5)$$

Internal friction is reduced because the normal pressure across the potential failure plane is lower in the amount of the pore pressure, not because the coefficient of friction is affected.

The graphical method of Mohr (1882) can be used to solve for the shear stress and effective normal stress on an arbitrary plane, given the extreme effective principal stresses σ_1 and σ_3 (Fig. 2). The point $(\sigma_1+\sigma_3)/2$ is the origin of a radius vector of length $(\sigma_1-\sigma_3)/2$, making an angle 2θ with the positive σ_n axis. The end of this vector marks the coordinates τ and σ_n, satisfying equations 4.

From the data of a series of triaxial compression tests for increasing values of effective confining pressure, one can plot a Mohr stress circle for each pair of values of σ_1 and σ_3 at failure to obtain a family of circles whose centers lie at successively larger distances out on the normal stress axis and whose radii become progressively longer. The Mohr (1900) envelope curve tangent to all these circles is the locus of points satisfying equation 5. The slope of the curve is $\tan \phi$ and the intercept on the ordinate at zero normal stress gives the value of τ_0.

Inspection of Figure 2 reveals that

$$\theta = \pm 45° + \frac{\phi}{2}. \tag{6}$$

There is a conjugate pair of potential failure planes, the acute included angle of which is bisected by the direction of maximum principal stress, and the intersection of which is parallel with the direction of intermediate principal stress. (For details of this method see, for example, Hubbert, 1951, p. 358–364; Handin and Hager, 1957,

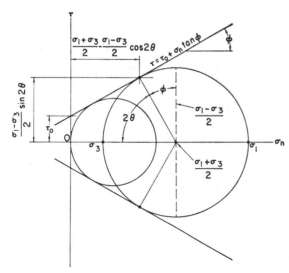

FIG. 2.—Mohr's graphical solution for shear stress τ and normal stress σ_n on plane of inclination θ, given extreme principal stresses σ_1 and σ_3. Pair of Mohr envelope curves for failure superimposed.

p. 26; Hubbert and Rubey, 1959, p. 123–125, 138–139).

PREVIOUS WORK

From the inception of experimental work on rock deformation under high hydrostatic confining pressure 50 years ago, it has been recognized that a jacketed specimen behaves differently from an unjacketed one which is exposed to the confining fluid. Griggs (1936, p. 567) noted that the ultimate strength of jacketed Solenhofen limestone under 10 kilobars kerosene pressure exceeded that of the unjacketed rock by 40 per cent. Goguel (1948, p. 168–206) found that the strengths of gypsum, limestone, sandstone, and shale were lower when these rocks were subjected directly to 500 bars oil pressure than when the pressure was transmitted to the specimens through a rubber sheath. Moreover, the unjacketed specimens remained brittle, whereas the rubber-encased rocks acquired ductility under confining pressure. Handin (1953, Fig. 24) observed that the strength of unjacketed rock salt exposed to kerosene under pressures from 250 to 1,000 bars was but 60 per cent of that of the copper-jacketed rock. All too frequently, the writers, and doubtless other workers as well, have inadvertently measured the properties of the unjacketed material simply because the jacket leaked. For porous materials at least, enhancements of strength and ductility under confining pressure were invariably less than in the properly jacketed specimens.

These do not appear to be surface effects. Thus, the properties of practically impermeable metals and of certain single cyrstals, for example, diamond, sapphire, and spinel, were found to be uninfluenced by the presence of a jacket in triaxial compression tests (Bridgman, 1941, p. 467–469). Goguel (1948, Fig. 71) measured the same strength for jacketed and unjacketed specimens of halite. The effects seem to be associated rather with the penetration of the confining fluid into the interstices of porous materials and are most probably purely physical in nature; that is, they are observed even though the confining fluid is chemically inert relative to the constituents of the material under the test conditions imposed.

Also, in extension tests the properties of impermeable ductile metals were discovered to be uninfluenced by a jacket[5] (Bridgman, 1952, p.

[5] Extension tests can be made on unjacketed materials, provided they are impermeable, but of course

106–107; 1953, p. 564–565). However, the breaking strengths of impermeable brittle materials which failed by extension fracturing were found to be sensitive to the composition and thickness of the jacket or to the nature of the confining fluid if a specimen was unjacketed (Bridgman, 1947, p. 250–251; 1952, p. 111–117; 1953, p. 566–568). Extension fracturing is thought to be initiated at a surface imperfection, probably a very small crack; therefore, here the confining medium probably does have a surface effect.

If, except to initiate an extension fracture, the fluid must invade a specimen to be effective, then the consequences of exposing the specimen to the fluid must depend on the permeability of the material, that is, on the extent to which the material is permeated during the time that a given confining pressure is applied. McHenry (1948) showed that whereas the breaking strength of jacketed concrete under 100 bars confining pressure exceeded the crushing strength severalfold, the strength of the unjacketed material confined by nitrogen was not enhanced. The strength of the unjacketed concrete confined by kerosene was observed to be about midway between the extreme values with zero pore pressure (jacketed) and full pore pressure (nitrogen-saturated). This suggested that under the low pressures applied, the relatively viscous kerosene could not pervade the concrete.

Bredthauer (1957, Fig. 7) observed no increase of strength upon application of 750 bars oil pressure to an initially water-saturated, unjacketed calcareous sandstone. On the other hand, the strength of the initially dry rock was augmented even though it was exposed to the confining oil.

Several experiments have been conducted on previously water-saturated, jacketed specimens. Griggs et al. (1951, Fig. 1; 1953, Fig. 2) studied the effect of interstitial fluids on the strength of Yule marble. Although the tests were made at 150°–300°C. and with carbonated water and magnesium chloride solution, as well as distilled water, there was no microscopic evidence that the effect of the solutions was anything but mechanical. Under confining pressures from 5 to 10 kilobars, the ultimate strength was reduced roughly in proportion to the amount of liquid present in the jacketed marble specimens. The greatest re-

duction exceeded 50 per cent for the maximum water content of about 3 per cent. Tests on water-saturated Navajo sandstone (Balmer and Hanson, 1951, p. 6–7) and on calcareous sandstone (Bredthauer, 1957, Fig. 7) showed that the ultimate strengths of the wet rocks were not so great as those of the dry ones at a given confining pressure. In none of these experiments was the magnitude of the pore pressure known.

Prior to the work of Handin (1958), there were few experiments with known pore pressures. McHenry (1948) had tested jacketed concrete samples subjected to confining pressures to 100 bars and interstitial pressures of nitrogen to 80 bars. His data are consistent with the idea of effective stress.

Subsequently, work on this problem has been done by Robinson (1959) on limestone, sandstone, and shale at pressures to 700 bars, by Heard (1960) on Solenhofen limestone to 5 kilobars and temperatures to 150°C., and by Serdengecti and Boozer (1961) on Berea sandstone to 1.4 kilobars. Their results will be compared with those of this report in the discussion to follow.

Apparatus and Procedures

The apparatus, methods, and procedures for conducting triaxial compression tests on dry samples at room temperature have been described in earlier papers (Handin, 1953, p. 319–321; Handin and Hager, 1957, p. 10–12). The additional equipment required for experimenting at high temperatures is discussed by Handin and Hager (1958, p. 2895). The only innovations are those needed to apply pore pressures independently of confining pressures.

The new apparatus is a simple press consisting of two platens joined by two tie rods (Fig. 3). A commercial 20-ton hydraulic ram is threaded into the right platen. The ram, activated by a 5000-psi. piston-type oil pump, drives a steel piston into the 0.5-inch (1.3-cm.) bore of the steel pore pressure cylinder (bomb) threaded into the left platen.

The hydrostatic pressure generated in the bomb is measured by a 50,000-psi. (3.5-kilobar) Baldwin pressure gauge, the electrical output of which is coupled to a Baldwin Type L "strain indicator." The smallest reliable reading is about 2 bars. The null indicator microammeter of the potentiometer circuit of the instrument has been replaced by a control meter of the contact type. If the pressure

---

the confining fluid must be denied access to the ends of a specimen.

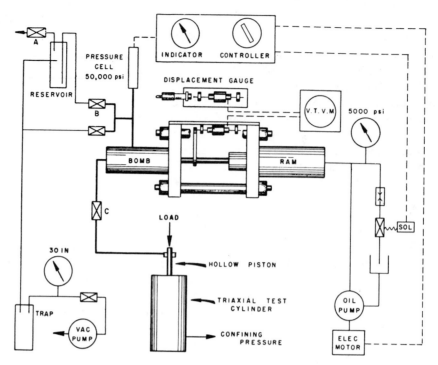

Fig. 3.—Schematic diagram of apparatus for triaxial compression tests with independently controlled pore pressures.

drops, the motor-driven pump is actuated. If the pressure increases, a solenoid valve is opened in the oil supply line to the ram. The pressure can be held within 1 per cent of any desired setting to 3 kilobars.

The motion of the piston is transmitted to the armature of a linear variable differential transformer, the output of which is connected in series with that of a similar transformer whose armature is adjusted by a micrometer screw. The piston displacement is remotely indicated by the micrometer reading when the null indicator voltmeter shows the two transformers to be in balance. Displacements can be measured to about ∓ 0.0005 inch (0.0013 cm.).

The bomb is coupled through valves B to a vacuum system and through valve C to the hollow piston of the triaxial test chamber. Fluids can then be injected through the piston into the pores of a jacketed specimen. The pore pressure can never exceed the confining pressure, since the thin-walled copper jacket has virtually no bursting strength. However, the pore pressure can have any value equal to or less than the confining pressure outside the jacket.

Since the pore pressure cylinder must be absolutely without leak at all pressures, use is made of

a simple but effective piston packing employed first by Adams (1936, p. 174). An ordinary rubber stopper is cut to a length of about 0.5 cm. and is forced into the bore of the cylinder, large end first. The ram is then allowed to advance the loosely fitting piston until it has just entered the chamber.

With valve A closed and valves B and C open, the entire pore pressure system, including the test sample, is evacuated. Confining pressure is then applied. Valve C and lower valve B are closed. Valve A is opened to atmospheric pressure, which forces liquid from the reservoir into the pore pressure cylinder. Upper valve B is then closed, and the pore pressure is raised to its desired value. The displacement gauge reading is recorded. Valve C is opened, permitting ingress of fluid into the high-pressure tubing between the valve and the triaxial test chamber, the hole in the piston, and the specimen itself. The pore pressure is returned to its former value, and a new displacement reading is taken. The difference in the readings multiplied by the area of the bore of the pore pressure cylinder gives the volume of fluid injected at a given pressure. The contribution of the tubing and the hollow piston to the total volume is known from a calibration experiment with a steel rod inserted in place of a porous specimen. The void volume of the rock specimen can therefore be measured at given confining and pore pressures.

When it is desired that the pore pressure just equal the confining pressure, an experiment is merely conducted on an "unjacketed" specimen. Since it would be difficult to assemble the apparatus without the support of the copper sleeve surrounding the specimen, use is made of a jacket with several pinholes punched through it. Because the "unjacketed" rocks are very weak, it is necessary to correct their stress-strain curves for the contribution of the jacket to the total load. The required stress-strain relations of copper were determined previously (Handin and Hager, 1958).

The initial dry bulk density ρ_b of a specimen is computed from the weight and the initial bulk volume, which is measured by calipering the undeformed cylinder. The average grain density ρ_s of the mineral constituents is determined by immersion of the crushed rock in toluene. The initial fractional porosity is then

$$f_0 = \frac{\rho_s - \rho_b}{\rho_s}. \tag{7}$$

Because confining pressure is applied to a specimen before the pore fluid (water in these experiments) is injected, the porosity at the beginning of an experiment differs from that measured under atmospheric conditions. To be certain that the water is, in fact, pervading a specimen, it is necessary to measure independently the compactions of the rocks and to determine their porosities under pressure. The bulk volume changes ΔV_b under appropriate confining pressures are measured by a method fully described elsewhere (Borg et al., 1960, p. 139–142). The compressibility of rock-forming minerals is only about 0.1 per cent per kilobar (Birch, 1942, p. 54–58). In computing porosity changes, it is therefore reasonable to assume that all the bulk volume change results from reduction of void volume, that is, the solid volume remains essentially constant. The new porosity is then

$$f = \frac{f_0 - \Delta V_b}{1 - \Delta V_b}. \tag{8}$$

In practice, confining pressure is applied first, and the pore water is then injected. The pore pressure is maintained until water ceases to enter the specimen, or up to a maximum of 24 hours. Because the test materials are not perfectly homogeneous, and because the porosity under confining pressure varies with pore pressure, the void volume of each specimen must be determined under particular test conditions from the amount of water introduced. The solid volume V_s, assumed constant, is known from the weight and average grain density; therefore, the new porosity is given by

$$f' = \frac{V_v}{V_v + V_s}, \tag{9}$$

where $V_v =$ void volume = volume of water injected.

In high-temperature tests the pressures are applied first, and the volume of water entering the specimen is measured. The temperature is then raised to the desired value and allowed to come to equilibrium. The volume of the pore pressure system changes because of expansion of the apparatus, the water occupying heated spaces within the apparatus, and the interstitial water in the specimen. Attempts to correct for these thermal expansions were not wholly successful. The porosity changes measured during the deformation of heated specimens were not regarded as reliable; the final porosities were therefore determined after release of pressure and removal of the specimens from the apparatus.

A triaxial test is conducted in the usual manner on a copper-jacketed specimen $\frac{1}{2}$ inch in diameter and 1 inch long. Throughout an experiment the pore pressure is held constant. The volume changes required to do this are determined from pore pressure piston displacement readings recorded every minute or so during the deformation. It is assumed that these volume changes result from variations of void volume only so that new porosities f'' can be calculated by equation 7. These porosities are then plotted as a function of longitudinal strain.

The final porosity of the deformed specimens can be checked roughly by an independent method after the specimens are removed from the apparatus. The final porosity of a dry sample is also determined in this manner, since volume changes during deformation are not measured. The value is calculated from the final bulk volume (at atmospheric pressure) V_b''' and the already determined solid volume, or

$$f''' = \frac{V_b''' - V_s}{V_b'''}. \tag{10}$$

Owing to the irregular shape of a permanently deformed cylinder, it is necessary to measure the final bulk volume by immersion in mercury. The final porosity f''' differs, of course, from f'' measured under pressure at maximum deformation because of the recovery of the specimen after release of confining pressure.

The median pore size and permeability of a deformed specimen can be determined by the capillary pressure method (Purcell, 1949, p. 39–41). In brief, the per cent of bulk volume occupied by the mercury entering the pores of the rock ("effective" porosity) is plotted against the injection pressure. If it is assumed that the pores are cylindrical capillary tubes, the pore size can be calculated approximately by the equation

$$P_c = \frac{2s \cos \alpha}{r}, \qquad (11)$$

where P_c = capillary pressure (dynes/cm.²), the minimum pressure required to inject a non-wetting liquid into the capillary of radius r (cm.); α = interfacial angle (140° for mercury against solid); and s = surface tension of mercury (480 dynes/cm.). The median pore size diameter (microns) is computed for that pressure at which 50 per cent of the effective pore space is filled. The permeability is given by

$$K = 0.66 F f''' \int_0^{100} \frac{dv_m}{P_c^2}, \qquad (12)$$

where K = permeability (millidarcys) as ordinarily determined in an air permeameter, f''' = final porosity (per cent), v_m = per cent of total void volume occupied by mercury, P_c = capillary pressure (atm.), and F = lithology factor, which depends primarily on the structure of the pore space and which is determined from the air permeabilities of the undeformed rocks. Since the lithology factor may be altered by a large deformation, the final value of F is not certainly known, and the value of the final permeability must be regarded as qualitative.

RESULTS

In Table I are listed the air permeabilities (K), grain densities (ρ_s), and initial porosities (f_0, equation 6) of representative undeformed specimens of the five rocks investigated. Also tabulated are the bulk volume changes (compactions) and porosities (f, equation 8) at 2 kilobars confining pressure, together with the final (permanent) bulk volume changes and final porosities (f''', equation 10) measured after release of confining pressure. None of the compactions is very large, and most of the bulk volume changes are recoverable; that is, the permanent compactions are small (less than 1 per cent).

A summary of triaxial compression tests is given for each rock (Tables II, IV–VII). The data listed include the experimental conditions of confining pressure, pore pressure, and temperature. When these two pressures are equal, it is to be understood that the specimen was unjacketed and exposed to the confining kerosene. Also tabulated are the total longitudinal strain achieved during an experiment, the ultimate strength (maximum ordinate of the stress-strain curve), and the shear fracture angle (measured relative to the maximum principal stress direction). The initial porosity f_0 is given for each specimen, together with the porosity f' (equation 9) after application of confining and pore pressures. The porosity f'' at the termination of the experiment

TABLE I. POROSITY AND PERMEABILITY DATA

Rock	At atmospheric pressure			At 2000 bars pressure			Final	
	Air permeability, k (md)	Grain density ρ_s (g/cm³)	Porosity, f_0 (percent)	Bulk volume change (percent)	Porosity, f (percent)	Porosity change (percent)	Bulk volume change (percent)	Porosity, f''' (percent)
Hasmark dolomite	0.71	2.91	3.5	1.8	1.8	50.0	0.2	3.3
Marianna limestone	<0.05	2.70	13.0	1.8	11.5	11.8	0.6	12.5
Berea sandstone	217	2.66	18.2	3.8	15.0	17.6	0.5	17.8
Muddy shale	<0.05	2.67	4.7	0.6	4.2	10.6	0	4.7
Repetto siltstone	<0.05	2.58	5.6	2.9	3.1	44.1	0.6	5.1

Table II. Summary of Experiments on Berea Sandstone

| | | | | | | Porosity (percent) | | | | | | | |
Conf. press. (kilobars)	Pore press. (kilobars)	Temp. (°C)	Total strain (%)	Ult. strength (bars)	Fracture angle (deg)	Predeformation Unconf.	Predeformation Conf.	After deformation Conf.	After deformation Change	After deformation Unconf.	After deformation Change	Pore Size (μ)	Permeability (md)
0	0 (dry)	24	2.2	490	26	18.2				16.7	-8.2	13.0	206
0	0 (wet)	24	19.4	720	30	18.5				16.9	-8.6	9.4	138
0.5	0	24	12.3	1590	27	17.3				16.6	-4.0	10.7	153
0.5	0.5	24	19.9	820	35	19.6				28.2	+44.0	17.8	1190
-	0	24	32.8	2480	34	22.7				20.7	-8.8	1.1	3.6
-	0.5	24	30.8	2420	*	19.2	16.4	15.5	-5.5	18.0	-6.2	2.7	42.2
-	0.5	24	18.0	1900	40	20.3	12.3	10.2	-17.1	18.9	-6.9	3.3	70.3
-	-	24	11.9	1670	36	20.7				28.2	+37.2	18.4	1020
1.5	1.5	24	22.0	780	35	18.5				25.3	+37.8	13.4	448
-	0	24	23.7	750	35	21.1				16.5	-21.8	0.8	1.3
2	0	24	30.4	4180	26	18.8				16.2	-13.8	0.6	1.2
2	0.5	24	29.2	4320	29	18.0	13.9	8.1	-41.4	13.1	-27.3	0.6	0.7
0.5	-	24	27.5	3410	30	20.1	16.5	11.8	-28.6	14.6	-27.1	1.3	2.2
-	1.5	24	27.1	2480	33	18.3	17.3	16.2	-6.2	16.9	-7.6	1.3	5.9
1.5	1.75	24	25.8	1740	34	18.8	16.7	17.7	+5.7	20.1	+6.9	16.9	490
1.75	1.75	24	9.1	1300	34	18.5	17.3	18.4	+6.0				
2	2	24	12.8	1490	32	17.9							
2	2	24	25.4	600	*	18.4							
0	0	24	24.3	640	35	18.8				19.2	+24.4	14.3	534
0	-	300	25.9	4250	31					15.9	-15.4	0.7	1.0
-	-	300	26.5	4250	*								
-	-	300	25.4	2340	32	17.8	14.4	16.0	+11.2	15.6	-12.3	1.0	2.7
-	-	300	25.4	2380	32	19.0				15.9	-16.3	0.9	1.8
2†	-	24	2.5	730	15	18.6				16.7	-10.2	11.6	205
5	5	24	15.8	630	32	18.8				23.8	+26.6	14.7	504

†Extension test.

*Stress-strain curve falling. Probably incipient shear.

TABLE III. EFFECTIVE PRINCIPAL STRESSES AT ULTIMATE STRENGTH OF BEREA SANDSTONE

Confining pressure (bars)	Pore pressure (bars)	Axial load (bars)	Effective stresses	
			σ_1 (bars)	σ_3 (bars)
0[1]	0	490	490	0
0[2]	0	720	720	0
500	0	1590	2090	500
500	500	820	820	0
1000	0	2450	3450	1000
1000	500	1790	2290	500
1000	1000	780	780	0
1500	1500	750	750	0
2000	0	4250	6250	2000
2000	500	3410	4910	1500
2000	1000	2480	3480	1000
2000[3]	1000	730	1000	270
2000	1500	1740	2240	500
2000	1750	1400	1650	250
2000	2000	640	640	0
5000	5000	630	630	0

[1]Dry.   [2]Wet.   [3]Extension test.

but before release of confining and pore pressures is listed together with the final porosity f''' measured under atmospheric conditions. Finally presented are median pore sizes (equation 11) and permeabilities (equation 12) calculated from capillary pressure data.

SANDSTONE

The Berea sandstone (Mississippian) sample, donated by R. A. Cunningham, was collected from the Cleveland quarry in Ohio. It is a medium-grained, low-rank graywacke consisting of about 60 per cent quartz, 20 per cent quartzite rock fragments, 5 per cent potash feldspar, 5 per cent calcite cement, 7 per cent argillaceous material, and small amounts of shale fragments and plagioclase feldspar. X-ray diffractometer analysis reveals that kaolinite is the only clay mineral present. The rock is moderately well cemented, and a hand specimen appears homogeneous. However, the porosity ranges from 17 to 22 per cent within a volume of a few cubic inches (see Table II).

The deformational behavior of the dry Berea sandstone is consistent with that of other previously tested, moderately well cemented sandstones (Handin and Hager, 1957, Figs. 19, 20, 24; 1958, Fig. 15). Ultimate strength at 2 kilobars confining pressure exceeds the crushing strength by a factor of about six. At atmospheric pressure the rock is brittle. At 2 kilobars, specimens can be shortened at least 30 per cent without rupture, the maximum achievable in the apparatus. Faulting occurs at low and intermediate pressures at

from 26° to 34° to the maximum principal stress direction. Heating to 300°C. at 2 kilobars confining pressure reduces the yield stress by about 10 per cent but affects ultimate strength and ductility very little (Table II).

Confirming the results of earlier workers, the data on the unjacketed sandstone show that ultimate strength is independent of confining pressure over the wide range of 0 to 5 kilobars (Table II). At a strain of about 1 per cent, all wet specimens fail suddenly at a stress difference of 600 to 800 bars and are regarded as brittle, in contrast to the dry rock deformed at 1 kilobar or more.

Fig. 4 shows the influence of pore pressure on the stress-strain relations of this sandstone. In the lower portion are stress-strain curves for a series of specimens compressed under a constant confining pressure of 2 kilobars but with different pore water pressures of 0, 0.5, 1, 1.5, 1.75, and 2 kilobars. Note two important effects of increasing the pore pressure (or reducing the effective pressure): The stress-strain curve is lowered and changes shape. Both ultimate strength and ductility are clearly affected.

That these properties are functions of effective stresses and not of the absolute values of confining and pore pressures is evident from comparisons of the three practically identical curves drawn in the upper portion of Figure 4. These are derived from tests at confining pressures of 0.5, 1, and 2 kilobars and pore pressures of 0, 0.5, and 1.5 kilobars. The rock behaves in a similar manner in every test in which the effective confining pressure is the same, namely, 0.5 kilobar.

The ultimate strength at a fixed confining pressure of 2 kilobars decreases nearly linearly with increasing pore pressure or decreasing effective pressure, $p_c - p_p$ (Fig. 5). If the concept of effective stress (equations 1) is valid, the effective principal stress relation at ultimate strength should be independent of pore pressure (Table III).

For the dry rock the relation between $\sigma_1 = S_1$ and $\sigma_3 = S_3$ is nearly linear (Fig. 6), approximately

$$\sigma_1 = a + b\sigma_3 = 0.7 + 2.8\sigma_3. \quad (13)$$

On the same graph are plotted the effective principal stresses at ultimate compressive strength for $p_c = 2$, $p_p = 0.5$, 1.0, 1.5, 1.75, and 2, $p_c = 1$, $p_p = 0.5$, and $p_c = 2$, $p_p = 1$ in an extension test. All points fall near the line defined by equation 13 which therefore expresses the principal stress rela-

TABLE IV. SUMMARY OF EXPERIMENTS ON MARIANNA LIMESTONE

						Porosity (percent)							
						Predeformation		After deformation					
Conf. press. (kilobars)	Pore press. (kilobars)	Temp. (°C)	Total strain (%)	Ult. strength (bars)	Fracture angle (deg)	Unconf.	Conf.	Conf.	Change	Unconf.	Change	Pore size (μ)	Permeability (md)
0	0	24	0.5	440	10	11.7				12.4	+6.0	.19	.17
0.5	0	24	5.7	1170	28					8.8		.22	.09
1	0	24	29.8	2640	**					14.9		.17	.11
1	0	24	30.4	2500	27								
2	0.5	24	25.8	1240	32	13.5	11.8	10.3	-12.5	14.4	+6.7	.27	.27
2	0	24	26.2	4120	**					9.7		.13	.01
2	0	24	28.0	4650	**	11.2				9.1	-18.7		
2	0.5	24	25.6	3330	*	11.6	9.0	6.1	-32.5	11.2	-3.5	.15	.03
2	1	24	24.5	2640	35	12.3	9.4	7.9	-16.1	11.9	-3.3	.14	.03
2	1.5	24	24.5	2140	35	12.3	12.2	12.2	0	12.4	+1.6	.42	.62
2	1.75	24	26.0	1240	34	14.0	12.1	13.4	+10.6	14.0	0	.26	.29
2	2	24	23.7	410	30	14.6				17.5	+19.8	.30	.27
2	0	300	25.5	4050	25					4.6			
2	0	300	25.7	3760	20	5.7				2.3	-59.5		
2	1	300	19.3	1910	23	11.5				11.2	-2.6	.15	.02
2†	1	24	1.5	850	15	11.3				10.0	-8.9	.16	.04
5	5	24	29.4	700	**		8.0	8.4	+5.0			.21	.07

*Terminated before fracture. Stress–strain curve rising.

**Stress–strain curve falling. Probably incipient shear.

†Extension test.

TABLE V. SUMMARY OF EXPERIMENTS ON HASMARK DOLOMITE

| 1 | 2 | 3 | 4 | 5 | 6 | Porosity (percent) | | | | | |
| | | | | | | Predeformation | | After deformation | | | |
Conf. press. (kilo-bars)	Pore press. (kilo-bars)	Temp. (°C)	Total Strain (%)	Ult. strength (bars)	Fracture angle (deg)	Unconf.	Conf.	Conf.	Change	Unconf.	Change
0	0	24	9.4	1160	29	2.9				3.5	+20.6
0.5	0	24	4.5	3380	24	2.7				5.8	+115.0
1	0	24	4.1	4150	25	2.3				8.7	+278.0
1	0	24	7.3	4050	28						
1	0.5	24	5.8	3760	**	1.9	1.4	0	-100.0	6.0	+216.0
2	0	24	16.6	5650	**						
2	0	24	18.9	5450	**					13.3	
2	0.5	24	17.3	5900	**	2.4	1.2	1.5	+25.0	13.9	+480.0
2	1	24	12.6	5650	33	2.6	1.4	2.6	+86.0	9.7	+274.0
2	1.5	24	19.2	5500	29	2.9	2.4	3.4	+42.0	11.6	+322.0
2	1.75	24	16.6	5400	**	3.4	3.3	3.5	+6.3	8.9	+132.0
2	2	24	2.0	745	5	3.2				6.0	+84.5
2	0	300	8.7	4680	32	2.2				7.9	+25.8
2	1	300	5.5	4430	34	3.4				4.8	+41.5

**Stress–strain curve falling. Probably incipient shear.

TABLE VI. SUMMARY OF EXPERIMENTS ON REPETTO SILTSTONE

| 1 | 2 | 3 | 4 | 5 | 6 | Porosity (percent) | | | | | | 13 |
| | | | | | | Predeformation | | After deformation | | | | Pore size (μ) |
Conf. press. (kilo-bars)	Pore press. (kilo-bars)	Temp. (°C)	Total strain (%)	Ult. strength (bars)	Fracture angle (deg)	Unconf.	Conf.	Conf.	Change	Unconf.	Change	
0	0	24	1.0	280	0-24	5.0				1.0	-80	.33
1	0	24	22.3	1790	35	5.8				1.2	-79	.52
1	0	24	27.8	1700	*							
1	0.5	24	25.4	1790	38	6.4	4.2	2.6	-38.2	1.5	-77	.25
2	0	24	22.7	2480	*							
2	0	24	28.0	2680	39	6.4				2.8	-56	.27
2	1†	24	25.4	1900	34	6.1	5.8	4.2	-27.6	8.1	+33	.17
2	1	24	22.6	2840	36	6.9	3.8	3.2	-15.7	3.0	-56	.27
2	1.5	24	22.9	2780	35							
2	1.75	24	23.8	3060	40	7.8	6.4	5.0	-21.9	2.9	-63	.21
2	2†	24	30.0	410	40							
2	0	300	20.4	1670	31	4.4				2.2	-50	.24
2	0	300	32.2	1160	*							

†Kerosene used for pore pressure liquid.

*Stress–strain curve falling. Probably incipient shear.

tion regardless of the pore and confining pressures applied.

The effect of pore pressure on ultimate strength can also be demonstrated by utilizing the graphical solution of Mohr (Fig. 2). One-half of the Mohr stress circles representing the extreme principal stresses at the ultimate strength of the dry rock under confining pressures of 0, 0.5, 1, and 2 kilobars are drawn in the lower half of Figure 7. The curve that is tangent to these circles is the Mohr envelope, the equation of which expresses the Coulomb-Mohr failure criterion (equation 2). The linear envelope for the Berea sandstone is

$$\tau = 0.2 + 0.55\sigma_n, \qquad (14)$$

where τ_0 (0.2 kilobar) is cohesive resistance at zero normal pressure, and ϕ (29°) is the internal friction angle (slope of the envelope, here constant).[6]

[6] In the calculation of ϕ, the diameter of the stress circle (the differential stress $\sigma_1 - \sigma_3$) at failure ought strictly to be corrected for volume changes. The total work done by the piston is $\Delta\sigma A(l_0 - l_1)$, where A is the area and l_0, l_1, are the original and final lengths of a specimen. Part of this work, $\Delta\bar\sigma A(l_0 - l_1)$, produces a shear strain. A volume strain is produced by the remainder, which must just equal $\pm\Delta V(p_c - p_p)$. Hence per unit volume

$$\Delta\sigma = \Delta\sigma + \frac{\Delta V}{\epsilon V}(p_c - p_p),$$

where ϵ is longitudinal strain, and where the sign is positive for a negative volume change, and vice versa. The greatest correction is needed when $p_p = 0$ and $\Delta V/V$ is large, say, 5 per cent (see Table II and equation 8). Here the corrected ultimate strength would be about 3.9 instead of 4.2 kilobars. From equations 2, 3, and 6, one can derive

$$\sigma_1 = 2\tau_0 \tan(45 + \phi/2) + \sigma_3 \tan^2(45 + \phi/2),$$

which gives $\phi = 29°$ for $\Delta\sigma = 4.2$, and $\phi = 26°$ for $\Delta\sigma = 3.9$. This correction would change θ (equation 6) by only 1.5°, which would not be measurable.

According to equation 6, faults should be inclined at 45° − 14.5° = 30.5° to the maximum principal stress (to the long axis of the cylindrical specimen in a compression test). This is, in fact, close to the values actually measured, except in the extended specimen (Table II).

If the presence of a fluid does not affect the mechanical properties of the rock,[7] then the superposition of the hydrostatic pore pressure should not vitiate equation 14. Indeed, the envelope tangent to the principal stress circles representative of ultimate compressive strengths of the saturated rock (upper half, Fig. 7) has an equation identical with equation 14. The center of a stress circle merely moves toward the origin along the normal stress axis by an amount equal to the value of the pore pressure p_p, as predicted by equation 5. The stress circle for the extension test (dashed half-circle, Fig. 7) is not tangent to the compression envelope curve. It will be recalled that the intermediate stress equals the least stress in compression and the maximum stress in extension (equations 1). The evidence indicates that the Mohr criterion (equation 2) is not independent of intermediate stress, as assumed.

The stress-strain curves (Fig. 4) reflect an influence of pore pressure not only on ultimate strength but also on the shapes of the curves which most probably are related to the deformation mechanisms involved. At effective confining pressures of 1.5 kilobars or more, the curves rise continuously, and the ductile rock appears to work-harden. At 1 kilobar the curve is nearly horizontal; the material is in the transitional state. At effective confining pressures below 1 kilobar, the curves rapidly attain a peak, then fall off to become roughly horizontal at total strains beyond about 3 per cent. The rock is regarded as brittle.

Ductility may be defined as total per cent strain before fracture, with the understanding that *fracture* implies total loss of cohesion and resistance to stress difference, separation into two or more parts, and release of stored elastic strain energy (Griggs and Handin, 1960, p. 348). In this

[7] The fault angles in the wet specimens (Table II) tend to be 3° to 5° smaller than those in the dry rock. This probably means that the angle of internal friction is in fact somewhat altered by the presence of water at intergranular contacts as it is in soils (Bishop and Eldin, 1953, p. 101). These differences are not detectable, however, in the Mohr construction (Fig. 6).

TABLE VII. SUMMARY OF EXPERIMENTS ON MUDDY SHALE

1 Conf. press. (kilobars)	2 Pore press. (kilobars)	3 Temp. (°C)	4 Total strain (%)	5 Ult. strength (bars)	6 Fracture angle (deg)	Porosity (percent)						13 Pore size (μ)	14 Permeability (md)
						7 Predeformation Unconf.	8 Conf.	9 Conf.	10 Change	11 After deformation Unconf.	12 Change		
0	0 (dry)	24	0.6	400	10	4.1				6.0		.13	.01
0	0 (wet)	24	4.2	830	0–40					4.4	+7.3		
0.5	0.5	24	2.5	1400	—	4.2				4.7	+11.9	.21	.05
0.5	0.5	24	5.4	1460	35					8.3		.24	.07
—	0	24	14.1	2460	33								
—	0	24	16.1	2500	33	4.9				5.4	+10.2	.23	.01
1.5	1.5	24	24.3	1810	36	5.1				5.4	+5.9	.33	.10
0	0	24	23.6	1920	35					10.9		.33	.10
2	0	24	23.3	3960	27								
2	2	24	26.8	4170	33	4.8				5.5	+14.5	.71	.29
2	0	300	18.2	2000	—					7.2		.26	.11
2	0	300	12.3	1950	34								
			10.5	2690	33								

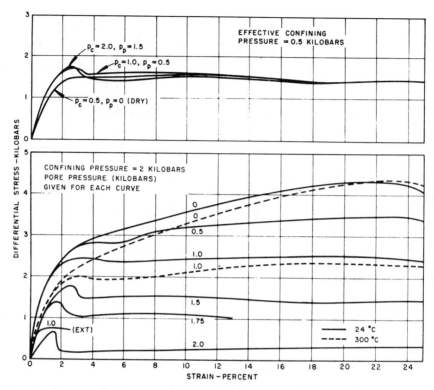

FIG. 4.—Stress-strain curves for Berea sandstone at different pore water pressures (kilobars). Below: all at 2 kilobars confining pressure and 24° or 300° C.; all in compression except curve marked *Ext* (for extension). Above: at confining pressures (p_c) of 0.5, 1, and 2 kilobars at 24° C.; at pore pressures (p_p) of 0, 0.5, and 1.5 kilobars; all at same effective pressure of 0.5 kilobar.

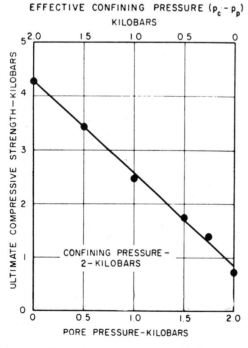

FIG. 5.—Ultimate compressive strength of Berea sandstone at 24°C., 2 kilobars confining pressure as function of pore pressure and effective confining pressure.

sense the sandstone at effective pressures below 1 kilobar has ductility, since strains of 25 per cent or more are achieved without total loss of cohesion or fracture in the ordinary sense. However, if the state of the rock is regarded as brittle when the deformation is characterized by a peaked stress-strain curve, then the brittle-to-ductile transition is also a function of effective confining pressure. At effective pressures below 1 kilobar, failure (without total loss of cohesion) occurs at about 3 per cent.[8]

Generally the form of a stress-strain curve cannot be associated with a specific mode of deforma-

[8] The sharp peak in the stress-strain curves for low effective confining pressures reflects the relatively high differential stress needed to break intergranular bonds, that is, to overcome the cohesive resistance. Further shearing resistance, albeit small, is probably due to two causes. Surface tensions in the wet rock provide some cohesive resistance, and the copper jacket applies a small confining pressure to the specimen, once the jacket is expanded. This latter effect can be estimated by regarding the jacket as a thin-walled cylinder under an internal pressure $p = 2h\sigma_y/d$, where h is wall thickness $= 0.01$ inch, σ_y is circumferential stress (yield stress of the copper) $= 7,000$ psi., and d is the diameter $= 0.50$ inch. p is about 280 psi., or 20 bars.

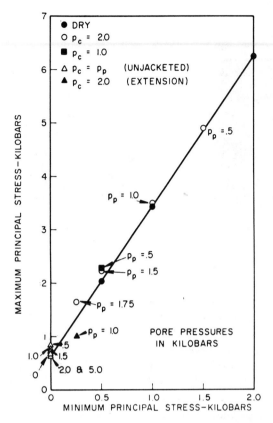

Fig. 6.—Effective maximum principal stress versus effective minimum principal stress at ultimate strength of Berea sandstone at 24°C.

tion. However, these sandstone curves are related to the macroscopic behavior of the rock (Fig. 8). The peaked curve typical of low effective pressures reflects the shear fracture of the brittle specimen on the left. The flat transitional curve obtained at intermediate pressures characterizes the middle sample, which has shortened mostly by localized shear without loss of cohesion along the through-going fault. The fault zone tends to broaden with increasing pressure until the deformation is distributed uniformly throughout the rock on the right. The stress-strain curve rises continuously. (Because of end constraints the deformation is not strictly homogeneous.)

The nature of the porosity change accompanying the deformation of the sandstone depends strongly on the effective confining pressure (Fig. 9). The porosity-strain curves do not have a common origin because (1) the initial porosities f_0 of the several Berea specimens are not the same, and (2) a portion of the void volume reduction (compaction) is recovered upon application of pore pressure; that is, $f \neq f'$ (equations 8 and 9). However, there is evidently a trend for f' (porosity at zero strain) to increase with pore pressure for a given confining pressure.

The 1.5- and 1-kilobar curves reflect a large reduction of porosity. At 0.5 kilobar the porosity

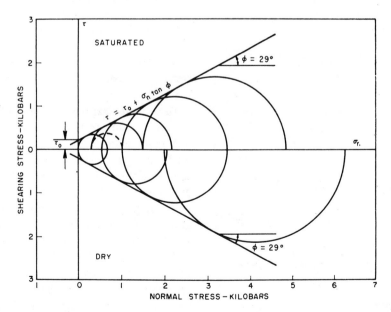

Fig. 7.—Mohr envelope curves (identical) for ultimate strength of Berea sandstone at 24° C. Lower curve for dry jacketed specimens in compression at 0, 0.5, 1, and 2 kilobars confining pressure. Upper curve for 2 kilobars confining pressure and 0.5, 1, 1.5, 1.75, and 2 kilobars pore pressure; all in compression except dashed half-circle for extension at 1 kilobar pore pressure.

FIG. 8.—Berea sandstone specimens compressed at different effective confining pressures (kilobars) at 24° C. From left to right: brittle (shear fracture), transitional (faulting), and ductile (flow).

remains essentially constant; there is a small decrease during the first 2 per cent of deformation. At 0.25 kilobar the porosity *increases*—the rock is dilatant. The void volume also enlarges as the result of a 3-per cent extension. Heating to 300°C. does not appear to influence the 1-kilobar curve.

The porosity-strain curves, as well as the stress-strain curves, appear to depend on effective confining pressure. Thus the shapes (though not the origins) of the curves for $p_c = 2$, $p_p = 1.5$ and $p_c = 1$, $p_p = 0.5$ are nearly alike.

The permanent pore volume change of the compacted dry rock under 2 kilobars confining pressure was small (Table I). It was therefore suspected that neither the stress-strain curve nor the porosity-strain curve would differ if all the confining pressure was applied first and then all the pore pressure, or if the confining and pore pressures were raised together gradually. This supposition was confirmed by an experiment in which the confining pressure was increased in increments of about 100 bars, at each of which the pore pressure was raised about 50 bars. The test was finally conducted at $p_c = 2$, $p_p = 1$ kilobar, and the results were practically identical with those presented here.

In order to explain the related effects of pore pressure upon strength, ductility, and porosity, one must acquire an understanding of the deformation mechanisms involved. Accordingly, thin sections of selected specimens, all deformed in compression at 2 kilobars confining pressure, were prepared and examined with the petrographic microscope. Fig. 10 shows a photomicrograph of the undeformed Berea sandstone.

The most conspicuous features of the deformed specimens as viewed in thin section are grain fracturing and macroscopic shear zones (faults) which transect the entire aggregate. Fracturing is most intense in the shear zones, where a "fault gouge" was developed.

The 30-per cent deformation of a dry specimen

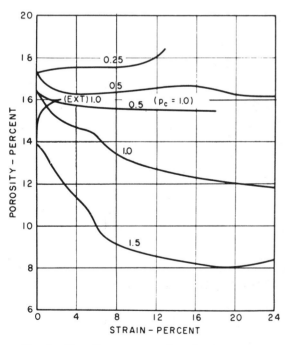

FIG. 9.—Porosity versus total strain of Berea sandstone at different effective confining pressures (kilobars) at 24°C. All at $p_c = 2$ kilobars except curve denoted $p_c = 1$. All in compression except curve marked *Ext* (for extension).

(750) is macroscopically homogeneous. Microscopic examination reveals one poorly defined through-going shear with indistinct boundaries (Fig. 11). Except for a few grains which appear to be protected by a coat of carbonate cement, the grains are generally fractured, most intensely along the shear zone. The total strain tends to be distributed throughout the specimen, and the macroscopic shearing occurs late in the deformation history.

In specimens 810 and 801, both shortened about 27 per cent at pore pressures of 0.5 and 1 kilobar, respectively, grain fracturing is general, except in grains entirely enclosed by cement. Each specimen contains several shear zones filled with quartz gouge and smeared-out cement. These do not necessarily transect the specimen.

Specimen 804, shortened 26 per cent under 1.5 kilobars pore pressure, is characterized by numerous, small conjugate shear zones along which grain fracturing is intense. Most grain fracturing is restricted to these zones, and much of the deformation occurs by slippage along these shears rather than by general crushing and subsequent rearrangement of grains and fragments.

Specimen 808 could be deformed only 9 per cent under 1.75 kilobars pore pressure before the large offset on the macroscopic through-going shear led to rupture of the jacket, after which pore pressure immediately rose to equal confining pressure. Most of the permanent strain is a result of slippage along one major shear zone which has very sharply defined boundaries and beyond which there is little or no grain fracturing (Fig. 12). Some of the porosity increase observed in this specimen may be due to granulation along the fault. However, the gouge zone is very narrow, less than 1 mm., and much of the increase of void volume is most probably ascribable to dilatancy.

Specimen 805 was shortened 25 per cent with pore and confining pressures equal. Under the microscope it is indistinguishable from the undeformed sandstone. The cement is undeformed; the grains are no more highly fractured than those in the starting material; and there is no shear zone. The deformation appears to be due solely to intergranular movements resulting in an increase of void volume.

Specimen 832 was shortened 25 per cent at 2 kilobars confining pressure and 1 kilobar pore

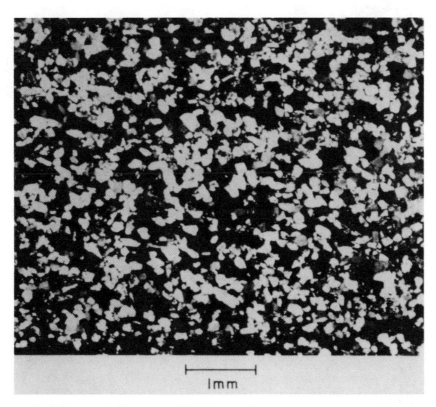

Fig. 10.—Photomicrograph of undeformed Berea sandstone. Thin section cut normal to bedding.

pressure at 300°C. Thin-section study reveals no difference in this sample and that which was also deformed at $p_p/p_c = 0.5$, but at room temperature. Heating does not appear to influence the deformation mechanisms nor to affect the inclinations of faults.

The amount of fracturing can be expressed numerically by a *fracturing index* used previously in studies of deformed unconsolidated sand aggregates (Borg *et al.*, 1960, p. 159). Each of 200 grains examined is assigned a number of 1 through 5 as it is unfractured, slightly, moderately, or highly fractured, or demolished. The fracturing index is computed by multiplying the per cent of grains in each category by its number, then summing these values. The index will vary from 100 to 500 as all the grains are unfractured or all are demolished. This method is subjective, but when the same worker examines all thin sections, the *relative* amount of fracturing from specimen to specimen is significant.

For the experimentally undeformed Berea sandstone the fracturing index is 114. The indices for the series of specimens deformed under 2 kilobars confining pressure and pore pressures of 0, 0.5, 1, 1.5, 1.75, and 2 kilobars are plotted against the effective pressure (Fig. 13). The index decreases with pressure until at $p_p = p_c$ the amount of fracturing is about the same as that in the experimentally undeformed sandstone. There is a marked break in the curve at about 0.8 kilobar, above which the fracturing index is nearly constant, and below which it drops rapidly.

Let us compare the measure of fracturing with the values of final porosity change $(f_0 - f''')/f_0$ (Fig. 13). There is a nearly constant reduction of porosity above 0.8 kilobar, at which there is a marked break in the curve corresponding to the break in the fracturing-index curve. Below 0.8 the porosity change decreases rapidly until at about 0.4 the porosity increases as a result of deformation, and this effect is clearly associated with the reduction of grain fracturing.

LIMESTONE

Of the several limestones tested previously, the Marianna was chosen because of its relatively high porosity (12 to 15 per cent) and permeability

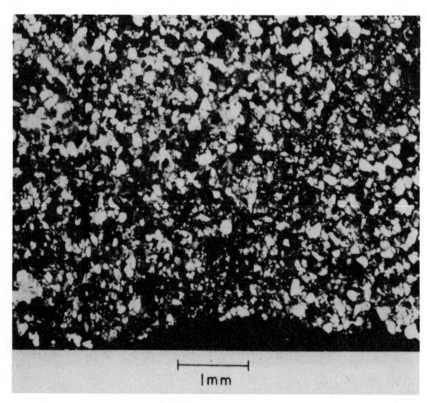

FIG. 11.—Photomicrograph of dry Berea sandstone specimen 750 compressed 30 per cent at 2 kilobars confining pressure, 24° C. Thin section cut parallel with load axis and normal to bedding.

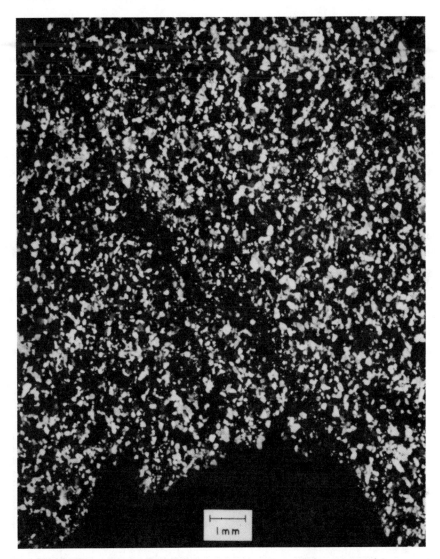

FIG. 12.—Photograph of water-saturated Berea sandstone specimen 808 compressed 9 per cent at 2 kilobars confining pressure, 1.75 kilobars pore pressure, 24° C. Thin section cut parallel with load axis and normal to bedding.

Samples of this Oligocene rock were cored from a geographically unoriented block collected in Jackson County, Florida. The rock is a massive fine-grained, white, rather friable limestone made up largely of clastic grains and small fossils. The average median pore size is of the order of 0.2 micron.

The stress-strain relations of the dry rock were determined in the earlier studies (Handin and

⫸→

FIG. 13.—Maximum porosity change and fracturing index of Berea sandstone specimens compressed at 2 kilobars confining pressure, 24°C., as functions of effective confining pressure. Porosities (f''') measured after release of pressure.

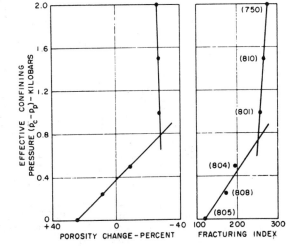

Hager, 1957, Fig. 14; 1958, Fig. 10). An increase of confining pressure from 0 to 2 kilobars results in tenfold increases of ultimate strength and ductility. Heating to 300°C. at 2 kilobars lowers the ultimate strength about 10 per cent.

Tests on unjacketed specimens reveal that ultimate strength is largely independent of pressure up to 2 kilobars (Table IV). At a strain of about 1 per cent, failure occurs suddenly at a differential stress of about 400 bars. Deformations of 20 per cent or more can be attained without total loss of cohesion, but the wet samples are regarded as brittle. At 5 kilobars the strength is greater—about 700 bars.

The influence of pore pressure on the stress-strain relations (Fig. 14) is similar to that observed for the sandstone. Again the ultimate strength at a fixed confining pressure of 2 kilobars decreases nearly linearly with effective confining pressure (or increasing pore pressure). The three stress-strain curves (Fig. 14) for the same effective pressure of 0.5 kilobar do not agree as well as the corresponding set for the sandstone, but they do reflect essentially similar behaviors of limestone

specimens compressed at the same effective pressure but at different confining and pore pressures.

The diagram (Fig. 15) of extreme effective principal stress relations at ultimate strength is also similar to that of the sandstones. All points fall near the line defined by the equation

$$\sigma_1 = 0.45 + 2.9\sigma_3 \qquad (15)$$

for the failure of the dry rock, so the relation is again nearly independent of pore pressure. Mohr's graphical solution (not reproduced here) would give a straight-line envelope, the equation of which would be

$$\tau = 0.2 + 0.55\sigma_n, \qquad (16)$$

and these equations would be approximately correct for any of the combinations of confining and pore pressures. The inclination of faults relative to the direction of maximum effective principal pressure (equation 6) should be $(90° - \tan^{-1} 0.55)/2 = 30°$. This angle should be independent of normal pressure (at room temperature), since equation 16 is linear. All measured angles are indeed close to 30°, with two exceptions (Table IV).

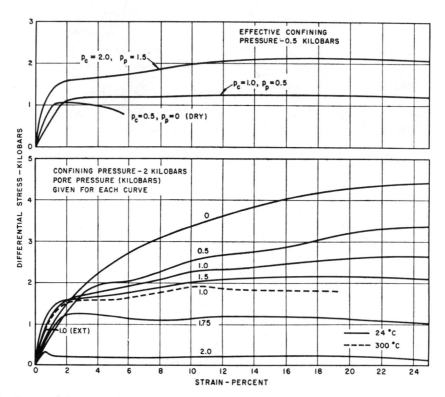

Fig. 14.—Stress-strain curves for Marianna limestone at different pore water pressures (kilobars). Below: all at 2 kilobars confining pressure and 24° or 300°C.; all in compression except curve marked *Ext* (for extension). Above: at confining pressures (p_c) of 0.5, 1, and 2 kilobars at 24°C.; at pore pressures (p_p) of 0, 0.5, and 1.5 kilobars; all at same effective pressure of 0.5 kilobar.

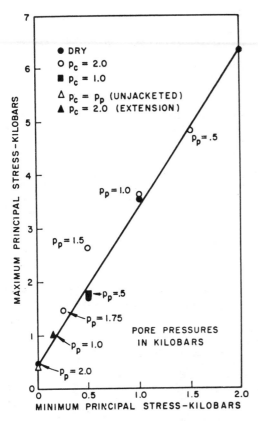

FIG. 15.—Effective maximum principal stress versus effective minimum principal stress at ultimate strength of Marianna limestone at 24°C.

The angle is less than 10° in the dry specimen deformed under atmospheric conditions. An angle of 15° was measured in the specimen which was extended rather than compressed. The fault angle tends to be steeper in specimens tested at 300°C.

It appears that the effects of pore pressure and temperature on ultimate strength at a given confining pressure can be superimposed. Thus at 2 kilobars, heating reduces the strength of the dry rock about 10 per cent. At the same confining pressure the strength of the dry rock at room temperature exceeds that of the rock with 1 kilobar pore pressure by about 40 per cent. The combined effect of the two variables would then be to lower the strength by 50 per cent. The strength of specimen 844, deformed at 300°C., 1 kilobar pore pressure, is actually 45 per cent of that of samples tested dry at room temperature (Table IV).

Since the behavior of the Marianna depended strongly on the effective confining pressure, meaningful results could not be obtained unless specimens were completely saturated so that the pore pressure was uniform throughout. It was discov-

ered that the pore pressure had to be maintained for about 24 hours to attain pervasion of water. It did not matter, however, whether the full confining pressure was applied first or whether it was raised gradually along with the pore pressure.

The notion of effective confining pressure appears also to apply to the ductility of this limestone. The dry rock is essentially brittle at 0.5 kilobar. The 1-kilobar stress-strain curve becomes horizontal; the material is in the transitional state. At 2 kilobars there is apparent work-hardening, and the rock is clearly ductile. For the wet rock the transition pressure is somewhat lower. Thus at an effective confining pressure of 1 kilobar the stress-strain curve rises continuously, but at 0.5 kilobar the curve reaches a peak and then decreases gradually ($p_c = 2$, $p_p = 1.5$) or is horizontal ($p_c = 1$, $p_p = 0.5$), as shown in Fig. 14. At 0 and 0.25 kilobar ($p_c = 2$, $p_p = 1.75$, 2) the curves are peaked. One may regard the limestone as brittle, for failure without total loss of cohesion has occurred early by faulting.

Although the effects are not so great as those observed in the sandstone, the porosity changes accompanying deformations of the limestone have similar trends. At high effective pressures the porosity decreases, but when p_p reaches about 1.6 kilobars, the porosity increases (Fig. 16). Like the sandstone, the limestone becomes dilatant. A plot of the final porosity changes against the effective pressure of a series of specimens deformed at 2 kilobars confining pressure and at pore pressures of 0, 0.5, 1, 1.5, 1.75, and 2 kilobars would be similar to like plots for the sandstone, although the magnitudes of the porosity changes would not be so great, and the points would be more scattered (see Table IV). Heating has little or no effect on the magnitude of the porosity changes for otherwise similar test conditions.

The Marianna is too fine grained for a petrographic study, but from the macroscopic viewpoint this limestone behaves much as the sandstone, that is, a nearly cohesionless, porous aggregate. Both ultimate strength and ductility are predictable on the basis of effective stresses.

DOLOMITE ROCK

All the dolomite rocks tested previously had very low porosities and permeabilities. Of these, the Hasmark (Cambrian), collected from a shallow mine shaft near Philipsburg, Montana, was chosen for this investigation because of its purity

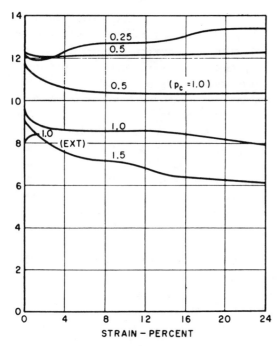

FIG. 16.—Porosity versus total strain of Marianna limestone at different effective confining pressures (kilobars) at 24°C. All at $p_c = 2$ kilobars except curve denoted $p_c = 1$. All in compression except curve marked *Ext* (for extension).

and relatively large grain size. Also, detailed microscopic studies had been made of experimentally deformed samples (Handin and Fair-

bairn, 1955). The initial porosity ranges from 2.3 to 3.4 per cent. The permeability is less than 1 millidarcy; the median pore size is below 1 micron.

Earlier studies reveal a fivefold increase in ultimate strength as confining pressure is raised from 0 to 2 kilobars. The dry rock is moderately ductile under pressure. At 2 kilobars, heating to 300°C. reduces the ultimate strength by only about 7 per cent and does not affect ductility (Handin and Hager, 1958, Fig. 3).

The stress-strain curves for the wet rock (Fig. 17) reflect the tendency of the material to become a little weaker as the pore pressure increases from 0 to 1.75 kilobars at 2 kilobars confining pressure. All the curves have much the same shape. The rock is in the transitional state, for at $p_c = 1$ kilobar (Handin and Hager, 1957, Fig. 9) it is clearly brittle, whereas at 5 kilobars there is work-hardening (Handin and Fairbairn, 1955, Fig. 5). When $p_p = p_c = 2$, the Hasmark is very much weaker and has also become brittle.

The principal stresses at the ultimate strength of the dry rock at 24°C. are plotted for confining pressures of 0, 0.5, 1, and 2 kilobars, together with the effective stresses in the saturated rock at $p_c = 2$ and $p_p = 0.5$, 1, 1.5, 1.75, and 2 kilobars (Fig. 18). Note that the principal stress relations differ in two respects from those of the sandstone and limestone. The curve for the dry rock is not

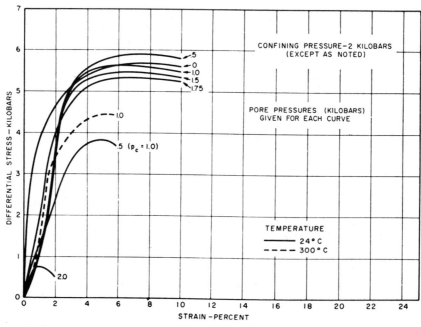

FIG. 17.—Stress-strain curves for Hasmark dolomite at different pore water pressures (kilobars). All in compression at 24° or 300° C.; all at 2 kilobars confining pressure except curve denoted $p_c = 1$.

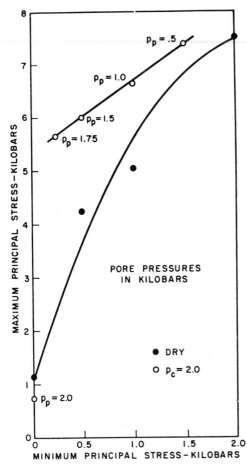

FIG. 18.—Effective maximum principal stress versus effective minimum principal stress at ultimate compressive strength of Hasmark dolomite at 24°C.

linear, and the points for the saturated rock do not fall on or even near that curve. Clearly, the notion of effective stresses is not efficacious here.

The reduction of strength due to heating to 300°C. appears to be greater at $p_p = 1$ than at $p_p = 0$. However, the reproducibility of the stress-strain curves for the wet rock has not been determined. As a first approximation one can probably superimpose temperature and pore pressure effects.

The fault angles in specimens deformed at room temperature are all close to 30°, whatever the magnitude of the pore pressure, except for 834 ($p_p = p_c$) with an angle of 15° (Table V). At 300°C. the values of the angles average out to about 35°, and this is regarded as significantly larger than the average room-temperature value.

Because the initial porosity of the Hasmark is very small, the precision of the measurements of void volume changes during deformation is poor. However, the porosity-strain curves (Fig. 19) do reveal that the higher the pore pressure, the larger the initial porosity under confining pressure (2 kilobars). This is regarded as good evidence that all available pore space, albeit small, is being filled with water. For small strains (less than about 3 per cent) the porosity decreases, nearly to zero at low pore pressures. As deformation proceeds, the porosity increases until at strains of about 14 per cent it exceeds the initial porosity of the undeformed rock.

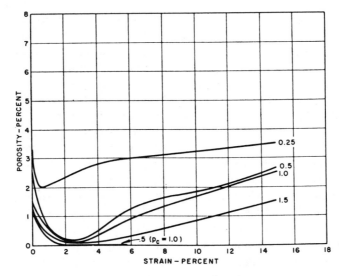

FIG. 19.—Porosity versus total strain of Hasmark dolomite at different effective confining pressures (kilobars) at 24° C. All in compression at 2 kilobars confining pressure except curve denoted $p_c = 1$.

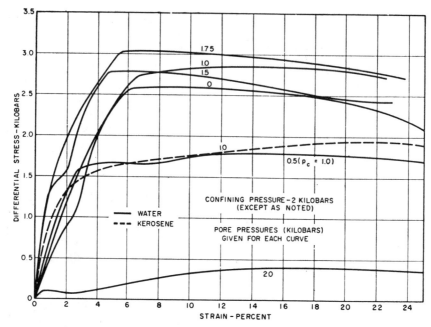

FIG. 20.—Stress-strain curves for Repetto siltsone at different pore pressures (kilobars) of water or kerosene. All in compression at 2 kilobars confining pressure, 24° C., except curve denoted $p_c = 1$.

SILTSTONE

Specimens of Repetto siltstone (Pliocene) were cut from well cores taken at a depth of about 12,500 feet in Ventura County, California. This rock is well indurated, dark gray, and fissile and contains alternating light and dark beds a few millimeters thick. It consists of about 13 per cent sand size (>60 microns) material and 63 per cent silt size (<4 microns) material. This is mostly quartz. The remaining 24 per cent is clay size material, of which half is montmorillonite. Twelve per cent of the rock is composed of about equal quantities of illite, kaolinite, and chlorite. The Repetto is heterogeneous even on the scale of a hand specimen. It has been difficult to obtain specimens sufficiently alike to permit legitimate comparisons of experimental results. For this reason fewer tests have been conducted than would be desirable. Loads were applied parallel to the well core about 45° to the bedding.

The stress-strain relations for the dry rock are available from the earlier reports (Handin and Hager, 1957, Fig. 30; 1958, Fig. 20). Increasing the confining pressure from 0 to 2 kilobars raises the ultimate strength by a factor of at least 10 and permits the attainment of permanent shortenings of 25 per cent or more. Heating to 300°C. at 1 or 2 kilobars reduces the ultimate strength by nearly 50 per cent.

The behaviors of the water-saturated specimens were not anticipated. Strengths appear to increase with pore pressure (Fig. 20). In the light of data for all other rocks tested, this result can certainly be regarded as anomalous. This increase of strength may be illusory, since its magnitude is of the same order as the reproducibility of the 2-kilobar stress-strain curves for the dry rock. Still there is no strength reduction, which would be expected if the pore pressure were even partially effective.

The failure of water pressure to influence mechanical properties in the manner expected may be due to one or both of two causes: (1) complete saturation is not achieved because the water enters the montmorillonite clay, which expands to reduce the permeability near the surface of the specimen, and (2) the effect of water on the expandable clay is actually to alter the mechanical properties of the aggregate. To test these ideas, an experiment was conducted with kerosene rather than water as the pore fluid. At $p_p = 1$, $p_c = 2$, the strength is drastically reduced (Fig. 20), and the stress-strain curve is very similar to that for $p_p = 0$, $p_c = 1$.

The principal stress relation at ultimate strength of the dry rock is nearly linear at confining pressures of 0.5, 1, and 2 kilobars at room temperature (Fig. 21).

$$\sigma_1 = a + 1.9\sigma_3. \qquad (17)$$

Points representing the effective principal stresses in specimens subjected to water pressures would fall on an essentially horizontal line, far from the line defined by equation 17. However, the point for the kerosene-saturated specimen at $p_c = 2$, $p_p = 1$ does plot on this principal stress curve. This is consistent with the results for sandstone and limestone, as are the stress-strain relations for the unjacketed siltstone (Table VI).

Mohr's graphical construction (not reproduced here) yields an envelope for ultimate strength which is a straight line, except very near the origin. $\phi = 17°$, so the predicted fault angles should be about 36° and should be independent of confining pressures above a few hundred bars. This is close to the values actually observed (Table VI), and these appear to be independent of both water and kerosene pressures. The angle is slightly steeper (31°) in specimen 720 heated to 300°C.

There is some evidence that complete saturation is not, in fact, attained with water. Thus at $p_p = 1$, $p_c = 2$ kilobars, the apparent porosity of specimen 850 is 3.8 per cent; the initial porosity was 6.9 per cent. On the other hand, under the same conditions specimen 969, filled with kerosene, has a porosity of 5.8 per cent compared with an initial value of 6.1 per cent (Table VI). Because it is unlikely that specimens are pervaded by water, the measurements of porosity changes

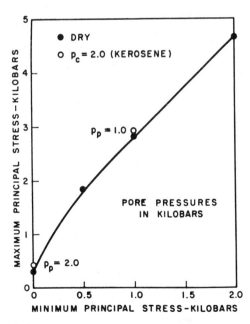

Fig. 21.—Effective maximum principal stress versus effective minimum principal stress at ultimate compressive strength of Repetto siltstone at 24°C.

accompanying deformations are not regarded as reliable. Final porosities measured after release of confining pressure have no consistent relationships to effective confining pressures. This further suggests that the water is ineffective. Note that the final porosity of kerosene-saturated specimen 969 exceeds the initial porosity, whereas in all other samples the bulk density has increased.

There is little doubt that the siltstone behaves as a sand-like aggregate, that is, pore pressure is fully effective, provided the interstitial fluid is inert relative to the clay mineral constituents of this rock.

SHALE

Specimens of Muddy Shale (Cretaceous) were cut parallel with a well core taken at a depth of 4,900 feet in Logan County, Colorado. This rock is dark gray, well indurated, homogeneous, and well laminated. It consists almost entirely of mixed-layer clay with subordinate amounts of illite, kaolinite, and chlorite, but no montmorillonite. The median grain size is about 10 microns, the pore size is 0.1 micron, and the permeability is less than 0.05 millidarcy.

The earlier work demonstrates large enhancements of ultimate strength and ductility of the dry rock with increasing confining pressure and a 50-per cent reduction of strength with increasing temperature from 24° to 300°C. (Handin and Hager, 1957, Fig. 29; 1958, Fig. 17). The nearly linear principal stress relation at confining pressures from 0 to 2 kilobars is given by

$$\sigma_1 = 0.4 + 2.8\sigma_3. \qquad (18)$$

Attempts to inject water or kerosene into specimens of the shale were unsuccessful. Even at the highest pore pressures the permeability is too low to permit pore pressure to come to equilibrium within a period of 24 hours. It may be argued that an effort should have been made to saturate with gas because of lower viscosity and surface tension. Unfortunately, this procedure would preclude void volume measurements, and one would never be certain that complete filling was attained. A more serious difficulty encountered in experiments on materials of low permeability is the time required for pore pressure to re-establish equilibrium throughout the material, once it is perturbed by a deformation. Thus whatever fluid is employed, and whether porosity determinations are made or not, one may never know the true pore pressure during a short-time triaxial

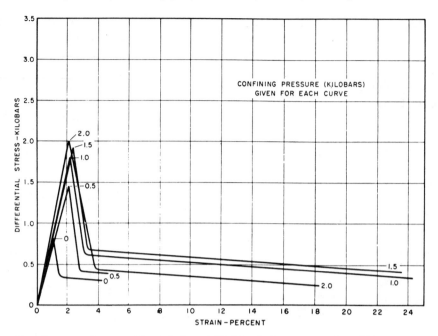

Fig. 22.—Stress-strain curves for unjacketed Muddy shale compressed at different confining pressures (kilobars) of kerosene at 24° C.

test. Since the pore pressure is measured externally to the specimen, one must assume that the permeability is sufficient to allow the interstitial pressure to attain a uniform magnitude throughout. Deformations must be slow enough to permit this; otherwise the results are meaningless. The strength of most, and probably all, materials differs with the rate of strain. Thus, if long-time tests are to be made to insure uniform pore pressure, then the time effects on the dry material must first be determined. Until this is accomplished, meaningful pore pressure experiments on ductile shales must be postponed.

However, in an effort to gain some insight into possible pore pressure effects on shales, a number of tests have been conducted on unjacketed samples. The stress-strain curves all peak at a strain of about 2 per cent, then drop off rapidly (Fig. 22). There is a tendency for the ultimate strength to increase with confining pressure, but although cohesion is not entirely lost at pressures above 0.5 kilobar, all the specimens must be regarded as brittle, in contrast to dry specimens under similar pressures (Handin and Hager, 1957, Fig. 29).

The strength of unjacketed, highly permeable materials such as the sandstone is independent of confining pressures up to 2 kilobars at least. This implies that $p_p = p_c$, that is, the effective confining pressure is always zero. The strength of the unjacketed shale does increase, but not nearly so rapidly as that of the dry rock (Fig. 23). This implies that full pore pressure is not applied because of the low permeability, but it must also mean that pore pressure is partially effective. The

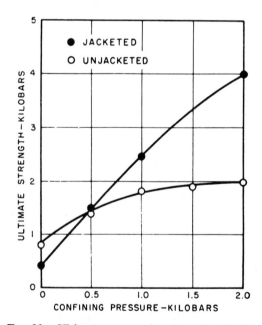

Fig. 23.—Ultimate compressive strengths of jacketed (dry) and unjacketed Muddy shale as functions of confining pressure at 24°C.

shale tends to behave in the same manner as the sandstone, in that high pore pressure favors low strength and brittleness. The magnitude of the interstitial kerosene pressure in the shale is unknown, but it now seems reasonable to suppose that it does not depart far from that of the confining pressure.

From the Mohr envelope of the ultimate strength of the dry shale (Fig. 24), one measures an angle ϕ of 30° to 22° over a range of confining pressures from 0 to 2 kilobars. The predicted fault angles would then be 30° to 34°. These are about what is observed at both 24° and 300°C. (Table VII), except for the 10° angle in specimen 217 compressed under atmospheric conditions.

The envelope determined from tests on unjacketed specimens is very non-linear, and ϕ varies from 35° to 3° as the pressure rises from 0 to 2 kilobars. The predicted fault angles would range from 28° to 44°. The observed values show no tendency to decrease with increasing pressure but remain nearly constant.

Porosity changes during deformation can not be measured. The values of final porosities determined after release of confining pressure are consistently greater than initial values (Table VII). The mechanism responsible for the dilatancy is uncertain, since the microscopic flow mechanisms in shale are unknown. Note, however, that the permeability and median pore size of specimen 898, deformed at the highest confining pressure (2 kilobars), are greater than those of samples compressed dry or at lower pressures. These properties are measured on fragments of the rock which do not contain visible faults. It is therefore suggested that the shale becomes dilatant (if effective pressure is small enough) in the same sense as does the sandstone. (Unfortunately, this concept cannot be verified by microscopic examination.)

Although the data are insufficient to prove the contention, it is considered probable that a shale would be affected by pore pressure much as a sand-like aggregate is affected. That is, the concept of effective stresses can be validly applied, provided that the pore pressure effects are purely mechanical and the magnitude of the interstitial pressure is known throughout the shale body.

DISCUSSION

EFFECTIVE STRESS

The data suggest, then, that the concept of effective stress (equations 1) can adequately account for the influence of pore pressure on both ultimate strength and ductility of sandstone,

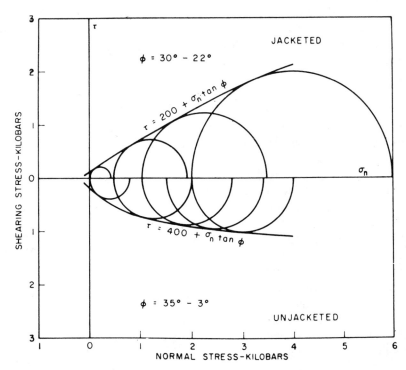

FIG. 24.—Mohr envelope curves for ultimate compressive strengths of jacketed (dry) and unjacketed Muddy shale at 24° C.

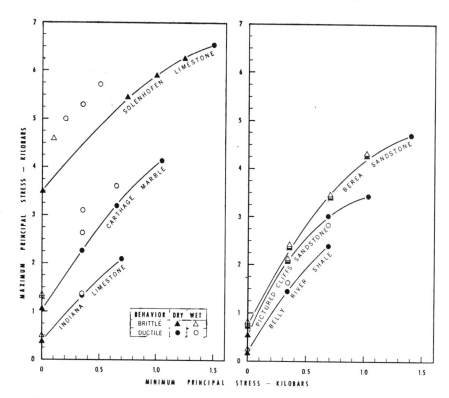

FIG. 25.—Effective maximum principal stress versus effective minimum principal stress at ultimate compressive strengths of jacketed rocks with pore pressures of water at 24° C.

limestone, siltstone, and probably shale, provided that (1) the interstitial fluid is inert, and (2) the permeability is sufficient to allow initial pervasion of the fluid and maintenance of constant uniform pore pressure throughout the deformation. Tests on dolomite, however, do not confirm this concept. What is the experience of other recent workers?

The significance of the results on other rocks is also demonstrated best by plotting the effective principal stresses at ultimate strength (as in Figs. 6, 15, 18, and 21). The stresses in the dry rock will be represented by filled triangles for the brittle state and filled circles for the ductile state. If the principal stress relation, defined by the curve drawn through these points, is independent of the absolute values of confining and pore pressures, then the open triangles and circles, denoting effective stresses in the saturated rock, must fall on the curve. Also, the brittle-to-ductile transition must occur at the same effective confining pressure. Fig. 25 shows plots of data derived from the triaxial compression tests of Robinson[9] (1959,

Figs. 7, 8, 9) on Pictured Cliffs sandstone (Cretaceous, New Mexico), "Indiana" limestone, "Carthage marble," and Belly River shale[10] (Cretaceous, Alberta); of Heard on Solenhofen limestone (1960, Fig. 11); and of Serdengecti and Boozer (1961, Fig. 3) on Berea sandstone.

Both ultimate strength and ductility of the sandstones, shale, and "Indiana" limestone are clearly functions of effective confining pressure at least to 1.5 kilobars. On the other hand, points representing the effective stresses in the rocks of low porosity—Solenhofen limestone (about 2 per

_____

required permanently to shorten the rock by 0.2 per cent when the yield stress is not well marked. This parameter corresponds with ultimate strength at the low effective confining pressures associated with peaked curves but not at the higher pressures associated with horizontal or continuously rising curves. Robinson's data are not therefore directly comparable with those of the writers and the other investigators.

[10] This shale has very low permeability and requires several days for complete saturation with brine. However, the permanent strains achieved prior to rupture were very small, and Robinson (personal communication, 1958) believes that the pore pressure must have remained essentially constant, since the porosity would be little affected. In any event the results are certainly consistent with those of the other rocks tested by him.

_____

[9] Robinson actually reports "yield strength," defined as the "maximum load supported by the rock" when the stress-strain curve is peaked but as the load

cent) and "Carthage marble" (porosity specified only as small)—do not correspond with the principal stress relations at ultimate strength, even though Heard's and Robinson's techniques were carefully designed to insure initial water saturation and maintenance of constant pore pressure during deformation. (The tests were "drained" in the usage of soil mechanics.) This is consistent with the behavior of the dolomite (porosity, 3.5 per cent). Note also that the dry Solenhofen does not acquire ductility until the confining pressure attains a value of 1.5 kilobars, whereas the wet rock is still ductile at an effective confining pressure of only 200 bars.

The data of all these investigators, as well as those of the writers, demonstrate that the strengths and ductilities of unjacketed permeable rocks are largely unaffected by confining pressure. This is expected, since the effective confining pressures are always zero. However, the early experiments by Griggs (1936, Fig. 4) on unjacketed Solenhofen limestone show that strength and ductility can be enhanced by much higher pressures. At about 8 kilobars kerosene pressure the limestone becomes ductile, and its ultimate strength is three times its crushing strength in air.

Empirically then, pore pressure does not always seem to be fully effective in consolidated rocks. What is the reason for this? Nearly forty years ago, Terzaghi (1923) showed from experiments on essentially cohesionless, water-saturated soils that the effective normal stress σ_n controlling internal friction was given by

$$\sigma_n = S_n - p_p, \tag{4}$$

where S_n is the total normal stress, and p_p is the pore fluid pressure or "neutral stress." Later he concluded that "the strains in clay and in concrete exclusively depend on the differences between the total stresses and the neutral stresses. In every point of the saturated material the neutral stresses act in every direction with equal intensity and they are equal to the pressure in the water at that point" (Terzaghi, 1936, p. 875).

Equation 4 has subsequently been confirmed empirically from many triaxial tests on soils and concrete by Terzaghi (1925; 1932; 1943), McHenry (1948), Bishop and Eldin (1950; 1953), and Bishop and Henkel (1957), among others. However, the theoretical basis for this equation has troubled most workers in the fields of soil mechanics and foundation engineering, including Terzaghi (1945) himself. Their problem, reviewed recently by Skempton (1961), is as follows.

An imaginary slip plane of area A through a water-saturated porous aggregate will cut through some grains and pass through some fluid-filled pore space. Whereas the total applied shearing force is balanced by an equal and opposite shearing force acting only in the projected area of the solids A_s, the total normal force is supposed to be balanced by two normal forces, one acting across the solid area, and the other acting across the area occupied by the interstitial fluid. This implies that the pore pressure is effective only in the area $(A - A_s)$ and that equation 5 becomes

$$\tau = \tau_0 + (S_n - f_b p_p) \tan \phi, \tag{19}$$

where $f_b = (A - A_s)/A$, the "boundary or surface porosity."

Since the measured values of f_b in equation 19 are in fact always equal to unity, or very nearly so, one must imagine a sinuous surface which passes through pore spaces and grain contacts, and then postulate that the total intergranular area A_s is almost vanishingly small. This reasoning might account for the behavior of, say, cohesionless sand composed of spherical grains of equal size, but it can not explain the fact that $f_b \approx 1$ even for consolidated rocks with volume porosities of less than 10 per cent.

In the writers' opinion, the fallacy of this concept of effective stress has been demonstrated unequivocally by Hubbert and Rubey (1959, p. 132–137; 1960, p. 617–628), who extend the Archimedes principle to the case of the permeable rock with interconnected pore space filled with water. They show that whenever the pore or "neutral" pressure is transmitted fully throughout the solid framework, then it must always be fully effective, regardless of porosity. If the interstitial fluid pressure is not fully effective, as in the rocks of low porosity, the configuration of the pore space must be such that this pressure is not uniformly distributed throughout the solid phase. The exact state of stress at all points in the body could then be computed only for very simple idealized situations; thus for practical purposes the influence of pore pressure must be determined empirically if equation 5 does not hold.

DEFORMATION MECHANISMS

The effects of pore pressure on strength, ductility, and porosity are clearly related and are now at least qualitatively explainable in terms of the deformation mechanisms which are important at the pressures, temperatures, and strain rate of these experiments.

Being essentially cataclastic, the deformation of the sandstone is most readily understood and is also best documented. At high effective confining pressures (about 1.2 kilobars or more) the grains are pressed together tightly, and the frictional resistance to slippage of grains past one another is very great. On the other hand, the stress concentrations at grain contacts are also large. Although some flow of cementing material and argillaceous matrix may occur, the deformation is largely cataclastic because the predominant mineral constituents (quartz and feldspar) do not flow—they fracture. Grain breakage is general, and local intergranular adjustments of grains and fragments tend to reduce pore space (Fig. 11). As the average grain size and porosity decrease with progressing deformation, the mean contact area increases, and hence the stress concentrations are reduced on the average. Further grain fracturing becomes increasingly difficult, and the differential stress required to deform the aggregate rises. From the standpoint of the stress-strain curve (Fig. 4), the rock is ductile and appears to work-harden. Eventually the stress difference becomes high enough to overcome internal friction. Macroscopic faulting may then occur, much as predicted by equations 5 and 6. Still further deformation is localized along faults oriented parallel to the most "dangerous" planes, until ultimately cohesion may be lost entirely.

Since the breaking strength of quartz (Griggs et al., 1960, Table 3), and presumably also of other minerals in the brittle state, is largely unaffected by heating, the ultimate strength and ductility of sandstone are insensitive to temperature changes in the range of 25° to 300°C.

At intermediate effective confining pressures (about 400 to 1,200 bars) the frictional term $\sigma_n \tan \phi$ is smaller because the effective normal stress is lower. Macroscopic shear failure occurs early in the deformation history. Its occurrence is marked by the peak in the stress-strain curve (Fig. 4), and once initiated, the faulting progresses with ever greater facility, so the differential stress falls continuously as the deformation proceeds. The friction is, however, still relatively high. Granulation occurs in the fault zone to which most of the grain fracturing is restricted. The porosity decreases little if at all (Fig. 13), and the rock is no longer ductile.

When the effective pressure reaches a value of the order of 200 bars, frictional resistance is low, macroscopic shear failure occurs readily at low differential stresses, and there is no grain breakage beyond the limits of the narrow, sharply defined shear zone (Fig. 12). The high pore pressure effectively "cushions" the grains. Stress concentrations at grain contacts are insufficient to cause rupture, but intergranular movements are relatively easy. The deformation is due to both early faulting and intergranular movements. Porosity increases as a result of the deformation (Fig. 13).

When the effective confining pressure is zero, the principal resistance to deformation is the low cohesive shearing strength (equation 5). The rock is very weak and brittle (Fig. 4). It can not deform cataclastically, and once the cohesive resistance is surmounted, deformation proceeds by intergranular movements. Since the rock is closely packed, these movements must result in increased void volume. The rock becomes dilatant (Fig. 9). Shear failure may occur when the deformation becomes large, but little or no fault gouge is produced, since there is virtually no friction.

The changes in median pore size and permeability (Table II) also fit into this picture of the sandstone deformation. For the state in which the material is dilatant ($p_p \geq 1,600$ bars), the final median pore size is about the same as that of the undeformed sandstone (15–20 microns). At lower pore pressures the pore size is drastically reduced, presumably because of grain fracturing, as is the permeability. In the dilatant state the permeability actually increases along with porosity.

The porous limestone behaves much as the sandstone, that is, a nearly cohesionless, granular aggregate, but the deformation differs, in that the constituent calcite grains can flow as well as fracture. At low temperatures the critical resolved shear stress for mechanical twinning is very low, about 50 bars, though the critical stress for translation gliding is rather high, about 1 kilobar (Turner et al., 1954, Fig. 7). Twinning would be expected at low effective confining pressures, but some translation gliding also occurs even at a pressure of only 20 bars (Paterson, 1958, p. 473). There is ample evidence from previous work that the brittle-ductile transition in limestones at room temperature occurs at an effective confining pressure of about 1 kilobar and is associated with the change of predominant deformation mechanism from cataclasis to intracrystalline gliding (Kármán, 1911; Robertson, 1955; Handin and Hager, 1957). Turner et al. (1956, p. 1265), Handin and Hager (1958, Figs. 10, 11, 13), and Heard (1960, Fig. 7) all show that heating enhances ductility and markedly reduces the yield stress of limestone. In the ductile range the strength reduction can be associated qualitatively with the lowered critical resolved shearing stresses for twin and translation gliding of the individual calcite grains (Turner et al., 1954, Fig. 7). Turner et al. (1956, p. 1290) found no evidence of recrystallization in wet or dry marble deformed at temperatures below about 300°C.

At high effective confining pressures (1 kilobar or more), the Marianna limestone is ductile, and the stress-strain curves (Fig. 14) rise continuously. This is probably due both to actual work-hardening of individual permanently deforming calcite crystals and to the porosity reduction (Fig. 16), which again tends to lower the average stress concentrations at grain contacts. At high strains the stress difference may attain a magnitude sufficient to overcome internal friction, and macroscopic faulting may eventually occur.

At intermediate pressures (about 500 bars), frictional resistance is lower, and the deformation is evidently due principally to faulting, since there is virtually no porosity change (Fig. 16) and no work-hardening (Fig. 14). However, some intragranular flow probably occurs, and some grain fracturing must be present, because the final median pore sizes and permeabilities after release of confining pressure have increased (Table IV).

At low pressures (0 to 250 bars) the deformation is accompanied by increases of porosity, permeability, and pore size; the rock is dilatant, as is the sandstone, and for the same reason.

The microscopic deformation mechanisms in the siltstone and shale can not be identified optically because of the very fine grain sizes. However, there is little doubt that the nature of the deformations is similar to that of the sandstone, except for the possibility of intracrystalline gliding in micaceous minerals at high effective confining pressures.

In addition to the nature of the constituent grains (brittle or ductile), one must consider the construction of the rock. The sandstones, Marianna limestone, siltstone, and shale behave as nearly cohesionless, sand-like aggregates in which the interstitial fluid has access to a very large fraction of the internal surface of the solid framework. On the other hand, the dolomite, marble, and Solenhofen limestone (as well as all other crystalline rocks) are aggregates of interlocked grains with very little pore space which may be irregularly distributed throughout the rock. The pore pressure may not be transmitted uniformly throughout the solid phase and hence may not be fully effective in opposing the external loads on the aggregate in bulk.

The deformation of the dolomite rock, for example, is due principally to macroscopic faulting without loss of cohesion. The flow mechanism in dolomite crystals in the temperature range of 25° to 300°C. is basal translation gliding, but the critical resolved shear stress is high, about 1 kilobar (Higgs and Handin, 1959, Fig. 12). One can expect the rock to be essentially brittle at the maximum effective confining pressure of these experiments (2 kilobars). Even at 5 kilobars a shortening of the order of 10 per cent is accounted for mostly by faulting, though there is small concomitant intracrystalline gliding (Handin and Fairbairn, 1955).

The porosity-strain curves (Fig. 19) show a reduction almost to zero at small strains (about 3 per cent), followed by an increase, until at strains of about 14 per cent the porosity exceeds that of the undeformed rock. One might regard the dolomite as dilatant, but this porosity change is not ascribable to the mechanisms suggested for the deformation of the porous sandstone and limestone. This is more readily apparent from an inspection of the values of the final porosity measured after release of confining pressure (Table VI). These are several-fold greater than initial values and are most probably associated with faulting which begins after voids are closed. This faulting occurs without permanent loss of cohesion, so long as confining pressure is maintained. Once the pressure is released, there is expansion in the sheared regions of a specimen. Complete separation may occur, especially in samples deformed at effective pressures of 1 kilobar or less. These specimens can be justifiably characterized as "fractured." Samples compressed at 2 kilobars may retain cohesion after release of confining pressure, even though large displacements have occurred on one or more macroscopic faults. For the same total strain, the effect on porosity tends to be smaller at 2 kilobars than at 1 kilobar, suggesting that cataclasis is less important and intracrystalline flow is more important at the higher effective confining pressures. Since pore pressure is not fully effective, it does not greatly reduce the normal stress across faults. Frictional resistance and hence ultimate shear strength (equation 2) are little influenced.

Why should the mechanical properties of unjacketed rocks be affected at all if the effective confining pressure must always be zero (e.g., Marianna limestone at 5 kilobars and Solenhofen limestone at 8 kilobars)? A truly cohesionless aggregate would possess no strength at all when confining and pore pressures were equal. However, rocks do have a finite cohesive strength (equation 2), and differential loading can develop intergranular stress concentrations, even if the effective confining pressure is zero. Furthermore, an individual mineral grain may be impermeable and thus under its own hydrostatic pressure, even if there is no effective pressure in the aggregate as a whole. Thus the breaking strength of the individual crystal is enhanced according to equation 2, but because the cohesive strength of quartz is exceedingly high, the frictional term becomes important only at very high normal pressures, and any effect is undetected in the sandstones. However, as the breaking strength of calcite (and doubtless dolomite) increases, the permissible stress difference eventually reaches the yield stress of the crystal. An aggregate like limestone can then deform by intracrystalline

gliding and so become ductile. It can work-harden until the bulk shear stress attains the limiting value set by equation 2 for faulting through the aggregate as a whole.

Terzaghi (1945) associated the increase of shear strength with confining pressure with the failure of intergranular bonds, because he supposed that the resistance to shear across grains would be independent of normal pressure. He envisaged a rough surface of failure. When the rock was sheared, a grain on one side of this surface was supposed to override the humps formed by a series of grains and voids on the opposite side. The enhancement of strength was ascribed to an increase of (1) resistance to shear in the bonding material because of high local mean pressure arising from stress concentrations at grain contacts, and (2) intergranular contacts which resulted in higher frictional resistance to slippage. The state of stress within the grains was regarded as irrelevant. This theory can not apply to the failure of consolidated rocks under high effective confining pressures, because shear occurs predominantly through grains. The breaking strength and the relative ductility of a mineral grain both depend upon the state of stress in that grain.

GEOLOGICAL IMPLICATIONS

Some engineering applications of the pore pressure data have been discussed previously (Handin, 1959). Some important geological implica-

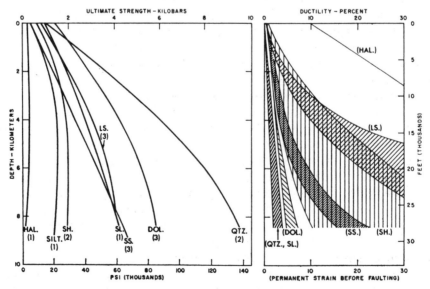

FIG. 26.—Ultimate compressive strengths and ductilities of dry rocks as functions of depth. Effects of confining (overburden) pressure and temperature (30° C./km.) included.

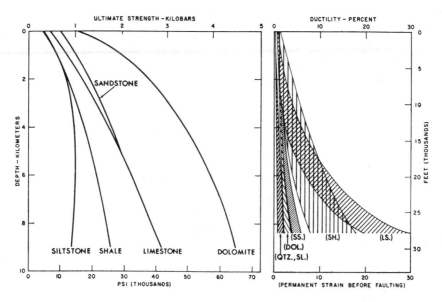

FIG. 27.—Ultimate compressive strengths and ductilities of water-saturated rocks as functions of depth. Effects of confining (overburden) pressure, temperature (30° C./km.), and "normal" formation (pore) pressure included.

tions will be emphasized here. The effects of pore pressure on ultimate strength and ductility can be illustrated with better geological perspective by plotting these parameters against depth of burial (Figs. 26, 27). This requires that the experimental variables be related to their natural counterparts. Pore pressure and temperature can be regarded as directly analogous to formation fluid pressure and earth temperature, respectively, but for the following reasons confining pressure can be associated only with a ·highly idealized state of stress in the crust.

1. The state of total stress in the jacketed laboratory specimen is initially hydrostatic since the liquid confining pressure effectively acts on all external surfaces. Because pore pressure is also hydrostatic, all principal effective stresses in the solid rock are equal. During an experiment the two radial principal effective stresses remain constant while the axial stress is increased or decreased until failure occurs. Both confining and pore pressures are independent experimental variables which are fixed throughout any single test. Strength and ductility are therefore determined as functions of effective confining pressure $(p_c - p_p)$ which must somehow be expressed in terms of depth.

2. Below flat country one principal total stress must be nearly vertical; the other two must be nearly horizontal. The magnitudes of the horizontal principal total stresses are not known a priori. However, the vertical total stress S_z must be equal to the total overburden pressure, that is, the weight per unit area of the superincumbent water-saturated rock. To calculate the vertical effective stress σ_z in the absence of knowledge of the hydrodynamic conditions on the ground water, one must assume that the pore water is in hydrostatic equilibrium. For this special case

$$\sigma_z = S_z - p_p = \rho_b g z - \rho_w g z, \qquad (20)$$

where z is depth below the free surface, ρ_b is bulk density of saturated rock, and ρ_w is density of water. From this relation one can estimate the magnitude of one of the principal effective stresses as a function of depth.

3. The quantity σ_z from equation 20 must still be related to the experimental variables. This can be done by assuming a natural state of stress in which all principal total stresses are the same and equal to the total overburden pressure, the "standard state" of Anderson (1951, p. 13). Since the formation pressure is hydrostatic, all principal effective stresses in the rock are also equal or "lithostatic." If confining and pore pressures are regarded as analogous to total overburden and formation pressures, respectively, then from equation 20

$$\sigma_1 = \sigma_2 = \sigma_3 = \sigma_z = p_c - p_p. \qquad (21)$$

4. Little is known about the absolute magnitudes of the principal effective stresses associated with natural rock deformations, but the state of stress would generally be truly triaxial, that is, all three principal stresses would be different. In the laboratory only two states are obtained ($\sigma_1 > \sigma_2 = \sigma_3$ and $\sigma_3 < \sigma_1 = \sigma_2$). None of the experiments can be expected to represent the natural phenomena exactly. Furthermore the essentially two-dimensional Coulomb-Mohr theory is not in fact independent of intermediate principal stress as supposed; that is, the Mohr envelopes for compression and extension tests do not coincide (see Fig. 7 and Handin and Hager, 1957, Fig. 43). The effect of this stress on mechanical properties has not yet been evaluated.

To investigate which natural situations may be approximated experimentally, let us consider Anderson's (1951) simplified theory of faulting. This is based on the Coulomb-Mohr criterion, but does at least suggest the probable directions of the principal stresses. Let us begin with the standard state, represented experimentally by equation 21, and assume that σ_z remains essentially constant.

When $\sigma_z = \sigma_1$, the fault is normal and dips about 60°. If only one horizontal principal stress is reduced, $\sigma_3 < \sigma_1 = \sigma_2$, analogous to an extension test. If both are reduced equally, $\sigma_1 > \sigma_2 = \sigma_3$ (compression test). When $\sigma_z = \sigma_3$, a thrust fault forms with a dip of about 60°. If one horizontal principal stress is increased, $\sigma_1 > \sigma_2 = \sigma_3$ (compression test), and if both are raised equally, $\sigma_3 < \sigma_1 = \sigma_2$ (extension test). When $\sigma_z = \sigma_2$, a vertical wrench fault develops. The magnitude of σ_2 might approximate that of σ_1 (extension test) or of σ_3 (compression test).

Which of these limiting values of σ_2, if either is the more geologically realistic, is conjectural for all three types of faulting. The curves of Figures 26 and 27 are valid only for the state $\sigma_1 > \sigma_2 = \sigma_3 = \sigma_z$, a limiting condition for thrust faulting and a condition possibly approached for wrench faulting. For normal faulting, strength and ductility at a given depth would be less than predicted because the horizontal principal effective stresses σ_2 and σ_3 would be less than σ_z. The effective confining pressure (equation 21) would therefore be lower than assumed for that particular depth.

Although the laboratory conditions must always depart in some degree from natural conditions, the curves are nevertheless useful in assessing the *relative* strengths and ductilities of different rocks at given depths or of the same rock at different depths. In constructing these curves the following specific assumptions are made with respect to the variations of natural environmental conditions with depth and to the procedures of data plotting.

1. The total overburden pressure gradient is computed from the density-depth curve of Nettleton (1934, Fig. 1) for sedimentary rocks. This is very nearly 230 bars/km. (1 psi./ft.) to a depth of 4.5 km. (15,000 ft.). Beyond, the pressure rises more rapidly than linearly to 2 kilobars (29,000 psi.) at 8.4 km. (27,600 ft.).

2. The choice of appropriate temperatures is more arbitrary. The heat flow across the continental surface of the earth is uniformly 1.2 microcal./cm.²-sec., but measured geothermal gradients vary from 5° to 70°C./km. (Birch, 1954, p. 646–647). This large spread of values is associated with differences in thermal conductivity. Shallow, porous sedimentary rocks have relatively low conductivities and hence high gradients. The fluid content of porous rocks strongly influences the conductivity. For thick sections of sedimentary rocks an average gradient of 30°C./km. (100°F./3,800 ft.) seems reasonable, though one must expect this figure to vary widely with rock type, geological structure, and hydrodynamics of underground fluids. On the other hand, errors of 10–20 per cent in estimating the temperature at a particular depth are insignificant, because the mechanical properties of the rocks involved are not sensitive to temperature changes of a few degrees.

3. The so-called "normal" formation pressure is equivalent to the hydrostatic head in a column of water from the surface to any particular depth, so the pressure gradient is about 100 bars/km. (0.45 psi./ft.). In fact this quantity is not constant because of small density changes due to variations of pressure (compressibility), temperature (thermal expansion), and salinity (concentration of dissolved solids), but errors in estimating the magnitude of the formation pressure are unlikely to be large. Dickinson (1953, p. 413) states that an average gradient of 0.465 psi./ft. has been established from the surface to a depth of 16,000 feet in the Gulf Coast region.

4. The formation pressure is everywhere either zero (Fig. 26) or "normal" (Fig. 27), and any pore pressure effects must be purely mechanical.

5. The curves for the dry rocks (Fig. 26) are derived from the stress-strain relations as functions of confining pressure and temperature determined previously (Handin and Hager, 1958). When data are available for more than one rock of a particular type, the strength-depth curve represents the average values, and the accompanying figures show the number of different rocks tested. The ductility-depth curves indicate the spread of measurements within each category, since this parameter varies widely, especially among the shales.

6. The curves for the water-saturated rocks (Fig. 27) are constructed as follows. Room-temperature strengths and ductilities of Berea sandstone, Marianna limestone, Hasmark dolomite, and Repetto siltstone at effective confining pressures appropriate to depths of 0, 2.2, 4.4, and 8.4 km. are read directly from the stress-strain curves (Figs. 4, 14, 17, 20). The Muddy shale is supposed to behave as though pore pressure were fully effective, and its curves are estimated from the stress-strain relations of the dry rock (Handin and Hager, 1957, Fig. 29). The influence of temperature on the deformation of the sandstone is negligible. Temperature effects on the other four rocks are taken from the earlier report (Handin and Hager, 1958, Figs. 3, 10, 17, 20). As the data suggest, pressure and temperature effects are superposable (e.g., Figs. 14, 17).

Comparison of Figs. 26 and 27 shows that pore pressure tends to reduce both strength and ductility at all depths, because these parameters depend principally upon effective confining pressure, and formation pressure lowers the effective overburden pressure in porous rocks. Of the materials tested previously, halite can not, of course, be subjected to water pressures, and the quartzite and slate are too impermeable for pore pressure experiments. However, the curves for dry specimens (Fig. 26) would probably be meaningful even if interstitial fluids were present, because the influence of "normal" formation pressure would probably be negligible. For halite, heat is far more important than effective pressure, and the rock salt in domes is characteristically "dry" anyway. The quartzite and slate (and other crystalline rocks as well) should behave much as the dolomite, so one would expect any pore pressure effects to be small.

Previous studies (Handin and Hager, 1957; 1958) have revealed the important fact that the deformational behaviors of all homogeneous rocks within each of relatively few categories are essentially similar. These are dolomite, halite, limestone, sandstone, siltstone and shale, and crystalline rocks like quartzite and slate. Thus one can use these curves (Figs. 26, 27) with considerable confidence to estimate the strengths and ductilities of all rocks of these types, whether tested or not, at any realistically simulated depths to about 9 km. (30,000 ft.).

The results of these short-time experiments can be applied directly to certain engineering problems. However, although Heard (1961) shows that a variation of 10^7 in strain rate does not much affect the strength of limestone below 300°C., geological interpretations must be fashioned with caution. Until the influence of time is fully evaluated, *absolute* values will not be meaningful, but again *relative* values can be useful in problems of structural geology which involve deformations of stratified rocks with very different mechanical properties. In this regard the concept of "competence" can be ambiguous and should be carefully defined by any geologist who feels compelled to use the term. The behavior of different beds during the folding process may depend upon both relative strength and ductility, and these parameters can be independent. For example, the strengths of sandstone and limestone are nearly identical, but the ductilities and hence the tendencies to fault or to flow uniformly are very different (Figs. 26, 27). Quartzite is stronger and more brittle than limestone, but this in turn is stronger but also more ductile than some shales.

Whenever the pore pressure is "normal," the formation must in effect be coupled to the atmosphere through a permeable conduit so that attainment of hydrostatic pressure is possible. Any system in which the formation pressure is "abnormal," that is, in excess of about 0.5 of overburden pressure, is not in equilibrium and must be isolated by some kind of permeability barrier. At least two important situations are being encountered more frequently as wells are being drilled deeper.

1. If the fluid pressure is initially "normal," deformation would tend to reduce pore space and so to raise the pressure if the fluids could not escape. If the rocks have low permeability and the deformation is rapid, the pressure could become "abnormal," even though the formation involved might eventually intersect the surface. An ex-

ample may be the Ventura anticline in Calfornia, where young, nearly impermeable siltstones are intensely folded (Watts, 1948). How long such disequilibrium pressures could be retained would depend on the effective permeability along and across the bedding.

2. Perhaps the commonest cause of high pressure is the very rapid deposition of thick sections of fine-grained sediments which are so young and impermeable that their connate waters have not yet been fully expelled. These sediments are simply undercompacted. The interstitial fluid supports most of the weight of the superincumbent material, and very high disequilibrium pressures can be retained even for geological times. Ratios of pore to overburden pressure as high as 0.9 have been reported (Rubey and Hubbert, 1959).

In regions of rapid deposition the connate water in the sands may be trapped below relatively impermeable layers of fine-grained sediments. In any event water is expelled from the surrounding compacting clays into the permeable sands as long as there is any potential gradient across the interface. The pore pressure measured in the sandstones must be essentially the same as that in the adjacent fine-grained rocks.

Whatever their cause, "abnormal" formation pressures will strongly affect the deformations of rocks in which they occur. Consider what might happen to the strength of a sandstone in an actual situation. The monotonically increasing strength-depth curve for Berea sandstone with "normal" pore pressure (Fig. 27) is replotted to 5 km. (15,000 ft.) in Figure 28. Superimposed is the predicted curve derived for "abnormal" pressures observed in isolated sandstones penetrated by a well in the Gulf Coast region (Dickinson, 1953, Fig. 6). The situation is "normal" to a depth of about 7,000 feet, below which the formation pressure increases rapidly to approach that of the overburden at about 12,000 feet. In this interval the curve would reverse as effective confining pressure drops, and the strength would become lower than that of the superjacent rock. Ductility would also be reduced. Both effects would tend to favor localized deformation in the high-pressure zone. Recognition of this situation, which can be expected for any rock in which the "abnormal" pore pressure is fully effective, may enhance our understanding of certain problems in structural geology.

1. The abnormally high-pressured, uncom-

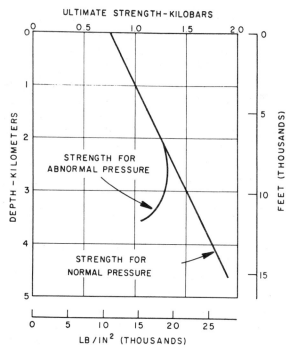

FIG. 28.—Ultimate compressive strength of water-saturated Berea sandstone as function of depth for a "normal" formation (pore) pressure gradient ($p_p \sim 0.5 p_c$) and for an "abnormal" gradient measured in Gulf Coast Louisiana. Sandstone is much weaker in high-pressure zone.

pacted rock can have a bulk density lower than that of the overlying material and hence can be gravitationally unstable. It may behave as rock salt in the sense that buoyant forces would tend to displace it upward relative to the denser normally pressured rocks. The nature of the resulting structure would seem to depend on the mechanical properties of the rock involved—diapirs of relatively weak and ductile salt, broad uplifts of moderately strong and ductile shale, and dikes of relatively strong and brittle sandstone.

2. It is at sites of rapid deposition of fine-grained sediments that one would expect to encounter low effective pressures and hence localized faulting. The so-called growth faults along the northern margin of the Gulf of Mexico may be one example. Hubbert and Rubey (1959) show how large blocks may be displaced many tens of kilometers on nearly horizontal overthrusts, where the frictional resistance may approach zero because high pore pressures lower the effective normal stresses across the faults (equation 5). These features are often associated with belts of thick, relatively young sediments bordering geo-

synclines (Rubey and Hubbert, 1959).

3. In sandstone, limestone, siltstone, and probably in shale, a compressional deformation reduces porosity when $p_p/p_c \leq 0.6$, and water must escape to maintain constant pore pressure. At ratios of 0.6 to 0.8 the void volume remains unchanged, but at 0.8 or more the rocks are dilatant, and water must flow into the system to hold the interstitial pressure steady.[11] The deformation of a rock element, effectively isolated by its impermeable surroundings, or impermeable itself relative to the rate of strain, would tend to raise formation pressures for initial values of p_p/p_c below about 0.8 and to lower them for ratios above. A temporary equilibrium (exclusive of gravitational compaction) might be established. In any event the deformation tending to increase pore pressure would be autocatalytic, since the strength would decrease progressively with effective pressure.

Conclusions

The following conclusions can be drawn from the results of tri-axial compression tests on Berea sandstone, Marianna limestone, Hasmark dolomite, Repetto siltstone, and Muddy shale at different constant confining pressures (to 2 kilobars), pore pressures of water (to 2 kilobars), and temperatures (to 300°C.) which realistically simulate the natural conditions of total overburden pressure, formation pressure, and earth temperature to a depth of about 9 km. (30,000 ft.).

1. The results are qualitatively consistent with those of soil mechanics from tests on loose aggregates at pressures of a few bars (for example, Bishop and Eldin, 1953); with those obtained earlier by McHenry (1948) for concrete at about 100 bars; and with those determined subsequently by Robinson (1959) for limestone, marble, sandstone, and shale at 1 kilobar, by Heard (1960) for Solenhofen limestone at 1.5 kilobars, and by Serdengecti and Boozer (1961) for sandstone at 1.4 kilobars. All suggest that the important mechanical properties—ultimate strength and ductility—are functions of the *effective stresses*,

[11] These values are only approximate, because the threshold of stress difference for intragranular flow or fracture, and hence porosity reduction, depends on the strengths of constituent minerals and bonding material, as well as effective confining pressure. However, they are probably qualitatively useful in problems involving porous sedimentary rocks to depths of the order of 25,000 feet.

provided that (*a*) the interstitial fluid is inert relative to the mineral constituents of the rock so that pore pressure effects are purely mechanical, (*b*) the permeability is sufficient to allow pervasion of the fluid and furthermore to permit the interstitial fluid to flow freely in or out of the rock during the deformation so that the pore pressure remains constant and uniform throughout (the test is "drained"), and (*c*) the rock is a sandlike aggregate with connected pore space, the configuration of which insures that the pore ("neutral") pressure is transmitted fully throughout the solid phase.

With these conditions, one can calculate the effective principal stresses (equation 1) by simply subtracting the pore pressure from the external loads. Thus the stress-strain relations of a dry rock deformed at, say, 1 kilobar confining pressure are the same as those of the saturated rock at 2 kilobars confining pressure and 1 kilobar pore pressure. In both instances the effective confining pressure is 1 kilobar. This concept holds for the sandstones and porous limestones, for the siltstone when provision (*a*) is satisfied (the interstitial fluid is kerosene instead of water), for the shales if provision (*b*) applies, but not for the crystalline rocks of low porosity (Hasmark dolomite, Solenhofen limestone, and marble), presumably because provision (*c*) does not hold.

2. At the outset of their laboratory investigations of the mechanical properties of rocks, the writers had to choose between two experimental methods. One approach would simulate the natural environment of deeply buried rocks as realistically as possible in every test, and this would have the advantage that a wealth of empirical data could be collected relatively rapidly. However, because of the complex interrelations among the significant variables, the favored method was to study the effects of these variables separately in order to acquire at least a qualitative understanding of how each contributed to the deformational behaviors of the rocks.

The influence of confining pressure alone on the properties of 23 dry rocks was examined first. Ultimate strength, ductility, and mode of deformation as functions of confining pressure were discovered to be much alike for all rocks within each of relatively few categories—dolomite, limestone, sandstone, siltstone and shale, and crystalline rocks including quartzite and slate (Handin and Hager, 1957). This encouraged the belief that the heterogeneity of crustal materials would not vitiate inferences as to the probable behavior of untested rocks under similar conditions and that the testing of fewer rocks (11) would be sufficient at elevated temperatures. The effects of temperature in combination with confining pressure also turned out to be consistent (Handin and Hager, 1958).

The number of experiments, and hence the cost and time of the work, grow rapidly with the addition of new interdependent variables. Pore pressure tests were made on only 5 rocks, but the advantage of the method of research decided upon becomes clear now. Since the pertinent properties of porous sedimentary rocks depend on effective confining pressure, since effects of heating can be superposed on pore pressure effects, and since all rocks within each category behave similarly, one can confidently estimate the short-time ultimate strength and ductility of any of these rocks, whether tested or not, under any conditions of overburden pressure, formation pressure, and temperature appropriate at depths to 9 km. (30,000 ft.).

3. Whenever the formation pressure is "normal" (equivalent to the hydrostatic head of a column of water to the surface), one can predict strengths and ductilities as functions of depth from the curves of Figures 26 and 27. However, the pressure can be abnormally high in undercompacted rocks in which the interstitial fluid is trapped by impermeable barriers. The internal (Coulomb) friction, and hence the strength and ductility of the rock, is low relative to that of the surrounding normally pressured rocks. Deformation tends to be localized within the high-pressure zone. Here the rocks also have abnormally low bulk densities and therefore tend to migrate upward.

4. Since the interstitial water affects the angle of internal friction little, if at all, the fault angles in most rocks should be inclined at 25° to 35° to the maximum principal compressive stress (essentially vertical and horizontal for normal and thrust faulting, respectively), as previously predicted (Handin and Hager, 1957; 1958).

5. The permanent shortening of porous sedimentary rocks is accompanied by a reduction of porosity whenever the ratio of pore pressure to confining pressure is of the order of 0.6 or less. At ratios of 0.6 to 0.8 the pore volume remains essentially constant, but above 0.8 the porosity increases, and the rocks are dilatant.

LITERATURE CITED

Adams, L. H., 1936, A simplified apparatus for high hydrostatic pressures: Rev. Sci. Instruments, v. 7, p. 174–177.

Anderson, E. M., 1951, The dynamics of faulting: London, Oliver and Boyd, 206 p.

Balmer, G. G., and Hanson, J. A., 1951, Strength and elastic properties of Navajo sandstone core from Glen Canyon Dam site, Arizona: U. S. Bur. Reclamation, Structural Lab. Rept. SP-30.

Birch, F., 1942, Handbook of physical constants: Geol. Soc. America Special Pub. 36, 325 p.

———— 1954, The present state of geothermal investigations: Geophysics, v. 19, p. 645–659.

Bishop, A. W., and Eldin, G. A. K., 1950, Undrained triaxial tests on saturated sands and their significance in the general theory of shear strength: Geotechnique, v. 2, p. 13–32.

———— and Eldin, G. A. K., 1953, The effect of stress history on the relation between ϕ and porosity in sand: 3d Internatl. Conf. Soil Mechanics and Foundation Engineering Proc., v. 1, p. 100–105.

———— and Henkel, D. J., 1957, The measurement of soil properties in the triaxial test: London, Arnold, 190 p.

Borg, I., Friedman, M., Handin, J., and Higgs, D. V., 1960, Experimental deformation of St. Peter sand: A study of cataclastic flow: Rock Deformation, p. 133–191, Geol. Soc. America Mem. 79.

Bredthauer, R. O., 1957, Strength characteristics of rock samples under hydrostatic pressure: Am. Soc. Mech. Engineers Trans., v. 79, p. 695–706.

Bridgman, P. W., 1941, Explorations toward the limit of utilizable pressures: Jour. Applied Physics, v. 12, p. 461–469.

———— 1947, The effect of hydrostatic pressure on the fracture of brittle substances: Jour. Applied Physics, v. 18, p. 246–258.

———— 1952, Studies in large plastic flow and fracture: N. Y., McGraw-Hill, 362 p.

———— 1953, The effect of pressure on the tensile properties of several metals and other materials: Jour. Applied Physics, v. 24, p. 560–570.

Coulomb, M., 1776, Sur une application des règles maximis et minimis à quelques problèmes de statique, relatif à l'architecture: Acad. Sci. Paris, Math. Phys. Mem., v. 7, p. 343–382.

Dickinson, G., 1953, Geological aspects of abnormal reservoir pressures in Gulf Coast Louisiana: Am. Assoc. Petroleum Geologists Bull., v. 37, p. 410–432.

Goguel, J., 1948, Introduction à l'étude mécanique des déformation de l'écorce terrestre: Paris, Imprimerie Nationale.

Griggs, D. T., 1936, Deformation of rocks under high confining pressure: Jour. Geology, v. 44, p. 541–577.

———— and Handin, J., 1960, Observations on fracture and a hypothesis of earthquakes, in Rock deformation: Geol. Soc. America Mem. 79, p. 347–364.

———— Turner, F. J., and Heard, H. C., 1960, Deformation of rocks at 500° to 800°C., in Rock deformation: Geol. Soc. America Mem. 79, p. 39–104.

———— Turner, F. J., Borg, I., and Sosoka, J., 1951, Deformation of Yule marble. Part IV—Effects at 150°C: Geol. Soc. America Bull., v. 62, p. 1386–1406.

———— Turner, F. J., Borg, I., and Sosoka, J., 1953, Deformation of Yule marble. Part V—Effects at 300°C: Geol. Soc. America Bull., v. 64, p. 1327–1342.

Handin, J., 1953, An application of high pressure in geophysics: Experimental rock deformation: Am. Soc. Mech. Engineers Trans., v. 75, p. 315–324.

—— 1958, Effects of pore pressure on the experimental deformation of some sedimentary rocks (abs.): Geol. Soc. America Bull., v. 69, p. 1576–1577.

—— 1959, Discussion of "Laboratory study of effect of overburden, formation, and mud column pressures on drilling rate" by R. A. Cunningham and J. G. Eenink: Jour. Petroleum Technology, v. 11, no. 1, p. 15–17.

—— and Fairbairn, H. W., 1955, Experimental deformation of Hasmark dolomite: Geol. Soc. America Bull., v. 66, p. 1257–1273.

—— and Hager, R. V., Jr., 1957, Experimental deformation of sedimentary rocks under confining pressure: Tests at room temperature on dry samples: Am. Assoc. Petroleum Geologists Bull., v. 41, p. 1–50.

—— and Hager, R. V., Jr., 1958, Experimental deformation of sedimentary rocks under confining pressure: Tests at high temperature: Am. Assoc. Petroleum Geologists Bull., v. 42, p. 2892–2934.

Heard, H. C., 1960, Transition from brittle fracture to ductile flow in Solenhofen limestone as a function of temperature, confining pressure, and interstitial fluid pressure, in Rock deformation: Geol. Soc. America Mem. 79, p. 193–226.

—— 1961, The effect of time on the experimental deformation of rocks (abs.): Jour. Geophys. Research, v. 66, p. 2534.

Higgs, D. V., and Handin, J., 1959, Experimental deformation of dolomite single crystals: Geol. Soc. America Bull., v. 70, p. 245–277.

Hubbert, M. K., 1951, Mechanical basis for certain familiar geologic structures: Geol. Soc. America Bull., v. 62, p. 355–372.

—— and Rubey, W. W., 1959, Role of fluid pressure in mechanics of overthrust faulting: Geol. Soc. America Bull., v. 70, p. 115–166.

—— and Rubey, W. W., 1960, Role of fluid pressure in mechanics of overthrust faulting, a reply: Geol. Soc. America Bull., v. 71, p. 617–628.

Kármán, T. von, 1911, Festigkeitsversuche unter allseitigem Druck: Zeits. Ver. deutsch. Ingenieure, v. 55, p. 1749–1757.

McHenry, D., 1948, The effect of uplift pressure on the shearing strength of concrete: Internatl. Cong. on Large Dams, 31 p.

Mohr, O., 1882, Ueber die Darstellung des Spannungszustandes eines Körperelements: Civilingenieure, v. 28, p. 113–156.

—— 1900, Welche Umstände bedingen die Elastizitätsgrenze und den Bruch eines Materiales?: Zeitschr. Ver. deutsch. Ingenieure, v. 44, p. 1524–1530, 1572–1577.

Nettleton, L. L., 1934, Fluid mechanics of salt domes: Am. Assoc. Petroleum Geologists Bull., v. 18, p. 1175–1204.

Paterson, M. S., 1958, Experimental deformation and faulting in Wombeyan marble: Geol. Soc. America Bull., v. 69, p. 465–475.

Purcell, W. R., 1949, Capillary pressures—their measurement using mercury and the calculation of permeability therefrom: Am. Inst. Min. Metall. Engineers Petroleum Trans., v. 186, p. 39–48.

Robertson, E. C., 1955, Experimental study of the strength of rocks: Geol. Soc. America Bull., v. 66, p. 1275–1314.

Robinson, L. H., Jr., 1959, The effect of pore and confining pressure on the failure process in sedimentary rock: Colo. School Mines Quart., v. 54, no. 3, p. 177–199.

Rubey, W. W., and Hubbert, M. K., 1959, Role of fluid pressure in mechanics of overthrust faulting: Geol. Soc. America Bull., v. 70, p. 167–205.

Serdengecti, S., and Boozer, G. D., 1961, The effects of strain rate and temperature on the behavior of rocks subjected to triaxial compression: Proc. 4th Symposium on Rock Mechanics, Pennsylvania State Univ., p. 83–97.

Skempton, A. W., 1961, Effective stress in soils, concrete and rocks: Pore pressure and suction in soils, p. 4–16, London, Butterworths.

Terzaghi, K., 1923, Die Berechnung der Durchlässigkeitsziffer des Tones aus dem Verlauf der Hydrodynamischen Spannungserscheinungen: Sitz. Akad. Wiss. Wien, v. 132, p. 105–124.

—— 1925, Principles of soil mechanics: Eng. News-Record, v. 95, p. 987–996.

—— 1932, Tragfähigkeit der Flachgründungen: Prelim. Pub. 1st Cong. Internatl. Assoc. Bridge and Structural Engineers, p. 659–683.

—— 1936, Simple tests determine hydrostatic uplift: Eng. News-Record, v. 116, p. 872–875.

—— 1943, Theoretical soil mechanics: N. Y., Wiley and Sons, 510 p.

—— 1945, Stress conditions for the failure of saturated concrete and rock: Am. Soc. Testing Materials Proc., v. 45, p. 777–801.

Turner, F. J., Griggs, D. T., and Heard, H. C., 1954, Experimental deformation of calcite crystals: Geol. Soc. America Bull., v. 65, p. 883–934.

—— Griggs, D. T., Clark, R. H., and Dixon, R. H., 1956, Deformation of Yule marble. Part VII—Development of oriented fabrics at 300° to 500°C.: Geol. Soc. America Bull., v. 67, p. 1259–1293.

Watts, E. V., 1948, Some aspects of high pressures in the D-7 Zone of the Ventura Avenue field: Am. Inst. Min. Metall. Engineers Petroleum Trans., v. 174, p. 191–205.

American Association of Petroleum Geologists Continuing Education Course Notes Series 4, *Pore Pressure: Fundamentals, General Ramifications and Implications for Structural Geology*, by P. E. Gretener, copyright 1977, p. 59-88.

From
*Pore Pressure: Fundamentals, General
Ramifications and Implications for
Structural Geology*
by P. E. Gretener

4.0 Structural Implications of Pore Pressure

It is evident from the previous discussions that pore pressure affects a number of physical properties profoundly. Therefore, the conclusion that pore pressure plays a key role in many structural processes is a logical one. In the following the possible, probable and certain effects of pore pressure on various types of rock deformation are assessed. Please note that the list is growing and this presentation reflects the state-of-the-art of 1976 and is presumably incomplete at that.

4.1 Geopressures and Thrust Faulting

Overthrusts were first described by A. Escher in 1841 (Staub,1954) for the Glarus area in Switzerland. Naturally such fantastic structures were not readily accepted by the geological fraternity. The controversy lasted for many decades. Alternate explanations such as Albert Heim's double fold were advanced for the undeniable observation of 'older over younger'. However, by the turn of the century overthrusts had become accepted, not in the least because by then their global distribution had been recognized. Mechanically the emplacement of such large relatively undisturbed plates remained an enigma.

It took until 1959 when Hubbert and Rubey in their now classic companion papers for the first time put forward a possible explanation for the unquestionably low friction that must exist at the base of thrust plates during movement. Abnormally high pore pressures were called upon to essentially float the thrust plates during the process. The idea was borrowed from Terzaghi (1950) who had previously suggested this mechanism for the smaller scale landslides. The Hubbert and Rubey hypothesis has proved an extremely fruitful concept far exceeding the original intentions of the authors as is demonstrated by the voluminous literature on the topic in the last decade and a half.

Overthrust belts in many parts of the world are now well explored and it is possible to compile their vital statistics (Gretener,1972). In its simplest form a thrust plate is shown in Figure 4.1-1. The thickness of such plates ranges from 3 to 6 kilometres, the advancement of individual major plates can be measured in tens of kilometres and the lateral extent may be in excess of 150 kilometres.

In reality thrust faults are not as simple as shown in Figure 4.1-1. A characteristic aspect of these faults is their 'stepping' up through the layered sequence. It seems that Rich (1934) was the first to call attention to this fact. It has been further substantiated by Douglas (1950) and others and has become a generally accepted observational fact. Dahlstrom (1970,p. 344, Fig. 9) has statistically analysed this phenomenon for the southern Canadian Rockies. Thrust faults tend to 'linger' in incompetent strata and step through the competent ones at a relatively steep angle. The fact is often somewhat obscured on modern cross sections (Bally et al.,1966; Royse et al.,1975) where the shape of the faults seems to be dictated by the availability of French curves. However, close inspection of these cross sections reveals that the authors fully adhere to the concept of stepping. Thus depending on the relationship between step spacing and total displacement, thrust sheets tend to become deformed as shown in Figures 4.1-2 and 4.1-3. One immediately notes that it is the aspect of step-

121

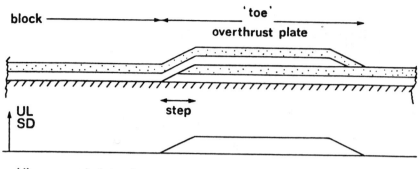

UL = uneroded load
SD = stratigraphic displacement

FIGURE 4.1-1

Schematic drawing of an overthrust plate.

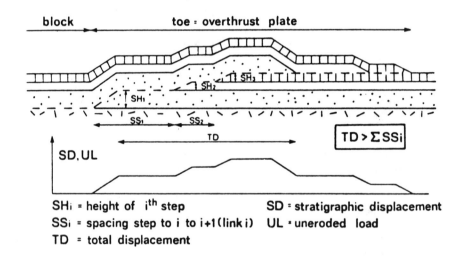

SH$_i$ = height of ith step   SD = stratigraphic displacement
SS$_i$ = spacing step to i to i+1(link i)   UL = uneroded load
TD = total displacement

FIGURE 4.1-2

Deformation of a thrust plate where the total displacement exceeds the sum
of the links

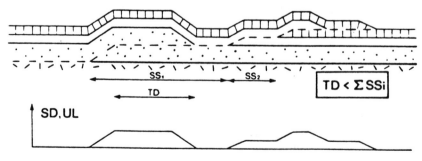

FIGURE 4.1-3

Deformation of a thrust plate where the total displacement is smaller than
the sum of the links

ping that is the trap forming mechanism in thrust faulting. Only in the region of the step is there a vertical component of movement leading to an anticlinal deformation in the thrust sheet. In the region of the link, movement is subparallel to bedding and only lateral translation occurs.

Unresolved questions surrounding the 'pore pressure theory':

1. The often relatively undistorted nature of major thrust plates (particularly in the southern Canadian Rockies, Bally et al.,1966) suggests that the old postulate of the geologists for low basal friction is a justified one. High pore pressures can resolve this enigma. Such pressures can be envisaged as caused by loading and aquathermal pressuring in the links usually comprising rocks of low permeability. The mechanism is difficult to accpet for the areas of the steps.

2. It is currently thought that thrust sheets move at average rates of a few centimetres per year (Hsü,1969). Thus the emplacement of a single major thrust plate is a matter of about one million years. At the same time the sheet moves over distances counted in tens of kilometres. Thus in order to have merit the conditions of high pore pressure must be maintained both in time and space. Gretener (1972) has suggested that one possible way out of this dilemma may be the fact that both pore pressures and movement on the thrust faults are discontinuous.*

3. The pore pressure theory in its present form does not account for the stepping nor does it provide any clues as to the step spacing.

Goguel (1969) indicates that the production of frictional heat at the base of overthrust plates may be sufficient to vaporize the formation waters and cause the high fluid pressures needed for easy gliding. Attempts to demonstrate such heating at the base of thrusts by analysis of the degree of carbonization of plant fragments by the author have yielded negative results so far. Goguel's concept is therefore viewed with some scepticism.

Thrust faulting leads to a duplication of the sequence and therefore constitutes a case of tectonic loading. In view of the currently accepted rates of advance (for detailed discussion se Gretener,1972,p. 592/93) the rate of loading must be termed fast. Thus high pore pressures may be induced in the overridden sequence by way of fast loading and aquathermal pressuring. This will lead to a series of events as shown in Figure 4.1-5. At a certain stage of advancement a lower thrust fault will become activated in preference to the initial fault. Where the lower fault steps through the sequence the initial fault, which is carried piggyback, will become folded. This succession of events which mechanically seems very plausible is the one generally accepted for the southern Canadian Rockies (Bally et al.,1966; Dahlstrom,1970; Jones,1971).

In addition one must not forget that during thrusting large loads are shifted from the inner portion to the outer area of the thrust belt. The isostatic response to this action cannot be ignored. From the observations on ice unloading

----

*Average rates for geological processes may be very useful to establish for certain purposes. They can, however, be extremely misleading in as much as they imply slow and continuous change. Where such change in fact occurs in spurts (Figure 4.1-4) the mechanism of the change is not at all explained by the average rate of change.

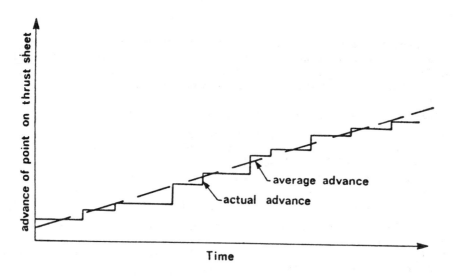

FIGURE 4.1-4

The average rate of change can be very misleading in trying to understand
the mechanism of a process that proceeds in 'jumps' such as fault movement,
erosion by flashflood, sedimentation by turbidity currents or through storm
action etc.

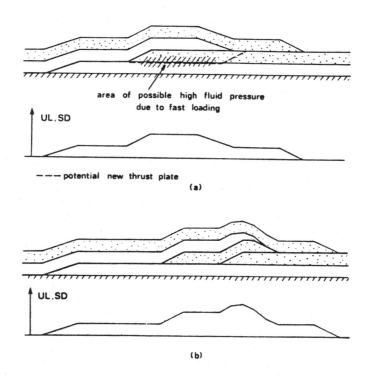

FIGURE 4.1-5

Overriding causes high pore pressures in the lower sequence (a) due to
fast loading and aquathermal pressuring. A lower thrust is activated and
the old thrust carried piggyback (b).

(Canadian Archipelago; Scandinavia) one can conlcude that this response is geologically speaking instantaneous. It is also substantial. If one assumes a sediment density of 2.5 g/cm^3 and a mantle density of 3.3 g/cm^3 one obtains the
following value for the resulting subsidence (S) :

$$3.3 \text{ x } S = 2.5 \text{ x } H$$

or $$S = (2.5/3.3)\text{x}H \simeq 0.75 \text{ H} \qquad\qquad 4.1/1$$

where H is the thickness of the overthrust sheet

Thus an extensive thrust plate of 5 km thickness can produce a maximum subsidence of about 3.8 km. Since isostatic compensation is a regional rather
than local phenomenon, the foredeep is a logical consequence to the advancing
load of the thrust sheets.

Combining isostatic adjustment and expected abnormal pressuring due to advancing
thrust plates Gretener (1972) has suggested that thrusting once initiated may
well become a self-perpetuating process, which is finally arrested in the following
manner. Usually the sedimentary skin involved in thrust faulting has the shape
of a wedge (see Bally et al,1966,p. 365, Fig. 10) thinning towards the craton.
As the thickness of the wedge reaches a critically low value, deformation terminates presumably because sufficient compressional stresses cannot be transmitted through such a sequence. A contributing factor may be the mechanical weakness of these youngest rocks of the thrust belt (foredeep sediments). Cross sections through various thrust belts suggest that the critical thickness may be
in the order of 3 to 4 km. In the case where the tip of the wedge rests on a
basement arch such as shown by Royse et al. (1975, plate IV) for the Idaho-Wyoming thrust belt this thickness may be considerably larger. Obviously the numerical value given above should not be taken too seriously at this time since
this aspect of thrust belts is not yet well understood.

The commercial trap is found where the reservoir rock rides up over a step.
Thus the hunt for hydrocarbons in a thrust belt amounts to the chase of the
leading edge of the reservoir as described for the southern Canadian Rockies
by Bally et al. (1966) and Keating (1966). In the case of the Canadian Rockies
where the reservoir rocks are carbonates of Paleozoic age with a strong velocity
contrast to the overlying Mesozoic clastics, special geophysical methods have
been evolved. Figure 4.1-6 shows such a case. The leading edge is usually ill defined since it is the locus of a well developed diffraction pattern. The involvement of the high velocity Paleozoics in the thrusting does, however, produce a pronounced velocity uplift on the deeper reflections such as the 'near-
Basement' reflection of Bally et al. Once this is recognized the velocity uplift can be used to determine the extent of the Paleozoic 'sliver'. While the
geometry of hydrocarbon traps is bound to be similar in all thrust belts, exploration methods are likely to differ. The exploration in the southern Canadian
Rockies is strongly tied to the large velocity contrast (15,000 to 20,000 ft/s;
4,500 to 6,000 m/s) between the Paleozoic carbonates and the Mesozoic clastics.
One cannot expect this basically favorable condition to exist everywhere. Also
Canadian experience in terms of structural style is not universally applicable.
The strong beams in the stratigraphic succession such as the Devonian Palliser
limestone and the Mississippian Rundle limestone have kept the deformation more
simple than elsewhere and suppressed the occurrence of incidental (Dahlstrom,
1970) disturbances.

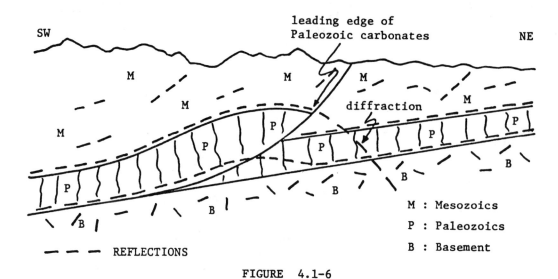

FIGURE 4.1-6

The 'Exploration Situation' in the southern Canadian Rockies

The acceptance of the succession of events as shown in Figure 4.1-5 also opens the possibility for stacked reservoirs. Indeed several of the fields in the Alberta foothills are of this type (Bally et al.,1966, plate 5). The deepest step forms the 'primary' trap and the deformation of the overlying sheets results in what one may term 'secondary' traps. Note that there is no time connotation implied, primary and secondary traps form simultaneously as the plate rides up the ramp of the deepest step.

If the theory of high fluid pressures at the base of thrust plates is indeed a viable one, as it seems at this time, then this has also consequences for the fluid migration during thrusting. Geopressure zones are perfect barriers to vertical fluid migration (Chapman,1973,p. 66). Thus one would assume that during deformation individual thrust plates would be largely closed systems with only minor leakage from one to the other in the areas of the steps. Whether or not such contemplations are important in terms of the present distribution of hydrocarbons depends on:

a) the occurrance of late, i.e. post-thrusting, migration
b) the continuous or discontinuous nature of the geopressures during thrusting

To this date in the southern Canadian Rockies only gas and condensate has been found with the exception of the Turner Valley field. This can be explained in terms of the Liquid Window Concept assuming that the hydrocarbons all originated in the deeper portions of the thrust belt, migrating updip into their present position. An alternative explanation would have to rely on the concept of differential entrapment of oil and gas as suggested by Gussow (1954).

In conclusion one can say the following :

1. In a thrust belt it is the peculiar behaviour of thrust faults, <u>stepping</u> through the layered sequence, that is the <u>trap forming process</u>.

2. High pore pressures existing along the links of thrust faults will affect the early migration of hydrocarbons and tend to isolate individual thrust sheets in terms of fluid movement. This may be obscured by later (post-thrusting) leakage. The argument will also become untenable if the discontinuous nature of the high pore pressures can be demonstrated.

3. The outward progression of thrusting, as it is generally accepted for the southern Canadian Rockies, seems mechanically most attractive. It also provides an easy explanation for folded faults (Fig. 4.1-5) and may result in stacked reservoirs. The backlimb thrusts of Douglas (1950) are one of the exceptions to this rule of outward progression.

4. The depth of the major sole fault (décollement) decreases towards the craton. The outer limit of the 'overthrust hydrocarbon province' occurs where the reservoir rocks lie below the décollement and are no longer involved in the thrusting. The question of the 'first sliver' (Fig. 4.1-6).

5. Present observations suggest the late, listric normal faults to be an integral part of thrust belts.

6. The unavoidable isostatic adjustments during thrusting have so far been much overlooked. There is a definite paucity of reference to that aspect of thrusting in the literature.

7. In the case of limited observational information (as is always the case) such concepts as: links-in-incompetent-units versus steps-in-competent-units; thrusting progressively younger towards the craton; displacement on faults must be constant and displacement on splays must be added; etc. provide a useful aid in arriving at a 'reasonable' interpretation.

4.2 Geopressures and Differential Compaction

Compaction usually occurs in response to the applied effective overburden stress. It is thus a continuous process proceeding as the load is applied in a regular fashion. Geopressures are generally indicative of a state of compaction disequilibrium. Under such conditions compaction is delayed and thus may proceed after the application of the load (sedimentation) has ceased. In the case of a fully and permanently sealed unit compaction may be delayed indefinitely.

Differential compaction occurs in areas where certain stratigraphic units display a lateral compactability contrast. This is the case where highly compactable sedimentary units, such as shales, envelope such bodies as reefs, lenticular sands, or highs on an erosional unconformity involving such non- or little compactable lithologies as, igneous and/or metamorphic rocks or well consolidated carbonates. These examples are shown in Figure 4.2-1. As compaction proceeds under a growing overburden load, structure is induced in the overlying beds. Drape structures formed in this manner are not known to be the locus of giant oil fields but some of them are commercial traps. In particular in central Alberta the upper Devonian Nisku as well as some Cretaceous sands draping over upper Devonian Leduc reefs have yielded commercial quantities of oil and gas.

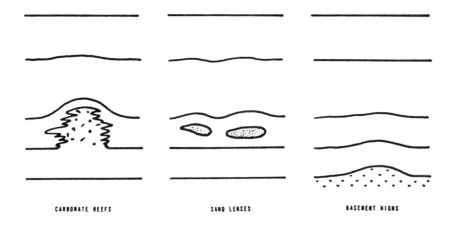

CARBONATE REEFS   SAND LENSES   BASEMENT HIGHS

FIGURE   4.2-1

Examples of differential compaction features.

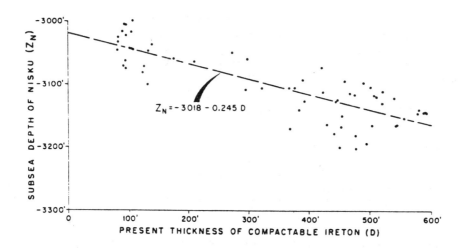

$Z_N = -3018 - 0.245\ D$

FIGURE   4.2-2

In this graph the structural elevation for a particular horizon as
found in a well is plotted against the thickness of the compactable
unit for the same well. The slope of the regression line is a measure
of the structure induced due to differential compaction. In this graph
the two units are the upper Devonian Nisku and Ireton Formations of
central Alberta.

The closure of such drape structures is a maximum immediately above the interval where the differential compaction occurs. It diminishes upward in the section and is always zero at the depositional surface. The latter requirement is a must since compaction occurs in response to loading only. This decay of the compaction moving upwards in the sequence can be analysed in the following manner (Labute and Gretener, 1969). For a number of wells (seismic data do not provide sufficient precision) the structural elevation of a given horizon is plotted versus the thickness of the compactable envelope as found in the same well. The latter will be greatly reduced or even zero in the area of the 'core' (reef, sandlens etc.). If structure is present the points will define a sloping line. The slope of this line is directly proportional to the structural relief on the particular horizon. An example of this is shown in Figure 4.2-2. The points where the compactable interval is thinnest are located directly over the non-compactable (or less compactable) body such as the reef. Points to the right in the diagramme represent the off-body position where the compactable unit is thickest. The scatter of the points may be attributed to a number of incidentals, such as superimposed later structure, interference due to an erosional unconformity, poor picks on logs etc. Plotting the slopes (as a measure of structural relief) for a number of horizons results in what has been termed the 'compaction curve'. This is a smooth curve with a maximum immediately above the differentially compacting interval. Such a curve is shown in Figure 4.2-3.

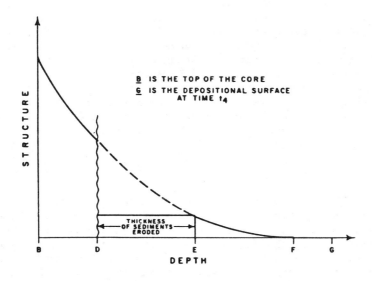

FIGURE   4.2-3

Structure induced by differential compaction diminishes uniformly upwards. Structural relief is a maximum at the top of the interval in which differential compaction occurs (B) and vanishes at level F. This indicates that compaction is complete under the load B-F. Erosion eradicates part of the compaction curve and results in the isostructure line D-E (solid line).

Labute and Gretener (1969) also investigated the effect of an unconformity on the compaction curve. The expected sequence of events is shown in Figure 4.2-4. The non- or less compactable body is designated as the 'core'. At the time t_o the core is embedded in a compactable unit. In the case of central Alberta, the Leduc reefs are enveloped by the Ireton shale. At time t_1 sedimentation reaches level 'C' and is followed by erosion to level 'D' at time t_2. Rebound in a large shale mass is considered to be negligible or non-existent during this period. Renewed sedimentation reaches the previous level of load at time t_3 and exceeds it at time t_4. Renewed compaction occurs in response to the excess load (E to F) at which time compaction is complete and further loading (F to G) produces no further effect. The result of erosion is a partial eradication of the compaction curve. The interval of maximum erosion (D-E) is represented by an isostructure line, due to the effect of the load above E. This is shown schematically in Figure 4.2-3. Labute and Gretener (1969) have applied the concept to the central Alberta situation where lower Cretaceous rocks rest directly on Mississippian and/or Devonian strata. The result is shown in Figure 4.2-5. The length of the isostructure line indicates that in the order of 2,000 feet of sediment were removed at the pre-Cretaceous unconformity. Later investigations by O'Connor and Gretener (1974b) lend further support to this result.

At a later date O'Connor and Gretener (1974a) contemplated the various causes that might interfere with the theoretical and smooth compaction curve. In particular they evaluated the effect of temporary or permanent isolation of the compactable unit. The predicted results are shown in Figure 4.2-6. Figure 4.2-6a shows the case where the compactable interval is restricted and compaction is incomplete. This produces a downward shift of the compaction curve and the drape closures are unexpectedly low. Figure 4.2-6b gives a case where compaction proceeds normally for a while and then is arrested by isolation of the compactable unit. As a results structures are developed only over a very limited depth range and their size is again much less than expected. Figure 4.2-6c presents the case of temporary isolation followed by sudden reopening. This leads to an isostructure line on the compaction curve. In contrast to the isostructure line produced by erosion this line ia above the line for normal compaction. Note also that sudden reopening results in what properly must be termed collapse rather than compaction. Figure 4.2-6d shows a situation where reopening of the isolated system is delayed until sedimentation is complete. In this case the isostructure line of 4.2-6c extends to the surface and such conditions have a surface manifestation.

O'Connor and Gretener (1974a) also investigated the effect of restriction on the isostructure line produced by erosion. The results of their findings are given in Figure 4.2-7. Depending on the severity of the restriction the isostructure line may be only reduced in length (Figure 4.2-7b) or completely eliminated (Figure 4.2-7d). In either case restriction of the compactable unit is another factor that will contribute to erroneous results when trying to evaluate the thicknes of eroded intervals by this method. In extreme cases the investigation may be totally invalidated.

The effects of geopressures in the realm of differential compaction may be grouped as beneficial and detrimental to the accumulation, preservation and discovery of hydrocarbons in such traps.

"CORE"

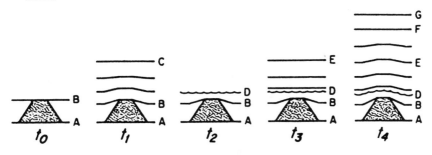

FIGURE 4.2-4

Development of drape structure due to differential compaction
when deposition is disrupted by erosion.

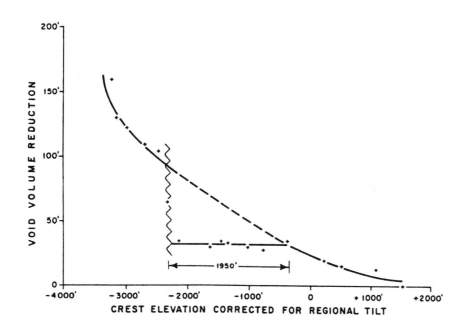

FIGURE 4.2-5

Actual compaction curve affected by erosion for central Alberta.
Void volume reduction is equal to closure on a particular horizon.

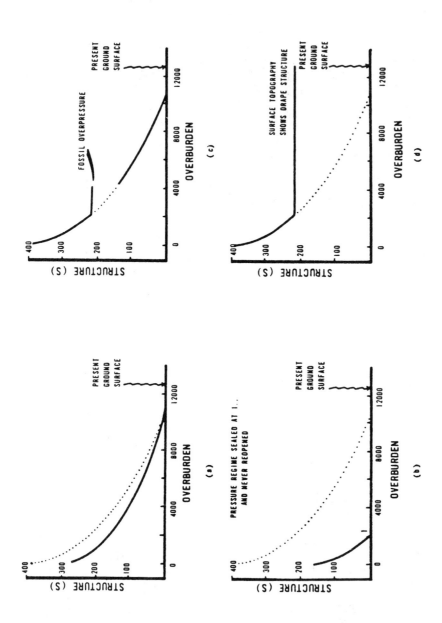

FIGURE 4.2-6

Effect of temporary or permanent restriction or isolation on the compaction curve (details see text).

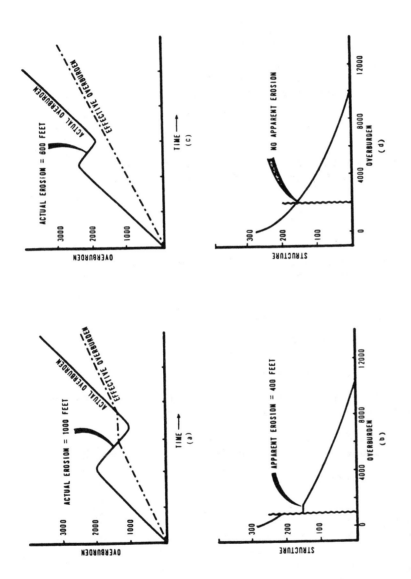

FIGURE 4.2-7

Effect of restriction on the isostructure line produced
by erosion (for details see text).

a) Beneficial Effects:

1. In 'cool' areas the main fluid migration may be retarded until the layers have moved into the liquid window.

2. Late opening of an isolated system may produce surface structures (Figure 4.2-6d) and thereby make the discovery of such traps easier.

b) Detrimental Effects:

1. Trap formation is late. Ergo trap filling is late. This may have a negative effect on the trap porosity since early presence of hydrocarbons tends to preserve porosity (Füchtbauer,1972).

2. Analysis of the compaction curve in terms of the maximum eroded interval will produce less reliable results and may even become impossible.

3. In areas of a high geothermal gradient the source rocks may be below the liquid window when the main fluid movement occurs.

4.3 Geopressures and Diapirism

The importance of diapiric structures for the oil explorationist needs no further elaboration. The literature on the subject is extremely voluminous to put it mildly.

The necessary requirements for diapirism are:

1. A density reversal, i.e. a low density material covered by a high density material.

2. High mobility (low viscosity) of the low density material.

3. Uneven loading or an initial perturbation on the surface of the low density material to provide a lateral pressure gradient in this material.

The best known sedimentary material to fulfill these requirements is salt and its behaviour has been well understood for many decades. The literature on the subject of salt diapirism is legion but one of the classic papers is the one by Nettleton (1934).*

More recently one has come to recognize that overpressured sediments are also excellent candidate for diapiric materials. Because of their abnormally high porosity they tend to be of low density. Due to the high porosity AND the low effective overburden stress they are also mechanically weak. Thus they are predestined to form diapirs. Indeed the overpressure theory offers a logical explanation for the sandstone dikes that have been observed in all parts of the world and that have baffled geologists for many years. Yet an even better prospect are the overpressured shales. Indeed today shale diapirism is a well ac-

---

*One must also keep in mind that the rise of granitic and basaltic magmas is governed by exactly the same principles. On a yet larger scale the formation of astenoliths is a must, the astenosphere (weak and light) being the diapiric material.

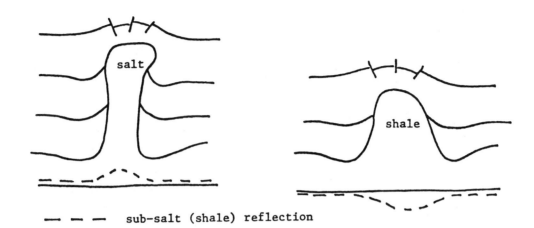

— — — sub-salt (shale) reflection

EFFECT OF DIAPRIC SALT AND SHALE MASSES ON DEEP REFLECTIONS

FIGURE 4.3-1

cepted feature of all continental margins (Short and Stäuble,1967; Bruce, 1973; Parsons,1975). Hedberg (1974) has pointed out that the production of methane in these shales, which are generally rich in organic material, may further enhance their diapiric properties. The Hedberg concept seems to be an entirely reasonable one in view of the fact that gas is associated with many mud volcanoes (Gansser,1960), the latter being no more than the surface expression of mud diapirs. In table 4.3-1 some of the physical properties of overpressured shales (with and without gas) are compared to those of salt.

Table 4.3-1

Some Physical Properties of Salt and Geopressured Shale

	Salt	Geop. Shale	G.S. + gas
thermal conductivity	high	low	very low
sonic velocity	high	low	very low
electrical resistivity	very high	very low	low
strength	low	low	very low
density	low	low	very low
caprock	present	absent	absent

It can be seen that while both materials possess low density and low strength, which they must in order to qualify as diapiric materials, they differ markedly in other respects. The sonic velocity of salt is high (at least compared to near surface clastic sediments) while the one of shale is low and in the case of the presence of free gas is very low. Domenico (1974) has shown that a rock in regards to sonic velocity will behave as fully gas saturated even though the free gas may occupy less than 10% of the pore space. Many 'bright spot fans' have found this to be true to their dismay. In view of the velocity contrast existing between salt and shale one can predict that sub-salt or sub-shale reflections should be characterized by velocity pull-ups or velocity sags as shown in Figure 4.3-1. Musgrave and Hicks (1966) have demonstrated that this is indeed one method to outline diapiric shale masses.

With the exception of basic intrusions that have a large excess hydrostatic head due to their deep origin and therefore can penetrate high into the low density section (Gretener,1969b), all diapirs are characterized by gravity lows. In the case of salt diapirs these lows are usually modified by the presence of the dense, near-surface caprock as shown by Nettleton (1962). This central high on Bouguer gravity maps is absent in the case of shale diapirs.

A typical feature of diapirs is the overhang that develops at shallow depth. Due to their deep origin many diapirs acquire sufficient hydrostatic excess pressure to actually penetrate into the shallow section where the sediment density is lower than that of the diapiric material. Due to the low strength the pressure in a diapir is always near hydrostatic. As a result in this shallow section the pressure in the diapir is sufficient to allow lateral spreading resulting in the well known and troublesome overhang. In this respect one does not expect salt and shale to behave differently.

An interesting case of a diapir is shown in Figure 4.3-2*. The caption indicates that the intrusive material may be of igneous origin. This can be ruled out on the basis that a rim syncline is present. The latter clearly indicates that the motherbed is at a shallow depth. It is after all the lateral depletion of the motherbed that gives rise to the rim syncline. Thus on the basis of structural considerations one can rule out what is obviously the author's preferred interpretation at the time of writing. This leaves salt or shale diapirism as viable alternatives. On the basis of the information given in table 4.3-1 one recognises that the quickest and easiest way for a decision is to drop an oceanic heat flow probe on top of the structure. For the case of a salt diapir the heat flow will be above normal (see Figure 1.5-15). The fact that salt domes are characterized by higher than normal heat flow is well demonstrated on the heat flow map for Germany (Creutzburg,1964). For a shale dome one would expect the heat flow to be abnormally low. Of course it is now well known that the structures shown in Figure 4.3-2 are indeed salt diapirs.

Other interesting features of shallow origin are the so-called mudlumps described by Morgan et al. (1968) and Shepard et al. (1968). They are of some significance in terms of the theories on overpressures since they are unquestionably pure loading features. Temperature and phase changes play no role at depths which measure at best a few hundreds of feet. In practical terms these shallow mud diapirs are of importance since they constitute a hazard to shipping and pipelines. They do penetrate to the surface, affecting the seafloor topography and in fact they may form at least temporary islands.

----

*This early unprocessed section taken from Leenhardt (1967) clearly demonstrates the need for geologists to be able to recognize diffraction patterns and multiples for what they are: fake features. Familiarity with such fundamental geophysical publications as e.g. Tucker and Yorston,1973¦Pitfalls in Seismic Interpretation¦ is a must for any modern structural geologist.

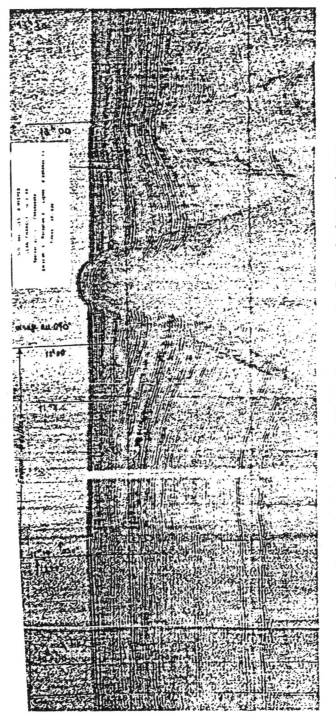

Fig. 10. *Another bump south of Toulon*. Same material as fig. 7 and 9, but a much more noisy ship (fig. 7: ship speed was 6 Knt; fig. 9: 7 Knt, and in the present case, 4 Knt). This structure seems to be of volcanic origin, but we have still no magnetic results.

FIGURE 4.3-2

For details see text. From Leenhardt (1967) with permission by the European Association of Exploration Geophysicists.

Summary od diapiric structures and their associated hydrocarbon traps :

1. The most important diapiric materials of sedimentary nature are : SALT and OVERPRESSURED SHALES.

2. There are 3 types of hydrocarbon traps associated with diapiric structures :

   a) The supra diapir traps, due to bending of the strata above a piercement dome or a pillow (flowage must not necessarily reach the piercement stage to create economic traps of this type). In the case of supra piercement traps the bending of the immediately overlying beds is extreme. This strong extension is usually relieved by a complex system of normal faults. As a result individual traps are small (relatively) and present problems to both exploration and exploitation. Deep seated piercement domes or pillows (salt or shale ridges) with moderate uplift will produce the most favorable, large unbroken structures, in beds well above the diapiric mass.

   b) The lateral diapiric traps, produced by the upward drag of the sedimentary layers pierced by a diapir. Such structures are confined to actual piercement features.

   c) The sub diapir traps, due to uplifting of the layers beneath a diapir in response to the relief of overburden pressure under such a structure. This type of trap is at this time largely hypothetical.

3. Diapirs are always characterized by negative gravity anomalies. In the case of salt structures a central positive anomaly may exist, depending on whether or not caprock is present. Shale diapirs will always produce simple gravity lows.

4. Overhangs at shallow depth are a typical feature of salt diapirs. Such overhangs make the exploration for and the exploitation of the underlying lateral traps difficult. At this time it is uncertain whether pure shale diapirs develop such overhangs. On purely theoretical grounds it is to be expected.

5. In terms of seismic exploration SALT must be considered a HIGH VELOCITY material ($\pm$15,000ft/s; 5km/s) at least in contrast to clastics at shallow depth. Overpressured SHALES are always a LOW VELOCITY material. Sub diapir reflections are correspondingly uplifted or depressed in time.

6. Salt is the best heat conductor amongst the common sedimentary rocks. The upper parts of salt domes therefore carry positive heat anomalies of up to $50°F+$ ($30°C$) according to model studies and field observations. Overpressured shales are extremely poor thermal conductors. Shale diapirs should have halos of abnormally low temperature associated with their upper terminations.

7. In rare cases salt and shale diapirs may have associated magnetic anomalies. In the Sverdrup basin of Arctic Canada diapirs puncture and upturn diabase sills. The near vertical edges of these basic sills result in magnetic anomalies of several hundred nanoTesla (1nT = 1γ) surrounding the diapirs.

4.4 Geopressures and Listric Normal Faults (Growth Faults)

The concept of a listric normal fault is shown in Figure 4.4-1. The curvature (flattening at depth) of such faults requires that the involved sequence be rotated as shown. It is in fact on the basis of this rotation that one can infer a flattening of the fault plane at depth. To the writer's knowledge no attempt has been made so far at a quantitative analysis of the geometry involved. One is tempted to conclude that the curvature of the rollover (the term used for the curvature by the Gulf Coast geologists) should provide a guide to the depth at which the fault parallels bedding. However, intuition is a poor guide and possibly no such correlation exists. Where the strata are subject to regional dip the rollover is an important trap forming process, as is the case in many coastal areas (Bruce,1973; Weber & Daukoru,1975).

Normal faulting is basically an extension phenomenon (Figure 1.1-3). The fact that listric normal faults flatten at depth and assume an orientation parallel to bedding indicates that the extension is confined to the skin above the flat portion of the fault. The listric normal faults are related to a plane or zone of weakness in the layered sequence and are gravity induced similar to many landslides. In fact listric normal faults differ only in size from the headwalls of common landslides. As in some cases of compressional folding the plane of weakness serves as a décollement and the structures associated with the listric normal faults cannot be traced past a critical depth. The stretching which the beds have to undergo in order to accomodate themselves to the curved fault plane may in part be provided by antithetic faults. Note also that the crest of the rollover is shifting down dip with depth, another dictate of the fault geometry.

An inclined layered sequence containing a zone of weakness is shown in Figure 4.4-2. In the case where the fluid pressure in the zone equals the total overburden stress the plate is in a state of floatation and solely supported by its toe. In a less extreme case a small residual shear stress may exist across the plane of weakness and lend additional support. When the strength of the toe is exceeded buckling and/or faulting occurs in the toe area and normal faulting will take place in the upper regions. In practice the normal faults will not turn abruptly into the gliding plane but join it along a gently curved path, giving rise to the typical listric normal fault geometry. Note that the displacement is mainly vertical near the surface where the fault plane is steep, often near vertical (Hamblin,1965, Fig. 1), and totally lateral at the depth where the fault parallels the bedding.

Where such faults form contemporaneous with sedimentation the hole on the surface acts as a sediment trap. The result is pronounced thickening of lithological units on the down thrown side. Loading thus tends to perpetuate the fault movement and leads to further enhancement of the rollover. It is these listric faults that are commonly referred to as growth faults. They differ from the simple listric normal faults by thickening of the units on the down thrown side and by a variable displacement dying out upwards as shown schematically in Figure 4.4-3. The gliding horizon for growth faults are usually overpressured shales.

Short and Stäuble (1967), Bruce (1973) and others have called attention to the fact that the pattern of growth faults is related to,and often complicated by, shale diapirism. Once growth faulting occurs on top of a thick overpressured shale mass the necessary requirements for diapirism are met. When such shale diapirs form, secondary growth faults may be initiated on their flanks where the dip

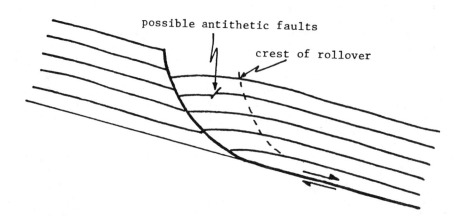

FIGURE 4.4-1

Schematic cross section of a listric normal fault.

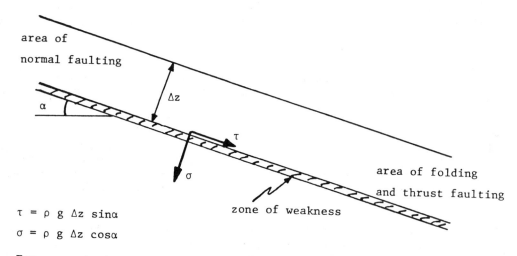

$\tau = \rho \ g \ \Delta z \ \sin\alpha$

$\sigma = \rho \ g \ \Delta z \ \cos\alpha$

For $p = \rho g \Delta z$ in the zone of weakness the normal stress (σ) vanishes and the plate is solely supported by its toe.

FIGURE 4.4-2

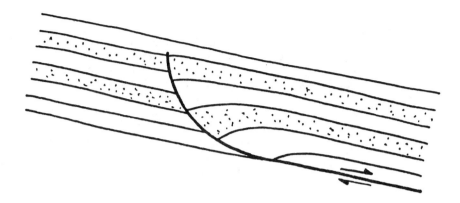

FIGURE 4.4-3

Schematic cross section of growth fault. Note thickening
of units and variable displacement.

of the uptilted beds exceeds the critical value for gravity sliding. The
experiments of Rettger (1935), now unfortunately almost forgotten, form a
beautiful portent to the whole discussion and investigation of deep shale
diapirism, mudlumps, and growth faults in the environment of a sloping sequence under an advancing load, a discussion which erupted 30 years later.
His experiments clearly let one expect that somewhere in the toe the extension
of the growth fault belt is compensated by compression. This, however, occurs
in deep water off the coast and has not yet received much attention.*

Most recently Weimer and Davis (1976) have described an ancient system of
growth faults in the Cretaceous rocks of the Denver basin. Appreciable
thickening of sandstone and coal beds is observed on the down thrown side
in the near surface. Gliding is restricted to the uppermost 5,000 feet with
the upper 4,000 feet of the Pierre shale being locally overpressured and acting
as a gliding horizon.

A different case yet belonging to the class of listric normal faults has been
reported by Bally et al. (1966) and Dahlstrom (1970). In their interpretation
the Flathead fault bordering the Rocky Mountain Trench in the Cranbrook-Flathead
area on the east side is such a feature. During the late post-orogenic uplift the
Rockies outermost skin was extended. As a result normal faulting took place.
These normal faults are of the listric type, flattening at depth to join with
former thrust planes. This would indicate that these thrust fault planes still
constituted planes of weaknesses much after thrusting had ceased. Displacement on
such faults consists of a positive part during active thrusting and a negative
part (back sliding) during the post-orogenic uplift. Mechanically this is by
far the most acceptable explanation for the Rocky Mountain Trench, which is
essentially a purely morphological feature. When viewing the wedge of Eocene-

----

*A thorough discussion of the topic of growth faulting and shale diapirism is
 given by Dailly (1976).

Oligocene Kishenehn Formation as shown by Jones (1969,Fig. 5) one wonders whether the term growth fault would not be appropriate. However, one can carry things too far.

Hamblin (1965) describes listric normal faults from the Colorado Plateau area. He refers to the rollover as 'reverse drag'. The Muav Canyon and the Hurricane faults are clearly of this type. The rollover demands flattening of these faults at depth and this is indeed what can be observed over the rather considerable elevation differences existing in the Grand Canyon area. Thus Hamblin (1965, p. 1155) reports that the Hurricane fault is essentially vertical on the Kaibab Plateau and flattens to about 60° at the bottom of the Grand Canyon. Note that these are listric normal faults but not growth faults. No thickening of the units is observed as one crosses from the up- to the down thrown side (see e.g. Stanley,Jr.,1970, p.307, Fig. 8).

The case of listric normal faults may be summarized as follows :

1. These faults are most prominent in the depocentres of coastal areas such as the Mississippi, the Niger and the Mackenzie Deltas. In those areas faulting and loading (deposition) occur contemporaneously. The result is a variable displacement on these faults (dying out upwards) and a pronounced thickening of equivalent units on the downthrown side. Because of these characteristics such listric normal faults are referred to as growth faults.*

2. The listric nature of this type of normal fault demands a décollement. The extension as manifested by the normal faults is therefore a skin effect and limited in depth.

3. As in thrust faulting the concept of a décollement requires a horizon of easy gliding. In coastal areas this is usually provided by overpressured shales. Since the latter fulfill all the requirements for diapirism, growth faulting and shale diapirism are frequently intimately related (Bruce,1973;Dailly,1976).

   In thrust belts late listric normal faults are thought to converge into the previous thrust planes at depth (Bally et al.,1966;P.B.Jones,1969). This interpretation implies that the extensional movement is surficial. It is dated as postorogenic and related to the late vertical uplift. This interpretation clearly postulates that the thrust planes were still acting as planes of weakness long after they ceased to function as gliding planes for the thrusting process.

4. The listric nature of such faults demands that the layers on the downthrown side be rotated, since the acceptance of a surficial open gap obviously does not constitute a viable interpretation. Superimposed on the regional dip it is this rotation which is the trap forming process and results in what is commonly called the roll-over.

---

*Growth faults are really no more than the headwalls of giant landslides, triggered and kept in motion by constant sedimentary loading of the head during a cycle of regression,and at depth controlled by internal planes of weakness.

5.0 References

AAPG: American Association of Petroleum Geologists
CSPG: Canadian Society of Petroleum Geologists
GSA: Geological Society of America
JPT: Journal of Petroleum Technology

Anderson,E.M.,1951, The Dynamics of Faulting and Dyke Formation; 2nd ed.,
        Oliver and Boyd, London, 206 p.

Anderson,R.A.,Ingram,D.S. and Zanier,A.M.,1973, Determining Fracture Pressure
        Gradients from Well Logs; JPT, Nov., p. 1259-1268.

Atwater,G.I. and Miller,E.E.,1965, The Effect of Decrease in Porosity with Depth
        on Future Development of Oil and Gas Reserves in South Louisiana;
        AAPG, 49/3, pt. I, p. 334 (Abstr.).

Bally,A.W.,Gordy,P.L. and Stewart,G.A.,1966, Structure, Seismic Data, and Oro-
        genic Evolution of Southern Canadian Rocky Mountains; CSPG, 14/3,
        p. 337-381.

Barker,C.,1972, Aquathermal Pressuring - Role of Temperature in Development of
        Abnormal-Pressure Zones; AAPG, 56/10, p. 2068-2071.

Brace,W.F.,1968,The Mechanical Effects of Pore Pressure on Fracturing of Rocks;
        Geol. Sur. Can. Paper 68-52, p. 113-124.

Brace,W.F. and Martin,III,R.J.,1968, A Test of the Law of Effective Stress for
        Crystalline Rocks of Low Porosity; Int. J. Rock. Mech. Min. Sc.,
        5/5, p.415-426.

Bradley,J.S.,1975, Abnormal Formation Pressure; AAPG, 59/6, p. 957-973.

Bruce,C.H.,1973, Pressured Shale and Related Sediment Deformation: Mechanism
        for Development of Regional Contemporaneous Faults; AAPG, 57/5,
        p. 878-886.

Brückl,E.,Roch,K.H. and Scheidegger,A.E.,1975, Significance of Stress Mesure-
        ments in the Hochkönig Massif in Austria; Tectonophysics, 29,
        p. 315-322, in Recent Crustal Movements, eds. N. Pavoni & R. Green.

Bullard,E.C.,1939, Heat Flow in South Africa; Proc. Roy. Soc. (London), Ser. A,
        v. 173, no. 955, p. 474-502.

Chapman,R.E.,1973, Petroleum Geology - A Concise Study; Elsevier, N.Y., 304 p.

Coates,D.F. and Grant,F.,1966, Stress Measurements at Elliot Lake; Can. Min. &
        Met. Bull., 59/649, p. 603-613.

Creutzburg,H.,1964, Untersuchungen über den Wärmestrom der Erde in Westdeutsch-
        land; Kali und Steinsalz, 4/3, p. 73-108.

Dahlstrom,C.D.A.,1970, Structural Geology in the Eastern Margin of the Canadian
        Rocky Mountains; CSPG, 18/3, p.332-406.

Dailly,G.C.,1976, A Possible Mechanism Relating Progradation, Growth Faulting, Clay
        Diapirism and Overthrusting in a Regressive Sequence of Sediments;
        CSPG, 24/1, p. 92-116.

Dickey,P.A.,1976, Abnormal Formation Pressure: Discussion; AAPG, 60/7,
          p. 1124-1127, with reply by Bradley, p. 1127-1128.

Dickinson,G.,1953, Geological Aspects of Abnormal Reservoir Pressures in
          Gulf Coast Louisiana; AAPG, 37/2, p. 410-423.

Domenico,S.N.,1974, Effect of Water Saturation on Seismic Reflectivity of
          Sand Reservoirs Encased in Shale; Geophysics, 39/6, p. 759-769.

Douglas,R.J.W.,1950, Callum Creek, Langford Creek, and Gap Map-Areas, Alberta;
          Geol. Sur. Can., Mem. 255, 124 p.

Drummond,J.M.,1963, Carbonates and Grade Size; CSPG, 11/1, p. 33-53.

Eisbacher,G.H. and Bielenstein,H.U.,1970, Interpretation of Elastic-Strain-
          Recovery Measurements near Elliot Lake, Ontario; Can. J. E. Sc.,
          7/2, p. 576-578.

Evans,D.M.,1966, The Denver Area Earthquakes and the Rocky Mountain Arsenal
          Disposal Well; The Mountain Geologist, 3/1, p. 23-36.

Evans,C.R.,McIvor,D.K. and Magara,K.,1975, Organic Matter, Compaction History
          and Hydrocarbon Occurrence - Mackenzie Delta, Canada; Proc. 9th
          World Petr. Congr., v. 2, p. 149-157.

Fabian,H.J.,1955, Carbon-Ratio-Theorie, geothermische Tiefenstufe und Erdgas-
          lagerstätten in Nordwestdeutschland; Erdöl und Kohle, Jahrg. 8,
          No. 3, p. 141-146.

Fertl,W.H.,1976, Abnormal Formation Pressures; Elsevier N.Y., 382 p.
          (with contributions by G.V. Chilingarian and H.H. Rieke,III).

Friedman,M. and Heard,H.C.,1974, Principal Stress Ratios in Cretaceous Lime-
          stone from Texas Gulf Coast; AAPG, 58/1, p. 71-78.

Füchtbauer,H.,1972, Diagenesis of Arenaceous Deposits; in 'Arenaceous Deposits:
          Sedimentation and Diagenesis; Nat. Conf. on Earth Sc., Banff, Alta.,
          Dept. Ext. Univ. Alta. & CSPG, p. 205-286.

Gansser,A.,1960, Ueber Schlammvulkane und Salzdome; Vierteljahrschrift der Natur-
          forschenden Gesellschaft Zürich, Jahrg. 105, p. 1-46.

Goguel,J.,1962, Tectonics; W.H.Freeman San Francisco, 384 p. (French text, 1952)

---------,1969, Le rôle de l'eau et de la chaleur dans les phénomènes tecto-
          niques; Revue de géographie physique et de géologie dynamique,
          11/2, p. 153-163.

Greiner,G.,1975, In-Situ Stress Measurements in Southwest Germany; Tectonophysics,
          29, p. 49-58, in Recent Crustal Movements, eds. N. Pavoni & R. Green.

Gretener,P.E.,1969a, Fluid Pressure in Porous Media - Its Importance in
    Geology: A Review; CSPG, 17/3, p. 255-295.

------------,1969b, On the Mechanics of the Intrusion of Sills; Can. J. E. Sc.,
    6/6, p. 1415-1419.

------------,1972, Thoughts on Overthrust Faulting in a Layered Sequence;
    CSPG, 20/3, p. 583-607.

Grossling,B.F.,1959, Temperature Variations due to the Formation of a Geosyn-
    cline; GSA, 70/10, p. 1253-1281.

Gussow,W.C.,1954, Differential Entrapment of Oil and Gas: a Fundamental Prin-
    ciple; AAPG, 38/5, p. 816-853.

Gysel,M.,1975, In-Situ Stress Measurements of the Primary Stress State in the
    Sonnenberg Tunnel in Lucerne, Switzerland; Tectonophysics, 29,
    p. 301-314, in Recent Crustal Movements, eds. N. Pavoni & R. Green.

Hamblin,W.K.,1965, Origin of 'Reverse Drag' on the Downthrown Side of Normal
    Faults; GSA, 76/10, p.1145-1164.

Hamilton,D.H. and Meehan,R.L.,1971, Ground Rupture in the Baldwin Hills; Science,
    172/3981, p. 333-344.

Handin,J. and Hager,R.V.,Jr.,1957, Experimental Deformation of Sedimentary
    Rocks under Confining Pressure: Tests at Room Temperature on Dry
    Samples; AAPG, 41/1, p. 1-50.

Handin,J.,Hager,R.V.,Jr.,Friedman,M. and Feather,J.N.,1963, Experimental De-
    formation of Sedimentary Rocks under Confining Pressure: Pore Pres-
    sure Tests; AAPG, 47/5, p. 717-755.

Hanshaw,B.B. and Zen,E-An,1965, Osmotic Equilibrium and Overthrust Faulting;
    GSA, 76/12, p. 1379-1386.

Hast,N. and Nilsson,T.,1964, Recent Rock Pressure Measurements and their Impli-
    cations for Dam Building; Trans. 8th Int. Congr. on Large Dams,
    Edinburgh, v. 1, Ques. 28, p. 601-610.

Heard,H.C. and Rubey,W.W.,1966, Tectonic Implications of Gypsum Dehydration;
    GSA, 77/7, p. 741-760.

Hedberg,H.D.,1974, Relation of Methane Generation to Undercompacted Shales,
    Shale Diapirs and Mud Volcanoes; AAPG, 58/4, p. 661-673.

Hottman,C.E. and Johnson,R.K.,1965, Estimation of Formation Pressures from
    Log-Derived Shale Properties; SPE Trans. AIME, v. 234, p. 717-722.

Hsü,K.J.,1969, A Preliminary Analysis of the Statics and Kinetics of the Glarus
    Overthrust; Ecl. geol. helv., 62/1, p. 143-154.

Hubbert,M.K.,1951, Mechanical Basis for certain Familiar Geologic Structures;
    GSA, 62/4, p. 355-372.

------------,1953, Entrapment of Petroleum under Hydrodynamic Conditions;
    AAPG, 37/8, p. 1954-2026.

Hubbert,M.K. and Willis,D.G.,1957, Mechanics of Hydraulic Fracturing; Trans. AIME, 210, p. 153-168.

Hubbert,M.K. and Rubey,W.W.,1959, Role of Fluid Pressure in Mechanics of Overthrust Faulting, I Mechanics of Fluid-Filled Porous Solids and its Application to Overthrust Faulting; GSA, 70/2, p. 115-166.

------------------------------,1960, Role of Fluid Pressure in Mechanics of Overthrust Faulting - A Reply; GSA, 71/5, p. 611-628.

Jaeger,J.C. and Cook,N.G.W.,1969, Fundamentals of Rock Mechanics; Methuen London, 513 p.

Jam,P.L.,Dickey,P.A. and Tryggvason,E.,1969, Subsurface Temperature in South Louisiana; AAPG, 53/10, p. 2141-2149.

Jones,P.H.,1969, Hydrodynamics of Geopressure in the Northern Gulf of Mexico Basin; JPT, July, p. 803-810.

Jones,P.B.,1969, The Tertiary Kishenehn Formation, British Columbia; CSPG, 17/2, p. 234-246.

----------,1971, Folded Faults and Sequence of Thrusting in Alberta Foothills; AAPG, 55/2, p. 292-306.

Keating,L.F.,1966, Exploration in the Canadian Rockies and Foothills; Can. J. E. Sc., 3/5, p. 713-723.

Kennedy,G.C. and Holser,W.T.,1966, Pressure-Volume-Temperature and Phase Relations of Water and Carbon Dioxide; GSA. Mem. #97, p. 371-384.

Kukal,Z.,1971, Geology of Recent Sediments; Academic Press, N.Y., 490 p.

Klemme,H.D.,1972, Heat Influences Size of Oil Giants; Oil & Gas J., 70/29, p. 136,141-144.

Labute,G.J. and Gretener,P.E.,1969, Differential Compaction around a Leduc Reef - Wizard Lake Area, Alberta; CSPG, 17/3, p. 304-325.

Laubscher,H.P.,1961, Die Fernschubhypothese der Jurafaltung; Ecl. geol. helv., 54/2, p. 221-282.

Lee,W.H.K.,1963, Heat Flow Data Analysis; Rev. Geophys., 1/3, p. 449-479.

Leenhardt,O.,1967, Topics on Seismic Research at the Monaco Oceanographic Museum; Geophys. Prosp., 15/3, p. 516-526.

Lewis,C.R. and Rose,S.C.,1970, A Theory Relating High Temperatures and Overpressures; JPT, January, p. 11-16.

Lohr,J.,1969, Die seismischen Geschwindigkeiten der jüngeren Molasse im ostschweizerischen und deutschen Alpenvorland; Geophys. Prosp., 17/2, p. 111-125.

Mackay,J.R.,1963a, The Mackenzie Delta Area, N.W.T.; Canada Dept. Min. and
        Techn. Sur., Geogr. Br. Mem. #8, 202 p.

-----------,1963b, Pingos in Canada; Proc. Int. Conf. Permafrost, Nat. Acad.
        Sc., Nat. Res. Council Publ. no. 1287, Washington, D.C., p. 71-76.

Magara,K.,1975a, Reevaluation of Montmorillonite Dehydration as Cause of Ab-
        normal Pressure and Hydrocarbon Migration; AAPG, 59/2, p. 292-302.

---------,1975b, Importance of Aquathermal Pressuring Effect in Gulf Coast;
        AAPG, 59/10, p. 2037-2045.

Mathews,W.H. and Mackay,J.R.,1960, Deformation of Soils by Glacier Ice and the
        Influence of Pore Pressure and Permafrost; Trans. Roy. Soc. Can.,
        3rd ser., v. 54, sec. 4, p. 27-36.

Maxwell,J.C.,1964, Influence of Depth, Temperature and Geologic Age on Porosity
        of Quartzose Sandstone; AAPG, 48/5, p. 697-709.

Miller,B.M.,1974, Geothermal and Geopressure-Relations as Tool for Petroleum
        Exploration; AAPG, 58/5, p. 916, (Abstr.).

Morgan,J.P.,Coleman,J.M. and Gagliano,S.M.,1968, Mudlumps: Diapiric Structures
        in the Mississippi Delta; AAPG, Mem. #8, p. 145-161.

Moses,P.L.,1961, Geothermal Gradients now known in Greater Detail; World Oil,
        May, p. 79-82.

Mostofi,B. and Gansser,A.,1957, The Story Behind the 5 Alborz; Oil & Gas J.,
        Jan., p. 78-84.

Müller,F.,1959, Beobachtungen über Pingos; Meddelelser om Grønland, Bd. 153,
        no. 3, 127 p.

Musgrave,A.W. and Hicks,W.G.,1966, Outlining of Shale Masses by Geophysical
        Methods; Geophysics, 31/4, p. 711-725.

Nettleton,L.L.,1934, Fluid Mechanics of Salt Domes; AAPG, 18/9, p. 1175-1204.

--------------,1962, Gravity and Magnetics for Geologists and Seismologists;
        AAPG, 46/10, p. 1815-1838.

Nichols,E.A.,1947, Geothermal Gradients in Mid-Continent and Gulf Coast Oil
        Fields; Trans. AIME, 170, p. 44-50 (reprinted 1956).

Obert,L.,1967, Determination of Stress in Rock - A State-of-the Art Report;
        ASTM, Spec. Techn. Publ. No. 429, 56 p.

O'Connor,M.J. and Gretener,P.E.,1974a, Quantitative Modelling of the Processes
        of Differential Compaction; CSPG, 22/3, p. 241-268.

-------------------------------,1974b, Differential Compaction within the
        Woodbend Group of Central Alberta; CSPG, 22/3, p. 269-304.

Parker,C.A.,1974, Geopressures and Secondary Porosity in Deep Jurassic of Mississippi; AAPG, 58/10, p. 2212 (Abstr.).

Parsons,M.G.,1975, The Geology of the Laurentian Fan and the Scotia Rise; CSPG. Mem. #4, p. 155-167.

Pavoni,N.,1968, Uber die Entstehung der Kiesmassen im Bergsturzgebiet von Bonaduz-Reichenau (Graubünden); Ecl. geol. helv., 61/2, p. 494-500.

Poland,J.F.,1972, Subsidence and its Control; AAPG. Mem. #18, p. 50-71.

Powers,M.C.,1967, Fluid-Release Mechanisms in Compacting Marine Mud Rocks and their Importance in Oil Exploration; AAPG, 51/7, p. 1240-1254.

Pusey,III,W.C.,1973, Paleotemperatures in the Gulf Coast using the ESR-Kerogen Method; Trans. Gulf Coast Assoc. Geol. Soc., 23, p. 195-202.

Raleigh,C.B.,1972, Earthquakes and Fluid Injection; AAPG Mem. #18, p. 273-279.

Raleigh,C.B. and Paterson,M.S.,1965, Experimental Deformation of Serpentinite and its Tectonic Implications; J. Geophys. Res., 70/16, p. 3965-3985.

Raleigh,C.B.,Healy,J.H. and Bredehoeft,J.D.,1972, Faulting and Crustal Stress at Rangely, Colorado; AGU Mem. #16, p. 275-284.

Ranalli,G.,1975, Geotectonic Relevance of Rock-Stress Determinations; Tectono-physics, 29, p. 49-58, in Recent Crustal Movements, eds. N. Pavoni and R. Green.

Rettger,R.E.,1935, Experiments in Soft-Rock Deformation; AAPG, 19/1, p.271-292.

Reynolds,E.B.,1970, Predicting Overpressured Zones with Seismic Data; World Oil, Oct., p. 78-82.

Reynolds,E.B.,May,J.E. and Klaveness,A.,1971, The Geophysical Aspects of Abnor-mal Fluid Pressures; Abnormal Subsurface Pressure; A Study Group Report, Houston Geol. Soc., p. 31-47.

Rich,J.L.,1934, Mechanics of Low-Angle Overthrust Faulting as Illustrated by Cumberland Thrust Block, Virginia, Kentucky and Tennessee; AAPG, 18/12, p. 1584-1596.

Rikitake,T.,1959, Studies of the Thermal State of the Earth, Part 2 - Heat Flow Associated with Magma Intrusions; Bull. Earthquake Res. Inst. Tokyo, 37/2, p. 233-243.

Rothé,J.P.,1970, Seismes artificiels; Tectonophysics, 9, p. 215-238.

Royse,Jr.,F.,Warner,M.A. and Reese,D.L.,1975, Thrust Belt Structural Geometry and Related Stratigraphic Problems, Wyoming-Idaho-Northern Utah; Rocky Mountain Assoc. Geol., 1975 Symposium, p. 41-54.

Rubey,W.W. and Hubbert,M.K.,1959, Role of Fluid Pressure in Mechanics of Over-thrust Faulting, II Overthrust Belt in Geosynclinal Area of Western Wyoming in Light of Fluid-Pressure Hypothesis; GSA, 70/2, p. 167-205.

Schmidt,G.W.,1973, Interstitial Water Composition and Geochemistry of Deep
Gulf Coast Shales and Sandstones; AAPG, 57/2, p. 321-337.

Secor,D.T.,1965, Role of Fluid Pressure in Jointing; Am. J. Sc., 263, p. 633-646.

Selig,F. and Wallick,G.C.,1966, Temperature Distribution in Salt Domes and
Surrounding Sediments; Geophysics, 31/2, p. 346-361.

Shepard,F.P.,Dill,R.F. and Heezen,B.C.,1968, Diapiric Intrusions in Foreset
Slope Sediments off Magdalena Delta, Columbia; AAPG, 52/11,
p. 2197-2207.

Short,K.C. and Stäuble,A.J.,1967, Outline of Geology of Niger Delta; AAPG,
51/5, p. 761-779.

Shouldice,D.H.,1971, Geology of the Western Continental Shelf; CSPG, 19/2,
p. 405-436.

Skempton,A.W.,1960, Significance of Terzaghi's Concept of Effective Stress;
'From Theory to Practice in Soil Mechanics', Wiley & Sons, N.Y.,
p. 42-53.

Sprunt,E.S. and Nur,A.,1976, Reduction of Porosity by Pressure Solution: Experimental Verification; GSA, Geology, 4/8, p. 463-466.

Stanley,Jr.,T.B.,1970, Vicksburg Fault Zone, Texas; AAPG, Mem. #14, p. 301-308.

Staub,R.,1954, Der Bau der Glarneralpen und seine prinzipielle Bedeutung für
die Alpengeologie; Glarus Verlag Tschudi & Co., 187 p.

Stuart,C.A.,1970, Geopressures; Supplement, Proc. 2nd Symposium on Abnormal
Subsurface Pressure, Louisiana State Univ. 121 p.

Suman,Jr.,G.O.,1974, Casing Buckling in Producing Intervals; Petr. Eng., 46/4,
p. 36,38,40,42.

Terzaghi,K.,1950, Mechanics of Landslides; GSA, Berkey Volume, p. 83-123.

Timko,D.J. and Fertl,W.H.,1971, Relationship between Hydrocarbon Accumulations
and Geopressure and its Economic Significance; JPT, August,
p. 923-933.

Tucker,P.M. and Yorston,H.J.,1973, Pitfalls in Seismic Interpretation; Soc.
Expl. Geophys., Monograph #2, 50 p.

Vorob'eva,K.I.,1962, Geothermal Features of the Ozek-Suat Oil Field and of other
Regions of the Tersko-Kuma Plain; Geologiya Nefti i Gaza; Petr.
Geol., 4/6-B, p. 359-362.

Weber,K.J. and Daukoru,E.,1975, Petroleum Geology of the Niger Delta; Proc.
9th World Petr. Congr., v. 2, p. 209-221.

Weimer,R.J. and Davis,T.L.,1976, Overpressured Shale Masses and Growth Faulting, New Tectonic Style in Cretaceous Foreland Basin, Colorado; AAPG, 60/8, p. 1413-1414 (Abstr.).

Zoback,M.D. and Byerlee,J.D.,1975, Permeability and Effective Stress; AAPG, 59/1, p. 154-158.

Special Publications on Abnormal Fluid Pressure :

AAPG Reprint Series No. 11 : Abnormal Subsurface Pressure (1974)

Fertl,W.H. : Abnormal Formation Pressures, Elsevier N.Y. (1976)

Houston Geol. Soc. : Abnormal Subsurface Pressure (1971)

Stuart,C.A. : Geopressures (1970)

Reprinted by permission of the Geological Society of America
from W. Hafner, *Geological Society of America Bulletin*, v.
62 (1951), p. 373-398.

BULLETIN OF THE GEOLOGICAL SOCIETY OF AMERICA

VOL. 62, PP. 373-398, 9 FIGS., 1 PL.   APRIL 1951

STRESS DISTRIBUTIONS AND FAULTING

By W. Hafner

Abstract

Tectonic deformations result from a condition of internal stress caused, in turn, by primary and secondary forces. In the geological literature, a great deal of discussion is based on a direct connection between forces and deformation, completely by-passing the concept of stress. This paper is a contribution in the intermediate field of stress relations. It presents the complete solutions of certain stress systems caused by various forms of boundary forces. Furthermore, the location and attitude of the fault surfaces likely to be associated with them is determined.

The basic concept of stress is briefly reviewed and some of the fundamental differences between the force-vector and the stress-tensor are pointed out. The fallacy of applying the familiar methods of vector-addition of forces to problems in stress is demonstrated.

For certain systems of external boundary forces acting on a portion of the earth's crust, the internal stress distribution can be calculated by means of the familiar equations of elasticity. Appropriate calculation methods for two-dimensional cases are shown and the basic equations applicable to a series of important boundary conditions are derived. The examples here presented include: (1) superposed horizontal compression with constant lateral and vertical gradients; (2) horizontal compression with exponential attenuation; and (3) sinusoidal vertical and shearing forces acting on the bottom of a block. The latter equations provide solutions for differential vertical uplift and for the important case of drag exerted on the bottom of the crust by convection currents in the substratum. Diagrams show configuration of the stress trajectories and distribution of the maximum shearing stress for the resulting stress systems.

A parallel series of diagrams shows the disposition between the relatively stable and unstable segments of the blocks and the probable attitude of the fault surfaces likely to be associated with the individual stress systems. The construction of the fault surfaces is based on the original stress distributions alone, the influence of local stress alterations due to the occurrence of fracture being disregarded. The full effect of this inter-action is not known, due to the extreme complexity of the problem. The fault patterns shown are strictly applicable only to the initial stages of fracture, but may also represent fair approximations during the more advanced stages, since the original stress remains the dominating influence and stress-alterations due to faulting diminish rapidly with distance.

CONTENTS

Illustrations

373

INTRODUCTION

The earth's crust is unceasingly undergoing deformations because of the interplay of forces acting upon it. The primary forces are sufficiently powerful to produce disturbances of enormous magnitude, such as mountain chains or the crustal down-buckling deduced from gravity observations; the secondary ones tend to bring about partial or complete restoration of equilibrium.

Many hypotheses have been advanced regarding the origin and nature of the deformative forces but, as yet, all these theories are speculative in varying degree and a wide divergence of viewpoint exists. Knowledge of the results of these processes is much more advanced, at least with respect to the uppermost portion of the crust. A vast and ever-growing body of observational data on the nature of the deformations is becoming available. In fact, the validity of the various hypotheses on the origin of the deformations rests largely upon the degree of agreement with this accumulated store of data, aside, of course, from the inherent soundness of the principles employed.

In all quantitative studies on the relationship between the original forces and the resulting deformations, an intermediate field of investigation enters, the condition of stress in the earth's crust. The original forces—whether primary or secondary, whether internal body or external boundary forces—set up a state of stress in the earth's crust which, in turn, produces the observed distortions. In the geological literature, we quite often find arguments attempting to explain the deformations *directly* in terms of the original forces. While this approach may be justified in certain simple qualitative considerations, it is an inherently fallacious procedure and more often than not leads to confusion, rather than elucidation, of the problem discussed. It is especially inappropriate in quantitative analysis.

This paper offers the results of some studies in the intermediate field of stress relations. It presents the complete solutions of the stress distributions produced by certain systems of external boundary stresses. The nature of the faulting most likely to be associated with them is analyzed, although certain complications for which no complete solutions are available have to be disregarded in this phase of the investigation. The stress distributions here analyzed were selected with the view of covering some of the geologically more important groups of boundary stresses acting on portions of the earth's crust, such as lateral compression, differential vertical uplift, and drag exerted on the bottom due to viscous currents in the substratum. No attempt is made to enter the more speculative field of the origin of these forces. However, the author believes that any advance made in our understanding of the relation between the observed tectonic phenomena and the causal stress systems will indirectly promote progress in the more important, but much more illusive, field of the ultimate causes of the earth's deformations.

ACKNOWLEDGMENTS

The writer wishes to thank Shell Oil Company, Incorporated, for permission to publish this paper. He further wishes to express his appreciation to Dr. M. King Hubbert for his stimulating influence concerning the quantitative approach to geological problems in general, for a critical reading of the manuscript, and for helpful suggestions during discussions of the fault problem. Thanks are also due to Miss B. J. King for assistance in the numerical calculations, and to C. B. Metcalfe for the drafting of the diagrams.

STRESS AND FORCE

In the geologic literature we encounter not infrequently a misunderstanding of the concept of stress. Many authors use the terms "force" and "stress" synonymously, in the sense of dynamical forces, and are not aware of the fundamental differences between the two concepts. In view of this fact, it appears advisable to review briefly the general properties of the

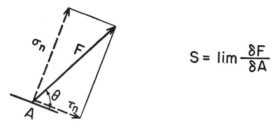

$$S = \lim \frac{\delta F}{\delta A}$$

FIGURE 1.—LIMITED DEFINITION OF STRESS

state of stress and its relation to forces. Such considerations, though of elementary character, may be helpful to readers not thoroughly familiar with these concepts.

Stress is force per unit area. A more precise definition (Timoshenko, 1934, p. 3) states that it represents the limiting value of the ratio of a surface force divided by the area upon which it acts (Fig. 1); in formula:

$$S = \lim \frac{\delta F}{\delta A}$$

This force may be resolved into a normal component σ_n, and a tangential component τ_n. Depending upon whether σ_n is directed outward or inward, it represents a tension or a pressure, respectively. The tangential component τ_n, which is acting parallel to the surface element, is a shearing stress.

This definition brings out one of the differences between stress and force: that the two quantities are dimensionally at variance. The dimensions of force are (m, l, t^{-2}); those of stress are (m, l^{-1}, t^{-2}). Hence, the statement that stress *is* force, as found in a well-known textbook on structural geology, is incorrect. However, the difference between stress and force goes much deeper. If we are dealing with dynamical forces only, it would always be possible by the process of resolution of forces to determine the normal and tangential components with respect to *any* direction of the surface element δA passing through the point. When dealing with stress, on the other hand, we know nothing whatever of the mag-

nitude of stress after rotation of the element into a different direction. To solve this simple problem, we need an expression for the state of stress which defines this quantity for *every* surface element leading through a given point, irrespective of direction. Mathematically, this more comprehensive expression is a tensor quantity and, more specifically, a symmetric tensor of second degree. This is in marked contrast to force, which is a vector quantity. A tensor requires for its description more than three components and is therefore a more complicated concept than a vector. The relation is thus similar to that between a vector and a scalar. It is as misleading to think of stress in terms of a vector as it would be to treat a problem involving force in terms of a scalar function. The specific properties of the stress tensor can be adequately described without recourse to the general theory of tensors. The most important fact is that a total of six quantities is needed for a complete determination of the state of stress at any one point in a solid. In contrast, force, being a vector, requires only three quantities for specification. This explains why the above simple problem is indeterminate if only the three parameters of a force-vector are given.

There are two convenient ways to describe the six parameters of the stress tensor (Fig. 2): (1) State the stress components acting on the surface of a cubic element whose faces are parallel to the coordinate axes or (2) Use the well-known stress ellipsoid. The stress ellipsoid is defined by three mutually perpendicular directions, called the principal directions, and the intensities of the stresses in these directions, called the principal stresses. Each method requires six independent quantities; if known at each point, the condition of stress throughout the body is completely determined. The stress systems analyzed in the present paper will be restricted to two-dimensional cases of plane deformations. These are characterized by the fact that variable displacements occur only in parallel planes, namely those planes which are oriented perpendicular to the axes of the structures. Elongation in the third, or longitudinal, direction may or may not take place; if it does, it is constant over the entire cross-section. In such systems, two of the independent shear stress components are automatically zero, and the normal stresses parallel to the longitudinal axes do not vary in this direction. The solution of stress problems is then reduced to the determination of three components instead

simultaneously at a point, the resultant force may be obtained by the well-known method of vector addition. Not infrequently we find this method applied to stresses too, sometimes quite specifically, more often in rather vaguely

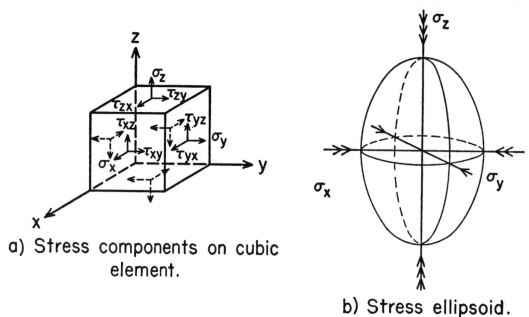

a) Stress components on cubic element.

b) Stress ellipsoid.

FIGURE 2. COMPLETE EXPRESSIONS OF STRESS TENSOR

of six, as in the general three-dimensional case. Taking the x-axis as the horizontal direction *in* the cross-section, the y-axis vertically downward (the z-coordinate being in the longitudinal direction), the three components then are σ_x, σ_y, and τ_{xy}, and the stress ellipsoid is reduced to a stress ellipse in the xy-plane (Fig. 3).

The stress ellipse provides a simple and convenient means of illustrating some of the fundamental differences between the stress-tensor and the force-vector. Two of them concern the methods of resolution and composition of forces and stresses. To illustrate the first case, assume a surface force acting in the direction of σ_x (Fig. 3). Its vector component at right angle, in the y-direction, is zero, whereas the corresponding stress component σ_y may be of any magnitude. This again shows that the process of resolution of forces for the purpose of determining the stress components acting upon a surface in any direction other than that specified in the limited definition of stress is inapplicable. A similar situation exists with respect to the composition of forces and stresses. When several forces act

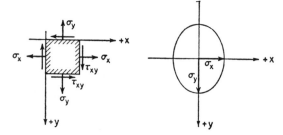

FIGURE 3.—TWO-DIMENSIONAL STRESS COMPONENTS AND STRESS ELLIPSE

phrased statements. Such deductions are basically wrong and lead often to complete confusion in the process of reasoning. When a solid is subjected to several simultaneous stresses, they can, of course, also be combined into a single stress system, but the procedure is quite different from the method of vector addition. The same is true for the respective results. The combination of stresses is based on the principle of superposition. It states that addition—in the form of linear functions—of the stress components from two, or several, valid stress systems, produces a new one which is likewise valid, provided the resulting deformation is still within the elastic limit. An important example of this principle will be given later. Again the stress ellipse is useful to illustrate the difference between the methods of combining stresses and forces. Let us assume a uniform and uniformly directed pressure σ_x acting in the direction of the x-axis and a second similar pressure σ_y acting in the y-direction. Each of these systems satisfies the required equations and, hence, represents a permissible stress system. The method of superposition yields a two-dimensional stress system characterized by the stress ellipse shown in Figure 4. Its principal directions coincide with the co-ordinate axes, and the intensity of the stress acting on any surface element at right angles to the xy-plane is represented by one of the radius vectors of the stress ellipse. It is to be noted, however, that the direction of the normal to the surface element and that of the corresponding radius vector are not identical (except for the principal planes); the exact relation between these directions is somewhat complicated but need not be considered here. If, on the other hand, we should apply the method of vector-addition to forces of equal intensity as σ_x and σ_y, the result would be totally different. Compare Figures 4a and 4b. The vector sum of f_x and f_y yields a force f_s which is substantially in excess of any radius vector in Figure 4a, regardless of direction. Hence this procedure cannot be applied to stress relations. It is, nevertheless, frequently employed in discussions on structural problems, and leads to serious errors, both in the quantitative results as well as the more general qualitative deductions.

A further difference between stress and force concerns statements of direction. One may speak of a force as acting in a certain direction, say towards the north. When applied to internal stresses, such statements have no sense. A stress component acting upon one side of a surface element exists only in co-ordination with a component of equal intensity but opposite direction, acting on the other side. This is true for normal, as well as shear stresses. Hence a pressure or a tension may exist in a north-south direction, but not towards the north or towards the south.

The co-existence of stress and counter-stress on opposite sides of a surface element is a general property existing on any imaginary internal surface of a body, as well as on all its boundary surfaces. One very important implication of this condition concerns its ap-

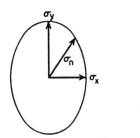 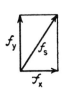

(a) Superposition of stresses   (b) Vector addition of forces

FIGURE 4.—COMPOSITION OF STRESSES AND FORCES

plication to the surface of the earth. Since the air is incapable of sustaining a shearing stress, τ_{xy} must be zero everywhere along this surface. Furthermore, the vertical pressure is reduced to the normal atmospheric pressure and cannot exceed it. This pressure is a very small quantity when compared to the internal stresses in the crust in which we are here interested; hence, it can safely be neglected. These considerations impose two specific boundary conditions which *must* be satisfied by every stress distribution applicable to geological problems: that σ_y and τ_{xy} vanish for every point along the free surface of the earth.

METHOD OF CALCULATING STRESS SYSTEMS

The methods applied to derive complete solutions of geologically significant stress distributions will be briefly outlined. Only two-dimensional cases will be treated. This simplification is fully justified for numerous structural problems but is not applicable, of course, to those of essentially three-dimensional character. A further assumption is that the stressed medium is homogeneous. This second restriction is geologically much more undesirable; however, the mathematical treatment of non-homogeneous cases would be extremely complicated.

The internal state of stress in a block of the earth's crust is adequately portrayed by the following sets of curves:

(1) *The stress trajectories*, defined by the property that the directions of the principal stresses are the tangents to these curves at all points. One set determines the direction of the maximum principal stress and a second one that of the minimum principal stress. The two sets are orthogonal everywhere.

(2) *Lines of equal maximum shearing stress*, consisting of a series of lines each of which connects all points of equal magnitude of this quantity.

The intensities of the two principal stresses can be illustrated by similar curves; however, they are less important for our purposes and are omitted from the illustrations to avoid overcrowding. Conversely, the external forces, or boundary stresses, form an integral part of each analysis and are, therefore, plainly marked on the diagrams.

The method here employed for the calculation of the internal stress distributions is based on the use of the Airy stress function, which offers an especially convenient approach for two-dimensional cases. It can be shown that for plane stress distributions—as well as for plane deformations—all three stress components can be expressed as second partial derivatives of a single function Φ which is a scalar function of x and y. Thus its role is somewhat similar to the potential function as applied to gravity, magnetism or other force fields. As is well known, the three force-components of such fields can likewise be expressed as partial derivatives of a single scalar function—the potential function—which greatly facilitates the mathematical treatment of the respective fields. However, the vector-components of the force fields are the *first* partial derivatives of the potential function, whereas the three components of the stress-tensor are the *second* partial derivatives of the stress function.

Now the theory of elasticity shows that the equations of static equilibrium are automatically satisfied if we define the stress components by the expressions

$$\sigma_x = \frac{\partial^2\Phi}{\partial y^2}; \quad \sigma_y = \frac{\partial^2\Phi}{\partial x^2}; \quad \tau_{xy} = -\frac{\partial^2\Phi}{\partial x \partial y}; \quad (1)$$

when body forces are absent, and by the expressions

$$\sigma_x = \frac{\partial^2\Phi}{\partial y^2}; \quad \sigma_y = \frac{\partial^2\Phi}{\partial x^2} - \rho g y; \quad \tau_{xy} = -\frac{\partial^2\Phi}{\partial x \partial y}; \quad (1)'$$

when gravity represents the only body force. Here ρ is the density of the material, assumed to be constant, and g is the constant of gravitational acceleration. (When applied to engineering problems, expressions $(1)'$ are usually stated in a different form, the term $-\rho g y$ here added to σ_y being replaced by a term $-\rho g x$, attached to τ_{xy}. (See Timoshenko, 1934, p. 25). The variation here adopted likewise satisfies the equilibrium equations and is preferable for geological investigations because the physical interpretation of the gravity term is more directly apparent. The resulting loss of symmetry is of no consequence.)

In order to yield a valid stress system, another requirement must be fulfilled. This arises from the fact that the stress components are connected by Hooke's law with the strain components (elongations), which, in turn, are defined as functions of the displacements. In two-dimensional problems, there are three stress and three corresponding strain components but only two displacements; hence, the former are not independent but must be connected by an equation establishing a condition of internal consistency of the displacements. This equation is called the compatibility equation. When applied to the stress function, it can be shown that a necessary—but also sufficient—requirement is that this function satisfy the equation:

$$\frac{\partial^4\Phi}{\partial x^4} + 2\frac{\partial^4\Phi}{\partial x^2\partial y^2} + \frac{\partial^4\Phi}{\partial y^4} = 0 \quad (2)$$

If this requirement is fulfilled, the stress function provides a method for the direct determination of the three stress components, making it unnecessary to first determine the displacements. The method then consists in finding suitable functions Φ which satisfy equation (2) and simultaneously yield a system of boundary stresses which are of interest to geological investigations.

Having found a suitable function of this type, the solution of the complete stress system is simple and straightforward. The three stress components are given directly by differentiation of Φ, according to formulae (1). The remaining significant stress quantities are then obtained by the well-known equations of elasticity. The maximum shearing stress at any point is:

$$\tau_{max} = \pm \left\{ \left(\frac{\sigma_x - \sigma_y}{2} \right)^2 \tau_{xy}^2 \right\}^{\frac{1}{2}} \qquad (3)$$

The intensities of the two principal stresses are:

$$\sigma_1 \text{ or } \sigma_2 = \frac{1}{2}(\sigma_x + \sigma_y) \pm \tau_{max}, \qquad (4)$$

and the directions of the stress trajectories are given by:

$$\tan 2\beta = \frac{2\tau_{xy}}{\sigma_x - \sigma_y}, \qquad (5)$$

where β represents the angle between the outward-directed normal of the surface element and the positive x-axis, in the direction from the latter towards the positive y-axis. In the following calculations, the two normal stress components σ_x and σ_y are taken positive if they represent tensile stresses, negative if they are compressive stresses.

The relation between the internal stress distribution of a body of limited extent and the forces causing it need some further explanation. The stress components defined by equations (1)' include the effect of gravity, whereas those of equations (1) are due only to the surface forces acting upon the boundaries of a block. In contrast to many engineering problems where gravity is a negligible factor, it constitutes one of the primary forces in geological systems and hence requires special consideration. This will be discussed in more detail in the next section. First we will focus our attention on the surface forces. The stress systems to be analyzed are defined by analytical functions which, in most cases, extend to infinity. In geological problems, on the other hand, we wish to investigate finite portions of the earth's crust which are exposed to the action of surface forces along their boundaries. Since such blocks are essentially stationary, the surface forces cannot be chosen arbitrarily but only such combinations are admissible which, on the whole, satisfy the equations of static equilibrium. Now the relation between the internal stress distribution defined by equations (1) and the boundary forces responsible for it is directly given by a fundamental prin-

ciple. It states that any imaginary surface inside a stressed medium can be made a boundary surface without causing a disturbance in the internal stress distribution, provided a system of surface forces equivalent to the former internal stress components is applied to this boundary. By this process we can arbitrarily select any desired location and limits of a block within the confines of an analytically defined stress distribution and can also immediately state the system of surface forces which produces it. This principle will be used extensively in the later treatment of specific examples.

The method of determining stress systems here outlined is, in some respects, an indirect approach. The stress distribution is defined first and the causal boundary forces are then fixed automatically. This is obviously not an altogether ideal procedure. Yet it has decided advantages and in reality is much more flexible than one might expect it to be. First the system of boundary forces thus derived automatically satisfies the requirement of static equilibrium. This imposition might constitute a tedious complication in all but the simplest cases if the direct approach of starting with arbitrarily selected boundary conditions were used. Second, numerous boundary systems representing some of the geologically most interesting cases can easily be obtained by suitable choices in the selection of the Airy stress function. Another important factor greatly enhances the flexibility of this method. As seen from equation (2), the stress function Φ is obtained by a solution of a fourth order differential equation. Hence it contains four independent integration constants which will also enter the expressions for the stress components (1). Now, as previously pointed out, any stress distribution applicable to our problems *must* satisfy the boundary conditions at the surface of the earth. Due to this restriction, two of the four integration constants must be expended for this purpose. The remaining two, however, are unassigned by general theoretical requirements and, therefore, can be used for the modification of the boundary stresses. It will be seen later that this freedom provides the means for a considerable degree of variability in the selection of geologically interesting cases.

The Standard State

E. M. Anderson (1942) has introduced the term "standard state," defined (p. 137) as

"a condition of pressure which is the same in all directions at any point, and equal to that which would be caused by the weight of the superincumbent material, across a horizontal plane, at the particular level in the rock. It is assumed in this definition that the surface is flat, and that the strata are of uniform specific gravity."

It will be shown that the standard state—according to this definition—represents a stress system composed of two parts: (1) the effect of gravity, and (2) a superposed horizontal stress which is constant in any horizontal plane but increasing uniformly with depth. The manner in which gravity affects our investigations must be clearly understood, since it represents one of the major causes of stress. Being a body force, it plays a somewhat different role in the analytical treatment than do the surface forces. For this reason—and also to illustrate the discussed calculation methods—the standard state is briefly derived as follows:

We select the Airy stress function in the form of a polynomial of third degree

$$\Phi = k_1 x^3 + k_2 x^2 + k_3 x^2 y + k_4 y^3,$$

which satisfies equation (2). The stress components (1)′ then are:

$$\sigma_x = \frac{\partial^2 \phi}{\partial y^2} = 2k_2 x + 6k_4 y;$$

$$\sigma_y = \frac{\partial^2 \phi}{\partial x^2} - \rho g y = 6k_1 x + 2k_2 y - \rho g y;$$

$$\tau_{xy} = \frac{\partial^2 \phi}{\partial x \partial y} = -2k_2 x - 2k_3 y.$$

In view of the boundary conditions at the surface, k_1 and k_2 must be set zero; k_3 and k_4 can be chosen arbitrarily. We select $k_3 = 0$; $k_4 = -\frac{1}{6}\rho g$ and obtain:

$$\sigma_z = \sigma_y = -\rho g y; \\ \tau_{xy} = 0 \tag{6}$$

These are the stress components for the standard state. The normal stresses are equal in all directions and for all points on a horizontal plane ($y = $ constant) and the shear stress is

zero throughout the body. This stress system is, therefore, equivalent to the hydrostatic pressure in a liquid. Its physical significance for solid bodies is apparent from the following consideration. In a homogeneous solid body of infinite horizontal extent, uniform lateral extension is prevented by lack of space. Therefore, if the weight of the body is the only source of stress, the strain components (elongations) in the horizontal directions are zero. From this the relation

$$\sigma_{hor} = \sigma_{vert} \frac{\nu}{1 - \nu}$$

is derived, where σ_{hor} and σ_{vert} represent the horizontal and vertical stress components, respectively, and ν is Poisson's ratio. It is seen that $\sigma_{hor} = \sigma_{vert}$ only if $\nu = 0.5$. Hence, the hydrostatic state is produced by gravity alone only in liquids. In solid bodies the horizontal stress components due to weight are substantially smaller than the vertical component; they are only a third of the latter if Poisson's ratio is taken to be 0.25, which is a good approximation in most cases. The term $6k_4 y$ in the above derivation of the standard state thus includes two effects: roughly a third of it is due to gravity, while the remaining portion represents a superposed horizontal stress, constant in any horizontal plane, but directly proportional to depth.

The important result derived from these considerations is that the effect of gravity can be incorporated into a stress system which is essentially hydrostatic in character. This is a consequence of the fact that a stress system of the form:

$$\sigma_x = ky; \qquad \sigma_y = \tau_{xy} = 0$$

is likewise a valid one—as shown by the above derivation—and thus can be combined with the weight of the body to produce a hydrostatic system. It does not matter whether the complementary horizontal stress ky actually exists in any particular case; the standard state represents an idealization which is probably only seldom realized in nature. Its principal usefulness—aside from offering a convenient "standard of reference—"arises from an application of the already discussed principle of superposition of stresses, which can now be formulated more precisely. Let Φ_1 and Φ_2 be two stress functions, each of which represents a correct solution. Then the sum of the two is

$$\Phi = \Phi_1 + \Phi_2 \qquad (7)$$

also a valid solution. Now let Φ_1 be the standard state which includes the effect of gravity. Since no other body force is involved, equation (7) states that we may superimpose upon it any other valid stress system Φ_2 caused by surface forces alone. Anderson has named such systems "supplementary stresses." Their mathematical treatment is now divorced from the effects of gravity, and the stress components are given by the simpler expressions of equations (1). The physical implication of this procedure is, of course, merely a statement of the fact that the standard state contributes nothing to the shearing stress and has no influence on the configuration of the stress trajectories. Conversely, it forms a major component of the total stress and, therefore, must be taken into account in phenomena depending upon the magnitude of the confining pressure. One of these phenomena is faulting.

FAULTING ASSOCIATED WITH STRESS DISTRIBUTION

While the determination of the stress properties is amenable to exact mathematical analysis, the occurrence of faulting presents an intricate problem for which no complete theory has yet been worked out. It is, therefore, important to discuss its critical aspects and to point out the shortcomings of the present treatment.

The principal object of the fault analysis is two-fold: (1) to know the attitude of the fault surfaces and their variations in the calculated stress systems, and (2) to determine the location of faulting in the stressed medium, i.e., the distribution of the stable and unstable portions.

The first problem is readily solved if we accept a widely held theory on the relation between the fault surfaces and the directions of the principal stresses. It states that fracture occurs along two planes which have a specific angle θ with the direction of the maximum principal pressure. This angle is a material constant and, therefore, variable. It is, however, always less than 45°, and for most types of rock falls between 20° and 40°. A value of 30° appears to form a good overall approximation. In three-dimensional stress systems the direction of the two fault surfaces is parallel to that of the intermediate principal stress, the angles θ being measured in the plane containing the maximum and least principal stresses. In the two-dimensional cases here analyzed, we shall presume that this is the plane containing the variable stress components—that the third dimension is parallel to the intermediate principal stress. Then the attitude of the fault surfaces is obtained by drawing two sets of curves having everywhere a selected constant angle with the direction of maximum principal pressure. This is evidently a simple procedure, once the configuration of the stress trajectories has been completely determined. While the choice of θ is arbitrary within certain limits, a value in the neighborhood of 30° appears most appropriate. Corresponding to the plus and minus sign of θ, the method always yields two sets of potential fault surfaces, which intersect each other at an oblique angle. While there is no theoretical basis for a discrimination between the two systems in a homogeneous medium, geological observations indicate that in most areas only one of them has been utilized, or that it strongly predominates. Nevertheless, examples of the co-existence of both sets within the same structural belt are known for all classes of faults.

The validity of the above theory on the attitude of the fault surfaces has been questioned by some authors (Jeffreys, 1936), but it finds much support both from the results of experiments and from geological observations. It is based on the assumption that fracture in a stressed medium takes place along that plane which offers the least resistance to it. To produce faulting, the available energy represented by the condition of strain has to overcome not only the strength of the material but also the internal friction opposing the motion. The latter is proportional to the normal pressure across the fault surface. The shear stress is always greatest in a direction at 45° to those of the maximum and minimum principal stresses. Now if we rotate a plane from this position towards the axis of maximum principal pressure, both the shear stress—which tends to produce faulting—and the normal pressure—which opposes motion—become smaller. Depending upon the specific properties of the material, there will be one particular angle for which the combined effect of the two opposing factors becomes most favorable for the occurrence of faulting; this is the angle θ above referred to. A detailed mathematical formulation of this principle is found in Anderson (1942, p. 9–10).

The second problem of faulting accorded attention here deals with the determination of the boundaries between the stable and unstable portions of the stress systems. We will first examine the problem solely upon the basis of the statical theory as applied to the original stress distribution.

The basic principle is easily stated: faulting occurs where the shearing stress exceeds the strength of the material. The magnitude of strength depends, first, upon the material, falling in the range of approximately 500–3000 kg/cm^2 under atmospheric pressure for the most common rock types. Second, it varies widely for a given material as a function of other factors, such as the confining pressure, temperature, presence or absence of solutions, and even the time of application of the shearing stress. Only the first of these has been adequately investigated in laboratory experiments. It was generally found that the breaking pressure increases with the confining pressure and that in many cases the rate of increase is nearly linear over a wide pressure range. Since little information is available on the influence of the other factors, the boundary equation for the separation of the stable and unstable portions adopted in the present paper is:

$$\sigma_{min} \leq n\sigma_{max} + \sigma_0 \qquad (8)$$

where:

σ_{min} = maximum compressive stress (compressive stresses being given the negative sign, the maximum compressive stress is, algebraically, the smallest),

σ_{max} = minimum compressive stress,

σ_0 = breaking strength under atmospheric confining pressure,

n = material constant.

As pointed out, while the standard state contributes nothing to the shearing stress in a homogeneous medium, it adds a vital component to the total confining pressure; hence σ_{min} and σ_{max} in (8) represent total principal stress, including that due to the standard state. They are obtained by adding to the components from equation (4) the hydrostatic pressure defined by (6):

$$\left.\begin{aligned} \sigma_{min} &= \tfrac{1}{2}(\sigma_x + \sigma_y) - \tau_{max} - \rho g y \\ \sigma_{max} &= \tfrac{1}{2}(\sigma_x + \sigma_y) + \tau_{max} - \rho g y \end{aligned}\right\} \qquad (9)$$

For the calculation of numerical examples it will be necessary to assign the constants ρg, σ_0 and n specific values. The selections for the present paper are:

ρg = 250 kg/cm^2 per kilometer or 400 kg/cm^2 per mile,

σ_0 = 1000 kg/cm^2,

n = 4.

While in nature each of these quantities is variable, the figures chosen represent the right order of magnitude. The value for n is based mostly on geological evidence, but it is also substantially in agreement with the results of experimental data.

In the practical application, a simple method was used for the solution of equation (8) which is illustrated in Figure 5. The curve σ_{st} represents the standard state, to which the principal supplementary stresses derived from equation (4) are added algebraically (compressive components being negative). Then the curve: $(n\sigma_{max} + \sigma_0)$ is drawn (dashed line). Faulting occurs where the latter is to the left of σ_{min}. The intersections between the two curves mark the boundaries between the areas of stability and those in which the shear stress exceeds the strength. In the specific examples, the boundary lines were determined for not only one, but several, values of the significant constants appearing in the formulae for the external boundary stresses.

In the preceding discussion, the distribution and mode of faulting were examined solely in relation to the original stress system. With the first occurrence of faulting, an entirely new problem arises. The process of fracturing causes an instantaneous and drastic alteration in the pre-existing stress system, accompanied by the introduction of a new internal boundary surface. It is thus evident that any subsequent faulting must take place in accordance with the changed conditions, and the value of the original stress system as a guide to further deformation may rightfully be questioned. This interaction between repeated changes in the stress distribution and the successive occurrences of faulting introduces a complication which presents great difficulties to an adequate mathematical analysis. No attempt is made here to deal with this problem. However, some qualitative arguments are advanced which may shed some light on the degree of control which the original stress distribution still exercises during the more advanced stages of faulting.

A mathematical solution of the alteration of stress due to the occurrence of a fault is given by Anderson for a simple case (1942, p. 144–156); this is the only example known to the writer. The results demonstrate that pronounced changes in both direction and intensity of the stresses occur in the immediate neighborhood of the fault surface, but that they diminish rapidly with distance. This is especially true in a direction at right angle to the fault where the alterations soon become insignificant. Near the central portion of the fracture the shear stress is greatly reduced, which has the effect of preventing further faulting in this vicinity. Conversely, there is a substantial increase in stress at both ends of the fracture, indicating a tendency for lateral extension, once the process has been initiated. While the solution applies to a specific case only, it is probable that the basic results are similar in other, more complicated situations.

In the light of these data we may visualize the following sequence of events. Under the action of external forces, a condition of stress is produced in a block of the earth's crust. As long as the shearing stress is everywhere below the strength of the material, the stress distribution is static and can be determined by means of the discussed methods. The various examples given below are rigorously correct during this stage only. With further intensification of the stresses, a point is reached where the strength is exceeded at some particular place, and faulting will occur there.

The general location of this place is determinable from the initial conditions. How far the fracture extends itself laterally into regions where the original shear stress is still well below the strength is an unsolved problem. Theoretical results suggest the possibility of

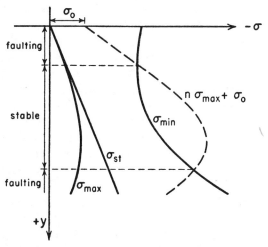

FIGURE 5.—DETERMINATION OF BOUNDARIES OF STABLE PORTIONS OF BLOCK

pronounced extension, whereas geological observations demonstrate that the majority of faults are of rather limited extent. Regardless of this apparent conflict, the principal result will be a drastic diminution of stress in the neighborhood of the origin of the fracture, while farther away the initial stress distribution remains substantially unaltered. The first fault thus provides local relief in the most intensively stressed portion of the block. After its occurrence, the further building-up of the stress system is presumed to go on, and additional portions of the block will gradually reach and exceed the strength limit. Renewed faulting then occurs some distance away from the first fracture, at a place where the stress-relief is of negligible amount. Hence, while no longer exactly the same, the conditions producing the new fracture should still conform closely to the original stresses. A further implication of this process of reasoning is the probability that the later faulting belongs to the same set as the initial one, since a secondary fracture utilizing the alternative complementary direction would lead straight into the protective zone of the preceding one. This, perhaps, is the explanation of the relative scarcity of complementary faults.

In this manner the process may continue until the maximum development of the system has been reached. At this stage the extent of the unstable areas has likewise reached its maximum expansion which, depending upon the ultimate intensities of the causal forces, will include a smaller or greater portion of the entire block. Due to the limited extent of the influence of previous fractures in adjacent areas, it appears probable that the initial faulting in each portion of the unstable segments is primarily controlled by the original stresses. The local relief of stress provided by the existing fractures will then gradually disappear again due to the tendency of the original stress system to re-establish itself. This can lead to renewed faulting in already fractured segments. The process thus becomes more and more complicated, on account of the introduction of a new internal boundary surface along each additional fault and the adjacent zone of pronounced stress alterations. In extreme cases, the stress system may be substantially changed during the final stages of deformation and the further course of events is then beyond the scope of theoretical analysis.

From this line of reasoning it appears justified to assume that the original stress system constitutes the controlling influence on the location and type of faulting at least during the initial stages of deformation in each faulted segment. The author, therefore, believes that illustrations showing the attitude of fault surfaces and the boundaries between stable and unstable segments based purely on the original stress system serve a useful purpose in studies of this nature.

EXAMPLES OF TWO-DIMENSIONAL STRESS SYSTEMS

Introduction

Using the methods outlined in the preceding sections, we will now present a series of examples of two-dimensional stress systems. One or several numerical examples for each group are illustrated graphically in Figures 6–8 and on Plate 1. In each figure, the upper diagram shows the complete solution of the internal stress distribution in the form of stress trajectories and lines of equal maximum shearing stress; the lower diagram shows the attitude of the fault surfaces and the distribution of the stable and unstable portions of the block. All graphs show the respective systems of boundary stresses.

Supplementary Horizontal Stress without Superposed Vertical Stress

In the first group to be analyzed, we assume the presence of a supplementary horizontal stress but absence of an associated vertical stress component, i.e., there is no pressure or tension across any horizontal plane in the body in addition to the normal hydrostatic component. Mathematically this is expressed by:

$$\sigma_y = \frac{\partial^2 \Phi}{\partial x^2} = 0 \quad \text{for all values of } y.$$

Integrating we obtain the stress function:

$$\Phi = cf(y)x + ax + bf_2(y) + d.$$

To satisfy equation (2)

$$\frac{\partial^4 \Phi}{\partial x^4} + 2\frac{\partial^4 \Phi}{\partial x^2 \partial y^2} + \frac{\partial^4 \Phi}{\partial y^4} = cxf_1^{\text{IV}}(y) + bf_2^{\text{IV}}(y) = 0$$

the fourth order derivatives of f_1 and f_2 must be zero. Hence the second order derivatives may be either linear functions of y, constants, or zero. The stress components then are (equations (1))

$$\left.\begin{array}{l} \sigma_x = \dfrac{\partial^2 \Phi}{\partial y^2} = cf_1^{\text{II}}(y)x + bf_2^{\text{II}}(y); \\[2mm] \sigma_y = \dfrac{\partial^2 \Phi}{\partial x^2} = 0; \\[2mm] \tau_{xy} = -\dfrac{\partial^2 \Phi}{\partial x \partial y} = -cf_1^{\text{I}}(y). \end{array}\right\} \quad (10)$$

The boundary conditions at the surface require that $f_1^{\text{I}}(y) = 0$ for $y = 0$. Keeping within the limits of the above restrictions we can set up the following three subgroups:

a) $f_1^{\text{I}}(y) = 0;$ $f_2^{\text{II}}(y) = y + d$

$$\left.\begin{array}{l} \sigma_x = by + d; \\ \sigma_y = 0; \\ \tau_{xy} = 0. \end{array}\right\} \quad (10a)$$

b) $f_1^{\text{I}}(y) = y;$ $f_2^{\text{II}}(y) = 0$

$$\left.\begin{array}{l} \sigma_x = cx; \\ \sigma_y = 0; \\ \tau_{xy} = -cy. \end{array}\right\} \quad (10b)$$

c) $f_1^{\text{I}}(y) = \tfrac{1}{2}y^2;$ $f_2^{\text{II}}(y) = 0$

$$\left.\begin{array}{l} \sigma_x = cxy; \\ \sigma_y = 0; \\ \tau_{xy} = -\dfrac{c}{2}y^2. \end{array}\right\} \quad (10c)$$

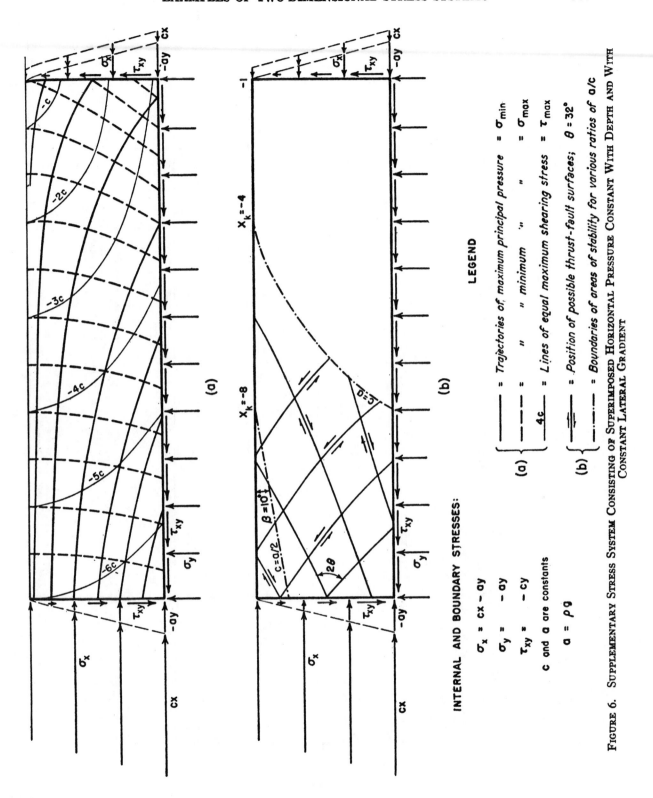

INTERNAL AND BOUNDARY STRESSES:

$\sigma_x = cx - ay$

$\sigma_y = -ay$

$\tau_{xy} = -cy$

c and a are constants

$a = \rho g$

LEGEND

$$
\left.\begin{array}{l}
\text{———} \\
\text{—·—·—} \\
\text{——} \quad 4c
\end{array}\right\} \text{(a)}
$$

= Trajectories of maximum principal pressure = σ_{min}

= "   " minimum "   " = σ_{max}

= Lines of equal maximum shearing stress = τ_{max}

$$
\left.\begin{array}{l}
\text{═══} \\
\text{—··—··—}
\end{array}\right\} \text{(b)}
$$

= Position of possible thrust-fault surfaces; $\theta = 32°$

= Boundaries of areas of stability for various ratios of a/c

FIGURE 6. SUPPLEMENTARY STRESS SYSTEM CONSISTING OF SUPERIMPOSED HORIZONTAL PRESSURE CONSTANT WITH DEPTH AND WITH CONSTANT LATERAL GRADIENT

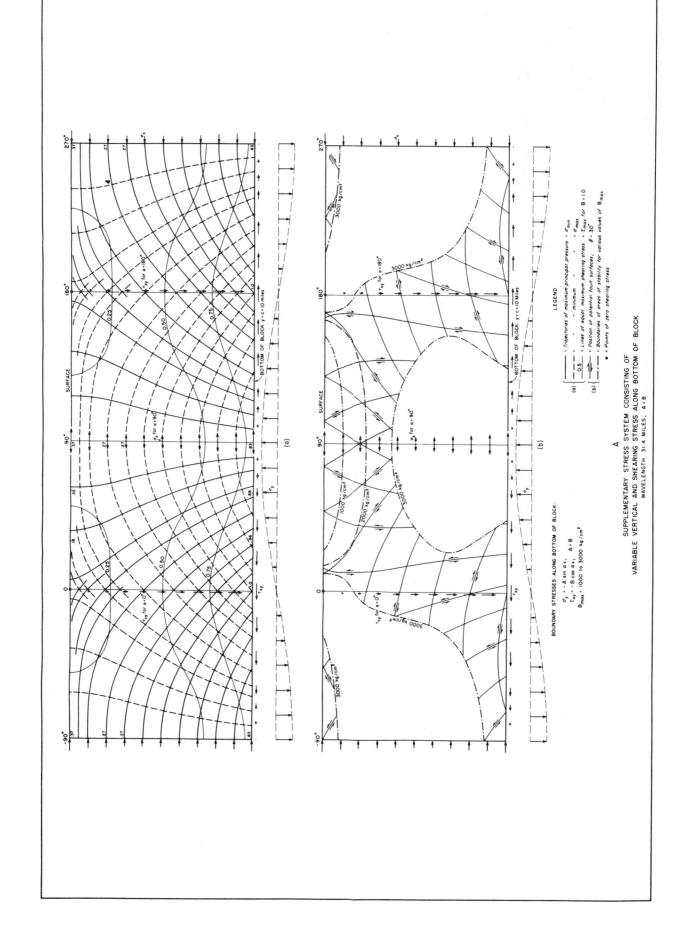

SUPPLEMENTARY STRESS SYSTEM CONSISTING OF
VARIABLE VERTICAL AND SHEARING STRESS ALONG BOTTOM OF BLOCK
WAVELENGTH 31.4 MILES; A = B

A

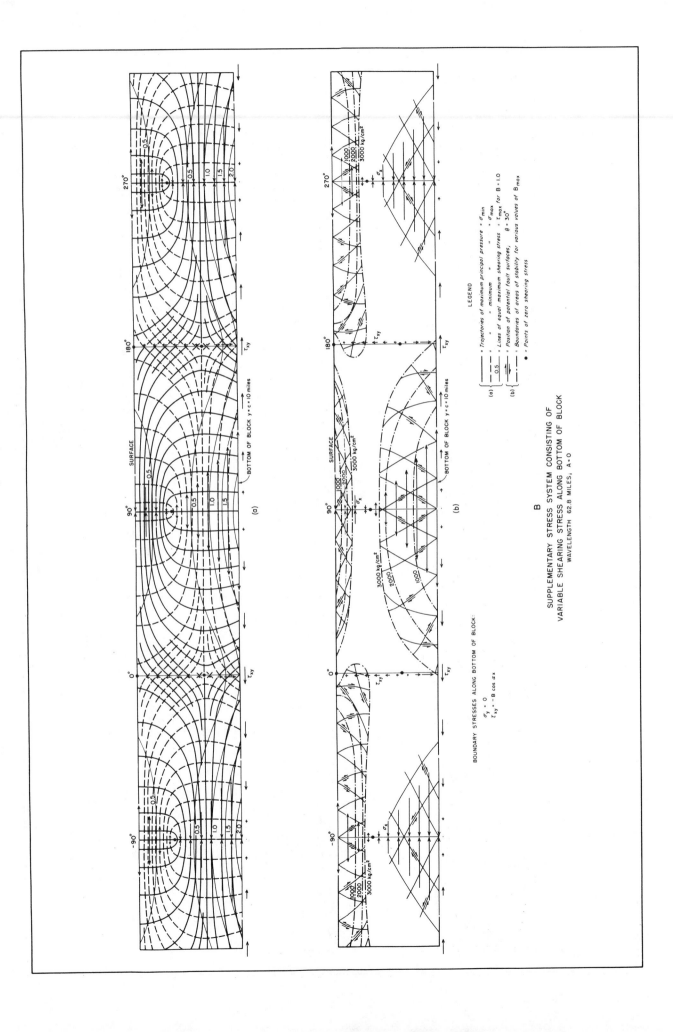

BOUNDARY STRESSES ALONG BOTTOM OF BLOCK:

$\sigma_y = 0$
$\tau_{xy} = -B \cos \alpha x$

LEGEND

(a) $\left\{\begin{array}{l} \underline{\hspace{1cm}} = \text{Trajectories of maximum principal pressure} = \sigma_{min} \\ \text{------} \quad " \quad " \quad \text{minimum} \quad " \quad " = \sigma_{max} \\ \underline{0.5} = \text{Lines of equal maximum shearing stress} = \tau_{max} \text{ for } B = 1.0 \end{array}\right.$

(b) $\left\{\begin{array}{l} \underline{\hspace{0.5cm}\longrightarrow\hspace{0.5cm}} = \text{Position of potential fault surfaces,} \quad \theta = 30° \\ \text{—·—·—} = \text{Boundaries of areas of stability for various values of } B_{max} \\ \bullet = \text{Points of zero shearing stress} \end{array}\right.$

B

SUPPLEMENTARY STRESS SYSTEM CONSISTING OF
VARIABLE SHEARING STRESS ALONG BOTTOM OF BLOCK

WAVELENGTH 62.8 MILES, A = 0

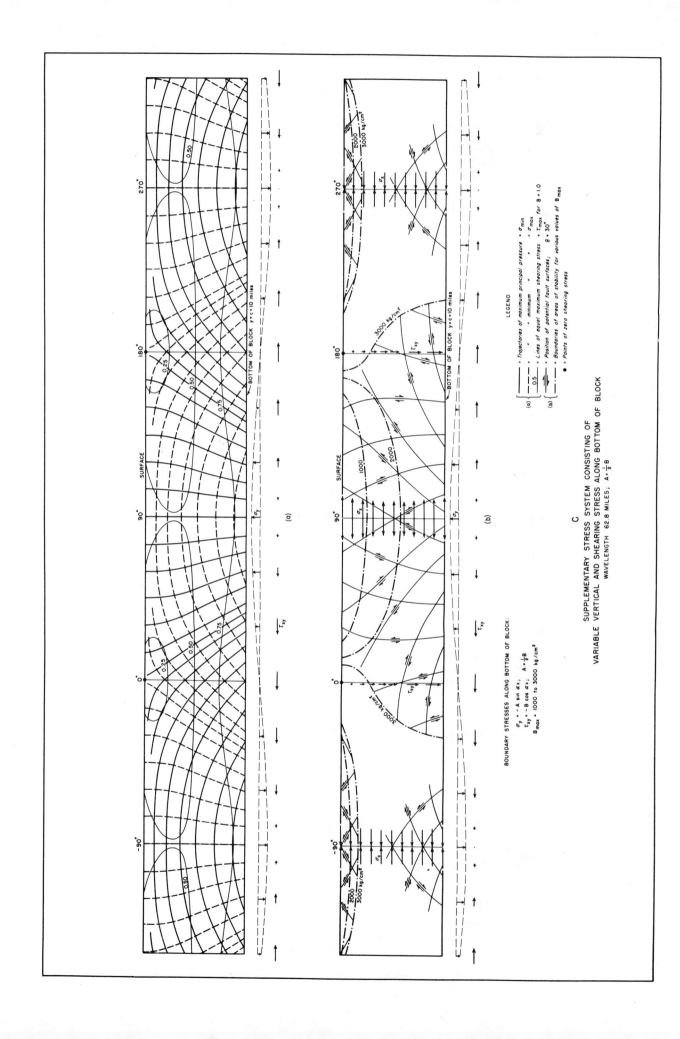

BOUNDARY STRESSES ALONG BOTTOM OF BLOCK:

$\sigma_y = -A \sin \alpha x;$ $A = \frac{1}{2}B$

$\tau_{xy} = -B \cos \alpha x;$

$B_{max} = 1000$ to 3000 kg/cm^2

LEGEND

(a) $\left\{ \begin{array}{l} \text{---} = \text{Trajectories of maximum principal pressure} = \sigma_{min} \\ \text{---} = \text{''} \text{''} \text{''} \text{minimum} \text{''} \text{''} = \sigma_{max} \\ \text{---} 0.5 = \text{Lines of equal maximum shearing stress} = \tau_{max} \text{ for } B = 1.0 \end{array} \right.$

(b) $\left\{ \begin{array}{l} \longleftrightarrow = \text{Position of potential fault surfaces;} \quad \theta = 30° \\ \text{---} = \text{Boundaries of areas of stability for various values of } B_{max} \\ \bullet = \text{Points of zero shearing stress} \end{array} \right.$

SUPPLEMENTARY STRESS SYSTEM CONSISTING OF
VARIABLE VERTICAL AND SHEARING STRESS ALONG BOTTOM OF BLOCK
WAVELENGTH 62.8 MILES; $A = \frac{1}{2} B$

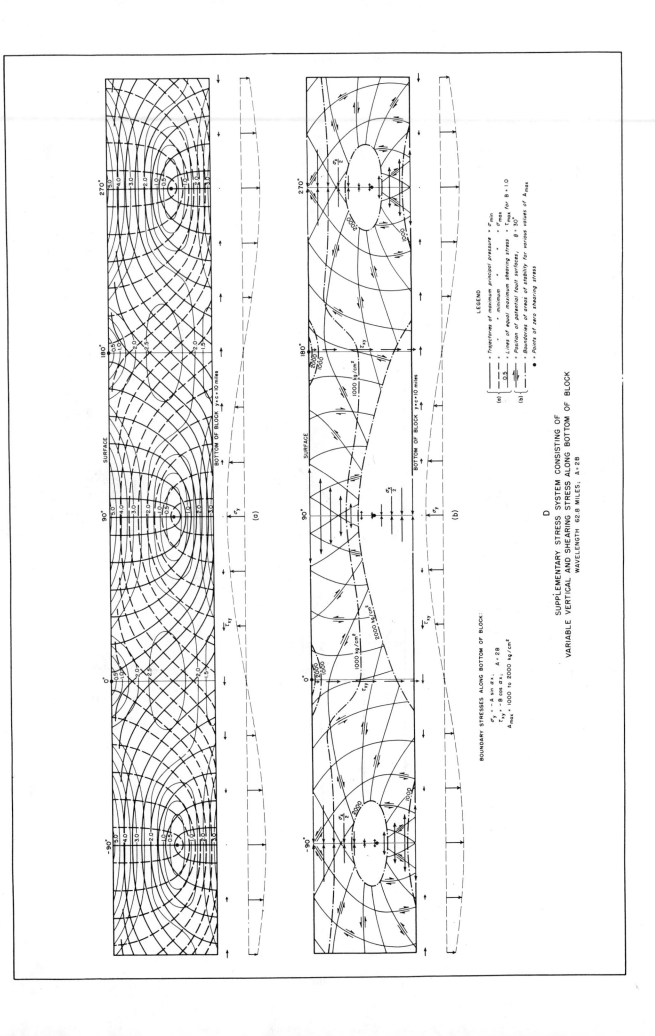

BOUNDARY STRESSES ALONG BOTTOM OF BLOCK:

$\sigma_y = -A \sin \alpha x$, $A = 2B$
$\tau_{xy} = -B \cos \alpha x$,
$A_{max} = 1000$ to 2000 kg/cm²

LEGEND

(a) $\left\{\begin{array}{l}\text{————} = \text{Trajectories of maximum principal pressure} = \sigma_{min} \\ \text{- - - -} \quad " \quad " \quad \text{minimum} \quad " \quad " = \sigma_{max} \\ \text{0.5} = \text{Lines of equal maximum shearing stress} = \tau_{max} \text{ for } B = 1.0\end{array}\right.$

(b) $\left\{\begin{array}{l}\text{———} = \text{Position of potential fault surfaces; } \theta = 30° \\ \text{—·—·—} = \text{Boundaries of areas of stability for various values of } A_{max} \\ \bullet = \text{Points of zero shearing stress}\end{array}\right.$

D

SUPPLEMENTARY STRESS SYSTEM CONSISTING OF
VARIABLE VERTICAL AND SHEARING STRESS ALONG BOTTOM OF BLOCK
WAVELENGTH 62.8 MILES; A = 2B

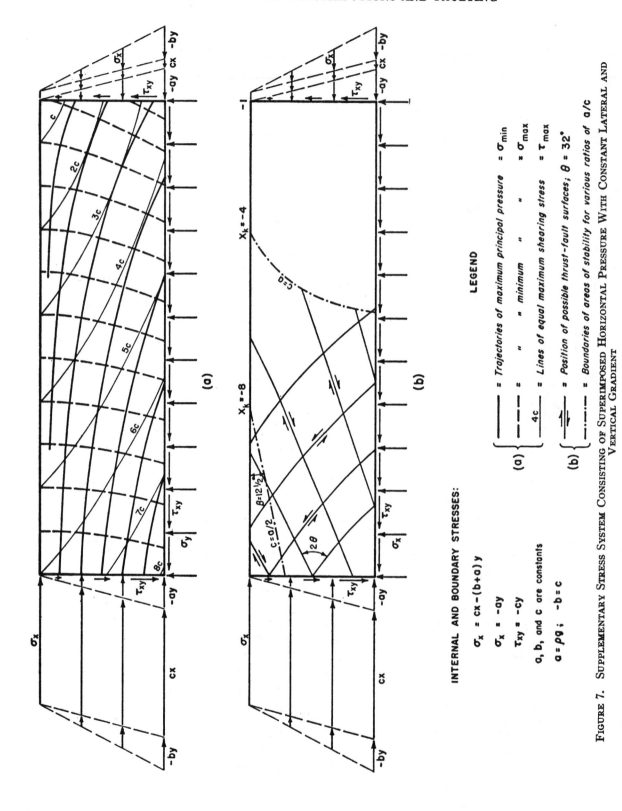

INTERNAL AND BOUNDARY STRESSES:

$\sigma_x = cx - (b+a)y$

$\sigma_x = -ay$

$\tau_{xy} = -cy$

$a, b,$ and c are constants

$a = \rho g; \quad -b = c$

LEGEND

$\left.\begin{array}{l} \rule{0pt}{12pt} \\ \rule{0pt}{12pt} \\ 4c \end{array}\right\}$ (a)
    = Trajectories of maximum principal pressure   = σ_{min}
    =   "   minimum   "   "   = σ_{max}
    = Lines of equal maximum shearing stress   = τ_{max}

$\left.\begin{array}{l} \rule{0pt}{12pt} \\ \rule{0pt}{12pt} \end{array}\right\}$ (b)
    = Position of possible thrust-fault surfaces; $\theta = 32°$
    = Boundaries of areas of stability for various ratios of a/c

FIGURE 7. SUPPLEMENTARY STRESS SYSTEM CONSISTING OF SUPERIMPOSED HORIZONTAL PRESSURE WITH CONSTANT LATERAL AND VERTICAL GRADIENT

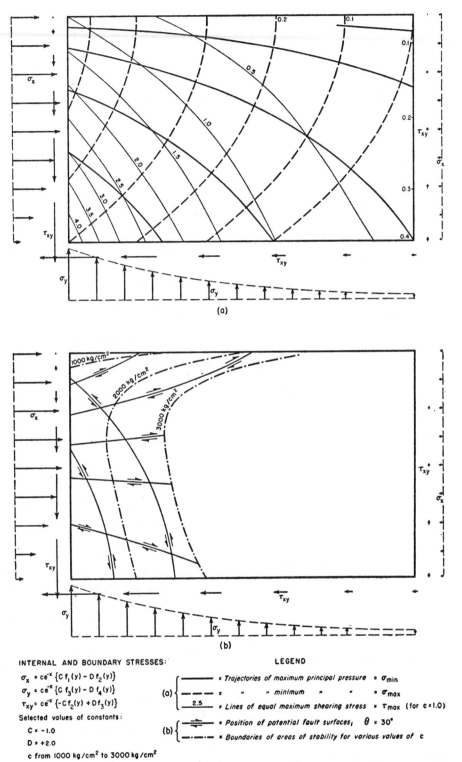

INTERNAL AND BOUNDARY STRESSES:

$\sigma_x = ce^{-x} \{C f_1(y) - D f_2(y)\}$

$\sigma_y = ce^{-x} \{C f_3(y) - D f_4(y)\}$

$\tau_{xy} = ce^{-x} \{-C f_2(y) + D f_3(y)\}$

Selected values of constants:

C = -1.0

D = +2.0

c from 1000 kg/cm² to 3000 kg/cm²

LEGEND

(a) {
— = Trajectories of maximum principal pressure = σ_{min}

—— = " " minimum " " = σ_{max}

—2.5— = Lines of equal maximum shearing stress = τ_{max} (for c = 1.0)
}

(b) {
⇄ = Position of potential fault surfaces; $\theta = 30°$

—·—·— = Boundaries of areas of stability for various values of c
}

FIGURE 8. SUPERPOSED HORIZONTAL STRESS DECREASING EXPONENTIALLY IN HORIZONTAL DIRECTION

In these equations, b, c, and d are arbitrary constants and may be assigned any value, including zero. Furthermore, any linear combination of equations (10a) to (10c) represents another allowable system.

The simplest case of a supplementary stress

system is given by (10a) if we let b equal zero. Then the superposed stress is restricted to a constant horizontal component $\sigma_x = d$, the other two components σ_y and τ_{xy} being zero everywhere. From (3) and (5) it follows that the maximum shearing stress is also constant— equal to half the horizontal stress—and that the stress trajectories are everywhere horizontal and vertical, respectively. The associated fault planes will be inclined at angles of about $\pm 30°$ to the horizontal (thrust faults).

We will next examine the stress systems of the first two sub-groups. The constant d is now selected to be zero since its contribution has already been discussed. The combination of (10a), (10b), and the standard state gives the total stress components:

$$\sigma_x = cx + by - ay,$$
$$\sigma_y = -ay,$$
$$\tau_{xy} = -cy;$$

where $a = \rho g$. The term $-ay$ represents the hydrostatic pressure and is the only one appearing in the vertical stress component. The horizontal component σ_x contains two additional terms, the first a linear function of x only and the second a linear function of y only. Hence σ_x has constant gradients in both the horizontal and vertical directions. The shear stress τ_{xy} is seen to be constant in any horizontal plane; further, it increases vertically at a constant rate which is equal to the lateral gradient of σ_x. The term ky in σ_x appeared previously as a component of the standard state; in the present formulae, it denotes the difference (excess or defect) between the actual magnitude of this stress and that portion of it which is absorbed in the standard state.

Figure 6[1] illustrates the case where the supplementary horizontal pressure is constant with depth ($b = 0$) and Figure 7 where it has a vertical gradient equal to the horizontal one ($b = c$). The external boundary stresses are shown on the end-surfaces and along the bottom of a rectangular block. On the two sides, the horizontal stress σ_x is broken down into its individual components. It is seen that addition of the term by does not change the equilibrium of the block in the lateral direction, a fact which

---

[1] Similar to Figure 25 in the companion paper by M. King Hubbert (1951).

explains that the shearing stress along the bottom is the same in both cases.

The trajectories of maximum principal pressure are curved lines, dipping downward away from the area of maximum compression. The curvature is stronger if the vertical gradient of σ_x is small. Since the trajectories are curved, so are the potential fault surfaces. The latter obviously belong to the class of thrust faults. The set dipping towards the area of maximum pressure is slightly concave upwards, the complementary set concave downwards. Thrust faults of the former type are very common in nature and the theoretically deduced curvature is frequently observed. The latter type appears to occur only rarely and little is known regarding the curvature.

The lines of equal maximum shearing stress are expressed in terms of multiples of the constant c. The shearing stress naturally increases towards the area of greater horizontal pressure but it also increases with depth, the latter effect being more pronounced if σ_x has a vertical gradient. To determine the boundary of the stable portion of the block, the constant c has to be fixed numerically. This is done best by selecting certain values of the ratio c/a, thus expressing the lateral gradient of the superposed horizontal stress in terms of a fraction of the vertical pressure gradient due to weight. In nature this ratio is evidently widely variable with place and time and, in fact, may assume any value from zero up to a magnitude large enough to cause thrusting. The most useful procedure, therefore, consists of assuming several ratios and calculating their respective boundary lines. In both Figures 6 and 7, two such lines are shown corresponding to the ratios $c = a$ and $c = a/2$.

These boundary lines are seen to be dipping towards the area of greater horizontal pressure. The rate of dip is primarily a function of the ratio c/a, being greater the larger this fraction is. It is steep only if the latter is near unity, when the horizontal pressure gradient approaches the magnitude of the vertical pressure gradient due to weight. For smaller values of c, the boundary lines are nearly straight and their inclination diminishes rapidly. Numerical examples on the dip β are: for $c = a/2$, β is about $10°$ in Figure 6 and $12\frac{1}{2}°$ in Figure 7, for c $= \alpha/10$, β is less than 2° in the former case.

These results suggest the following conclusions. If the supplementary horizontal pressure has only a small lateral gradient, say of the order of half the vertical pressure gradient or less, the potential area of thrusting is confined to a shallow, gently dipping wedge. The resulting deformation will consist of a series of slice-thrusts covering a broad belt but extending only to shallow depth. The presence of a vertical gradient slightly steepens the wedge. However, the boundary becomes very steep or nearly vertical only if the horizontal gradient approaches the magnitude of the vertical increase of the hydrostatic pressure. In that event, thrusting can take place throughout a thick, but probably only narrow, zone of the crust. Such a condition appears to have occurred, for instance, in the marginal belts of several ranges in the Rocky Mountain province.

The boundary stresses of the third sub-group (formulae 10c) are also characterized by the fact that the shearing stress is constant in all horizontal planes. Its vertical gradient, however, is now a function of the second order of y, instead of the first order. The shearing stress is balanced by a horizontal pressure, again having constant lateral and vertical gradients but increasing more rapidly in the diagonal direction. This case has not been further analyzed.

The most general expression for the stress systems satisfying the assumption of absence of a vertical stress component is given by the superposition of equations (10a), (10b) and (10c), as follows:

$$\left.\begin{array}{l} \sigma_x = c_1 xy + c_2 x + by + d; \\ \sigma_y = 0; \\ \tau_{xy} = -\dfrac{c_1}{2}y^2 - c_2 y \end{array}\right\} \quad (11)$$

It is seen that the stipulation $\sigma_y = 0$ is associated with two additional general properties of the internal stress system: (1) that the shearing stress is a function of y only, i.e., constant in all horizontal planes, and (2) that σ_x has linear gradients in both the horizontal and vertical directions. That the first two properties are reversible can readily be demon-

strated by deriving the stress system based on the assumption:

$$\tau_{xy} = -\frac{\partial^2 \Phi}{\partial x \partial y} = -k_1 f(y).$$

Using the same direct integration method as before and satisfying equation (2) and the boundary conditions at the surface, this leads again to equations (11). Hence, the reversed statement: "constancy of the shearing stress in all horizontal planes is associated with absence of a vertical supplementary stress throughout the body," is also true. Thus the heading of this Section could equally well read "Supplementary shearing stress constant in all horizontal planes."

Supplementary Horizontal Stress Decreasing Exponentially in the Horizontal Direction

We have investigated the condition of the presence of a supplementary horizontal stress coupled with absence of a superposed vertical component. In this section we again take a supplementary horizontal stress, but drop the assumption that σ_y be zero throughout the block. Instead, we now impose the condition that σ_x decrease exponentially in the horizontal direction.

For this purpose we take the Airy stress function in the form:

$$\Phi = ce^x f(y). \quad (12)$$

Applying equation (2) to this function yields the differential equation

$$f(y) + 2f^{II}(y) + f^{IV}(y) = 0$$

the general solution of which is

$$f(y) = A \sin y + B \cos y + Cy \sin y + Dy \cos y. \quad (13)$$

Substituting (13) in (12) and differentiating we obtain the stress components (1). The compulsory boundary conditions $\sigma_y = \tau_{xy} = 0$ for $y = 0$ can be taken care of by a proper disposition of the two constants A and B. After eliminating these constants, the stress components become:

$$\begin{array}{l} \sigma_x = ce^x\{C(2\cos y - y\sin y) - D(\sin y + y\cos y)\}; \\ \sigma_y = ce^x\{Cy\sin y \qquad\quad - D(\sin y - y\cos y)\}; \\ \tau_{xy} = ce^x\{-C(\sin y + y\cos y) + Dy\sin y\}. \end{array}$$

Let:

$$f_1(y) = 2\cos y - y\sin y; \quad f_2(y) = \sin y + y\cos y;$$
$$f_3(y) = y\sin y; \quad f_4(y) = \sin y - y\cos y;$$

then the stress components can be written in the form:

$$\begin{aligned}
\sigma_x &= ce^x\{Cf_1(y) - Df_2(y)\} = ce^x F_1(y);\\
\sigma_y &= ce^x\{Cf_3(y) - Df_4(y)\} = ce^x F_2(y);\\
\tau_{xy} &= ce^x\{-Cf_2(y) + Df_3(y)\} = ce^x F_3(y).
\end{aligned} \quad (14)$$

The direction of the stress trajectories is given by:

$$\tan 2\beta = \frac{2\,F_3(y)}{F_1(y) - F_2(y)}. \quad (15)$$

The functions F_1 to F_3 are constant in the horizontal direction since they are functions of y only. Therefore, equation (15) shows that all stress trajectories are parallel curves and the same applies to the fault surfaces under the assumption of constancy of angle between the latter and the directions of maximum principal pressure.

When writing equations (14) in the form:

$$\sigma = c\,\epsilon(x, y),$$

it is evident that the constant c has the dimension of stress. Its value can be selected arbitrarily and determines the absolute magnitude of the stress components, whereas the function $\epsilon(x, y)$ determines their areal variations. The two remaining integration constants C and D are dimensionless and, being likewise fully independent, permit the selection of numerous variations. The degree of variability is further enhanced by the fact that the position of the bottom of the block can be chosen at any desired lower limit of the y functions, the unit of length also being arbitrary. The range of opportunity thus provided can easily be exploited by drawing graphs of the four functions f_1 to f_4 and making visual estimates of the combinations obtainable after multiplication with various constants.

Figure 8 illustrates the case where $C = -1$ and $D = +2$. In this and all succeeding illustrations, only the supplementary boundary stresses are shown; the normal hydrostatic components are omitted to simplify the drawings. The shearing stress acting on the bottom of the block is no longer constant but, like the superposed horizontal stress, increases exponentially towards the pressure area. The same is true for the vertical component which now forms a part of the stress system. All three boundary stresses diminish rapidly away from the area of compression. The zone of potential faulting consists of a narrow, nearly vertical, belt and a shallow outwardly protruding wedge near the surface. Figure 8b indicates three boundary lines limiting the area of stability; they correspond to values of the constant c of 1000, 2000, and 3000 kg/cm^2, respectively.

The general fault system in the shallow portion of the unstable segment is similar to that obtained in the preceding stress systems, except for a slight change in attitude. The thrust faults dipping towards the pressure area are less steeply inclined. This divergence increases with depth where, because of a gradual clockwise rotation, the fault surfaces of this set become practically flat and finally even overturned. It appears doubtful, however, that the extreme stage is ever realized, because the sudden decrease of σ_x near the lower left-hand corner of the block does not represent a boundary condition which one would expect in nature. As already stated, we are at liberty to match the bottom of the block with any desired point of the (σ_x, y) curve, for instance, that for which σ_x reaches a maximum. In that case the lateral pressure increases throughout the entire thickness of the block, thus providing a stress system which fully satisfies our intuitive concept of geological probability. Figure 8 demonstrates that even within these limitations the thrust planes reach a practically horizontal attitude (in the vicinity of approximately half of the total depth to which calculations were carried out).

It is thus evident that stress systems of this type produce thrust faults which are only gently inclined at shallow depth and become nearly horizontal at greater depth. Such attitudes are frequently observed in nature and are usually interpreted to indicate subsequent tilting of the fault surfaces. It has been shown here that the gently inclined attitudes may equally well be explained as original features of the stress system.

The preceding two sections were principally concerned with various forms of horizontal compression acting on the two sides of a block. We now focus on stresses acting on the bottom of a block. These may be superposed vertical stresses, shearing stresses, or a combination of both. It is evident that a vertical component which is constant laterally is of no interest for geological purposes. Such a condition may exist within a limited width of the crust, but it then forms only a part of a broader system in which these stresses eventually die out sideways. Hence, only a laterally variable vertical component will be investigated. The most suitable form of lateral variation is that of a sinusoidal curve. This is the most convenient assumption from the mathematical standpoint and in all probability also forms the most satisfactory approximation to many actual cases of differential vertical uplift.

Examples of both constant and variable shearing stresses along the bottom of a block were already encountered in association with superimposed horizontal pressures; however, if variable, they had a strictly unidirectional gradient. This property renders them inapplicable for many important phenomena, such as the drag produced along the bottom of the crust by the action of convection currents in the underlying substratum. The treatment of this problem likewise leads to a sinusoidal form of the primary boundary stresses.

To obtain stress systems of the desired type we select the Airy stress function in the form

$$\Phi = \sin \alpha x \cdot f(y), \qquad (16)$$

where f is a function of y only and

$$\alpha = \frac{2\pi}{l}$$

Application of equation (2) to this stress function yields the following equation for $f(y)$:

$$\alpha^4 f(y) - 2\alpha^2 f''(y) + f^{IV}(y) = 0. \qquad (17)$$

The general solution of this differential equation is (Timoshenko, 1934, p. 45)

$$f(y) = C_1 \cosh \alpha y + C_2 \sinh \alpha y \\ + C_3 y \cosh \alpha y + C_4 y \sinh \alpha y. \qquad (18)$$

The stress components (1) are obtained by inserting this function of y into the stress function (16) and subsequent differentiation. They are:

$$\sigma_z = \sin \alpha x \{ C_1 \alpha^2 \cosh \alpha y + C_2 \alpha^2 \sinh \alpha y \\ + C_3 \alpha (2 \sinh \alpha y + \alpha y \cosh \alpha y) \\ + C_4 \alpha (2 \cosh \alpha y + \alpha y \sinh \alpha y) \}$$

$$\sigma_y = -\alpha^2 \sin \alpha x \{ C_1 \cosh \alpha y + C_2 \sinh \alpha y \\ + C_3 y \cosh \alpha y + C_4 y \sinh \alpha y \}$$

$$\tau_{xy} = -\alpha \cos \alpha x \{ C_1 \alpha \sinh \alpha y + C_2 \alpha \cosh \alpha y \\ + C_3 (\cosh \alpha y + \alpha y \sinh \alpha y) \\ + C_4 (\sinh \alpha y + \alpha y \cosh \alpha y) \}$$

The mandatory boundary conditions at the surface are satisfied if we let:

$$C_1 = 0; \qquad C_2 - \frac{C_3}{\alpha}$$

This imposition reduces the stress components to the form:

$$\sigma_z = \alpha \sin \alpha x \{ C_3 (\sinh \alpha y + \alpha y \cosh \alpha y) \\ + C_4 (2 \cosh \alpha y + \alpha y \sinh \alpha y) \};$$
$$\sigma_y = -\alpha \sin \alpha x \{ -C_3 (\sinh \alpha y - \alpha y \cosh \alpha y) \\ + C_4 \alpha y \sinh \alpha y \}; \qquad (19)$$
$$\tau_{xy} = -\alpha \cos \alpha x \{ C_3 \alpha y \sinh \alpha y \\ + C_4 (\sinh \alpha y + \alpha y \cosh \alpha y) \}.$$

The remaining two integration constants C_3 and C_4 are available for suitable selections of the boundary stresses. From the form of equations (19) the proper choice evidently consists of functions of the type

$$\sigma_y = -A \sin \alpha x; \\ \tau_{xy} = -B \cos \alpha x; \qquad (20)$$

where A and B are arbitrary constants. These boundary stresses are placed along the bottom of a block, the position of which is determined by $y = c$, c being likewise a constant. Substituting functions (20) for the left-hand sides of the corresponding equations (19) and replacing y by c, we obtain two equations from which the as yet unassigned integration constants C_3 and C_4 can be calculated in terms of the remaining constants α, c, A and B. They can then be eliminated from equations (19).

Performing these steps we arrive at the following final expressions for the stress components:

$$\left.\begin{aligned}
\sigma_x &= \sin \alpha x \{-k_1 f_1(y) + k_2 f_2(y)\} = \sin \alpha x F_1(y); \\
\sigma_y &= \sin \alpha x \{-k_1 f_3(y) - k_2 f_4(y)\} = \sin \alpha x F_2(y); \\
\tau_{xy} &= \cos \alpha x \{k_1 f_4(y) - k_2 f_1(y)\} \ = \cos \alpha x F_3(y);
\end{aligned}\right\} (21)$$

where:

$$\left.\begin{aligned}
f_1(y) &= \sinh \alpha y + \alpha y \cosh \alpha y; \\
f_2(y) &= 2 \cosh \alpha y + \alpha y \sinh \alpha y; \\
f_3(y) &= \sinh \alpha y - \alpha y \cosh \alpha y; \\
f_4(y) &= \alpha y \sinh \alpha y;
\end{aligned}\right\} (22)$$

and:

$$\left.\begin{aligned}
k_1 &= \frac{A \alpha c \cosh \alpha c - B \alpha c \sinh \alpha c + A \sinh \alpha c}{\sinh^2 \alpha c - \alpha^2 c^2}; \\
k_2 &= \frac{A \alpha c \sinh \alpha c - B \alpha c \cosh \alpha c + B \sinh \alpha c}{\sinh^2 \alpha c - \alpha^2 c^2}.
\end{aligned}\right\} (23)$$

Equations (21) represent the general solution for the internal stress components produced by sinusoidal boundary stresses acting on the bottom of a block.

It can easily be verified that equations (21) satisfy the imposed boundary conditions, both the mandatory ones at the surface, as well as the selective ones at the bottom. All three stress components are seen to vary in the horizontal direction according to sinusoidal functions, and in the vertical direction according to the more complicated hyperbolic functions f_1 to f_4 (equations 22). The latter are multiplied by two constant factors, k_1 and k_1 which, in turn, are expressions of four other constants, α, c, A, and B (equation 23). Since α is defined by $\frac{2\pi}{l}$, the terms αc, αy, and functions thereof, are dimensionless, whereas the constants A and B and, therefore, k_1 and k_2, have the dimensions of stress (m, l^{-1}, t^{-2}). The latter result is, of course, also directly obtainable from equations (20). The significance of the four constants is:

$c = y_c$ is the depth of the bottom of the block along which the boundary stresses are acting;

$l = \dfrac{2\pi}{\alpha}$ is the distance, in the horizontal direction, of a full wave length of the sinusoidal variations of the stress components;

A and B represent the maximum values of the supplementary vertical and shear components at the bottom.

All four constants are mutually independent and may be assigned any desired value. For instance, any combination of thickness and lateral extent (wave length) of the block is permissible. The same applies to the ratio of the amplitudes of the two supplementary boundary stresses. Letting A equal zero yields a case of pure drag along the bottom, whereas B equal zero corresponds to differential vertical uplift alone. Though it is questionable that these latter conditions exist in nature, or are even closely approached, it is of interest to examine the general behavior of these two extreme cases. On the whole, it is seen that equations (21) provide the solution for a substantial variety of situations.

The form of the stress trajectories is given by:

$$\tan 2\beta = \frac{1}{\tan \alpha x} \frac{2F_3(y)}{F_1(y) - F_2(y)}. \qquad (24)$$

This equation shows that, for values of x denoting uneven multiples of a quarter wave length ($x = \dfrac{l}{4}$, $\dfrac{3l}{4}$ etc.), the two principal directions are vertical and horizontal, respectively, and for values of x denoting even multiples of a quarter wave length ($x = 0$, $\dfrac{2l}{4}$ etc.), they are both inclined at 45°. For all other values of x, the directions of the trajectories are complicated functions of y. In the horizontal direction they are constant for those specific depths for which either F_3 or $F_1 - F_2$ are zero. The former case always includes the surface of the block, where the trajectories are horizontal and vertical; the latter, where the trajectories are at 45°, is only occasionally encountered. Of special importance in the construction of the stress diagrams are those points for which the right-hand side of equation (24) is indeterminate (for instance, because of the condition of zero over zero). These are singular points for which the directions of the principal stresses are undefined. They occur at places where the shearing stress is zero. If located internally they are associated with a 90-degree turn of the stress trajectories when passing the point in a vertical or horizontal direction. Owing to the complexity of the expressions for the stress components, the complete solution of equation (24) and the general equations (3) and (4), is best carried out by numerical and graphical methods.

A series of specific examples is shown on Plate 1. The boundary stresses along the bottom of the block are clearly illustrated in each case. In all instances, a full wave length of the sinusoidal stress variations is shown, but the lateral extent of the block is left undetermined. It should be remembered that we are at liberty to terminate the blocks at any desired place by the simple expedient of applying the appropriate boundary stresses. We might be interested, for instance, in a full wave length of the diagram, or in any one of several significant portions, such as the positive or the negative zone of the supplementary vertical stress. It will be seen later, however, that in some systems the associated horizontal stress component σ_x varies in the vertical direction in a manner which probably excludes it from consideration as a primary boundary stress. Its role should then be restricted to that of a purely internal stress. In such cases, the logical choice of the lateral limits of the block is restricted to the 0- and 180-degree positions (Pl. 1, B and D). To facilitate this aspect of the analysis, the horizontal and lateral shearing stress components have been added to the diagrams at the most important positions. From equations (21), it is seen that σ_x is zero for all points $x = 0$ or $l/2$ but reaches maximum values (positive or negative) for all points $x = l/4$, $3l/4$ etc. The relations are similar for the shear stress components, except for a phase change of one-quarter of a wave length.

The boundary lines segregating areas of stability from those subject to faulting were again calculated for several values of the maximum supplementary boundary stresses, the usual selections being 1000, 2000, and 3000 kg/cm^2. In this connection, a remark concerning the symmetry of the diagrams is in order. Equations (21) show that all quantities which depend solely upon the supplementary stress components have the full degree of symmetry characteristic of the trigonometric functions. These include the stress trajectories, the lines of equal maximum shearing stress and the attitudes of the potential fault surfaces. Conversely, the boundary curves of the stable areas do not display this symmetry between the

positive and negative halves of a full wave length. The reason can be seen from a study of Figure 5. For any two points half a wave length apart, σ_{min} and σ_{max} change sign, whereas the line representing the standard state always remains negative. Hence, there is a switch in position between the two principal stresses with respect to the line representing the hydrostatic pressure. This causes a completely different distribution between the stable and unstable segments in the positive and negative half wave length. The only exception is for the points $x = 0$ and $l/2$, for which the interchange is of no consequence due to the fact that $\sigma_{max} = -\sigma_{min}$. This lack of symmetry of the boundary lines represents a striking feature of the diagrams. It is another consequence of the fact that the occurrence of faulting is a function not only of the maximum shearing stress but also of the total confining pressure. Gravity once more exerts a decided influence upon the results.

Plate 1, A represents an example of a relatively narrow wave length. The selected constants are: $l = 31.4$ miles, $c = 10$ miles, and $A = B$. The wave length here chosen roughly equals the distance between adjacent mountain ranges in portions of the Basin-and-Range province, notably Nevada. This statement is made to facilitate a mental association of the selected width with certain well-known structures. Although the diagram covers the two adjacent negative quadrants, we shall limit our attention to the positive half wave length. The supplementary stress system is composed of a vertical pressure gradually decreasing from a maximum value at the center ($\alpha = 90°$) to zero at both ends ($\alpha = 0$ and $180°$), and of a bottom shear stress increasing from the center towards both ends. The two primary stresses set up a horizontal tensile component σ_x which is nearly constant with depth and has the same lateral distribution as σ_y. Only a downward-directed shearing stress acts on the two sides of the block. The necessity of its presence is immediately realized from a consideration of the equilibrium requirement in the vertical direction; it counteracts the upward push at the bottom due to the superposed vertical pressure.

The stress trajectories are vertical and horizontal in the central portion of the block, but gradually turn into a 45-degree position towards both ends, except within a thin zone immediately below the surface. This rotation is associated with corresponding changes in the attitude of the fault surfaces. In the central upper portion of the block, normal faults occur, having the characteristic dips for this class of faults. Away from the center, the dip of one of the two sets gradually increases, finally becoming vertical, or even slightly overturned, near the ends. The complementary set decreases in dip, changing from a normal 60° at the center to as low as 15° in the deeper end portions. These changes in attitude do not apply to the surface, where the directions of the stress trajectories remain unaltered, because of the required boundary conditions. There is, however, a rapid transition with depth from this normal position along the surface into the significant changes in orientation below.

During the gradual build-up of a stress system of this type (lower diagram of Pl. 1, A), instability with resultant faulting is first reached in a broad, but relatively thin, layer below the surface. With continued intensification of the primary stresses, this zone expands with depth and, to a lesser degree, sideways (at the surface it never reaches the two endpoints, since they represent singular points of zero shear stress). When their maximum value has attained a magnitude of the order of 3,000 kg/cm^2, it has spread out to embrace nearly the full depth of the outer portions of the block, but it does not yet cover the deeper central portion where the maximum shear stress is still below the strength. The process of this gradual expansion of instability is illustrated graphically by the three boundary lines corresponding to values of 1000, 2000 and 3000 kg/cm^2, respectively, of the maximum value of the uplifting force at the bottom.

The changes in attitude of the normal faults associated with this stress system are of considerable interest. Faults which are steeply dipping, vertical, or even slightly overturned, are frequently observed in areas where the dominating tectonic force appears to have been a differential vertical uplift. Numerous examples of this type have been described, for instance, from the Basin-and-Range province and from those portions of the Rocky Mountains in which vertical deformation predominates. If the dip is less than 90°, such faults are still assigned to the class of normal faults; if overturned, they are usually designated as reverse faults. Anderson has emphasized that faults of this nature belong in many instances to an entirely different class, the transcurrent or lateral class. This is often, but by no means always, true. An alternative explanation, applicable to those cases where lateral displacement is lacking, is provided by the present analysis. It shows that steeply dipping normal faults and reverse faults can be the direct product of an original stress system such as that here analyzed. They occur on the flanks of the uplift, beginning at a shallow depth. To be observable at the present surface, we merely have to assume that the uppermost layer has since been removed by erosion. The rather wide occurrence of such faults is not unexpected in view of the probability that the causal stress system is frequently realized in nature. However, this system is essentially of local character, which explains the fact that near-vertical normal faults and reverse faults form no part of Anderson's classification, based exclusively upon regional stress systems.

The complementary set of low-angle normal faults also deserves some attention. Faults of this type appear to be rare and are seldom mentioned in the literature. An outstanding case, however, has been described by C. R. Longwell from the Desert and Sheep ranges in southern Nevada (Longwell, 1945). Here Longwell found the presence of an extensive system of low-angle normal faults, all located on the flanks of a major anticline and dipping towards its axis. The range of the dips is mainly from 10° to 25°, and the form of the fault surfaces is further characterized by a slight upward concavity. This combination of facts agrees remarkably well with the general behavior of the low-angle normal faults emerging at the surface inside the positive quadrants. The corresponding dips in Plate 1, A are not quite as low as those described by Longwell from southern Nevada, but it can logically be explained by assuming a further flattening due to a subsequent rise of the central portion of the uplift.

Another characteristic feature shown on the diagram is the reversal in curvature of the low-dipping normal faults with depth. Its verification in nature would be contingent upon finding an example similar to the Desert-Sheep ranges, but exposed at a lower level. No clear-cut example of a combination of the near-vertical and low-angle, normal faults is known to the author, probably because of the protective influence of preceding fractures. The question of why very steep normal faults occur much more frequently than flat ones is less easily answered.

Plate 1, B–D presents a series of examples with a broader wave length. While the thickness of the block is the same as in the preceding case (c = 10 miles), the wave length is now twice as large, 62.8 miles. This distance is more comparable to that between adjacent ranges in certain portions of the Rocky Mountains. The difference between the three cases is in the ratios of the superposed vertical and shear-stress components. They are: $A = 0$ in Plate 1, B, $A = B/2$ in Plate 1, C and $A = 2B$ in Plate 1, D.

The results in the second case (Pl. 1, C) are quite similar to those in the preceding example. Evidently, the doubling of the wave length is roughly compensated by a reduction of the ratio A/B by half. The only notable difference lies in the relative proportions of the stable and the unstable segments. It is seen that for equal values of B_{max}, the depth penetration of the faulted portions is much greater in Plate 1, C than in Plate 1, A. For $B_{max} = 3,000 \ kg/cm^2$, it now embraces the entire positive half of the wave length, excepting two small areas in the immediate vicinity of the corners of zero shearing stress.

The negative half of the wave length illustrates the conditions due to the presence of a variable vertical tensile stress acting on the bottom of the block, in combination with a shear stress component of equal size but opposite directions as compared with the positive quadrants. There is, of course, no absolute tension at this depth, the supplementary tensile stress being always smaller than the standard state. It represents a partial relief of the pressure due to the weight of the overburden. The boundary stresses at the bottom set up a hori-

zontal compression which may produce thrust faults within a thin layer below the surface.

Major changes in the character of the stress system accompany alterations of the above ratio A/B in either sense. Although it is doubtful whether these cases still fall within the domain of geological reality, they demonstrate the nature of the changes in both directions. Plate 1, B represents the extreme case where the supplementary stress applied to the bottom of the block consists of a sinusoidal shear stress only, A being zero. Now the internal horizontal stress component is no longer approximately constant with depth, nor even of uniform sign. At the 90-degree position σ_x changes from a maximum compressive stress at the surface to a maximum tensile stress at the bottom; at 270°, the signs are reversed. At some particular depth along these positions, singular points of zero shearing stress occur, accompanied by a 90-degree turn in the direction of the stress trajectories. A second pair of singular points is found at the 0 and 180-degree positions. The latter points are further characterized by the fact that the horizontal plane containing them is, like the surface, a principal plane where all trajectories are vertical or horizontal. The 90-degree switch here takes place in the horizontal direction.

A peculiar and important feature of a stress system of this type is that it produces two layers of faulting, separated by an intervening stable zone. Corresponding to the described variations in σ_x in the first half wave length, there is a zone of normal faulting in the lower part of the block, underlying a thin zone of thrusting in the uppermost portion. In the second half wave length the distribution is reversed, but thrusting in the lower portion does not occur unless the boundary stress at the bottom reaches very high values, exceeding 3,000 kg/cm^2 (although these areas are essentially stable, the attitude of a few potential fault surfaces has been indicated in the diagram). Conversely, at the 90-degree position, the strength limit at the bottom is reached for comparatively small values of B_{max}, due to the large tensile stress component.

The effect of a change in ratio in the other direction is illustrated in Plate 1, D. Here the ratio A/B is equal 2, the vertical stress component now being the larger one. Although the shear stress component is still of appreciable magnitude, the resulting effects are equally pronounced as in the preceding case. They are also quite similar, though of opposite sense. The horizontal component σ_x again has a large vertical gradient, accompanied by a change in sign and by the occurrence of singular points. (They are absent, however, at the 0 and 180-degree positions.) The distribution of faulting is the reverse of that encountered in Plate 1, B. The upper portion of the first half wave length is a zone of normal faulting, whereas the lower portion is stable up to very large magnitudes of the boundary stresses. The second half wave length is roughly divided into three parts, namely, an upper layer of thrusting, an intermediate stable one, and a lower one of normal faulting. One important difference between the two cases—aside from their opposite character—lies in the relative proportions of the stable areas. This is especially marked in the vicinity of the 0- and 180-degree positions, where almost the entire thickness of the block is stable in Plate 1, B and unstable in Plate 1, D. In the latter case, the faulting here encountered is quite similar to that previously found in Plate 1, C, one set consisting of nearly vertical normal or reverse faults and the complementary set of gently dipping normal faults.

The stress system illustrated in Plate 1, D is obviously due to tectonic forces causing differential uplifts and depressions in adjacent belts of the crust. The former correspond to the positive, the latter to the negative half wave lengths of the diagrams. The primary vertical stresses applied at the bottom—representing an uplifting force in the positive belts and a partial relief of pressure in the negative zones—are probably accompanied by shear stresses of smaller but still appreciable magnitude. Possibly the ratio selected in Plate 1, D is already weighted too strongly in favor of the vertical component to be geologically significant. Diagrams B and D are nonetheless of interest due to the result that a departure in either direction from a certain "balanced" ratio produces stress systems in which a zone of thrusting may occur above a zone of normal faulting and vice versa, the two faulted layers being separated by a stable zone.

CONVECTION CURRENTS

Another important usefulness of the equations derived in the preceding section (equations 20 to 23) consists of their application to convection currents. Several authors (Griggs, 1939; Vening Meinesz and others, 1948), have demonstrated the likelihood of a periodic existence of large-scale convection currents in the weak substratum underlying the crust of the earth. Griggs shows that of all the primary forces so far suggested as the basic cause for mountain building, only two are potentially of sufficient magnitude to provide acceptable working hypotheses. These are (1) lateral compression due to thermal contraction, and (2) the viscous drag exerted on the bottom of the crust by convection currents in the substratum. He rejects the first one on account of other difficulties, but demonstrates the adequacy of the second on the basis of theoretical considerations, experimental data, and the results obtained from an ingenious scale model with dimensionally correct physical properties. In view of the results of these and related investigations, the effect of convection currents on the deformation of the crust has become of considerable geological interest.

The shearing stress exerted on the bottom of the crust by the viscous flow in the substratum is evidently best expressed in the form of a sinusoidal curve, the half wave length of which represents the width of the convection cell. Hence, it conforms to the second of the two primary boundary stresses analyzed in the preceding section (equations 20). Furthermore, it appears logical to assume that the region of the crust above the rising current endures a supplementary upward pressure, and the portion above the sinking column experiences a downward drag causing a reduction in the hydrostatic pressure. Therefore, the first of the two boundary stresses of equations (20) is probably also effective. This relationship between the location of the convection cells and the boundary stresses acting on the bottom of the crust is schematically illustrated in Figure 9. On the other hand, the numerical value of the ratio of the two stresses A/B is not safely determinable from qualitative considerations. Whatever it may be, equations (20) to (23) provide the basic solution for the determination of the stress distribution in the crust due to the action of convection currents in the substratum.

The selection of the critical constants is as yet highly speculative. In a recent paper on convection currents, Vening Meinesz (1948, p. 200) discusses an apparent example from the

pressure is small, say of the order of half, or less, of the vertical pressure gradient due to weight, then the potential zone of faulting is limited to a thin wedge below the surface in which numerous shallow slice thrusts are expected to occur. If the gradient is sufficiently

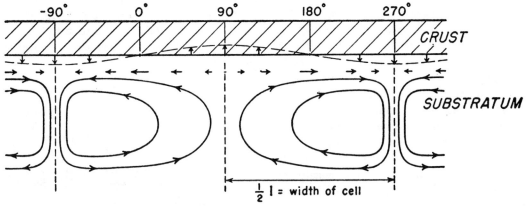

FIGURE 9. CONVECTION CURRENTS AND BOUNDARY STRESSES AT BOTTOM OF CRUST

Indonesian Archipelago and assumes the following constants: depth of crust $c = 30$ kilometers, distance between rising and sinking currents $l/2 = 340$ kilometers, and maximum shear stress at the bottom of the crust $B = 3,320$ kg/cm^2. This case, making the further assumption $A = 0$, has been calculated and was found to yield a stress system almost identical to that shown in Plate 1, B. It is not reproduced here since the only essential difference between the two pictures consists of a reduction in the ratio of thickness over wave length by the order of about three to one. A further reason for this omission lies in the uncertainty regarding the proper ratio A/B. The presence of an associated vertical stress component being surmised, the correct solution may correspond more probably to a flattened-out version of Plate 1, C or to another closely related system.

CONCLUSIONS

A superposed horizontal compression with constant lateral and vertical gradients is always accompanied by a shear-stress component being constant in all horizontal planes. The associated thrust-fault surfaces are slightly curved, concave upwards for the set dipping towards the region of increased compression. If the lateral gradient of the superposed horizontal

large (approximating the vertical gradient of the hydrostatic pressure), thrust faults may penetrate to great depths, but may then be confined to a comparatively narrow zone. The same is true if the superposed horizontal compression has an exponential—instead of a linear—rate of decrease. It is then associated with vertical and shear stress components likewise decreasing exponentially in the horizontal direction. The inclination of the resulting thrust-fault surfaces diminishes rapidly with depth. Hence, a gently dipping, or even horizontal, attitude may not necessarily imply a subsequent rotation but can also result directly from a local stress system of this type.

A laterally variable vertical boundary stress acting on the bottom of a block is most likely associated with a corresponding shear-stress component, the latter being 90° out of phase if the vertical component is assumed in the form of a sinusoidal curve. For certain "balanced" ratios of the two components, this combination causes a supplementary horizontal tensile stress in the belt subjected to the increased upward pressure and a corresponding compressive stress in the adjacent depressed belts. The boundary zones between these belts are characterized by absence of a superposed horizontal pressure and the occurrence of maximum shear-stress components. This causes a

45-degree rotation of the stress trajectories, which, in turn, leads to a pronounced change in the attitude of the fault surfaces. The standard normal faulting occurs over the central uplifted portion, but towards both flanks one of the two sets gradually changes to very steep, vertical, and finally overturned positions, whereas the complementary, but rarely realized, set of normal faults assumes unusually low inclinations. Thus the nearly vertical normal faults, so frequently encountered in tectonic provinces characterized by differential vertical uplift, are explained as a primary feature of such stress systems. The anomalous attitudes are caused by the described rotation of the stress trajectories. They cannot be ascribed to variations in the ratio of supplementary horizontal and vertical pressures, as is occasionally expressed in the literature, based on a false application of the methods of vector analysis. On the other hand, as Anderson has pointed out, vertical and reverse faults are in numerous instances not variations of normal faults but belong to the dynamically entirely different class of transcurrent faults.

A pronounced departure of the ratio of the two primary stress components in either direction from the discussed balanced condition produces theoretically interesting cases which may, or may not, be realized in nature. One of the principal features of these systems is the co-existence of two diverse zones of faulting at different levels in the crust, one situated above the other but separated from it by an intervening zone of stability.

Finally, the equations derived provide the basic solution for the stress distribution in the crust caused by convection currents in the substratum. A detailed application to this phenomenon must await a more secure knowledge of the pertinent constants.

REFERENCES CITED

Anderson, E. M. (1942) *The dynamics of faulting,* Oliver and Boyd, London, 183 pages.

Griggs, David (1939) *A theory of mountain building,* Am. Jour. Sci., vol. 237, p. 611–650.

Hubbert, M. King (1951) *Mechanical basis for certain familiar geologic structures,* Geol. Soc. Am., Bull., vol. 62, p. 355

Jeffreys, Dr. H. (1936) *Note on fracture,* Royal Soc. Edinburgh, Pr., vol. 56, pt. 2, p. 158–163.

Longwell, C. R. (1945) *Low-angle normal faults in the Basin and Range Province,* Am. Geophys. Union, Tr., vol. 26, pt. 1, p. 107–118.

Timoshenko, S. (1934) *Theory of elasticity,* Eng. Soc., Mon., 403 pages.

Vening Meinesz, F. A. (1948) *Major tectonic phenomena and the hypothesis of convection currents in the Earth,* Geol. Soc. London, Quart. Jour., vol. 103, pt. 3, p. 191–207.

SHELL OIL CO., SHELL BUILDING, HOUSTON, TEXAS
MANUSCRIPT RECEIVED BY THE SECRETARY OF THE SOCIETY, SEPTEMBER 23, 1949

FOLDING

Reproduced by permission of the Geological Society from Thoughts on the tectonics of folded belts by A. W. Bally, in *Thrust and Nappe Tectonics*, Geological Society Special Publication 9, 1981, p. 13-32.

Thoughts on the tectonics of folded belts

A. W. Bally

SUMMARY: Balanced cross sections and their palinspastic reconstruction in structurally simple external zones of folded belts suggest that a limited amount of continental lithosphere was subducted (Ampferer or A-subduction). This process appears to be associated with preceding or synchronous basement remobilization. Therefore, the rigid or ductile nature and the timing of basement deformation remain as some of the most important orogenic problems. Decoupling and ductility contrasts within the lower continental crust and within the overlying sedimentary sequences are responsible for varying structural styles in mountain ranges.

Gravity gliding as an important factor for mountain building is examined in some detail. Soft sediment gravity tectonics on passive continental margins are dominated by listric normal growth faults. This style contrasts with observed styles of deformation in folded belts. Gravity tectonics induced by stretching of the underlying basement area is commonly observed during the rifting phase of passive continental margins and in episutural basins associated with orogenic systems.

Three opposing schools of thought are today proposing their images of mountain ranges:

The *fixists* (Beloussov 1962, 1975, 1977) visualize mountain building as the product of asthenospheric and related lithospheric diapirism. In their view, widespread thrusting and folding in mountain ranges is explained in terms of either gravity gliding or else gravity spreading (as defined by Price 1971, 1973).

Adherents of an *expanding earth* (Carey 1975, 1977) view mountain building as the fixists do, but essentially as an ensialic process; they differ from fixists because they see the origin of the oceans not by a process of oceanization but instead by accretion and spreading processes along mid-ocean ridges that record the vicissitudes of an expanding earth.

Finally, *plate tectonic* devotees see mountains as the product of subduction processes on converging plate boundaries; overthrust and folded belts are, in essence, of compressional origin and represent excess sediments and slices of the underlying crystalline crust that have been scraped off and decoupled from subducting lithospheric slabs. A large number of observations in mountain ranges can be well fitted into a plate tectonic frame of reference. However, it is only fair to state that despite the brilliant early intuitions of Ampferer (1906), Argand (1924), Staub (1928), and other alpine geologists, plate tectonics are not easily directly deduced from observations that are limited only to mountain ranges. The plate tectonics hypothesis remains anchored mainly in geophysical and marine geological observations.

Deep-sea sediments and possible remnants of oceanic floor (ophiolites) occupy, areally, only minor portions of folded belts. Consequently, a plate tectonic origin of mountain ranges is not all that obvious to an unbiased observer. Much of what we see today in mountain ranges suggests widespread mobilization and 'ductilization' of an earlier rigid sialic lithosphere. In other words, while the continental lithosphere of cratons remains rigid, the continental lithosphere of mountain ranges shows pervasive remobilization during orogenic processes, suggesting repeated lithospheric 'softening'.

In this paper I attempt to provide a perspective of the evidence for, and the relative roles of, normal faulting, thrust faulting, and folding in mountain ranges. Such a perspective may help in judging the realism of some of the geotectonic images mentioned previously. It must be realized, however, that the phenomena on which I propose to concentrate are but the near-surface expression of deeper seated igneous and metamorphic processes, which are much more difficult to unravel but so much more important for an understanding of orogeny.

Although this paper deals mainly with compressional tectonics and thrust faulting, most figures illustrate normal faulting. It is assumed that specialists in thrust faulting and nappe tectonics are familiar and have ready access to the relevant illustrations. However, the seismic examples of gravity-induced normal faulting included in this paper may suggest to folded belt specialists what to look for when they attempt to differentiate gravity from compressional tectonics.

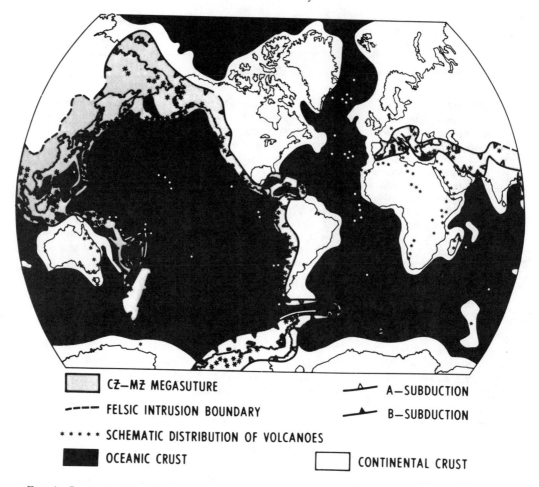

CZ–MZ MEGASUTURE

---- FELSIC INTRUSION BOUNDARY

* * * * * SCHEMATIC DISTRIBUTION OF VOLCANOES

OCEANIC CRUST

A–SUBDUCTION

B–SUBDUCTION

CONTINENTAL CRUST

FIG. 1. Cenozoic–Mesozoic megasuture of the world and its boundaries. B-subduction zones face oceans; A-subduction boundaries face continental cratonic areas. They are characterized by widespread décollement folding and thrust faulting. In China the boundary is an ill-defined envelope around Mesozoic and Cenozoic granitic intrusives (after Bally & Snelson 1980, with permission of Canadian Society of Petroleum Geologists).

Without entering into detailed descriptions, Fig. 1 may serve to show a number of orogenic configurations. In earlier publications I proposed the term megasuture for foldbelts *and* sedimentary basins that are included in them (Bally 1975; Bally & Snelson 1980). For instance, the Mesozoic–Cenozoic megasuture includes all regions of intensive Mesozoic–Cenozoic mountain building and sedimentary basins involved in these processes. In plate tectonics jargon, the megasuture is the integrated product of all subduction-related processes which form the counterpart of the Mesozoic–Cenozoic ocean-spreading processes. The megasuture was introduced primarily to help in classifying sedimentary basins. The term also emphasizes the point that the bottoms of sedimentary basins within the megasutures are typically as deep as the adjacent

mountains are high; and that the evolution of such basins has be be viewed as part of the total evolution of mobile belts (see Fig. 2).

Four types of megasuture boundaries are: B- (or Benioff) subduction boundaries where oceanic lithosphere is subducted; A- (or Ampferer) subduction boundaries where continental lithosphere is subducted; transform fault system boundaries; and—in China—a boundary which is an enevlope around felsic intrusives. This fourth boundary type is required because in China the continent facing boundary of the Mesozoic–Cenozoic megasuture is not associated with obvious external foldbelts involving former passive margin sequences and overlying foredeep sequences. Instead one can trace an ill-defined outline of Mesozoic and Tertiary intrusives which invade deeply into China and Mongolia.

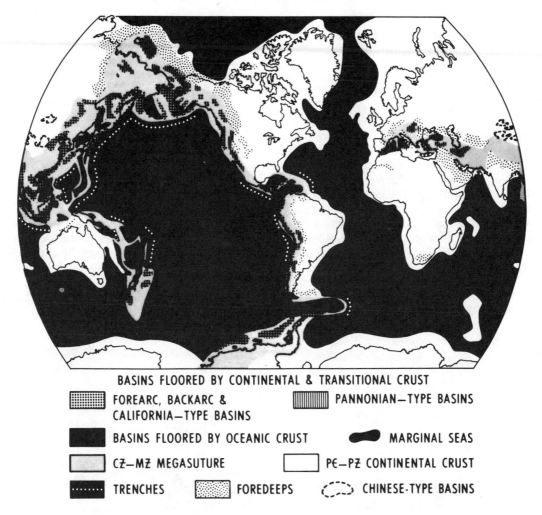

BASINS FLOORED BY CONTINENTAL & TRANSITIONAL CRUST

FOREARC, BACKARC & CALIFORNIA—TYPE BASINS		PANNONIAN—TYPE BASINS
BASINS FLOORED BY OCEANIC CRUST		MARGINAL SEAS
CZ—MZ MEGASUTURE		PЄ—PZ CONTINENTAL CRUST
TRENCHES	FOREDEEPS	CHINESE-TYPE BASINS

FIG. 2. Sedimentary basin families associated with the Cenozoic–Mesozoic megasuture (after Bally & Snelson 1980, with permission of Canadian Society of Petroleum Geologists).

With respect to the Mesozoic–Cenozoic megasuture, we differentiate four major 'orogenic' or megasuture types:

(1) *The SW Pacific type* contained between B-subduction and transform boundaries; a system of island arcs and marginal seas.

(2) *The NW Pacific type* contained between B-subduction and transform boundaries on the Pacific side and the felsic intrusion boundary of China. Within this orogenic type, marginal seas are opened and closed and continental fragments are captured (e.g. Indochina, South China platform, or the Lut Block of Iran).

(3) *The Cordilleran type* contained between a seaward B-subduction and/or transform boundary and a landward A-subduction boundary. The opening and closing of marginal seas appears to be less dominant in the history of the Cordilleras and it would appear that strike-slip rifting of continental fragments plays a major role.

(4) *The Alpine–Himalayan type* is contained within two A-subduction boundaries facing the Eurasian craton to the N and the African–Arabian and Indian continents to the S. This type is the end product of continental collisions.

Note that widespread overthrusting of former passive margin sediments and widespread regional basement remobilization and metamorphism is characteristically associated with A-subduction boundaries. B-subduction boundaries exhibit some imbricate thrusting, and deformation of oceanic sediments in the accretionary wedges of island arcs. Finally there is some thrusting associated with transform systems.

The case for compression and subduction in the external zones of foldbelts

External zones of foldbelts (the Externides of Kober 1928) are the accretionary wedges associated with Benioff subduction zones and more important, they are the folded belts that are associated with A-subduction zones.

The seismic record of Benioff zones favours extension and normal faulting in the peripheral bulge that is formed on the oceanward side of a deep-sea trench and compression in the shallow portions of the subduction zones and the associated accretionary wedge. Reflection seismic data on inner walls of deep-sea trenches suggest accretionary structures formed by 'offscraping' of oceanic sediments that overlie a gently dipping oceanic crust (Fig. 3). Scholl *et al.* (1977) noted that large volumes of pelagic sediments should be, but are not, exposed in Palaeozoic and Mesozoic Circum-Pacific moun-

tain systems. These authors suggest that accretionary wedges may represent mostly deformed slope deposits and that much of the pelagic sediments were subducted with the oceanic crust.

Of course, one is tempted to construct balanced cross-sections (Dahlstrom 1969, 1970) in such a setting. However, the premises for the method are not easily fulfilled, because in most cases, the sediments that form the accretionary wedges are not dated by drilling, and further, because penetrative fabrics and the style of deformation indicates dominance of ductile flow and shear (for details, see von Huene *et al.* this volume). High ductility is suggested by outcrop studies of mélanges and the high pore pressures reported in some wells that are associated with subduction zones (Shouldice 1971). The situation is further complicated by an overprint of gravitational sliding and normal growth faults that show up on a number of reflection lines (see Colombia section, Fig. 3).

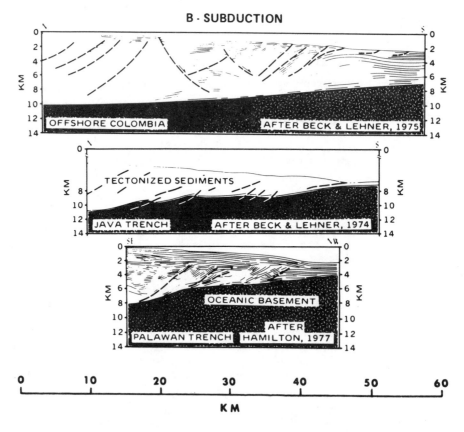

FIG. 3. Sketches of typical accretionary wedges associated with B-subduction zones. (a) Offshore Columbia (after Beck & Lehner 1975); (b) Java Trench (after Beck & Lehner 1974); (c) Palawan Trench (after Hamilton 1977); Drawings after Bally & Snelson (1980, with permission of Canadian Society of Petroleum Geologists).

Although one may rightly question the advisability of using balanced cross section techniques in Franciscan terranes, a recently published reconstruction of northern California by Suppe (1979) suggests some 175 km of shortening of Franciscan, a far cry from the shortening one would expect from offscraping of sediments during the subduction of an oceanic slab that was several thousands of kilometres wide.

One is forced to conclude that the current state of knowledge is inadequate to assess the amount of shortening associated with accretionary wedges of island arcs. The amount of underthrusting can only be derived from reconstructions that are based on magnetic stripe interpretations of the ocean floor. To understand structural deformation in accretionary wedges of island arcs and their ancient equivalents, it is increasingly more important to develope criteria that differentiate the effects of superficial gravitational sliding from the effects of compression and/or extension caused by the sinking oceanic lithospheric slab.

Shortening in folded belts associated with A-subduction boundaries has been studied for a number of years. In fact, the original concept of subduction in the Alps was based on such studies (Ampferer 1906; and later documented by more specific reconstructions: Spengler 1953–59; Trümpy (1969) has provided a detailed palinspastic reconstruction of the Glaronese Alps).

More accurate approximations to palinspastic reconstructions became possible with the aid of structural sections that were based on reflection seismic data. These data allow us to map the basement underlying the frontal folded belts in the Rocky Mountains (Bally *et al.* 1966; Price & Mountjoy 1970; Gordy *et al.* 1975; Royse *et al.* 1975) and in the Appalachians (Gwinn 1970; Roeder *et al.* 1978). A gentle mountainward dipping basement surface observed on numerous reflection seismic sections constrains possible interpretations and leads to more rigid 'balancing' of cross sections as outlined by Dahlstrom (1969), Gwinn (1970), and Roeder *et al.* (1978).

Following Dahlstrom, balanced cross sections are only justified in a 'concentric regime', and the method needs modification if applied in a 'similar fold' regime. Bearing all cautions in mind, shortening in the Canadian Rockies exceeds 160 km and may reach 270 km (Bally *et al.* 1966; Price & Mountjoy 1970; also see Price this volume). For the Helvetic Nappes of the Alps and allowing for the uncertainties due to internal deformation, the accurate nature of

tectonic units, and the estimation of eroded and buried parts, Trümpy (1969) estimates shortening in the order of 30–40 km. Spengler (1953–59) offers estimates typically in the order of 100–150 km for the northern Calcareous Alps.

Gwinn (1970) derives about 80 km for the Central Appalachians, a figure which is increased by Roeder *et al.* (1978) to about 140 km and which is to be further modified to in excess of 200 km in view of the new COCORP data in the area (Cook *et al.* 1979; Hatcher 1972).

For the Scandinavian Caledonides, which are at an erosional level that is particularly favourable for tracing the authochtonous foreland underlying higher thrust sheets, Gee (1975) postulates a total shortening in excess of 500 km. Similar amounts are postulated by Binns (1978) for the Caledonides of nothern Scandinavia. Note that Gee's figure is based mostly on credible inference from surface geology, without reflection seismic data. In the Caledonides like elsewhere, shortening estimates are critically dependent on resolving the penetrative strain recorded in allochthonous sequences, a point properly emphasized by Hossack (1978).

This author computed for the Jotun Nappe 65% vertical shortening and a transverse elongation of 160%. To restore this flattened nappe, its stratigraphic thickness needs to be multiplied by 2.15 to arrive at the original thickness. Even so, the displacement of the Jotun Nappe alone exceeds 290 km.

In other words, the amounts of shortening in external folded belts that are based on reasonably accurate geometric reconstructions and in some cases on reflection seismic data are typically in the range of 50–500 km. Even though in some folded belts of the world the amounts of shortening may be two or three times larger, the figures still are short of the several thousand kilometres that may be deduced from plate motions based on magnetic stripe reconstructions or else on palaeomagnetic data. This suggests either that plate tectonic reconstructions are basically incorrect or else that the subduction mechanism is most efficient in destroying much of the evidence which could be used for a quantitative kinematic check of reconstructions that are based on palaeomagnetics. Because of the convincing nature of the palaeomagnetic data, I favour the second alternative.

For an understanding of mountain building and the genesis of continental crust, it is desirable to examine the fate of the basement which was originally underlying the excess sediment now piled up in thrust sheets, nappes and folds. All authors who have actually carried out quantitative reconstructions in the external zones of folded belts agree that the crust underlying the excess sediments remained in the subsurface: that is, nowhere has a corresponding tectonically denuded surface been exposed, nor is there evidence for subaerial exposure of such a surface that was followed by subsequent burial. Consequently, gravitational gliding will have to be ruled out as a major factor in mountain building.

The concept of gravitational spreading as developed by Price & Mountjoy (1970), Price (1971, 1973), and Elliott (1976a, b) suggests that gravitational forces dominate the emplacement of thrust sheets and, following Elliott, that significant surface slopes are required to form thrusted and folded mountain belts. The linkage of folding in the external zones with the emplacement of metamorphic folds of the Pennine Nappe type is not clear and adequately documented because in a number of cases, such metamorphism precedes the deformation in the external foldbelts.

To conclude: the basement originally underlying excess sediments of external foldbelts remained at depth either to form a mountain 'root' leading to formation of a thickened lithosphere or else it was engulfed in the overall lithospheric subduction process which led to the formation of mountains. In all cases the apparent disappearance in depth of continental lithosphere (A-subduction) is limited to hundreds of kilometres and contrasts with B-subduction involving disposal of thousands of kilometres of oceanic crust. There is little doubt that surface slopes are increased during both A- and B-subduction processes. Such slopes may facilitate gravity spreading. However, the proof for gravity spreading rests on demonstrating that diapir-like metamorphic structures in the internal zones of folded belts formed at the same time as the foreland folds and thrusts which record substantial shortening, and also on a demonstration that extensional tectonics in more brittle overlying sequences are synchronous with the shortening in the foreland. To my knowledge, such conclusive proofs have yet to be published.

This paper is not concerned with the mechanics of the A-subduction process. Conceptual and mechanical difficulties of subduction of substantial portions of continental lithosphere are in part overcome by the simple observation that in all cases we deal with a presumably attenuated continental crust of the

lithosphere of a former passive margin. Nevertheless, until recently, there was a great deal of reluctance by plate tectonic experts to accept A-subduction. It has been argued that the high buoyancy of continental crust would prevent significant subduction of the continental lithosphere. Molnar & Gray (1979) calculate that significant fractions of the continental crust may be subducted if these could be detached from their upper part. Furthermore, these authors suggest that the gravitational force acting on sinking oceanic lithosphere may pull continental lithosphere into the asthenosphere. Such a pull is counteracted by the buoyancy of the light continental crust. Under these circumstances and using varying assumptions, typical values between a few kilometres and up to 330 km in length of subducted continental crust appear to be reasonable. These calculations depend on the thicknesses of lower continental crust that may be detached during the subduction process. If somewhat more extreme assumptions are used, much greater lengths of continental lithosphere may be subducted.

The model of Bird et al. (1975) and Bird (1978) provide a thermal and mechanical scenario which leads to the 'delamination' of sub-crustal lithosphere by insertion or wedging of less viscous asthenosphere between an upper crustal layer and the underlying denser subcrustal lithosphere.

While there is little to add to these interesting models, it is of some comfort to know that subduction of continental lithosphere (A-subduction) which for some time was repulsive to theoretical plate tectonicians has now been accepted in principle. Thus, as geologists, we may now continue to gather observations that may bracket the actual amount of shortening observed in mountain ranges.

The phenomenology of gravity tectonics of sedimentary sequences

Although all preceding considerations seriously limit the role and significance of gravity gliding processes for mountain building, it still may be useful to develop criteria for recognition of gravity tectonics by looking at some obvious examples.

Reviews of gravity gliding have been offered by de Sitter (1954) and North (1964). Various aspects of gravity tectonics are dealt with in a book dedicated to van Bemmelen (de Jong &

Scholten 1973). While many of these authors spend a great deal of effort to convince the readers of the correctness of their mountain building images, very little is offered by way of description of large regions that are unambiguously dominated by gravity tectonics. Two examples of such areas are the Niger Delta and the northern Gulf of Mexico.

The Niger Delta, as described by a number of authors (e.g. Delteil *et al.* 1976; Weber & Daukoru 1976; Lehner & de Ruiter 1977), is prograding on a foundation of high pore pressure shales that overlies an oceanic crust. It is characterized by extensive growth fault systems and shale diapirism. Particularly the toe of the Delta is characterized by imbrications that are similar to features often observed in folded belts (Lehner & de Ruiter 1977).

Like the Niger Delta, the Gulf Coast Tertiary is also prograding on a high pore pressure shale substratum that in turn appears to be underlain by Mesozoic carbonates and a basal salt-bearing sequence which provides an additional unstable base. Where penetrated by the drill, the basement appears to be continental

and an extension of the Palaeozoic mountain system of the Ouachitas and the Appalachians. However, moving towards the Gulf, the crust changes from continental to transitional and/or oceanic (Fig. 4). The structural evolution of the Gulf of Mexico is far from being unraveled, but a few characteristic details shown in reflection seismic lines may offer useful reminders for geologists interested in gravity tectonics in folded belts and may suggest what to look for in support of gravity-gliding concepts.

As suggested on Fig. 4, the edge of the salt mass appears an allochthonous glacier-like tongue overlying very young Tertiary sediments (Watkins *et al.* 1978; Humphris 1978). Widespread pre-Cretaceous listric normal growth faults are restricted to the sedimentary sequence and flatten out at the base of the salt (Figs 5, 6, & 7). Farther up in the sequence and towards the Gulf of Mexico, extensive listric normal growth faults dissect Tertiary clastics and flatten within the high pore pressure shale section at depth. Diapiric salt movement and questionable shale diapirs are intimately associated with these growth fault systems (Figs

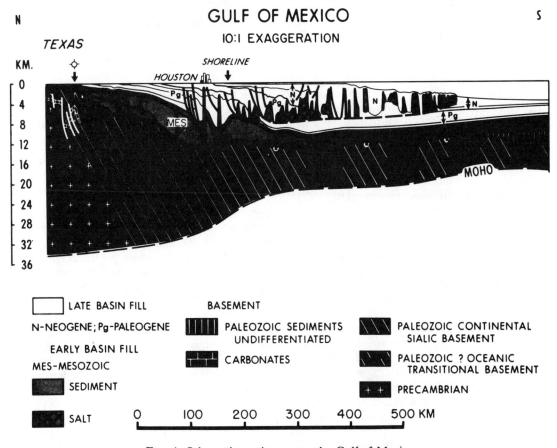

FIG. 4. Schematic section across the Gulf of Mexico.

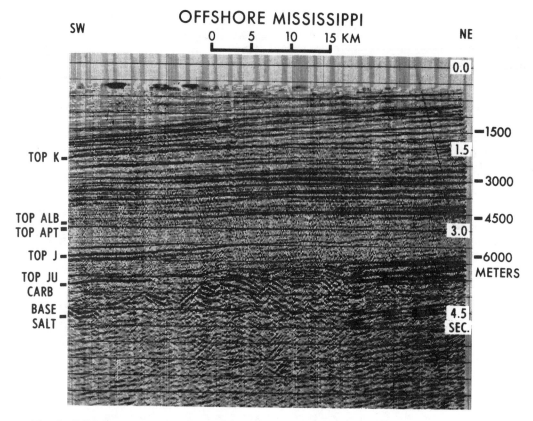

FIG. 5. Offshore Mississippi, reflection seismic section. Note listric normal faults separating salt rollers involving Jurassic carbonates.

8 & 9). The interaction of sedimentation and growth faulting in this region has been summarized by Curtis (1970) and Curtis & Picou (1978), and its relevance for the genesis of hydrocarbon deposits is well illustrated by Curtis (1979). Detailed studies of listric normal faulting and sedimentation are also known from western Ireland (Rider 1978) and Spitzbergen (Edwards 1976).

Gravity tectonics of the type indicated are typical in areas underlain by soft sediments, unstable high pore pressure shales and salt that are associated with high rates of subsidence. Listric normal growth fault systems can be observed in depths in excess of 20 000 ft (6000 m).

An example which shows listric normal faulting that is more or less synchronous with compressional folding is seen on Mexico's eastern offshore (Buffler *et al.* 1979; Watkins *et al.* 1976; Fig. 10). Although more data and calibration by drilling are needed, it appears possible that the amount of shortening represented in the folds of the Mexican ridges may correspond roughly to the amount of stretching by

listric normal faulting underlying the Mexican shelf. On the other hand, if the amount of shortening of the linear folds substantially exceeds the postulated stretching, we may interpret the data as a much younger equivalent (i.e. Pleistocene) of the Laramide Sierra Madre Oriental folds, onshore to the W.

Friends of gravity gliding will be quick to point out that—contrary to the Gulf of Mexico example—much of the deformation occurring in folded belts affects already lithified sediments and in several cases the underlying basement. Consequently, it is also of some importance to characterize clearcut gravity tectonics in 'hard' rocks.

Some very well documented examples of local superficial gravitational gliding have been given by Pierce (1957, 1963, 1973) for the Heart Mountain area of Wyoming and by Reeves (1946) for the Bearpaw Mountains of Montana.

Normal faults responding to extension of the upper crust are most common in the western Cordillera of the USA and the southernmost segment of the Canadian Rocky Mountains.

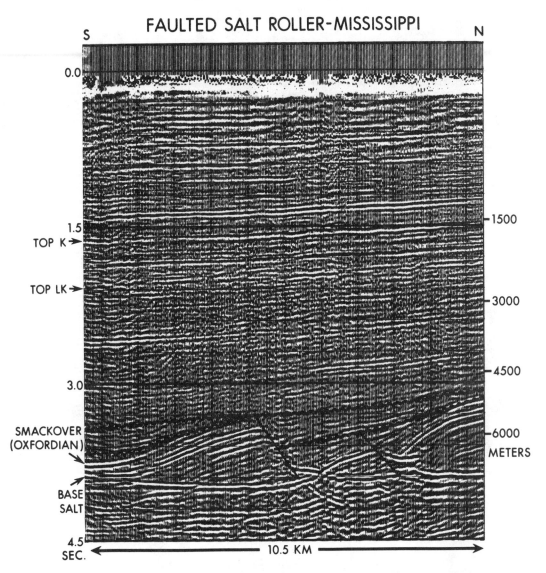

FIG. 6. Gulf Coast in Mississippi, reflection seismic section. Salt rollers, bounded by normal faults which flatten and merge with base of the salt.

Here again, reflection seismic data indicate the listric nature of the faults. The evidence for this is the presence of continuous reflections underlying obvious normal faults that are post-thrusting in age and can be mapped on the surface. The presence of such reflection data does not permit the direct straight-line projection into the subsurface of the large fault offsets seen at the surface; instead the normal faults have to flatten quickly at depth (see Fig. 11; Bally *et al.* 1966; McDonald 1976).

Another expression of the listric nature of these normal faults is the widespread rotation into the fault plane of beds that were deposited while the fault was active (Fig. 12). In the

western Cordillera the evidence for the listricity of normal faults ranges from superficial low-angle normal faults that simulate the base of a major landslide system to intermediate depth fault systems and exhumed, formerly deep fault systems that separate the more ductile deformation realms of metamorphic core complexes from the brittle overlying sediment cover (Davis & Coney 1979; Davis in press; Effimoff & Pinezich, 1980).

The normal fault systems in the western Cordillera are part of a very complex megashear system that—most unfortunately for 'gravity gliders'—clearly postdate the major overthrusting events of the western

A. W. Bally

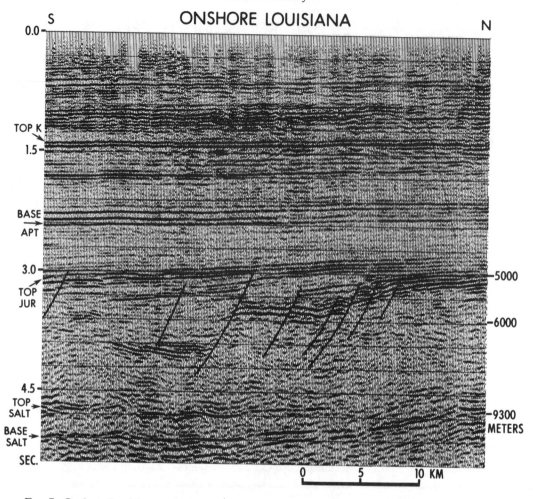

FIG. 7. Onshore Louisiana, reflection seismic section. Note pre-Cretaceous listric normal faults.

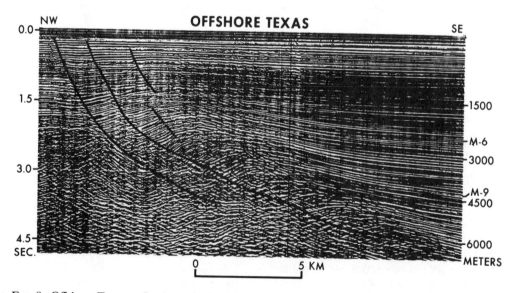

FIG. 8. Offshore Texas, reflection seismic section. Note extensive listric normal growth faults in Miocene section. M-6, M-9 are Miocene marker beds.

OFFSHORE TEXAS

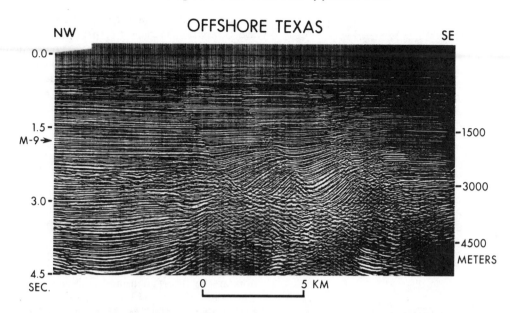

FIG. 9. Offshore Texas, reflection seismic profile showing the interaction of growth faults and diapiric structures. M-9 is a Miocene marker bed.

Cordillera. Note, however, that in a somewhat similar setting, extensional gravity tectonics and normal faulting are also common in the Vienna and the Pannonian Basins (Prey 1974). There, however, the normal faulting roughly occurs during the same time brackets as the last phase of overthrusting in the adjacent Carpathians. However, traditional gravity gliding is precluded by the simple fact that inner portions of the Carpathians are typically subsiding during the Tertiary, instead of furnishing an elevation from which nappes could glide towards the adjacent foredeep.

We may conclude that if we are to explain extensive thrusting and the formation of nappes by gravity gliding, we ought to search for extensive listric normal faulting that is synchronous and directly associated with the overthrust phenomena we observe. The amount of extension of these faults should be comparable to the amount of compressional shortening in the associated foldbelt. So far, I have failed to find in any mountain ranges evidence that would so link the emplacement of overthrust sheets with commensurate extensional fault systems in the inner portions of orogenic belts.

Thrust faulting involving the crystalline basement

So far, the discussion has been concerned with

the evidence for the relative role of gravity gliding versus shortening by subduction-related compression. Support has been obtained by studying surface data and reflection seismic data that illustrate décollement tectonics of sedimentary sequences. A few words concerning the involvement of continental crystalline basement in orogenic processes is now in order.

One group of examples of orogenic basement involvement includes cases where slabs of varying thicknesses of, more or less, rigid crystalline basement rocks form thrust sheets. Examples are the eastern Alpine thrust sheets, the Main Central Thrust of the Himalayas, the Blue Ridge of the Appalachians, the Caledonides of Norway, and many others. In this first group, the basement has not been pervasively remobilized and contrasts with a second group of examples where the crystalline basement was extensively remobilized as in the Penninic Nappes of the Alps or the Shuswap complex of the Canadian Cordillera. Both groups can be viewed as end members for transitions showing varying degrees of basement mobility have been mapped across many folded belts. Instead of reviewing in detail the structural deformation of crystalline basement in folded belts, only the following points will be re-emphasized:

The involvement of crystalline continental basement in folded belts is characteristic and

A. W. Bally

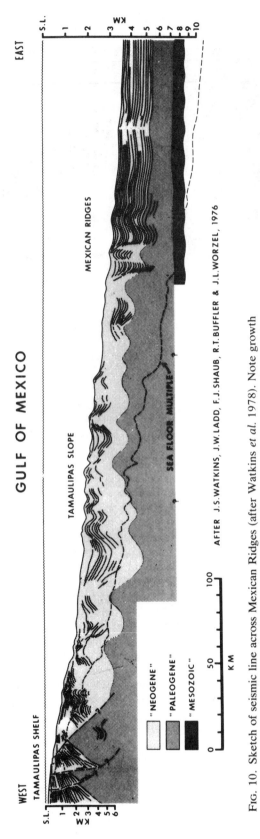

FIG. 10. Sketch of seismic line across Mexican Ridges (after Watkins *et al.* 1978). Note growth faults on left side of section and compressional décollement folds on the right side of the section.

constitutes the main reason why folded belts are often regarded as the product of ensialic orogenies. Some authors see a contradiction between a purely ensialic orogeny and orogenies that are related to plate tectonic processes. It is true that a number of mobile belts involve oceanic crust or ophiolites in suture zones. These permit us to postulate oceans of unknown width in plate tectonic reconstructions. However, in other cases, the evidence for such oceans is not so obvious, and at least in one case, folding and associated basement remobilization is entirely ensialic (Amadeus Basin of central Australia: Wells *et al.* 1970). Admittedly, the last example is unusual, because it affects a relatively small inverted basin located in the foreland of the Tasmanides. Nevertheless, the Amadeus Basin appears to offer an unambiguous case of ensialic deformation with décollement folding accompanied by the formation of small crystalline nappes.

Thrust faulting involving basement clearly indicates the existence of decoupling levels within the continental lithosphere. These may be mostly in the lower crust or may be in the upper mantle. Recent geophysical work in the Alps and Appennines (Angenheister *et al* 1972; Giese *et al.* 1973, 1978; Mueller *et al.* 1976; Mueller 1977) gives evidence for widespread low-velocity layers within the crust that extend well into the foreland. Thus the case for a layered crust with potential intracrustal decoupling levels is reinforced. The rheologic characteristics of these low-velocity layers are as yet poorly defined and consequently, the genesis of crustal low-velocity layers remains a matter for speculation. Some of the geological consequences of this problem are discussed by Hsü (1979).

There is also evidence for widespread crustal decoupling in the foreland of folded belts. A recent reflection line across the Wind River Mountains of Wyoming (Smithson *et al.* 1978, 1979; Brewer *et al.* in press) indicates that this basement uplift is underlain by a thrust fault which can be followed to a depth of about 35 km (see Fig. 13). If—as I believe—this line is characteristic for foreland block faulting, then it would follow that a major Laramide decoupling zone occurring at or near the base of the continental crust was underlying the Cordilleran foreland or the central and southern Rockies of the USA.

In a more speculative vein, attention should be called to Ziegler's (1978) observation that the well-known inversions observed in northwestern Europe (e.g. Wealden Anticlinorium,

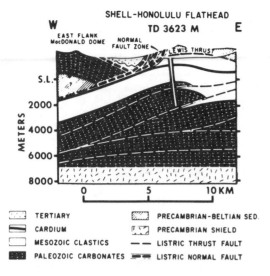

FIG. 11. Post-orogenic normal faults in Flathead area of British Columbia (from Bally *et al.* 1966, with permission of Canadian Society of Petroleum Geologists).

W Netherlands Basin, Lower Saxony Basin, and the Polish Anticlinorium, etc.) occur during the Meso-alpine deformation. This may well mean that with the incipient alpine collision, stresses are transmitted over more than 500 km across a rigid basement plate that overlies a deeper, less competent decoupling level. Such a level may well be located within the lower crust or the upper mantle. The concept is further supported by structural observations (horizontal stylolites, joints, minor re-

verse and strike-slip faults) in the sediment cover of the intervening platforms (Wagner 1974; de Charpal *et al.* 1974; Wunderlich 1974; for an updated review, see also Letouzey & Tremolières 1980).

There is a great deal of similarity between the foreland tectonics described in the preceding paragraphs and the continental collision processes that were described by Molnar & Tapponier (1978) and Tapponier & Molnar (1976) for Central Asia. There the late Palaeogene–Neogene collision of India with Eurasia caused extensive strike-slip faulting, thrust faulting, and normal faulting in Central Asia, which according to Molnar & Tapponier led to much of China being squeezed in an eastward direction. It should be noted that the collision also led to the emplacement of basement thrusts in the Himalayas, and it appears likely to me that some thrust faults associated with the Tien Shan and the Nan Shan uplifts involve thick segments of the continental crust in a manner comparable to the Rocky Mountain foreland.

It is concluded that decoupling within the lower crust and perhaps in the upper mantle is essential to explain thrust faulting of continental basement slabs. More geophysical information is needed to map such decoupling levels. These may well coincide with low-velocity layers that have been determined in a number of crustal studies. We are accustomed to explaining varying styles of deformation in folded and thrust-faulted sedimentary sequences as a function of ductility contrasts that are inherent

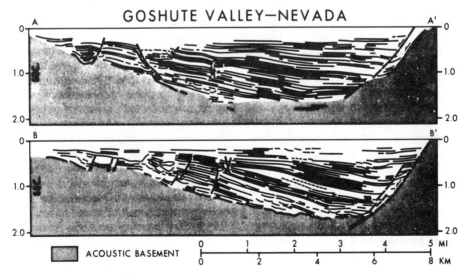

FIG. 12. Sketch of seismic line across Goshute Valley, NE Nevada (after Bally & Snelson 1980), with permission of Canadian Society of Petroleum Geologists).

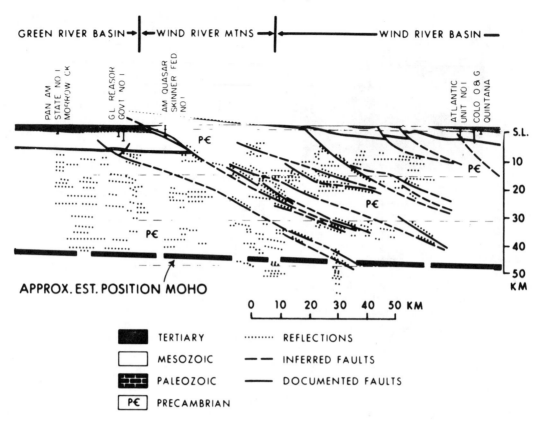

GREEN RIVER BASIN ➡◄━WIND RIVER MTNS━► ◄━━━━━━━WIND RIVER BASIN ━

FIG. 13. Interpretation of Wind River Mountains, Wyoming, based on COCORP line (Smithson *et al.* 1978).

in the stratigraphic layering of the deformed sequences. By analogy, the style of deformation of basement slabs is dependent on ductility contrasts within the crystalline basement.

A number of plate tectonic models imply crustal decoupling (Armstrong & Dick 1974; Oxburgh's flake tectonics 1972; Bird's delamination 1978; Molnar & Gray 1979). A significant increase in understanding requires that future work concentrates on getting more detailed geological and geophysical documentation of deep crustal decoupling.

Turning to basement mobilization in the inner folded belts, it may be said that the causes for widespread regional remobilization and metamorphism of the basement remain obscure and may be due to burial and loading by higher thrust sheets or possibly due to thermal uplifts.

Overthrusting and strike-slip faulting

In recent years a number of authors have discussed the nature of thrust faulting as-

sociated with strike-slip faulting (Lowell 1972; Wilcox *et al.* 1973; Harding 1973, 1974, 1976; Sylvester & Smith 1976; Harding & Lowell 1979). These authors document and systematize the phenomenology of wrench faulting with surface, subsurface examples and clay model studies. An important problem relates to the scale and importance of thrust faulting, associated with strike-slip faulting. Clearly *en échelon* faults and thrust faults occur in a wrench fault regime. The upthrust interpretation proposed by various authors needs more verification by reflection seismic data. There remains, however, the question of how much shortening of sediments can be taken up by strike-slip faulting.

A schematic cross section and reconstruction made by my colleague, R. E. Farmer, illustrates the problem (Fig. 14). The Taiwan foldbelt has a structural style that appears to be similar to the Rocky Mountain Foothills of Canada and the outer Carpathians of Rumania. Although no seismic data concerning the underlying basement are published, one may assume gentle eastward dip by the reconstruction of the stratigraphic wedge that

TAIWAN

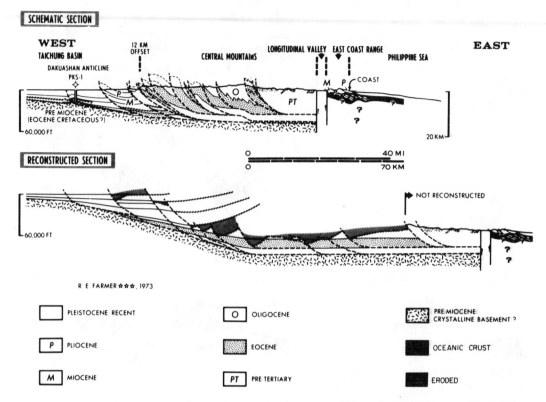

FIG. 14. Schematic cross section and reconstruction across Taiwan by R. E. Farmer (Shell Oil Company). *** Indicates quality.

is involved in the deformation. As a 'denuded' basement does not outcrop in the adjacent mountains, conventional gravity gliding may be precluded.

Accepting the admittedly shaky premises of the cross section, one has to conclude that a strip more than 50 km wide of pre-Tertiary basement apparently was subducted to form a deep 'root' under Taiwan. The crustal character of that basement is unknown. The subduction of the basement could have preceded the strike-slip displacement along the fault of the Longitudinal Valley or else the subduction process occurred during the deformation of that fault and during the folding in western Taiwan. Phases of subduction alternating with strike-slip faulting over short geological time spans can also be imagined.

The conclusion is that a substantial room problem may occur in palinspastic reconstructions of balanced cross sections across foldbelts that appear to be related to or later modified by strike-slip fault systems. Obviously, more reflection seismic data are needed to gain a better feeling for the dimension of the prob-

lem. Plate tectonic reconstructions based on palaeomagnetic data frequently suggest the location of orogenic systems in an overall strike-slip/shear context and therefore it becomes increasingly more important to differentiate and determine the scale of thrusting related to strike-slip faulting, A-subduction and B-subduction.

Folded belts, basement mobilization and plate tectonic reconstructions

A corollary of many of the preceding comments and the megasuture concept is that the crystalline basement of folded belts did not behave as part of a rigid lithosphere during orogenic processes. Consequently, it is of some importance to separate relatively more rigid lithospheric realms from basement that has been remobilized and subjected to regional metamorphism and from basement fragments that were overthrust or else were rifted by complex strike-slip movements.

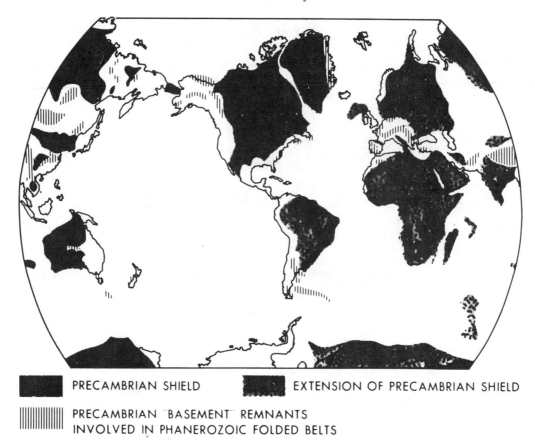

FIG. 15. Preserved Precambrian plate remnants. Precambrian 'basement' remnants that have been remobilized within Phanerozoic foldbelts do not qualify as 'rigid' microplates.

Superficial décollement folds and thrust sheets of Externides (Kober 1928) are probably in most cases underlain by a gently dipping basement ramp which represents unambiguous rigid lithosphere. The inner remobilized and metamorphosed portions of folded belts and small interior basement blocks captured during the orogenic process obviously represent either 'ductilized' lithosphere or else stray crustal fragments.

Plate tectonic reconstructions that are based on palaeomagnetic or stratigraphic points of control which are located within folded belts obviously have to be contrasted and related to lithospheric 'cratonic' segments that are now adjacent to these foldbelts. Therefore, there is an urgent need for reasonably accurate structural reconstructions of folded belts. At the same time, it is desirable to map the outlines of pre-Mesozoic and Precambrian continental lithospheric remnants to provide some of the building blocks for plate tectonic reconstructions. Outlines of the different types of building blocks are shown on Fig. 15 for the begin-

ning of the Palaeozoic and on Fig. 16 for the beginning of the Mesozoic. The extensive occurrence of Precambrian and Palaeozoic basement remnants in later folded belts again emphasizes the 'ensialic' aspects of mountain building.

Conclusions

Normal faulting in folded belts and their foreland may be: —listric normal faulting involving the basement and related to the passive margin phase preceding orogenic deformation (for analogue, see de Charpal *et al.* 1978); —surficial soft-sediment listric normal faulting related to the drifting phase preceding orogenic deformation; —listric normal faulting related to the genesis of synorogenic accretionary wedges of B-subduction zones; —synorogenic normal faulting in the foreland related to the foreland bulge associated with subduction zones (Buchanan & Johnson 1968; Hopkins 1968; Laubscher 1978); —syn or

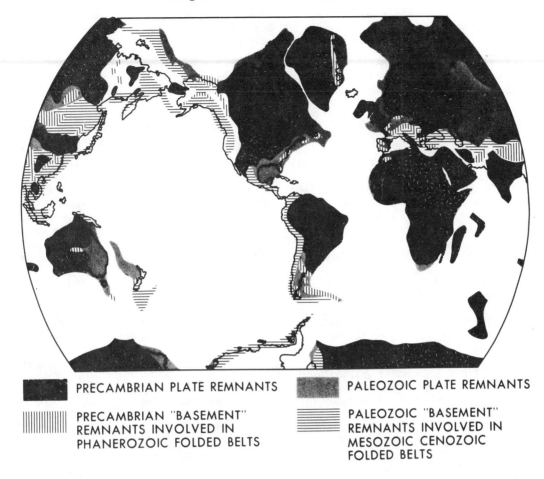

PRECAMBRIAN PLATE REMNANTS

PALEOZOIC PLATE REMNANTS

PRECAMBRIAN "BASEMENT" REMNANTS INVOLVED IN PHANEROZOIC FOLDED BELTS

PALEOZOIC "BASEMENT" REMNANTS INVOLVED IN MESOZOIC CENOZOIC FOLDED BELTS

FIG. 16. Preserved Precambrian and Palaeozoic plate remnants. Palaeozoic and Precambrian outcrops that have been remobilized within Phanerozoic foldbelts do not qualify as preserved "rigid" microplates.

postorogenic listric normal faulting associated with stretching and shearing of the orogenic system (e.g., Great Basin, Vienna and Pannonian Basins).

Thrust faulting in the folded belts may be:
—minor preorogenic thrust faulting at the toe of deltaic systems of passive margins (for analogue, see Lehner & de Ruiter 1977);
—synorogenic listric thrust faulting involving the continental basement within the mountain range (e.g. Himalayas and eastern Alpine Nappe) and in the foreland of mountain ranges (e.g. Wind River Mountains); —synorogenic thrust faulting or sedimentary sequences related to A-subduction processes (e.g. Canadian Rocky Mountains, Appalachians, and Externides of the Alpine system); —synorogenic thrust faulting of sedimentary sequences related to B-subduction processes (see Fig. 3);
—synorogenic thrust faulting related to strike-slip faults (Fig. 14).

All information on folded belts permits and supports their plate tectonic origin at subduction or else at transform plate boundaries. However, much of what we actually observe in mountain ranges involves décollement of and within sedimentary sequences, as well as significant decoupling within the deeper continental crust. Normal faulting and thrust faulting due to gravity gliding is not important for mountain building. This and the absence of wide-spread tectonically denuded basement argues strongly against a major role for gravity gliding.

Future geological and geophysical work should aim at defining and mapping such decoupling levels, particularly those that occur within the crust. Seismic crustal studies are a particularly promising aid in mapping intracrustal decoupling levels.

ACKNOWLEDGMENTS. I would like to thank my colleagues R. E. Farmer for providing the geological cross section of Taiwan, R. L. Nicholas for providing the Gulf Coast reflection lines, and K. Arbenz and S. Snelson for reviewing the paper and for their helpful suggestions. I also thank Shell Oil Company for permission to publish the seismic sections and the paper.

References

AMPFERER, O. 1906. Über das Bewegungbild von Faltengebirgen, Austria. *Jahrb. geol. Bundesanst*, **56**, 539–622.

ANGENHEISTER, G., BÖGEL, H., GEBRANDE, H., GIESE, P., SCHMIDT-THOME, R. & ZEIL, W. 1972. Recent investigations of surficial and deeper crustal structure of the eastern and southern Alps. *Geol. Rdsch.* **61**, 349–95.

ARGAND, E. 1924. La tectonique de l'Asie, *Compt. Rend IIIe. Congr. Int. geol. Liège*, Imprimerie Vaillant-Carmanne (transl. & ed. by Carozzi, A. V.). Hafner Press, New York, 218 pp.

ARMSTRONG, R. L. & DICK, H. J. B. 1974. A model for the development of thin overthrust sheets of crystalline rock. *Geology*, **2**, 35–40.

BALLY, A. W. 1975. A geodynamic scenario for hydrocarbon occurrences. *Proc. Ninth World Petrol. Congr. Tokyo*, **2**, Applied Science, Essex, 33–44.

——, GORDY, P. L. & STEWART, G. A. 1966. Structure, seismic data, and orogenic evolution of Southern Canadian Rocky Mountains. *Bull. Can. Pet. Geol.* **14**, 337–81.

—— & SNELSON, S. 1980. Realms of subsidence. *In*: *Facts and Principles of World Oil Occurrence*. Mem. Can. Soc. Pet. Geol. **6**, (in press).

BECK, R. H. & LEHNER, P. 1974. Oceans, new frontier in exploration. *Bull. Am. Assoc. Petrol. Geol.* **58**, 376–95.

—— & ——, with collab. of DIEBOLD, P., BAKKER, G. & DOUST, H. 1975. New geophysical data on key problems on global tectonics. *Proc. Ninth World Petrol. Congr. Tokyo*, **2**, Applied Science, Essex, 3–17.

BELOUSSOV, V. V. 1962. *Basic Problems in Geotectonics*. McGraw Hill, New York, 809 pp.

—— 1975. *Foundations of Geotectonics* (in Russian). Nyedra, Moscow, 260 pp.

—— 1977. Gravitational instability and the development of the structure of continents. *In*: SAXENA, S. K. & BATTACHARJI, S. (eds). *Energetics of Geological Processes*. Springer, New York, 3–18.

BINNS, R. E. 1978. Caledonian Nappe correlation and orogenic history in Scandinavia north of lat. 67°N. *Bull. geol. Soc. Am.* **89**, 1475–90.

BIRD, P. 1978. Initiation of intracontinental subduction in the Himalaya. *J. geophys. Res.* **83**, 4975–87.

——, TOKSÖZ, M. N. & SLEEP, N. H. 1975. Thermal and mechanical models of continent–continent convergence zones. *J. geophys. Res.* **80**, 4405–16.

BREWER, J., SMITHSON, S. B., KAUFMAN, S., & OLIVER, J. in press. The Laramide orogeny: evidence from COCORP deep crustal seismic profiles in the Wind River Mountains, Wyoming. *J. geophys. Res.*

BUCHANAN, R. S. & JOHNSON, F. K. 1968. Bonanza gas field—a model for Arkoma Basin growth faulting. *In*: CLINE, L. M. (ed). *A Guidebook to the Geology of the Western Arkoma Basin and Ouachita Mountains, Oklahoma*. Okla. City Geol. Soc. 75–85.

BUFFLER, R. T., SHAUB, F. T., WORZEL, J. L. & WATKINS, J. S. 1979. Gravity slide origin for Mexican ridges foldbelt, southwestern Gulf of Mexcio. *Bull. Am. Assoc. Petrol. Geol.*, **63**, 426.

CAREY, S. W. 1975. The expanding earth—an essay review. *Earth Sci. Rev.* **11**, 105–43.

—— 1977. The Expanding Earth: *Development in Geotectonics Series*, **10**. Elsevier, Amsterdam, 488 pp.

COOK, F. A., BROWN, L. D., HATCHER, R. D., KAUFMAN, S. & OLIVER, J. E. 1979. Preliminary interpretation of COCORP reflection profiles across the Brevard zone in northwest Georgia. *Abstr. Spring Meeting Am. Geophys. Un.*

CURTIS, D. M. 1970. Miocene deltaic sedimentation Lousiana Gulf Coast. *In*: MORGAN, JAMES P. (ed). *Deltaic Sedimentation Modern and Ancient*. Spec. Publ. Soc. Econ. Paleo. Miner. **15**, 293–308.

—— 1980. Source of Oils in Gulf Goast Tertiary. *J. Sediment. Petrol.* (in press).

—— & PICOU, JR., E. B. 1978. Gulf Coast Cenozoic: A model for the application of stratigraphic concepts to exploration on passive margins. *Trans. Gulf Coast Assoc. geol. Soc.* **XXVIII**, 103–20.

DAHLSTROM, C. D. A. 1969. Balanced cross sections. *Can. J. Earth Sci.* **6**, 743–57.

—— 1970. Structural geology in the eastern margin of the Canadian Rocky Mountains. *Bull. Can. Pet. Geol* **18**, 332–406.

DAVIS, G. A. in press. Problems of intraplate extensional tectonics, Western United States. *In*: *Nat. Res. Council Continental Dynamics publication*.

DAVIS, G. H. & CONEY, P. J. 1979. Geological development of the Cordilleran metamorphic core complexes. *Geology*, **7**, 120–4.

DE CHARPAL, O., GUENNOC, P., MONTADERT, L. & ROBERTS, D. G. 1978. Rifting, crustal attenuation and subsidence in the Bay of Biscay. *Nature*, **275**, 706–11.

——, TREMOLIÈRES, P., JEAN, F. & MASSE, P. 1974. Un example de tectonique de plate-forme: les Causses Majeurs, sud du Massif Central, France. *Rev. Inst. Fr. Pet. Paris* **29**, 641–59.

DE JONG, K. A. & SCHOLTEN, R. 1973. *Gravity and Tectonics*. Wiley-Interscience, New York, 502 pp.

DELTEIL, J. R., RIVIER, F., MONTADERT, L., APOSTOLESCU, V., DIDIER, J., GOSLIN, M. & PAT-

RIAT, P. H. 1976. Structure and sedimentation of the continental margin of the Gulf of Benin. *In*: DE ALMEIDA, F. F. M., (ed). *Continental Margins of Atlantic Type*. Ann. Acad. Brasil **48**, 51–66.

DE SITTER, L. U. 1954. Gravitation gliding tectonics, an essay in comparative structural geology. *Am. J. Sci.* **252**, 321–44.

EDWARDS, M. B. 1976. Growth faults in Upper Triassic deltaic sediments, Svalbard, *Bull. Am. Assoc. Petrol. Geol.* **60**, 314–55.

EFFIMOFF, I. & PINEZICH, A. A. 1980. Tertiary structural development of selected valleys based on seismic data—Basin and Range Province, Northeastern Nevada. *Paper for Roy. Soc. Meeting, London*. (in preparation).

ELLIOTT, D. 1976*a*. The motion of thrust sheets. *J. geophys. Res.* **81**, 949–63.

—— 1976*b*. The energy balance and deformation mechanisms of thrust sheets, *Philos. Trans. R. Soc. London*, **283**, 289–312.

GEE, D. G. 1975. A tectonic model for the central part of the Scandinavian Caledonides. *Am. J. Sci.* **275A**, 468–515.

GIESE, P., MORELLI, C. & STEINMETZ, L. 1973. Main features of crustal structure in central and southern Europe based on data of explosion seismology. *In*: MÜLLER, S. (ed), *Tectonophysics*, **20**, 367–79.

—— & RÜTTER, K. J. 1978. Crustal and structural features of the margins of the Adria Microplate. *In*: CLOSS, H. *et al.* (eds). *Alps, Appennines, Hellenides*, Schweizerbart'sche Verlagsbuchhandlung, Stuttgart, 565–87.

GORDY, P. F., FREY, F. R. & OLLERENSHAW, N. C. 1975. Structural geology of the foothills between Savanna Creek and Panther River SW Alberta, Canada. Guidebook Can. Soc. Petrol. Geol., *Can Soc. Explor. Geophys.*, *Explor. Update* 1975, Calgary.

GWINN, V. E. 1970. Kinematic patterns and estimates of lateral shortening, Valley and Ridge and Great Valley provinces. *In*: FISHER, G. W. *et al.* (eds). *Studies of Appalachian Geology, Central and Southern*. Wiley, New York, 460 pp.

HAMILTON, W. 1977. Subduction in the Indonesian Region. *In*: TALWANI, W. & PITTMAN, W. C. (eds). *Island Arcs, Deep Sea Trenches and Back-arc Basins*. Maurice Ewing Series **1**, Am. geol. Un., 15–31.

HARDING, T. P. 1973. Newport–Inglewood trend, California—an example of wrenching style of deformation. *Bull. Am. Assoc. Petrol. Geol.* **60**, 366–78.

—— 1974. Petroleum traps associated with wrench faults. *Bull. Am. Assoc. Petrol. Geol.* **58**, 1290–304.

—— 1976. Tectonic significance and hydrocarbon trapping consequences of sequential folding synchronous with San Andreas faulting, San Joaquin Valley, California. *Bull. Am. Assoc. Petrol. Geol.* **60**, 366–78.

—— & LOWELL, J. D. 1979. Structural styles, their plate tectonic habitats and hydrocarbon traps in petroleum provinces. *Bull. Am. Assoc. Petrol. Geol.* **63**, 1016–58.

HATCHER, R. D. 1972. Developmental model for the southern Appalachians. *Bull. geol. Soc. Am.* **83**, 2735–60.

HOPKINS, H. R. 1968. Structural interpretations of the Quachita Mountains. *In*: CLINE, L. M. (ed). *A Guidebook to the Geology of Western Arkoma Basin and Ouachita Mountains, Oklahoma*. Okla. City Geol. Soc. 104–8.

HOSSACK, J. R. 1978. The correction of stratigraphic sections for tectonic finite strain in the Bygdin area, Norway. *J. Geol. Soc. London*, **135**, 229–41.

HSÜ, K. J. 1979. Thin skinned plate tectonics during Neo-Alpine Orogenesis *Am. J. Sci.* **279**, 353–66.

HUMPHRIS, C. C. 1978. Salt movement on continental slope, northern Gulf of Mexico. *Am. Assoc. Petrol. Geol. Studies in Geology*, **7**, 69–85.

KOBER, L. 1928. *Der Bau der Erde*, Bornträger, Berlin. 500 pp.

LAUBSCHER, H. P. 1978. Foreland folding. *Tectonophysics*, **47**, 325–37.

LEHNER, P. & de RUITER, P. A. C. 1977. Structural history of the Atlantic margin of Africa. *Bull. Am. Assoc. Petrol. Geol.* **61**, 961–81.

LETOUZEY, J. & TREMOLIÈRES, P. 1980. Paleo-stress fields around the Mediterranean since the Mesozoic derived from microtectonics: Comparisons with plate tectonic data. *In*: SCHEIDEGGER, A. E. (ed). *Tectonic Stress in the Alpine–Mediterranean region*. Springer Verlag.

LOWELL, J. D. 1972. Spitsbergen Tertiary orogenic belt and the Spitsbergen fracture zone. *Bull. geol. Soc. Am.* **83**, 3091–102.

McDONALD, R. E. 1976. Tertiary tectonics and sedimentary rocks along the transition basin and range province to plateau and thrust belt province, Utah. *Rocky Mtn. Assoc. Geol. Guidebook*, 281–371.

MOLNAR, P. & GRAY, D. 1979. Subduction of continental lithosphere: some constraints of certainties. *Geology*, **7**, 58–62.

—— & TAPPONIER, P. 1978. Active tectonics of Tibet. *J. geophys. Res.* **83**, 5361–75.

MUELLER, S. 1977. A new model of the continental crust. *Monogr. Am. Geophys. Un.* **20**, 289–317.

——, EGLOFF, R. & ANSORGE, J. 1976. Struktur des tieferen undergrundes Entlang der Schwizer geotraverse. *Schweiz. mineral. petrogr. Mitt.* **56**, 685–92.

NORTH, F. K. 1964. Gravitational tectonics. *Bull. Can. Pet. Geol.* **12**, 185–225.

OXBURGH, E. Z. 1972. Flake tectonics and continental collision. *Nature*, **239**, 202–4.

PIERCE, W. G. 1957. Heart Mountain and South Fork detachment thrusts of Wyoming. *Bull. Am. Assoc. Petrol. Geol.* **41**, 519–626.

—— 1963. Reef Creek detachment fault, northwestern Wyoming. *Bull. geol. Soc. Am.* **74**, 1225–36.

—— 1973. Principal features of the Heart Mountain fault and the mechanism problem. *In*: DE JONG, K. A. & SCHOLTEN, R. (eds) *Gravity and Tec-*

tonics. Wiley-Interscience, New York, 457–71.

PREY, S. 1974. External zones. *In:* MAHEL, M. (ed). *Tectonics of the Carpathian Balkan Regions— Explanation to the Tectonic Map of the Carpathian–Balkan Regions and their Foreland.* Geol. Inst. Dionyz Stur. Bratislava, 75–84.

PRICE, R. A. 1971. Gravitational sliding and the foreland thrust and fold belt of the North American Cordillera: Discussion. *Bull. geol. Soc. Am.* **77,** 1133–38.

—— 1973. Large-scale gravitational flow of supra-crustal rocks, southern Canadian Rockies. *In:* DE JONG, K. A. & SCHOLTEN, R. (eds). *Gravity and Tectonics.* Wiley-Interscience, New York, 491–502.

—— & MOUNTJOY, E. W. 1970. Geologic structure of the Canadian Rocky Mountains between Bow and Athabasca Rivers: a progress report. *Spec. Pap. geol. Assoc. Can.* **6,** 7–25.

REEVES, F. 1946. Origin and mechanics of the thrust faults adjacent to the Bear-paw Mountains, Montana. *Bull. geol. Soc. Am.* **57,** 1033–48.

RIDER, M. H. 1978. Growth faults in carboniferous of western Ireland, *Bull. Am. Assoc. Petrol. Geol.* **62,** 2191–213.

ROEDER, D., GILBERT, JR., D. E. & WITHERSPOON, W. D. 1978. Evolution and macroscopic structure of Valley and Ridge thrust belt. Tennessee and Virginia. *Univ. Tenn. Geol. Sci. Studies in Geology,* **2,** Knoxville, 25 pp.

ROYSE, JR., F., WARNER, M. A. & REESE, D. C. 1975. Thrust belt structural geometry and related stratigraphic problems, Wyoming–Idaho–northern Utah. *Rocky Mtn. Ass. Geol. Symp.* Deep Drilling Frontiers in Central Rocky Mtns., 41–54.

SCHOLL, D. W., MARLOW, M. S. & COOPER, A. K. 1977. Sediment subduction and offscraping at Pacific Margins. *In:* TALWANI, M. & PITTMAN, W. C. (eds). *Island Arcs. Deep Sea Trenches and Back-Arc Basins.* Maurice Ewing Series **1,** Am. geophys. Un., 199–210.

SHOULDICE, G. H. 1971. Geology of the western Canadian continental shelf. *Bull. Can Pet. Geol.* **19,** 405–36.

SMITHSON, S. B., BREWER, J. A., KAUFMAN, S., OLIVER, J. & HURICH, C. 1978. Nature of the Wind River thrust, Wyoming, from COCORP deep reflections and gravity data. *Geology,* **6,** 648–52.

——, ——, ——, —— & —— 1979. Structure of the Laramide Wind River uplift, Wyoming, from COCORP deep reflection data and from gravity data. *J. geophys. Res.* **84,** 5955–72.

SPENGLER, E. 1953-59. Versuch einer Rekostruktion des Ablagerungsraumes der Decken der nördlichen Kalkalpen I-III. *Jahrb. geol. Bundesanst.* 1953, 1956, 1959.

STAUB, R. 1928, *Der Bewegungsmechanismus der Erde,* Bornträger, Berlin, 270 pp.

SUPPE, J. 1979. Cross Section of southern part of northern coast ranges and Sacramento Valley, California. *Geol. Soc. Am. Map & Chart Series* **MC–28B.**

SYLVESTER, A. G. & SMITH, R. R. 1976. Tectonic transpression and basement-controlled deformation in the San Andreas fault zone, Salton trough, California. *Bull. Am. Assoc. Petrol Geol.* **60,** 2081–102.

TAPPONIER, P. & MOLNAR, P. 1976. Slip-line theory and large-scale plate tectonics. *Nature,* **264,** 319–24.

TRÜMPY, R. 1969. Die Helvetischen Decken der Ostschweiz: Versuch einer palinspastischen Korrelation und Ansätze zu einer kinematischen Analyse. *Eclog. geol. Helv.* **62,** 105–38.

VON HUENE, R., ARTHUR, M. & CARSON, B. *this volume.* Ambiguity in interpretation of seismic data from modern convergent margins: an example from the IPOD Japan trench transect.

WAGNER, G. H. 1964. Druckspannungsindizien in den Sedimenttafeln des Rheinischen Schildes. *Geol. Rdsch.* **56,** 906–13.

WATKINS, J. S., LADD, J. W., SHAUB, F. J., BUFFLER, R. T. & WORZEL, J. L. 1976. Southern Gulf of Mexico, east–west section from Tamaulipas shelf to Campeche Scarp. *Am. Assoc. Petrol. Geol. Seismic sec.* **1.**

——, ——, BUFFLER, R. T., SHAUB, F. G., HOUSTON, M. H. & WORZEL, J. L. 1978. Occurrence and evolution of salt in deep Gulf of Mexico. *Am. Assoc. Petrol. Geol. Studies in Geology,* **7,** 43–65.

WEBER, K. J. & DAUKORU, E. 1976. Petroleum geology of the Niger delta. *Proc. Ninth World Petrol. Congr. Tokyo,* **2,** Applied Science, Essex, 209–221.

WELLS, A. T., FORMAN, D. J., RANFORD, L. C. & COOK, P. J. 1970. Geology of the Amadeus Basin, central Australia. *Bull. Bur. Miner. Resour. Geol. Geophys. Melbourne.* **100,** 222 pp.

WILCOX, R. E., HARDING, T. P. & SEELY, D. R. 1973. Basic wrench tectonics. *Bull. Am. Assoc. Petrol. Geol.* **57,** 74–90.

WUNDERLICH, H. G. 1974. Die Bedeutung der Süddeutschen Gross-scholle in der Geodynamik Westeuropas. *Geol. Rdsch.* **63,** 755–72.

ZIEGLER, P. A. 1978. Northwestern Europe: tectonics and basin development. *Geol. Mijnbouw.* **57,** 589–626.

A. W. BALLY, Shell Oil Company, P.O. Box 481, Houston, Texas 77001, U.S.A.

Reprinted with permission from *Journal of Structural Geology*, v. 8, p. 325-339, Steven E. Boyer, Styles of folding within thrust sheets: examples from the Appalachian and Rocky Mountains of the U.S.A. and Canada, Copyright 1986, Pergamon Press plc.

Styles of folding within thrust sheets: examples from the Appalachian and Rocky Mountains of the U.S.A. and Canada

STEVEN E. BOYER

Sohio Petroleum Company, Technology Center, 5400 LBJ Freeway, Suite 1200, Dallas, TX 75240, U.S.A.

(*Received* 18 *July* 1985; *accepted in revised form* 13 *August* 1985)

Abstract—Folds in a single thrust sheet can be classified as trailing edge (formed over footwall ramps), intraplate (due to shortening within the body of the sheet) and leading edge. Leading edge folds are usually removed by erosion at the front of a thrust sheet, but their unfaulted equivalents can be examined at the lateral termination of the thrusts. Most folds of these various classes possess a kink-band geometry with sharp hinges and long, planar limbs. The orderly nature of kink-band folds breaks down when thin incompetent units separate thicker competent members. The resulting fold geometry, ranging from simple disharmonic to hinge-collapse, is controlled by the thickness of interbedded incompetent materials. The Appalachian and Rocky Mountain thrust belts provide examples of these various classes and styles of folds.

INTRODUCTION

IN THE recent past, folds in thrust and fold belts were modelled as being cylindrical and concentric (Dahlstrom 1969, p. 220, fig. 3); that is, the fold axes lie in the plane of bedding and bedding units maintain constant thickness about the folds. Folds were drawn as sinusoidal and projected to depth from surface data using the methods of Busk (1929). However, Faill (1969, 1973), Laubscher (1976, 1977) and Roeder *et al.* (1978) have shown that folds in most fold and thrust belts possess kink-band geometries, with narrow hinges and long planar limbs. Since folds are rarely concentric, the Busk method cannot be applied to the construction of cross sections in kink-folded terranes (Faill 1973, p. 1291).

Nevertheless, many geologists persist in the use of concentric fold models. Therefore, I have chosen to demonstrate the kink-band nature of thrust-belt folds by presenting maps, cross-sections, field photos and sketches, well data, and seismic interpretations drawn primarily from the Rocky Mountains of the western United States (Fig. 1). Additional examples have come from the Canadian Rocky Mountains and the central and southern Appalachians of the eastern United States. The following discussion deals primarily with folds in two-dimensional cross-sections and does not attempt to describe complexities which occur in the third dimension.

TRAILING EDGE RAMP ANTICLINES

Folds in a single thrust sheet fall into three classes: trailing edge, intraplate, and leading edge (Fig. 2), all of which possess kink-band geometries.

Trailing-edge folds develop over footwall ramps in the thrust surface (Fig. 4a), and contain two kink-bands, one caused by the footwall ramp itself and the other resulting from the cutoff of hangingwall units against the upper thrust surface. Kink planes in the hangingwall units originate from the base of the ramp (1 in Figs. 4a & b), the top of the ramp (2 in Figs. 4a & c), and bound the hangingwall cutoffs (3 in Figs. 4a & d).

Not all ramp anticlines are as simple as the ones in Fig. 4. Late in the movement of a thrust sheet, imbricate

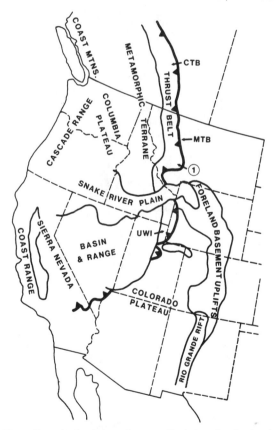

Fig. 1. Tectonic sketch map of western North America showing the Canadian thrust belt of Alberta and British Columbia (CTB), the Montana thrust belt (MTB) and the Utah–Wyoming–Idaho thrust belt (UWI).

325

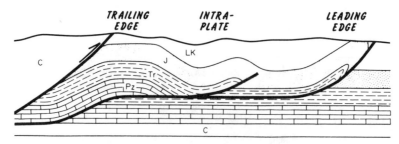

Fig. 2. Thrust anticlines are related to footwall ramps (trailing edge), shortening within the body of the sheet (intraplate), or fault propagation at the leading edge of a thrust (leading edge).

faults may initiate at the footwall ramp (Fig. 5a), resulting in a compound anticline. A cross-section through Carter Creek Field (Fig. 5b) by Lamerson (1982) shows an example of a compound ramp-anticline from the Utah–Wyoming thrust belt (location 1 in Fig. 3). The late imbrication on the Absaroka ramp may have occurred during propagation and movement on the lower and younger Darby thrust to the east.

INTRAPLATE FOLDS

Intraplate folds accommodate shortening strain within the body of a thrust sheet and are commonly cored by imbricate faults which propagate from the basal thrust (Fig. 2). Figure 6(a) shows, in simplified form, a typical fault-cored intraplate fold. Note that fold form is a function of position within the fold. The fold has a box-like configuration with two distinct kink planes at higher levels, but with depth the fold becomes a chevron

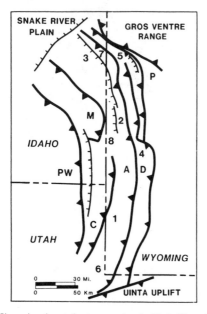

Fig. 3. Six major thrust sheets comprise the Utah–Wyoming–Idaho thrust belt: the Paris/Willard (PW), Meade (M), Crawford (C), Absaroka (A), Darby (D) and the Prospect (P). The Gros Ventre and Uinta uplifts, cored by Precambrian basement, were coincident with movement on the Prospect thrust and post-dated movement on the Darby thrust. Numbers are locations referred to in the text.

as the two kink-bands merge. A thrust-cored kink fold from the Appalachians (Fig. 6b), lacks the upper box-like form but demonstrates thrust displacement decreasing upward as it is replaced by shortening in the chevron fold.

Elk Horn anticline (Fig. 7) in the Montana thrust belt (Fig. 1, location 1) is a large version of an intraplate fold possessing a box-like geometry. It presumably tightens with depth, and seismic data and surface mapping suggest that folds of this type in this portion of the thrust belt are cored by imbricate faults above a basal thrust.

Bear Creek anticline in southeast Idaho (location 3 in Fig. 3) has been breached by erosion to reveal an imbricate fault and chevron fold of lower Paleozoic carbonate rocks in its core (Fig. 8a). At the Triassic level the enveloping fold has a kink-band geometry and plunges toward the south (Fig. 8b). The imbricate fault and fold are due to shortening within the Absaroka thrust sheet and the imbricate fault presumably joins the Absaroka thrust at depth.

As shown by Faill (1973, p. 1291), kink-band folds in the Pennsylvania Valley and Ridge Province have planar limbs and narrow hinges, a geometry which holds for fold wavelengths of a few centimeters to eighteen kilometers. An anticline 6 km northeast of Afton, Wyoming (Fig. 9a; location 2 in Fig. 3), developed in thick-bedded Paleozoic carbonates, and a small chevron fold in Cambrian shales and siltstones from the Appalachians (Fig. 9b) further demonstrate that kink-band folding occurs on all scales.

LEADING-EDGE FOLDS

Elliott (1976) suggested that folding precedes thrusting. A thrust propagates laterally and up-section with a fold at its tip. The forelimb of the advancing fold front is continuously being cut by the propagating thrust. Such 'fault-propagation folds' (Suppe & Medwedeff 1984) appear at the emergent edge of thrust sheets as 'leading-edge folds' (Fig. 2).

The Darby thrust in Wyoming (Fig. 3) clearly demonstrates the relationship between thrusting and leading-edge folds. From south to north, Darby thrust displacement at the Mississippian level decreases from >26 km (location 4 in Fig. 3) to 3 km (location 5) over a distance of 100 km, and at its northern surface termination

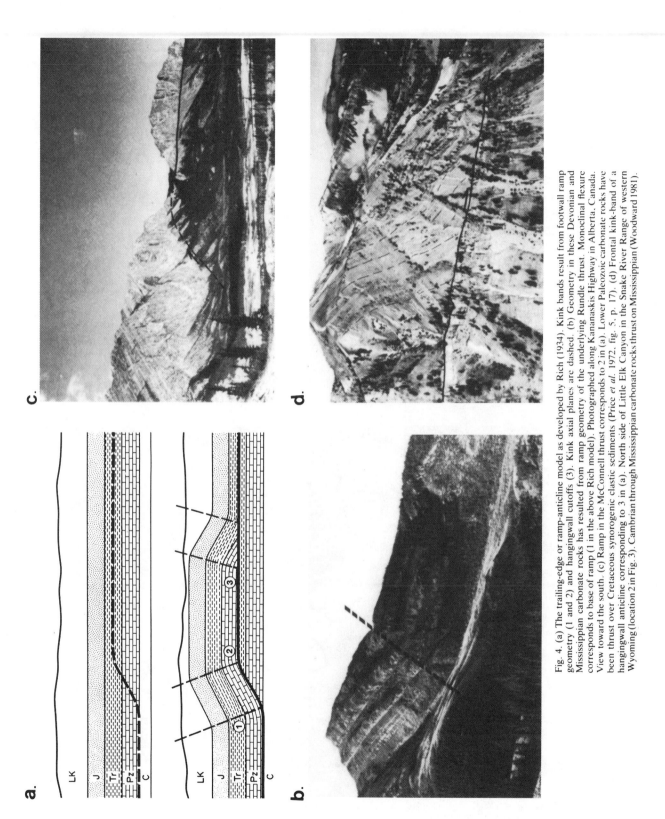

Fig. 4. (a) The trailing-edge or ramp-anticline model as developed by Rich (1934). Kink bands result from footwall ramp geometry (1 and 2) and hangingwall cutoffs (3). Kink axial planes are dashed. (b) Geometry in these Devonian and Mississippian carbonate rocks has resulted from ramp geometry of the underlying Rundle thrust. Monoclinal flexure corresponds to base of ramp (1 in the above Rich model). Photographed along Kananaskis Highway in Alberta, Canada. View toward the south. (c) Ramp in the McConnell thrust corresponds to 2 in (a). Lower Paleozoic carbonate rocks have been thrust over Cretaceous synorogenic clastic sediments (Price et al. 1972, fig. 5, p. 17). (d) Frontal kink-band of a hangingwall anticline corresponding to 3 in (a). North side of Little Elk Canyon in the Snake River Range of western Wyoming (location 2 in Fig. 3). Cambrian through Mississippian carbonate rocks thrust on Mississippian (Woodward 1981).

327

a.

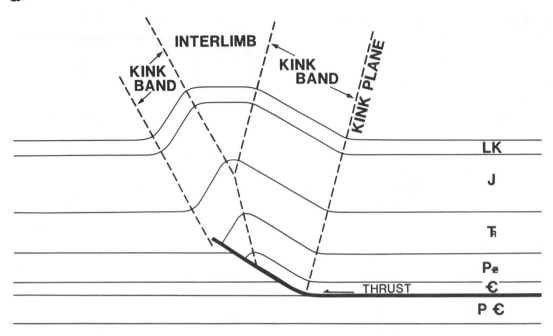

b.

Fig. 6. (a) In a thrust-cored kink anticline, fold shortening in the upper layers is balanced by faulting at lower levels. Fault displacement decreases upward as faulting gives way to folding. The lengths of all stratigraphic horizons are equal (adapted from figs. 6 and 9 of Faill 1973). (b) Minor thrust losing displacement into a kink fold. Upper Carboniferous Gizzard Group. Along Tennessee route 8 near Dunlap, Tennessee, in the Valley and Ridge Province. View towards the north.

Fig. 7. Elkhorn Anticline, viewed from the south. North side of Coal Gulch; Sections 15 and 16, Township 4 North, Range 7 East, Montana (Skipp 1977) (location 1 in Fig. 1). Mm, Mission Canyon Limestone; Pdt, Devils Pocket (dolomite) and Tyler (mudstone and siltstone with interbedded carbonates) Formations; Je, Ellis Group (sandstone, shale, and limestone); Jm, Morrison Formation (mudstone and shale interbedded with siltstone and sandstone).

a.

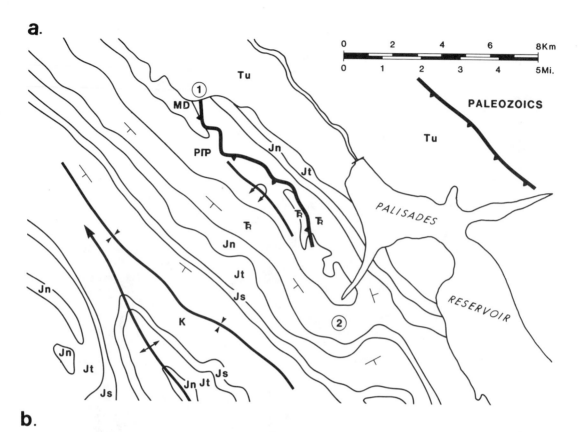

b.

Fig. 8. (a) Map of box fold (2), cored by unnamed thrust (1) (location 3 in Fig. 3). MD. Paleozoic carbonates; PP, Paleozoic clastics and carbonates; Tr, undifferentiated Triassic red-beds and carbonates; Jn, eolian sandstones (Nugget Formation); Jt, carbonates (Twin Creek Formation); Js, siltstones. shale. and sandstone (Preuss and Stump Formations); and K. Cretaceous clastics with minor carbonates (Gardner 1961). (b) Aerial photograph of box fold (2 in Fig. 8a) in Triassic red-beds. View towards the south.

330

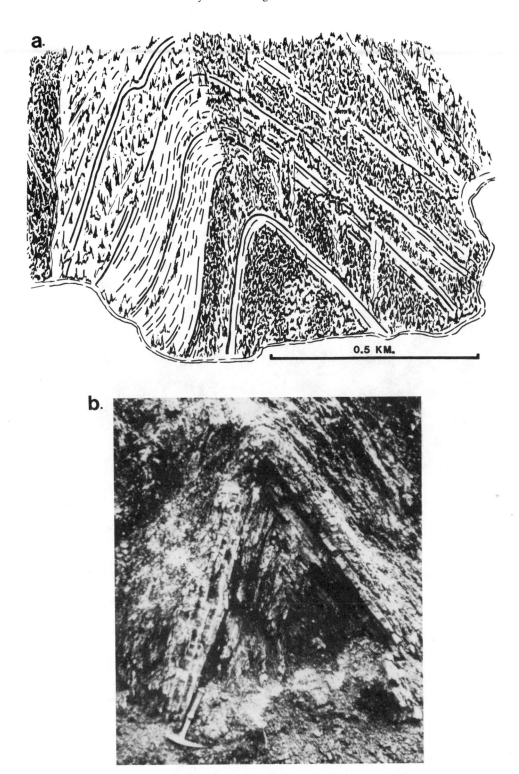

Fig. 9. (a) Chevron fold in Paleozoic carbonates. Northeast of Afton, Wyoming (Rubey 1973, location 2 in Fig. 3). Viewed from the north. Drawn from an aerial photograph. (b) Chevron fold in Cambrian shale and siltstone. Valley and Ridge Province of southwest Virginia.

331

Fig. 11. Chevron folds at the northern termination of the Lewis thrust. View towards the north. Photographed from the Kananaski's Highway, Alberta, Canada.

a.

b.

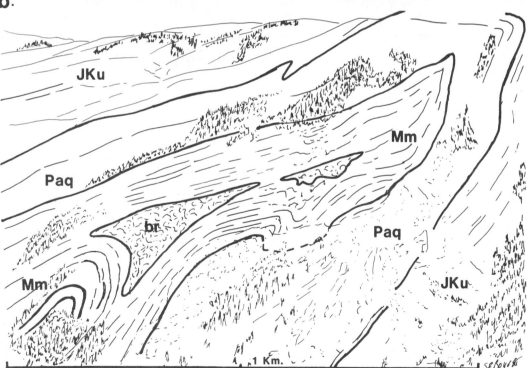

Fig. 15. (a) Aerial photograph of Middle Fork Anticline (Skipp & Hepp 1968), a collapse fold cored by Paleozoic carbonates. View toward the north-west. Location 1 in Fig. 1. (b) Sketch of Middle Fork Anticline as viewed from the south. Mm, Mission Canyon limestone; Paq, Quadrant sandstone and Amsden dolomite and claystone; JKu, undifferentiated limestone and clastics.

— This page intentionally blank —

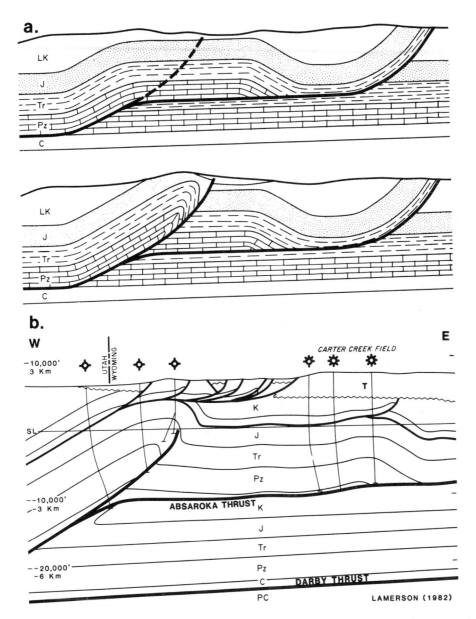

Fig. 5. (a) An out-of-sequence thrust imbricate (dashed) may initiate at a pre-existing thrust ramp. Movement on the later imbricate results in a compound ramp anticline. (b) Compound ramp anticline from the Wyoming thrust belt (location 1 in Fig. 3) (after Lamerson 1982).

(location 5) the Darby thrust passes into a box fold (Fig. 10a). A cross-section, constructed using seismic control and down-plunge projections of map data, shows that the thrust is dying up-section, as well as along strike, into a kink anticline (Fig. 10b). The Darby termination box fold, with its NW- and NE-plunging fold axes, demonstrates that such folds are usually noncylindrical. As a result, leading-edge folds can be expected to exhibit complex variations in geometry along a thrust trace.

Another example of a major thrust dying along strike into a fold or fold complex is the Lewis thrust of the Canadian and Montana thrust belts (Mudge & Earhart 1980, Price et al. 1972, p. 122 and fig. 72, p. 123,

Stockmal 1979). At Mt. Kidd the northern surface termination of the Lewis thrust is expressed as a series of chevron folds (Fig. 11).

DISHARMONIC FOLDING

The folds discussed up to this point have had orderly kink-band geometries with sharp hinges and planar limbs. Formational thicknesses are relatively constant in the limbs of these folds; but, as Ramsay (1974) has demonstrated, in chevron folds (a subset of kink-band folds) there may be a marked variation of thickness in hinge areas.

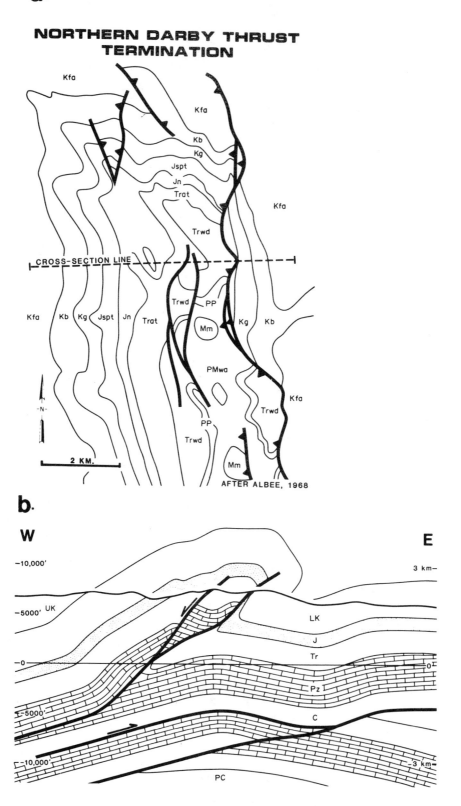

Fig. 10. (a) Map of the northern termination of the Darby thrust at Munger Mountain. Vicinity of location 5 in Fig. 3. Modified from Albee (1968). (b) Darby thrust dying upsection into fold. Thrust shortening in the Paleozoic and lower Mesozoic is balanced by fold shortening in the Upper Mesozoic.

a.

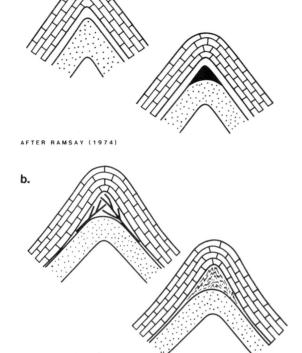

AFTER RAMSAY (1974)

b.

Fig. 12. (a) At left is a simple chevron model with uniform crestal thickening. In a real chevron fold (at right), with interbedded competent members (limestone and sandstone patterns) and incompetent shale or evaporites (unpatterned), competent units will maintain constant thickness and a void (black) will result. (b) This void may be filled by flowage, wedging or mineralization (not shown). After Ramsay (1974).

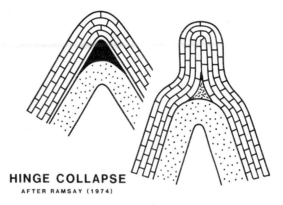

HINGE COLLAPSE
AFTER RAMSAY (1974)

Fig. 14. When an insufficient volume of incompetent material is available to fill void in the hinge area of a chevron fold, hinge collapse results (after Ramsay 1974).

In a simple chevron model all units maintain constant thickness in the limbs and undergo uniform thickening in the hinges (Fig. 12a). For interbedded sequences of competent and incompetent rocks, this model is unrealistic. Competent units will maintain constant thickness about the fold resulting in a void (Fig. 12a; Ramsay 1974, fig. 4, p. 1744). Such crestal voids become filled by (1) crystallization of mineral species which are mobile during deformation (Ramsay 1974, fig. 12, p. 1746), (2) flow of incompetent material from the limbs (Fig. 12b; Ramsay 1974, p. 1746 and fig. 14, p. 1747) or (3) wedging and flowage of interbedded competent and incompetent materials (Fig. 12b; Cloos 1961, fig. 12, p. 113).

When the incompetent material is of sufficient thickness extreme disharmonic folding results (Nickelsen

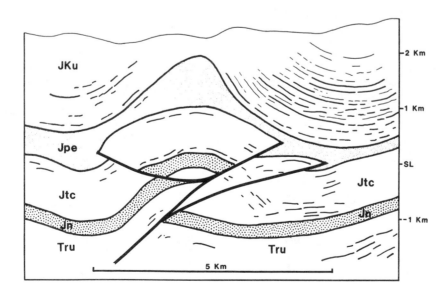

Fig. 13. Disharmonic folding in Mesozoic clastics and carbonates. Drawn from a seismic profile (location 8 in Fig. 3). JKu, Cretaceous synorogenic and Jurassic marine clastics; Jpe, Preuss evaporites; Jtc, Twin Creek argillaceous limestone; Jn, Nugget eolian sandstone; Tr, red-beds and carbonates. Traced from a seismic line.

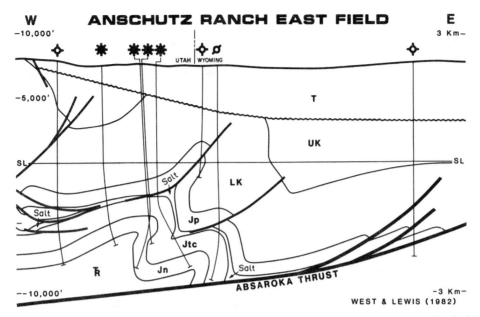

Fig. 16. Cross section through Anschutz Ranch East Field in the Utah–Wyoming thrust belt (location 6 in Fig. 3). Collapse fold has resulted from flowage of evaporites which separate Jurassic Twin Creek Limestone (Jtc) from Jurassic Preuss/Stump fine-grained clastics (Jp). Jn, Nugget sandstone; Tr, red-beds and carbonates; LK and UK, undifferentiated Upper and Lower Cretaceous synorogenic clastic sediments (after West & Lewis 1982).

1979, Nickelsen & Cotter 1983, p. 128 and fig. III-2, p. 130). In the Utah–Wyoming–Idaho thrust belt (Fig. 2) evaporites at the base of the Jurassic Preuss marine clastics are responsible for much of the disharmonic folding. The lower Mesozoic section is often folded into tight synclines and broad faulted anticlines, whereas the upper Mesozoic rocks above the evaporitic unit are independently folded into tight anticlines and broad synclines (Fig. 13).

As stated previously the crestal void between competent units can be filled by flowage or imbrication of interbedded incompetent materials (Fig. 12b). However, when there is insufficient incompetent material to fill the void, hinge collapse results (Fig. 14; Ramsay 1974, pp. 1746–7, fig. 13, p. 1747, and fig. 15, p. 1748).

Middle Fork anticline (Skipp & Hepp 1868) provides an excellent example of a large-scale hinge-collapse fold (Fig. 15). The fold, located northeast of the Bridger Mountains of Montana (1 in Fig. 1), is composed of the Mission Canyon Limestone of the Mississippian Madison Group. In Montana an anhydrite/gypsum unit, measuring 8–15 m in thickness, is located in the middle of the Mission Canyon Limestone. The evaporite zone provided a glide zone for disharmonic folding between the upper and lower Madison Group. However, the amount of slip was so great that the thin evaporite zone could not fill the resulting void in the hinge zone, leading to hinge collapse.

Drilling at the Anschutz Ranch East field in the Utah–Wyoming thrust belt (location 6 in Fig. 3) has revealed another collapse feature (Fig. 16). As in Fig. 13, the cause of the disharmony is an evaporite unit within the Jurassic sequence.

CONCLUSIONS

Most thrust-belt folds, including trailing-edge, intraplate, and leading-edge folds, and folds at the lateral terminations of major thrusts, possess kink-band geometries. The positions of kink-bands in trailing-edge ramp-anticlines are controlled by the geometry of the underlying thrust. Intraplate folds are due to bulk shortening within the body of a thrust sheet and are cored by imbricate faults which splay from the basal thrust. Leading edge anticlines form as the frontal part of a thrust as a result of thrust propagation. These frontal anticlines are often removed by erosion, but analogous folds can be observed at the lateral terminations of major thrusts.

Not all kink-band folds are orderly. Varying degrees of disharmonic folding may result from the presence of varying quantities of interbedded incompetent units. Hinge-collapse folds, an extreme case of disharmonic folding, result when the incompetent material is too thin to fill crestal voids, but is of sufficient thickness to allow interbed slip.

Acknowledgements—I wish to thank Jake Hossack, J. P. Platt and anonymous reviewers for their helpful reviews and SOHIO Petroleum Co. for permission to publish this paper. I first visited several of the structures described in this manuscript while employed with the Atlantic Richfield Co. I thank Atlantic Richfield Co. for the opportunity to conduct field work in the Rocky Mountains and would like to acknowledge helpful discussions with fellow ARCO field party members from 1978 to 1980. Mark Purington of SOHIO drafted the figures and numerous revisions of the text were typed by Kathryn Rechel.

REFERENCES

Albee, H. F. 1968. Geologic map of the Munger Mountain quadrangle, Teton and Lincoln Counties, Wyoming. *U.S. Geol. Surv.* Map GQ-705.

Busk, H. G. 1929. *Earth Flexures*. Cambridge University Press, London.

Cloos, E. 1961. Bedding slips, wedges, and folding in layered sequences. *Extrait C. r. Soc. géol. Finlande* **33**, 105–122.

Dahlstrom, C. D. A. 1969. The upper detachment in concentric folding. *Bull. Can. Petrol. Geol.* **17**, 326–346.

Elliott, D. 1976. The energy balance and deformation mechanisms of thrust sheets. *Phil. Trans. R. Soc.* **A283**, 289–312.

Faill, R. T. 1969. Kink band structures in the Valley and Ridge Province, Central Pennsylvania. *Bull. geol. Soc. Am.* **80**, 2539–2550.

Faill, R. T. 1973. Kink band folding, Valley and Ridge Province, Pennsylvania. *Bull. geol. Soc. Am.* **84**, 1289–1314.

Gardner, L. S. 1961. Preliminary geologic map of the Irwin quadrangle. *U.S. Geol. Surv.* Open File Report OF 61-53.

Lamerson, P. R. 1982. The Fossil Basin area and its relationship to the Absaroka thrust fault system. In: *Geologic Studies of the Cordilleran Thrust Belt* (edited by Powers, R. B.). Rocky Mountain Association of Geologists, 279–340.

Laubscher, H. P. 1976. Geometrical adjustments during rotation of a Jura fold limb. *Tectonophysics* **36**, 347–366.

Laubscher, H. P. 1977. Fold development in the Jura. *Tectonophysics* **37**, 337–362.

Mudge, M. R. & Earhart, R. L. 1980. The Lewis thrust fault and related structures in the Disturbed Belt, northwestern Montana. *Prof. Pap. U.S. Geol. Surv.* **1174**.

Nickelsen, R. P. 1979. Sequence of structural stages of the Alleghany orogeny at the Bear Valley Strip Mine, Shamokin, Pennsylvania. *Am. J. Sci.* **279**, 225–271.

Nickelsen, R. P. & Cotter, E. 1983. Silurian depositional history and Alleghanian deformation in the Pennsylvania Valley and Ridge. 48th Annual Field Conf. of Pennsylvania Geologists.

Price, R. A. *et al.* 1972. The Canadian Rockies and tectonic evolution of the southeastern Canadian Cordillera. *XXIV Int. Geol. Congress*, Excursion AC 15.

Ramsay, J. G. 1974. Development of chevron folds. *Bull. geol. Soc. Am.* **85**, 1741–1754.

Rich, J. L. 1934. Mechanics of low-angle overthrust faulting illustrated by Cumberland thrust block, Virginia, Kentucky, and Tennessee. *Bull. geol. Soc. Am.* **18**, 1584–1596.

Roeder, D., Yust, W. W. & Little, R. L. 1978. Folding in the Valley and Ridge Province of Tennessee. *Am. J. Sci.* **278**, 477–496.

Rubey, W. W. 1973. Geologic map of the Afton quadrangle and part of the Big Piney quadrangle, Lincoln and Sublette Counties, Wyoming. *U.S. Geol. Surv. Misc. Geol. Inv.* Map I-686.

Skipp, B. 1977. Geologic map and cross section of the Wallrock quadrangle, Gallatin and Park Counties, Montana. *U.S. Geol. Surv.* Map GQ-1402.

Skipp, B. & Hepp, M. 1968. Geologic map of the Hatfield Mountain quadrangle, Gallatin County, Montana. *U.S. Geol. Surv.* Map GQ-729.

Stockmal, G. S. 1979. Structural geology of the northern termination of the Lewis thrust, Front Ranges, southern Canadian Rocky Mountains. Unpublished M.Sc. thesis, The University of Calgary, Alberta.

Suppe, J. & Medwedeff, D. A. 1984. Fault-propagation folding. *Abs. with Programs geol. Soc. Am.* **16**, 670.

West, J. & Lewis, H. 1982. Structure and palinspastic reconstruction of the Absaroka thrust, Anschutz Ranch area, Utah and Wyoming. In: *Geologic Studies of the Cordilleran Thrust Belt* (edited by Powers, R. B.). Rocky Mountain Association of Geologists, 633–640.

Woodward, N. B. 1981. Structural geometry of the Snake River Range, Idaho and Wyoming. Unpublished Ph.D thesis, Johns Hopkins University, Baltimore, Maryland.

The American Association of Petroleum Geologists Bulletin
V. 62, No. 6 (June 1978), P. 984-1003, 16 Figs.

Fold Development in Zagros Simply Folded Belt, Southwest Iran[1]

S. P. COLMAN-SADD[2]

Abstract In the simply folded belt of the Zagros Mountains, a sequence of Precambrian to Pliocene shelf sediments about 12 km thick has undergone folding from Miocene to recent time. Much of the section (6,000 to 7,000 m), consisting of Cambrian to Miocene rocks, forms a single structural lithic unit, the Competent group. It is bounded above and below by detachment zones in evaporite deposits. Structures in the Competent group are typical of parallel folds formed by buckling and developed by a combination of flexural-slip and neutral-surface mechanisms. They include bedding-plane slickensides, extension structures on anticlinal crests, and congested anticlinal and synclinal fold cores. The neutral-surface component of folding has had an important influence on fluid migration. The asymmetry of Competent group folds reflects shearing in the lower detachment zone. The enormous size of the folds is the result of many factors acting together; chief among these is the great thickness of the structural unit.

Folding induced by salt movement may have occurred in the Competent group but is unrelated to the Cenozoic buckle folds; it provides a mechanism for salt diapirism through competent strata, and an explanation of how room was made for diapirs and why they rarely contain relics of country rocks. Preexisting diapirs have been reactivated in anticlines by the tectonic stresses causing buckling, but their movement generally has been halted in synclines. Diapirs are unlikely to have been initiated during buckle folding.

The basement has not taken part in the folding, but instead has been deformed by strike-slip faulting.

INTRODUCTION

The Zagros Mountains[3] trend southeast through northeastern Iraq and southwestern Iran. They are the topographic expression of an orogenic belt in which deformation began in the Late Cretaceous, but has been most extensive since the Miocene; recent earthquakes (Nowroozi, 1972; McQuillan, 1973a) are evidence that the orogeny continues to the present.

Falcon (1969) divided the Zagros orogene into three structural belts with strikes paralleling the mountain trend (Fig. 1). From southwest to northeast these are:

Simply folded belt—A thick sequence of late Precambrian to Pliocene shelf sediments, without any visible angular unconformity, has been folded into a series of huge anticlines and synclines; the folding has taken place since the Miocene and is reflected in the topography, which is dominated by anticlinal mountains and synclinal valleys. The anticlinal oil traps of Iran and northeast Iraq are in this belt.

Imbricated belt—Sediments similar to those in the simply folded belt have been deformed more

intensely with the development of a series of thrust faults; deformation initially took place in the Late Cretaceous and has continued since the Miocene.

Thrust belt—A wide variety of lithologies including crushed limestones, radiolarites, and ultrabasic and metamorphic rocks has been intensely thrust faulted in this belt.

The Zagros orogenic belt is bounded by two stable platforms. On the southwest is the Arabian platform, where shelf sediments laterally equivalent to those in the simply folded belt are virtually undeformed and overlie the metamorphic rocks of the Arabian shield. On the northeast is the Precambrian metamorphic basement of central Iran, overlain by Paleozoic and Mesozoic shelf sediments and a chain of Cenozoic and possibly Mesozoic volcanoes that parallel the main Zagros trend (Stöcklin, 1968a).

The Zagros deformation has been attributed variously to the interaction of crustal plates (e.g., Haynes and McQuillan, 1974) or to uplift in the

[1]Manuscript received, April 4, 1977; accepted, December 2, 1977.

[2]Department of Mines and Energy, Government of Newfoundland and Labrador, St. John's, Newfoundland, Canada.

Field observations were made while the writer was an instructor at Pahlavi University, Shiraz, Iran. Special thanks are due to S. Edgell, S. J. Haynes, and H. McQuillan for transport, guidance, and hospitality in Iran. The writer also thanks R. Gibbons and I. Knight for critical reading of the manuscript, and K. Byrne, W. Howell, and G. Martin for drafting the figures. Publication authorized by B. Greene, Director, Mineral Development Division, Newfoundland Department of Mines and Energy.

[3]Editor's note:

Because many of the geographic spellings used by Colman-Sadd are long familiar to western oil-oriented readers we have retained them, contrary to usual *Bulletin* practice which follows that of *National Geographic Atlas of the World*; thus "Zagros Mountains" rather than "Kūhha-ye Zāgros."

Article Identification Number
0149-1423/78/B006-0003$03.00/0

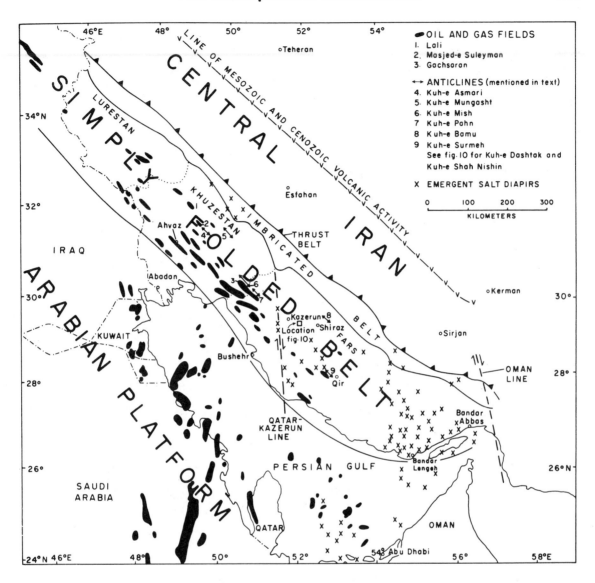

FIG. 1—Location map, southwest Iran and Persian Gulf.

final stages of the classic geosynclinal cycle (e.g., Kashfi, 1976). The object of this paper is to examine the way the rocks of the simply folded belt reacted to the deformation, the fundamental causes of which are not discussed. The intention is to show how the structures of the simply folded belt illustrate some of the general principles of structural geology, and to use these principles to interpret the structures at depth.

STRATIGRAPHIC AND STRUCTURAL DIVISIONS

The stratigraphy of the Zagros simply folded belt has been described in detail in numerous publications by oil geologists working in the re-

gion (e.g., Lees and Richardson, 1940; Dunnington, 1958; James and Wynd, 1965). For the purposes of this paper the generalized stratigraphic column published by the British Petroleum Company Ltd. (1956) is sufficient (Fig. 2). The stratigraphic column is divided into the five structural divisions of O'Brien (1950), which are crucial to an understanding of the Zagros folding. The divisions are:

Basement group—All information on the basement rocks is derived from fragments brought to the surface in salt diapirs. The fragments include granite, gabbro, basalt, amphibolite, and schist (Harrison, 1930; Kent, 1970; Haynes and Mc-Quillan, 1974).

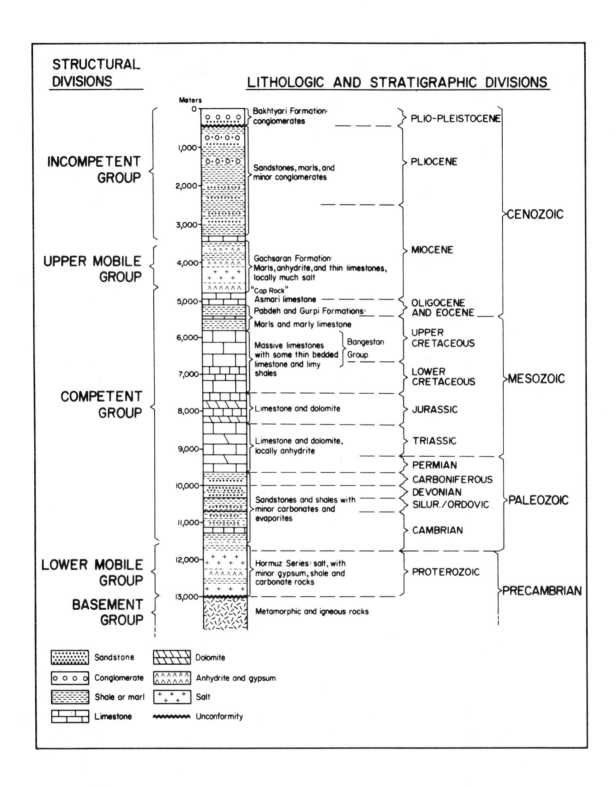

FIG. 2—Generalized structural, lithologic, and stratigraphic divisions, Zagros simply folded belt, southwest Iran (modified after O'Brien, 1950; British Petroleum Co. Ltd., 1956).

Lower Mobile group—This group consists of the Hormuz salt which is not exposed in place, but is present as numerous salt diapirs, particularly in Fars Province (Fig. 1). The salt is associated with gypsum, shale, and carbonate rocks, as well as the igneous and metamorphic rocks of the basement; the complete heterogeneous suite comprises the Hormuz Series (Blanford, 1872). For many years the series was considered to be Early Cambrian, but at least part of it is of late Proterozoic age (Stöcklin, 1968b; Kent, 1970).

Competent group—The thickest group in the stratigraphic sequence includes sediments ranging from Cambrian to Miocene in age. A comparatively thin development of Cambrian to Carboniferous rocks (in the order of 2,000 m) consists principally of shales and sandstones with subordinate carbonate rocks and evaporites. The Permian to Upper Cretaceous section is dominated by carbonate rocks with minor marls, shales, and evaporites; the thickness is generally 3,000 to 4,000 m. The upper part of the group consists of Upper Cretaceous to Oligocene shales and limestones, overlain by the massive oil-bearing Asmari limestone and its caprock of anhydrite (basal part of the Gachsaran Formation); the thickness of these rocks is about 1,000 m. The total thickness of the Competent group is 6,000 to 7,000 m.

Upper Mobile group—The Competent group is overlain by the Miocene salt, gypsum, anhydrite, and marls of the central part of the Gachsaran Formation. The structural mobility of these rocks is famous (Lees and Richardson, 1940; O'Brien, 1950; Hills, 1963; de Sitter, 1964) and it is difficult to measure their stratigraphic thickness; in the composite type section at the Gachsaran oil field the saliferous members (2 to 6) are thought to be about 2,000 m thick (James and Wynd, 1965), but this is exceptional.

Incompetent group—The remaining part of the stratigraphic sequence consists of a variety of thin-bedded marls, shales, sandstones, and conglomerates, with minor anhydrite (lower Miocene to Pliocene), unconformably overlain by the Pliocene to Pleistocene syntectonic conglomerates of the Bakhtyari Formation. The thickness of the Incompetent group varies, reflecting unstable tectonic conditions during its deposition; an average figure is between 3,000 and 4,000 m.

The general behavior of O'Brien's structural divisions during deformation is well established. The Basement group is thought either to have remained rigid or to have undergone block faulting (Lees, 1952; Falcon, 1969). The Competent group has become detached from the basement and has been folded independently; the Lower Mobile group forms the zone of detachment (Lees, 1950; O'Brien, 1950). The folds in the Competent group are huge, relatively simple structures (Figs. 3, 11) and thrust faulting is not extensive in the simply folded belt (although it is in the imbricated belt). The Incompetent group has been complexly folded and thrust (Fig. 4); its structures bear little relation to the simpler folds in the Competent group below. The two disharmonically folded groups are separated by a detachment zone in the Upper Mobile group (Lees and Richardson, 1940; O'Brien, 1950).

The main oil accumulations of the Zagros Mountains are in anticlinal traps in the Asmari limestone at the top of the Competent group. As a result of exploration for oil, the structures in the overlying Incompetent and Upper Mobile group and their relations to the Competent group have become well known. Interest in the Competent group, however, generally has been limited to the extreme upper parts of anticlines where the oil is trapped; the natures of the synclines and the anticlinal cores have remained largely obscure (Falcon, 1969, 1974; Hull and Warman, 1970). The following discussion therefore is concentrated on the development of structures in the Competent group, and the reader is referred to O'Brien (1950, 1957) and Falcon (1969) for further details of folding in the Incompetent group.

PARALLEL FOLDING

Definition

Van Hise (1896) distinguished two principal kinds of folds, parallel and similar. In parallel folds the folded layers ideally have parallel bounding surfaces and maintain a constant thickness perpendicular to these surfaces. In contrast the thickness of the layers in similar folds varies, commonly increasing on the hinges and decreasing on the limbs. The Competent group of the Zagros Mountains generally conforms to the parallel style; the massive carbonate members that dominate the group maintain a more or less constant thickness from fold limb to fold hinge. Incompetent members of the group, notably anhydrite beds, may vary considerably in thickness, but because they are quantitatively of minor importance they can be considered as little more than lubrication between the layers of carbonate rock.

Concentricity and Detachment

Ideal parallel folds can be depicted as a series of approximately circular arcs, hence the synonym, concentric folds. On any one fold the radius of curvature in the layers on the outer arc of the

FIG. 3—Looking southwest from Kuh-e Shah Nishin anticline. In foreground is horizontal Bangestan limestone on outer arc of anticline. Asmari limestone, forming escarpment in middle distance, dips into syncline shown in Figure 10 and reappears on dip slope of Kuh-e Dashtak in far distance. Distance to crest of Kuh-e Dashtak is about 10 km.

FIG. 4—Recumbent folding in Incompetent group, northwest of Bandar Lengeh. Structure is presumed to overlie disharmonically much simpler upright folds in Competent group.

fold is much greater than in the layers near the core (Fig. 5). In anticlines there is therefore a lower limit below which the folds no longer can follow the ideal parallel style; there is a corresponding upper limit in synclines.

Because parallel folds cannot continue indefinitely either upward or downward, they must change in style to similar or chevron folds, or the folded layer must become detached from the rocks both above and below. In the case of the Competent group of the Zagros Mountains, there is a lower zone of detachment in the Lower Mobile group, and an upper zone in the Upper Mobile group.

The level of erosion in the Zagros Mountains is such that only the upper part of the Competent group normally is exposed. Thus we see the outer arcs of the anticlines and the inner arcs or cores of the synclines (Fig. 5). This results in an impression of broad anticlines (Fig. 3) and narrow, pinched synclines (Figs. 10, 11, 12; Hull and Warman, 1970; McQuillan, 1973a; Haynes and McQuillan, 1974). A much deeper level of erosion would give the opposite effect of pinched anticlines and broad synclines; the latter situation is well illustrated by the Franklinian fold belt of the Canadian Arctic where a largely carbonate sequence of Ordovician to Devonian rocks has undergone detachment folding above a layer of Ordovician evaporites (Tozer and Thorsteinsson, 1964; Kerr, 1974).

Mechanisms of Parallel Folding by Buckling

Folds formed by compression parallel with the layering are known as buckle folds; those formed by differential forces perpendicular to the layering are bending folds (Ramberg, 1970). Discussion herein is restricted to buckle folds and shows how the northwest-trending folds in the Competent group conform to this mode of formation. Upright buckle folds are formed by a regional maximum principal stress normal to the fold axes and tangential to the earth's surface; the intermediate principal stress is parallel with the fold axes, and the minimum principal stress is vertical.

Flexural-Slip and Neutral-Surface Folding

Folding by buckling is achieved by two mechanisms, flexural-slip and neutral-surface folding (Hills, 1963; Ramsay, 1967; Fig. 6). Flexural-slip folding takes advantage of the planar inhomogeneity of most rocks, in this case sedimentary bedding. During buckling, individual layers slip over each other away from the constricted fold cores toward the outer arcs of folds. The maximum relative movements of the layers and the maximum strain of the whole multilayered se-

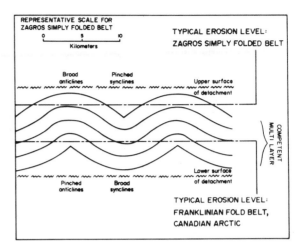

FIG. 5—Concentricity of parallel folds. At higher erosion level anticlines are broad and synclines pinched; at lower erosion level reverse is true. Potential faulting is omitted.

quence occur in the fold limbs; in the hinges there is no relative movement and the strain is minimal.

In ideal neutral-surface folding there is no slip parallel with layer boundaries. Therefore, the rocks in the core of a fold do not move outward toward the limbs as folding progresses. They are compressed by a local maximum principal stress, parallel with the layer boundaries and normal to the fold axis, whereas the rocks on the outer arc of the fold are stretched parallel with a similarly oriented minimum principal stress. The local maximum principal stress on the outer arc of an upright fold is usually gravity. Within the buckled layer is a surface of no finite strain, the neutral surface, which separates the stretched outer arcs of folds from the compressed cores. The neutral surface may be located anywhere within the layer, may cut across the layering at a small angle, and may change its position through time. In a buckle formed by neutral-surface folding, the strain is greatest in the fold hinges and least in the fold limbs, which is opposite to flexural-slip folds.

Parallel folding generally is achieved by a combination of the two mechanisms. The element of flexural-slip folding is likely to increase with decreasing cohesion of the internal layering. Neutral-surface folding is favored by a high wavelength to thickness ratio which allows greater amounts of shortening for a given amount of congestion and strain in fold cores.

Faults Associated with Folding

Flexural-slip folding is accomplished by "faulting" parallel with the boundaries of the buckled

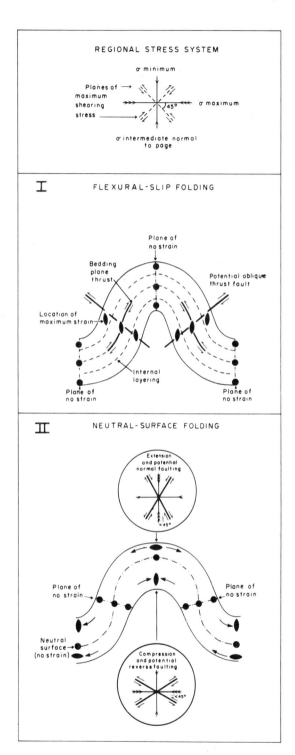

FIG. 6—Parallel folds developed by (I) flexural slip, (II) neutral-surface folding.

layer. The planes of maximum shearing stress, which are close to potential fault planes (Anderson, 1951), lie at 45° to the maximum and minimum principal stresses. As soon as the folding begins, the internal layering in each limb of the fold begins to rotate toward parallelism with one of these planes; the complementary plane on each fold limb crosses the limb obliquely (Fig. 6,I). In a multilayered series the initial weakness parallel with the layers causes shearing to take place preferentially along layer boundaries, resulting in flexural slip along bedding-plane thrusts; at first, oblique thrusts are unlikely to form. However, as folding progresses, the layering in each fold limb is rotated past the plane of maximum shearing stress and the incremental movement parallel with the layer boundaries diminishes (Ramsay, 1967, p. 393). Movement related to the complementary planes then may take place and thrust faults may cut through the fold limbs. In asymmetric folds the thrusts form across the steepest limbs, as these have been rotated farthest.

The greatest strain in a neutral-surface fold is in the hinge (Fig. 6,II). The outer arc of the fold is stretched, creating extension joints at shallow depths and, if the fold is upright, normal faults because the local maximum principal stress is usually gravity. In the core of the fold, compression occurs. The local maximum principal stress parallels the layering and is normal to the fold axis. The minimum principal stress is vertical, normal to both the fold axis and the layer boundaries. The stress pattern causes conjugate sets of reverse or thrust faults to form across the layering. Theoretically no faults will cross the neutral surface, provided it is spatially static, for it is a surface of no strain.

Excessive compression in fold cores may be accompanied by straightening of the limbs and yielding of the hinges to form chevron folds (Johnson and Honea, 1975). Such folds involve slip between competent layers and thickness variation in incompetent layers. They may continue indefinitely normal to the layering and, as folding becomes tighter, the chevron style may propagate upward in the anticlines and downward in the synclines.

Parallel Folding in Simply Folded Belt

The folding of the Competent group has been accomplished by a combination of the flexural-slip and neutral-surface mechanisms.

Evidence of Flexural-Slip Folding

Indications of movement between the layers of the Competent group are provided by numerous bedding-plane slickensides on fold limbs (Fig. 7).

FIG. 7—Reverse fault cutting Permian-Triassic dolomite in core of Kuh-e Surmeh anticline. Slickensides on bedding surface indicate degree of flexural-slip folding.

The slickensides are normal to the fold axes. Other evidence of bedding-plane movement is the variation in the thickness of individual, incompetent evaporite beds between massive carbonate members, especially in the more deeply eroded anticlinal cores (e.g., Kuh-e Surmeh), where folding has tended toward the chevron style and the evaporites have been squeezed from the limbs to the hinges of the folds.

Evidence of Neutral-Surface Folding

The outer arcs of anticlines in the Competent group are well exposed and provide numerous examples of extension structures. Compressive structures in fold cores can be observed in the few anticlines that have been eroded or drilled below the neutral surface, and can be seen in some of the synclines where they have not been hidden by

alluvium. None of the synclines have been eroded deeply enough to expose any of the extension structures that may be present in their lower parts beneath the neutral surface.

Extension structures—Most of the anticlines capped by the Asmari limestone have normal faults on or near their crests. The faults are parallel, normal, or oblique to the fold axes. Documented examples near the Khuzestan oil field belt are the grabens parallel with the fold axes of Kuh-e Pahn (McQuillan, 1973b) and Kuh-e Asmari (McQuillan, 1974); normal fault patterns in the Qir area of Fars Province are shown on the map of McQuillan (1973a); Figure 8 shows the normal fault at Dasht-e Arjan, which forms the northwestern wall of the graben crossing the Kuh-e Shah Nishin anticline west of Shiraz. The

FIG. 8—Normal fault cutting Asmari limestone. Fault forms northwest wall of Dasht-e Arjan graben on crest of Kuh-e Shah Nishin anticline, between Shiraz and Kazerun.

faults extend downward at least into the Creta-ceous Bangestan Group where they are exposed, for example, in the Tang-e Gurguda section through Kuh-e Mish.

The oil fields in the Asmari limestone depend for their excellent reservoir properties on wide-spread fracturing. McQuillan (1974) studied the distribution and orientation of fractures visible on 1:30,000 air photographs of the Kuh-e Asmari anticline; he found that the fractures had two dominant attitudes, parallel with and normal to the fold axis, and that they provided "a textbook example of structure-related tension-fracture sets." Smaller scale ground-observed fractures show a more random orientation and distribu-tion; McQuillan (1973b) concluded that they pre-dated the main development of the fold and were controlled by depositional and diagenetic factors. There is no marked development of shear joints (McQuillan, 1974). This can be attributed to the fact that the limestone lies at the top of the Com-petent group. On anticlines such as Kuh-e As-mari, it is located on the outer arc of the fold and has been subject to lateral extension rather than compression; structures such as shear joints can be expected in the neighboring synclines where the limestone is presumed to be in tightly com-pressed fold cores.

Compressive structures—A deep hole drilled into the Masjed-e Suleyman anticline to test pre-Asmari formations discovered a thrust fault that repeats the middle Jurassic section in the core of the fold (Baniriah et al, 1967). Only a few anti-clines in the simply folded belt have been eroded deeply enough to expose such compressive struc-tures. One of these is Kuh-e Surmeh in central Fars Province, where Paleozoic rocks are ex-posed; Figure 9 shows the northwest-plunging end of the anticline, with tightly folded Permian-Triassic dolomite and anhydrite near the core of the fold, and overlying more broadly folded Me-sozoic limestone in the middle distance. The Permian-Triassic dolomite is reverse faulted (Fig. 7) and locally a chevron style of folding is devel-oped with a sharp hinge and straight limbs; the room problem caused by this style of folding has been solved by the squeezing of anhydrite from the limbs to the hinge. By contrast the geometry of the anticline in the overlying Mesozoic and Ce-nozoic sediments is broad and concentric; the only sign of congestion is a thrust fault in part of the oversteepened southwest limb.

FIG. 9—Northwest plunging end of Kuh-e Surmeh anticline, showing tightly folded Permian-Triassic rocks in fold core overlain by more broadly folded Mesozoic limestone in middle distance. Three bedding planes are outlined.

Congestion in anticlinal cores is significant in indicating the type of folding in the simply folded belt. It was stated earlier that folding could be caused either by buckling (compression parallel with the layering) or by bending (differential forces perpendicular to the layering). Bending, caused for example by the uplift of basement blocks or by salt movement at depth, results in stretching of the entire overlying stratigraphic section, for in the absence of lateral shortening the surface area of the folded layer must be greater than that of the original horizontal layer. The prominent compressive structures in anticlinal cores can be explained only by lateral shortening caused by a regional maximum principal stress tangential to the earth's surface and normal to the fold axes.

Compression in synclinal cores above the neutral surface was illustrated by Hull and Warman (1970, Fig. 7). They show broad anticlines in the Asmari limestone and underlying Cretaceous rocks, separated by an isoclinal syncline in which the two limbs of Asmari limestone lie one on top of the other. There are examples of this type of

folding close to the imbricated belt at Kuh-e Bamu, north of Shiraz, and Kuh-e Mungasht northeast of the oil field belt (Fig. 1).

In most places in the simply folded belt, synclines in the Competent group are hidden by disharmonically folded rocks of the Upper Mobile and Incompetent groups; they are obscured further by deposits of alluvium. The detailed structure is therefore generally a matter of guesswork (Falcon, 1969, 1974; Hull and Warman, 1970). There are however a few examples where the structure can be determined, and from which the style of deformation in less exposed areas can be inferred. One such syncline lying between Kuh-e Dashtak and Kuh-e Shah Nishin on the Shiraz-Kazerun road is described here (Figs. 1, 10). The alluvial cover is broken just southeast of the main road across most of the width of the structure. The salt of the Upper Mobile group is poorly developed in this area and the sediments of the lower part of the Gachsaran Formation have been folded in harmony with the Competent group. The oldest rocks exposed in the syncline belong to the feature-forming Asmari limestone. They

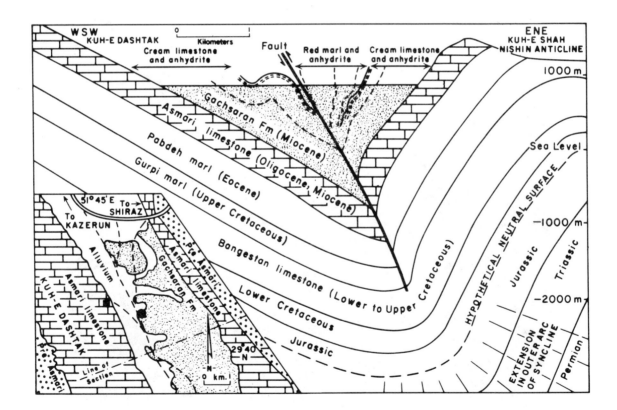

FIG. 10—Section through syncline between Kuh-e Dashtak and Kuh-e Shah Nishin, based on surface geology and extrapolated downward by analogy with syncline shown by Gansser (1964, photo 18). Neutral surface is assumed in Jurassic and congestion in fold core has been partly relieved by flexural slip. Solid squares indicate villages.

are overlain by approximately 600 m of thin-bedded, cream limestone and anhydrite, above which is at least 800 m of interbedded red marl and anhydrite with minor limestone. On the southwest limb of the syncline, the Asmari limestone and the limestone and anhydrite unit dip at angles of less than 40° (Figs. 10, 11); the red marl and anhydrite unit is present only locally and is absent on the line of section. On the northeast limb, the Asmari limestone has dips up to 60° (Fig. 12) and the dip gradually increases into the center of the syncline where the red marls and anhydrite are overturned. The two limbs are separated by a zone of brecciated limestone 50 m wide, which is interpreted as a reverse fault; there are no other structural discontinuities and there is no evidence of gravitational slumping. The interpretation of the syncline at depth is based on published photographs of parallel folds, and in particular on a syncline in the Himalayas shown by Gansser (1964, photo 18). The reverse fault is of a kind present in some fold cores (Ramsay, 1967, Figs. 6-2, 7-66); in its lower part the strata on the hanging wall are older than the strata on the foot wall, whereas in its upper part the opposite is true. The changing age relation between foot wall and hanging wall has been caused by different angles between bedding and the fault plane on the two limbs of the syncline and by flowage in the incompetent rocks of the Gachsaran Formation; it does not indicate any change in the sense of movement on the fault. In Figure 10 a neutral surface is assumed in the Jurassic, which is its approximate position in the Kuh-e Surmeh anticline; a generous allowance has been made for folding by flexural slip, as this is indicated by numerous bedding-plane slickensides.

Neutral-Surface Folding and Oil Migration

Normal faults and extension joints related to neutral-surface folding have played an important role in the migration of oil in the simply folded belt. In the principal oil fields of the belt, reservoir rocks in the Tertiary Asmari and Cretaceous Bangestan limestones are in fluid connection (the Lali oil field is a notable exception; British Petroleum Co. Ltd., 1956). Along the margins of the fold belt, however, in the less pronounced structures, the Bangestan Group has its own separate pressure system (e.g., Ahvaz oil field). It has been suggested that the oil originally collected in Cretaceous reservoirs and then migrated into the Asmari limestone during the late Cenozoic folding (Dunnington, 1958; Grieg, 1958; Kent and War-

FIG. 11—Asmari limestone forming northeast dip slope of Kuh-e Dashtak. Two villages shown in Figure 10 are outlined; in foreground are inverted beds of red marl and anhydrite forming northeast limb of syncline.

FIG. 12—Southwest limb of Kuh-e Shah Nishin anticline. In foreground is brecciated lime-stone of reverse fault of Figure 10; in middle distance are inverted beds of red marl and anhydrite; and in far distance is uninverted Asmari limestone.

man, 1972). In the tighter structures of the main oil-field belt, neutral-surface folding caused enough extension on the outer arcs of the anti-clines to form fractures as far down as the Ban-gestan Group. In the structures at the edge of the belt folding was not so tight, so there was less extension and the impermeable layer between the Asmari and Bangestan limestones was not breached.

Other factors also affect the amount of exten-sion on the outer arcs of the anticlines, the most notable of these being the extent to which folding is accomplished by the flexural-slip mechanism. An increase in the incompetent horizons in the stratigraphic column tends to reduce the internal cohesion of a folded sequence and therefore en-courages flexural-slip folding; this reduces the amount of stretching and discourages the devel-opment of extension structures on the outer arcs of folds. The increasing shaliness of the Compe-tent group northwestward (James and Wynd, 1965) may partly explain the separate Asmari and Bangestan reservoirs at the Lali oil field.

Extension fractures on anticlines die out down-ward toward the neutral surface. The amount of

fluid migration across bedding, therefore, is re-duced greatly near and below this surface, which is likely to separate different pressure systems just as it separates the extended and compressed parts of the anticline. The separation of reservoirs is illustrated by the deep hole drilled in the Masjed-e Suleyman oil field (Baniriah et al, 1967). The hole penetrated the Asmari and Bangestan lime-stones, which are in fluid connection in the ex-tended part of the anticline. It passed through the neutral surface into the compressed part where it encountered a thrust fault. Below the fault it en-tered a repetition of the Middle Jurassic section containing a reservoir of high-pressure gas. The difference in pressure between the Jurassic and Asmari-Bangestan reservoirs caused an under-ground blowout of gas into the Asmari limestone and a consequent lowering of the oil-gas interface in this reservoir.

Asymmetry of Folding

Most of the folds in the simply folded belt are asymmetric and, with a few exceptions, the steep-est limbs of the anticlines are on the southwest sides. There are several possible causes of asymmetric folding (de Sitter, 1964; Billings, 1972), but the mechanism most applicable to the regionally consistent asymmetry of the simply folded belt is shearing in the detachment zone of the Lower Mobile group. During folding the Competent group has moved southwestward relative to the Basement group and imperfect detachment has caused drag of one upon the other. The synclines in the Competent group are separated from the basement by a thinner mobile layer than are the anticlines (Fig. 13). Furthermore, beneath the synclines, where the Competent group has been forced downward toward the basement, the vertical stress is greater than beneath the anticlines (Johnson and Honea, 1975). This has had the effect of reducing the stress deviator and increasing the hydrostatic component of the stress system, which in turn has reduced the tendency of the rocks in the Lower Mobile group to fracture or flow (Ramsay, 1967, p. 289). The more perfect detachment beneath the anticlines has caused them to override the synclines as both have moved southwest relative to the basement.

The asymmetry of folding only implies relative movement between cover and basement, and not absolute movement. It is therefore equally consistent with: (a) the Competent group being forced southwestward across the static foreland of basement rocks (geosynclinal theory); or (b) the Basement group moving northeastward and being subducted along the thrust belt while the overlying sediments are deformed against the continental mass of central Iran (plate-tectonic theory).

Factors Other Than Gravity Affecting Wavelength

The Competent group of the simply folded belt is a multilayered "structural lithic unit" (Currie et al, 1962) in which folds of spectacular size are formed; the fold wavelengths are in the order of 10 to 20 km with amplitudes of 2 to 3 km (Kent and Warman, 1972). There are many stratigraphic and lithologic factors that influence the initial wavelengths of folds in multilayered sequences, and hence determine the eventual fold size. In the Competent group most of these factors are strongly biased toward large initial wavelengths:

1. Wavelength increases with increasing thickness of the multilayer, provided there is strong interaction between the layers (for examples, see Currie et al, 1962). The Competent group ranges in thickness from 6,000 and 7,000 m, which is nearly twice that of the thickest structural lithic unit quoted by Currie et al (1962); none of the incompetent members of the group is thick enough to allow significant structural independence in any of the competent layers.

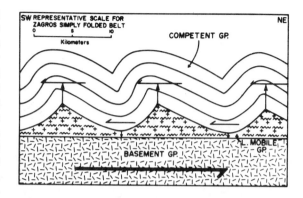

FIG. 13—Asymmetric folds caused by shearing in detachment zone. Arrows show relative movement, and upward and downward displacement in anticlines and synclines. Potential faulting is omitted.

2. Wavelength increases with increasing thickness of the individual competent layers, and with an increasing ratio between the thickness of the competent layers and that of the incompetent layers (Biot, 1961; Ramberg, 1964). The Competent group is dominated by thick, massive carbonate units occupying most of the Permian to Cretaceous part of the section (3,000 to 4,000 m) and represented at the top of the group by the Oligocene to Miocene Asmari limestone (300 m). Throughout most of southwest Iran the most important incompetent rocks in the Competent group are the Pabdeh-Gurpi marls, which are generally less than 1,000 m thick, and the shales of the Cambrian to Carboniferous succession which form only a part of about 2,000 m of sedimentary rock. Northwestward, in Lurestan, there is an increase in the shaliness of the Competent group and this may be a contributing factor in the shorter wavelengths in this province.

3. Wavelength increases as the competence (shear moduli) of the competent layers and incompetent interbeds increase, and as the competence of the enclosing medium decreases (Biot, 1961; Johnson and Honea, 1975). The competence contrast between the carbonate rocks of the Competent group and the medium, the salt of the Upper and Lower Mobile groups, is obviously substantial. The shear modulus for parts of the Asmari limestone (at 1 atm), calculated from the elastic constants of Richards (1933) using the relations of Birch (1966), is 0.24 megabars (24 $\times$ 10[6]kPa), compared with 0.15 megabars (15 $\times$ 10[6]kPa) for a quasi-isotropic aggregate of halite (Voigt, quoted by Birch, 1966). The competence contrast may be considerably greater at the base

of the Competent group as a result of the rise in temperature with depth (see, for example, Handin and Hager, 1958; Gussow, 1968; Heroy, 1968, on the effect of temperature on the ductility of salt). Within the Competent group, the incompetent interbeds consist mainly of marl, shale, and anhydrite, which are more competent than the salt of the Upper and Lower Mobile groups. According to Johnson and Honea (1975), this competence difference should have the particular effect of increasing the wavelengths on the outer arcs of the folds close to the boundaries of the multilayer. Thus it would tend to make the anticlines in the upper part of the Competent group broader and the synclines more pinched, and would reinforce the effect of concentric folding.

4. Wavelength increases with increasing confining pressure (Johnson and Honea, 1975). At the initiation of folding, the Competent group was overlain by 1 to 3 km of rocks of the Upper Mobile and Incompetent groups. Therefore, the confining pressure at its upper surface was in the region of 250 to 750 bars (25,000 to 75,000 kPa), whereas at its lower surface, with about 11 km of overburden, the pressure was about 2,500 to 3,000 bars (250 × 10³ to 300 × 10³ kPa).

5. Wavelength increases if the distance between the buckled multilayer and a rigid layer increases, provided that the space between the buckled and rigid layers is occupied by a less competent medium (Ramberg, 1963, 1970). Thus the thicker the Lower Mobile group (the less competent medium), which separates the Competent group (the buckled multilayer) from the Basement group (the rigid layer), the greater should be the wavelengths of the folds. The absence of salt diapirs in Khuzestan and Lurestan suggests a thinner Lower Mobile group than elsewhere and this may explain in part the narrower folds prevalent in these provinces (Ala, 1974).

GRAVITATIONALLY INDUCED STRUCTURES

The sediments of the simply folded belt have an inherent gravitational instability. A sequence consisting principally of carbonate and clastic rocks with an original thickness of nearly 12 km overlies the low-density Hormuz salt of the Lower Mobile group. The most obvious results of this instability are the numerous salt diapirs of Fars Province and offshore in the Persian Gulf (Figs. 1, 14; Harrison, 1930; O'Brien, 1957; Kent, 1958, 1970; Ala, 1974). Salt movement inevitably results in some folding in the overlying rocks, and conversely folding in these rocks due to orogenic processes must result in some salt movement. Salt-induced folds may occur on the Arabian platform and are inferred to have formed in the simply folded belt before being obscured by the

FIG. 14—Diapir of Hormuz salt intruding Incompetent group, northwest of Bandar Lengeh.

late Cenozoic orogeny. Salt movement during the orogeny is implied by the theory of detachment folding and is indicated by still active salt diapirs and "glaciers."

Folding Induced by Gravitational Instability

Ramberg (1971, model 13) investigated the folding of a viscous layer in a less viscous and less dense medium, restricted below but not above by a rigid layer. The model has a general correspondence with the Persian Gulf region: the viscous layer is represented by the Competent group, the medium by the salt of the Lower and Upper Mobile groups, and the rigid layer by the Basement group. Ramberg found that two orders of folds with different wavelength/thickness ratios should occur, one formed by bending due to gravitational instability and the other by buckling due to layer-parallel compressive stress. The ratios of the bending folds were about 100 whereas those of the buckle folds were about 10. These wavelength/thickness ratios cannot be applied directly to the Persian Gulf region, because the parameters of the theoretical model do not correspond with those of the natural example. The contrast between the wavelengths of the two types of folds is, however, still valid. Bending folds resulting from deep salt movements and the formation of salt anticlines should have considerably larger wavelengths than the buckle folds of the simply folded belt.

Possible examples of bending folds are on the eastern side of the Arabian Peninsula and offshore in the Persian Gulf (Greig, 1958; Falcon, 1967; Mina et al, 1967; Kent, 1970; Kent and Warman, 1972). They have been formed since the beginning of the Cretaceous and have been attributed variously to salt movement, basement faulting, or both; they have wavelengths in the order of 50 km (Kent, 1970, Fig. 5) and are either domal or have overall north-south axial trends. They are unrelated to the northwest-southeast buckle folds of the late Cenozoic Zagros orogeny, which have wavelengths of 10 to 20 km. The extension of salt-induced Cretaceous structures into the simply folded belt is suggested by the numerous salt diapirs of Fars Province, many of which are demonstrably older than the late Cenozoic folding (Kent, 1958). The diapirs have a crude north-south alignment (Ala, 1974) which may indicate the presence of low-amplitude salt anticlines that have been obscured by the later, more intense buckle folds.

The formation of salt-induced bending folds provides a mechanism for the diapiric piercement of the massive carbonate rocks of the Competent group. In the absence of lateral shortening caused by layer-parallel compressive stress, a folded layer must have a greater surface area than the original unfolded layer (Fig. 15). If the folded layer consists of brittle, competent material as does the Competent group, it will be fractured and forced apart rather than stretched. The vacant space can be occupied either by a downfaulted block creating a graben, or by the upwelling of salt. Because the salt flows into a space created by the breaking apart of the folded strata, there is no need for it to "punch a hole" (Kent, 1970) in these rocks. Also there is no "room problem," for the salt does not displace rock but merely fills an otherwise empty space. The rarity of relics representing the vent volume of the diapir (Kent, 1970; Ala, 1974) is explained similarly because the expected "relics" actually form the present vent walls.

In most cases salt diapirs are thought to emanate from the crests of salt anticlines (Trusheim, 1960; Murray, 1968) where the gravitational instability is greatest (Fig. 15,II). It is, however, just as likely that the competent layer will break apart in the synclines, especially as the greatest extension in these structures is at the base of the layer, adjacent to the salt. If this happens, synclinal diapirs similar to those offshore of Abu Dhabi may form (Kent, 1970).

The broad, low-amplitude, bending folds of the Arabian platform form the oil traps in that region. Similar structures may have provided pre-

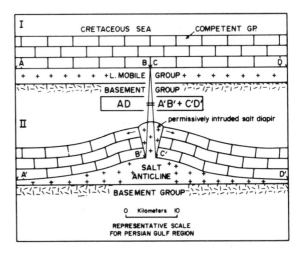

FIG. 15—Bending fold formed by salt movement *II* beneath initially horizontal competent strata *I*. No allowance has been made for flexural slip; lower surface of competent layer has maintained its original length, and upper surface has been compressed slightly in synclines; all extension has been consolidated in diapiric vent. Result of folding is permissive intrusion of salt diapir.

orogenic traps for the primary accumulation of hydrocarbons in the simply folded belt; during the late Cenozoic orogeny fracturing above the neutral surfaces of buckle anticlines would have allowed migration to the present Asmari limestone reservoirs (Dunnington, 1958; Greig, 1958; Kent and Warman, 1972).

Relation Between Buckle Folding and Salt Diapirs

Many of the salt diapirs of the Persian Gulf region must be attributed to the forces of gravity acting in a relatively stable tectonic environment (Nettleton, 1934), either because they predate the late Cenozoic orogeny or because they are outside its area of influence. Others, however, have been active during the orogeny and must have been influenced in one way or another by the associated tectonic stresses.

Diapir Initiation during Buckle Folding

The initial intrusion of salt diapirs from the Lower Mobile group into the Competent group during active folding is likely to be restricted to lines of major faulting (e.g., the Qatar-Kazerun line), where the regional stress system locally is disrupted. In areas where deformation has been

principally by folding, the tectonic stresses are such that they discourage the formation of significant new diapirs.

Within the Lower Mobile group the ductility of the salt has allowed a relatively uniform distribution of the deviatory stress responsible for the Zagros folding. By contrast the stresses at the base of the Competent group are distributed very unevenly (Fig. 6,II). In the congested cores of the anticlines very high layer-parallel compressive stresses are present and the hydrostatic-stress component, which is the mean of the three principal stresses (Ramsay, 1967, p. 39), is correspondingly high. Whereas salt can be expected to flow into the spaces beneath the Competent group vacated by the upward movement of the anticlines, it will not move from these zones of lower hydrostatic stress into zones of much higher hydrostatic stress actually within the anticlinal cores (Fig. 16, II). To do so, it would have to push aside the rocks of the Competent group against a layer-parallel maximum principal stress exceeding that prevalent in the Lower Mobile group. The initiation of diapirs at the base of anticlines during folding is therefore contrary to the principles that salt moves from zones of higher hydrostatic stress

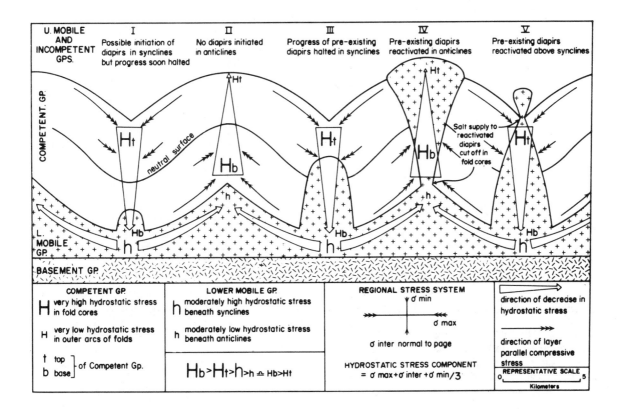

FIG. 16—Relation between salt diapirism and active buckle folding. Potential faulting is omitted.

to zones of lower hydrostatic stress, and that diapiric intrusions stretch and push aside the overlying rocks or pass through them permissively. It also can be concluded that the anticlines formed by buckling do not have salt cores, as shown, for example, by Falcon (1969, Fig. 2). Diapirs and salt anticlines formed in the stable preorogenic environment provide ample evidence that the Hormuz salt cannot support the small stress deviators created by inhomogeneities in only a part of 12 km of sedimentary rock. It is therefore most unlikely that it can support the much greater tectonic stress deviator that has been able to raise the full 12 km of sedimentary rock into huge folds during the Zagros orogeny.

In synclines, the base of the Competent group forms the outer arc of the fold and is therefore subject to extension normal to the fold axis, and to a relatively low hydrostatic stress component (Fig. 16,I). The hydrostatic stress within the adjacent salt of the Lower Mobile group is higher, reflecting the more evenly distributed tectonic stresses and the downward pressure of the syncline. There is therefore a suitable pressure gradient for the movement of salt from the Lower Mobile group into the Competent group; furthermore extension at the base of the Competent group can provide space for a diapiric intrusion. Once a diapir has penetrated into the Competent group, however, its growth is discouraged by an upward increase in the layer-parallel stress normal to the fold axis as the fold core is approached. This increase from the minimum principal stress to the extreme compressive stress of the synclinal core exceeds the relatively slight decrease in the vertical stress due to overburden; the hydrostatic stress, therefore, increases upward and the growth of the salt diapir is halted.

Diapir Reactivation during Buckle Folding

Diapirs that intruded the Competent group as a result of gravitational instability before the onset of buckle folding have been affected differently depending on whether they are in synclinal or anticlinal folds. Those that are in synclines and have not penetrated to the top of the Competent group (Fig. 16,III) have not been reactivated because the layer-parallel stress normal to the fold axis increases upward, and there is a consequent increase in the hydrostatic stress component as the fold core is approached. If a synclinal diapir penetrated into rocks above the Competent group before the onset of folding (Fig. 16,V), it may have undergone minor reactivation but only in its upper parts where salt has been squeezed upward out of the synclinal core. However, diapirs in anticlines (Fig. 16,IV) have been subject to an up-

ward decrease in both the layer-parallel stress and the vertical stress exerted by the overburden. They, therefore, have been squeezed rapidly upward toward the surface, but have been cut off from their supply of salt in the Lower Mobile group by the high stresses in the anticlinal cores at the base of the Competent group.

Distribution of Diapirs

Harrison (1930) found that, although most diapirs pierce anticlines, they are just as common on flanks or plunging ends as on crests; he also noted that a few are in synclines.

Although preexisting diapirs and salt anticlines almost certainly have affected the character of the late Cenozoic folding (Kent, 1970), it is a reasonable assumption that they have not preferentially localized either anticlines or synclines; they therefore should have a uniform distribution with respect to these structures. Diapirs on anticlines, however, more commonly are exposed because they have been selectively reactivated and the depth of erosion is greater. They do not occupy any special positions relative to the folds because their positions were determined before folding began. The upward decrease in hydrostatic stress that causes reactivation is greatest in the centers of the anticlines, but it also occurs on the flanks and plunging ends of the folds, and the principal requirement for reactivation is a preexisting diapir.

BEHAVIOR OF BASEMENT

Lees (1952) suggested that the basement was cut by a series of northwest-dipping thrust faults and thus was involved intimately in the folding of the Competent group. The absence of in-situ basement rocks anywhere in the Zagros Mountains southwest of the thrust belt, even where Cambrian rocks are thrust to the surface, makes this unlikely (Falcon, 1969). Furthermore the presence of the neutral surface of the Zagros folds in the middle of the Mesozoic, about half way down the sedimentary succession, suggests that these folds do not have a rigid core of basement rocks.

Although major thrust faults do not occur in the basement beneath the simply folded belt, the presence of strike-slip faults is well documented (Falcon, 1969; McQuillan, 1973a; Haynes and McQuillan, 1974). These faults, of which the most prominent are the Qatar-Kazerun and Oman lines, have dextral displacements and are oriented north-south at about 40° to the main Zagros trend. As pointed out by Falcon (1969), they coincide in orientation and sense of movement with the shear fractures that would be formed by a maximum principal stress normal to the Zagros fold trend. In the Phanerozoic sedimentary rocks, gravity is the minimum principal stress and the rocks have been deformed by folding and thrust faulting. In the basement, however, with its much greater mass, gravity is the intermediate principal stress and consequently deformation has been by strike-slip faulting (Anderson, 1951).

CONCLUSIONS

The main conclusions that have been drawn in this discussion of the folding of the Competent group in the Zagros simply folded belt are summarized in the following.

1. The Competent group can be considered as a structural lithic unit bounded above and below by detachment zones in the evaporitic Upper and Lower Mobile groups.

2. The contrast between broad anticlines and narrow pinched synclines is a reflection of the concentricity of the folds and the shallow level of erosion; the contrast can be expected to decrease and then be reversed as the folded layer is exposed more deeply by erosion.

3. The northwest-southeast folds of the simply folded belt are buckle folds formed by compression parallel with the layering, in contrast to the generally north-south folds of the Arabian platform which are bending folds formed by differential forces acting perpendicular to the layering.

4. The buckle folds of the simply folded belt have been formed by a combination of flexural-slip and neutral-surface folding.

5. Flexural-slip folding is indicated by bedding-plane slickensides.

6. Neutral-surface folding is indicated by (a) normal faults and extension joints on the outer arcs of anticlines, and (b) thrust or reverse faults and a chevron style of folding in the inner arcs of anticlines and synclines. Extension structures are inferred to be present on the unexposed outer arcs of synclines.

7. The approximate position of the neutral surface is in the Jurassic part of the sedimentary succession.

8. Extension structures caused by neutral-surface folding have facilitated the migration of oil across the bedding and have caused the fluid connection of the reservoir rocks in the tighter anticlines.

9. The neutral surface separates the extended and compressed parts of anticlines and as a result is a major divide between different fluid systems.

10. The asymmetry of folding is caused by shearing in the detachment zone of the Lower Mobile group. It implies relative movement between the Competent group (toward the south-west) and the Basement group (toward the north-east).

11. The very large wavelengths of the folds are the result of most of the factors affecting wavelength being strongly biased toward large size; the most important among these is the great thickness of the Competent group.

12. Bending folds formed by deep salt movement should have much greater wavelengths than the northwest-southeast buckle folds. Preorogenic, low-amplitude bending folds, similar to examples on the Arabian platform, are inferred to have formed in the simply folded belt.

13. Extension caused by bending folds has broken apart the Competent group strata and allowed the permissive intrusion of salt diapirs; it provides a solution to the "room problem" and explains why diapirs rarely contain relics of the rocks through which they pass.

14. Salt diapirs are unlikely to have been initiated during active buckle folding, except perhaps along major fault zones. The cores of buckle anticlines are not occupied by salt.

15. Preexisting salt diapirs have been reactivated by buckle folding in anticlines, but their progress has been halted in synclines except where they already have penetrated through the Competent group.

16. The Basement group has not taken part in the folding of the Competent group, but instead has undergone strike-slip faulting during the same period of deformation.

REFERENCES CITED

Ala, M. A., 1974, Salt diapirism in southern Iran: AAPG Bull., v. 58, p. 1758-1770.

Anderson, E. M., 1951, The dynamics of faulting, 2d ed.: Edinburgh, Oliver and Boyd, 206 p.

Baniriah, N., G. C. Beckman, and J. Birks, 1967, Repressuring of the Masjid-i-Sulaiman oilfield from a deep underground Jurassic gas leak and remedial killing operations: 7th World Petroleum Cong., Mexico, Proc., v. 3, p. 765-771.

Billings, M. P., 1972, Structural geology, 3d ed.: Englewood Cliffs, N.J., Prentice-Hall, 606 p.

Biot, M. A., 1961, Theory of folding of stratified viscoelastic media and its implications in tectonics and orogenesis: Geol. Soc. America Bull., v. 72, p. 1595-1620.

Birch, F., 1966, Compressibility; elastic constants, in Handbook of physical constants: Geol. Soc. America Mem. 97, Sec. 7, p. 97-173.

Blanford, W. T., 1872, Notes on the geological formations seen along the coasts of Baluchistan and Persia, from Karachi to the head of the Persian Gulf, and some of the Gulf Islands: India Geol. Survey Recs., v. 5, p. 41-45.

British Petroleum Company Ltd., 1956, Oil and gas in southwest Iran, in Symposium sobre yacimientos de petroleo y gas, v. 2: 20th Internat. Geol. Cong., Mexico, p. 33-72.

Currie, J. B., H. W. Patnode, and R. P. Trump, 1962, Development of folds in sedimentary strata: Geol. Soc. America Bull., v. 73, p. 655-674.

De Sitter, L. U., 1964, Structural geology, 2d ed.: New York, McGraw-Hill, 551 p.

Dunnington, H. V., 1958, Generation, migration, accumulation, and dissipation of oil in northern Iraq, in L. G. Weeks, ed., Habitat of oil, a symposium: AAPG, p. 1194-1251.

Falcon, N. L., 1967, The geology of the north-east margin of the Arabian basement shield: Adv. Sci., v. 24, p. 31-42.

———— 1969, Problems of the relationship between surface structure and deep displacements illustrated by the Zagros Range, in Time and place in orogeny: Geol. Soc. London Spec. Pub. 3, p. 9-22.

———— 1974, Southern Iran: Zagros Mountains, in Mesozoic-Cenozoic orogenic belts, data for orogenic studies: Geol. Soc. London Spec. Pub. 4, p. 199-211.

Gansser, A., 1964, Geology of the Himalayas: London, John Wiley and Sons, 289 p.

Grieg, D. A., 1958, Oil horizons in the Middle East, in L. G. Weeks, ed., Habitat of oil, a symposium: AAPG, p. 1182-1193.

Gussow, W. C., 1968, Salt diapirism: importance of temperature, and energy source of emplacement, in Diapirism and diapirs: AAPG Mem. 8, p. 16-52.

Handin, J., and R. V. Hager, Jr., 1958, Experimental deformation of sedimentary rocks under confining pressure; pt. 2, tests at high temperature: AAPG Bull., v. 42, p. 2892-2934.

Harrison, J. V., 1930, The geology of some salt plugs in Laristan (southern Persia): Geol. Soc. London Quart. Jour., v. 86, p. 463-522.

Haynes, S. J., and H. McQuillan, 1974, Evolution of the Zagros suture zone, southern Iran: Geol. Soc. America Bull., v. 85, p. 739-744.

Heroy, W. B., 1968, Thermicity of salt as a geologic function, in Saline deposits: Geol. Soc. America Spec. Paper 88, p. 619-630.

Hills, E. S., 1963, Elements of structural geology: New York, John Wiley and Sons, 483 p.

Hull, C. E., and H. R. Warman, 1970, Asmari oil fields of Iran, in Geology of giant petroleum fields: AAPG Mem. 14, p. 428-437.

James, G. A., and J. G. Wynd, 1965, Stratigraphic nomenclature of Iranian oil consortium agreement area: AAPG Bull., v. 49, p. 2182-2245.

Johnson, A. M., and E. Honea, 1975, A theory of concentric, kink, and sinusoidal folding and of monoclinal flexuring of compressible, elastic multilayers. III. Transition from sinusoidal to concentric-like to chevron folds: Tectonophysics, v. 27, p. 1-38.

Kashfi, M. S., 1976, Plate tectonics and structural evolution of the Zagros geosyncline, southwestern Iran: Geol. Soc. America Bull., v. 87, p. 1486-1490.

Kent, P. E., 1958, Recent studies of south persian salt plugs: AAPG Bull., v. 42, p. 2951-2972.

———— 1970, The salt plugs of the Persian Gulf region: Leicester Literary and Philos. Soc. Trans., v. 64, p. 56-88.

———— and H. R. Warman, 1972, An environmental review of the world's richest oil-bearing region—the Middle East: 24th Internat. Geol. Cong., Canada, Proc., Sec. 5, p. 142-152.

Kerr, J. W., 1974, Geology of Bathurst Island Group and Byam Martin Island, Arctic Canada: Canada Geol. Survey Mem. 378, 152 p.

Lees, G. M., 1950, Some structural and stratigraphical aspects of the oilfields of the Middle East: 18th Internat. Geol. Cong., Great Britain, Proc., pt. 6, p. 26-33.

———— 1952, Foreland folding: Geol. Soc. London Quart. Jour., v. 108, p. 1-34.

———— and F. D. S. Richardson, 1940, The geology of the oil-field belt of S.W. Iran and Iraq: Geol. Mag., v. 77, p. 227-252.

McQuillan, H., 1973a, A geological note on the Qir earthquake, SW Iran, April 1972: Geol. Mag., v. 110, p. 243-248.

———— 1973b, Small-scale fracture density in Asmari Formation of southwest Iran and its relation to bed thickness and structural setting: AAPG Bull., v. 57, p. 2367-2385.

———— 1974, Fracture patterns on Kuh-e Asmari anticline, southwest Iran: AAPG Bull., v. 58, p. 236-246.

Mina, P., M. T. Razaghnia, and Y. Paran, 1967, Geological and geophysical studies and exploratory drilling of the Iranian continental shelf—Persian Gulf: 7th World Petroleum Cong., Mexico, Proc., v. 2, p. 871-903.

Murray, G. E., 1968, Salt structures of Gulf of Mexico basin—a review, in Diapirism and diapirs: AAPG Mem. 8, p. 99-121.

Nettleton, L. L., 1934, Fluid mechanics of salt domes: AAPG Bull., v. 18, p. 1175-1204.

Nowroozi, A. A., 1972, Focal mechanism of earthquakes in Persia, Turkey, West Pakistan, and Afghanistan and plate tectonics of the Middle East: Seismol. Soc. America Bull., v. 62, p. 823-850.

O'Brien, C.A.E., 1950, Tectonic problems of the oilfield belt of southwest Iran: 18th Internat. Geol. Cong., Great Britain, Proc., pt. 6, p. 45-58.

———— 1957, Salt diapirism in south Persia (Iran): Geologie en Mijnbouw, n. s., v. 19, p. 357-376.

Ramberg, H., 1963, Fluid dynamics of viscous buckling applicable to folding of layered rocks: AAPG Bull., v. 47, p. 484-505.

———— 1964, Selective buckling of composite layers with contrasted rheological properties; a theory for simultaneous formation of several orders of folds: Tectonophysics, v. 1, p. 307-341.

———— 1970, Folding of laterally compressed multilayers in the field of gravity, I: Physics Earth and Planetary Interiors, v. 2, p. 203-232.

———— 1971, Folding of laterally compressed multilayers in the field of gravity, II, numerical examples: Physics Earth and Planetary Interiors, v. 4, p. 83-120.

Ramsay, J. G., 1967, Folding and fracturing of rocks: New York, McGraw-Hill, 568 p.

Richards, T. C., 1933, On the elastic constants of rocks, with a seismic application: Phys. Soc. Proc., v. 45, p. 70-81.

Stöcklin, J., 1968a, Structural history and tectonics of Iran: a review: AAPG Bull., v. 52, p. 1229-1258.

——— 1968b, Salt deposits of the Middle East, *in* Saline deposits: Geol. Soc. America Spec. Paper 88, p. 158-181.

Tozer, E. T., and R. Thorsteinsson, 1964, Western Queen Elizabeth Islands, Arctic Archipelago: Canada Geol. Survey Mem. 332, 242 p.

Trusheim, F., 1960, Mechanism of salt migration in northern Germany: AAPG Bull., v. 44, p. 1519-1540.

Van Hise, C. R., 1896, Principles of North American Pre-Cambrian geology: U.S. Geol. Survey 16th Ann. Rept., 1894-1895, pt. 1, p. 571-843.

VOLUME 10   MARCH, 1962   NUMBER 3

JOURNAL

of the

ALBERTA SOCIETY

of

PETROLEUM GEOLOGISTS

═══════════

FOLDING[1]

S. Warren Carey

University of Tasmania

Hobart, Tasmania

ABSTRACT

Folds disclose the components of deformation oblique to the bedding but there is additional latent deformation which may be large. Similar folding is simpler physically and geometrically than concentric. Similar folding is not just a more intense development from concentric folding; the two types are at opposite ends of a behaviour spectrum, the former developing under isotropic conditions where bedding is irrelevant, the latter under anisotropic conditions where the difference between adjacent beds is great. Real folding is distributed between these extremes. Transport is normal to the axis in all folds but similar folds imply transport in the plane of the axial surface, whereas concentric folds imply transport at a large angle to this surface. Similar folds persist in depth, whereas concentric folds imply a décollement at a depth of the same order as the fold amplitude. Bedding surface area shows great expansion in similar folding, but crustal shortening is not implied. Bedding surface area remains constant in concentric folding which implies superficial shortening vanishing on a décollement, but over-all crustal shortening is not implied.

Superposed similar folds of great apparent complexity may be analysed into their components. In general the original thickness of any stratigraphic unit is not less than the maximum orthogonal thickness of the same bed in the folded and refolded condition. Diapiric folds (salt domes, mantled gneiss domes or orogenic axial zones) have an over-all toroidal pattern analogous to the circulation of convection cells.

All of the characteristic structures of an orogen—axial zone of intense similar folding, basement horsts, serpentinite belts along steep inner thrusts, tracts of outwardly transported nappes and recumbent folds, and apparently autochthonous folds of the frontal belt—may develop in the absence of crustal shortening, and may develop even during progressive secular extension of the orogenic zone, transverse to its folds.

─────────────────────────────────

[1]Fourth Annual Honorary Address, Alberta Society of Petroleum Geologists. Presented May 24, 1960.

95

96

Structural geology began a century and a half ago when Hall recognized folding as deformation. From the outset it was assumed that folds were produced by shortening, on the analogy of crumpling of piles of paper. From that idea it was assumed that folding came from orogenesis, and in general implied compression. This gradually became a law of geology, and then an axiom. Tonight I challenge that axiom.

Role of Bedding in Folding.—The idea of folding begins with some set of parallel or near parallel reference surfaces (which for simplicity we will call bedding, though aware that reference surfaces other than bedding may be used). Without such bedding there may be deformation, but there is no discernible folding. Strata such as in Fig. 1 we would regard as strongly folded. By contrast, we would say that the beds in Fig. 2 are little folded. This might be true, but it could be quite false. Bedding can only disclose folding which is transverse to itself. The beds in Fig. 2 could be just as strongly folded as those of Fig. 1 even though they look innocent. Fig. 3 could be the extension of Fig. 2. The curve on the end of Fig. 3 is identical with that of Fig. 1. In other words the amount of deformation in Fig. 3 is exactly the same in degree and in form as that in Fig. 1. Herein lies an important lesson. Folding does not disclose the whole deformation. It can only reveal that part of the deformation which is transverse to the bedding. The component of the deformation

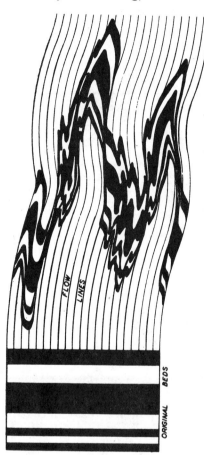

Fig. 1.—Similar folding

parallel to the bedding in any element of a fold is not disclosed. It is our habit to neglect this latent component — and assume that the fold we see gives the whole deformation. There is no reason why this unseen component should not be large. On the contrary there *is* reason why it should be large since such deformation proceeds most rapidly in the most ductile layers, whereas the transverse component we do see must deform the least ductile beds. Of this, more anon.

Similar and Concentric Folding. — Geologists have recognized two contrasting forms of folding (Fig. 4 and Fig. 1) known respectively as concentric and similar folding. In the former the orthogonal thickness of each bed remains constant and a normal to

Fig. 2.

Fig. 3.

one bed is also normal to the beds above and below it. As the centre of curvature at any point lies on the normal, and as the thickness of each bed is constant, the beds are everywhere concentric surfaces. In similar folding each folded surface has the same shape but successive bedding surfaces would coincide with each other if displaced along parallel flow lines.

The terms concentric and similar folding were introduced by van Hise (1896). He also used parallel folding as a synonym for concentric folding in the sense that in such folding a normal to any stratum is normal to other strata above and below it. Hence the normals see all beds as parallels. This usage has crossed directly into French (Goguel, 1952, p. 40). However Stoces and White (1935, figs. 203 and 205) misquote van Hise and use parallel folding as a synonym for similar folding, their idea being that if one folded stratum is displaced parallel to itself in the direction of shear it will coincide with the next stratum. In view of this contradictory usage, and as van Hise defined concentric and similar folding unambiguously and merely added parallel folding as a synonym, it would be best to drop the term parallel folding. More recently, the terms flexure folding and shear folding have been substituted for concentric and similar folding respectively and have gained wide currency. However as both types of folding involve shear to a comparable degree, and as both types of folds are flexures etymologically and in ordinary English, these substitutes should be dropped on grounds both of priority and precision. Flexure was defined by Powell (1876, p. 10-11) for bends caused primarily by vertical movement, in contrast to folds due to horizontal pressure. This usage was followed by Bailey Willis (1934, p. 77). Nevin (1949, Chapter III) and Hills (1939, p. 75) use flexure in a quite general sense with its ordinary English meaning. If the term flexure is used at all it should be used in this general sense.

Fig. 4.—Concentric folding.

Concentric folding is commonly regarded as simple folding, and similar folding as more complex. In fact the reverse is true. Similar folding as in Fig. 1 is physically simple. It is the kind of folding which occurs in isotropic materials. Concentric folding as in Fig. 4 is physically complex. It can only develop in the presence of great inequality of material. All folding involves change of shape. A reference sphere to measure strain at a point in the material becomes a triaxial ellipsoid during the folding. Since the volume of the reference sphere does not alter, at least one diameter must lengthen, at least one must shorten and there must be at least one diameter which remains the same. In isotropic material the stress field alone determines the orientation of the lengthening, shortening and unchanged diameters. If the bedding has no function other than reference surfaces, (if for example the beds are simply marked by dye and have no other physical difference), then the directions which shorten and lengthen and remain constant are determined by the stress field, and do not have any necessary relation to where the bedding happens to be. Concentric folding, which requires that the thickness measured normal to the bedding should remain constant, can only develop if the stress field is governed by the bedding, and this can only remain true if the physical properties of the beds are very different, that is, if the material is very anisotropic.

SIMILAR FOLDING

Simple Similar Folding.—The logical development of a theory of folding is to begin with the simplest possible case, the deformation of physically isotropic material, and having understood that, to proceed to introduce other variables. Hence we begin by analysis of similar folds such as Fig. 1 which turn out to be physically and geometrically the simplest kind of folding.

Folding implies displacement, which in turn means that there must be flow lines, a flow line being the path of movement of reference points in the material. In the simplest case these flow lines are parallel. It is quite easy to set up conditions where the flow lines converge or diverge, but let us deal first with the parallel case. Consider the material between the pair of flow lines on the extreme right of Fig. 1. The segment of each bed before deformation is a rectangle, but as the bed deforms it becomes a parallelogram. Neither the volume nor density changes apart from mineralogical changes which we may neglect in this discussion and by definition there is no movement across flow lines, hence the areas of these parallelograms must be the same as the areas of the corresponding equivalent rectangles. As the bounding flow lines are parallel the intercepts of all beds on the flow lines must remain constant, for the areas of the parallelograms equal these intercepts multiplied by the normal distance between flow lines, which is constant. Here lies the geometrical law of similar folding: Provided the flow lines are parallel, the thickness of a bed *measured in the direction of flow* remains constant. If the flow lines converge or diverge, the thickness of a bed measured in the direction of flow increases or decreases in inverse proportion to the normal distance between flow lines.

Reverting to Fig. 1, the folded beds in the upper part may be broken up into a series of parallelograms each of which has the same area as the equivalent rectangle in the undeformed beds below. The cross-section of each bed in the fold is exactly equal to the area of the corresponding undeformed bed below. The total area of the whole fold above is exactly equal to the whole rectangular segment of undeformed beds below. Despite the appearance of extreme attenuation and rapid thickness variation, the thickness of each bed remains absolutely constant throughout the fold, when measured in the direction of flow.

In contrast with concentric folds, where the apparent thickness of beds in sections has to be corrected for the plunge of the fold (and in such cases is always greater than the original thickness of the bed) no such plunge correction is necessary in drawing sections across similar folds in any plane which contains the flow lines, for in all such cases the thickness of beds is constant in the direction of flow. However if the section plane is oblique to the flow lines and intersects the bedding at a larger or smaller angle than do the flow lines, the intercept of the bed in the section is respectively less than or greater than the original thickness of the bed.

Flow lines in similar folds are easy to find. For every fold axial trace and every axis of inflection is a flow line. It must be so, because a fold axial trace is a line along which the angle of shear changes, and that can only happen along a flow line. Similar folding is usually characterised by large numbers of little folds and crenulations and these are found to be remarkably persistent down their axial surfaces, for that indeed is an essential property of similar folding. Hence it is very easy to identify the flow lines in similar folds on any level in a mine. Although the folding may look complicated, the flow lines are always much simpler and may be projected towards the next level, and all of the bedding surfaces projected down using the law of similar folding, namely that thicknesses remain constant in the direction of flow.

In uniform material in a stress field with a uniform gradient flow lines remain parallel, and hence they can be projected using the method of tangential circular arcs described by Busk (1929) for the projection of bedding. Large variations of flow properties are necessary to cause much convergence of flow lines so the projection by this method will never be far wrong and the convergence will be apparent and the projection can be corrected accordingly.

Superimposed Similar Folds with the Same Strike.—Fig. 5c shows a slightly more complicated case. Many of us who have worked among the crystalline metamorphic rocks have seen structures like this, but we have been inclined to dismiss them as too hard to resolve. In fact this folding is still very simple physically. It obeys the rules of isotropic deformation. The original beds 5a were first deformed to the similar folds 5b and subsequently by further similar folding to form 5c. The two fold templates involved are shown in 5d and 5e. The area of the folded beds shown in 5c is exactly equal to the corresponding areas in 5b and 5a, because thicknesses in the direction of flow were held constant in each transformation.

Fig. 5f is added to prove that overprinted similar folding may not be commutative; *AB* may not be the same as *BA*. Fig. 5c and Fig. 5f are both produced from folding the beds 5a by the same similar folds 5d and 5e, but in Fig. 5c the 5e fold was impressed before 5d and in Fig. 5f, 5d was first. Superposed folds are not commutative unless they have the same directrix. (See classification table on p. 114).

Fig. 6 shows another pattern produced in a single bed by successive development of similar folds which have the same strike but opposed attitudes. By

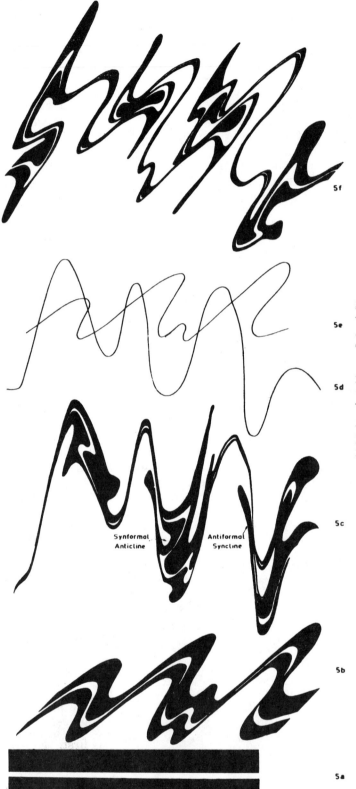

5f

5e

5d

Superposed similar folds with the same strike but different attitude. (5a) The beds before folding. (5b) The beds after folding according to the pattern 5e. (5c) The same beds after the additional fold 5d has been superposed. (5f) The pattern resulting from folding the beds in the reverse order, i.e. 5d before 5e.

Synformal Anticline

Antiformal Syncline

5c

5b

5a

Fig. 5.

repeating such overprintings with a wide variety of random folds, patterns emerge, recognition of which in geological exposures suggests the form of the

basic folds involved. The hook structures so prominent in Fig. 5c producing antiformal synclines (structures geometrically like anticlines but having the younger strata in the core) and synformal anticlines, are characteristic of overprinted similar folds which have identical strike but significantly different attitude.

For example it is not uncommon in strongly folded rocks to find a hooked outcrop pattern such as Fig. 7a, which might be the plan of a body of dolomite or amphibolite in schists. This outcrop pattern might suggest that the dolomite body has been dragged and cut off by a fault as in 7b. This might be right, but it might be wrong. The pattern could

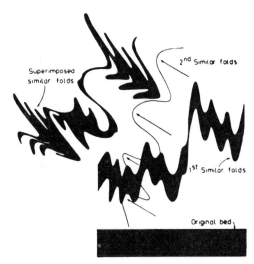

Fig. 6. — Superposed similar folds with parallel strike but strongly divergent attitudes.

be the outcrop (7c) of the kind of hooked structure shown in Fig. 5c. If the geological map contains a number of such hooked patterns the probability grows

that the second interpretation may be correct. In that event the hooks disclose the strike, dip of axial surface, and plunge of the second folds, as well as the strike of the first folds, and the relative attitude of the two generations.

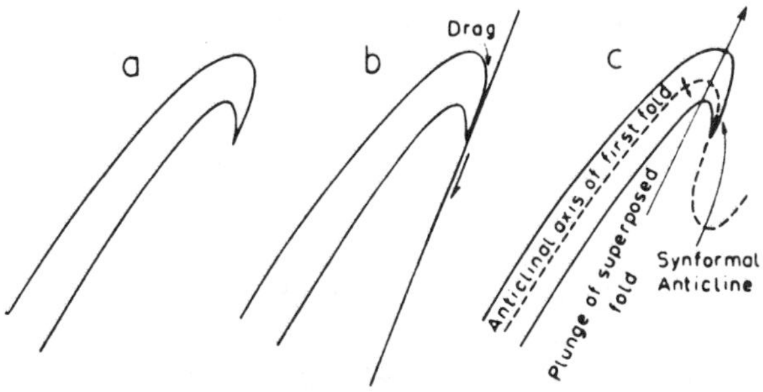

In the absence of plunge to produce the hooked outcrops, structures of this type are still more likely to be misinterpreted

Fig. 7.—Alternative interpretations of outcrop patterns.

in field mapping. Take for example the five repetitions of the one bed shown on Fig. 8a, where the facing and dip of each outcrop is known with certainty and fold axes are also known. The obvious interpretation which suggests itself is Fig. 8b which shows ordinary overturned anticline and syncline. However the true section may be Fig. 8c which is a refolded anticline. This might not be easy to discover if the axes were horizontal but the correct solution would show up if the structure could be followed down-pitch, which would bring to the surface the tell-tale hook. Most structural geologists know that if they look at a geological map with their sight line in the direction of strike and inclined to the map at the angle of axial plunge what they see is the structural section across the area. Thus the folds of Fig. 5c would outcrop as in Fig. 9 if plunging at 15°.

Analysis of Superimposed Similar Folding. — Although physically simple, folds may appear quite obscure when met in field exposures (e.g., Fig. 10 which represents a vertical section). However a systematic attack may pro-

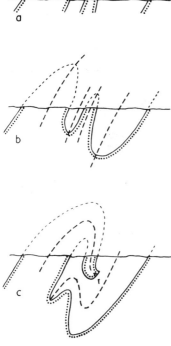

Fig. 8. — Alternative interpretations of outcrop patterns.

duce order from a seemingly hopeless confusion. Three kinds of analysis are possible: (a) analysis of the inflected symmetry surfaces drawn with respect to the bedding, (b) statistical analysis of bedding using plots of poles, traces or intersections, as developed by Weiss, McIntyre and others, (c) statistical analysis of the bedding curvature. Of these the most direct is the symmetry analysis.

Begin the analysis of Fig. 10 by tracing off the bedding surfaces, then numbering them in sequence (Fig. 11). At this stage it is not known which is oldest or youngest, and continuity may be lost in the finely attenuated zones, but these areas can usually be solved by working around them. If they can't a second or even a third sequence may be started using Roman or Greek letters, or capitals and lower case. Once this is done several places will be found where the numbers are symmetrical, e.g. at X where the numbers run 9 10 11 10 9 or at Y where the numbers run 9 8 7 6 5 6 7 8 9. These points are clearly on fold axes, and X and Y are of different kinds, one having the highest number in the core and the other the lowest.

These symmetry axes may now be drawn in (Fig. 12) using firm continuous lines where the axis is narrowly confined, broken lines where the axis is certainly present but it is not closely defined, and dotted lines where the general course of the axis can be inferred, but its actual position cannot be drawn with any precision. These axes are marked with a zero or a cross-bar sign according to whether the core bed carries a minimum or a maximum index number. Where the core formation makes an apex the axis must pass into the adjacent formation, and the obvious place to draw the line is through the apex tip. No great error will ensue if this is done although actually in overprinted folding the emergence of an axis is more commonly away from the tip.

From Fig. 12 emerges the fact that two kinds of fold axes can be drawn. One set consists of nearly straight parallel lines but groups of zero and cross-

Fig. 9.—Outcrop pattern of Fig. 5e if plunging at 15°. Look in the direction of plunge (left to right) with 15° angle of the sight line onto the page, and the cross-section of the fold (as in Fig. 5c) will be seen.

Fig. 10.—Section of simple overprinted similar folds with parallel strikes, but different attitudes.

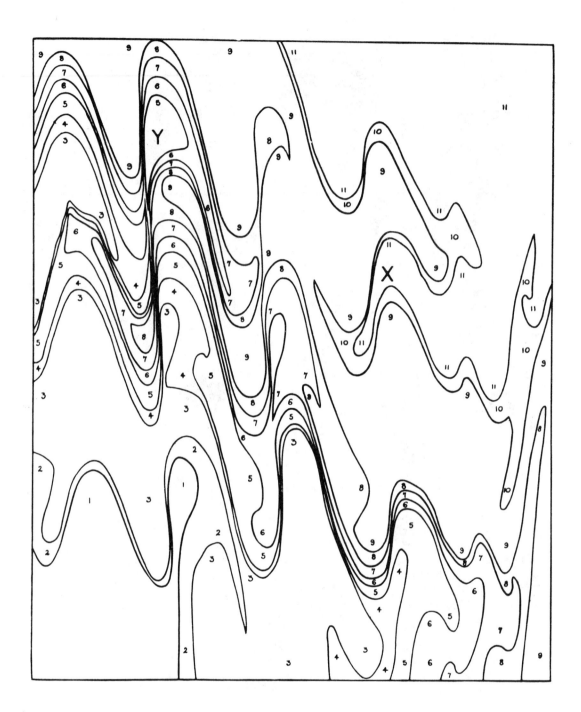

Fig. 11.—Sequence of rock units derived from Fig. 10.

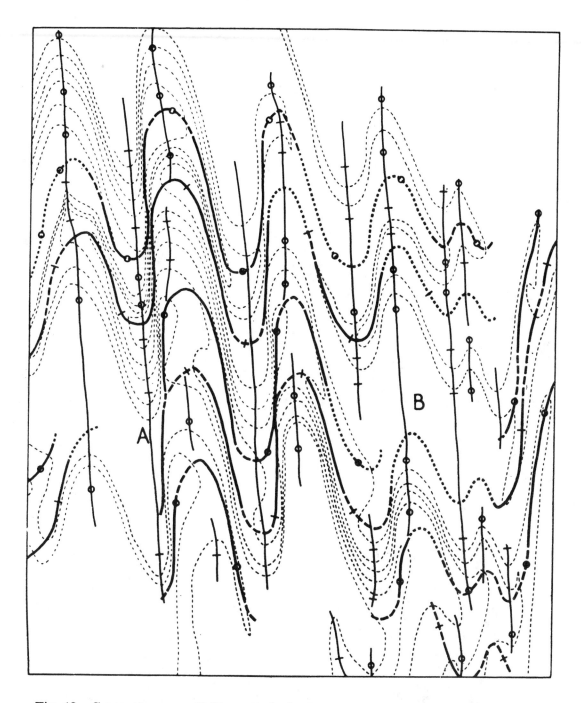

Fig. 12.—Symmetry axes (fold axes) derived from Fig. 11. The original "bedding" here shown in light dotted lines is receding and the involuted axial surfaces of the first folds which act as "bedding" for defining the second folds, become more prominent.

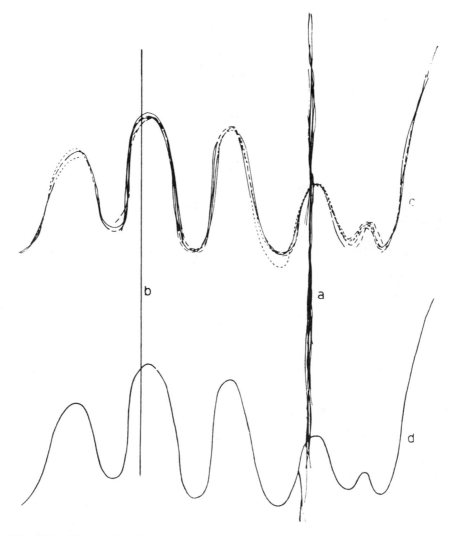

Fig. 13.—Determination of mean directrix and mean fold template of
superposed folds.

bar signs alternate along them. The other set has strongly inflected courses
but these axes have all zeros or all cross-bars and they are cut by the first set,
which forms axes for their inflections as well as for the bedding from which
they were drawn. Further, whenever the straight set crosses one of the inflected set, the sign of the straight axis changes. It is clear that two sets of
fold axial traces have been filtered out, the inflected set belonging to the early
folding and the straight set to the later folding. In addition it is now possible
to complete the early set of axial traces across strongly attenuated zones such
as at A by joining axial traces of like sign, and across broad poorly defined
areas such as B, using the three clues that (a) axial traces of like sign must
be joined, (b) the early axial surfaces will be inflected on the later axial
surfaces according to the pattern shown further up or down the later axial
traces, and (c) the number of earlier axial traces inserted across a gap must
be such that change of sign occurs along the later axial traces at each earlier
axial trace crossed, and only at such crossings.

The two sets of fold axial traces are taken off on Fig. 13. This can now be resolved further by using the law of similar folding (that thicknesses measured in the direction of flow remain constant), in order to wipe out the later folding and regain the first folding. After the early folding, the fold axial surfaces of the first folding were a set of planes or near-planes and the deformation of these (Fig. 13) is the later folding.

We proceed as follows. First check the later axial traces for parallelism. They are flow lines of the later folding and if they are not parallel the intercepts on them need correction for convergence or divergence as previously explained. In the present case the axial traces are essentially parallel. Next trace off the longest of the axes of later folds, and moving the tracing parallel to itself, to get the best approximation to congruency, trace off all the other axes (Fig. 13a). The mean direction of these is then adopted as directrix (Fig. 13b). Next trace off the most continuous of the inflected early fold axes, then move the tracing parallel to the directrix until the best fit is obtained against the next folded axis, and trace it off on top of the first. Where the axis was drawn in continuous line indicating that it was closely defined it should be so traced, and likewise where it was shown in broken or dotted line indicating more uncertainty it should be traced thus because this part of the fold may be better known from the other axes. When all the axes are traced off (Fig. 13c), a single mean curve is drawn (Fig. 13d) to represent the second folding, giving preference to the closely defined lines.

In proceeding through the sequence of Figs. 10, 11, 12, and 13, the bedding has been progressively filtered from the picture to yield the skeleton of the second folding, namely its form and its directrix. We now return to Fig. 10 and use these elements to isolate the first folding (Fig. 14). Fig. 14a is the original exposure of Fig. 10. Fig. 14b is constructed from this by displacing all parts of this figure in the direction of the directrix (Fig. 13b) by amount indicated by the second fold (Fig. 13d). The deformation of AB and CD, the boundaries of the rectangles ABCD, show the overall movement. In Fig. 14b the first folding shows through for the first time — albeit crudely and unrefined, but none-the-less distinctly. The irregularity of the fold limbs may come from three causes: (a) the drafting uncertainty in determining the points where bedding and directrix intersect acutely; (b) some irregularity of directrix, and other minor departures from strictly similar folding; and (c) real deformation in the original fold limbs. The first two of these, which are artificial, can be filtered out by repeating the process of Fig. 13, which is carried out in Fig. 14c, from which the smoothed shape of the first folding is shown in Fig. 14d. If we smooth all the folds in Fig. 14b to conform with Fig. 14d we get back to the condition of the rectangle ABCD after the first folding (Fig. 14e).

It is now a simple matter, by repeating this process, to wipe out the first folding and get back to the original beds before deformation (Fig. 14f) and establish a stratigraphic succession with thicknesses of all formations. The figure ABCD of Fig. 14f bounds the material which ends up as the rectangle ABCD of Fig. 14a. Alternatively an original rectangle would deform to a figure reciprocal to this, with sides shaped like Figs. 13d and 14d.

Some very fundamental facts about similar folding emerge from Fig. 14.

Fig. 14.—Resolution of superposed similar folds to the original beds.

14a.—Final superposed pattern. 14b.—Initial reversal of later folds. 14c.—Successive beds of 14b compared by superposition. 14d.—Mean fold from 14c. 14e.—Result of "refining" 14b by 14d. 14f.—Removal of 14d from 14e to produce original beds.

First, the original stratigraphic thickness of any bed which has undergone one of more superposed foldings by the similar folding law, is never less than the *maximum* orthogonal thickness of that bed in its final state. This should be axiomatic if we pause to visualize what happens during similar folding. Yet text-books almost invariably talk of thickening of the fold crests, whereas the real process is one of thinning of the limbs (as measured orthogonally). The universality of *separation* of boudins wherever a thin more viscous forma-

Fig. 15a.—Alternative form of Fig. 14e if simple shear is added. 15b.—Alternate form of Fig. 14f if simple shear is added.

tion is present to act as a marker· of the movement, confirms this generalisation.

Second, any chosen point on Fig. 14a can be identified precisely on Fig. 14f, and hence stratigraphic relations can be determined. Thus a stratigraphic anomaly, which is not immediately apparent on Fig. 14a shows up immediately on Fig. 14f, as well as on 14e. It is obvious that bed 10 on Fig. 14f which runs out of the diagram at E should continue across through F and G. This is also clear on Fig. 14e, and when we study Fig. 14a it is clear enough that bed 10 which runs out across AB at E should be brought back again into the diagram at E F and G by the sharp fold (of the second group) immediately to the left of E. Nevertheless I had worked a good deal with Fig. 14a and did not realize that there was anything anomalous about it until the analysis was done. The analysis indicates that the exposure should be re-examined to see what happens to bed 10.

The length of the black tongue extending to the right of FG on Fig. 14f is a measure of the distance formation 10 and 11 should project into 14a between F and G. A complete topological identity exists between Figs. 14f and 14a. For example if we follow the boundary from B towards C on each of these figures, we start at B high in formation 11, then rise a little stratigraphically, then descend quickly down through the sequence almost to 10, then rise stratigraphically a little across a first generation syncline, then on to its other limb descend quickly across bed 10 almost to the base of bed 9 in the second generation anticline, whence we rise stratigraphically in bed 9 to point C. It is instructive to follow out each boundary of ABCD step by step in the two diagrams.

Third, an inescapable element of indeterminancy remains in all of these figures. At the outset of this address I pointed out that bedding cannot disclose deformation which is parallel to itself, nor can it disclose rotations of the whole. Our induction that the beds were originally horizontal, and hence must have been rotated does not arise from any consideration of the geometry of the folding but from a quite independent condition. The transformations of Fig. 14 are equally applicable to the involution of a set of parallel lodes as to the folding and refolding of a set of beds which were originally horizontal.

The arbitrary factor entered between Fig. 14a and Fig. 14b where the fold of Fig. 13d was displaced by a series of translations parallel to the directrix, sufficient to reduce 14d to a straight line normal to the directrix. It is in fact much more probable that the straight line should have been at an acute angle to the directrix rather than normal to it. (The normal case involves the arbitrary assumption that the axial surfaces of the second folding were normal to the axial surfaces of the first). If a different angle had been adopted, Fig. 14e would have been replaced by some form such as Fig. 15a, related to Fig. 14d by a simple shear parallel to the second directrix (13b). However this would not have affected the next result since the angle of simple shear introduced into Fig. 15a would be cancelled by a different angle for the first directrix implied by the new pattern. There is, then, an arbitrary angle of simple shear reciprocally distributed between the two sets of folding, which remains undefined by the final fold pattern. If we knew for certain that the beds were originally horizontal and that no foldings or rotations of the whole are involved other than these two folds, then the attitude of the intermediate foldings can in fact be determined.

However there is still an indeterminate bedding shear before the first folding. Thus Fig. 14a could equally well be derived from Fig. 15b as from 14f, the two latter differing only in simple shear parallel to the bedding.

Superposed Similar Folding with Transverse Strikes.—Overprinted folding is not recognisable morphologically if the axial surfaces of the two generations are the same. Clues may be found in the mineralogy or metamorphic grade but not in the form of the folds. If the axial surfaces of the two fold generations are not parallel, two different patterns develop according as the foldings have the same strike, differing only in attitude (such for example are the folds shown in Figs. 5c, 5f, 6 and 10), or have transverse strikes. Folding superposed with transverse strikes shows up more conspicuously in plan than in section since most sections are dominated by one or other fold system. Again the pattern differs markedly according to whether the first folding is strongly overturned or not.

O'Driscoll (this volume) using ingenious card-pack models has studied the forms produced when the two generations of folds intersect orthogonally or obliquely but share a common directrix. The orthogonal case produces domes and basins. The resulting shapes of a single horizon are shown in his Plate IIIB, Text-fig. 2 and Plate IVA, the outcrop in plan is shown in his Plate IVA,

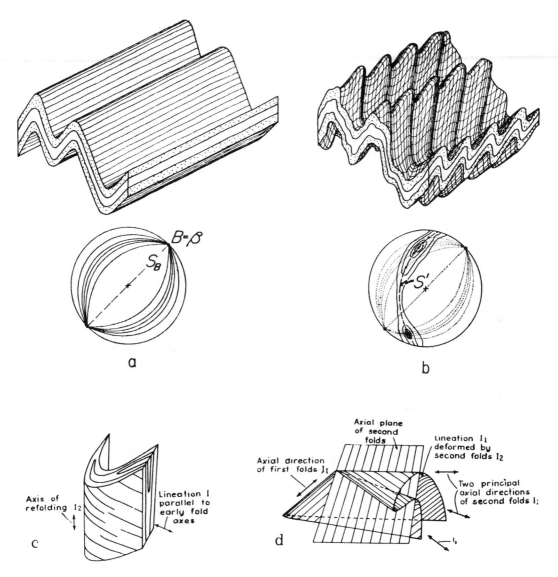

Figs. 16a and 16b.—Superposed cylindroidal folding with oblique axes, as drawn by Weiss, showing stereographic projections of the bedding and fold axes (from Weiss, 1959, Fig. 5). 16c and 16d.—Superposed deformation of early lineated folds (from Ramsay, 1960, Figs. 1 and 2). Axes and directrices are oblique.

B, C, the sections parallel to the respective fold systems are shown in his Plate IVA and traces on oblique surfaces are shown in his Plate V. Oblique intersection of superposed folds is shown in his Plate IC, IIB, C and IIIA,C and Text-figs. 1, 4, 5 and 6, which show that the resulting domes and basins have sigmoid fold axes which have right or left echelon according to the relative angles of shear of the two interfering fold limbs. It is commonly assumed that such echelon sigmoid anticlines indicate wrenching transcurrent movements. However O'Driscoll emphasizes that nothing but vertical transport is involved in producing these sigmoid echelon patterns. Offset occurs whenever the interfering fold axes are parallel or have a common component of movement (i.e. they are not orthogonal). These offsets cause the sigmoid patterns (offset in opposite senses on either side of the composite crest) and may cause migration of crests or culminations below an unconformity (see O'Driscoll, Text-fig. 9).

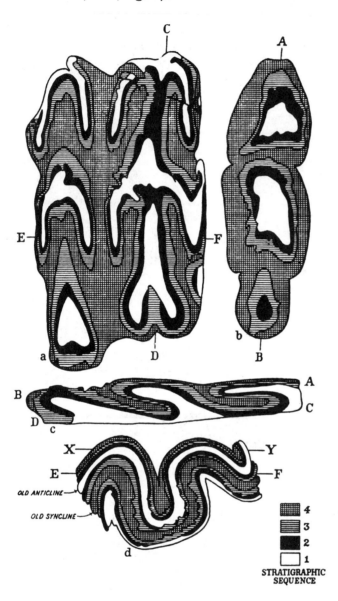

Fig. 17.—Superposition of similar folds with orthogonal strikes and divergent directrices (from Reynolds and Holmes, 1954, Text-fig. 13). EF is the axial direction of the first folding and CD (or AB) that of the second. (a) Surface outcrops after slicing. (b) Surface outcrops on the right-hand side only, as seen at a higher level than that of (a). (c) Section of the model, before slicing off its top, cut along CD of (a) and AB of (b). AB and CD represent the levels at which (b) and (a) were respectively sliced. (d) Section of the model, before slicing off its top, cut along EF of (a). XY and EF represent the levels at which (b) and (a) were respectively sliced.

Weiss (1959) has studied the same kind of cases as O'Driscoll by statistical analysis of S-planes (see Fig. 16).

Reynolds and Holmes (1954) have investigated by field mapping and petrofabric analysis, confirmed with plasticene models, a case of overprinted folding where the strikes are orthogonal but the directrices are oblique. They have shown that under these conditions the traces on a horizontal surface (corresponding to Figs. 19 and 20 of O'Driscoll) develop trident-, heart-, stirrup- and anchor-shaped outcrops (Fig. 17).

When the orthogonal axes of Reynolds and Holmes are made oblique, with the fold directrices also divergent, the same patterns are produced in a skewed form, as shown in Fig. 18. The obliquity also results in sigmoidal axial traces for the same reasons as in O'Driscoll's common directrix case.

In the left hand column of Fig. 18 the fold axes are at right angles to each other. In the right hand column they intersect at 30°. In Fig. 18a both folds are symmetrical but one has twice the wave length of the other. In Fig. 18b the first is overturned. In 18c the first folds are overturned but crestlines are at different heights, although the axial lines for all folds are horizontal and wave lengths are equal. Fig. 18d shows conditions identical to

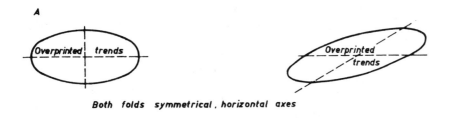

A

Overprinted | trends

Overprinted
trends

Both folds symmetrical, horizontal axes

B

Older rocks if
antiform
younger if
synform

First trend

Axial dip of first folds

Second trend

Crests of first folds similar height

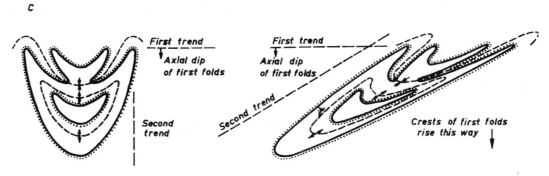

C

First trend

Axial dip
of first folds

Second
trend

First trend

Axial dip
of first folds

Second trend

Crests of first folds
rise this way

First folds have progressively rising crests

D

First folds one quarter wave - length of 2ⁿᵈ folds

Fig. 18.—Comparison of orthogonal and oblique intersection of superposed fold axes with variations of fold patterns. These are all plan views.

18c except that the first folds have one quarter of the wave length of the second folds. All these closed patterns in the left hand column have bilateral symmetry; the plane of symmetry is the axial surface of the second folding and lines connecting corresponding points across this plane indicate the axial directions of the first folding. The points of the tridents (or hearts or anchors) point up dip on the overfolding of the first set of folds.

If the fold axes cross obliquely, similar patterns develop but they are skewed, but even though skewed by 60° as in Fig. 18 the patterns are easily recognised. Where the axes are so close in direction that these patterns do not show up even when foreshortened by oblique line of sight, the strikes are near enough to parallel to be analyzed as parallel strikes. Whenever such trident patterns appear on plans of complexly folded rocks they should be recognised as indicative of transverse superposed folding and the successive fold axes and their symmetries filtered out.

In anticlinal structures the younger strata are outside the closed structures and vice versa in synclines, but within such a closed structure a stratigraphic symmetry surface (axial surface of first folds) may outcrop as an inner closed structure. Within this closed structure reversed structures occur (antiformal synclines and synformal anticlines) in which the relations are the opposite to the above rules (see Fig. 18c). It is tedious but not otherwise difficult to reconstruct structure contours on theoretical folds of this type, using the principles worked out for similar folding. The same bed may be repeated at two or three levels vertically above one another so colours are necessary to differentiate the contours. With such a structure contour map, outcrop patterns may be superimposed on topographic relief, in the usual way of combining structure and topographc contours.

Possible Categories of Superposed Folding.—We can now summarise in a exhaustive list the possible categories of superposed folding, according to the relations the respective strikes and directrices bear to each other.

CLASSIFICATION OF SUPERPOSED SIMILAR FOLDS

Strikes	Directrices	Composite Symmetry	Figured Example	Pattern
1. Parallel	Common	Orthorhombic	Fig. 1	Overprinting not deducible topologically
2. Parallel	Oblique	Monoclinic	Figs. 5c, 5f, 6, 12, 13, 14	Hooked patterns, antiformal synclines & synformal anticlines
3. Orthogonal	Common	Orthorhombic	O'Driscoll's, Plates IIIB, IVA, B and C, and Text-fig. 2	Orthogonal domes and basins
4. Oblique	Common	Monoclinic	Fig. 16 (from Weiss; O'Driscoll's Plates IC, 11B, C IIIA, C, & Text-figs, 1, 4, 5, 6	Echelon domes and basins, sigmoidal fold axes
5. Orthogonal	Oblique	Monoclinic	Figs. 17 (from Reynolds and Holmes), 18	Heart, trident and anchor patterns
6. Oblique	Oblique	Triclinic	Fig. 18, 16c and 16d (from Ramsay 1960)	Skewed heart, trident and anchor patterns

Patterns (1), (3) and (4) are commutative, that is, the same pattern results irrespective of which fold is impressed first. This is true because the flow lines (directrix) are identical. Patterns (2), (5) and (6) are not commutative.

The plane of shear of the later folding tends to dominate the overall attitude of the folding, because surfaces already nearest to this attitude after the first folding are rotated least by the second folding, whereas surfaces departing most from this attitude after the first folding are rotated most by the second (compare for example Fig. 5c and 5e). Original a-lineations belonging to the early folding remain unchanged in categories (1), (3) and (4), (that is when the two foldings have a common directrix), but in categories (2), (5) and (6) the early a-lineations are dispersed in the plane containing the two directrices (see Ramsey, 1960). Original b-lineations belonging to the first folds are dispersed in all categories, and form a girdle in the great circle representing the plane of the early fold axis (the original b) and the superposed directrix. These girdles show a maximum at the second directrix, and higher density in the minor arc of the great circle between the early axis and the late directrix.

This list exhausts the possibilities for systematic overprinting. It might be thought that transcurrent shear with the directrix horizontal or oblique, and not contained in the plane of the directrix or axis of the first folds might introduce another category but this is not so. It is clear from Plate IIIB of O'Driscoll that patterns (3) and (4) or two variants of (4) may be produced from each other by giving O'Driscoll's card pack a simple shear with a horizontal directrix. A simple shear in the direction of the other fold axis may be imposed as well by tilting the base of the card pack, which is done in O'Driscoll's Plate IIIA. If these two shears are added simultaneously (e.g. by raising one corner of the pack shown in O'Driscolls' Text-fig. 2) the effect is the same as that of a dipping simple-shear axis. All these forms are topologically continuous and do not introduce additional categories.

Apparent Compression in Similar Folding. —Folds such as in Fig. 1 are commonly assumed to be caused by intense crustal compression — the basic dogma of structural geology. But these folds have the same width normal to the strike before folding and after; the amount of shortening is exactly nil. Such folding does not prove compression —it merely proves flow, and flow can occur in many stress environments. Flow implies stress difference, no more and no less, and

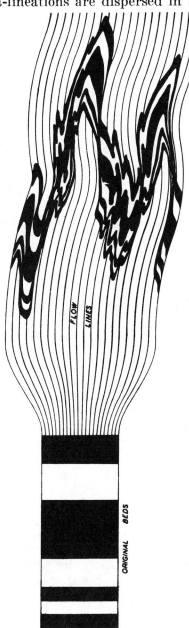

Fig. 19.—Similar folds showing dilation to twice width during folding.

stress difference may result as well from extension as from compression. In fact the folds of Fig. 1 could have developed during crustal dilation, as is shown by Fig. 19 where an identical set of folds has developed while the crustal width normal to the strike has doubled. In this figure the flow lines have diverged to twice their interval and hence the intercepts in the direction of flow are reduced to half. The overall volume of each of the folded beds shown in the upper part of Fig. 19 is identical with the volume of the corresponding unfolded bed below. Such circumstances arise wherever flow lines diverge as they move upwards as they do for example in many diapirs, and in fact, as I will contend later, as they do in most orogenic axial zones.

Certainly I could have drawn this same set of folds produced from a set of beds suffering lateral shortening to half their width. But therein lies the point. Folds of this kind when met in the field do not have any unique implication either of compression or dilatation. The folds imply differential transport and the assumption that intense folding necessarily implies intense or even mild compression is a geological fallacy we should now outgrow.

Geologists in the field, mapping folds such as in Figs. 5c, 5f, 6 and 10, have almost invariably taken it for granted that these were compressional structures. Nothing but a complex history of dire compression could produce folds such as those in Fig. 10. Such has been the creed. But in all of these particular examples the amount of shortening is precisely none, and every one of them could have been produced in a dilatating environment.

CONCENTRIC FOLDING

Deformation Anistropy.—Concentric folding such as Fig. 4 obeys the law that the orthogonal thickness of each bed remains constant through the fold. This implies that a normal to any bed at any point is also normal to the beds above and below it and that the centre of curvature is the same for all beds at the points intersected by such a normal to them, that is, the beds are continuously concentric. The restriction that the deformation does not change the orthogonal thickness can only be satisfied by marked anisotropy parallel to the bedding.

Let us start from simple isotropic folding in which all beds have identical flow properties and as a consequence all deformation is governed solely by stress field irrespective of how this field crossed the bedding. Retaining simplicity in all the other variables, let us replace each alternate bed by a bed with different flow properties such that the alternate beds flow ten times as fast under the same load as did the original material. Now under a uniform stress field the amount of deformation in a specified time will be ten times as great in the more yielding material as in the other material. This means that flow lines will be much closer together in the yielding material, just as magnetic flux lines are crowded more closely in layers of more highly permeable material, or as hydraulic flow lines are more closely spaced in alternate beds which are ten times as permeable as the intervening beds. As the disparity in rate of flow increases in the fold case, or magnetic permeability or hydraulic permeability increases in the analogues, so the flux lines become increasingly concentrated in the more receptive layers until they become virtually parallel to the bedding. In the limit, as the disparity becomes very great, they are parallel to the bedding and the result is concentric folding, where the deformation is entirely by shear flow parallel to the bedding.

Geologists used to be taught that similar folding was a more extreme form which followed on from greater intensity of the same kind of phenomenon which produced concentric folding in the simpler cases. This notion is false. Concentric and similar folding are the ideal cases at opposite ends of a spectrum. Similar folding is the deformation of isotropic material where bedding has no physical significance. Concentric folding is the other extreme where bedding so controls the deformation that all deformation is by bedding slip and all flow is parallel to the bedding. This can only occur if the resistance to flow on bedding partings is vanishingly small in comparison with flow transverse thereto. Neither of these extreme ideal conditions is commonly satisfied in nature. Real folding is spread along the spectrum in between — some of it approximating towards the similar end, and much of it towards the concentric end with gross concentric folding of "competent" beds owing to the much more rapid yielding of thin incompetent beds between them.

Our empirical experience of the distribution of similar and concentric folding is wholly in accord with this physical picture. Where do we find concentric folds? In the sedimentary basins resting on stable floors or platforms, or flanking the orogenic piles which spill over onto these platforms. And where do folds approximate most towards the similar pole? In the axial orogenic zones, among the crystalline schists, as well as in ice, salt and gypsum, where the material is inherently isotropic to deformation, or nearly so. The sediments on the platforms consist of shales, sandstones, conglomerates and limestones, which have a very wide spread of rate of flow under given load — a range in viscosity of perhaps a million.

It is wholly to be expected that folds in an interbedded series of such materials where the amplitude of the folds is an order or more greater than the thickness of the beds, should conform reasonably well with sections projected by tangential concentric arcs which preserve orthogonal thickness. But as these rocks become buried deeply in geosynclines and temperatures and fluid pressures rise, so their folding departs more and more from the concentric mode and enters the similar field. The viscosity of all crystalline materials declines exponentially with rising absolute temperature. All rocks flow more rapidly under the same load as burial increases, but this does not in itself produce similar folds. The important point is that rising temperatures and fluid pressures carve bigger slices from the viscosities of the more "competent" rocks than from those whose viscosities were already low, leading thus to a convergence of viscosities in orogenic axial zones. Such convergence means physical isotropy for deformation, and hence similar folding. In their classic study of the intensely deformed and mineralised Broken Hill lode Gustafson, Burrell and Garretty (1950) emphasized that as the various formations trend into the attenuated zone, it availed nothing that the rock be garnet granulite, sillimanite schist or greenstone, whatever rock was near squirted up where the yield was greatest, irrespective of rock type or fancied competence. The Broken Hill tectonic style is therefore akin to Fig. 5—typical similar folding.

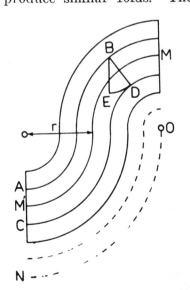

Fig. 20.—Arc lengths round concentric folds.

Length around Concentric Folds. — Provided the centres of curvature of all parts of the fold all lie outside the folded plate, the length of any concentrically folded bed is the same as its original length and hence the length along all beds are the same if a complete fold flank is measured from crest to trough. That this is true is seen from Fig. 20 where only two centres of curvature are shown and the medial horizon M consists of two equal quadrants. The succeeding horizons gain in arc length as they approach one centre what they lose as they move away from the other centre. The interpolation of any number of additional centres of curvature between the two which make the crest and trough of the fold does not alter this equality. Since length round the fold is constant, the area of the folded surface is also constant. If a shorter segment ABCD is taken, the fold trace AB is shorter than the fold trace CD by the amount of the arc DE where BE is drawn parallel to AC (see Goguel, 1952, p. 118). The length of the

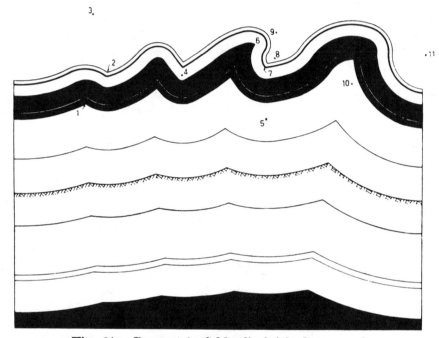

Fig. 21.—Concentric folds diminish downwards.

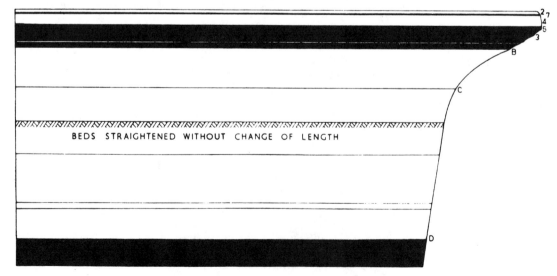

BEDS STRAIGHTENED WITHOUT CHANGE OF LENGTH

Fig. 22.—Lengths of the beds involved in concentric folds of Fig. 21.

double arm MM′ is πr and the length of the arc NO is also πr. But if another horizon is drawn an equal interval below 0, the equality fails because the relation is only true while the centres of curvature lie outside the folded plate. Accordingly, any horizon below the centre of curvature of an anticlinal bend becomes cuspate, and has shorter length within the fold if the fold remains concentric.

The concentric folds of Fig. 4 are continued downwards in Fig. 21 which illustrates two important rules governing concentric folds:

(1)   All concentric folds die out in depth.

(2)   The length along beds through one or a series of folds is constant while the centres of curvature fall outside the section, but become substantially shorter downwards as each centre of curvature (indicated by numbers) is reached.

Fig. 22 shows the result of straightening all the beds in the fold of Fig. 21. The numbers of the fold centres which correspond to the changes in the lengths of beds are indicated at the right hand end of the figure. The only part of these folds which has developed freely is the part between 4 and 6. All the beds below 6 are shorter, and each additional centre of curvature reached causes an additional inflexion in the length profile curve. This is true both upwards and downwards from the master formation which controls the shape of the fold. These facts require substantial adjustments below. In Fig. 22 the main adjustment has occurred between B and C, which is the weak zone which permitted this type of folding. The strata below C have remained relatively inert while the folding slid by above them. Concentric folds are essentially skin folds (*plis de couverture*) and have no persistence in depth. Although he realised that a décollement is implied, Goguel met this problem by crumpling the lowest bed, ingeniously maintaining the length of each bed and its local orthogonal thickness (Fig. 23). He has not solved the space problem by crumpling the lowest bed, he has merely swept it under the carpet. The décollement is just outside his diagram. The very next bed must part company altogether. Not only is there a décollement under at least one flank of the fold where the lowest bed slides a long distance on the next surface below, but these two surfaces must separate by the full amplitude of the fold. Moreover the shortening implied by the fold does not affect at all the underlying formations or the basement below them.

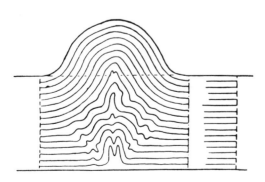

Fig. 23.—Goguel's suggested crumpling of the anticlinal core to preserve bedding length and orthogonal thickness.

De Sitter (1956, p. 199) has tried to meet this dilemma by suggesting that the lack of space in the core of the fold (Fig. 24) results in higher pressures there, which causes break-thrusts and tectonic extrusion of the core strata. But this device does not eliminate the décollement. It may solve the problem so far as it concerns the strata shown overlapping in Fig. 24, but if we are to believe that crustal shortening is the cause of folding, the full prism of basement corresponding to the shortening involved in the fold has to be annihilated. The décollement is still present

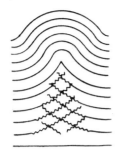

Fig. 24. — Lack of space in the core of a concentric fold (after Goguel, 1952, fig. 60. and de Sitter, 1956, fig. 137).

between the fold which shortens and the underlying basement which does not.

Because the fold amplitude diminishes rapidly in depth the length adjustment between the upper and lower beds must be in the direction of bedding surface. This may take the form of continuously distributed bedding slip in a very weak formation with disharmonic folding above and below it, or a single décollement such as in Fig. 25. If we assume that the beds on the extreme left have not moved with respect to the basement, then the amount of slippage on the décollement surface under the right limb of the first fold is the differential shortening in the upper and lower folded beds, perhaps 500 metres in a typical fold. The décollement under the second fold is greater — perhaps a kilometre, and under the right hand part of the fold, perhaps something of the order of two or three kilometres. That adds up to a substantial overthrust, but here we find it latent under a set of ''normal'' folds of not very great intensity. The implied presence of such blind overthrusts at comparatively shallow depth where no sign of faulting or thrusting is exposed at the surface is somewhat startling, and has very important implications in petroleum geology. It is none-the-less valid, as examples below demonstrate.

The curve at the right-hand edge of Fig. 22, which can be plotted from the surface dips, indicates the depth to the main décollement zone. The folded profile is determined absolutely by the surface dips, as in Busk (1929) and this in turn discloses the depth of décollement provided the folds are strictly concentric. In fact cause and effect is the reverse of this. The sequence of viscosities in the sediments determines which beds will act competently with respect to the others, and which beds will yield. This viscosity sequence determines the form of the folds, the thickness and amplitude of folding, and the depth of décollement.

Burning Springs Anticline, West Virginia. — The Sandhill Well on the Burning Springs Anticline in West Virginia was completed in 1955 in Precambrian basement at a depth of 13,331 feet. According to the Director of the West Virginian Geological Society, ''probably few wells, certainly not one in the Appalachian area, has received the critical analysis given by specialists to the Sandhill Well.'' Figures 26 to 29 are reproduced from (Woodward 1959) in ''A symposium on the Sandhill deep well, Wood County, West Virginia.''

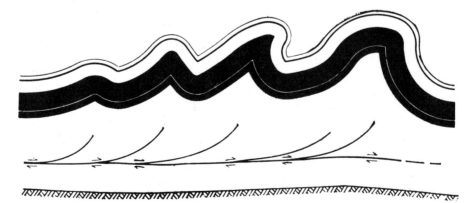

Fig. 25.—Décollement implied by concentric folds of Fig. 4.

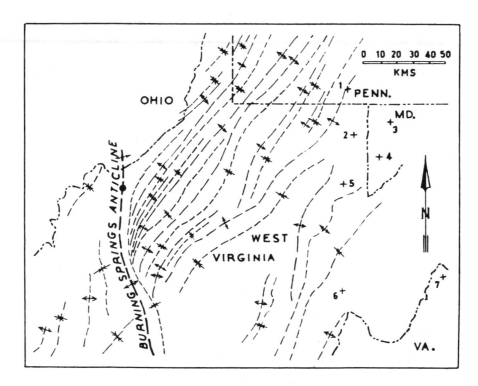

Fig. 26.—Structural setting of the Burning Springs Anticline (after Woodward, 1959). Numbered gas fields with disturbance on Oriskany décollement horizon are: 1, Summit; 2, Terra Alta; 3, Accident; 4, Mt. Lake Park; 5, Etam; 6, Glady; 7, Bergton.

Fig. 26 shows the regional setting of the Burning Springs Anticline. West of it are flat-lying unfolded strata dipping towards the Appalachian geosyncline at less than 20 feet to the mile. East of it lie a bundle of gentle concentric folds each of which dies out en echelon on the Burning Springs Anticline, which is the front fold of the fold system and much more sharply folded. Such a pattern proclaims a décollement thrust surface under all the folds to the east, which diminishes in magnitude with each successive anticline and finally rises into the core of the Burning Springs Anticline where it terminates. The strata west of this anticline are all autochthonous and not displaced with respect to the basement below. The strata east of the Burning Springs Anticline are all displaced westwards for distances exceeding a kilometre with respect to the basement below, a displacement which increases eastwards. The Burning Springs Anticline is sharply folded (Fig. 27), and is believed to be strike-faulted at the surface (Woodward, p. 11) but not seriously so, so this quite substantial overthrust is essentially blind. The south-eastern trending wing of the Burning Springs Anticline must pass down in depth to a sinistral tear fault, even though this too expressed itself at the surface only as echelon folds. The axes of the echelon folds striking into and terminating on the Burning Springs Anticline are normal to the direction of transport, which is normal to the Appalachian front, but they swing round towards the drag direction of the sinistral couple along the Burning Springs Anticline.

All this — the necessity for a large blind overthrust, the buried sinistral tear, and the steeply overthrust core of the Burning Springs Anticline were deducible from and implied by the surface outcrop map, when the principles of concentric folding enunciated above are applied. Yet when these structures were encountered in the drilling they caused surprise and puzzlement.

Fig. 28 summarizes the drilling experience as presented and interpreted by Woodward. The first 4012 feet down to the Onondaga Chert (Unit A) behaved normally. There was no unwonted disturbance, the beds were nearly horizontal on the crest of the anticline and the surface closure (1650 feet) continued. The next 1843 feet (Unit B) was quite abnormal. All manner of drilling and water troubles were met. Stratigraphic units repeated themselves as often as three times, and in 1858 feet of drilling only 185 feet of additional

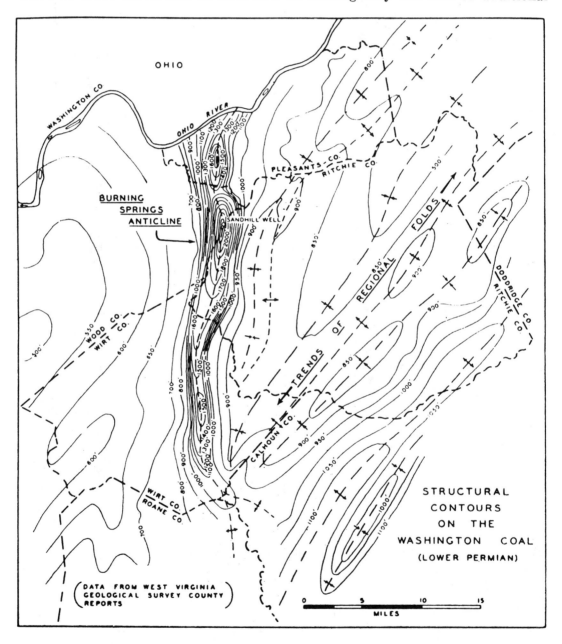

Fig. 27.—Structure of the Burning Springs Anticline, West Virginia reproduced from Woodward 1959 (with the permission of the West Virginia Geological Survey).

COMPOSITE STRUCTURE
SANDHILL WELL

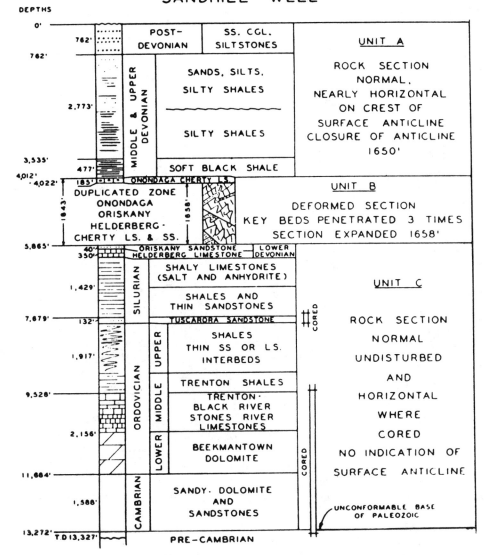

Fig. 28.—Structure—stratigraphic interpretation of Sandhill well by Woodward (1959, p. 162). (Reproduced with the permission of the West Virginia Geological Survey).

strata were drilled. The extra thickness is the amplitude of the anticline, which completely vanished in the décollement zone, and below the blind décollement, did not exist at all. Once the décollement was passed the remaining 8407 feet (Unit C) was drilled without trouble, the beds were horizontal, and there was no sign of the Burning Springs Anticline.

That the anticline really does vanish at the décollement is made quite clear by Woodward from the comparative sections in the Sandhill well and in wells 10 miles to the west and 10 miles to the east respectively:

COMPARATIVE ELEVATIONS ON THE ORISKANY SURFACE (FROM WOODWARD)

Intervals or elevations	10 miles west	Sandhill Well	10 miles east
Elevation, Washington Coal _____	525' A.T.	2400' A.T.	850' A.T.
Elevation, Berea Sandstone _____	-1618' A.T.	288' A.T.	-1218' A.T.
Berea-Huntersville interval _____	2900'	3260'	3600'
Elevation, top Huntersville _____	-4518' A.T.	-2972' A.T.	-4818' A.T.
Thickness, Huntersville _____	175'	185'	195'
Excess thickness, Segment B _____	—	1658'	—
Elevation, top of Oriskany _____	-4693' A.T.	-4815' A.T.	-5013' A.T.

The Lower Permian Washington Coal has a regional dip west of 16 feet to the mile for the 20-mile interval and the Burning Springs Anticline rises 1713 feet above this slope at the Sandhill Well. The Berea Sandstone has a regional dip west of 20 feet to the mile for the 20-mile interval and the anticline rises 1708 feet above this slope. The top of the Huntersville has a regional dip east of 15 feet to the mile (owing to regional thickening towards the geosyncline axis) and the anticline rises 1696 feet above this slope. The Huntersville formation, which should be 185 feet thick at the Sandhill Well, is repeated through 1843 feet; this is an excess thickness of 1658 feet, which is approximately the full height of the anticline, and at the base of the Huntersville (top of the Oriskany) the anticline has vanished on the décollement. This horizon has a regional dip of 16 feet to the mile to the east over the 20-mile interval and its altitude as drilled in the Sandhill Well is only 38 feet above the mean gradient. The amplitude of the anticline declines from 1713 to 1708 feet, that is five feet in the Washington-Berea stratigraphic interval of 2112 feet, and from 1708 to 1696 feet, that is 12 feet in the Berea-Huntersville stratigraphic interval of 3260 feet, and from 1696 feet to 38 feet, that is 1658 feet in the Hunterville stratigraphic interval of 185 feet. In the 5372 ft. stratigraphic interval above the Huntersville the anticline loses one foot of amplitude for every 316 feet of strata. This of itself implies that some minor décollement slippage is present in the weaker zones of the main arch-forming strata. But in the Huntersville interval the loss of amplitude is 1658 feet in 185 feet of strata or six thousand times as much per foot as in the strata above. This is a real décollement, although there is no physical sign of it at the surface. It is of the greatest importance to petroleum geology that such décollements be predicted in advance of drilling. The décollement must continue east under all the folds of West Virginia increasing in displacement with every surface fold. Woodward (1959, p. 168) states that one of the curious features that has been discovered in the commercial development of the 'mountain type' Oriskany gas fields (as at Terra Alta, Accident, Summit, Glady, Beraton, Etam and others) is that there is a zone of acute deformation at the general level of the Onondaga-Oriskany-Helderberg horizon, which is not visible at the surface (see Fig. 26). This disturbed zone is clearly the décollement surface which turns up into and terminates in the Burning Springs Anticline. The Oriskany gas fields mentioned lie about 150 km. east of the Burning Springs Anticline and extend for some 150 km. along the strike (Fig. 26). In this area the décollement is about halfway between its emergence on the Burning Springs Anticline and its root in the Appalachians. At the Glady-Accident line the westward displacement on the décollement must be several kilometres, being the sum of the surface shortening of all the folds across West Virginia.

In the décollement zone in the Sandhill well the drillers found very heavy flows of salt water which rose more than 3000 feet in the well and defied

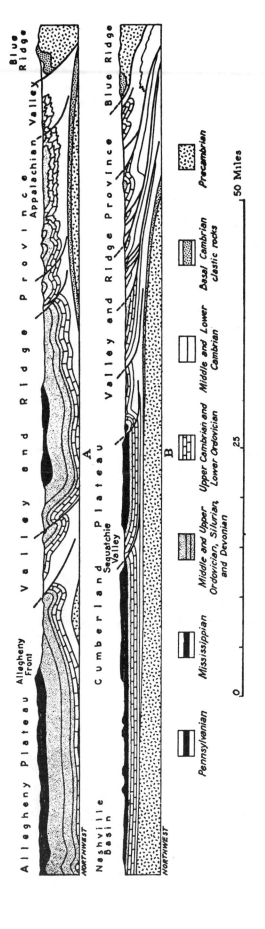

Fig. 29.—Décollements in Cambrian shales of the Rome and Conasauga Formations in Valley and Ridge province of Tennessee (from King, 1959, following Rodgers 1953).

attempts at lowering (Corbett, 1959, pp. 171-172). The presence of this décollement which has clearly been an important channel for fluid migration could have played a dominant role in the migration of petroleum in this region, and may have been crucial, not only in the presence of gas and oil in the Oriskany, but in the somewhat peculiar distribution of oil in the West Virginia area.

The thrust surface finally turns down into the root zone of the Appalachian orogen. Its overall pattern is very similar to the outermost thrust in the tectonic profile shown in Fig. 44, which is based on a section through the Cumberland Plateau further south along the Appalachian front. Nearer the orogen axis décollements appear in higher strata and produce the more steeply dipping overthrusts and the disharmonic folding so clearly portrayed by Darton's sections of the Pennsylvanian anthracite belt (Darton, 1940). All this is closely in accord with the general section given in Fig. 44.

Allegheny and Cumberland Plateaus.—Fig. 29 which is reproduced from King (1959, fig. 25), following Rodgers (1953), shows cumulative décollement surfaces developed under the concentric folds of the Valley and Ridge province in West Virginia and Tennessee where the décollements are developed in Middle Cambrian shales of the Rome and Conasauga Formations. Here overthrusts peel upwards into the axial surface of most anticlines although the main décollement is mostly blind. The relation of these décollement surfaces to the orogen as a whole is shown in Fig. 44. Large latent displacement parallel to the bedding in the Cambrian shales in this region offers a supreme example of the false innocence of bedding referred to in the introduction to this address.

This Valley and Ridge province was the classic area which inspired Bailey Willis to establish the geometry of concentric folding. The other type area for this kind of folding is the Jura, which is also skin folding with décollement, here in the Triassic and Permian salt formations (Fig. 30). It is not chance that concentric folding should dominate here. The tectonically weak formations so ordain.

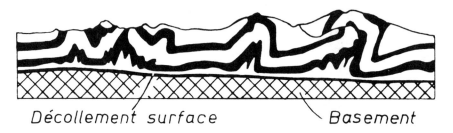

Fig. 30.—Jura-type décollement.

The Burning Springs Anticline, the Valley and Ridge folds and thrusts and the Jura folds have all been assumed to imply crustal shortening. This does not follow. Such folds necessarily die out in depth and the crust below them is not shortened. These folds do imply translation of the superficial layers by simple shear about a horizontal axis of shear. This translation can be produced in other ways beside compression, so it is false to deduce from the presence of folding, either concentric or similar, that the crust has been shortened.

REVIEW OF SIMILAR AND CONCENTRIC FOLDING

Similar and concentric folds are compared and contrasted.

SIMILAR FOLDS	CONCENTRIC FOLDS
Isotropic deformation	Anisotropic deformation
Shear direction not related to bedding	Shear parallel to bedding
Thickness of beds constant in direction of flow	Thickness of beds constant orthogonally
Length of bed around fold increases	Length of bed round fold constant
Surface area of bed increases	Surface area of bed constant
Persistent down axial surface (plis de fond)	Dies out down axial surface (plis de couverture)
Implies absence of décollement below	Implies décollement below
Implies transport in direction of axial surface which proceeds from considerable depth	Transport in direction of axial surface in main part of fold proceeds from transport parallel to bedding below main part of fold
No implication of shortening or of extension	Implies superficial shortening but not crustal shortening

Analogue Models of Folding.—Early structural geologists conceived folding by mental models of stacks of paper buckled by lateral compression. Such models produced folds. But these models involved limitations and restrictions which are inherent in the model but not necessarily in nature. The paper model constrains the length round the bed and the area of surface of the bed to remain constant during folding. It also involves zero cohesion between beds, leading to easy bedding surface slip, and ordains constancy of orthogonal thickness of beds. The models necessarily involve shortening of the width of the folded slab, which is thought of as crustal shortening, for although the model certainly requires a décollement surface below the lowest bed, this was overlooked since it is outside the limits of the model.

All these properties are properties of concentric folding which therefore acquired status as normal or simple folding, whereas other modes of folding were regarded as extreme cases or aberrant cases, or strain beyond some normal undefined limit.

A different model could have been chosen, by taking layers of coloured plasticine and forming folds by raising rams (of varied shapes) from below, or by injecting honey through elongated apertures below. This model would also have produced folds, but they would be folds in deformationally isotropic material, and there would be no lack of cohesion on bedding surfaces. The boundary area of the folded slab would remain unchanged during deformation, the length around the fold and the bedding surface area would greatly increase, orthogonal bedding thicknesses would be less in the limbs, all orthogonal thicknesses would be equal to or less than the original orthogonal thickness, the bedding in the direction of flow would be constant, and there would be no décollement below the fold, and no implication of crustal shortening. All these properties are properties of similar folds.

Experiments on salt domes have followed this pattern, but salt domes have not been regarded as compressional structures anyway. More recently the Russian group led by Beloussov has produced experimentally a variety of fold types by vertical motions or rams and basement blocks. This work approaches nearer to reproducing the conditions in the real earth than any other experimental tectonic work. No crustal shortening is involved. The "ram" need not be something more rigid* or more viscous. Salt is more mobile than the materials it intrudes, and likewise the driving pressure in the hearts of folds in the axial zones of orogens may be rheid materials.

The same mental straight-jacket of folding stacks of paper leaves the notion that folds are long in the direction of their axes — and that in the ideal case the folded surface may be generated by a straight line moving parallel to itself —the so-called cylindroidal folding. A great deal of folding is not like this, but is produced by a tongue rising steeply in one place. A salt dome is the obvious example, but mantled gneiss domes and many other structures are of this type, where each surface is shaped like a cap, which rests over a similar cap below it, and so on down. This is *paraboloidal* folding. The axis of the paraboloidal cap is the direction of flow, and all beds maintain their thickness measured in this direction. The section of the paraboloid normal to the flow axis may be circular, or elliptical, or have more involute shapes. The ellipse may be so eccentric that the folded surface is a flattish tongue whose axis is the direction of flow. The individual flow lobes within a salt dome are commonly of this type (e.g. the salt shown in black in Fig. 34) as are also many lobes in migmatites, gneisses and the hypothermal ore bodies. Where the cross-section increases in the direction of flow the surface stretches, but when the flow lines converge, the surface does not shrink in again like elastic, but converges only approximately in its own surface; after a series of such stretchings and constrictions the surface has many wrinkles large and small, so characteristic of the folds in salt and ice and schists, or anything that has flowed a great distance. The axes of these wrinkles are the flow direction, not the normal to it as some students of petrofabrics insist, and these are not drag folds in the commonly accepted sense. An outcrop such as Fig. 31 (a) might be interpreted as indicating a sinistral couple in the plane of the paper. But if it is a case of paraboloidal folding the motion is normal to the paper and a considerable distance down the flow line the position of A and B might be as in Fig. 31 (b) and still further down as in 31 (c). Similar folding involves great extension of the bedding surface and once extended it would require a very long chance to contract it again precisely in the same line. Hence rocks which have flowed long distances usually develop the wrinkled surface of crepe, with long flute- or mullion-like minor folds in the direction of flow.

All of this is easy to comprehend if we do our thinking with a realistic model, but it is hard to imagine paraboloidal folding in paper. There is an important lesson here. In geology we think in terms of models — either physical analogues or models in the form of algebraic symbols adopted to represent the properties of the earth. Many of these models involve greater or less generality than the real earth, and hence thinking in terms of the model leads to conclusions valid for the model but not necessarily valid for the real earth.

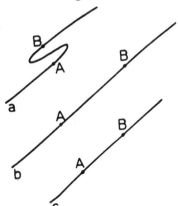

Fig. 31.—Successive sections up the flow lines of a paraboloidal fold, showing development of minor folds by weaving of flow lines.

---

* In fact rigidity is quite irrelevant. It is the viscosity which is significant here (see Carey, 1954).

THE TECTONIC IMPLICATIONS OF FOLDING

Interpretation of a Fold Pattern.—Fig. 32 is a map of complexly folded tectonite. The wavy lines are all bedding. The gaps are places of concealed outcrop. All dips are steep. All deformation has been by flow without fracture.

Fig. 32.—Plan of a folded tectonite (part of a level plan of a mine).

A map like this could be matched in many places in the Canadian shield. There is a marked ENE trend in the southern part of the map swinging more north-easterly in the south-east corner. But in the northern part of the map the trend is NNW. This clearly suggests an area of over-printed folding with intense compression both from the NNW and the NE. Fig. 33 is a regional map of the same area and its surroundings on a smaller scale. This broader view confirms the ENE and NNW axes but there is now an additional clear N-S folding trend. Dips are consistently steep. The plain view shows intense

Fig. 33.—Regional setting of Fig. 32 (area marked B).

crenulation on tiny as well as large scale. There is no doubt, we might conclude, that these are strongly over-printed structures, which have suffered intense horizontal compression in not less than three orogenies.

But how wrong can we be! The horizontal compression in this area is nil. The direction of transport is along the fold axes — normal to the paper — not transverse to them in the plane of the paper. These maps are traced directly from the map of the Grand Saline Salt dome in Texas (Muehlberger, 1958).

Lessons from Salt Domes.—Fig. 34 is a section through the Heide salt dome in Germany (after Bentz, 1949) which shows repeated injections of salt following each other up the dome. The salt shown stippled is the highest stratigra-

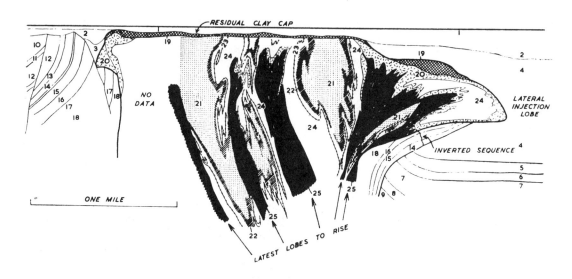

Fig. 34.—Successive intrusion lobes of salt revealed by potash mining, Heide Salt Dome, Germany (after Bentz). (1) Diluvium; (2) Tertiary; (3) Upper Cretaceous; (4) Senonian; (5) Turonian; (6) Cenomanian; (7) Albian; (8) Aptian; (9) Hauterivian; (10) Lias; (11) Rhaetic; (12) Middle Keuper; (14) Upper Muschelkalk; (15) Middle Muschelkalk; (16) Lower Muschelkalk; (17) Red Beds; (18) Bunter; (19) Residual clay cap; (20) Anhydrite cap; (21) Youngest rock salt; (22) Red salt clay; (23) Potash bed; (24) Lower young rock salt; (25) Older rock salt.

phically and the salt shown black is the lowest stratigraphically. To the right of the salt are the Triassic strata [Bunter (18) to the Muschelkalk (14)] inverted and overturned to dip at about 30° wrongside up, overthrust over the Cretaceous [Hauterivian (9) to Senonian (4)], with still older salt overthrust over inverted Triassic. If this heavily overthrust section were met in field mapping, it would certainly be concluded that strong crustal shortening had occurred. Yet the crustal shortening across this area is nil. The salt has injected vertically in a complex diapiric fold through the non-folded strata and tilted them back then finally over-ridden them in one direction. If the structure had been perfectly symmetrical it would have over-ridden them in all directions, but such perfection is rare and any asymmetry tends to accentuate itself. The moral of this example is that we must resist the temptation to interpret overthrusting as due to crustal shortening, because such a conclusion does not necessarily follow.

Salt dome folding is a fertile field for tectonic contemplation. Nettleton (1936) and many other authors have emphasised the applicability of fluid mechanics to their analysis. Fig. 35 reproduced from Carey (1954) shows

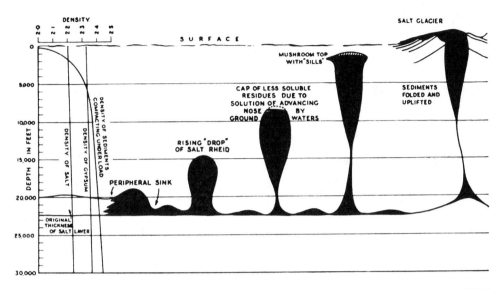

Fig. 35.—Density inversion causing the rise of salt domes (from Carey, 1954).

how the density of average geosynclinal sediments is less than that of salt (2.2) when deposited but increases with burial load to values significantly greater than that of salt under the same load. The situation is mechanically similar to a layer of dense syrup overlying a layer of lighter viscous oil, provided we take a time scale long enough for both silt and sediments to behave as a rheid. The salt therefore rises like a tear drop penetrating the sediments. At a depth of about 500 metres the density of salt and adjacent sediments is equal, and above this less work is done against gravity if the salt intrudes the sediments laterally, lifting the lighter sediments above.

An even more instructive analogy comes from comparing the layer of lighter salt underlying a layer of denser sediments with a layer of warmer (and hence less dense) fluid below a layer of colder denser fluid. This results in a series of convection cells, whose circulation is driven mechanically simply by the density difference. In both the salt dome and the convection cell the energy driving the motion is gravitational potential energy. This salt dome —convection cell analogy can be taken a stage further. The convection circulation has a toroid (doughnut) form — upwards in the central trunk, radially outwards at the top, and downwards to the base then radially inwards to the central trunk. The salt dome transport has precisely this form.

In the trunk of the salt dome the flow is upwards. At the top the flow is radially outward, symmetrically in the ideal case but more commonly regional dip or some other inequality tips the flow in one direction as in the right hand side of Fig. 34. In the rim syncline the flow is downward as the supporting salt flows away from beneath it, and this centripetal salt flow completes the toroid circuit.

The analogy with a convection toroid is close, but not perfect. In general the viscosity of a convecting medium varies only slightly round the circuit.

In the salt-sediment toroid the viscosity of the salt is an order or two less than that of the sediments. There are three consequences of this. (a) The cross-section area of the salt sections are an order or more smaller than the cross-section area of the corresponding sediment portions. (b) The salt travels an order or more further than do the sediments (the rise of the salt is at least ten times the amount of subsidence of the rim syncline). (c) The salt flows without fracture whereas fractures directly related to the toroid motion are numerous in the sediments. Stated another way, the salt is a rheid for this time scale whereas the sediments are intermediate between rheids and brittle solids (see Carey, 1954). The probability of fracture increases as the dome develops, not because of increasing strain, but because of increasing rate of strain. In the early stages the stress difference in the materials is infinitesimal. But with each increment to the dome the weight of the salt column becomes less and the weight of the sediment column becomes greater, so the dome rises at an accelating rate, and faulting makes an increasing contribution to the total transport of the sediments, commencing where the flow lines are closest, and extending its domain as the flow rate accelerates.

The analogy with the toroidal circuit of a convection cell gives a complete explanation of the tectonic pattern of the various parts of a salt dome (Fig. 36).

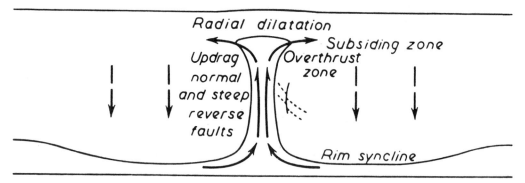

Fig. 36.—Salt dome toroidal circuit arising from the same kind of density inversion as a convection toroidal circuit.

(1) Above the salt dome cap the circuit is radially outwards. Hence this area suffers radial dilatation and produces characteristic tensional patterns with grabens (Fig. 37) which have been beautifully reproduced by Parker and McDowall (1951) in dynamically scaled models. These tension fractures are usually ascribed to stretching of the rising dome, and whereas this explanation is valid, tension is inevitable in this centrifugal area anyway.

(2) Immediately adjacent to this dilatation zone the overhang region displays overthrust tectonics approaching horizontal nappes, with steeply overturned strata (Fig. 34). These overthrusts are in the direction and sense of the toroid flow lines, and adjust the motion of the more rapidly flowing salt to the less rapidly moving sediments. The motion of these overthrusts is continuous back-flow with the boundary faults of the salt around the trunk. In this they are mechanically analogous to the overthrusts shown in Fig. 34 (near number 14 on that figure) although in Fig. 36 the viscosity difference is less marked than in Fig. 34, and hence the overthrust zone is more distributed.

(3) The sediments surrounding the trunk show the strong updrag appropriate to that part of the circuit, with steep reverse faults with peripheral strikes.

(4)   The cylindrical contact surface between sediments and salt is one of strong slippage owing to the very different flow parameters of sediments and salt.

(5)   The trunk salt has most complex poly-paraboloidal folding with intense crenulation owing to the great distance of transport.

Salt domes are small enough and simple enough and develop entirely above an inert floor so that it is easy for us to visualize the whole structure.   Orogens are larger and more complex and have ill-defined roots.   But the integral understanding of salt domes takes us a long way along the road to the understanding of orogens.

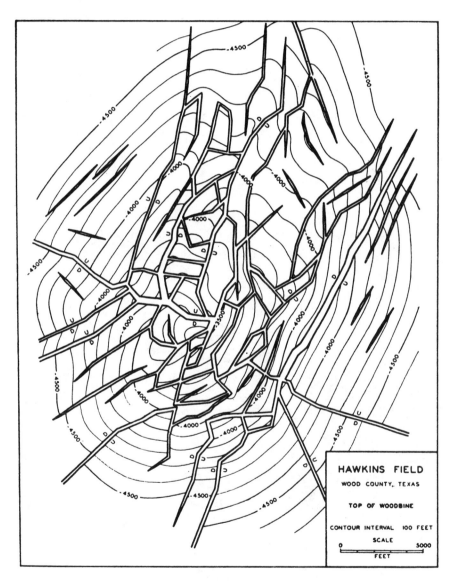

Fig. 37.—Tension structures above Hawkins salt dome, Texas. (From Parker and McDowall, 1951).

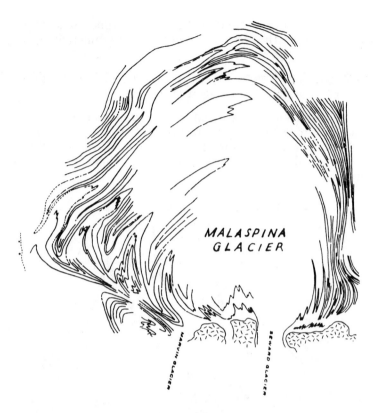

Fig. 38.—Plan showing flow fold pattern of Malaspina Glacier, Alaska. (From Sharpe, 1958).

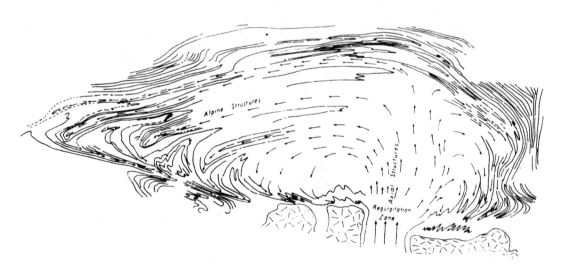

Fig. 39.—Section showing analogous flow fold pattern where material extrudes upwards and the lobe flattens under its own weight.

The Lesson of the Malaspina Glacier.—Fig. 38 is a map compiled by Sharpe (1958) showing the folding in plan as the Malaspina Glacier pushes out onto the piedmont, finding relief by spreading laterally wherever there is space occupied only at a lower level. The arrows show the direction of flow of the various lobes. The width of the main lobe is some thirty kilometres. Fig. 39 for comparison illustrates the kind of structures found in the Alps. In the axial zone is a zone of roots where all the nappes rise steeply out of the pre-orogenic trough of the Alps. One by one they turn over and flow out towards the NW as great nappes, some over-riding clearly above, others digging into the backs of others ahead of them. The outermost structures in the pre-Alps are the farthest travelled. Some lobes—fewer in number turn over towards the SE. If not true in detail this diagram certainly contains the fundaments of the Alpine structures. Yet this diagram has been made from the Malaspina map simply by changing the ratio of the vertical scale to the horizontal in the ratio of three to one, to allow for the difference that in the case of the Alps the upward extruding lobe is weighed down by gravity whereas in the case of the Malaspina Glacier the restraint to outward movement is no greater than the restraint laterally.

Apart from this ratio change, the two diagrams correspond completely line for line, fold for fold. Yet Alpine geologists have told us that the Alps represent great shortening, and that the overthrust belt was formerly eight times as wide as it is now, that it has been compressed to one-eighth of its width. Yet we *know* that the Malaspina lobe produced similar geometrical and structural patterns while it was being dilated to 30 times its original width. Can there be any question of the conclusion that the structures of the Alps could be produced by the upward extrusion of the contents of the geosyncline, spreading laterally where the pile lacked lateral support? The jaws of the extruding zone may have remained fixed while material was squeezed out from within (as in the case of the Malaspina Glacier) or they could even have themselves dilated during the process. In the former case the overall length of the crust would have remained unchanged, in the latter case the length of the crust would have *increased* during the formation of the Alps.

Haller (1956) has drawn sections like Fig. 40 representing the deformation of migmatites during the Caledonian orogeny in East Greenland. He correctly interpreted these structures as extrusion flow and concluded that no crustal shortening is implied.

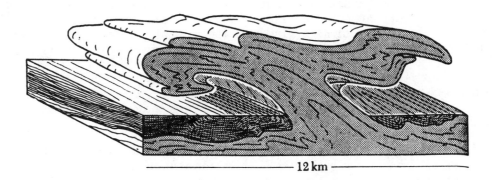

Fig. 40.—Migmatite gneiss regurgitation diapirs in Greenland. (From Haller, 1956.)

The structure of an Orogen.—Orogens do not all fit a single mould, yet they have several common characters. Away to the flank there is usually a basement shield. Lapping on to this in carbonate-quartzite-argillite facies are strata which thicken towards and dip regionally towards the geosyncline, forming a syncline (e.g. Allegheny Syncline) which is asymmetric in thickness, facies and tectonics. The shield side of this syncline is simple and thin. The orogenic side shows concentric folds (e.g. Valley and Ridge province), then overthrusts, with thickening strata exposed at progressively lower stratigraphic horizons, until finally the rim is reached in the form of the pre-orogenic basement horsts overthrust over the fore-syncline (e.g. Blue Ridge, Green Mountains, etc. in the Appalachians and Mt. Blanc, Aar, etc., in the Alps). Beyond this rim lies the greywacke belt (schists lustrés), the orogenic axial zone of vertical tectonics — nappe roots, paraboloidal folds, mantled gneiss domes, migmatitic diapirs and granitic diapirs.

The types of movements in these respective zones differ. The shield is mildly positive. The shield limb of the fore-syncline subsides gently throughout with rate increasing towards the orogen. The Valley and Ridge belt suffers concentric folding with buried décollements, and overthrusting away from the orogen. This involves no real crustal shortening, but only the gravity overflow of the extruded axials over the rim of the syncline which is itself dragged upwards and outwards from the orogen, overthrusting the sediments. This zone is completely equivalent to the overthrust zone on the right hand side of Fig. 34 (but on a larger scale), and equivalent to the lateral thrust zone of Fig. 39. The axial zone of the orogen proclaims vertical motion from every lineament, as in the trunks of salt domes. Fig. 41 shows a plan of the axial zone in North Carolina, showing circular swirling trends due to vertical motion normal to the paper as in the Grand Saline salt dome plan (Fig. 33). In both figures (41 and 33) the flow is upward, normal to the plane of the figure. Dips are all steep. Strikes may box the compass. Mantled gneiss domes of the axial zone show fir-tree structure in their associated minor folds and drag folds(Fig. 42) which indicate by their couples that the domes are diapiric, vertically rising structures. The notion that drag folds should face the opposite way is of course derived from the assumptions of concentric folding (constant length round fold, accommodated by bedding slip) and has no necessary validity in folds of other kind. Drag folds do however indicate the sense of the local simple shear couple.

Greywacke sediments, cherts and spilites accumulated in deep oceanic trenches outcrop six to eight km. above their cradle. Metamorphosed cores commonly reach the kyanite zone which requires burial to some 18 to 20 km. for its genesis. The presence of such grades at the present surface in the metamorphic kernels of orogenic belt does not mean a misfit between field and laboratory results, but only that great vertical transport is involved in the orogenic axial zones. This is not quite the same as Billings' conclusion (1960, p. 380) that "whenever we see kyanite in the field at least ten miles of rock has been eroded away," for the removal of the overlying material might have been partly by tectonic outflow over the fore-syncline.

Fig. 43a and 43b reproduce a comparison made by Macgregor (1951, fig. 6A and 6B) of "gregarious batholiths" which form diapiric migmatitic domes in Rhodesia and Canada in rocks of similar age and on similar scale. Macgregor believes that motion is upward flow rather than tangential compression and states (p. xli): "the dominant force causing the folding of the Rhodesian base-

Fig. 41.—Appalachian axial zone in North Carolina. The motion is upwards, normal to this map. Black, granite; zebra, ultramafic rocks; stippled, younger strata. The curving lines show the strikes of the geosynclinal strata.

Diapir
Motion

Fig. 42. — Fir-tree drag folds on mantled gneiss domes, implying upward motion.

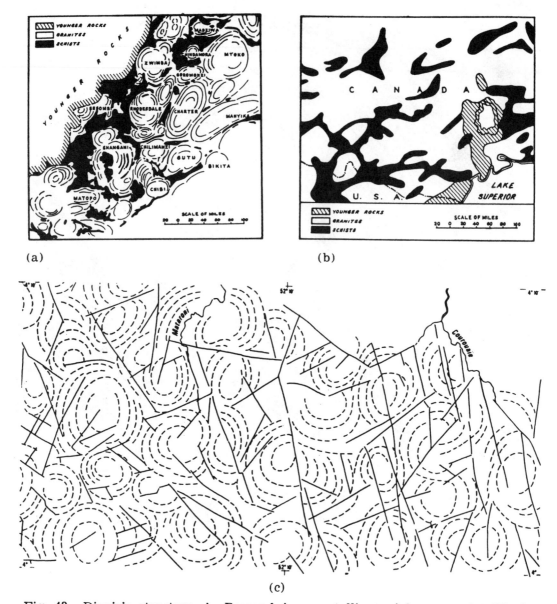

(a)

(b)

(c)

Fig. 43.—Diapiric structures in Precambrian crystalline axial zones. (a) Rhodesian structures, interpreted as gregarious diapiric batholiths (from Macgregor, 1951). (b) Canadian structures, similar in pattern and scale (*ibid*). (c) Similar migmatite dome pattern in French Guiana (from Choubert, 1960).

ment complex appears to be gravity." Fig. 43c, for comparison, shows migmatitic domes in French Guiana, reproduced from Choubert (1960). In his summary, Choubert states (p. 801) that "these structures may be perfectly circular. No preferred direction is visible, even if, locally, secondary events may flatten the concentric forms into ellipses. This isotropy shows clearly that tangential forces have played no part in these tectonics. We have there an example of deformation due uniquely to vertical forces" (translation S.W.C.). The granitic bathyliths and stocks of orogenic belts are diapiric with sympathetic gross doming in the deeper zones where elevated temperatures and reduced viscosities are general, but are impatient, transgressive and through-cutting

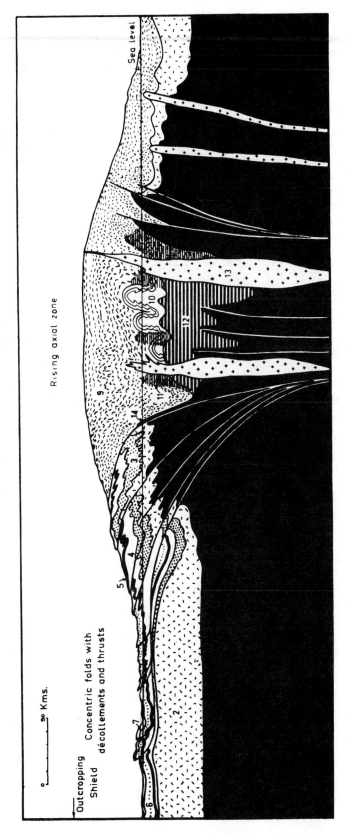

Fig. 44.—Section through the Appalachians with schematic restoration.

1.	Mantle.	6. Middle Palaeozoic.	11. Granitised sediments.
2.	Sialic basement.	7. Carboniferous.	12. Altered mantle.
3.	Lower Cambrian.	9. Geosynclinal greywackes, slates	13. Magmatic granite.
4.	Lower and Middle Cambrian.	and schists.	14. Serpentine belt.
5.	Lower Palaeozoic carbonates.	10. Gneiss domes.	

in the superficial zones where the viscosities of magma and invaded sediments are of vastly different order. The spectrum of viscosities in the orogenic zone produces an hierarchy of diapiric injection, schists diapiric through greywackes, migmatites through schists, and magmatic granites and porphyries piercing the whole symposium.

A section through half an orogen (the Appalachians) is drawn in Fig. 44. This has been held rigidly to fit all available data. It is drawn to natural scale. (Most sections drawn of orogens grossly exaggerate the vertical scale). It includes the Mohorovicic discontinuity. (It is much easier to draw plausible sections of orogens if this is left out.) The surface structure is held firmly to the known structure through the Allegheny Plateau, across the Blue Ridge into North Caroline as presented in the Geological Society of America guidebooks for the North Carolina meeting. The westernmost 160 km. of the profile is identical with Fig. 29 which shows this portion on larger scale. The décollements shown there are simply the upward and outward flow lines, completely analogous with the upward and outward flow lines and overthrusts in Fig. 34.

The altitude of the regurgitating axial zone is shown in Fig. 40 to rise to nearly 50 km. This is of course fictitious since orogens are rapidly eroded as they rise and great clastic wedges of recycled sediment accumulate round the rising zone. Hence if the fold structures are projected to completion they inevitably produce altitudes greater than those which existed at any one time.

While the regurgitation of the axial zone is in process, the altitude of the orogenic crest reaches a dynamic equilibrium even though the upward extrusion of the orogenic heart continues. If for example a viscous material like honey is extruded through a gap between rigid or more viscous materials (Fig. 45) the static pressure at point A equals the load of the overburden AC, so there is a pressure gradient from A to B which causes flow in this direction at a rate determined only by the viscosity and this stress gradient. If the honey is being squeezed up at a faster rate then the material will thus flow away, then the crest C must rise until the height AC is great enough to cause the material to flow out towards B at the rate at which it comes in from below. Thereafter the growth is mainly by outflow of B and the height of C rises only at a rate sufficient to maintain the stress gradient over the increasing distance AB. Fig. 45 sketches flow lines and isopotential surfaces during the flow. The flow lines are the directions in which the thrust surfaces will run, each flow line over-riding the one beneath it. These should be compared with the attitudes of flow and fault surfaces in Figs. 34, 36, 39 and 44. Note that these thrust surfaces turn downwards into the diapiric root whence they spring and hence differ fundamentally from the more superficial type of gravity thrust whose sole surface turns upwards to an open outcrop like a giant landslide.

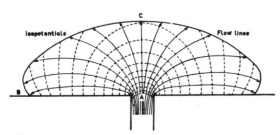

Fig. 45.—Flowlines (arrows) and isopotentials (dotted) in extruding mass.

If the left half of Fig. 44 is straightened out, something like Fig. 46 results. This could equally well be a section across the Gulf of Mexico, where J, K, P, N and Q represent respectively the maximum thicknesses of the Jurassic, Cre-

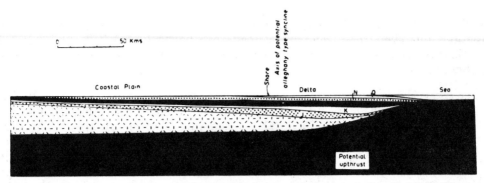

Fig. 46.—Section through the coastal plain and delta of the Gulf of Mexico, a contemporary geosyncline.

taceous, Palaeogene, Neogene and Quaternary. The Moho rises from some 35 km. depth under the Ozarks to a depth under the Gulf of less than 10 km., overlain by basalt, on which the Quaternary sediments are still accumulating. The pre-Mesozoic basement descends from the surface in the Ozarks and Ouachitas to a depth of at least 12 km. along the present shore, and must wedge out altogether before the edge of the delta is reached. The sediments are therefore in an asymmetric trough at least 12 km. deep resting on a rising floor of continental basement on the one side and on a rising simatic floor on the other. The sediments at J, K, P, N, and Q are all likely to be in the greywacke facies but on the foreland side these grade into carbonates and more cratonic facies.

If regurgitation now commenced at the place indicated on Fig. 42 by the double arrow, an emergent anticlinorium would appear off shore, which would supply a great clastic wedge of flysch-like sediments probably becoming continental, and gravity creep in these young sediments would produce the incipient flank folds. Ultimately the thinning wedge of basement would reach the surface and would thrust over the deep asymmetric non-volcanic syncline, with an axis not far from the present shore line. The structures produced from nothing but vertical motion on this site would approximate to those of Fig. 44 and would resemble those of the right hand side of Fig. 34.

In Fig. 44 the pre-geosynclinal basement emerges as thrust slices over-riding the miogeosynclinal trough. However, other belts of pre-geosynclinal basement may also be found further in, in the heart of the eugeosynclinal zone. If the early (taphrogenic) stage the geosyncline develops anastomosing grabens with intervening horsts (like the Basin and Range horsts of Nevada, or on a larger scale the Ruwenzori horst of Central Africa) these basement horsts become massifs in the subsequent axial zone and may be transported with the main nappes. Some of the Alpine nappes appear to have such a history.

In comparing Figs. 44 and 46 it is apparent that the greywackes rest directly on a simatic floor, or nearly so. It has always seemed to me that this was inescapable as the axial zones of geosynclines sink consistently for tens or even hundreds of millions of years. Much smaller loads have made a substantial degree of isostatic adjustment in the few thousand years since the melting of the Labrador and Scandinavian ice sheets, and this in regions of highly stable non-volcanic crust. No substantial gravity load could persist for anything like such times, so the geosynclinal piles must have grown and developed while remaining essentially in isostatic equilibrium throughout, as indeed the Gulf Coast is today. Narrow trenches may be negative loads and still be

142

subsiding if the rate of crustal extension is greater than the rate of local correction, but even these temporary loads are borne by positive anomalies in the adjacent crust so that transverse sections of 100 km. or more are close to equilibrium.

Some very significant features of the Appalachian orogen are shown in Fig. 47, reproduced from King (1959). The front against the miogeosyncline is bowed into a series of arcs. The over-thrusting is most intense where the arcs bow outward. Where the arcs bow inward towards the crystalline eugeosyn-clinal core the folding and overthrusting is less severe. This is the opposite from what would be expected if the Appalachians were produced by inward compression, but it is exactly what one would expect if the folding was due to upward and outspreading processes such as are suggested by Figs. 39 and 45. Even though the regurgitation process commenced by extruding equally all along the trough (which is not very probable) this equality is unstable in that any slight inequality is favoured, and increases because warmer rocks are higher there, and ratio of flow under the same load increases exponentially with the temperature (Carey, 1954, p. 69). Hence an initial uniformity of upward movement all along the trough tends to develop into maximum extrusion at one point, with an outward spreading with arcuate overthrusting from that focus.

Fig. 47.—Arcs and clastic wedges of the Appalachians (from King, 1959).

It is clear from Fig. 47 that such an outflowing tumour developed during the Middle Ordovician in North Carolina and spread detrital products as a clastic wedge across Kentucky and Tennessee. In the Upper Ordovician a similar sedimentary tumour bulged upwards in New England and its tectonic and detrital outwash spread as overthrust sheets (Taconic Klippe) and a thick clastic wedge across New York State. In the Upper Devonian a sedimentary tumour developed in Maryland and spread its clastic products across Pennsylvania. Finally in the Pennsylvanian a bulge appeared in Georgia and spread its waste across Alabama. At the time of each such bulge, less pronounced upward extrusion may have occurred all along the trough. Wherever the bulge was greatest, tectonic spreading causing outward overthrusting and folding was greatest, and the surrounding apron of recycled clastic material was also greatest. A modern equivalent of these upward bulges is found in the string of islands south-west of Sumatra, which rose from the main trough line of the Java deep, and lie on the main negative gravity anomaly axis. They are no doubt at present supplying a thick clastic wedge of sediments onto the surrounding sedimentary pile.

The pattern of arcs, overthrustings and clastic wedges of the Appalachians follows simply from the regurgitation extrusion picture, but is quite anomalous when an explanation is attempted according to the obsolete compressional vice theory.

The continental basement thins away under the geosyncline which must therefore be a stretching zone. Far indeed from the doubled thickness of continental rocks contemplated for the geosynclinal floor by the moribund tectogene concept, the actual thickness of continental basement approaches zero. However this implies that the orogenic root must develop from the mantle. There are at least two ways in which this might come about: (a) There may be an accession of new magmatic granite derived from deep differentiation within the mantle. Van Bemmelen and Rubey have both suggested mechanisms for this, although my own induction leads to a still different mechanism associated with the expansion of the earth's deep interior. (b) The geosyncline is a dilating zone of the earth's crust and commences as a rift valley zone in the craton with marginal fault systems which gradually become buried as the crust thins and sedimentation runs its course. The marginal faults which frame the Gulf province, or the north side of the Aquitanian basin, the Lisbon scarp of Portugal and the Darling Scarp and associated faults of Western Australia, are of this type. This dilating zone extends deep into the mantle. The Moho, commencing at 35 km. works it way up to the sea floor (see Carey, 1958, fig. 3) and upward movement at great depth is implied. Hot water and other fluids may lead to extensive serpentinisation of the mantle over such a zone and this could provide the root which is only known from its density and average seismic velocity. Very extensive outcrops of serpentinite in orogenic belts such as in New Caledonia may perhaps be horst-like blocks of the central region of a structure like (12) in Fig. 44. Fig. 44 shows both solutions (a) and (b) above, which are not mutually exclusive and could both be correct. It also shows a tectonically injected tongue of serpentinite (14 on Fig. 44), derived from the mantle, intruded along a steep border at the boundary between the axial zone and the outward spreading zone of thrust sheets. Serpentine belts of this kind are common in orogenic systems.

Clark and Fyfe (1961) have shown that in this environment ultrabasic liquids are also possible, and since the geosynclinal zone as herein interpreted is one of excess temperature gradient throughout its history, the probability of Clark's and Fyfe's mechanism is enhanced. The Steinmann trinity is a natural consequence.

An orogen such as I have drawn does not imply compression or crustal shortening — only outflow. The geosyncline commences as a stretching zone in the crust, and may continue to widen throughout the subsiding, sedimentary and diastrophic phases. There is nothing in the geometry or fabric of an orogen to deny the proposition that the foreland cratonic blocks on opposite sides of an orogen moved apart even during the climax of orogensis.

CONCLUSION

In this address I have not dealt with block-folding which Beloussov in Russia has so ably developed, and which has its dominant place in the early stage of orogens and is part of the story in the Wyoming pattern of folding. Nor have I touched on folding in simple shear where the shear axis is not horizontal as in concentric folding, but vertical, conditions which are dominant in California, Papua, and is a second part of the story in Wyoming. I have omitted all reference to the commonest structure on the earth's surface, the cratonic basin, and with it Brock's embossed shield pattern of Africa.

But what I have done is to try to demonstrate that the time-honoured axiom that folding and orogenesis are necessarily due to crustal shortening, is a myth. We looked in similar folds for clear evidence of compression but it was not

144

there. We looked in concentric folds but it was not there either, and axial zones of the orogenic heart know great vertical translation but no compelling evidence of compression, beyond the reverent faith that it is there.

REFERENCES

Bentz, A., 1949, "Erdöl und Tektonik in Nordwestdeutschland," Amt. für Boden-forschung, Hannover-Celle.

Billings, M., 1960, "Diastrophism and Mountain Building," Geol. Soc. Am. Bull. Vol. 71, p. 363-398.

Busk, H. G., 1929, "Earth Flexures," Cambridge U. Press.

Carey, S. W., 1954, "The Rheid Concept in Geotectonics," Jour. Geol. Soc. Aust. Vol. 1, p. 67-117.

Carey, S. W., 1958, "The Tectonic Approach to Continental Drift," Continental Drift—a symposium, University of Tasmania, Hobart.

Choubert, B., 1960, "Le Problème des Structures Tectoniques Surimposées en Guyane française," Soc. Géol. France, 7S, Vol. II, No. 7, pp. 855-861.

Clark, R. H., and Fyfe, W. S., 1961, "Ultrabasic Liquids," Nature, Vol. 191, No. 478A, p. 159.

Corbett, D. M., 1959, "Drilling Appalachian Area Deep Well to Basement Rock," a Symposium on the Sandhill Deep Well, Wood Country, West Virginia. Rept. of Investigations No. 18, W. Virg. Geol. Surv.

Darton, N. H., 1940, "Some Structural Features of the Northern Anthracite Coal Basin, Pennsylvania," U.S.G.S. Prof. Paper No. 193D, pp. 779-802.

de Sitter, L. U., 1956, "Structural Geology," McGraw-Hill, New York.

Goguel, J., 1952, "Traité de Tectonique," Masson, Paris.

Gustafson, J. K., Burrell, H. C., and Garretty, M. D., 1950, "Geology of the Broken Hill, N.S.W.," Bull. Geol. Soc. Am., Vol. 61, 12, pt. I.

Haller, J., 1956, "Probleme der Tiefentektonik Bauformen in Migmatit-Stockwerk der Ostgrönländischen Kaledoniden," Geol. Rundschau, 45, pp. 159-167.

Hills, E. S., 1939, "Outlines of Structural Geology," Methuen, London.

King, P. B., 1959, "The Evolution of North America," Princeton Univ. Press, Princeton, New Jersey, U.S.A.

Macgregor, A. M., 1951, "Some milestones in the Pre-Cambrian of Southern Rhodesia," Trans. Proc. Geol. Soc. S. Africa, Vol. 54, XXVII.

Muehlberger, W. R., 1958, "Internal Structure of the Grand Saline Salt Dome, Van Zandt County, Texas," Univ. Texas Bur. Ec. Geol., Rept. of Investigations No. 18.

Nettleton, L. L., 1936, "Fluid Mechanics of Salt Domes," Gulf Coast Oilfields—a Symposium," Am. Assoc. Petrol. Geol., pp. 79-108.

Nevin, C. M., 1949, "Principles of Structural Geology," Wiley, New York.

O'Driscoll, E. S., 1962, "Experimental Patterns in Superposed Similar Folding," Jour. Alberta Soc. Petr. Geol., Vol. 10, No. 3, pp. 95-144.

Parker, T. J., and McDowall, A. N., 1951, "Scale Models as a Guide to Interpretation of Salt-Dome Faulting," Bull. Am. Assoc. Petrol. Geol., Vol. 35, No. 9, pp. 2076-2094.

Powell, J. W., 1876, "Geology of the Uinta Mountains," U.S. Geog. and Geol. Surv. of the Terr., pp. 10-11.

Ramsay, J. G., 1960, "The Deformation of Early Linear Structures in Areas of Repeated Folding," Jour. Geol., Vol. 68, No. 1, pp. 75-93.

Reynolds, D. L., and Holmes, A., 1954, "The Superposition of Caledonoid Folds on an Older Fold System in the Dalradians of Malin Head, Co. Donegal," Geol. Mag., Vol. XCI, No. 6, pp. 417-433.

Rodgers, J., 1953, "The Folds and Faults of the Appalachian Valley and Ridge Province," Kentucky Geol. Surv., No. 1, pp. 150-166.

Sharp, R. P., 1958, "Malaspina Glacier, Alaska," Geol. Soc. Am. Bull., Vol. 69, No. 6, pp. 617-646.

Stoces, B., and White, C. H., 1935, "Structural Geology, with Special Reference to Economic Deposits," Macmillan, London.

van Hise, C. R., 1896, "Principles of North American Pre-Cambrian Geology," 16th Ann. Rept. U.S. Geol. Surv., Pty. 1, pp. 581-843.

Weiss, L. E., 1959, "Geometry of Superposed Folding," Geol. Soc. Am. Bull., Vol. 70, No. 1, pp. 91-106.

Willis, B. and Willis, R., 1934, "Geologic Structures," McGraw Hill, New York.

Woodward, H. P., 1929, "Structural Interpretation of the Burning Springs Anticline," a Symposium on the Sandhill Deep Well, Wood County, West Virginia, Rept. of Investigations No. 18, West Virginia Geol. Surv.

Reprinted by permission of the Geological Society of America from J. B. Currie, H. W. Patnode, and R. P. Trump, *Geological Society of America Bulletin*, v. 73 (1962), p. 655-673.

J. B. CURRIE   *University of Toronto, Toronto, Ontario, Canada*
H. W. PATNODE   *Gulf Research & Development Co., Pittsburgh, Pa.*
R. P. TRUMP   *Gulf Research & Development Co., Pittsburgh, Pa.*

Development of Folds in Sedimentary Strata

Abstract: Information concerning the mechanics of rock folding based on theoretical considerations, laboratory experiments, and field observations is presented in this paper. Theoretical considerations are based on the theory of elastic stability. Photoelastic model experiments that illustrate the theory and qualitatively describe complex cases are included. The wave lengths of folds are compared with the dimensional arrangement of sedimentary strata. A plot of this relationship shows the wave length to be approximately equal to 27 times the thickness of the dominant member. The importance of the stratification of sedimentary rocks in determining the geometry of folds is emphasized. The concept of describing a sedimentary section in terms of structural lithic units is discussed.

CONTENTS

1. INTRODUCTION

Folds in stratified rocks represent one of the most common expressions of geologic deformation. Interpretation and prediction of their position in a structural pattern constitute problems of both scientific and economic interest.

The general problem of rock folding covers a broad geological field with respect to the areal size of folds and their structural type. This discussion examines only a segment of the problem, namely the folding of sedimentary rocks under tangential loads, as in the structural environment of a deformed sedimentary basin.

Thus, flexures due directly to vertical uplift, to differential compaction, or to tectonics involving a major part of the earth's crust are not considered.

The object of the paper is to present information concerning mechanics of fold development from theoretical considerations, laboratory experiments, and field observations. A theoretical approach is limited by the restriction that only idealized geologic systems can be treated readily and by a lack of knowledge regarding physical properties of rock strata. Experimental work with models investigates the process of deformation in more complex systems. Analysis of field structures is

Geological Society of America Bulletin, v. 73, p. 655–674, 9 figs., 4 pls., June 1962

655

hampered by difficulty in obtaining adequate measurements of structural geometry and by inability to observe structure at more than one stage of development. Thus, the three directions of study are necessary.

Other geologic variables also complicate an analysis of folding. Assumptions must be made concerning the physical properties of sedimentary rocks at the time of deformation, the thickness and reaction of material overlying competent members, and the nature of applied forces and structural boundaries.

With these limitations in mind, the following discussion is written in generalized form. Elementary theory is employed to derive some possible physical relationships having geologic significance; experimental results are cited that demonstrate the processes under discussion; field observations are used to illustrate the occurrence of these relationships in folded sedimentary strata.

2. THEORETICAL CONSIDERATIONS

2a. Elastic Stability

The observed periodicity of folds and the apparent horizontal shortening in many regions of structural deformation constitute important reasons for considering the possibility of buckling as a mode of deformation of sedimentary rocks. Also, large compressional loads are necessary to restrict tensile stresses within the relatively low tensile strength of rocks. This is especially evident in regions of intense folding. Further, groups of minor folds with steeply dipping limbs, commonly observed in the field, represent flexures developed as a result of forces parallel rather than perpendicular to the bedding.

The importance of a buckling concept to analysis of fold structures is derived from the geometric control that occurs in the earliest stages of this type of deformation. This control is expressed by development of a particular and persistent fold wave length at low amplitudes of buckling. In later stages any subsequent failure may be related more closely to the fold geometry than to the original stress system.

Study of instability of beams provides information on conditions under which buckling may occur and on relationships obtained during the buckling process. In this discussion the energy method is used to present basic relationships that have been treated by Timoshenko (1936), Biot (1957; 1959a; 1959b) and Jeffreys (1952). Geologic applications of elastic buckling have been treated previously

by Smoluchowski (1909), Goldstein (1926), and Gunn (1937). The theory presented is given to provide the necessary outline of elastic considerations.

Initially, the problem of a weightless beam in a nonrigid medium is solved by finding the load required to make the work input to the beam equal to the energy required to bend the beam. The beam is assumed to fulfill the linear elastic assumption for small deflections. The work input (W) is equal to the applied load (P) multiplied by the shortening (δ) of the bar that is accomplished by a change in the shape of the beam. The strain energy input is practically cancelled by the strain energy required at all stages of deformation.[1]

The geometric shortening (δ) is given approximately by:

$$\delta = \frac{1}{2} \int \left(\frac{dy}{dx}\right)^2 dx \quad \ldots \ldots \ldots \text{(a)}^2$$

and hence

$$W = \frac{P}{2} \int \left(\frac{dy}{dx}\right)^2 dx . \quad \ldots \ldots \ldots \text{(1)}$$

The energy required to bend the beam can be found by using an approximation of the moment in terms of the beam deflection. For small curvatures this relationship is:

$$M = EI \left(\frac{d^2y}{dx^2}\right) \quad \ldots \ldots \ldots \ldots \text{(1a)}$$

where: M = the moment; E = modulus of elasticity; I = moment of inertia; and y = deflection at x.

The energy required to bend the beam is:

$$E_r = \int \frac{M^2}{2EI} dx. \quad \ldots \ldots \ldots \text{(1b)}$$

Substituting equation (1a) into (1b):

$$E_r = \int \frac{EI}{2} \left(\frac{d^2y}{dx^2}\right)^2 dx. \quad \ldots \ldots \ldots \text{(1c)}$$

---

[1] There is a slight increase in the load required at higher amplitudes. This increase occurs until the proportional limit is exceeded, after which there is a rapid decrease in the required load.

[2] The length of a curved bar is equal to $\int [1 + dy/dx)^2]^{\frac{1}{2}} dx$. The approximation given is based on the fact that $(1+\alpha)^{\frac{1}{2}}$ is nearly equal to $(1+\alpha/2)$ if α is very small. Thus the shortening is approximately equal to $\frac{1}{2} \int (dy/dx)^2 dx$. This approximation is only valid for the low amplitudes considered in the early stages of buckling. A curve showing the values for the shortening at high amplitudes is given in Figure 7.

Because cross sections of the beams under consideration are all rectangular, the moment of inertia can be written in terms of the thickness and width of the beam:

$$E_r = \frac{b\,ET^3}{24} \int \left(\frac{d^2y}{dx^2}\right)^2 dx \quad \ldots \ldots \ldots \quad (2)^3$$

known to be a sine curve. Under the end conditions of Figure 1 it can be written:

$$y = a \sin\left(\frac{2\pi x}{L}\right) \quad \ldots \ldots \ldots \ldots \quad (6)$$

where: a = amplitude of the curve; L = wave length of the curve.

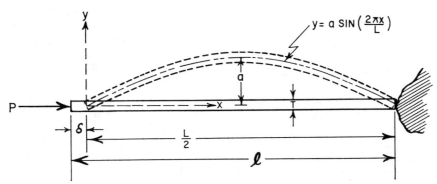

Figure 1.   Sketch of the buckling of a weightless beam with hinged ends

where: b = width of beam; T = thickness of beam.

Both equations (1) and (2) may be divided by the width of the beam for simplification. Equation (1) becomes:

$$W_i = \frac{P_w}{2} \int \left(\frac{dy}{dx}\right)^2 dx \quad \ldots \ldots \ldots \ldots \quad (3)$$

where: W_i = work input per unit width; P_w = load per unit width.

Equation (2) becomes:

$$E_w = \frac{ET^3}{24} \int \left(\frac{d^2y}{dx^2}\right)^2 dx \quad \ldots \ldots \ldots \ldots \quad (4)$$

where: E_w = bending energy required per unit width.

The beam loses its elastic stability when the work input equals the energy required to buckle the beam. Thus

$$\frac{P_w}{2} \int \left(\frac{dy}{dx}\right)^2 dx = \frac{ET^3}{24} \int \left(\frac{d^2y}{dx^2}\right)^2 dx. \quad \ldots \ldots \quad (5)$$

The deflection curve is needed to solve this equation. The shape of the deflection curve is

---

[3] The stiffening effect of wide beams or plates is neglected in this equation and in later discussion of an elastic material surrounding a beam. The stiffening increases the required energy by a factor of $(1/1-\mu^2)$ where μ is Poisson's ratio for the material. However, this factor is small in geological cases when compared to the uncertainty concerning the modulus of elasticity.

Equation (5) can be rewritten for the case shown in Figure 1 as:

$$\frac{P_w}{2} \int_0^{\frac{L}{2}} \left[\frac{2\pi a}{L} \cos\left(\frac{2\pi x}{L}\right)\right]^2 dx$$

$$= \frac{ET^3}{24} \int_0^{\frac{L}{2}} \left[-\frac{4\pi^2 a}{L^2} \sin\left(\frac{2\pi x}{L}\right)\right]^2 dx \quad \ldots \ldots \quad (6a)$$

$$\int_0^{\frac{L}{2}} \cos^2\left(\frac{2\pi x}{L}\right) dx = \int_0^{\frac{L}{2}} \sin^2\left(\frac{2\pi x}{L}\right) dx = \frac{L}{4}.$$

$$P_w \left(\frac{2\pi^2 a^2}{L^2}\right) = \frac{2\pi^4 a^2 ET^3}{3L^4} \quad \ldots \ldots \ldots \quad (6b)$$

$$P_w = \frac{\pi^2 ET^3}{3L^2} \quad \ldots \ldots \ldots \ldots \ldots \quad (6c)$$

The length (l) shown in Figure 1 is approximately equal to $L/2$; therefore:

$$P_w = \frac{\pi^2 ET^3}{12l^2}. \quad \ldots \ldots \ldots \ldots \quad (7)$$

Equation (7) (the familiar Euler formula to be found in the elementary strength of materials texts) gives the critical load per unit width for the case shown in Figure 1, namely that of a beam with hinged ends that are constrained to move parallel to the original centroid of the beam. If any load greater than the critical load is applied to such a beam, it will buckle, *i.e.*, it will deflect laterally. If the applied load is less than the critical pressure, the beam will only shorten without buckling.

Experiments show that short columns can withstand a characteristic amount of compressive stress and remain within the proportional limit of the material. This limit (σ_p) can be substituted in equation (7).

$$P_w = \sigma_p \, T = \frac{\pi^2 E T^3}{12 l^2} \, ; \, \sigma_p = \frac{\pi^2 E}{12}\left(\frac{T}{l}\right)^2 \quad . \quad . \quad . \quad (8)$$

or

$$\left(\frac{l}{T}\right)^2 = \frac{\pi^2 E}{12 \sigma_p} \, ; \, \frac{l}{T} = \sqrt{\frac{\pi^2 E}{12 \sigma_p}} \, . \quad . \quad . \quad . \quad . \quad . \quad (9)$$

Thus, whenever the length-to-thickness ratio is greater than the value given by the right side of equation (9), a beam will shorten elastically and without deflection until the critical load is reached. Under a greater load, the beam will deflect sinusoidally, and this sinusoidal deflection will continue until curvature becomes great enough to cause a fiber stress in excess of either the tensional or compressional proportional limit. Should the length-to-thickness ratio be less than that given by equation (9), the beam will either buckle nonelastically or crush.

2b. Elastic Stability of Members with Lateral Restraint

The foregoing theory provides information concerning relationships between two of the three factors necessary for the study of structural instability in sedimentary rocks. These factors comprise the work input to a member and the energy required to bend it. The remaining factor constitutes the energy required to overcome any lateral restraint that is present. This, of course, lumps several unrelated or apparently unrelated unknowns into a single term for the purpose of simplicity. Geologically, this lateral restraint may be due to the weight of a competent sedimentary unit or the weight and rigidity of surrounding strata.

The presence of a continuous lateral restraint changes the emphasis of the stability problem from that of determining the load required for buckling to one of establishing the probable wave length in the buckled member. Equation (7) indicates that when a member without lateral restraint is considered, the critical buckling load should decrease as the length of the member is increased. Thus, theory suggests the conclusion that an infinitely long member without lateral restraint would be unstable at any compressional load and would buckle into an infinitely long half-wave. Such a case can be approximated only under special conditions, for example, those of a buckling member surrounded by a fluid medium of equal density. The weight of the member itself will, in general restrict the wave length to some finite value.

To evaluate theoretical relationships concerning lateral restraints in geologic cases, it is necessary to develop an approximate or idealized picture of the structurally important parts of a sedimentary section and of the loading conditions to which a section is subjected.

One possible loading system comprises a simple horizontal shortening of the entire section. The total amount of shortening and the rate at which shortening takes place are not considered here. The former must be taken into account if the final shape of the structure is to be analyzed; the rate of shortening becomes important if the compression modulus of the material is time-dependent. For the present discussion, it is assumed that a linear stress-strain relationship exists which is independent of time.

There are two other types of loading that may be important in geologic instability problems. These comprise loads transmitted to the competent member from its surrounding material by shear or friction and gravitational load that develops when beds are tilted. However, a solution of simple shortening provides an important part for the study of other loading conditions.

The sedimentary section may be represented ideally as a series of individual rock sheets or plates, each of which has homogeneous strength properties, is essentially flat-lying before deformation, and has areal dimensions far in excess of its thickness or buckling wave length. Although the individual sheets may have homogeneous strength properties, there is a geologically significant difference of strength properties among the various sheets that make up the section. This characteristic is commonly referred to as the relative competency of rock units. In general, limestones, dolomites, and sandstones are considered to be competent rocks, whereas shale and evaporites constitute the incompetent units of stratigraphic sections. In this paper, Young's modulus is used as a measure of the competency of a rock unit. The question of the properties of rocks is being avoided as much as possible to eliminate complete dependency on small specimen measurements of rock properties. The attempt here is to provide the reader with opportunity to substitute his own values for rock properties.

Thus, the problem is that of determining the structure that may develop when a sequence of strata having essentially infinite areal dimensions and characteristic Young's modulus is subjected to simple shortening. A

transmit the maximum load and if the boundaries between the member and the surrounding medium are assumed to be frictionless, equation (5) can be rewritten:

$$\frac{P_w}{2}\int\left(\frac{dy}{dx}\right)^2 dx = \frac{ET^3}{24}\int\left(\frac{d^2y}{dx^2}\right)^2 dx + R_L \quad . \quad .(10)$$

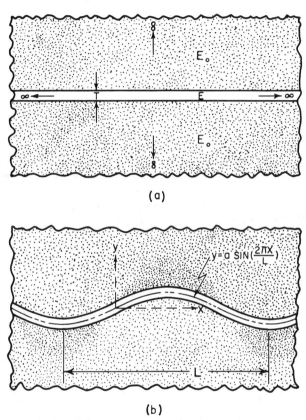

(a)

(b)

Figure 2. Sketch of the buckling of a beam in a continuous, infinite medium

simple system may be chosen first, namely, a unit of thickness T embedded in a homogeneous medium of infinite thickness (Fig. 2a). Additional assumptions, that the plate does not change its thickness during the early stages of deformation and that folding will be symmetrical around the line of the original centroid, permit one to forego direct consideration of the load due to the weight of the member and its surrounding material.[4]

If the embedded member is required to

where: R_L represents the energy required to overcome the lateral restraint.

Biot (1937) gives a solution for the load necessary to produce a sinusoidal deflection in the plane of a semi-infinite plate. For a plate of unit width this solution can be written (Fig. 2b):

$$F = y\frac{E_o\pi}{L} \quad . \quad . \quad . \quad . \quad . \quad . \quad . \quad . \quad . \quad . \quad . \quad (10a)$$

where: F = force necessary to produce deflection y at x; E_o = elastic modulus of the medium; and L = wave length of deflection

$$y = a\sin\frac{2\pi x}{L} \; .$$

---

[4] The question concerning the influence of the weight of the member and stress conditions in the surrounding medium has been discussed by Biot (1959a; 1959b).

When the member lies within an infinite plate, the required force is twice that given by expression (10a). Should the elasticity of the material above and below the member have different values, their sum becomes the appropriate value.

The energy required to produce deflection is

$$R_L = \int \left(\frac{2y \, E_o \pi}{L} \right) \frac{y}{2} \, dx. \quad \ldots \ldots \quad (11)$$

Equation 10 can then be written

$$\frac{P_w}{2} \int \left(\frac{dy}{dx} \right)^2 dx = \frac{ET^3}{24} \int \left(\frac{d^2y}{dx^2} \right)^2 dx$$

$$+ \frac{E_o \pi}{L} \int y^2 dx \, . \quad \ldots \ldots \ldots \quad (12)$$

The assumed deflection curve is

$$y = a \sin \left(\frac{2\pi x}{L} \right);$$

and as in the development of equation (7),

$$P_w \left(\frac{2\pi^2 a^2}{L^2} \right) \left(\frac{L}{4} \right) = \frac{2\pi^4 a^2 ET^3}{3L^4} \left(\frac{L}{4} \right)$$

$$+ \frac{a^2 E_o \pi}{L} \left(\frac{L}{4} \right).$$

$$P_w = \frac{\pi^2 ET^3}{3L^2} + \frac{E_o L}{2\pi}. \quad \ldots \ldots \ldots \quad (13)$$

Equation (13) indicates the load per unit width that is required to produce instability in the embedded member. To find the wave length for which this critical load is a minimum, the first derivative of P_w with respect to L is set equal to zero.

$$\frac{2\pi^2 ET^3}{3L^3} = \frac{E_o}{2\pi}, \, L^3 = \frac{4\pi^3 ET^3}{3E_o},$$

or

$$L = 2\pi T \sqrt[3]{\frac{E}{6E_o}}. \quad \ldots \ldots \ldots \ldots \quad (14)$$

Hence, buckling of the member produces a sinusoidal shape, at least within the range of low amplitude, and the fold wave length will be governed by relationships expressed in equation (14).

The restrictions of wave length to member thickness discussed earlier with respect to equation (8) apply also to a member embedded within an elastic medium. To gain a measure

of this limitation in the present problem, one can substitute the wave length of equation (14) in equation (13) and divide the resulting expression by the member thickness (T). Then,

$$\sigma = \frac{1}{2} \sqrt[3]{\frac{EE_o^2}{6}} + \sqrt[3]{\frac{EE_o^2}{6}}.$$

$$\sigma = \frac{3}{2} \sqrt[3]{\frac{EE_o^2}{6}} \quad \ldots \ldots \ldots \quad (15)$$

where: σ = axial compressional stress in the member.

In terms of wave length and thickness, expression (15) becomes

$$\sigma = \frac{\pi^2 ET^2}{L^2}. \quad \ldots \ldots \ldots \ldots \quad (16)$$

If approximate values are assumed for the elasticity and proportional limit of an embedded member, a range of elastic stability applied to folding can be evaluated approximately. Using the value $E = 10^6$ psi and $\sigma_p = 40,000$ psi and the relationships in equation (15),

$$E_o = \sqrt{\frac{16\sigma_p^3}{9E}} < 10^4 \text{ psi} .$$

Using these values, the elastic modulus of the surrounding medium must be less than 1/100 of that exhibited by the embedded member to allow elastic instability. The particular values of E and σ_p are employed to emphasize the critical importance of the ratio E/E_o to the buckling process rather than to suggest that these values should be used to represent rock strength over a wide range of conditions.

Introducing this ratio in equation (14) gives

$$\frac{L}{T} = 2\pi \sqrt[3]{\frac{E}{6E_o}} > 16,$$

and indicates that for the chosen values of E and σ_p, the lowest possible length-to-thickness ratio is 16.

To approximate boundary conditions other than those of an infinite surrounding medium, a system may be treated in which a member of thickness T lies between two elastic sheets of thickness J, against which are placed two infinitely long rigid boundaries. Also, the two sheets on each side of the dominant member are assumed to comprise a series of individual elastic columns of width (dx) that are not connected to each other (Fig. 3a). This viewpoint requires that equation (12) be modified to account for effects due to a surrounding medium of limited thickness.

The force required to shorten a column is

$$F = \frac{\delta EA}{l}$$

and equation (10) can be written

$$\frac{P_w}{2} \int \left(\frac{dy}{dx}\right)^2 dx = \frac{ET^3}{24} \int \left(\frac{d^2y}{dx^2}\right)^2 dx$$

$$+ \frac{E_o}{J} \int y^2 \, dx \, . \quad \ldots \ldots \ldots \ldots \quad (18)$$

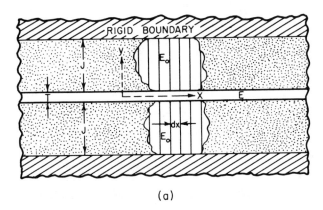

(a)

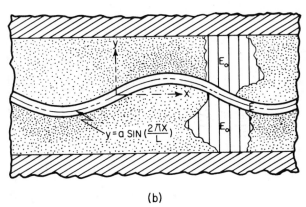

(b)

Figure 3. Sketch of the buckling of a beam in a medium made up of vertical columns bounded by rigid members

where: F = force necessary to shorten column a distance of δ; E = modulus of elasticity of column; A = cross-sectional area of column; and l = length of column.

Using the symbols of Figure 3b, this is written:

$$F = y \frac{E_o}{J} dx \, .$$

The energy required is

$$R_L = \int \left(2y \frac{E_o}{J} dx\right) \frac{y}{2} \, , \quad \ldots \ldots \ldots \quad (17)$$

The deflection curve is again taken to be

$$y = a \sin\left(\frac{2\pi x}{L}\right) .$$

Thus,

$$P_w = \frac{\pi^2 ET^3}{3L^2} + \frac{E_o L^2}{2\pi^2 J} \, . \quad \ldots \ldots \ldots \quad (19)$$

Following the procedure used to obtain equation (14),

$$\frac{2\pi^2 ET^3}{3L^3} = \frac{2E_o L}{2\pi^2 J} \, ; \, L^4 = \frac{2\pi^4 ET^3 J}{3E_o} \, . \quad \ldots \ldots \quad (20)$$

Substituting (20) into (19) ($P_w = \sigma T$) and dividing by (T) the axial compressional stress becomes

$$\sigma = 2\sqrt{\frac{EE_oT}{6J}} \quad . \qquad \ldots \ldots \ldots \ldots \quad (21)$$

Comparison of equations (14) and (20) indicates that they are almost numerically

Although this approximation could be improved by using the stress function, the present form is considered adequate in view of unknowns of the field cases. A somewhat analogous problem has been treated by Gough and others (1940) in a study of composite beams.

Another approximation of geological interest is obtained by replacing a single member of thickness T by n members, each of equal

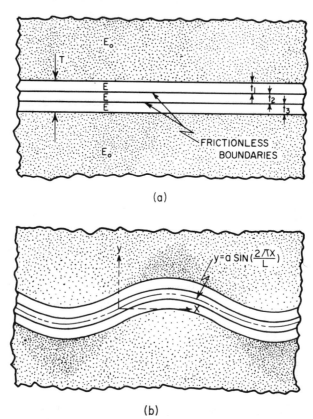

(a)

(b)

Figure 4.   Sketch of the buckling of multiple members in an infinite medium

equivalent when J is equal to the wave length L. Both the stress function given by Biot (1937) and the experimental examples of Plate 2 show that the stresses are greatly reduced at a distance L from the member. This suggests that equation (14), the solution for buckling of a dominant member in a medium with infinite boundaries, is an acceptable approximation when the thickness of the surrounding medium is equal to, or greater than, the predicted wave length. Equation (20) is valuable when the thickness of the medium is less than the predicted fold wave length (note the experimental examples illustrated by Plate 2).

thickness t, and having a total thickness T (Figs. 4a, 4b). Equation (12) can then be written:

$$\frac{P_w}{2}\int\left(\frac{dy}{dx}\right)^2 dx = \frac{Et_1^3}{24}\int\left(\frac{d^2y}{dx^2}\right)^2 dx$$

$$+ \frac{Et_2^3}{24}\int\left(\frac{d^2y}{dx^2}\right)^2 dx + \ldots \frac{Et_n^3}{24}\int\left(\frac{d^2y}{dx^2}\right)^2 dx$$

$$+ \frac{E_o\pi}{L}\int y^2 dx, \qquad \ldots \ldots \ldots \ldots \quad (22)$$

and when

$$t = t_1 = t_2 = \ldots \ldots t_n,$$

$$\frac{P_w}{2} \int \left(\frac{dy}{dx}\right)^2 dx = \frac{nEt^3}{24} \int \left(\frac{d^2y}{dx^2}\right)^2 dx$$

$$+ \frac{E_o\pi}{L} \int y^2 dx \ldots \ldots \ldots \ldots (23)$$

As before,

$$y = a \sin\left(\frac{2\pi x}{L}\right)$$

$$P_w = \frac{n\pi^2 Et^3}{3L^2} + \frac{E_o L}{2\pi} \ldots \ldots \ldots (24)$$

$$L = 2\pi t \sqrt[3]{\frac{nE}{6E_o}} \ldots \ldots \ldots (25)$$

$$L = 2\pi T \sqrt[3]{\frac{E}{6n^2 E_o}} \ldots \ldots \ldots (26)$$

$$\sigma = \frac{3}{2} \sqrt[3]{\frac{E E_o^2}{6n^2}} \ldots \ldots \ldots (27)$$

The above case, the most favorable to elastic instability, is highly dependent on the assumption that the boundaries of individual competent members are frictionless. Geologically, this assumption is not strictly valid. However, the incompetent beds that usually separate dominant members do allow the small amount of adjustment required between these members. The effect, provided the separating member is not greater than two or three times the thickness of the competent members, is equivalent to the case of an essentially frictionless boundary. It is illustrated by experimental examples described in the following section (Pl. 2, fig. 11).

Before the experimental studies are discussed, the influence of gravity and the possible time-dependent properties of rocks warrant further comment. Biot (1957; 1959a; 1959b) has examined these problems in terms of generalized material properties of which the linear modulus, employed in this paper, can be considered a special case.

The weight of the sedimentary section has an effect which must be considered when rocks at the surface are subjected to differential vertical movements due to the actual buckling of the surface strata or as a reaction by conforming members to near-surface buckling. This effect can be evaluated qualitatively by extending equation (12) with the term given by Jeffreys (1952, p. 311–314).

3. EXPERIMENTAL WORK

Experimental studies serve to illustrate the process of buckling outlined theoretically in the foregoing discussion. They serve also to test assumptions used to simplify theoretical analysis and to suggest the types of geometry and strain distribution that may occur in advanced stages of structural development.

A series of qualitative experiments have been devised to examine, photoelastically, the buckling of competent members embedded within an optically active medium of less competency. Gelatin specimens, the incompetent material, were prepared from a 2-2-3 parts (by weight) mixture of gelatin, glycerin, and water poured into molds (Pl. 1, fig. 1) after the mixture had been heated for 2 hours at 70°C. These gelatin models are experimentally advantageous in that they can be readily prepared and molded at slightly above room temperature without pressure. They have high optical sensitivity and little or no molding stresses. This mixture can be prepared within a range of Young's moduli of 1–10 psi (measured by elongating a gelatin bar with a weight and measuring the short time elongation) and the specimens are adequately transparent. Although gelatin models exhibit a compression modulus that is time dependent, their departure from a linear stress-strain relationship is not severe if the rate of loading is reasonably rapid and the total loads applied are not excessive. The examples shown were completely loaded within half a minute, and the average axial stress of the total sample was less than 1 psi.

Gum rubber strips were employed to simulate competent members embedded within the gelatin molds. The strips have a measured Young's modulus of 100 psi. Hence, the ratio of Young's moduli of competent members to surrounding incompetent material ranged from 10 to 100 and lay within the upper portion of this range in most experiments. Plaster of Paris, metal foil, paper, and plastic sheets have been tested qualitatively to evaluate deformation to be expected when high ratios of elastic moduli are used. To obtain low ratios of elastic moduli, specimens have been prepared having a high elasticity gelatin embedded within a mold of weaker gelatin. Although these experiments were not originally intended to be quantitative, all tests of an isolated member gave the wave length-to-thickness ratio predicted by equation (14) within the accuracy of the determination of the ratio of elasticities.

Plate 1 demonstrates four stages in the preparation and study of the rubber-in-gelatin models. Figure 1 of Plate 1 shows the mold with rubber strips in place, prior to placing its cap in position and pouring the liquid gelatin mixture. The undeformed model is pictured under ordinary light in Figure 2 of Plate 1 with rubber members 1/64 inch and 1/16 inch in thickness, cast within the gelatin. Figures 3 and 4 of Plate 1 illustrate the deformation of the competent members of different thickness, both independently and under the influence of adjacent members. Figure 3 of Plate 1 (polarized light) shows °0–90° isoclinics (Frocht, 1948) developed within the gelatin after the competent members began to buckle. Dark areas and lines within the model indicate positions at which the directions of principal stresses are either horizontal or vertical. The development of isoclinics in the earliest stages of buckling permits measurement of fold wave length before the fold curvature is visible in the competent member. Figure 4 of Plate 1 shows the model in a late stage of the deformation and again under ordinary light.

Four series of experiments are illustrated by Plate 2. In Figures 1, 2, and 3 of Plate 2, the gray bands arranged symmetrically about the buckled rubber members are isochromatic fringes (loci of constant shear stress) in the surrounding gelatin. In successive experiments (Figs. 1, 2, 3), increased thickness of the competent member is accompanied by a resultant increase of the fold wave length ($\frac{L}{T} = 10$ for $\frac{E}{E_0} = 20$) in accordance with equation (14). The apparent thickening of the rubber members suggested by several photographs in Plate 2 is due to viewing parallax rather than a change in thickness of these members during deformation.

The upper parts of Figures 4, 5, and 6 of Plate 2 show the specimens in ordinary light; the lower parts of each photograph again illustrate isochromatic fringes. These examples illustrate the effect of spacing on the buckling mode of adjacent members. The example shown in Figure 4 of Plate 2 indicates, by its pattern of isochromatic fringes, that adjacent members are sufficiently far apart to deform independently. In Figure 5 of Plate 2, the shape and wave length of the outer members (1/64 inch thick) are significantly influenced by the configuration of the central member

(1/32 inch thick). When the competent members are brought still more closely together (Pl. 2, fig. 6) they combine to give a single wave length as suggested by the relationships of equation (25).

A relative increase of fold wave length (Pl. 2, figs. 7, 8, 9) develops when the spacing of two competent members of equal thickness is decreased and finally is replaced by a single member having their combined thickness (Pl. 2, fig. 9). A more complex condition is illustrated by Figures 10 and 11 of Plate 2. The spacing of members in Figure 10 of Plate 2 is sufficient to permit them to deform almost independently, whereas the decreased thickness of incompetent material in Figure 11 of Plate 2 restricts the independent action of the competent members and an increased wave length results from their combined action, again as suggested by equation (25). These photographs also indicate a geometrical requirement not included in the development of equation (25), namely, that the outer members of such a series cannot develop as sinusoidal curves.

The experiment illustrated by Figure 5 provides some information concerning high-amplitude folds. The wave length of the gelatin bar is restricted by the weight of the bar. The trajectories of maximum shear stress, sketched on the outline of the buckled member, are derived from photoelastic study of the deformed gelatin bar. The figure indicates that the neutral axis, or dividing line between areas of tension and compression, approaches the center of the fold only during the high amplitude stage. The entire cross section of the member is in compression until a certain curvature is reached. A measure of the flank dips that may be achieved in geologic structures before important tensile stresses are developed can be obtained from the axial stress of equation (16) and from an estimate of the fiber stress given by expression (1a). It should be emphasized that this is using a low-amplitude approximation for a high-amplitude flexure and can only be used as a guide. $\tan \alpha = 2\pi a/L$ represents the flank dip or maximum slope of a sine curve. The fiber stress and axial loads cancel one another at $a = T/2$. By substitution

$$\tan \alpha = \frac{\pi T}{L}. \quad \ldots \ldots \ldots \ldots (28)$$

Using the value of $L/T = 27$ from field examples (Fig. 6) the angle α becomes 5° and represents the flank dip below which no major

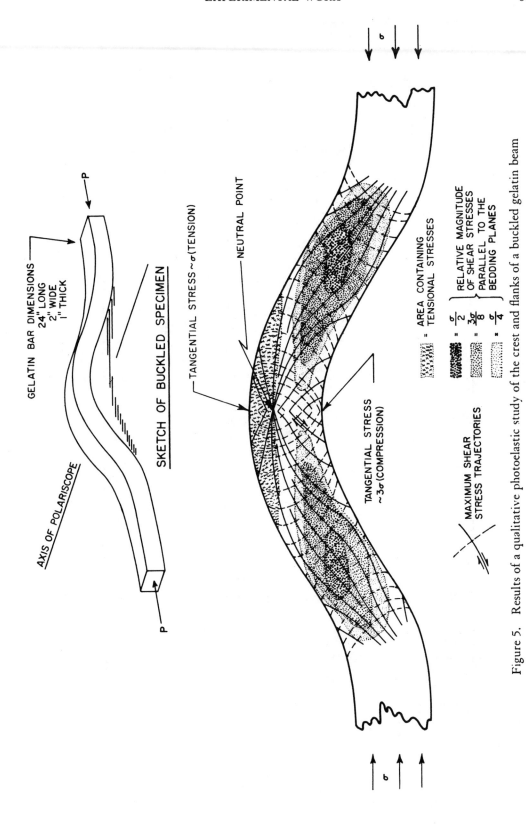

Figure 5.   Results of a qualitative photoelastic study of the crest and flanks of a buckled gelatin beam

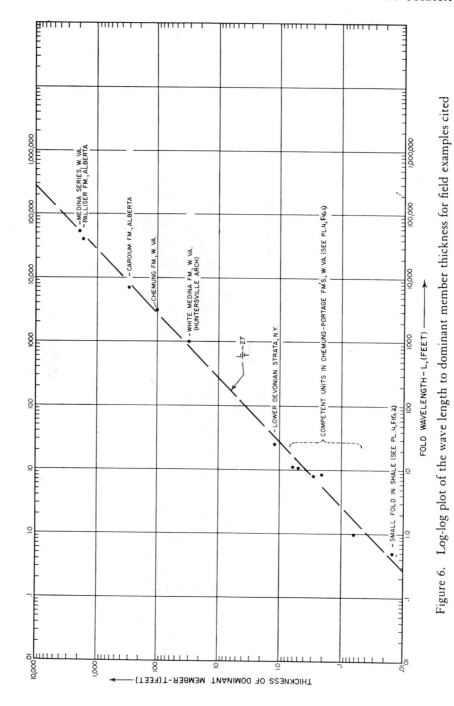

Figure 6.   Log-log plot of the wave length to dominant member thickness for field examples cited

tensile stresses will exist in the controlling or dominant member of the fold. Thus, one line of evidence suggests that the first radical deviation of the competent folding member away from a sinusoidal shape will not begin until maximum flank dips, due to folding, are in excess of 5°. The point at which a buckling failure can occur in a brittle column has been discussed by Chapman and Slatford (1957). Although such progressive changes in fold shape are to be expected in advanced stages of folding, the initially established wave length constitutes a persistent property that remains unaltered.

Plate 3 illustrates an experiment in which the embedded member is comprised of segments of rubber strips simulating a beam of low tensile strength and relatively high compressive strength. Figure 1 of Plate 3 shows that the rubber strip in the photographs (1/16 of an inch by half an inch by five inches) is made up of segments which are approximately 1/16 inch long. Figure 2 of Plate 3 is a photograph (polarized light) of the model before deformation. The fold begins to develop as continuous flexures (Pl. 3, figs. 3, 4) analogous to those of the previous experiments and continues without discontinuities until the flank angles exceed those expected (15°) from equation (28). The vertical lines on the left side of each photograph mark the boundaries of gelatin strips poured and allowed to set successively. The alteration of the normal fringe pattern is due to built-in stresses caused by the successive pourings. The right side (this extends just past the left edge of the rubber strip) is made by a single pouring of gelatin.

4. FIELD EXAMINATION OF FOLDS

The sub-equal spacing of anticlinal crests or synclinal troughs is a geometrical characteristic commonly apparent where a series of folds is developed either on the scale of large structures within a sedimentary basin or on the much smaller scale of minor folds and drag flexures. From the viewpoint outlined in the foregoing discussion of theoretical and experimental work, one would expect the wave length of folds to be controlled by individual competent members or groups of such members in the sedimentary section.

In field practice the wave length is commonly measured from one anticlinal or synclinal axis to the next and would thus, in its measured length, be shorter than that predicted theoretically, the theoretical wave length being more nearly equal to the arc length of geologic structure developed beyond a low amplitude. For the present comparison of field observations with theory and experiments, the distance between fold axes will be taken as the wave length since the graphical plot in Figure 7 suggests that, for folds having flank dips of less than 25°, the error in wavelength determination introduced by this procedure does not exceed 5 per cent.

The commonly complete exposure of small folds in outcrops allows study of the relationship between competent and incompetent members and observation of dominant members that may control fold wave length. The uniform thickness and well-defined boundaries of interbedded siltstone and shale strata in the Portage and Chemung Formations of Pennsylvania and West Virginia (Pl. 4, fig. 1) provide excellent examples for study.

The relationship of wave length to dominant member thickness for five groups of folds measured in these Chemung-Portage rocks is illustrated by Figure 6. The controlling member was assumed to be the thickest competent member (siltstone) within the sequence of folded beds. Thus for a range of dominant member thickness from 1 inch to 8 inches, the fold wave length varied from 1 foot to 10 feet approximately.

Thin, controlling members are illustrated by a fold in shale (Pl. 4, fig. 2). The thickness of these dominant members is seemingly designated not by an obvious lithologic change but rather by the spacing of the shale laminae (1/8 inch) between which a continuous surface of bedding-plane slip appears to have occurred during folding.

The Huntersville arch, west of Marlington, West Virginia (Price, 1929), also suggests the important role of continuous slip surfaces in defining the boundaries and hence the thickness of a dominant member. The arch crops out on the west flank of the Brown's Mountain anticline. Rocks exposed in the fold are quartzites of the Medina Formation; its wave length is approximately 1000 feet. Two quartzite members, each 10 feet thick, the margins of which are defined by continuous, slickensided surfaces, lie at the base of the fold. A 30-foot interval of quartzite lies above these beds in which numerous bedding surfaces are apparent but none have continuity along the bed. This observation suggests that the entire 30-foot interval acted as a dominant member which controlled development of a fold having a 1000-foot wave length (Fig. 6).

The relationship of fold wave length to dominant member thickness approximates that

members is uncertain. Folds developed in Upper Cretaceous beds within the Foothills Belt of western Alberta possibly indicate the relationship of dominant member thickness to fold wave length in this size of structure.

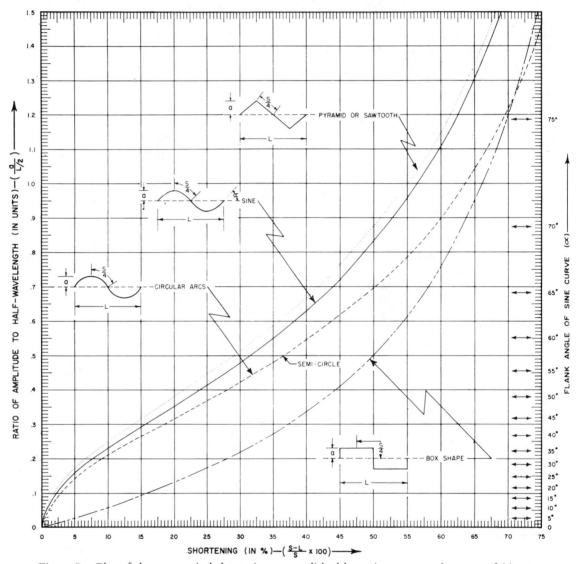

Figure 7.   Plot of the geometrical shortening accomplished by a sine curve and extreme fold types

predicted by equation (14). In this case the ratio of elastic moduli for the competent to incompetent members would amount to about 500.

Delineation of dominant members in folds having a wave length greater than those already discussed is made difficult by the impossibility of observing the entire lithologic section. Thus, a determination of the controlling member or

The Cardium formation within the Foothills in the vicinity of the North Saskatchewan and Clearwater rivers is approximately 300 feet thick. It is underlain by 1200 feet of Blackstone shale and overlain by 1500 feet of Wapiabi shale. The Cardium maintains a substantially uniform thickness in the area; its upper and lower sandstone members are 20–50 feet thick; a third member of similar thickness is com-

Figure 1. Mold with rubber strips in place, prior to placing cap and pouring gelatin mixture

Figure 2. Undeformed model with rubber members 1/64 inch and 1/16 inch in thickness, cast within gelatin

Figure 3. Model undergoing deformation, polarized light

Figure 4. Highly deformed model, ordinary light

STEPS IN THE PREPARATION AND STUDY OF RUBBER-IN-GELATIN MODELS

Figure 1.   1/64-inch rubber strip

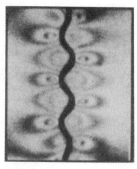

Figure 2.   1/32-inch rubber strip

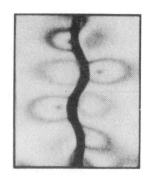

Figure 3.   1/16-inch rubber strip

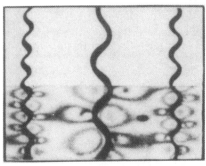

Figure 4.   Separated 1/64-inch and 1/32-inch strips

Figure 5.   Interfering 1/64-inch and 1/32-inch strips

Figure 6.   Combining 1/64-inch and 1/32-inch strips

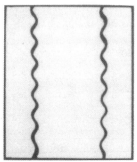

Figure 7.   Separated 1/64-inch strips

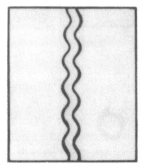

Figure 8.   Interfering 1/64-inch strips

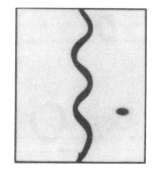

Figure 9.   Single 1/32-inch strip

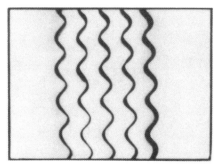

Figure 10.   Separated 1/64-inch strips

Figure 11.   Interfering 1/64-inch strips

STUDY OF THE BUCKLING OF CONTINUOUS RUBBER STRIPS IN GELATIN

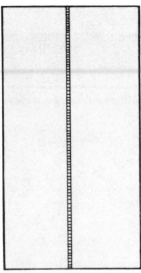

Figure 1. Sketch of 1/16-inch segmented rubber strip

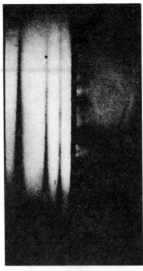

Figure 2. Undeformed strip, polarized light

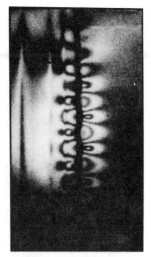

Figure 3. Early deformation

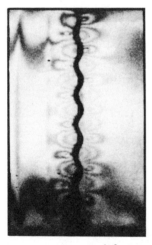

Figure 4. Late deformation, strip continuous

Figure 5. Fault developing

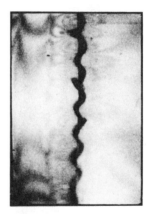

Figure 6. Late stage, several faults

STUDY OF THE BUCKLING OF A SEGMENTED RUBBER STRIP IN GELATIN

Figure 1.   Thin continuous stringers in the Portage-Chemung formations, West Virginia

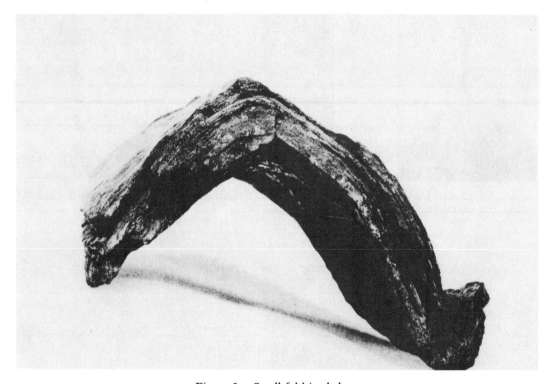

Figure 2.   Small fold in shale

STRATA SHOWING COMBINED COMPETENT MEMBERS

monly present. Thus the Cardium formation may represent a dominant member bordered above and below by an incompetent medium several times its thickness. A question arises as to whether the Cardium sandstone units have acted during folding as a single member or as three separate members. This problem was discussed from a theoretical standpoint in the derivation of equation (25). Two points suggest that selection of the single, 300-foot member is preferable: (1) the shale interval between sandstone members is not uniform and continuous; and (2) the elastic modulus calculated from the seismic velocity of this shale interval is one-tenth of the modulus obtained in like manner for the sandstone members. The calculated modulus for shales such as the surrounding Blackstone and Wapiabi shales is about one-hundredth that of these sandstone members.

The folds in which the Cardium formation can be observed in this area of Alberta have a wave length of 1–1½ miles. Thus the Cardium thickness to fold wave length ratio approximates the expected relationship illustrated in Figure 6.

Uncertainty in definition of dominant member thickness becomes greater when major structures within a sedimentary basin are examined. Two cases will serve to illustrate, however, that even on this large scale of deformation, it is possible that certain lithologic members within the section control the periodic spacing of folds in which they are involved. The wave length of major folds along the Allegheny Front in the Central Appalachian Mountains of West Virginia is about 10 miles. The seemingly most competent zone in the post-Ordovician sedimentary rocks comprises the Silurian and lowest Devonian section which includes the quartzite beds of the Medina Series at its base. Dominantly shale sections of Devonian strata lie above this zone, and the Martinsburg Formation that consists of shale interbedded with thin limestones lies below it. Thus the dominant member of these folds comprises a sequence of beds approximately 2000 feet thick whose lithology is primarily limestone, sandstone, and quartzite. This thickness is plotted against a fold wave length of 10 miles (Fig. 6).

Another example, of similar scale, comprises major structures involving Paleozoic rocks of the Foothills and eastern part of the Rocky Mountain Front Range in central Alberta. The spacing of anticlinal crests of the major

fold structure ranges between 6 and 10 miles. The competent and massive Palliser limestone section in the Devonian sequence exhibits a thickness of approximately 1500 feet (Fig. 6) and may be the dominant member that controls the wave length in structures of which it is a part.

The Rundle formation, which consists of massive carbonate units, may also represent a dominant structural member. The total thickness of these massive members is in excess of 1000 feet. Less competent rocks within the Fernie and Banff Formations occur above and below the Rundle respectively. The Palliser and Rundle intervals may provide an example of the multiple dominant units discussed in the derivation of equation (25). Their relationship is further complicated by the probability that the section underlying the Palliser is more competent than that overlying the Rundle. Thus it gives rise to a possible deviation from the simple sinusoidal relationship assumed in equation (25). The Banff formation may not only reduce the coupling between the Palliser and Rundle but may also provide a zone of adjustment which allows development of higher amplitude folds in the Rundle sequence than in the underlying Palliser structures.

Thrust faults are commonly associated with fold structure in these areas, and their possible influence on development of periodic folds is important. Experimental studies suggest that, because the ratio of elasticities for competent to incompetent members is low in geologic cases, there is not only the possibility that thrusts may develop as a result of advanced folding, but also the likelihood that a thrust of minor displacement may cut the dominant member in the very early stages of folding without materially influencing the course of fold development at low amplitudes. The latter situation is anticipated, because a fault which breaks the dominant member provides only localized relief. To alter the course of fold development it must extend its rupture through the entire rock section associated with the structural system whose periodicity is controlled by that dominant member.

Although the relationship of fold wave length to dominant member thickness (Fig. 6) supports the concept that folds develop through elastic instability, it does not preclude the possible effect of nonelastik rock properties in folding. The conditions of loading in geologic folds, the rates at which they are applied, and the physical properties of rocks at their time of deformation constitute factors that may necessitate consideration of nonelastic rock properties even in the initial stages of fold development.

5. STRUCTURAL LITHIC UNITS

Thus, we see the control that individual

fluence the final structural geometry of a folded area.

If the competent and incompetent beds within a sedimentary basin are considered in terms that combine their geometrical and strength properties, then a particular stratigraphic column is divisible into units having

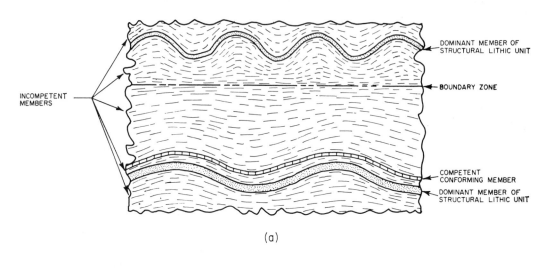

(a)

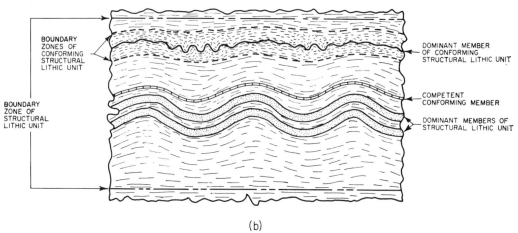

(b)

Figure 8.   Idealized sketches of structural lithic units showing the terminology used

dominant members exert on the wave length of structures developed during folding. However, the presence of many competent beds within a sedimentary section, separated by zones of incompetent strata, suggest that in sedimentary basins the interaction and control of folding by these dominant members will be complex. Folds developed in different beds of a stratigraphic column displaying wave lengths that range from a few feet to several miles indicate clearly that many dominant members in-

their own characteristic reaction to deformation. These are termed structural lithic units.

A structural lithic unit contains a dominant member, or members, whose thickness and physical properties (relative to the surrounding material) determine the wave length of major folds within the unit. Other competent members of lesser thickness may also be included in the unit, together with incompetent strata (Fig. 8). These zones conform to the shape imposed on them by folding of the dominant member. The margins of the unit consist of incompetent rocks that serve as boundary zones. Field examples of structural lithic units can be readily observed in small-scale features but are less apparent when major geologic structures are considered.

Figure 9a illustrates diagrammatically the relation of minor folds to a larger flexure in Lower Devonian limestone strata of the northern Appalachian Basin near Catskill, New York. The boundaries of this small structural lithic unit are indicated by the letters A and B. The dominant members that control the period of the minor fold are presumably the 12-18-inch-thick limestone beds denoted by the letter C. The relationship of wave length to dominant member thickness in this fold is shown in Figure 6.

An example of large structural lithic units (Fig. 9b) is afforded by structures in the

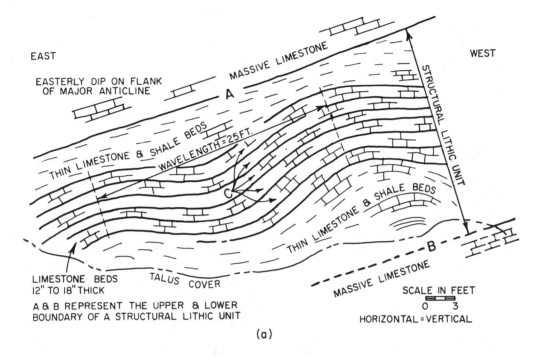

(a)

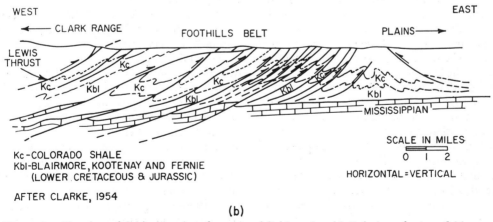

(b)

Figure 9.   Sketches of field examples of structural lithic units. (a) Relation of minor folds to a large flexure in lower Devonian limestone strata of northern Appalachian Basin near Catskill, New York; (b) structures in southern foothills of Rocky Mountains in southern Alberta

southern foothills of the Rocky Mountains in southern Alberta. The eastern portion of a cross section through the Clarke Range and Foothills belt (Clarke, 1954) suggests that the Jurassic and basal Lower Cretaceous beds form a transition zone between structural geometry developed in Cretaceous rocks above them and in Mississippian-Devonian beds below. Thrust faults that cut Paleozoic strata have produced fault segments 4–6 miles long within these rocks.

In the near-surface Cretaceous rocks, the spacing of thrust-fault slices is considerably less than that within the Paleozoic beds at depth. However, the geometry of structure in the lower structural lithic unit is recognizable in the upper unit by the notably large displacement evident on the thrust faults that cut upward from the Paleozoic limestones and dolomites.

Hence the Jurassic and basal Lower Cretaceous beds appear to act as an incomplete boundary between two major structural lithic units. This boundary is incomplete to the extent that the large spacing of fault segments within the lower unit comprising Mississippian and Devonian strata is superimposed on, and reflected in, the closely spaced imbricate structure of the Cretaceous beds at the surface.

6. CONCLUSION

Field observation and laboratory study indicate that the stratification of sedimentary rocks is significant in determining how a sequence of rocks within a basin will respond to deformation. The behavior of these rocks as competent or incompetent members, described here by the relative elastic moduli of interbedded rock types, and the limits to which strata can be deformed before they cease to act as continuous members are each dependent, to a degree, on the specific arrangement of sedimentary units in the stratigraphic column.

By combining field information with study of theoretical and experimental systems that seem mechanically analogous, one gains information that contributes to an understanding of factors controlling the development of folds in sedimentary rocks. This line of thought leads to the conclusion that much of the folding and faulting of strata is related to deformation best described in terms of a buckling process.

In this process, the physical properties and thickness of a dominant member control the fold wave length that develops in the early stages of deformation. Quantitative values deduced to describe the relationships between dominant member thickness and fold wave length are dependent not only on concepts regarding the dominant member's response to applied loads, but also on assumptions concerning the properties and initial geometry of the material around it.

The field examples of fold spacing cited in this study suggest that a linear relationship exists between fold wave length and dominant member thickness. This relationship, although tested in a limited number of cases, is approximated by structures whose spacing ranges from a foot to several miles; the thickness of the dominant member that controls this wave length ranges from a fraction of an inch to about 1500 feet.

Experimental work adds useful information to the study of folding in that it constitutes a method of simulating, in part, the complex structural interaction of competent and incompetent members. The qualitative photoelastic experiments employed here support the concept of fold development through buckling of a dominant member within a relatively less competent medium. They demonstrate the importance of dominant members and of subordinate or conforming members to a gradually developing fold geometry.

If one extends the results from a study of folding, in which only idealized cases can be treated quantitatively, to prediction of actual structure, one encounters difficulties imposed by the great variability of folded rocks in field occurrences. These difficulties may be lessened by dividing the sedimentary column into structural lithic units.

Because the division of a sedimentary section into its structural lithic units requires that thought be given to the manner in which the rock members may respond to deformation, this division serves to focus critical attention on individual rock units from a structural viewpoint. Those groups of strata that act as dominant members may be outlined; the sedimentary units that constitute conforming members may be designated; and the incompetent beds important to the process of deformation either within the structural lithic unit or its margins may be studied.

Prediction of geologic structures must also take into account the progressive sequence of structural events in the development of fold systems. These events may influence the nature of structural lithic units, the effectiveness of a dominant member, and the significance of incompetent beds to the structural process. For example, two competent members of nearly equal thickness may act together within a single structural lithic unit in the early stages of deformation, but at a later stage they may begin to act independently and modify the original unit toward a major element that contains a minor lithic unit within it. In other cases, a dominant member that controls deformation may fold initially, but in later stages it may serve as the locus of displacement on a fault. Also, a sequence of structural events that places an increasing restriction on incompetent beds at the core of a fold in the dominant member may finally curtail the folding process and require that further relief be obtained by fault displacement.

REFERENCES CITED

Biot, M. A., 1937, Bending of an infinite beam on an elastic foundation: Jour. Applied Mechanics, v. 4, p. A1–A7

—— 1957, Folding instability of a layered viscoelastic medium under compression: Royal Soc. London Proc., Ser. A, v. 242, p. 444–454

—— 1959a, Folding of a layered viscoelastic medium derived from an exact stability theory of a continuum under initial stress: Applied Mathematics Quart., v. XVII, no. 2, p. 185–204

—— 1959b, The influence of gravity on the folding of a layered viscoelastic medium under compression: Jour. Franklin Inst., v. 267, no. 3, p. 211–228

Chapman, J. C., and Slatford, J., 1957, The elastic buckling of brittle columns: Inst. of Civil Engineers Proc., Paper no. 6147, v. 6, p. 107–125

Clarke, L. M., 1954, Cross section through the Clarke Range of the Rocky Mountains of southern Alberta and southern British Columbia: Alberta Soc. Petroleum Geologists, Guidebook of Fourth Ann. Field Conf., p. 105–109

Frocht, M. M., 1948, Photoelasticity: New York, John Wiley and Sons, Inc., v. 1, p. 99–214

Goldstein, S., 1926, The stability of a strut under thrust when buckling is resisted by a force proportional to the displacement: Cambridge Philos. Soc. Proc., v. 23, p. 120–129

Gough, G. S., Elam, F. C., and de Bruyne, N. A., 1940, The stabilization of a thin sheet by a continuous supporting medium: Jour. Royal Aeronautical Soc., v. 44, p. 12–43

Gunn, Ross, 1937, A quantitative study of mountain building on an unsymmetrical earth: Jour. Franklin Inst., v. 224, no. 1, p. 19–53

Jeffreys, H., 1952, The earth: Cambridge, Cambridge Univ. Press, p. 311–314

Price, P. H., 1929, Pocahontas County: W. Virginia Geol. Survey County Repts.

Smoluchowski, M., 1909, Folding of the earth's surface in formation of mountain chains: Acad. Sci. Cracovie, Bull. 6, p. 3–20

Timoshenko, S., 1936, Theory of elastic stability: New York and London, McGraw-Hill Book Co., Inc., p. 64–169

Manuscript Received by the Secretary of the Society, October 14, 1960

Reprinted by permission of the Geological Society of America from M. Friedman, J. Handin, J. M. Logan, K. D. Min, and D. W. Stearns, *Geological Society of America Bulletin*, v. 87 (1976), p. 1049-1066.

Experimental folding of rocks under confining pressure: Part III. Faulted drape folds in multilithologic layered specimens

M. FRIEDMAN
J. HANDIN
J. M. LOGAN   } *Center for Tectonophysics, Texas A&M University, College Station, Texas 77843*
K. D. MIN
D. W. STEARNS

ABSTRACT

Drape folds and reverse faults are produced experimentally at confining pressures to 2.0 kb and shortening rates of 10^{-3} to 10^{-6} sec^{-1} by displacing a block of brittle sandstone (2 by 3 by 12.6 cm) along a lubricated saw cut into one to five initially intact layers (0.2 to 1.0 cm thick and as much as 12.6 cm long) of limestone, sandstone, and rock salt. The saw cut is inclined at from 30° to 90° to the layer boundary. The deformation is characterized from studies of fault geometry, displacements and sequence, bedding-plane slip, layer-thickness changes, and the development of fault gouge, fold hinges, microfractures, calcite twin lamellae, and dimensional orientations of grains (in the rock salt). Stress trajectories are inferred from faults, microfractures, and calcite twin lamellae, and strains are calculated from layer-thickness changes and from calcite twin lamellae.

Reverse faults curving concave downward propagate upward from the saw cut in the forcing block. With increasing displacement along the precut faults, the faults and associated gouge zones in the layer steepen and become progressively younger toward the upthrown block as displacement increases. The faults are preceded by swarms of extension microfractures that form throughout the deformation and that are the best clues to the stress trajectories. The downthrown layers are thickened by uniform flow and by repetition caused by the faulting. They are displaced away from the faults by bedding-plane slip. Trajectories of the greatest principal compressive stress (σ_1) are inclined at low angles to the layer boundaries near the faults and become perpendicular to these boundaries away from the fault. The maximum deformation of the downthrown block occurs when the saw cut is inclined at about 65° to the layering. The upthrown layers are all extended parallel to the layering and perpendicular to the fold axes, as indicated by extension fractures, thinned layers, and cal-

cite twin lamellae and the development of graben zones and low-angle normal faults that are conjugate to the reverse faults. The layers are translated by bedding-plane slip away from the fault zone. Trajectories of σ_1 are inclined from 45° to 90° to the layering.

The fabric data are internally consistent, and inferred stresses are in good agreement with those calculated from an elastic solution of the experimental boundary conditions. Principal strains calculated from calcite twin lamellae are within an average of 0.01 of those calculated from layer-thickness changes and permit clear resolution of individual events in domains of superposed deformations. *Key words: structural geology, experimental rock deformation, confining pressure, multilithologic layered specimens, drape folds, reverse faults, dynamic petrofabrics, stress trajectories, strain analyses.*

INTRODUCTION

Our program of experimental folding of rocks under confining pressure dealt first with the buckling of monolithologic, single-layer rock beams (Handin and others, 1972) and then with the buckling of multilithologic layered rock specimens (Pattison, 1972; Handin and others, 1976). In this paper we describe a third set of experiments in which faulted "drape folds" are formed by forcing an essentially rigid "basement" block of sandstone (2 by 3 by 12.6 cm) along a lubricated saw-cut surface or "reverse fault" into an intact "sedimentary veneer" consisting of one to five layers of limestone, sandstone, and rock salt at confining pressures to 2 kb, shortening rates of 10^{-3} to 10^{-6} sec^{-1}, and room temperature (Fig. 1, Table 1). Folding of this type is stable in contrast to unstable buckling. The geometry of the fold is controlled primarily by the shape, size, and displacement of the forcing member rather than by the aspect ratio and material properties of the folded layers as in buckling. Drape folds, therefore, are here referred to as stable or forced

folds. (Excluded are the supratenuous or compaction folds that some workers also call drape folds.)

Drape folding is an important structural style that is well developed in the Wyoming province of the western United States (Berg, 1962; Prucha and others, 1965; Stearns, 1971; among others) and in many other forelands throughout the world (Lees, 1952). The structural problems related to the formation of these folds (Stearns, 1971) are the stimulus for our experiments. For example, (1) What are the mechanisms of deformation during drape folding? (2) Are the hinges of the fold fixed, or do they migrate during structural growth? (3) How is extension of the layers over the uplifted forcing member manifested — by thinning through cataclastic or uniform flow, by faulting, or both? (4) If thinning does not occur, does detachment along bedding planes permit the fold to develop as postulated by Stearns (1971)? (5) If the layers fault, what types of faults form, and what is their sequence in space and time? (6) How are large rigid-body rotations accomplished during brittle deformations? (7) What are the boundary conditions on the layers, and how might the fold geometry change with boundary conditions? (8) Will dynamic petrofabric analysis be useful in recognizing and detailing this structural style? (9) Are drape folds amenable to detailed numerical modeling?

The reader may find that our experiments provide insight into many of these questions. One must be cautious, however, in extrapolating between the laboratory and the field. There is a scale difference of about 10^5 between the 10-cm specimens and prototype structures like Rattlesnake Mountain near Cody, Wyoming. Moreover, the specimens by no means exactly simulate the natural stratigraphic sequence of a sedimentary veneer in lithologic variation, relative thickness of beds, or multiplicity of bedding planes along which displacements may occur. Nor do the physical conditions of these tests necessarily reflect those for a

Geological Society of America Bulletin, v. 87, p. 1049–1066, 16 figs., July 1976, Doc. no. 60711.

1049

313

specific natural structure. Nonetheless, the experiments do treat real rocks deformed under confining pressure, and the results do shed light on the kinematics, mechanisms, sequences, and relative positions of the deformations. Moreover, they demonstrate how large rigid-body rotations in one domain can be obtained from cataclastic flow in adjacent domains. In addition, microfractures in quartz and calcite and calcite twin lamellae prove to be reliable criteria for mapping principal stresses and strains. Details of the microscopic fabric and inferred stress trajectories are in excellent agreement with a previously studied natural counterpart (Friedman, 1969), with regard to faults within the specimens, with previous model studies, and with numerical modeling (Min, 1974; Min and others, 1975).

In this paper the experimental results are followed by discussions of the petrofabric

analyses, the final results of which are illustrated in stress-trajectory diagrams. The essential results of the calcite-fabric study, including derivations of both stress (Turner, 1953) and strain (Groshong, 1972, 1974) are given, but the data are given in detail elsewhere (Friedman and others, 1976). Correlations with numerical modeling of these experiments (Min, 1974) are outlined. Tentative extrapolations to natural structures are mentioned briefly; they are discussed in detail elsewhere (Stearns and Weinberg, 1975).

EXPERIMENTAL RESULTS

Our high-pressure rock-deformation apparatus, techniques for sample preparation and jacketing, and starting material were described fully in our first paper on experimental folding of rocks (Handin and others, 1972). Below we describe features that are unique in drape-fold experiments.

Specimen Configuration

Each specimen consists of an essentially rigid forcing member containing a saw-cut surface that simulates a basement fault and one or more intact layers that simulate the sedimentary veneer (Fig. 1). The forcing member is a block of Coconino Sandstone (2 by 3 by 12.6 cm) with a saw cut inclined at 30° to 90° to the layering (Table 1). This surface is lubricated with molykote to reduce friction. Experiments with the saw cut at 90° to the layering require a different specimen configuration, and results for such specimens will be reported elsewhere. Most of the tests reported here are done with 65° saw cuts. In most experiments the sharp leading edge of the forcing block is retained, but in a few tests it is beveled to provide a blunter contact (Fig. 2, d). The resulting notch is filled with modeling clay.

The veneer is varied to investigate the effects of ductility contrasts, lithologic variations, and loading conditions. As listed in Table 1, the veneer consists of (1) monolayers (1 cm thick) of rock salt or of Indiana Limestone, (2) three layers (each 0.3 cm thick) of Coconino Sandstone, Indiana Limestone, or different combinations of both, and (3) five layers (each 0.2 cm thick) of alternating sandstone and limestone. All the layers are as wide as the forcing block (3.0 cm) and vary in length as stated below. In specimen 328, a layer of lead separates the forcing member from the overlying layers of limestone and sandstone. All surfaces of the forcing block and the layers are prepared with a surface-grinder 60-grit wheel. There is negligible deformation of surface grains (even of calcite) due to grinding.

Once the specimen is assembled, a 2-mm grid pattern (ink stamp pad) is placed on all four sides. Lead strips are attached then, and molykote is applied to the ends that will be in contact with the piston (Fig. 1). Tests with and without the lead strips show that the lead does not influence the deformations; these strips merely facilitate handling the specimen after deformation. The assembly is jacketed with two layers of heat-shrink polyolefin tubing (Fig. 1, e).

Experimental Conditions

All tests are run on room-dry samples at room temperature. Most are at 1-kb confining pressure and a nominal axial shortening rate of 10^{-4} sec^{-1}. A few tests are run at other confining pressures in the range of 0.34 to 2.0 kb and at rates of 10^{-3} to 10^{-6} sec^{-1} to evaluate in a cursory way the influence of these parameters (Table 1). The lengths of the layers are changed to vary the end-loading conditions (Fig. 1, a–d; Table 1). For condition I, the layers are as long as the forcing member (12.6 cm), so that they

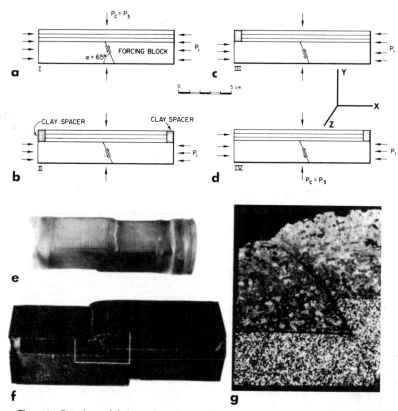

Figure 1. Experimental design and specimen configuration: a–d, Schematic diagrams show loading conditions I to IV, where P_1 is greatest principal compressive stress across boundaries and $P_3 = P_c$ is confining pressure; clay spacers transmit confining pressure only; XZ plane is parallel to layering; X is parallel to P_1, and Z is parallel to eventual fold axis and "strike" of precut surface in forcing block. e, Typical specimen in polyolefin jacket. f, Oblique view of specimen 295 with lead strip parallel to XY surface removed; change in magnitude of throw along "strike" of fault and "graben" zone on upthrown block can be seen in lead; thin sections parallel to XY contain area outlined in white. g, Photomicrograph of XY thin section for specimen 295 shows displacement of forcing block and upthrust in monolayer of Indiana Limestone; fault zone is marked by calcite gouge and en echelon fractures.

are end loaded by the piston as well as transversely loaded by the confining pressure and movement along the saw cut in the forcing member. For condition II, the layers are only 11.5 cm long, and they are separated from the pistons by modeling clay, which transmits only the confining pressure to the ends. For conditions III and IV, the layers are 12.0 cm long, and they contact the piston only at one end, the other end having the clay spacer. The end against the piston is on the eventual "up" and "down" blocks of the forcing member for conditions III and IV, respectively. Most of the experiments done to date have been under conditions I and II.

Displacement along the saw cut in the forcing member, a function of total shortening, is the other controlled parameter (Table 1). The maximum displacement of about 0.9 cm is limited by the flexibility of

TABLE 1.   EXPERIMENTAL DATA

Specimen no.	Loading condition (Fig. 1)	Layer lithology (top to bottom)*	Angle of saw cut in forcing block (°)	Confining pressure (kb)	Axial displacement rate (cm/sec)	Displacement along saw cut (cm)	Maximum differential axial force (10^10 dynes)
Forcing block alone							
332		None	65	1.0	10^{-4}	0.49	0.653
333		None	65	1.0	10^{-4}	0.68	0.844
339		None	65	1.0	10^{-4}	1.02	0.891
Single layer †							
278	I	Salt	39	1.0	10^{-4}	..	..
288	II	Salt	39	1.0	10^{-4}	0.38	..
289	II	Salt	39	1.0	10^{-4}	0.53	..
290	II	Salt	60	0.35	10^{-4}	0.56	..
292	II	Salt	60	0.7	10^{-4}	0.50	0.872
294	II	Salt	60	1.0	10^{-4}	0.50	0.962
291	II	Ls	60	0.35	10^{-4}	..	..
293	II	Ls	60	0.7	10^{-4}	..	0.814
295	II	Ls	60	1.0	10^{-4}	0.47	1.051
Three layers							
328§	II	SS-LS-Lead	65	1.0	10^{-4}	0.44	0.850#
303	II	SS-SS-SS	65	1.0	10^{-4}	0.77	1.144#
302	II	LS-SS-SS	65	1.0	10^{-4}	0.57	0.822#
300	II	SS-LS-SS	65	1.0	10^{-4}	1.05	1.314#
301	II	SS-LS-LS	65	1.0	10^{-4}	0.68	..
296	II	LS-SS-LS	75	1.0	10^{-4}	0.06	2.301
297	II	LS-SS-LS	70	1.0	10^{-4}	0.8	1.622
298	II	LS-SS-LS	65	1.0	10^{-4}	0.63	1.131#
314	II	LS-SS-LS	65	0.5	10^{-4}	0.59	0.667
315	II	LS-SS-LS	65	2.0	10^{-4}	0.73	1.917
316	II	LS-SS-LS	65	1.0	10^{-3}	0.50	1.160
318	II	LS-SS-LS	65	1.0	10^{-6}	0.54	0.987
329	II	LS-SS-LS	65	1.0	10^{-4}	0.49	0.912#
330	II	LS-SS-LS	65	1.0	10^{-4}	0.30	1.231#
331	II	LS-SS-LS	65	1.0	10^{-4}	0.18	0.841#
334	II	LS-SS-LS	65	1.0	10^{-4}	0.49	1.214#
335	II	LS-SS-LS	65	1.0	10^{-4}	0.40	1.381#
336	II	LS-SS-LS	65	1.0	10^{-4}	0.52	1.137#
337	II	LS-SS-LS	65	1.0	10^{-4}	0.38	1.240#
338	II	LS-SS-LS	65	1.0	10^{-4}	0.30	1.449#
299	II	LS-LS-LS	65	1.0	10^{-4}	0.70	1.346#
348	II	LS-SS-LS	30	1.0	10^{-4}	..	.. (leak)
349	II	LS-SS-LS	50	1.0	10^{-4}	0.96	1.032
350	II	LS-SS-LS	30	1.0	10^{-4}	0.73	0.974
353**	II	LS-SS-LS	90	1.0	10^{-4}	0.73	1.231
354**	II	LS-SS-LS	75	1.0	10^{-4}	0.83	1.519
Three layers, forcing block beveled††							
18	II	LS-SS-LS	65	1.0	10^{-4}	0.23	1.423
19	II	LS-SS-LS	65	1.0	10^{-4}	0.28	1.763
20	IV	LS-SS-LS	65	1.0	10^{-4}	0.42	1.359
21	IV	LS-SS-LS	65	1.0	10^{-4}	0.63	1.494
Five layers							
304	II	LS-SS-LS-SS-LS	65	1.0	10^{-4}	0.44	0.928#
305	II	LS-SS-LS-SS-LS	65	1.0	10^{-4}	0.43	1.415#

* Sequence of layering is specified as top to bottom, with bottom layer adjacent to forcing block; SS = sandstone; LS = limestone.
† Unless otherwise indicated, total thickness of layer(s) above forcing member is 0.9 ± 0.1 cm.
§ Total thickness of layers in specimen 328 is 0.73 ± 0.1 cm.
# Frictional effect of forcing block subtracted.
** Length of layer is 5.5 ± 0.1 cm.
†† See Figure 2, d.

the jacket. The displacement is varied to provide specimens taken to just short of the maximum allowed and also to lesser amounts for studies of the sequential development of the fold and associated faults. The reproducibility of the deformations has also been tested by study of four specimens (specimens 334–337) experimentally deformed identically. Three tests without sedimentary layers are made under the same conditions to obtain a knowledge of the frictional characteristics along the fault in the forcing block (specimens 332, 333, and 339). During the tests, differential axial force and displacement are recorded on an X-Y recorder. Curves of differential axial force versus axial displacement are then ob-

tained by subtracting the frictional effect of the forcing block from the recorded differential axial forces of the tests (Fig. 3).

Force-Displacement Curves

The record of differential axial force versus axial displacement for a series of tests (Fig. 3) shows that the initial parts of the curves are roughly linear to about 0.08-cm axial displacement. For tests at 0.5 and 1.0 kb, the curves then bend concave downward or they finally become nearly horizontal with slight oscillations or they exhibit a small downward concavity until the tests are terminated. Pronounced strain hardening occurs at 2.0-kb confining pressure.

These curves are markedly different from those for the thin-beam buckles (Handin and others, 1972, Fig. 23; Pattison, 1972, Fig. 3; Handin and others, 1976, Fig. 3). There are no sharp peaks followed by pronounced work softening as for the unstable folds. No geometric instability is evident, and the axial force must be kept essentially constant or increased to continue the deformation.

The small oscillations beyond the yield points, particularly at 1.0-kb confining pressure, may be related to the fracturing and faulting. Observations show that faults are initiated from the saw cut in the forcing block and grow toward the upper layers. As the faults develop, layers are bent, hinges develop, and folds grow by uniform flow and by rigid-body rotation. This sequential deformation must occur in the range of steady-state or work-hardening parts of the curves.

PETROFABRIC RESULTS

The faulted drape folds are carefully examined petrographically with the view toward their mechanical behavior in order to describe their deformations, to provide structural detail for comparison with natural counterparts and with numerical models, and to evaluate further certain petrofabric techniques as reliable measures of stress and strain. The locations, senses, and magnitudes of "bedding-plane slip," bedding-thickness changes, and faults are delineated. Strain and (or) stress patterns are inferred from microfracture and calcite-fabric analyses. Hinges and gouge zones are described. From these data, the mechanisms and sequences of deformation are established, principal stresses are inferred from the faulting, microfracturing, and development of calcite twin lamellae, and strains are calculated from bedding-thickness changes and the twin-lamellae analyses. Results of the dynamic analysis are illustrated in stress-trajectory diagrams.

Most of the detailed observations described below are made on specimens deformed at 1.0-kb confining pressure and at a shortening rate of 10^{-4} sec^{-1} with 65° saw cuts in the forcing block. Variation in the intensity of deformation and in structural detail at these standard conditions is due primarily to differences in the magnitude of displacement along the saw cut and to the beveling of the leading edge of the forcing block. The four specimens deformed to the same displacement show essentially identical features. Cursory examination of specimens deformed at other confining pressures and displacement rates (Table 1) do not show any significant structural differences when compared to those deformed at standard conditions. This obtains primarily because the brittle-ductile con-

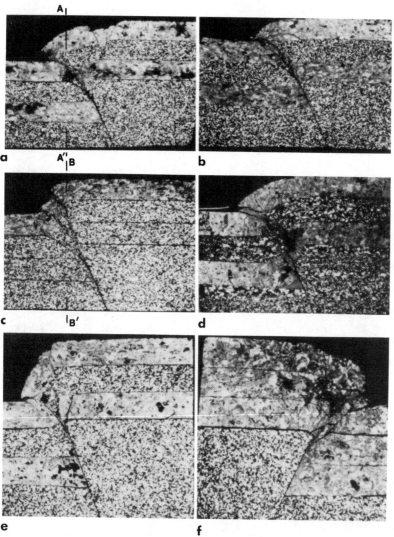

Figure 2. Photomicrographs of XY thin sections for specimens 298, 336, 302, 21, 297, and 299 (a–f, respectively). Each layer is 0.3 cm thick. Cross-polarized light. Lines A–A′ and B–B′ refer to locations of thin sections cut parallel to fold axes, as shown in Figure 10, a and c.

trast between the limestone and sandstone does not change over the range of conditions investigated. Variations in deformational intensity, but not in overall style, do occur with changes in saw-cut angle between 30° and 90°, and these will be described. The following descriptions of the specimens deformed at the standard conditions, therefore, lead to valid generalizations for the whole suite of experiments.

Observational Procedures

After deformation, the heat-shrink jacket is removed along with one of the lead strips along the XY plane (Fig. 1). The remaining strips constrain the specimen and prohibit differential movements of the layers, fault zones, or forcing blocks. Epoxy is applied to the exposed surface and allowed to saturate the specimen. Once the epoxy hardens,

the other lead strip parallel to the XY plane is removed, and the specimen is again cemented. This treatment preserves the integrity of the specimen and facilitates measurements of fault displacements and of offsets of the grid pattern and the preparation of thin sections.

Once the surface measurements are completed, a thin section is prepared parallel to the XY surface. This section contains the area of interest around the upper part of the forcing member and the faulted and drape-folded veneer (see, for example, Figs. 1, f and g, and 2). Sections parallel to the YZ plane are also cut from selected specimens. Most of the sections are made while the lead strips on the upper XZ surface are still in place. Optical studies are made with the aid of the universal stage.

Sandstone and Limestone Veneers

Bedding-Plane Slip. The grid pattern inked on each specimen prior to deformation is used primarily to determine the location, magnitude, and sense of shear displacement along the bedding surfaces. Data for a representative number of specimens (loading condition II) indicate that the displacements are consistent at locations a, a', c, and c' (Table 2, Fig. 4). At other places, the displacements are either nil or inconsistent, or there are too few data to warrant generalization. Some of the inconsistencies may arise from randomized offsets that occur from specimen to specimen; however, that the senses of displacement are consistent at certain locations indicates an overriding effect that must be meaningful — namely, the net displacement of particles along the lower boundary of the veneer is away from the fault zone (Fig. 4, b). It will be demonstrated later that the magnitudes of the displacements (as much as 0.47 mm) are large compared to the strains and there-

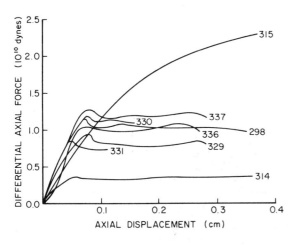

Figure 3. Typical curves of differential axial force displacement, with frictional effect of forcing block subtracted from recorded differential axial force. Numbers at ends of each curve are specimen numbers (Table 1).

TABLE 2. DATA FOR SENSES AND MAGNITUDES OF BEDDING-PLANE SLIP

Specimen no.	Location and sense of shear displacement*	Magnitude of shear displacement† (mm)
296	c ⇄, c' ⇄	c, 0.47; c', 0.09
297	a ⇄, a' ⇄, c ⇆, c' ⇆	a, 0.09; a', 0.09; c, 0.28; c', 0.28
299	a ⇄, a' ⇄, b ⇄, b' ⇄, c .., c' ⇆	a, 0.47; a', 0.46; b, nil; b', 0.19; c, nil; c', 0.37
300	a .., a' ⇄, c ⇆, c' ⇆	a, nil; a', 0.19; c, 0.28; c', 0.28
301	a ⇄, a' ⇄, c ⇆, c' ⇆	a, 0.19; a', 0.19; c, 0.28; c', 0.28
302	a ⇄, a' ⇄, c ⇆, c' ⇆	a, 0.19; a', 0.19; c, 0.09; c', 0.19
303	a ⇄, a' ⇄, c ⇆, c' ⇆	a, 0.09; a', 0.28; c, 0.47; c', 0.19
305	a .., a' .., c ⇆, c' ⇆	a, nil; a', nil; c, 0.09; c', 0.09
314	b ⇄, b' .., c ⇆, c' ⇆, d ⇆, d' ..	b, 0.09; b', nil; c, 0.19; c', 0.09; d, 0.09; d', nil
315	a ⇆, a' ⇄, b ⇆, b' ⇆, c ⇆, c' ⇆	a, 0.28; a', 0.28; b, 0.28; b', 0.28; c, 0.28; c', 0.19
316	a ⇄, a' ⇄, c ⇆, c' ⇆, d .., d' ⇄	a, 0.19; a', 0.19; c, 0.19; c', 0.19; d, nil; d', 0.19
318	a ⇄, a' ⇄, b ⇄, b' ⇄, c ⇆, c' ..	a, 0.19; a', 0.28; b, 0.19; b', 0.19; c, 0.09; c', nil
328	b .., b' .., d .., d' ..	b, nil; b', nil; d, nil; d', nil
330	a ⇄, c .., c' ⇆, d ⇆, d' ⇄	a, 0.28; a', 0.28; c, nil; c', 0.09; d, 0.28; d', 0.19
331	a ⇄, a' .., c ⇆, c' ⇆	a, 0.09; a', nil; c, 0.28; c', 0.37
335	a ⇄, a' .., b .., b' .., c ⇆, c' ⇆ d .., d' ..	a, 0.09; a', nil; b, nil; b', nil; c, 0.09; c', 0.19 d, nil; d', nil
336	a ⇄, a' ⇄, b .., b' ⇄, c ⇆, c' ⇆ d .., d' ..	a, 0.38; a', 0.28; b, nil; b', 0.19; c, 0.28; c', 0.28 d, nil; d', nil
337	a ⇆, a' .., b ⇆, b' ⇆, c ⇆, c' ⇆ d .., d' ⇆	a, 0.19; a', nil; b, 0.19; b', 0.09; c, 0.19; c', 0.28 d, nil; d', 0.09
338	a ⇄, a' ⇄, b .., b' .., c ⇆, c' ⇆ d .., d' ..	a, 0.19; a', 0.19; b, nil; b', nil; c, 0.09; c', 0.09 d, nil; d', nil
19	a ⇄, a' ⇄, b ⇄, b' .., c ⇆, c' ⇆ d ⇆, d' ⇆	a, 0.09; a', 0.09; b, 0.09; b', nil; c, 0.09; c', 0.09 d, 0.19; d', 0.19

* Locations and senses of displacement are with respect to inclination of fault in forcing member, as shown in Figure 4. Absence of data for any of the locations means slip displacement is essentially zero.
† Displacement of upper layer with respect to underneath layer. Magnitudes of slip are measured to nearest small division of an eyepiece-micrometer binocular microscope, 13× magnification.

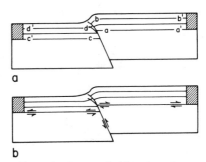

Figure 4. Data on bedding-plane slip: a, Sketch of typical specimen shows locations a through d and a' through d' where slip is measured. b, View of consistent senses of shear displacement (Table 2). In b, absence of displacement arrows at localities b, d, b', c', and d' means that there are too few data for valid interpretations.

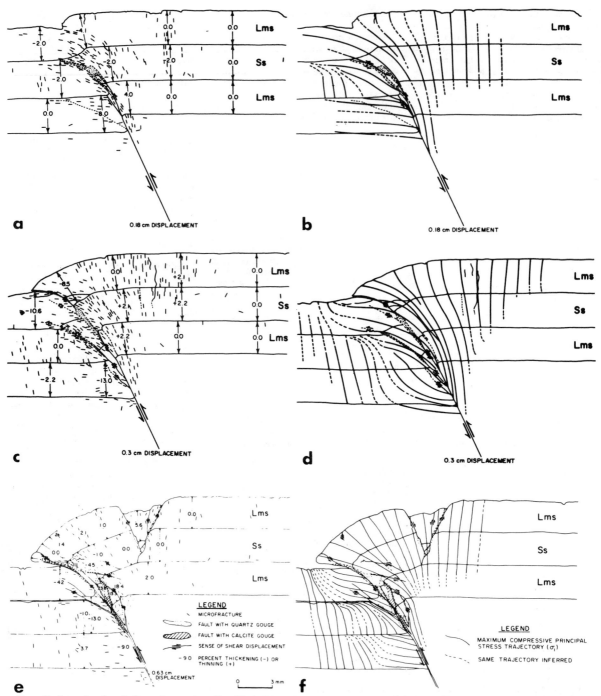

Figure 5. Examples of results from microscopic studies of faulted drape folds with displacements of 0.18 cm (a, b), 0.3 cm (c, d), and 0.63 cm (e, f) along basement fault for specimens 331, 338, and 298, respectively. Diagrams a, c, and e show deformation features observed in thin section. Microfractures are shown schematically as if all are same length; however, their locations and orientations are accurate. Arrows oriented perpendicular to bedding indicate strain calculated from bedding-thickness change (thinning is counted positive). Diagrams b, d, and f show stress trajectories inferred from all microscopic data from a, c, and e.

318

fore are primarily rigid-body motions. This is clearly true of the slip at localities a' and c', where the strain is zero, and at c, where the strain produces a thickening of the layers adjacent to the main fault. That is, the particle motion associated with the strain parallel to bedding is opposed to that of the bedding-plane slip. On the other hand, the thinning of the beds at locality a conforms to the sense of the bedding-plane slip.

There is no correlation between the magnitude of bedding-plane slip at each locality and displacement along the main fault (compare Tables 1 and 2). Perhaps the initial offsets in specimen assembly, the superposition of strains and rigid-body translations, and small differences in the fault configuration mask what otherwise is expected to be a direct correlation.

Bedding-Thickness Changes. Changes in bedding thickness provide the only macroscopic measure of the strains in the various domains of each fold (Figs. 5, 6). Initially we planned to use the distortion of the inked grid pattern, but we found that some of the domains of homogeneous deformation are too small relative to grid size (2.5 mm), and that the width of the grid line relative to spacing introduces errors that are large compared to the strains. On the other hand, the thickness of the layers within the veneer is accurately measurable in thin section to better than ± 0.01 mm. Strains normal to bedding are calculated from measurements of layer-thickness changes ($\Delta t = t_f - t_o$) at specific locations (t_f) with respect to the starting thickness of each layer (t_o), that is, $\Delta t / t_o$. Limestone layers in representative specimens are thinned in domains 1, 3, and 5 (see Fig. 11 for locations of domains) on the upthrown block, and they are thickened in domains 2 and 4 in the downthrown block (Figs. 5, 6). In domains 2 and 4 the thickening is the result largely of uniform intragranular flow (Fig. 5) as well as repetition along reverse faults (Fig. 5, a and c). The thickening increases toward the main upthrust. In domains 3 and 5, the beds are consistently thinned with increasing attenuation toward the fault. In comparable specimens, the thinning in domain 5 increases with increasing displacement along the main fault (Fig. 7). The strains in domain 1 are complicated by bending and adjacent thrust faulting. Maximum thinning occurs in parts of the domain where layer-parallel extension due to bending is greatest. Thickening occurs in a few specimens because of repetition along small thrusts and (or) because of uniform flow (Figs. 5, a and c, 6, a). Changes in sandstone-layer thickness are due to cataclasis (Figs. 5, 6). As we show below, the strains determined from bedding-thickness changes correlate well with those calculated from data on calcite twin lamellae and with the orientations of the principal stresses inferred from orientations of microfractures and calcite twin lamellae.

Comparison in thin section of the areas occupied by the layers before and after deformation shows significant area (volume) changes. These require either large bulk-volume changes or, more likely, mass transport of material along the fold axes. This aspect of the work is currently being studied.

Faults. The word "fault" here implies a discontinuity at least several grain diameters in length, along which shear displacement has occurred. When the displacement is relatively large, there is readily visible offset of layer boundaries, and quartz or calcite gouge is well developed (Figs. 2, 5, 6, 8, 9). For small or incipient displacements, the fault trace is marked by coalesced precursive microfractures, microscopic feather fractures formed after faulting (Conrad and Friedman, 1975), and rigid-body rotations of fragments (Fig. 8).

Depending on location with respect to the basement upthrust, some faults tend to shorten and others to extend the veneer parallel to bedding. Shortening is primarily due to the high-angle reverse fault that curves upward into a low-angle thrust fault (the main upthrust) and to smaller reverse faults in the downthrown block (Figs. 2, c, 5, c and e, 6, a). The extensional faults form only in the upthrown block. They include the normal faults, which form a graben, and the low-angle normal faults, which are conjugate to the upthrust (Figs. 2, a, 5, e).

The sequence of faulting is determined from studies of several specimens, identical except for the magnitude of the displacement along the saw cut in the forcing block (Fig. 5). Initially, incipient or small reverse faults occur low in the veneer of the downthrown block. These tend to terminate at layer boundaries; some curve asymptotically into these boundaries. The faulting then shifts to higher levels in the downthrown block. Each small fault has its own gouge zone, which, although it is a "zone of weakness," is largely ignored by subsequent deformation. Deformation in the downthrown block ends with the propagation of the main upthrust through a zone of precursive microfractures (Fig. 5, a, c, and e). Finally, if layer-parallel extension in the upper part of the upthrown block is large enough, normal faults occur to form a graben (Fig. 5, a, c, and e). The normal faults provide the major compensatory extension required by the low-angle thrusting. Lesser thrusting is accommodated by extension microfracturing and thinning of beds. Comparison of the strain parallel to layer-ing and perpendicular to the fold axis calculated from calcite twin lamellae (described below) in domain 3 of specimen 298 (5 percent, with graben; Figs. 2, a, 5, e, 8, b, 11) and specimen 295 (2 percent, incipient graben; Figs. 1, g, 12) indicates that the graben forms when this extension is between 2 and 5 percent.

A view of the typical upthrust assemblage along "strike" is provided by thin sections cut parallel to the fold axes (Fig. 10). Differential fault displacements along strike produce slivers of varying thickness and spoon-shaped gouge zones.

Gouge Zones. Quartz and calcite gouge develop in all fault zones in sandstone and limestone layers (Figs. 1, 2, 5, 6, 8, 9). There is an outward transition from very fine grained gouge through cataclastic rock to simply fractured or twinned material. Along the main upthrust, calcite gouge is commonly dragged into juxtaposition with sandstone layers and vice versa (Figs. 2, a, 5, a, c, and e). As noted previously, the existence of one gouge zone does not preclude the development of other nearly parallel zones a few grain diameters apart and located higher in the veneer as displacement of the forcing block increases.

In addition, cataclastic sandstone — highly fractured material transitional to finely comminuted gouge — occurs in zones through which faults would certainly have propagated had the displacement of the forcing member been greater (Fig. 5, a and c). Cataclastic flow in these zones and in the adjacent gouge, along with differential displacement on bounding faults, causes overlying layers to undergo rigid-body rotations (Figs. 2, c and e, 9).

Hinges. Hinge zones, where the rate of change of dip is large, are sites of intense cataclasis, gliding flow, and faulting (Figs. 2, 5, 6, 8). Accordingly, they are often sites of pronounced change in bed thickness. At least three types of hinges are recognized: (1) the type created by discrete faults such as a graben (Fig. 2, a) or fractures such as in the downthrown syncline (Fig. 6, g); (2) the type that occurs in layers that undergo rigid-body rotation because of cataclastic flow and differential fault displacements in an underlying cataclastic zone — for example, an upper limestone layer (Figs. 2, c and e, 9); and (3) the type that within a layer is produced by uniform flow (cataclastic and [or] gliding) within that layer (Figs. 2, a, b, d, and f, 8, c).

It is possible to gain some insight into the fixed or migrating nature of the hinges from the distribution of deformation within the layer, sequence of faulting, development of accompanying cataclastic zones, distribution of thinning of the layers, and comparison of experiments with different displace-

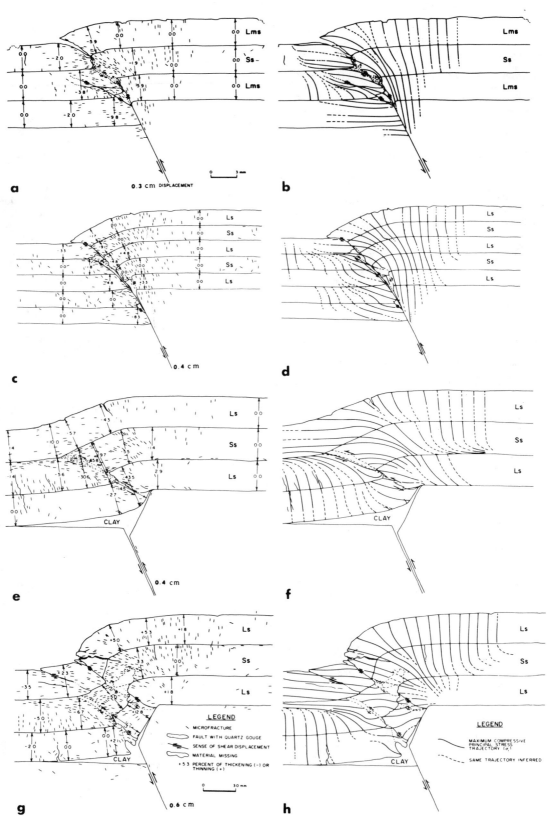

Figure 6. Examples of results from microscopic studies of faulted drape folds, specimens 330, 305, 20, and 21 in a and b, c and d, e and f, and g and h, respectively. Explanation otherwise similar to that for Figure 5.

320

ments along the saw cut. The synclinal hinges in the downthrown block of specimens 20 and 21 (Figs. 6, e and f, 8, c) migrate downward and away from the fault with increasing displacement on the precut fault. Anticlinal hinges associated with discrete faults are fixed, but in those specimens in which two or more such hinges form, the sequence of hinge formation migrates toward the upthrown block (see, for example, Figs. 2, c, 9). Hinges caused by flow within layers seem to migrate away from the fault (compare Fig. 5, a, d, e, and f).

Microfractures. Microfractures in sandstone develop within individual detrital grains (Fig. 8). Quartz grains contain single microfractures or sets of two or more roughly parallel ones; the maximum length of these is limited only by grain size. Chains of fractures can coalesce to form larger scale fractures or faults. In the Indiana Limestone, microfractures often transect fossil, sparry-calcite cement, and micrite boundaries (Fig. 8, a and b). Most are <0.8 mm long. Microfracturing is the major mechanism of deformation in the sandstone. Microfractures in the limestone occur along with twin gliding in most domains (Figs. 5, 6, 8) where differences in the relative abundances of microfractures and twin lamellae clearly delineate the degree of brittle versus ductile deformation. The deformation mode can be associated with the superposition of stresses that lead to changes in mean stress — the lower the mean stress, the greater the degree of microfracturing for the same stress difference.

Microfractures form even at small displacements along the underlying basement fault and continue to form with further displacement (compare Fig. 5, a, c, and e). For loading condition II (Fig. 1), the microfracturing tends also to migrate upward into the extended upthrown block with increasing displacement. In some specimens the microfractures are best developed in the upthrown block (Fig. 5, c), but in others they are most abundant in the zones of reverse faulting (Fig. 5, e). In specimens 20 and 21 (beveled leading edge of forcing block, loading condition IV; see Table 1) the microfracture density is as high in the downthrown block as it is along the principal faults or in the upthrown block (Fig. 6, e and g).

The microfractures exhibit a strong preferred orientation within each domain of statistically homogeneous stress and strain (Figs. 5, 6, 8, 9). Indeed, these domains can be readily delineated by microfracture studies. Orientations are presented on cross sections rather than in conventional equal-area fabric diagrams, because orientations are strong, the fracture planes are all inclined at nearly 90° to the cross section, and the positions of the microfractures within the specimen can also be illustrated. In the upthrown block, the microfractures are, with few exceptions, oriented nearly perpendicular to the layering. In the zone of reverse faulting, they are either inclined at 15° to 30° to faults or they occur in swarms subparallel to faults that would have developed had the displacements been greater along the underlying fault (compare Figs. 5, a and c, and 7, a and c). The general pattern of microfractures throughout the zone of faulting is to curve (concave downward) similar to the upthrusts. In the downthrown block, in regions unaffected by incipient reverse faults, the microfractures are parallel

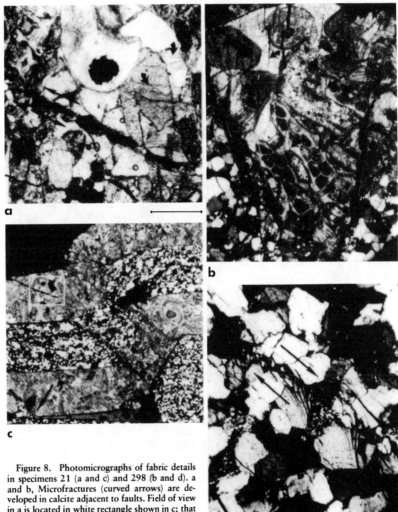

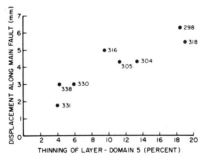

Figure 7. Plot of percentage of thinning of limestone layer, domain 5, versus displacement along main fault in forcing block. Specimen number given for each data point. Absolute lengths are measured optically with a calibrated eyepiece micrometer and are accurate to ± 0.01 mm.

Figure 8. Photomicrographs of fabric details in specimens 21 (a and c) and 298 (b and d). a and b, Microfractures (curved arrows) are developed in calcite adjacent to faults. Field of view in a is located in white rectangle shown in c; that in b covers graben zone in specimen (compare Fig. 2, a). c, Photomicrograph illustrates cataclasis of middle sandstone layer and hinge development in limestone layers. Note development of incipient "synclinal" hinge (straight arrow). d, Microscopic feather fractures are rotated along low-angle normal fault in specimen 298. Photomicrographs are taken in crossed-polarized light; scale line = 0.4 mm in a, 0.6 mm in b, 3.8 mm in c, and 0.2 mm in d.

to the layers, and at other locations they lie at high angles to the layer boundaries.

Deformation of the Forcing Block. Deformation of the unbeveled leading edge or corner of the forcing block increases with increasing displacement along the saw cut (see Fig. 5, a, c, and e) and with decreasing saw-cut angle. Initially, microfractures form inclined at about 80° to the upper

boundary of the forcing block, and they are parallel to and continuous with those in the lowest layer of the veneer. With increasing displacement, shear takes place along the microfractures such that the upper boundary of the forcing block is rotated along a series of down-to-the-left steps. At still larger displacement along the saw cut, the microfractures coalesce to produce a small,

nearly vertical reverse fault in the forcing block (Fig. 5, e).

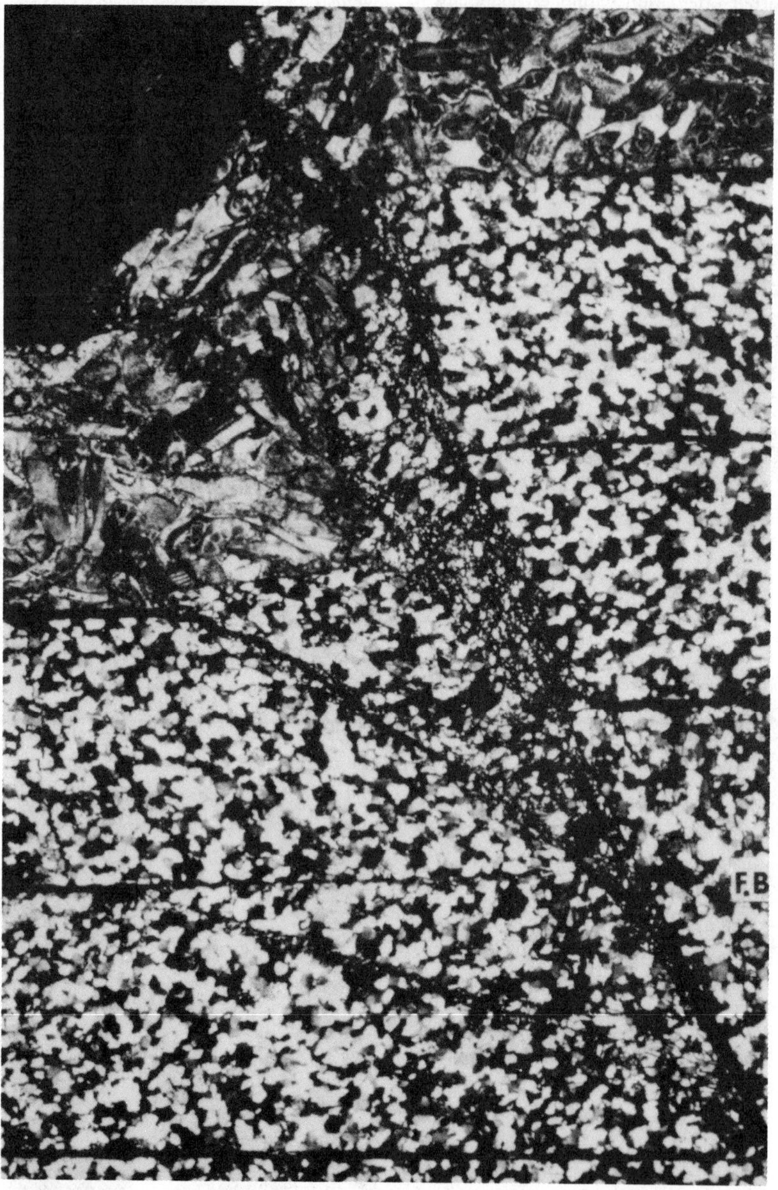

Calcite Fabrics. Dynamic analyses of calcite twin lamellae in selected specimens include determinations of the orientations and relative magnitudes of the principal stress axes after Turner (1953) and the principal strains by the least-squares solution of Groshong (1972, 1974). The undeformed Indiana Limestone is satisfactory experimental material in that it is essentially free of twin lamellae, and the c-axes of calcite-cement crystals and fossil fragments are randomly oriented. In hindsight, however, it is somewhat unsatisfactory, owing to high porosity (10 to 15 percent) and large mean grain size (about 0.3 mm). Initial collapse of pores under confining pressure and further collapse during deformation probably generate twin lamellae in adjacent grains. These are unavoidably sampled during analysis, and so they contribute scatter to the data — that is, they add to the apparent heterogeneity of the stress and strain fields. The large grain size causes a problem because the domains in which the deformation is statistically homogeneous are themselves small. Accordingly, sample population per domain is smaller (8 to 70) than desirable, even though data are obtained from several parallel thin sections for some specimens. Most of the results come from studies of thin sections cut perpendicular to fold axes. However, examination of mutually perpendicular thin sections from specimen 298 (second section cut parallel to the fold axis) reveals that no significant data are lost through use of a single section.

Compression and extension axes are inferred from calcite twin lamellae in specimens 295, 296, 299, 300, 328, 20, and 21. The relatively small numbers of calcite grains per domain limit somewhat the power of this tool; nevertheless, certain significant trends are readily apparent. The data for specimens 298, 295, and 299 are illustrated (Figs. 11, 12, 13) because the numbers of data points are the largest available. As in many previous studies (for example, Friedman, 1963; Carter and Friedman, 1965; Handin and others, 1972), the compression axes in this study adequately show the significant features of the stress field.

In specimen 298 the compression axes exhibit a definite, nonrandom pattern in each of the five domains (Fig. 11). In the downthrown block, domains 2 and 4, the maximum principal compressive stress σ_1 inferred from each point diagram is subparallel to the layering. The pattern shifts 90° in domains 3 and 5 of the upthrown block, where compression axes are distributed along broad great-circle girdles, with many axes inclined at 60° to 90° to the layering. Thus the layer-parallel σ_1 in the

Figure 9. Photomicrograph shows fabric details in specimen 302, composed of one limestone and two sandstone layers (horizontal layer boundaries in black). Reverse faults steepen and get progressively younger with increasing displacement along major fault. Development of microfractures near faults, of gouge zones, of microfractures in leading edge of forcing block (F.B., lower right) and of rotated blocks of limestone layer are discussed in text. Crossed-polarized light; layer thickness is 3.0 mm.

downthrown block and the layer-parallel extension of the upthrown block, including the graben formation, are clearly reflected in the distributions of compression axes. In domain 1 the pattern is again rotated, reflecting both the dip of the extended layer (axes nearly normal to bedding) and possibly the horizontal orientation of σ_1 associated with the thrust fault. That is, domain 1 is not one of homogeneous, single-phase deformation. A second suggestion of superposed deformation is evident in domains 3 and 5, where compression axes are at low angles to the fold axis as well as at high angles to the layering. It will be shown later that these patterns agree with the strain analyses of calcite twin lamellae, the orientations of microfractures in this specimen, and the layer thickening and thinning.

Similar patterns are recorded for specimens 295 and 299 (Figs. 12, 13). Domains in the downthrown block (2, 4, and 7) all show compression axes that reflect a layer-parallel σ_1. Domains in the upthrown block (Fig. 12, domain 3'; Fig. 13, domains 1, 3, and 5) exhibit axes inclined at high angles to bedding and reflect layer-parallel extension. In specimen 295, domain 1 also seems to show compression axes related both to thrust faults and to curvature of the layer. The compression-axis patterns in domain 3, specimen 295, and in domain 6, specimen 299, are poorly defined.

Strain analyses of calcite twin lamellae are based on optical measurements of the total thickness of twinned material per twin set per grain (Groshong, 1972, 1974; Friedman and others, 1976). Each set of twin lamellae provides one equation for the shear strain in a specific orientation. Deviatoric strains are calculated from simultaneous solution of at least five such equations in which shear strains and twin-set orientations are the known quantities, and the strain components are the unknowns. In principle, for perfectly homogeneous strain, five equations and the condition of constant volume, $\epsilon_z = -(\epsilon_x + \epsilon_y)$, suffice. In practice, however, at least twenty are needed; the overdetermined condition is solved by the least-squares method to provide an estimate of the precision of the resulting strain components and principal axes. We gained confidence in this technique by finding excellent agreement with surface strains determined independently in experimental thick-beam folds of Lueders Limestone (Teufel and Friedman, 1974). The results for specimens 298 and 295 are summarized below; a more detailed account is given elsewhere (Friedman and others, 1976). The axis of greatest shortening is ϵ_1 (shortenings counted positive), ϵ_2 is the intermediate principal strain, and ϵ_3 is the axis of least shortening — that is, greatest elongation.

The orientations of the principal strain axes and their magnitudes vary between domains just as do the inferred stresses (Figs. 11–13; Table 3). In specimen 298, the principal strain axes agree with the stress axes in domain 2 but disagree in domain 4 (Fig. 11). In domains 1 and 5, two sets of principal strain axes (1' and 5' versus 1″ and 5″) are recognized on the basis of the sense of calculated expected shear strains. That is, from the least-squares strain tensor, an expected shear strain for each twin set can be calculated. Because twin gliding in calcite occurs only with a positive sense of shear, all of the expected shear strains should be positive for a single homogeneous deformation. As Groshong (1974) pointed out, a negative shear strain indicates that the host grain was not properly oriented for twin gliding with respect to the computed strain tensor. A few small negative values might occur if the least-squares solution has a large standard error or if the deformation is not exactly homogeneous, but many and large negative values indicate either gross heterogeneity or multiple homogeneous deformations during which twin gliding occurs with respect to two or more different stress systems. Segregation of the data on the bases of positive or negative expected shear strains thus permits recognition of superimposed deformations. This approach has been verified in study of right-circular cylinders experimentally shortened sequentially along two axes 90° apart (Teufel, 1975). In domain 1, the two sets of principal strains reflect the nearby thrust fault (ϵ_1, the axis of maximum shortening, is nearly horizontal, 1') and the thinning of the layer (ϵ_1 is normal to and ϵ_3 parallel to the layering, 1″). Further, calcite crystals located close to the fault yield the data in diagram 1'. In domain 5, the thinning of the layer is indicated (5″ and 5*), while shortening parallel to the fold axis is suggested (5'). In domain 3, ϵ_1 is also normal to the layers, and ϵ_3 is horizontal, in agreement with the strain plan of

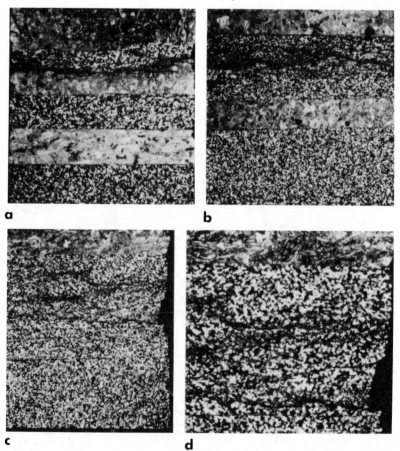

Figure 10. Photomicrographs of "strike" sections cut parallel to fold axes (ZY plane). a, Specimen 298; line of section is A–A' in Figure 2, a. b, Specimen 316, same line of section. c, Specimen 302; line of section is B–B' in Figure 2, c. d, Detail of c. Thickness of limestone layer is 3.0 mm. Crossed-polarized light.

layer thinning and graben formation. For domains 1 and 5, the strains determined from the two perpendicular thin sections agree (see Fig. 11, 1* and 1', 5* and 5"). In general, the strain magnitudes are <5 percent (Table 3). Strains calculated normal to the layering from the calcite twin lamellae and from layer-thickness changes differ on the average by only 0.01 (Table 3; uncertain values for domain 3 omitted). Similar agreement (to ±0.01 strain) is found for the thick-beam fold of Lueders Limestone between strains calculated from the twin lamellae and those calculated from distortion of a surface grid pattern (Teufel and Friedman, 1974; Friedman and others, 1976).

In specimen 295, the principal-strain axes agree with the pattern of compression axes in domains 2 and 3' where the latter are well defined (Fig. 12). In domain 2, ϵ_1 is parallel to the layer boundary, and ϵ_3 is normal to it; in domain 3', the axes are rotated 90° so that ϵ_1 is normal and ϵ_3 is parallel to the layering. These are identical to the strain patterns in similarly located domains in specimen 298 (compare Figs. 11 and 12). In domains 3 and 4' the compression axes are scattered, but the principal strains are consistently oriented even though they are calculated from only 8 and 14 sets of lamellae, respectively. In each domain the axis of greatest shortening is inclined at low angles to the fold axis. Shortening parallel to the fold axis is also recognized from the strain analysis of the thick-beam fold of Lueders Limestone (Teufel and Friedman, 1974; Friedman and others, 1976), and it is suggested in specimen 298 (Fig. 11, domain 5'). It is clear that in the folded beam this shortening is caused by "anticlastic" bending — the "rubber-eraser" effect (Ramsay, 1967, p. 402). Principal strains throughout specimen 295 are less than 3 percent. Correlation with layer-thickness change is impossible because the domains are too small compared to the thickness.

Salt Veneer

Specimens with monolayers of salt are deformed at confining pressures of 0.35, 0.7, and 1.0 kb and loading conditions I and II, and with saw cuts of 40° and 60° (Table 1, Fig. 14). The layer (1.0 cm thick) is prepared from a cattle salt block, which texturally consists of clear halite crystals 1 to 3 mm long surrounded by a fine-grained matrix of halite and cattle nutrients (Fig. 14). The salt is ductile under all confining pressures and loading conditions tested, although after deformation the layer contains as many as three parallel macroscopic extension fractures oriented parallel to the fold axis and normal to the layering (Fig. 14, a, c). Strains calculated from length changes are conspicuously different for

loading conditions I and II. For both loading conditions, the salt is thickened in the domains immediately above the underlying saw cut — more than 40 percent in specimen 278 (Fig. 14, a, b, and e). When the salt is both end and transversely loaded (condition I), there is conspicuous elongation parallel to the fold axis — 13 percent in specimen 278 (Fig. 14, c). The strain in this direction is negligible, however, when the layer is not end loaded (Fig. 14, d).

Microscopically, the halite "phenocrysts" are highly elongated locally (Fig. 14), presumably owing to gliding flow. In the domain of greatest layer thickening (domain 2 and part of domain 3), the crystals are elongated normal to the layer boundary, and their length-to-thickness ratios are >2.0, whereas elsewhere the ratios are <2.0. In domain 1 the rock is essentially undeformed. Similar grain elongations are recorded for quartz grains in small folds subject to bending stresses (Hara, 1966). The region of negligible shape change is used to define the neutral zone along the axial plane of the fold. By analogy, the neutral zone in the salt layer is in domain 1, high in the layer.

DISCUSSION

Petrofabric Analyses

Dynamic petrofabric analyses of the faulted drape folds are important because they provide (1) documentation of the details of the deformations for correlations with large-scale natural counterparts, (2) a test of the petrofabric techniques themselves, not only in axially symmetric de-

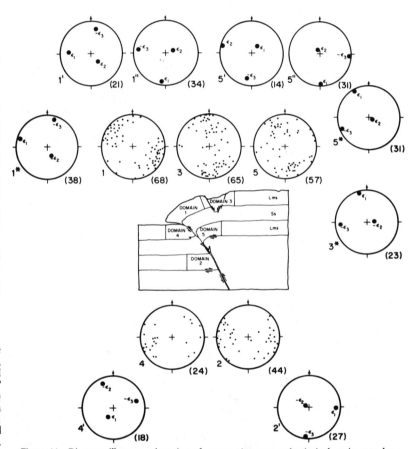

Figure 11. Diagrams illustrate orientation of compression axes and principal strain axes determined from calcite twin lamellae in domains 1 through 5, specimen 298. Data are plotted in lower-hemisphere equal-area projection; plane of each diagram is perpendicular to fold axis. Number of sets of ϵ_1 twin lamellae from which data are obtained is given in parentheses for each diagram, and numbers 1 to 5 correspond to domains 1 through 5. Compression axes are shown in diagrams 1 through 5. Principal strain axes are shown in diagrams with superscript symbols. Diagrams 1', 1", 5', and 5" represent two different solutions based on segregated data in domains 1 and 5, respectively (see text). Diagrams 1*, 3*, and 5* represent strain analyses from thin sections cut parallel to fold axis; otherwise data are from sections cut perpendicular to fold axis.

formed right cylinders, as in most previous work, but also in rocks deformed under complex boundary conditions that give rise to superposed deformations, and (3) maps of the stress and strain distributions as inferred from known deformations; the latter enhances our understanding of the mechanics of folding and affords a basis on which to judge the applicability of analytical boundary-value solutions (Min, 1974; Min and others, in prep.). Toward these ends, petrofabric analyses of faulting, bedding-plane slip, microfracture orientations, bedding-thickness changes, and calcite twin lamellae have proved to be remarkably consistent.

Deformation Patterns. Petrofabric analyses of the five domains (1, 3, and 5 in the upthrown block and 2 and 4 in the downthrown block) of typical drape-folded specimens loaded under condition II (Figs. 2, 5, 6) illustrate the nature and reproducibility of the deformation.

Domain 1 is one of superposed deformation involving thrust faulting and layer-parallel extension (thinning) due to both bending and thrusting. Sequential tests (Figs. 5, 6) suggest that the layer-parallel extension develops first, followed by low-angle thrusting in some specimens. Distant from the thrust fault, the layer is thinned and microfractures are normal to bedding, as are σ_1 and ϵ_1 determined from the calcite twin lamellae. The sense of bedding-plane slip (Fig. 4) agrees with the strain plan. Close to the main fault, the microfractures are nearly parallel to layering. In some specimens the layers are thickened (Fig. 5, c), and ϵ_1 derived from the calcite twin lamellae is correspondingly parallel to the layering.

Domain 3 is characterized by layer-parallel extension. The beds are thinned; faults are normal and form grabens when thrust displacement in domain 1 and bending produce layer-parallel extension of 2 to 5 percent; microfractures are normal to bedding; and both σ_1 and ϵ_1 inferred from the calcite twin lamellae are at high angles to bedding. Some suggestion of superposed deformation is seen in the calcite data, which indicate a tendency for ϵ_1 to lie at low angles to the fold axis at some stage of the deformation (Figs. 11, 12).

Domain 5 also is characterized by layer-parallel extension. The beds are thinned; faults along lateral domain boundaries are high-angle reverse faults; microfractures are inclined at from 60° to 90° to the layering (nearly normal to layering distant from fault); and σ_1 and ϵ_1 inferred from the twin lamellae are inclined at high angles to the layering. The sense of shear along the lower boundary agrees with the strain plan (Fig. 4).

Domain 2 is one of single-phase layer-parallel shortening. σ_1 positioned at 30° to small reverse faults is parallel to layering; microfractures are parallel to layering; beds are thickened; σ_1 and ϵ_1 inferred from twin lamellae are parallel to layering. The sense of bedding-plane slip is such as to transport material in the layer *away* from the fault and thus is opposed to the strain plan. This distinguishes between the events of rigid-body translation and strain.

Domain 4 is one mainly, but not wholly, of single-phase, layer-parallel shortening. With the exception of the calcite data, the orientations of fabric elements and derived quantities are the same as in domain 2. The calcite data, however, are not straightforward. In specimen 298 (Fig. 11) the compression axes are grouped subparallel to bedding, but the strain axes do not conform. Instead ϵ_1 is inclined at low angles to the fold axis, although the strain normal to the layering is an extension of 0.019 compared to an extension of 0.042 computed from the thickness change (Table 3). The same conflicting results occur in domain 4', specimen 295 (Fig. 12). Why the stress and strain axes derived from twin-lamellae data should differ is dealt with elsewhere (Friedman and others, 1976). That calcite twin lamellae should be the first, and in this case the only, features to reflect superposed deformations is understandable on the basis of the low resolved shear stress for twin gliding.

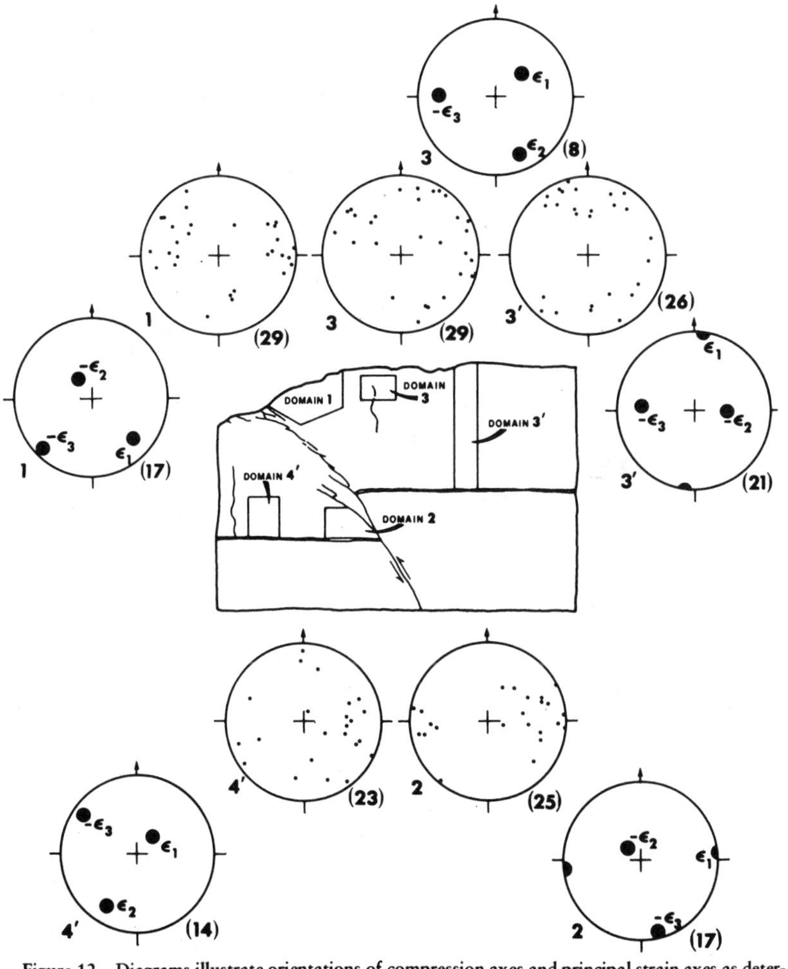

Influence of Loading Conditions

The purpose in varying the boundary loading conditions (Fig. 1) is to determine the nature of accompanying changes, if any, in structural detail. If significant differences result and if they can be correlated with those in natural counterparts, new insights

Figure 12. Diagrams illustrate orientations of compression axes and principal strain axes as determined from calcite twin lamellae in domains 1 through 4, specimen 295. See caption of Figure 11 for explanation.

into natural boundary conditions may be gained. Although this aspect of the work has received only cursory attention to date, the results from a few comparisons seem significant. For example, in salt specimens, extension parallel to the fold axis is large under condition I (Fig. 14, c) but negligible for condition II. For the limestone-sandstone-limestone veneer deformed under condition IV, and with the leading edge of the forcing block beveled (specimens 20 and 21, Fig. 6, e–h), microfractures are more abundant and synclinal bending is conspicuous in the downthrown veneer as compared to similar specimens deformed at condition II (compare Figs. 5 and 6). Variations in the saw-cut angle

from 30° to 90° produce different intensities of deformation but not changes in the overall style. In general, for similar displacement along the saw cut, deformation in the downthrown block decreases whereas extension in the upthrown layers increases as the saw-cut angle is changed from 65° to 30°. Although detailed studies of specimens with 75° and 90° saw cuts have just been initiated, it is clear that the deformation in the layers is reduced (fewer faults and microfractures, smaller strain) as the angle is steepened from 65° to 90°. It is our intention to test these trends in field studies.

Another result of the loading condition is the brittle behavior of the limestone in domains 1 and 3, particularly, and its pre-

dominantly ductile behavior in domain 4. Under the conditions of our experiments, we expect the sandstone always to be brittle, but in triaxial compression tests the limestone is ductile under 1-kb confining pressure and brittle in extension tests at the same confining pressure (Handin and Hager, 1957). Thus, microfracturing of the limestone indicates superposition of stresses such as to reduce the mean pressure and thus enhance brittle behavior. The abundance of fractured calcite relative to twinned calcite is a measure of this transition, and fracturing is greatest in domains 1 and 3, where layer-parallel extension caused by bending is superposed on the stresses across the boundaries of the layers to lower the mean pressure.

Stress Trajectories. The final stage in the petrofabric analysis is the preparation of stress-trajectory diagrams derived from the orientation of faults, microfractures, and calcite twin lamellae (Figs. 5, b, d, and f, 6, b, d, f, and h). The trajectories conform also to the strains calculated from bedding-thickness changes and the calcite twin lamellae. The dynamic significance of a fault is well known. Shear displacements relative to bedding tend to extend the layers (normal faults) or to shorten them (reverse and thrust faults), and σ_1 is positioned relative to the fault at about $\theta = 30°$ ($\theta = 45° - \phi/2$, where ϕ is the angle of internal friction of the Coulomb-Mohr criterion), commensurate with the sense of shear, σ_2 is parallel to the fault plane and at 90° to the slip line, and σ_3 is mutually perpendicular to σ_1 and σ_2. The significance of calcite twin lamellae was reviewed earlier. The existence of bedding-plane slip proves that contacts between beds are not welded, and the relative slip is a measure of the net motion of particles across layering, a motion due primarily to rigid-body displacement. Bedding-thickness changes are a measure of finite strain perpendicular to layering. They help define the strain plan for a given domain in the deformed body, and they provide an independent check of strains computed from the data on twin lamellae.

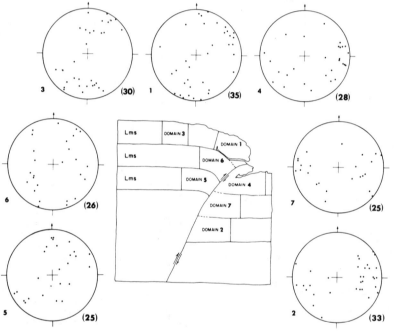

Figure 13. Calcite fabric data for specimen 299. Compression axes are derived from calcite twin lamellae in each of seven domains. See caption of Figure 11 for explanation.

TABLE 3. STRAINS CALCULATED FROM CALCITE TWIN LAMELLAE COMPARED WITH THOSE FROM LAYER-THICKNESS CHANGES, SPECIMEN 298

Domain no.	Strain from layer-thickness change	Principal strains from calcite twin lamellae			Strain normal to layering, calcite twin lamellae	Number of twin sets in analysis	Difference in strains normal to bedding
		ϵ_1	ϵ_2	ϵ_3			
1	0.021*	0.022	0.004	−0.026	0.016	55	0.005
2	−0.037	0.035	−0.008	−0.028	−0.023	28	−0.014
3	0.056†	0.030	0.001	−0.041	0.014	19	0.042†
3§	0.056†	0.036	−0.001	−0.026	0.016	23	0.040†
4	−0.042	0.081	−0.038	−0.043	−0.019	18	−0.023
5	0.020	0.041	0.001	−0.052	0.036	60	−0.016
5§	0.020	0.039	0.018	−0.056	0.017	31	0.003

* Shortenings (for example, layer thinning) are counted positive.
† Value is probably too large because material at top of graben was lost in preparation of thin section (Fig. 2, a).
§ From thin section cut parallel to fold axis. Other data are from thin sections cut perpendicular to fold axis.

Microfracturing is an important phenomenon because (1) it serves as a measure of relative ductility, (2) microfracture orientations are a clue to the strain plan, and (3) these orientations can be used to map principal-stress orientations, provided that the type of fracturing — shear or extension — is known. We argue that essentially all the microfractures in our specimens are of the extension type (Griggs and Handin, 1960) for the following reasons: (1) No shear displacements are visible even under high magnification. (2) Photoelastic studies of granular aggregates have shown that extension fractures form from stress concentrations at grain boundaries (Gallagher and others, 1974). (3) Microfractures immediately adjacent to faults (microscopic feather fractures) are extension features (Conrad and Friedman, 1975). (4) Microfractures precursory to the development of macroscopic shear fractures or faults are probably extension fractures — that is, they are oriented subparallel to the greatest compressive stress axis across the boundaries of test specimens (see, for example, Brace and others, 1966; Bieniawski, 1967; Hoshino and Koide, 1970; Friedman, 1975). Indeed, this latter fact is better illustrated in our specimens (for example, Fig. 5, a, c, and e) than in the previous work cited. (5) Microfractures are well developed normal to the layers where the layers are extended or thinned and parallel to them where they are shortened. (6) In fractured and twinned calcite grains, the orientation of σ_1 favorable for twin gliding is parallel to the fractures, and the statistically determined ϵ_3, the greatest extension, tends to be oriented perpendicular to the traces of the microfractures. Accordingly, each microfracture is interpreted as an extension fracture; as such, σ_3 and ϵ_3 are normal and σ_1 and ϵ_1 are parallel to the microfracture plane.

The stress-trajectory maps (Figs. 5, 6) are made by using all the fabric data. Neglecting the instances of superposed stress states (discussed above), the predominant orientation of σ_1 changes little throughout the displacement history of the underlying basement fault. Trajectories in the upthrown block are essentially identical in all specimens. Those in the downthrown block are more complex and appear to depend on the end-loading condition and (or) the shape of the leading edge of the forcing member (compare Figs. 5, b, d, and f, and 6, b and d,

with Fig. 6, b and h). Discontinuities in the trajectories at layer boundaries indicate detachment and relative slip. Bent-beam neutral zones develop in the middle layer(s) (Figs. 5, b–f, 6, b–h), most conspicuously when the leading edge of the forcing block is beveled (Fig. 6, f and h).

Comparison with Numerical Models

Min (1974; Min and others, 1975) has obtained elastic boundary-value solutions, using the geometric and material parameters of the experiments. Here we illustrate only one of his solutions for comparison with the observed results. In this solution (Fig. 15) the three veneer layers are treated as a single one for which Poisson's ratio is 0.25, and the boundary between the forcing block and the lowermost layer is regarded as frictionless. The orientations of the trajectories depend on the shape of the layer (ratio of length to thickness) and Poisson's ratio. They are independent of displacement on the forcing fault, Young's modulus, and the modulus of rigidity as noted by Sanford (1959). The magnitudes of the differential stresses do depend on displacements.

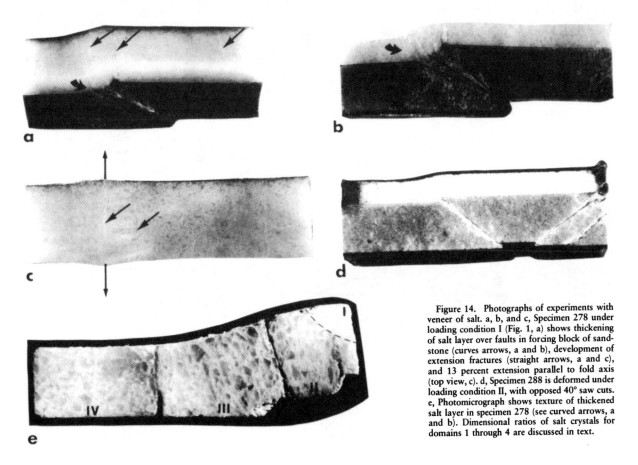

Figure 14. Photographs of experiments with veneer of salt. a, b, and c, Specimen 278 under loading condition I (Fig. 1, a) shows thickening of salt layer over faults in forcing block of sandstone (curves arrows, a and b), development of extension fractures (straight arrows, a and c), and 13 percent extension parallel to fold axis (top view, c). d, Specimen 288 is deformed under loading condition II, with opposed 40° saw cuts. e, Photomicrograph shows texture of thickened salt layer in specimen 278 (see curved arrows, a and b). Dimensional ratios of salt crystals for domains 1 through 4 are discussed in text.

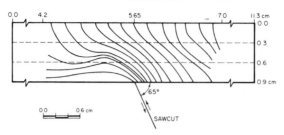

Figure 15. Trajectories for greatest principal compressive stress resulting from elastic boundary-value solution for experimental loading condition II (Fig. 1, b) after Min (1974). Poisson's ratio (0.25) is same for all layers, and interface between forcing member and lowermost layer is frictionless.

Comparison of computed stress trajectories (Fig. 15) with those inferred from the petrofabric analyses (Figs. 5, 6) shows that the two patterns are very similar. In detail, note that both sets of σ_1 trajectories are subparallel to the layers in the lowermost layer of the downthrown block, they are inclined at from 30° to 60° to the layer boundaries in the zone of faulting, and they are inclined from 60° to 90° to bedding in the upthrown block. Further comparisons have been discussed by Min (1974).

Correlations with Natural Structures

The experiments are not scale models of natural forced folds. Composed of real rocks (as opposed to modeling clay or unconsolidated sand), they are at best directly comparable to similarly deformed natural structures of the same size. There are many aspects of the experiments, however, that have a direct bearing on natural-scale structures. These deal with comparable mechanisms of deformation, kinematics, fabrics, and deformational history. For instance, many features resulting from brittle basement deformation in the Rocky Mountain foreland, which have puzzled geologists for years, can now be understood in light of the mechanisms operative within the brittle unit of the experiments (Stearns and Weinberg, 1975).

The geometrical similarities between experiment and field occur over several scales. In some instances even the gross structural features in cross section compare too well to be ignored. For example, the geometry produced in specimen 298 (Fig. 16, a) compares very closely with that of a cross section through the Owl Creek Mountains of Wyoming (Fig. 16, b). The normal-fault zone behind the curved, high-angle reverse fault is produced in the experiment in precisely the same relative position as it is in nature — immediately above the sharpest curvature in the curved fault. This same phenomenon is also typical of other mountain systems in Wyoming, like the southwest front of the Wind River Mountains, the southern Granite Mountains, and the east side of the Rock Springs uplift.

Closer geometrical correlation occurs on a smaller scale. In many places in the Rocky Mountain foreland province, immediately to the basinward side of the large mountain-boundary reverse faults in the basement, the Cambrian-Precambrian contact dips toward the basin at angles of as much as 60°. This contact is flat and planar. That is, the contact has not been folded. It has been difficult, if not impossible, to resolve at the field scale how rigid-body rotations of the upper basement of as much as 60° can occur in these brittle materials without producing curved contacts. Such

occurrences are common in Wyoming — for example, along the southeast corner of the Beartooth Mountains, in the Seminoe Mountains, and particularly around the flanks of the Hanna Basin. Similar features have developed in the experiments. For example, in specimen 302 (Figs. 2, c, 9) the two sandstone layers below the limestone become part of the forcing block. They are faulted, and the faults are accompanied by zones of gouge and cataclastic sandstone. The upper contact with the limestone is still planar, yet it has been rotated as much as 60°. Similar rigid-body rotation can also be observed in the small frontal blocks in other experiments (see, for example, Fig. 2, a and e). It would appear that the rotation results from cataclasis and differential displacement along the faults below the limestone. The exact mechanisms by which these rotations occur are currently being studied. Although the phenomena are not fully understood, it is encouraging that such rotations do occur under controlled conditions and that the large rotated frontal blocks found in nature have a rational counterpart in the experiment.

The experiments also suggest features to look for in the field. For example, the faults in specimen 302 referred to above are surrounded by zones of intense cataclasis. With this in mind, we studied a Precambrian block at Rattlesnake Mountain, near Cody, Wyoming, and similar cataclastic zones were found bounding large rotated blocks of intact material. Similarly, in the Seminoe, Beartooth, and Bennett Mountains, there are without exception numerous narrow cataclastic zones where the basement surface is rotated and an absence of these zones where it is not rotated.

The deformation along the corners of brittle basement blocks is being restudied in a broad field program in which the experiments are serving as a basis for a better un-

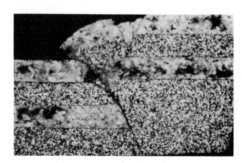

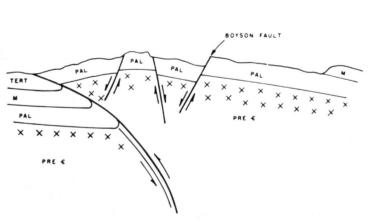

Figure 16. Comparison of gross structural features in specimen 298 (left) and in cross section through Owl Creek Range, Wind River Canyon, Wyoming (right). Both show curved reverse fault and extension zone with graben developed in upthrown layers.

derstanding of the natural deformations. The insights provided by the experiments are allowing completely new explanations of field relationships.

The drape folds in the layered sedimentary rocks above the Precambrian basement in the Rocky Mountain foreland are perhaps best duplicated by specimen 21 (Figs. 2, d, 8, c). Here the clay that filled in the beveled leading edge of the forcing member flowed freely during the deformation and thus probably is comparable to some of the movements in the Cambrian shales that separate the Precambrian crystalline basement from the overlying Paleozoic carbonate rocks in Wyoming. Note that the best unfaulted draping of the layers occurs in this type of experiment. It is also interesting to note that the incipient synclinal hinge is forming some distance from the forcing block, as illustrated by the injected clay (Fig. 8, c). That part of the layer between this new hinge and the well-developed hinge at the base of the forcing member is uplifted as a nearly rigid member of the veneer. This is precisely what is observed in natural drape folds in the Rocky Mountain foreland (Stearns, 1971).

Another example of correlation between experimental and natural structures is the sequence and orientation of reverse faulting and the associated microfracture fabric in the downthrown layers. Recall that in the experiments, small low-dipping reverse faults (frontal faults) develop early in the displacement history along the saw cut and that younger and more steeply dipping faults occur toward the high block as displacement along the saw cut increases (see Figs. 5, a, c, and e; 9). The overall microfracture pattern adjacent to these faults exhibits concentrations parallel and conjugate (up to 45°) to the faults. This fault geometry and sequence and the microfracture pattern is essentially identical to that developed at depth in the Tertiary sandstone in the downthrown block of the Oak Ridge reverse fault, Saticoy field, Ventura County, California (Friedman, 1969). Recognition of these faults as an integral part of this structural style early in the history of drilling would have led to more efficient development of the field.

Finally, it should be emphasized that the microscopic deformation mechanisms operative in the experiments (microfracturing and twin gliding) are also the principal ones in natural counterparts. The best correlation of microfractures is that in the Saticoy field discussed above. In addition it has been shown that calcite twin lamellae in naturally folded Madison Limestone (Mississippian) at the Teton anticline reflect some of the same stress states as nearby macrofractures (Friedman and Stearns, 1971). Moreover, calcite twin lamellae in

the Madison Limestone at Rattlesnake Mountain, near Cody, Wyoming, reflect the stresses associated with synclinal bending in the downthrown layers and with the major faulting (Friedman, unpub. data). More detailed microscopic petrofabric analyses of various parts of natural drape folds are currently in progress.

CONCLUSIONS

Faulted drape folds are produced experimentally in multilithologic layered specimens under known, yet complex, loading conditions. Observational studies provide knowledge of the mechanisms of deformation, kinematics, fabrics, and structural history of the specimens. In full consideration of scale differences, there are significant correlations with natural counterparts that improve understanding of the natural structures and that suggest ideas to be pursued in field studies. Results are summarized in the Abstract; here we state certain general conclusions that we regard as particularly significant:

1. The observational data and force-displacement curves demonstrate that drape folds are controlled primarily by the shape and displacement of the forcing block rather than by the aspect ratio and mechanical properties of the folded layers, as in buckling.

2. Fabric data are internally consistent, and inferred stresses are in good agreement with those calculated from an elastic solution for the experimental boundary conditions. They demonstrate that the major features of the deformation are established early in the history of displacement of the forcing blocks.

3. Principal strains calculated from calcite twin lamellae after Groshong are within an average of 0.01 of those calculated from layer-thickness changes. The strain technique permits clear resolution of individual events in domains of superposed deformations.

4. Observational studies provide structural details of the extension of the upthrown layers, the reverse fault zone, and the complexities in the downthrown layers, which should aid in the recognition of this structural style in the field and in the subsurface.

5. Some of the experiments demonstrate how large rigid-body displacement of a layer boundary can result from cataclastic flow and faulting in an adjacent domain.

6. Although these specimens are not scale models of natural prototype structures in the Wyoming province and elsewhere, our results do shed light on the mechanics of deformation of this structural style, especially with regard to sequences and relative locations of certain structural features.

ACKNOWLEDGMENTS

We thank several graduate students in Friedman's course in structural petrology for their help in the fabric studies and C. E. Corry for use of his specimens incorporating the beveled leading edge of the forcing block. Our research on folding has been generously supported by National Science Foundation Grants GA-23332, GA-36127X, and DES74-22954.

REFERENCES CITED

Berg, R. R., 1962, Mountain flank thrusting in Rocky Mountain foreland, Wyoming and Colorado: Am. Assoc. Petroleum Geologists Bull., v. 48, p. 2019–2032.

Bieniawski, Z. T., 1967, Mechanism of brittle fracture of rock. Part II, Experimental studies: Internat. Jour. Rock Mechanics and Mining Sci., v. 4, p. 407–423.

Brace, W. F., Paulding, B. W., Jr., and Scholz, C., 1966, Dilatancy in the fracture of crystalline rocks: Jour. Geophys. Research, v. 71, p. 3939–3954.

Carter, N. L., and Friedman, M., 1965, Dynamic analysis of deformed quartz and calcite from the Dry Creek Ridge anticline, Montana: Am. Jour. Sci., v. 263, p. 747–785.

Conrad, R. E., II, and Friedman, M., 1975, Microscopic feather fractures in the faulting process: Geol. Soc. America Abs. with Programs, v. 7, p. 153–154.

Friedman, M., 1963, Petrofabric analysis of experimentally deformed calcite-cemented sandstones: Jour. Geology, v. 71, p. 12–37.

——1969, Structural analysis of fractures in cores from the Saticoy Field, Ventura County, California: Am. Assoc. Petroleum Geologists Bull., v. 53, p. 367–389.

——1975, Fracture in rock: Am. Geophys. Union Geophys. Rev., v. 13, p. 352–358.

Friedman, M., and Stearns, D. W., 1971, Relations between stresses from calcite twin lamellae and macrofractures, Teton anticline, Montana: Geol. Soc. America Bull., v. 82, p. 3151–3161.

Friedman, M., Teufel, L. W., and Morse, J. D., 1976, Strain and stress analyses from calcite twin lamellae in experimental buckles and faulted drape-folds: Royal Soc. London Philos. Trans. (in press).

Gallagher, J. J., Jr., Friedman, M., Handin, J., and Sowers, G. M., 1974, Experimental studies relating to microfracture in sandstone: Tectonophysics, v. 21, p. 203–247.

Griggs, David, and Handin, John, 1960, Observations on fracture and a hypothesis of earthquakes, in Griggs, D., and Handin, J., eds., Rock deformation: Geol. Soc. America Mem. 79, p. 347–364.

Groshong, R. H., Jr., 1972, Strain calculated from twinning in calcite: Geol. Soc. America Bull., v. 83, p. 2025–2038.

——1974, Experimental test of least-squares strain gage calculation using twinned calcite: Geol. Soc. America Bull., v. 85, p. 1855–1864.

Handin, J., and Hager, R. V., Jr., 1957, Experimental deformation of rocks under confining pressure; Tests at room temperature on dry samples: Am. Assoc. Petroleum Geologists Bull., v. 41, p. 1–50.

Handin, J., Friedman, M., Logan, J. M., Pattison, L. J., and Swolfs, H. S., 1972, Experimental folding of rocks under confining pressure: Buckling of single-layer rock beams: Am. Geophys. Union Mon. 16, p. 1–28.

Handin, J., Friedman, M., Min, K. D., and Pattison, L. J., 1976, Experimental folding of rocks under confining pressure: Part II. Buckling of multilayered rock beams: Geol. Soc. America Bull., v. 87, p. 1035–1048.

Hara, Ikuo, 1966, Dimensional fabric of quartz in a concentric fold: Japanese Jour. Geology and Geography, v. 37, p. 123–139.

Hoshino, Kazuo, and Koide, Hitoshi, 1970, Process of deformation of the sedimentary rocks: Internat. Soc. Rock Mechanics Cong., 2nd, Beograd 1970, Proc., v. 1, paper 2-13.

Lees, G. M., 1952, Foreland folding: Geol. Soc. London Quart. Jour., v. 108, p. 1–35.

Min, K. D., 1974, Analytical and petrofabric studies of experimental faulted drape folds in layered rock specimens [Ph.D. dissert.]: College Station, Texas A&M Univ., 90 p.

Min, K. D., Gangi, A. F., and Logan, J. M., 1975, Analytic and petrofabric studies of experimentally faulted drape folds in multilithologic layered rock specimens: Geol. Soc. America Abs. with Programs, v. 7, p. 217.

Pattison, L. J., 1972, Petrofabric analysis of experimentally folded multilayered rocks [M.S. thesis]: College Station, Texas A&M Univ., 71 p.

Prucha, J. J., Graham, J. A., and Nickelson, R. P., 1965, Basement-controlled deformation in Wyoming province of Rocky Mountain foreland: Am. Assoc. Petroleum Geologists Bull., v. 49, p. 966–992.

Ramsay, J. G., 1967, Folding and fracturing of rocks: New York, McGraw-Hill Book Co., 568 p.

Sanford, A. R., 1959, Analytical and experimental study of simple geologic structures: Geol. Soc. America Bull., v. 70, p. 19–52.

Stearns, D. W., 1971, Mechanisms of drape folding in the Wyoming province: Wyoming Geol. Assoc., 23rd Ann. Field Conf., Guidebook, p. 125–144.

Stearns, D. W., and Weinberg, D. M., 1975, A comparison of experimentally created and naturally formed drape folds: Wyoming Geol. Assoc., 27th Ann. Field Conf., Guidebook, p. 159–166.

Teufel, L. W., 1975, Strain analysis of twinned calcite for two experimentally superposed deformations of Indiana Limestone: Geol. Soc. America Abs. with Programs, v. 7, p. 240.

Teufel, L. W., and Friedman, M., 1974, Strain analysis of calcite twin lamellae in a thick buckled beam of Lueders Limestone [abs]: EOS (Am. Geophys. Union Trans.), v. 55, p. 420.

Turner, F. J., 1953, Nature and dynamic interpretation of deformation lamellae in calcite of three marbles: Am. Jour. Sci., v. 215, p. 276–298.

MANUSCRIPT RECEIVED BY THE SOCIETY MARCH 19, 1975
REVISED MANUSCRIPT RECEIVED JULY 10, 1975
MANUSCRIPT ACCEPTED JULY 17, 1975

Reprinted by permission of Elsevier Science Publishers from
Tectonophysics, v. 17, no. 4 (1973), p. 299-321.

Tectonophysics, 17 (1973) 299-321

© Elsevier Scientific Publishing Company, Amsterdam – Printed in The Netherlands

LAYER SHORTENING AND FOLD-SHAPE DEVELOPMENT IN THE BUCKLING OF SINGLE LAYERS

PETER J. HUDLESTON and OVE STEPHANSSON

Department of Geology, University of Minnesota, Minneapolis, Minn. (U.S.A.)
Institute of Geology, University of Uppsala, Uppsala (Sweden)

(Accepted for publication June 5, 1972)

ABSTRACT

Hudleston, P. J. and Stephansson, O., 1973. Layer shortening and fold-shape development in the buckling of single layers. *Tectonophysics*, 17: 299–321.

The progressive development of folds by buckling in single isolated viscous layers compressed parallel to the layering and embedded in a less viscous host is examined in several ways; by use of experiments, an analogue model to simulate simultaneous buckling and flattening and by an application of finite-element analysis.

The appearance of folds with a characteristic wavelength in an initially flat layer occurs in the experiments for viscosity ratios ($\mu_{layer}/\mu_{host} = \mu_1/\mu_2$) of between 11 and 100; progressive fold development after the initial folds have appeared is similar in the experiments and in the finite-element models. Except for the finite-element model for $\mu_1/\mu_2 = 1,000$ layer-parallel shortening occurs in the early stages of folding and a stage is reached where little further changes in arc length occur. The amount of layer-parallel shortening increases with decreasing viscosity contrast, and becomes relatively unimportant after the folds have attained limb dips of about $15°-25°$.

Thickness variations with dip are only significant here for the finite-element model with $\mu_1/\mu_2 = 10$, and in experiments for $\mu_1/\mu_2 = 5$ where the layer is initially in the form of a moderate-amplitude sine wave. The variations range from a parallel to a near-similar fold geometry, and in general depend on the viscosity contrast, the degree of shortening and the initial wavelength/thickness ratio. They are very similar to the variations predicted by the analogue model of combined buckling and flattening. The difference between the thickness/dip variations in a fold produced by buckling at low viscosity contrast and one produced by flattening a parallel fold is marked at high limb dips and very slight at low limb dips.

Many natural folds in isolated rock layers or veins show thickness/dip relationships expected for a flattened parallel fold, and some show relationships expected for buckling at low viscosity contrasts. Studies of the wavelength/thickness ratios in natural folds have suggested that competence contrast is often low. Many folds in isolated rock layers or veins whose geometry may vary between parallel and almost similar, and may be indistinguishable from those of flattened parallel folds, have probably developed by a process of buckling at low viscosity contrasts.

INTRODUCTION

Natural fold profiles are observed to vary from parallel to nearly similar in shape for competent rock layers either isolated in a uniform rock mass or forming part of a multi-layered sequence. Various attempts have been made to explain these observations in terms of models of folding processes. For example, similar folds have been interpreted as being the result of differential simple shear deformation of passive layers (e.g., Turner and Weiss, 1963, p. 480), or as being the result of "flattening" (the superposition of a finite homogeneous strain) gently buckled layers (e.g., Mukhopadhyay, 1965). Parallel folds have a geometry that may be consistent with flexural slip or flexural flow folding (e.g., Hills, 1963, p. 227) or with internal deformation taken up by tangential—longitudinal strain (Ramsay, 1967, p. 397). The properties of these models have been discussed in detail by Ramberg (1963), Ramsay (1967) and Hudleston (1973a). The results of the analytical and experimental models of buckle fold development discussed in this paper enable many natural folds in isolated competent layers, with shapes varying from parallel to nearly similar, to be interpreted in terms of a single theory of buckling in viscous materials where viscosity contrast is low and layer-parallel shortening important. It should be noted that we are concerned here with small folds where effects of gravity may be ignored.

De Sitter (1968) considered that the maximum amount of shortening that could be accommodated by flexural slip folding was about 36% for a concentric parallel fold model. With a more general fold shape considerably more shortening can be accommodated in a parallel fold. Still further shortening can be developed by flattening, the effect of which is discussed in detail by Ramsay (1967, pp. 411—415) and Hudleston (1973a). A result of flattening noted by Mukhopadhyay (1965) is that for a given degree of flattening and size of error in measuring thickness in the fold profile, there is a critical limb dip below which a fold is indistinguishable from a similar fold. Studies of natural fold shapes have indicated that many folds in competent layers have geometric forms almost identical to those produced by uniformly flattening a parallel fold to various degrees (e.g., Ramsay, 1962), that many folds are almost similar in form (Ramsay, 1967, p. 421), and that for folds of nearly similar geometry, systematic but slight divergences from a true similar geometry exist that can be related to the relative competence of the layers (Hudleston, 1973c).

A better understanding of the mechanism of single-layer fold formation came from the theoretical analyses of viscous buckling made by Biot (1957, 1961) and Ramberg (1959, 1961). From the early results of buckling theory, Flinn (1962, p. 424) concluded that "as soon as buckling is present it will be acted on by the homogeneous strain and further folding will be a mixture of similar folding due to homogeneous strain and concentric folding due to the buckling. Which type predominates depends on the competence difference." The idea of two contrasted kinds of shortening in buckling was treated theoretically by Ramberg (1964a). For a low-amplitude fold he distinguished "buckle shortening", in which the wavelength decreases while the arclength remains constant and "layer shortening" representing the uniform change in length of the layer in response to the compressive stress. A more detailed review of pertinent aspects of buckling theory is given in the following section.

In this paper we examine the progressive changes in various parameters of fold geometry, including arclength, limb dip, amplitude and thickness variations with dip in experimentally produced folds, in a simple analytical model combining buckling and flattening increments of deformation and in models produced by finite-element analysis. We find that these three approaches give similar results and that these are consistent with limited data on natural folds.

REVIEW OF RELEVANT ASPECTS OF BUCKLING THEORY AND FINITE-AMPLITUDE FOLDING

Most of the theoretical work on buckling of layered systems done by Biot (1961, 1965) and Ramberg (1961, 1964a) concerns only the instantaneous development of folds from initial sinusoidal configurations or from initial irregularities in the layers that can be considered to be the sum of a harmonic series of sine and cosine waves of different wavelengths and amplitudes. One assumption of the theory is that the amplitudes of the initial sinusoidal waves or components of a harmonic series must be small (usually limb dips must be less than about 10–15°). One of the most important results of the buckling theory is the prediction of a dominant wavelength that is most likely to appear in a given system, the dominant wavelength being that of the sinusoidal wave which is initially amplified the most. Both Biot (1961) and Ramberg (1961) show that for a single viscous layer isolated in a less viscous medium, the ratio of dominant wavelength to thickness of the layer depends only on the ratio of the viscosities of the layer and the enclosing medium.

Ramberg (1964a, 1971) has derived expressions for the rate of overall shortening due to buckling and for that due to layer shortening. He was able to show quantitatively that the ratio of buckle to layer shortening increases both with viscosity contrast and with amplitude. Thus, in the development of a single fold shortening by buckling will become dominant at a certain fold amplitude (except at very low viscosity contrasts). The expressions given by Ramberg (1964a, 1971) are valid as long as the ratio amplitude/wavelength is less than about 0.1. The effects of layer shortening as a function of viscosity contrast and amplitude are also apparent from the equations for amplification developed by Biot (1961) and are most clearly seen in graphed form (see Ramsay, 1967, fig. 7–37; Hudleston, 1973a, fig. 15). The simple expression for the dominant wavelength derived by Biot and Ramberg is not applicable at low viscosity contrasts, because no finite shortening of the layer is allowed for in their derivations. Sherwin and Chapple (1968) modified Biot's basic theory to take layer-parallel shortening into account. They showed that the dominant wavelength changes progressively as the layer is shortened and thickened and they derived expressions relating dominant wavelength, viscosity contrast, amplification and layer shortening (see Sherwin and Chapple, 1968, fig. 3). This modified theory allows more realistic estimates of viscosity contrast to be obtained for natural folds, and allows total shortening in the profile plane of the folds to be estimated (Sherwin and Chapple, 1968; Hudleston, 1973c). It is still only valid in so far as the folds are of fairly low amplitude (with limb dips less than about 20°, see Hudleston, 1973c, p. 124).

Recently, mathematical studies of fold development to high amplitudes with large finite deformations in viscous materials have been made by use of numerical methods of finite-difference analysis (Chapple, 1968) and finite-element analysis (Dieterich and Carter, 1969; Stephansson and Berner, 1971). Chapple (1968) analysed the development of folding in a thin (thickness/wavelength ratio of 0.025) inextensible viscous layer embedded in a less viscous medium, where the layer was initially in the form of a low-amplitude sine wave. Fold shape, and strain rate and finite-strain fields were computed for successive stages of fold development. Chapple showed that the fold shape depended only on the ratio of the actual wavelength (L), to the dominant wavelength (L_d) given by the basic theory of Biot (1961) or Ramberg (1961). He also showed that a probable limit to the wavelength-selection process, which operates during the initiation of folding according to the infinitesimal buckling theory (Biot, 1961), is reached when the folds attain limb dips of about $15°$.

Dieterich and Carter (1969) and Dieterich (1969) studied the development of folds to large amplitudes in viscous layers set in a less viscous medium by the application of finite-element analysis. An advantage that this technique affords over the finite-difference method is that restrictions of inextensibility and linear bending strain distribution across the layer need not be imposed. Computations were made with viscosity contrasts of 42/1 and 17.5/1 between layer and matrix. Finite-strain and stress fields were derived throughout the folding history for folds in layers that were initially in the form of low-amplitude sine waves at the dominant wavelength/thickness ratio. Layer shortening occurred in the early stages of folding and was greatest in the case of the lower viscosity contrast. At the higher viscosity contrast very slight variations in thickness with dip were detectable at high amplitudes, and at the lower viscosity contrast very pronounced thickening in the hinges and thinning in the limbs of the folds were apparent. Dieterich (1969) was able to show that the pattern of slaty cleavage found in many natural folds is more closely related to the finite-strain distribution than to the stress distribution at any stage. The cleavage appears to lie perpendicular to the maximum finite-shortening direction in the rock.

Stephansson and Berner (1971) studied the development of folds in a multilayered system using finite-element analysis. The multilayer consisted of five layers with relative viscosities 1/100/1/1,000/1 and thicknesses 9/1.5/5/1/9. The dominant wavelength for this system was derived by applying the theory developed by Ramberg (1970). Initially the two more competent layers had the form of low-amplitude sine waves at the dominant wavelength. Progressive fold shape and stress distribution were found to depend on the compressibility of the materials. For incompressible materials the competent layers were very slightly thickened in the hinges at high amplitudes, whereas for compressible materials ($\dot{e}_2/\dot{e}_1 = 0.33$) the competent layers actually show a thinning in the hinges.

AN APPROXIMATE ANALYTICAL MODEL TO SIMULATE FLATTENING DURING BUCKLING

Layer-parallel shortening during the early stages of buckling and subsequent non-uniform changes in arclength and thickness around a developing fold are an integral part of the buckling process at low viscosity contrasts. To gain some idea of the nature of these changes in thickness and limb dip, without taking other factors into account, a simple analogue model has been designed that considers buckling to consist of two processes, alternately adding small increments to the fold development. We consider a "buckling" process in which limb dip increases with no change in thickness or arclength (which is an unknown factor) and a "flattening" process in which the fold is subjected to a uniform compressive strain. The model is approximate because the two processes are interdependent components of a single folding mechanism (Bayly, 1971) and are therefore simplified. In the model, described in

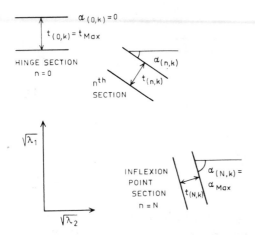

Fig. 1. Analogue model for the simultaneous buckling and flattening of a fold; three of the $N+1$ discrete sections of the fold after k increments of buckling and flattening. The quadratic elongation directions are those of both incremental and finite strains in the hinge section, and of incremental strains only in the other sections.

detail by Hudleston (1973a), the fold is represented by a large number $(N+1)$, of discrete sections between hinge and inflexion points (Fig. 1), each section defined solely by its dip, α, and its thickness, t. For each buckling increment, k (k odd), and for the n-th section ($1 \leqslant n \leqslant N$), α changes according to:

$$\alpha_{(n,k)} = \alpha_{(n,k-1)} \left[1 + \frac{\theta}{\alpha_{(N,k-1)}} \right] \tag{1a}$$

with a special case at the initiation of folding:

$$\alpha_{(n,1)} = \frac{\theta n}{N} \tag{1b}$$

where θ is the maximum rotation of the limb for each buckling increment. Thickness is kept constant so that:

$$t_{(n,k)} = t_{(n,k-1)} \qquad (0 \leqslant n \leqslant N) \tag{2a}$$

and

$$t_{(n,1)} = 1.0 \qquad (0 \leqslant n \leqslant N) \tag{2b}$$

For each flattening increment (k even) α and t change according to:

$$\alpha_{(n,k)} = \tan^{-1} \left[(\lambda_1/\lambda_2)^{\frac{1}{2}} \tan \alpha_{(n,k-1)} \right] \tag{3}$$

$$t_{(n,k)} = \frac{t_{(n,k-1)} (\lambda_1)^{\frac{1}{2}} \cos \alpha_{(n,k)}}{\cos \alpha_{(n,k-1)}} \tag{4a}$$

and

$$t_{(0,k)} = t_{(0,k-1)} (\lambda_1)^{\frac{1}{2}} \tag{4b}$$

where λ_1 and λ_2 are the incremental quadratic elongations (see Ramsay, 1967, eq. 3–34 and 7–30). In the calculations N was taken as 50 and a range of values of θ and $(\lambda_1/\lambda_2)^{1/2}$ were used to vary the relative amounts of buckling to flattening. Eq. 1–4 were solved for successive increments until the maximum limb dip reached about $90°$; the total number of increments, k, the finite strain at the hinge and final values of t and α were recorded. Input data are shown in Table I and results in Fig. 2.

The resultant fold geometry varies between parallel (almost no flattening) and similar (large components of flattening) and differs systematically from the geometry of flattened parallel folds, very slightly at low limb dips and most markedly at high dips. These differences are apparent on plots of t'_α against α (Fig. 2).

TABLE I

Data for model simulating flattening during buckling. Input data for the analysis are listed in columns 2 and 3, results in columns 3–5. Curve numbers in column 1 refer to Fig. 2.

Curve no.	Increment of flattening $((\lambda_2/\lambda_1)^{1/2})$	Maximum increment of buckling in degrees (θ)	Number of increments (k)	Total contribution to limb rotation by buckling in degrees	Finite strain at the hinge $((\lambda_2/\lambda_1)^{1/2})$
1	0.998	1	173	86	0.84
2	0.990	2	83	80	0.66
3	0.990	1	153	75	0.47
4	0.900	5	27	60	0.25
5	0.990	0.25	437	54.5	0.113
6	0.990	0.125	655	41	0.038
7	0.900	0.5	95	23	0.007

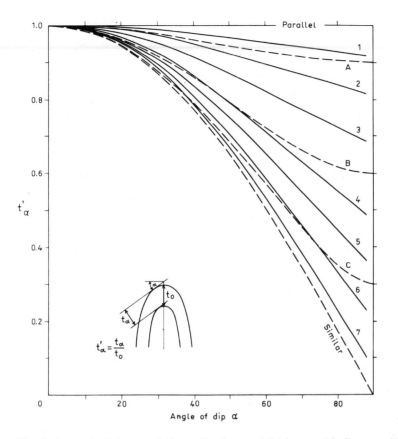

Fig. 2. Curves 1–7 show variations of orthogonal thickness with dip according to the analogue model of simultaneous buckling and flattening (see Table I). Curves *A, B* and *C* are for an initially parallel fold flattened by strains of $(\lambda_2/\lambda_1)^{1/2}$ = 0.75, 0.5, and 0.25 respectively.

EXPERIMENTAL RESULTS

Hudleston (1973b) described a series of buckling experiments using single viscous layers embedded in a less viscous matrix, with variable viscosity contrast (expressed as the ratio μ_1/μ_2) between layer (viscosity μ_1) and matrix (viscosity μ_2). The materials used were solutions of ethyl cellulose in benzyl alcohol, and the experiments were carried out under conditions of plane strain and pure shear. For viscosity contrasts between 24/1 and 100/1 significant folds formed by buckling at values of overall shortening well within the limits imposed by the apparatus (about 60%). The folds had a characteristic wavelength which depended on the viscosity ratio. The mean maximum limb dips of folds for individual experimental runs varied between about 60° (for μ_1/μ_2 = 24) to about 80° (for μ_1/μ_2 = 100). In all experiments layer-parallel shortening took place until folds attained limb dips of about 15–25° after which only very slight changes in overall arclength occurred, and no significant variations of thickness with dip could be detected in any of the experiments even at the most advanced stages of fold development (Hudleston, 1973b). For experiments at a viscosity contrast of 11/1, folds developed by buckling after substantial layer-parallel shortening but did not attain amplitudes that were high enough for possible variations of thickness with dip to become apparent at the limits of shortening fixed by the apparatus. Mean maximum limb dips of about 15–25° were recorded in these experiments with total layer shortening of about 50–60%. Variations of arclength and limb dip with overall shortening are shown in Fig. 3 for selected experiments at each of the four viscosity contrasts. It was not found possible to calculate amplification because the initial amplitudes were too small to be measured accurately. Infinitesimal-amplitude buckling theory predicts that fold amplitude should increase exponentially with time (Biot, 1961) or with shortening if an expression for this replaces time in the theoretical expressions (Sherwin and Chapple, 1968). However, it is clear that amplitude and limb dip are both limited by the geometry of the folding at advanced stages of fold development. This geometrical constraint may be seen qualitatively in Fig. 3. At the two lower viscosity contrasts the slope of the graph of limb dip against shortening increases continuously with shortening, whereas at the two higher viscosity contrasts the slope of this graph has a maximum value at some stage of fold development and then decreases. This effect is most pronounced for $\mu_1/\mu_2 = 100$.

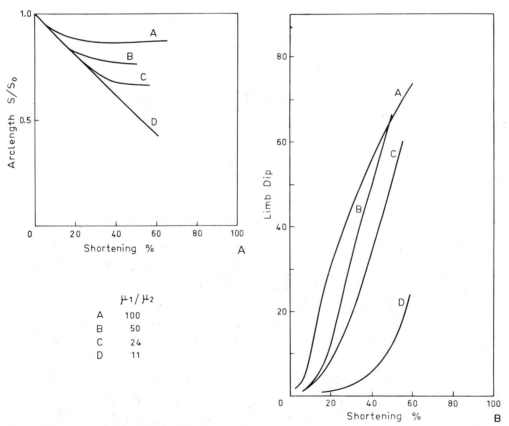

$$\begin{array}{cc} & \mu_1/\mu_2 \\ A & 100 \\ B & 50 \\ C & 24 \\ D & 11 \end{array}$$

Fig. 3. Changes of arclength, S, and limb dip with total shortening in selected experiments at different viscosity contrasts. S_0 is the initial arclength.

For a viscosity contrast of about 5/1 fold generation by buckling was barely apparent at the limit of shortening of the apparatus, and it was not found possible to check whether a characteristic wavelength was appearing. Fold development to higher amplitudes was studied by starting deformation with fairly open folds instead of a flat layer. The initial wavelengths were not constant and were considerably greater than the predicted dominant wavelength for this viscosity contrast. The initial and final configurations of a layer in one experiment are shown in Fig. 4. Initially the folds were parallel and almost sinusoidal in shape; during deformation both arclength and thickness changed around the folds to produce thickening in the hinges and thinning in the limbs. On plots of t'_α against α a close similarity between the geometry of the experimentally produced folds and flattened parallel folds is

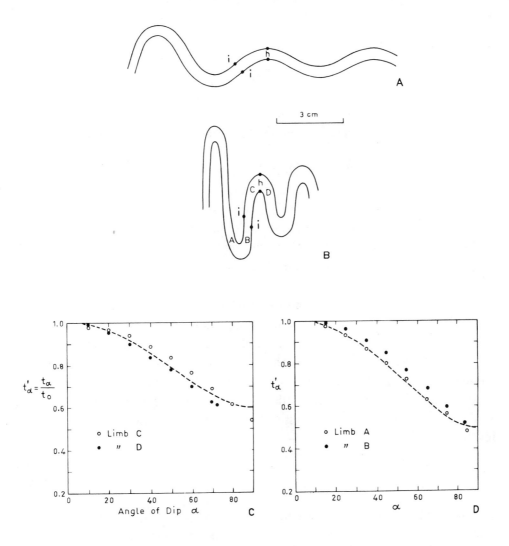

Fig. 4. Experimental deformation at a viscosity contrast of ca. 5/1. A. Initial configuration. B. Final configuration after a total shortening of about 50%. C. $t'_\alpha = t_\alpha/t_0$ plotted against angle of dip, α, for fold limbs C and D. D. t'_α against α for fold limbs A and B.

apparent, and an even closer similarity between the geometry of these folds and the geometry of folds formed by flattening during buckling predicted by the analytical model described in the previous section.

It is of interest here to briefly mention the results of some multilayer buckling experiments. It appears that significant folding may develop in laterally compressed multilayered models where rheological contrast (usually a contrast of viscosities) is very slight between layers. Folding may occur at fairly low values of total shortening (e.g., Ghosh, 1968) compared, for example, with the amount of layer shortening undergone before significant folds appear in a single isolated viscous layer ten times more viscous than the surrounding medium. At low viscosity contrasts, ease of slip between layers and layer anisotropy appear to be important factors controlling fold development (Ghosh, 1968; Bayly, 1971). Thickening in fold hinges in both competent and incompetent layers alternating in a multilayered sequence is commonly observed in experiments (see Ghosh, 1968, fig. 14, 4). Away from the ends of the multilayered unit the overall fold geometry is usually similar, and the geometry of individual layers within the unit may approach the similar model too. Bayly (1971) argues that nearly perfectly similar folds may develop by buckling in a multilayered system with the right amount of layer anisotropy. It would thus appear to be considerably simpler to produce similar folds in a multilayered system than in isolated single viscous layers.

FOLD-SHAPE DEVELOPMENT STUDIED BY MEANS OF THE FINITE-ELEMENT METHOD

The finite-element method may be used to determine stresses and either displacements in elastic bodies or displacement rates in viscous bodies for which exact analytical solutions cannot be obtained. Descriptions of the method and applications to geological problems are given by Dieterich and Onat (1969), Dieterich and Carter (1969), Voight and Dahl (1970), Stephansson and Berner (1971) and Berner et al. (1972).

The body to be analysed is partitioned into a finite number of interconnected elements. All forces and displacements (boundary conditions) applied to the body are replaced by statically equivalent forces and displacements acting at the nodal points of the elements. The nodal point forces and displacements are related by the "stiffness" of each element (Stephansson and Berner, 1971). The stiffness is a constant which depends on the geometry and material nature of the element and upon whether the analysis is performed in plane stress or plane strain. In the present study the stress and strain rate tensors are related by the viscous flow law:

$$\sigma_{ij} = 2\mu\dot{\epsilon}_{ij} + \lambda\dot{\epsilon}_{kk}\delta_{ij} \tag{5}$$

where σ_{ij} are the components of the stress tensor, $\dot{\epsilon}_{ij}$ the components of the strain rate tensor and δ_{ij} is the Kronecker delta. μ and λ are viscous constants analogous to Lame's parameters for elastic solids. λ/μ can be written in terms of $\dot{e}_2/\dot{e}_1$ which in effect expresses the compressibility of the viscous material, where $\dot{e}_2$ and $\dot{e}_1$ are the rates of lateral extension and longitudinal contraction in the material undergoing uniaxial compression (cf. Poisson's ratio for an elastic material).

The initial configuration of all the models studied here consisted of a viscous layer, in the form of a low amplitude (limb dip 3°) sine wave of one half wavelength, embedded in a less viscous medium. The boundaries of the model were two adjacent axial surfaces (which remain plane and parallel throughout deformation for reasons of symmetry), and two free surfaces initially normal to the axial surfaces at a distance of one wavelength either side of the layer. The fold model was subjected to a constant shortening strain rate across the axial surfaces. To satisfy symmetry requirements, nodal points situated along the axial surfaces were only free to move along that surface. The nodal point at the inflexion point in the middle of the stiff layer was kept fixed in position. In contrast to the models studied by Dieterich and Carter (1969), the full thickness of the buckling layer was studied. The model system consisted of 384 quadrilateral elements (425 nodal points) each split up into four triangular elements for the computations. The nodal points were more closely spaced near the boundaries between the layer and the embedding medium, and each layer boundary was represented by lines connecting 17 nodal points.

The finite continuous deformation is treated as a series of small quasi-static incremental deformations. At each successive configuration the model is subjected to a small shortening increment between the fold axial surfaces. The displacements of all the nodal points may then be found that minimises the strain energy in the system for this increment of shortening. These small displacements give rise to a new configuration of the nodal points and the process is repeated.

In the various models the initial wavelength and amplitude were kept constant and the thickness of the buckling layer and viscosity ratio μ_1/μ_2 were varied. Computations were made for compressible materials, compressibility expressed by $\dot{e}_2/\dot{e}_1 = 0.4$, corresponding to a value of 4 for λ/μ in the constitutive equation (5).

Results for a viscosity ratio $\mu_1/\mu_2 = 10$

Three models were studied for a viscosity contrast between layer and matrix given by $\mu_1/\mu_2 = 10$, where μ_1 is the viscosity of the layer. In one model the wavelength/thickness ratio, L/h, was 7.4, that predicted for the dominant folds at this viscosity contrast by the theoretical expression of Biot (1961) and Ramberg (1961). In the other two models the L/h ratios were taken as 4.0 and 15.1. Progressive finite strain in the models was built up as the sum of a series of small incremental deformations, with increments of 2.5% overall shortening for the dominant wavelength model and 5% for the other models.

The progressive growth of the folds is shown in Fig. 5, from which it is apparent that fold shape depends upon the L/h ratio. Flattening dominates the thick-layer model, whereas a large amount of buckle shortening accompanies the deformation of the thin layer. Plots of t'_α against α at different values of shortening for the dominant wavelength model are shown in Fig. 6A. Here the fold geometry changes systematically from parallel to become more similar as the shortening increases. The curve relating t'_α to α for the fold at 80% shortening is almost identical with curve 5 of the analogue model of simultaneous buckling and flattening (see Fig. 2). Thickness–dip relationships for the thin-layer and thick-layer folds at 80% shortening are shown in Fig. 6B; these are quite similar to those produced by uniformly flattening a parallel fold by strains of $(\lambda_2/\lambda_1)^{1/2}$ = 0.7 and 0.2 respectively; and are also similar to members of the family of curves derived for the analogue model of buckling and flattening (see curves 3 and 7 in Fig. 2). It is difficult to see which gives a better fit. The changes in limb dip, amplitude and arclength with shortening in the three models are recorded in Fig. 7. The increases in limb dip are nearly the same in the dominant wavelength and thin-layer models, and much less in the thick-layer model, where the limb dip at any stage is not much greater than that in a passive fold (for which μ_1/μ_2 = 1) of the same initial amplitude/wavelength ratio. Changes in arclength are initially identical to those of a flat plate; such changes would be represented on the graph in Fig. 7 by a straight line joining the points (0,1) and (100,0). In the dominant wavelength and thin-layer folds arclength changes diverge from this straight line and become very slight after shortenings of about 50 − 60%. In the thick-layer fold changes in arclength depart much less from those of a

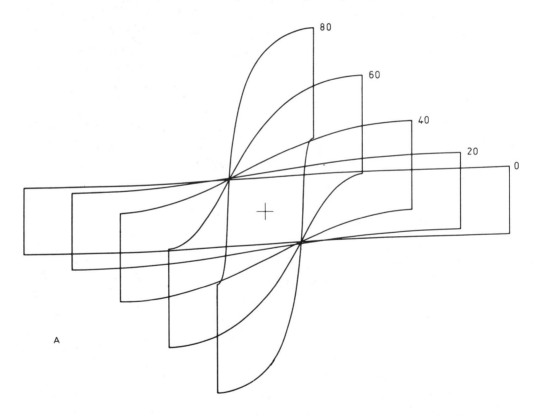

Fig. 5. Single-layer buckle folds produced by finite-element analysis. Progressive development of fold shape with overall shortening for models with viscosity ratio, μ_1/μ_2 = 10. A. Wavelength/thickness ratio L/h = 15.1. B. Dominant wavelength/thickness ratio, L_d/h = 7.4. C. Wavelength/thickness ratio less than the value given by the basic theory of Biot (1961) and Ramberg (1961), L/h = 4.

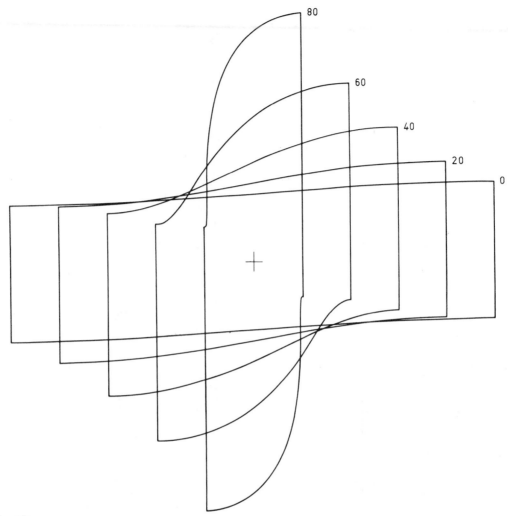

Fig. 5B.

flat plate. It is apparent that some mutually compensatory changes in length of arc in the hinge and limb regions of the dominant wavelength and thin-layer folds must be occurring because relative thickening in the hinges and thinning in the limbs is recorded (Fig. 6) for only slight changes in overall arclength (Fig. 7). It is apparent from Fig. 7 that the thick-layer fold undergoes changes in geometry not very different from those undergone by a passive fold. The thin-layer fold has behaved in the most "competent" fashion and the thick-layer fold in the least "competent" fashion.

The use of compressible materials in the analyses results in volume changes which are roughly proportional to the amount of shortening, and which vary with wavelength/thickness ratio of the competent layer, due to differences in the finite strain within the layer at any stage of shortening. The decreases in the volume of the buckling layer at 80% shortening in the three models studied here with L/h = 4, 7.4 and 15.0 are 49, 37 and 25% respectively. The effects of compressibility on fold geometry will be discussed later.

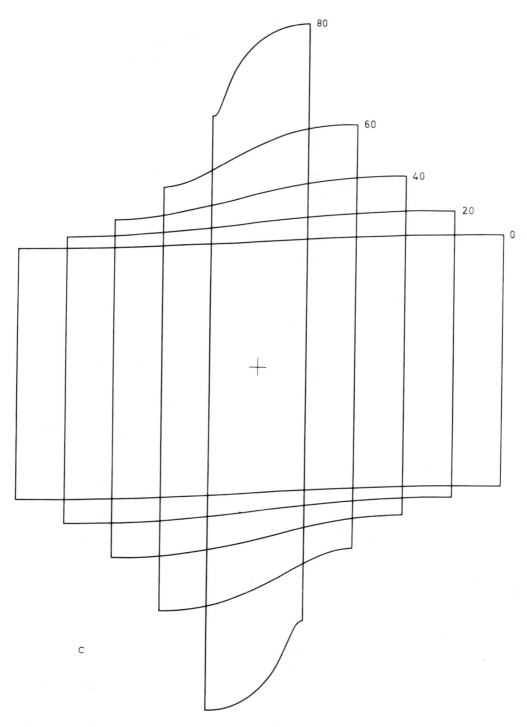

Fig. 5C. For legend see p. 310.

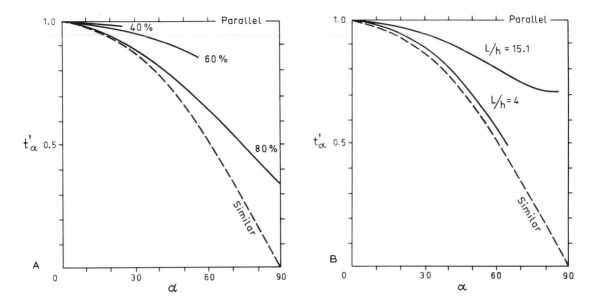

Fig. 6. Variation of orthogonal thickness t'_α with dip α for the finite-element models with $\mu_1/\mu_2 = 10$. A. For $L_d/h = 7.4$ for various values of shortening. The curve for 80% shortening is almost identical with curve 5 in Fig. 2 of the analogue model of simultaneous buckling and flattening. B. For $L/h = 4$ and 15.1 at 80% shortening.

Results for viscosity ratios $\mu_1/\mu_2 = 100$ and 1000

Keeping the initial wavelength and amplitude the same as for the previous models, the thickness of the buckling layer was adjusted to carry out analyses of fold development at the dominant wavelength for viscosity contrasts given by $\mu_1/\mu_2 = 100$ and 1,000. The progressive growth of folds in these two models is shown in Fig. 8. In both cases the folds maintain an almost parallel shape and follow a path of development similar to that traced by the dominant wavelength folds in a numerical study made by Chapple (1968) on fold development in viscous materials. A slight relative thinning in the hinge regions of the fold in both models is apparent, and is due to the use of a compressible relationship in the constitutive equations. Changes in arclength, amplitude and limb dip are recorded in Fig. 7. The geometrical constraint (noted in the experimental work on buckling) imposed on the increases in limb dip and amplitude is apparent in Fig. 7, in which the slopes of the curves recording changes of limb dip or amplitude with shortening attain a maximum value at an early stage of shortening and then decrease. For the model with $\mu_1/\mu_2 = 100$ layer-parallel shortening occurred until the fold attained a limb dip of about 35°, at a shortening of 20%, after which very slight increases in arclength occurred. In fact, only slight changes in arclength occurred after the fold had attained a limb dip of about 25°. For the model with $\mu_1/\mu_2 = 1,000$ no stage of arclength decrease was noted, and very slight increases accompanied shortening.

One factor that should be taken into account before any further discussion of the results

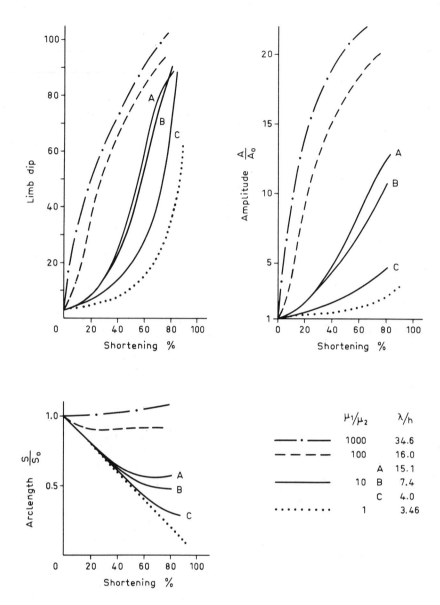

Fig. 7. Changes of limb dip, amplitude and arclength with overall shortening for finite-element models at different viscosity contrasts and wavelength/thickness ratios. A_0 and S_0 are the initial amplitude and arclength.

of the finite-element analyses is the effect on these results of having employed a compressible viscous flow law in the computations. Such a law was utilized for the sake of simplifying the computational procedure (see Dieterich and Onat, 1969); it would be more realistic to assume that rock undergoing deformation behaves in an essentially incompressible fashion. Dieterich and Onat (1969) note that if λ/μ is large with respect to unity the rate of change of volume caused by any given average normal stress is much smaller than the rate of shear strain caused by a shear stress of the same magnitude. Thus approximate solutions to incompressible problems may be obtained by using a value for λ/μ of much greater than unity. Dieterich and Onat use a value of 4 for λ/μ, and this was also the value employed in this study. The effect of employing compressible materials will be to give rise to a rate of volume change which will depend on the value of the mean normal stress, and to finite-volume changes which will vary with the finite strain. In these fold models, the variations in volume change throughout any model and hence the effects of compressibility on fold geometry will decrease with the viscosity contrast between layer and matrix. The most important effect observed in the models studied here is an apparent relative thinning in the hinge regions of the folds compared with the limbs, for $\mu_1/\mu_2 = 100$ and $1,000$. Folds developed in incompressible materials show either no variation in thickness or slight thickening in the hinge regions compared to the limbs. The least effects of compressibility on fold geometry is for $\mu_1/\mu_2 = 10$. Changes of arclength, amplitude and limb dip with shortening will be

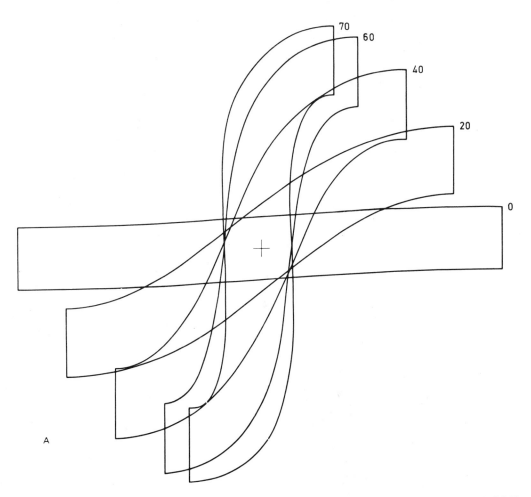

Fig. 8. Single-layer buckle folds produced by finite-element analysis. Progressive development of fold shape with overall shortening. The initial wavelength/thickness ratio is taken from the basic equation of Biot (1961) and Ramberg (1961). A. Viscosity ratio $\mu_1/\mu_2 = 100$. B. Viscosity ratio $\mu_1/\mu_2 = 1,000$.

similar to changes undergone by incompressible materials, and thickness variation with dip for $\mu_1/\mu_2 = 10$, will be similar for compressible and incompressible cases.

NATURAL FOLDS

Nature provides many examples of folded veins or layers embedded in a fairly uniform host rock, such as folded sandstone layers in a shale sequence, folded quartz veins in a uniform sedimentary sequence or ptygmatic veins in a granite gneiss or migmatite. The periodic nature of these folds is well known, and a linear relationship between the arclength of one complete wave and thickness for folds in isolated layers has been noted by several authors (e.g., Ramberg, 1963; Ramsay, 1967). It has been argued that the wavelength of the folds at the time of their initiation is preserved as the arclength of the mature folds at high amplitudes. Single-layer buckling theory predicts the appearance of a dominant wavelength in a layer of uniform thickness and a linear relationship between wavelength (of the low-am-

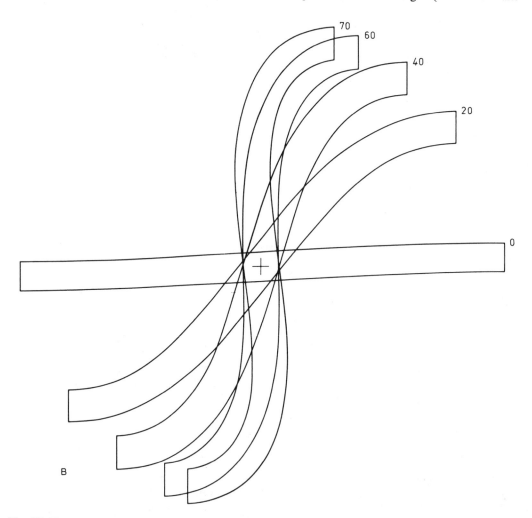

Fig. 8B. For legend see p. 315.

plitude folds) and thickness for viscous materials. Interpretation of small-scale natural folds in terms of this theory have been made by Ramberg (1963), Ramsay (1967), Sherwin and Chapple (1968) and Hudleston (1973c). By modifying the basic buckling theory to account for layer shortening Sherwin and Chapple (1968) were able to estimate the viscosity contrast between layer and host rock and the total shortening within the profile plane of the folds for quartz veins in phyllite. At the low viscosity contrasts frequently met with in rocks, it is apparent from the work of Sherwin and Chapple (1968) and Hudleston (1973b, 1973c) that the arclength observed in the mature folds in rock is not that initially amplified the most according to the basic Biot (1961) or Ramberg (1961) theory. Hudleston (1973b) argued that the modified wavelength-selection process described by Sherwin and Chapple (1968) operates until the folds attain limb dips of about $15 - 20°$, after which for practical purposes, layer shortening ceases and the arclength is "fixed", so that the arclength in mature folds at high amplitudes represents the wavelength at the $15 - 20°$ limb dip stage.

In many examples of folds in isolated competent layers or veins no systematic variations of thickness with limb dip are apparent, and the folds are approximately parallel in form (see Ramberg, 1963; Sherwin and Chapple, 1968). Hudleston (1973c) described ptygmatic folds in pegmatitic veins in which variations of thickness with limb dip were apparent. Part of a folded vein is shown in profile section in Fig. 9, and thickness variations with dip are recorded for several of the folds shown. Nearly all the folds are geometrically similar to a parallel fold that has been flattened by a strain of between $(\lambda_2/\lambda_1)^{1/2} = 0.4$ and $(\lambda_2/\lambda_1)^{1/2} = 0.5$. Surface irregularities in the veins are probably responsible for some of the variation in fold geometry apparent in plots of t'_α against α (see Fig. 9, fold E, limb 6). In some folds thickness variations with dip are almost identical to those produced by flattening a parallel fold (Fig. 9, fold C, limb 4; fold C, limb 3). Several limbs (Fig. 9, limbs $1, 2, 5$) show thickness/dip relations very similar to those predicted by the simple analogue model of flattening during buckling described above (see Fig. 2) and observed in experiments where flattening accompanies buckling (see Fig. 4). The folds were probably formed by a process of buckling at a low viscosity contrast. It is possible that a fairly uniform flattening affected the folds at a late stage of development when the folds had become tightly appressed and resistance to buckling was high as material was being squeezed out from between the limbs (see Chapple, 1968).

DISCUSSION AND CONCLUSIONS

The experimental results, the simple analogue model of buckling and flattening and the finite-element model results are all consistent in their records of progressive changes in geometry during fold development by single-layer buckling. In contrast to the situation observed in the experiments, where no significant variations of t'_α with α were detected in any of the folds at viscosity contrasts varying between $\mu_1/\mu_2 = 100$ and $\mu_1/\mu_2 = 11$, the finite-element analyses indicated that for $\mu_1/\mu_2 = 10$ variations of t'_α with α appear as folding develops, and that their nature depends on the amount of shortening and on the wavelength/

thickness ratio. The thickness variations with dip are similar to those predicted by the simple analogue model of simultaneous buckling and flattening, and to those observed in folds developed in experiments with $\mu_1/\mu_2 \approx 5$ from initial sinusoidal deflections of large L/h. These variations appear to be characteristic of buckling at low viscosity contrasts between layer and matrix. A comparison between thickness/dip variations produced in this way and those produced by uniformly flattening a parallel fold shows that differences are marked at high limb dips and very slight at low limb dips, such that it may be impossible to distinguish between the two (see Fig. 2 and 6). For the experiments at $\mu_1/\mu_2 = 11$ it seems probable that fold development was not mature enough for t'_α variations with α to become apparent. It also seems probable that thickness/dip variations during folding are not important for $\mu_1/\mu_2 > 24$. Layer-parallel shortening before significant folding was apparent in all the experiments and in the finite-element models, except for $\mu_1/\mu_2 = 1,000$. Where initial layer-parallel shortening has taken place, the folding history may be

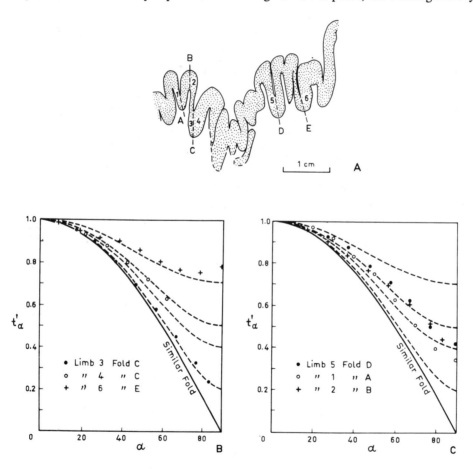

Fig. 9. Ptygmatic fold profile in a pegmatitic vein in biotite schist from Monar, Scotland. Dashed lines on the profile are datum lines for measuring orthogonal thickness, t, and dip, α. $t'_\alpha = t_\alpha/t_0$ is plotted against α for 6 fold limbs of the profile. Dotted lines in these graphs are plots for parallel folds flattened by strains of $(\lambda_2/\lambda_1)^{1/2} = 0.2, 0.4, 0.5$ and 0.7.

approximated by a stage where the layer shortens as a flat plate followed by a stage where only slight changes in arclength occur and, for high viscosity contrasts, the fold retains the geometry of a parallel fold to high amplitudes (see Hudleston, 1973b). This is not the case for $\mu_1/\mu_2 = 10$. For thin-layer and dominant-wavelength folds the second stage of fold development, marked by little overall change in arclength, is less well pronounced and is accompanied by mutually compensatory changes in arclength in the hinge and limb regions of the fold and the appearance of thickness variations with dip. With further shortening, greater than that attained in the experiments or the finite-element models it seems clear that the arclength must eventually begin to increase considerably as elongation of the limbs becomes dominant over continued shortening in the hinge regions. This is a stage of fold development not considered here, although presumably important at very great shortenings.

It should be noted that the form of the curves recording limb dip, arclength and amplitude changes with shortening (Fig. 3, 7) are dependent on the size of the initial irregularities or initial sinusoidal deflections in the layer. For example, with a much lower amplitude initial sine wave in the finite-element models the near-horizontal segments of all the curves relating arclength to shortening would be depressed and all the curves recording increase of limb dip with shortening would be pushed to the right. Although amplification (A/A_0) depends only on shortening for a particular viscosity contrast in the infinitesimal-amplitude buckling theory (Biot, 1961), the form of the curves drawn in Fig. 7 will change with the amplitude of the initial sine wave due to departure from the constraints of the theory when limb dips of more than a few degrees are attained. With sufficiently small irregularities, amplification curves would approach the theoretical curves (see Hudleston, 1973a). Curve B for the dominant wavelength fold in Fig. 7 is fairly similar to the theoretical curve for amplification according to the infinitesimal-amplitude theory (see Ramsay, 1967, fig. 7–37) for low values of shortening. As μ_1/μ_2 is increased the departure of the observed amplification curves from the theoretical curves becomes more pronounced. It is apparent from Fig. 7 that the thin-layer fold for $\mu_1/\mu_2 = 10$ grows more rapidly than does the "dominant wavelength" fold. The reason for this is that the dominant wavelength continuously varies with shortening (Sherwin and Chapple, 1968) until some finite stage of folding is reached (perhaps when the folds have attained limb dips of $15 - 25°$), when the process of selection of the dominant wavelength effectively ceases (see Chapple, 1968; Hudleston, 1973b). At quite an early stage in the shortening history, the amplification of the fold with L/h initially 15 overtakes that of the fold with L/h initially 7.4, the instantaneous dominant wavelength.

The experiments and various models described in earlier sections of the paper indicate that thickness variation with dip may be expected in natural folds produced by buckling at low viscosity contrasts between layer and host rock. The resultant fold geometry may vary between a parallel and a near-similar fold model, depending upon the viscosity ratio μ_1/μ_2, the degree of shortening and the wavelength/thickness ratio. The thickness/dip variations should be similar to but in general distinguishable from those produced by flattening a parallel fold, especially at high limb dips. Studies of natural folds (Sherwin and Chapple, 1968; Hudleston,

1973c) suggest that viscosity ratios are often rather low in situations where folds develop in isolated single layers, and the geometry of some of these folds closely resembles that of a flattened parallel fold or that of a fold developed both experimentally and in finite-element models by buckling at low viscosity contrasts. If indeed it is possible for parallel folds to become uniformly flattened, it is not clear how this might be effected (see Hudleston, 1973a). The development of folds in single layers by buckling at low viscosity contrasts probably accounts for many of the "flattened parallel folds" that are commonly observed in nature, including some that are almost similar folds.

ACKNOWLEDGEMENTS

We would like to thank J.G. Ramsay and H. Ramberg for critically reading the manuscript. Part of this work was supported by the Swedish Board for Technical Development under contract 70-135/U797.

REFERENCES

Bayly, M.B., 1971. Similar folds, buckling and great-circle patterns. *J. Geol.*, 79: 110–118.
Berner, H., Ramberg, H. and Stephansson, O., 1972. Diapirism in theory and experiments. *Tectonophysics*, 15: 197–218.
Biot, M.A., 1957. Folding instability of a layered viscoelastic medium under compression. *Proc. R. Soc. (Lond.), Ser. A*, 242: 444–454.
Biot, M.A., 1961. Theory of folding of stratified viscoelastic media and its implication in tectonics and orogenesis. *Geol. Soc. Am. Bull.*, 72: 1595–1620.
Biot, M.A., 1965. *Mechanics of Incremental Deformations*. Wiley, New York, N.Y., 504 pp.
Chapple, W.M., 1968. A mathematical theory of finite-amplitude rock-folding. *Geol. Soc. Am. Bull.*, 79: 47–68.
De Sitter, L.U., 1968. *Structural Geology*, McGraw–Hill, New York, N.Y., 2nd. ed., 551 pp.
Dieterich, J.H., 1969. Origin of cleavage in folded rocks. *Am. J. Sci.*, 267: 155–165.
Dieterich, J.H. and Carter, N.L., 1969. Stress-history of folding. *Am. J. Sci.*, 267: 129–154.
Dieterich, J.H. and Onat, E.T., 1969. Slow finite deformations of viscous solids. *J. Geophys. Res.*, 74: 2081–2088.
Flinn, D., 1962. On folding during three-dimensional progressive deformation. *Q. J. Geol. Soc. Lond.*, 118: 385–433.
Ghosh, S.K., 1968. Experiments of buckling of multilayers which permit interlayer gliding. *Tectonophysics*, 6: 207–249.
Hills, E.S., 1963. *Elements of Structural Geology*. Methuen, London, 483 pp.
Hudleston, P.J., 1973a. Fold morphology and some geometrical implications of theories of fold development. *Tectonophysics*, 16: 1–46.
Hudleston, P.J., 1973b. An analysis of "single-layer" folds developed experimentally in viscous media. *Tectonophysics*, 16: 189–214.
Hudleston, P.J., 1973c. The analysis and interpretation of minor folds developed in the Moine rocks of Monar, Scotland. *Tectonophysics*, 17 (in press).
Mukhopadhyay, D., 1965. Effects of compression on concentric folds and mechanism of similar folding. *J. Geol. Soc. India*, 6: 27–41.

Ramberg, H., 1959. Evolution of ptygmatic folding. *Nor. Geol. Tidsskr.*, 39: 99–155.

Ramberg, H., 1961. Contact strain and folding instability of a multilayered body under compression. *Geol. Rundsch.*, 51: 405–439.

Ramberg, H., 1963. Strain distribution and geometry of folds. *Bull. Geol. Inst. Univ. Upps.*, 42: 1–20.

Ramberg, H., 1964a. Selective buckling of composite layers with contrasted rheological properties; a theory for simultaneous formation of several orders of folds. *Tectonophysics*, 1: 307–341.

Ramberg, H., 1964b. Note on model studies of folding of moraines in piedmont glaciers. *J. Glaciol.*, 5: 207–218.

Ramberg, H., 1970. Folding of laterally compressed multilayers in the field of gravity, I. *Phys. Earth Planet. Inter.*, 2: 203–232.

Ramberg, H., 1971. Folding of laterally compressed multilayers in the field of gravity, II. Numerical examples. *Phys. Earth Planet. Inter.*, 4: 83–120.

Ramsay, J.G., 1962. The geometry and mechanics of "similar" type folds. *J. Geol.*, 70: 309–327.

Ramsay, J.G., 1967. *Folding and Fracturing of Rocks*. McGraw-Hill, New York, N.Y., 568 pp.

Sherwin, J.-A. and Chapple, W.M., 1968. Wavelengths of single layer folds: a comparison between theory and observation. *Am. J. Sci.*, 266: 167–179.

Stephansson, O. and Berner, H., 1971. The finite-element method in tectonic processes. *Phys. Earth Planet. Inter.*, 4: 301–321.

Turner, F.J. and Weiss, L.E., 1963. *Structural Analysis of Metamorphic Tectonites*. McGraw-Hill, New York, N.Y., 545 pp.

Voight, B. and Dahl, H.D., 1970. Numerical continuum approaches to analysis of nonlinear rock deformation. *Can. J. Earth Sci.*, 7: 814–830.

Reprinted by permission of the Geological Society of America
from John G. Ramsay, *Geological Society of America Bulletin*, v. 85, no. 11 (1974), p. 1741-1754.

Development of Chevron Folds

JOHN G. RAMSAY   *Department of Earth Sciences, University of Leeds, Leeds LS2 9JT England*

ABSTRACT

Multilayered rock complexes with regular alternations of competent and incompetent layers of thickness t_1 and t_2, respectively, and with high ductility contrast form folds of the chevron style when subjected to compression along the layering. The geometric forms of progressively developing chevron folds are analyzed using a model whose properties are based on the geometric forms of naturally deformed rock layers. It is found that the chevron fold style is only stable where no strongly marked variations in competent layer thickness t_1 exists. The thickness of the incompetent layer exerts no influence on fold model stability. Slight variations of competent layer thickness can be accommodated by local modifications of the fold style, such as limb faults, bulbous hinge zones, or layer boudinage, but if any strongly marked variation exists, the fold limbs become curved. The chevron model involves dilation at the hinge zones, and saddle reef formation, incompetent layer flow into the hinge, or slow hinge collapse generally results from actual or potential dilation. The speed of development of chevron folds is calculated under conditions of constant stress and of constant load. Folding starts slowly but rapidly accelerates; the later stages are characterized by a progressive slowing in shortening rate and fold growth, leading either to a stage of locking up of the fold or to modification of its geometry by limb thinning and hinge thickening toward a more similar geometric style. The strains in the hinge zone region are related to the rates of shear taking place in the fold limbs; the geometric model is not completely stable throughout the fold development, and complex progressive strain increments can
occur in the hinge zones and lead to the development of superposed small-scale structures indicating reversals of principal axes of incremental strain. *Key words: structural geology, deformation, fractures, foliation, mathematical geology.*

INTRODUCTION

In recent years, the study of folds has been developed on a sound mechanical basis. Studies of simple situations with single layers of competent ductile material surrounded by a less competent matrix have now been extended into more complex situations involving multilayered complexes of different layer competence. The works of Ramberg (1959, 1961, 1964), Biot (1964, 1965), Chapple (1968, 1969), and Cobbold and others (1971) have been outstanding in this field. They have mostly analyzed the mechanics of the initial instability during fold formation. These studies have made a great impact on the understanding of folding, but structural geologists are also interested in the later stages of fold development and in how the geometric forms of the initial folds become modified as deformation proceeds and more shortening is taken up across the ends of the layers. Although mathematical solutions for the geometric forms of the initial fold forms are now available, there are, as yet, none available for the later stages of fold growth. However, numerical solutions to these problems have evolved from experiments: first, with layered model materials with the model used as a scaled analog computer (Currie and others, 1962; Ghosh,

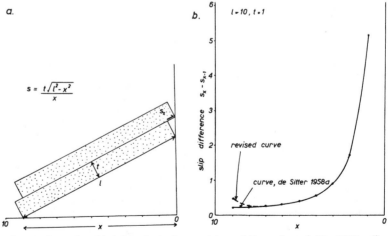

Figure 1. Geometric model of progressive development of chevron folds according to de Sitter (1958). a. Slip *s* that takes place between two adjacent layers of thickness *t* and limb length *l*, which are situated on limb of developing chevron fold. b. Amount of incremental slip for 10 percent increments of shortening.

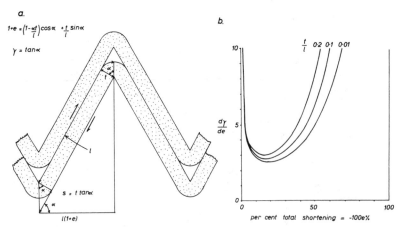

Figure 2. Flexural slip model of chevron folds (Ramsay, 1967). a. Slip s taking place between two adjacent competent layers of original limb length l as result of chevron folding to limb dip α with total layer shortening of $1 + e$. b. Changes in slip with shortening at different amounts of total shortening.

1966, 1968; Cobbold and others, 1971); and second, with numerical analysis using finite element techniques (Dieterich and Carter, 1969; Dieterich, 1970; Stephansson and Berner, 1971), which employ a digital computer to grow folds in incremental stages according to given laws of material flow. The geometric results of these model experiments can be compared with those of naturally deformed rock layers. If the cross section layer geometry and finite strain states are closely similar, then we might deduce that the rock behaved like the materials of the model, but it is possible to get the same approximate geometry from different materials undergoing different deformation histories. However, to many who have made detailed studies of naturally deformed rock layers, the correspondences between the model work and natural process are extremely close. I, for one, am convinced that this geometric correspondence is not just by chance.

Although in general it is not possible to find the exact equations that account for progressive fold growth, field geologists know that certain styles of fold structure keep recurring, and some of them are geometrically fairly simple. Because of their simplicity, one is tempted to try to discover the functions that might express their evolutionary development. The chevron fold style is such an example, and in this paper, I attempt to make a geometric analysis of the progressive development of such folds.

The concept behind this study is to set up a geometric model based on controls imposed by observations on naturally deformed, layered rocks, and then to investigate the properties of this model to see if any of its special geometric features help to aid an understanding of the geometric characteristics of naturally deformed rocks. De Sitter (1956, 1958) was one of the first to successfully develop this approach to the

study of chevron folds. His model consists of layers of constant thickness, sliding one over the other without internal deformation but losing cohesion along the fold axial surface, thereby producing a series of dilation openings along the hinge zone (see Fig. 1a). His investigations of the development potential of this model showed that, for progressive shortening increments of 10 percent along the direction of the initial layering, the slip increments increased greatly with the amount of total shortening (Fig. 1b). This deduction led him to consider the possibility that the frictional resistance along the slip surfaces on the fold limbs might be too large to accommodate the progressively large increase in slip necessary for the structures to continue developing; the folds would become progressively more difficult to amplify, and there would be some critical interlimb angle (he suggested 60°) where the fold would become locked.

Replotting his graphical data shows that the first points on the increment curve (Fig. 1b) calculated from the amount of slip over the first 10 percent of shortening should be higher than that of the next increment between 10 and 20 percent shortening (Fig. 1b, revised curve). Differentiation of the function used to plot Figure 1b to determine the rate of increase of slip with shortening shows that the slope of the slip function is negative during the first stages of folding, and that a minimum value exists at some finite shortening, depending upon the values of t and l. At higher shortenings, the slip derivative has a positive slope that progressively increases with shortening, as is apparent from de Sitter's graph. As a result of this recalculation, I decided to investigate further the change in slip rate with shortening, and I modified de Sitter's model before doing this (Ramsay, 1967, p. 440–447). The geometric model chosen (Fig. 2a) was made to accord more exactly with the conditions generally observed in

natural chevron folds, so that the layer thickness of the competent beds around the fold hinge is approximately constant. This feature suggested that the competent layers on the fold limb were subject to some internal deformation by simple shear parallel to the bedding surfaces. Analysis of this model showed the following features:

1. Straight-limbed chevron folds are not geometrically stable where pronounced thickness variations occur in folds having constant limb length. This explains why chevron folds are generally localized in regularly bedded multilayer sequences such as are found, for example, in typical turbidite flysch deposits, and it also leads to an explanation of why the ratio of limb length to thickness of the competent layers generally exceeds 10/1.

2. An analysis of the changes of shear strain with progressive shortening (Fig. 2b) confirmed de Sitter's ideas of increasing slip during the latter part of the fold development, and an evaluation of the values of shear stress acting along the folded layers showed that the stress decreased as the fold reached the later stages of formation. The fold must therefore lock up at some angle determined by the frictional properties of the material separating the competent layers.

3. Inside the competent layers, the rates of shear strain are very high at the start of folding and subsequently decrease. There is, therefore, a critical threshold that must be overcome before folding of the chevron style can begin to develop. For any given compressive stress deviator acting along the layer, the threshold is a function of dip; certain minimum dip values must be attained (probably by buckling instability) before folding goes on in the manner suggested by the model analysis. The threshold of folding is also controlled by the internal flow properties of the competent layers and is not dependent upon the frictional effects along surfaces separating the competent layers. As the folds evolve, they become progressively easy to form; easiest growth occurs when the interlimb angles have values between 140° and 100°, the exact value depending upon the ratio of thickness to limb length. Taken with the conclusions in 2 above, these features explain why open chevron folds are very uncommon and why it is usual to find natural chevron folds with interlimb angles locked around 60°.

The following discussion extends these previous studies, but its philosophy of approach is the same: to choose what appears to be a geologically realistic model of natural chevron folds, to deduce properties inherent in this model but that were not primary conditions for setting up the model, and to relate these deduced properties to actually observed situations.

The geometric model used below is slightly more elaborate than those used previously, in order to accord more exactly with situations observed in nature. In particular, the effects are calculated of variably thick layers of incompetent material situated between the competent layers. The geometric predictions from this new model will be developed to analyze the following problems: (1) stability of the chevron fold style, with particular reference to variations in layer thickness; (2) dilation problems around the hinge zone, and how these might modify the fold style in this region; (3) strain history and speed of development of chevron folds, based on the flow properties of the materials; (4) relation of flow in the fold limbs to flow pattern at the hinge, and in particular the nature of the structures formed at the hinge zone during progressive fold development.

GEOMETRIC BASIS FOR A NEW MODEL FOR CHEVRON FOLDS

The following features usually shown by naturally formed chevron folds were used to set up the conditions for the model:

1. The fold limbs are straight, and the length of the layers involved in the hinge zone is generally short compared with the layer length in the fold limbs.

2. Chevron folds occur in regularly bedded sequences containing two alternating rock types. The more competent layers in the alternating sequence show a fairly constant thickness, and the ratio of thickness of competent layers to limb length is generally less than 1/10 and often very much less than this ratio. The fairly common occurrence of shear and tensile fractures in the more competent layers that are infilled with fibrous crystalline vein material during deformation suggests that the strain rates involved during the folding were enough to bring the more competent rock into the region of brittle-ductile transition. The more ductile layers may or may not show these features.

3. The folded competent layers have a constant thickness throughout the structure, even around the hinge zone, and individual bedding plane markers within a competent unit keep a consistent orthogonal spacing around the fold hinges (Fig. 3). The component layers within a competent bed, therefore, accord closely with the parallel or concentric fold model (Ramsay, 1967, p. 365 — convergent isogon fold with style 1B). This feature suggests that internal deformation within the competent layer is probably taken up mostly by shear parallel to the layer boundaries and not by tangential longitudinal strain.

4. The less competent layers usually have a constant thickness along the fold limbs but often show marked variations of thickness in the hinge zones (Fig. 3), generally of the divergent isogon fold style (Ramsay, 1967, p. 365).

5. The less competent layers frequently show high internal strains by ductile flow, but the two-dimensional strains within the bedding surfaces situated on the fold limbs are generally low; these features suggest that the predominant deformation of these less competent layers on the fold limbs is one of simple shear. Slickensides and growth fiber veins (Durney and Ramsay, 1973) are common on the fold limbs and indicate that differential displacement has predominantly taken place in a direction perpendicular to the fold hinge lines. The competent layers often show only very low internal deformation.

6. Saddle reefs are common in the incompetent material located at the fold hinges.

The geometric model of the fold profile that best incorporates these features is shown in Figure 4. The model is one of plane strain with no displacements perpendicular to the profile section. Competent layers are folded parallel to symmetric folds with limb dips of $\pm\alpha$. The arc lengths within the competent layers remain of unchanged length l between two adjacent fold hinges. The thickness t_1 remains constant throughout the fold, and internal deformation is developed entirely by displacement parallel to bedding surfaces, and therefore distortion is by simple shear. The incompetent layers are also assumed to be folded without change in their initial thickness (t_2) on the fold limbs. Dilation occurs in both competent and incompetent layers as a result of the folding, and the model assumes that dilation is accommodated at the fold hinge zone.

PROPERTIES OF THE MODEL

After a shortening along the direction of the original layering (expressed as a negative extension, e), the layers situated on each fold limb take up a dip, α. On the fold limb, the total slip between the top of one competent layer and the top of the adjacent competent layer (that is, the total slip across a competent plus incompetent pair) is given by

$$S_T = (t_1 + t_2) \tan \alpha , \quad (1)$$

and this may be expressed as an average shear strain parallel to the layering:

$$\gamma_T = \tan \alpha . \quad (2)$$

On the fold limbs, the slip across the boundaries of any competent layer is equivalent to half the length of the curved outer arc of the hinge zone:

$$S_{t1} = \alpha t_1 , \quad (3)$$

and this can also be expressed in terms of a shear strain parallel to the layering:

$$\gamma_{t1} = \alpha . \quad (4)$$

The slip across the bounding surfaces of the incompetent layer on the fold limbs is the

difference between (1) and (3):

$$S_{t2} = (t_1 + t_2) \tan \alpha - t_1 \alpha , \quad (5)$$

which can be expressed as a shear strain by dividing by the layer thickness t_2:

$$\gamma_{t2} = (t_1/t_2 + 1) \tan \alpha - t_1 \alpha/t_2 . \quad (6)$$

From trigonometrical relations of triangle ABC, it follows that

$$(l - t_1\alpha + t_1 \tan \alpha) \cos \alpha = l(1 + e)$$

or

$$1 + e = (1 - \alpha t_1/l) \cos \alpha + t_1 \sin \alpha/l. \quad (7)$$

STABILITY OF THE CHEVRON FOLD STYLE WITH VARIATIONS IN LAYER THICKNESS

For a given t_1/l ratio, it follows from (7) that there is a simple relation between the shortening produced by folding and the angle of dip of the fold limb. This relation has been graphically recorded in Figure 5, and the geologic consequences of variations of angle of dip with t_1/l ratio will now be examined.

Providing that the thickness of successive competent layers (t_1) remains fairly constant, the model illustrated in Figure 4 is geometrically stable: that is, for a given total shortening and constant fold wave length, the limb dips remain constant. Variations of thickness of the incompetent material (t_2) do not affect this stability in any way. If, however, the competent layers have variable thickness, the graphed relations of Figure 5 indicate that it is not possible to accommodate the same shortening with the same limb dip. Figure 6 illustrates the effect of taking up 50 percent shortening by chevron folding in sequences that have t_1/l ratios of 1/10 and 1/5, respectively. Provided that these two sequences are separated, as in this diagram, one could probably make them fit together by some sort of accommodation at their interface, but if thick competent layers are interbedded with thinner competent layers, there is a much more difficult compatibility problem.

We will examine simple solutions to this by first considering isolated, anomalously thick or anomalously thin beds intercalated in a more regular sequence.

The first problem — an isolated thick layer in a generally thin bedded sequence of competent layers — is illustrated in Figure 7. It is clear that changes must be made in our theoretical model, because it is generally impossible to have one layer in a chevron fold with a different angle of dip than the others, and it is necessary for its dip to conform with those of the general system around it. When it does conform, its limb length will be too long. There are three possibilities:

1. The extra slack is taken up somewhere near the hinge zone, and the dips of the folded surfaces around the hinge zone change in the manner shown in Figure 8a.

Figure 3. Synclinal chevron fold hinge, illustrating characteristic geometric features employed in model of Figure 4. Culm Series, Longpeak, Hartland Quay, southwest England.

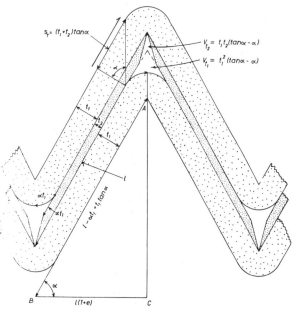

Figure 4. Chevron fold model in multilayered sequence of layers of different properties and thicknesses. Description in text.

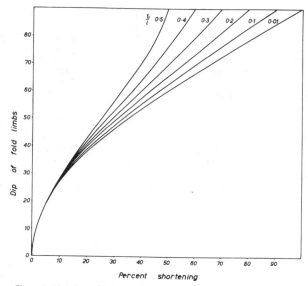

Figure 5. Variations of limb dip with percentage shortening for systems with differing t_1/l ratios.

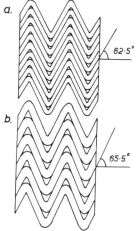

Figure 6. Chevron folds in sequences with t_1/l ratios of (a) 0/1 and (b) 0/2, both with total shortening of 50 percent.

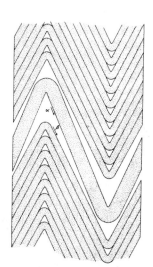

Figure 7. Geometric problem posed when single thick competent bed is incorporated into succession of folded thinner layers. For same total shortening, dip in thick layer (α) is always greater than that in thin layers (β).

358

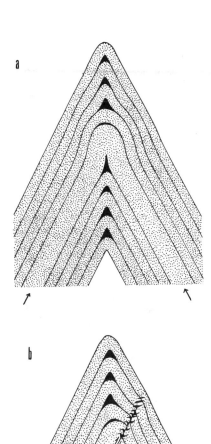

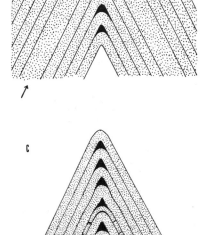

Figure 9. Bulbous hinge structure in graywacke sequence of Culm, Hartland, southwest England.

Figure 10. Bulbous hinge structure controlled by anomalously thick competent layer in experimentally deformed multilayered plasticene model.

Bulbous hinge structures of this type are especially common in the graywacke sequence of the flyschlike Culm series of North Devon (Fig. 9), and this effect can also be verified by simple model experiments (Fig. 10).

2. The extra limb length can override the material on the opposite limb, producing a limb thrust, in which the fault surface lies parallel to the bedding on the limb situated on the inner arc of the competent layer and cuts across the bedding planes of that layer on the opposite fold limb (Figs. 8b, 11). The movement on faults of this type shows no systematic overthrust sense; sometimes the left-hand limb is thrust over the right-hand limb, and at other localities the arrangement is the opposite. Zones of en echelon tension fractures are common in the shear zones passing through the ruptured competent layer (Fig. 8b).

3. The extra-thick limb might undergo a change of shape of internal ductile flow by shortening parallel to the bedding surface with increasing thickness.

The second problem concerns an isolated, thin competent layer in a sequence of uniformly thick competent strata. If the thin layer conforms with the dip of the surrounding chevron fold limbs, its length must increase. It can do this in two ways: (1) development of boudinage structure or extensive fissures oriented perpendicular to the bedding surfaces (Fig. 8c); (2) extension by ductile flow, with consequent reduction of the orthogonal thickness of the layer.

From a study of the graphs of Figure 5, it is apparent that the geometric problems posed by thickness variations are not severe during the early stages of fold development, when the interlimb angles are greater than 100°, but they become greater as the fold tightens. The effects described above are therefore likely to develop during the later stages.

If thin and thick competent layers are developed in a mixed fashion throughout the structure, it is generally the thicker, stronger layers that lead to the dominant limb dip in the system; lengths of the thinner competent layers increase by fracture or flow. The structural effects are often controlled by the relative thickness of incompetent material acting as padding between the competent strata. If the ratio t_2/t_1 is high — say, greater than 3 — there may be sufficient less competent material to adjust the shape changes required by any anomalously thick competent layer, and the curvature of the competent layers thereby can change to take up the extra limb length

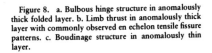

Figure 8. a. Bulbous hinge structure in anomalously thick folded layer. b. Limb thrust in anomalously thick layer with commonly observed en echelon tensile fissure patterns. c. Boudinage structure in anomalously thin layer.

without severely upsetting the overall chevron style. However, when large variations of competent layer thickness exist in situations where the incompetent material is thin, the straight-limbed chevron style cannot exist, particularly if the proportion of competent layer thickness to limb length is high (greater than 1/10). The graphs indicate that the problems of variable dip with competent layer thickness are not great where the t_1/l ratios are less than 0.02; with such values, doubling the competent layer thickness leads to only small increases in the potential limb length of this extra-thick layer.

DILATION IN THE FOLD HINGE ZONE

Because the model is one of plane strain, the dilation in the system is given by the area change in the cross section. In unit

Figure 11. Limb thrust controlled by anomalously thick competent sandstone layer in generally thin bedded succession, Culm series, Hartland Quay, southwest England.

Figure 12. Quartz-filled saddle reef structure, Hartland Quay, southwest England.

depth of material, the volume changes are given by

$$V_T = t_1(t_1 + t_2)(\tan \alpha - \alpha)$$
$$V_{t1} = t_1^2 (\tan \alpha - \alpha)$$
$$V_{t2} = t_1 t_2 (\tan \alpha - \alpha) , \qquad (8)$$

where V_T, V_{t1} and V_{t2} refer to the total volume change, volume change in the competent layer of thickness t_1, and incompetent layer of thickness t_2, respectively. These volumetric effects can also be expressed as dilations per unit volume of original material (Δ):

$$\Delta_T = \Delta_{t1} = \Delta_{t2} = l_1 (\tan \alpha - \alpha)/l . \qquad (9)$$

All dilations increase with dip and shortening, and the change of dilation with increase in dip is very marked indeed; it is the same for both competent and incompetent materials and is given by

$$\frac{d\Delta}{d\alpha} = \frac{t_1 \tan^2 \alpha}{l} . \qquad (10)$$

In the model shown in Figure 4, the dilation spaces are localized in the fold hinge zones, as one sees in naturally produced structures. The dilation spaces in the competent layer occur on the outer arcs of the folded competent layers. In the incompetent layer, the predicted spaces occur in the layer as two triangular wedges cutting through the beds and pointing toward the inner arcs of the competent layer. This arrangement seems geometrically most realistic when one takes into account the drag exerted on the surfaces of the incompetent material by the sliding motion of the adjacent competent layers. It is important to note that the dilation on the incompetent material is a function of the competent layer thickness and does not depend upon the thickness of the incompetent layers.

True *saddle reef structures* might develop in the dilation spaces by the crystallization of material from solutions passing through the rock structure, and possibly also by diffusion through the walls of the cavities (Fig. 12). The crystalline material in such cavities is often locally derived by removal of the most soluble (or most easily diffusable) components from the rocks surrounding the cavity. Solution generally proceeds most readily in regions of high differential stress and high mean pressure, as might be set up on the inner arcs of the competent strata, and precipitation most readily occurs in zones of low differential pressure or low mean pressure, as is typically found in the dilation spaces. The size of the saddle reef structures formed in this way is a function of thickness/length ratio of the competent layers, whereas the actual morphology of the saddle reef depends upon the proportional thickness of competent to incompetent layers. Examples of predicted saddle shapes for ratios t_1/t_2, of 3/1, and of 1/3 are illustrated in Figure 13.

The *flow of incompetent material* into the potential hinge spaces is a method of adjusting the geometry of the fold to potential dilation where no space-filling material is available. This is particularly favored if the ductility of the incompetent material is low. The material flowing into the hinge zones is derived from the fold limbs. The two sides of a competent rock sandwich squeeze together, enabling the less competent sandwich filling to flow away in both directions from a central point on the fold limb. The thickness of the incompetent layer is reduced, and additional and rather complex displacements and strains are set up by the flow and superposed on the strains being produced by shear of the incompetent layer between its retaining walls of competent rock (Fig. 14). The finite strain trajectory pattern will be complicated as a result of the complexity of the displacement; it is possible that cleavage induced by the deformation will show variations in intensity and orientation to accord with the XY planes of the finite strain (Fig. 14). From the geometric models of potential dilation spaces shown in diagrams a and c of Figure 13, it is clear that considerable flow must be imposed on the incompetent material if the t_1/t_2 ratio is high, whereas much less flow is necessary where this ratio is small.

Hinge collapse is a very common feature of chevron folds and is related to the dilational effects at the hinges. If the two processes described above do not occur, there is likely to be a progressive inward displacement of the competent layers toward the hinge zone to keep pace with the physical cavity trying to develop in this region (Figs. 13, 15). The geometric forms of the component layers resulting from this progressive slow collapse depend upon the t_1/t_2 ratio. Examples of structures that could de-

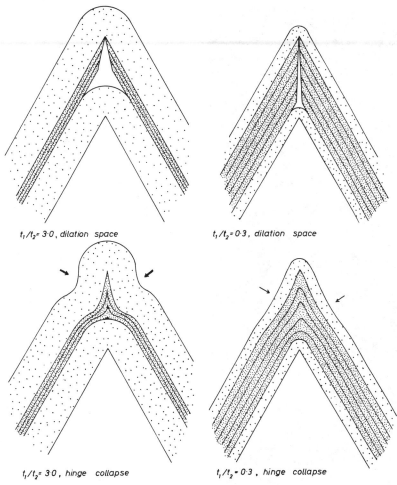

$t_1/t_2 = 3 \cdot 0$, dilation space

$t_1/t_2 = 0 \cdot 3$, dilation space

$t_1/t_2 = 3 \cdot 0$, hinge collapse

$t_1/t_2 = 0 \cdot 3$, hinge collapse

Figure 13. Potential dilation spaces giving rise to saddle reef structure or hinge collapse.

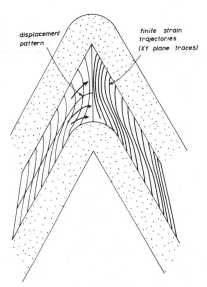

displacement pattern

finite strain trajectories (XY plane traces)

Figure 14. Displacement and strain patterns caused by flow of less competent material into dilation regions at fold hinge zone.

velop with ratios of 3/1 and 1/3 are shown in Figure 13, b and d. The distortional effect on the competent layer is very marked when the proportion of competent to incompetent rock is high. The bulbous fold nose that develops is somewhat like that form previously discussed, in which an anomalously thick layer is folded in with a uniformly thin layered competent sequence (Fig. 8a). With hinge collapse, the displacements of the competent layer are toward the inner arc of that layer, whereas the extra-thick competent layer produces a displacement away from the inner arc. Both these effects lead to complex variations of curvature, and the now-standard practice of subdividing a folded layer into domains on the recognition of the positions of maximum, minimum, and zero curvature (Fleuty, 1964; Ramsay, 1967) leads to a complex nomenclature for these structures.

All the three possible dilational effects described above are likely to be insignificant during the early stages of folding, but they become progressively more important as the fold develops and the interlimb angle decreases.

STRAIN HISTORY OF THE FOLD

As shortening of the layers goes on in a direction perpendicular to the axial surface, the interlimb angle decreases and the state of internal deformation of the layers changes. At each stage of fold development, the shear in the limbs changes, and it changes by different amounts in the competent and incompetent layers within the fold. The average shear increment parallel to the bedding surfaces can be determined for a competent-incompetent layer pair and for each layer individually. At any angle of dip, the rate of increase of shear strain parallel to the layer can be compared with the rate of change of dip. A more interesting calculation, however, is a comparison of the shear strain with the shortening perpendicu-

lar to the axial plane. This is obtained from the relations:

$$\frac{\partial \gamma}{\partial e} = \frac{\partial \gamma}{\partial \alpha} \frac{\partial \alpha}{\partial e}, \qquad (11)$$

and from equations (2), (4), (6), and (7), we obtain the functions

$$\frac{\partial \gamma_T}{\partial e} = \frac{\sec^2 \alpha \, \csc \alpha}{\alpha t_1/l - 1} \qquad (12)$$

$$\frac{\partial \gamma_{t1}}{\partial e} = \frac{\csc \alpha}{\alpha t_1/l - 1} \qquad (13)$$

$$\frac{\partial \gamma_{t2}}{\partial e} = \frac{[(t_1/t_2 + 1) \sec^2 \alpha - t_1/t_2] \csc \alpha}{\alpha t_1/l - 1} \qquad (14)$$

I have graphed functions (12) and (13) and have discussed their geologic significance (Ramsay, 1967, Figs. 7-114, 7-116, p. 433–447); the main conclusions have been summarized in this paper. Function (14) involves for the first time the effect of changing the thickness ratio t_1/t_2. The incremental shear is a function of three variables, α, t_1/l, and t_1/t_2. I discussed (Ramsay, 1967, p. 446) the influence of the first two of these, using a slightly simpler model than that used in this paper, and showed that the shear strains increased consistently with shortening. This feature led to the suggestion that the properties of the material between the competent layers was especially important in controlling the locking up of the chevron structure. For any given shortening, the incremental shears are always greatest for folds having high t_1/l ratios, and this feature probably leads to a locking up of such structures at a lower total shortening than in those where the ratio is low. The effect of the third variable (t_1/t_2) is to in-

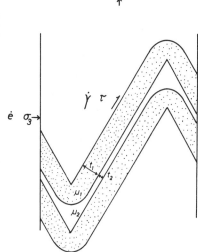

Figure 15. Anticlinal fold with hinge collapse area in upper part, Culm series, Hartland Quay, southwest England.

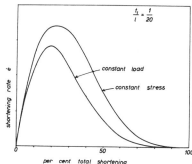

Figure 16. Control of contraction rate $\dot{e}$ by shear strain rate $\dot{\gamma}$ on fold limbs.

Figure 17. Speed of shortening in chevron fold under conditions of constant stress and constant end load.

crease consistently the incremental shears as the proportion of competent to incompetent layers increases; it therefore leads to a locking up of multilayer systems having a high proportion of competent strata at a lower total shortening (and lower interlimb angle) than those systems of similar materials but with a low proportion of competent rock.

SPEED OF GROWTH OF CHEVRON FOLDS

Multilayer systems with chevron folds are characterized by having small ratios of layer thickness to limb length, and values of 1/50 are not uncommon. Because most of the rock material is therefore situated in the fold limbs, it seems reasonable to assume that the incremental deformations going on in the limbs are particularly important in controlling the speed of development of the fold. This implies that the strain rate $\dot{e}$ normal to the axial surface will be controlled by the shear strain rate $\dot{\gamma}_T$ on the limbs (Fig. 16). Equation (12) relates the increment of shear in the fold limbs to the increment of shortening, and therefore it can be modified to describe the strain rates of these variables:

$$\dot{e} = \dot{\gamma}_T(\alpha t_1/l - 1) \cos^2\alpha \sin\alpha. \quad (15)$$

We will now make the assumption that, to a first approximation, the layers involved in the fold have linear stress-strain rate prop-

erties connected by a parameter of Newtonian viscosity $\bar{\mu}$ according to

$$\tau = \bar{\mu}\dot{\gamma}_T \quad (16)$$

where $\bar{\mu}$ is the *mean* viscosity across the competent-incompetent layer pair and τ is the shear stress. This mean viscosity clearly describes the average stress–strain rate properties. If the layers of thickness t_1 and t_2 have real viscosities μ_1 and μ_2, respectively, the mean viscosity may be determined:

$$\bar{\mu} = \frac{\mu_2(\mu_1 t_1/\mu_2 t_2 + 1)}{1 + t_1/t_2}. \quad (17)$$

If principal normal stresses σ_1 and σ_3 are acting along lines parallel to and perpendicular to the fold axial surface, the shear stresses along any surface inclined at an angle α to these axes is assumed to be homogeneous and is given by

$$\tau = (\sigma_1 - \sigma_3) \sin\alpha \cos\alpha. \quad (18)$$

Substituting (16) and (18) in (15),

$$\dot{e} = \frac{(\sigma_1 - \sigma_3)}{\mu}(\alpha t_1/l - 1)\sin^2\alpha\cos^3\alpha . \quad (19)$$

This function can be combined with (7) to express the speed of shortening of a given system at any stage in its development in terms of the total shortening already accomplished. The variations of shortening rate in a chevron system with $t_1/l = 1/20$ has been plotted graphically in Figure 17 in terms of arbitrary rate units (controlled by the mean viscosity and principal stress difference) for conditions of constant stress difference in directions parallel to and perpendicular to the axial plane, and also for conditions of constant load perpendicular to the axial plane. The constant load curve lies under the constant stress curve because the area of the end faces on which the load is applied increases during folding, and therefore the stress difference falls. These curves give a clear picture of the changes of rate of development of a chevron fold. The beginning stages are sluggish, but fold development rapidly accelerates as the interlimb angle approaches 110°, and at this point the rate of folding progressively decelerates as the structure tightens. It is also possible to integrate the strain rate function and compute the changes of shortening and shape with reference to time. This has been done for a fold with $t_1/l = 1/20$ under conditions of constant load, and the results are plotted in Figure 18. This graph brings out rather dramatically the variations in progressive changes in the folds. The initial stages of fold development are slow but do not last long, and they are followed by an explosive development of the fold. The late stages are of decelerating growth and a very long period of slow tightening. If the material has plastic properties, it is likely to lock up somewhere in these later stages of fold development, particularly if the stress difference decreases to a value below the yield points of the materials, either by geometric changes of the area over which the applied load acts or by relief of orogenic stress by the actual development of the folds. If the materials have truly viscous properties, the fold will never actually become locked up unless the orogenic stresses are completely removed. If the orogenic stress continues to act, layer-parallel shear in the limbs no longer dominates the kinematics, and some other mechanism probably takes over to continue shape changes in the rock mass. The mechanism most likely to occur during these last stages of chevron fold development will probably involve extension of the limb length compensated by a decrease in thickness of the layers. This mechanism is the well-known effect that produces "flattened flexural slip" or "flattened paral-

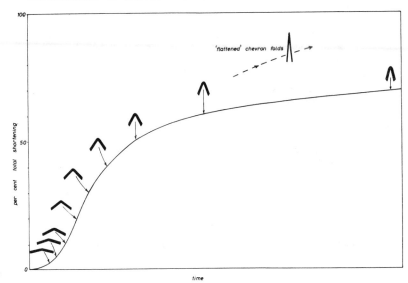

Figure 18. Progressive development of chevron folds under constant end load with $t_1/l = 0.05$.

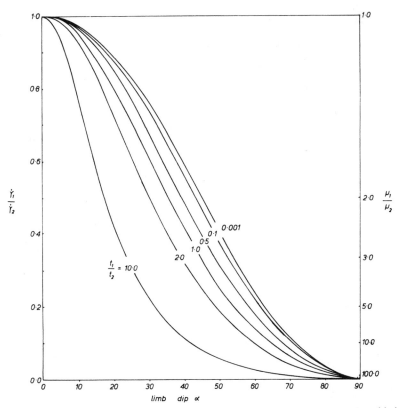

Figure 19. Variations in proportion of shear strain rates in adjacent layers according to geometric model of Figure 4.

lel" folds (de Sitter, 1958; Ramsay, 1962, 1967, p. 411–415). The fold shape in both competent and incompetent layers is modified toward the similar fold form, but it never actually reaches that style (Ramsay, 1967, p. 433–434).

STRUCTURES FORMED AT HINGE ZONES OF CHEVRON FOLDS

The model predicts that, at every stage of development of the structure, it is always possible to determine the shear increments taking place in the fold limbs. The shear strain rates $\dot{\gamma}$ on the fold limbs can always be expressed in terms of the shortening rate $\dot{e}$ by reorganizing equations (13) and (14) into the forms:

$$\gamma_{t1} = \frac{\dot{e} \csc \alpha}{\alpha t_1/l - 1},   (20)$$

$$\gamma_{t2} = \frac{\dot{e}[(t_1/t_2 + 1) \sec^2\alpha - t_1/t_2]}{\alpha t_1/l - 1} \csc \alpha .   (21)$$

The ratios of the shear strain rates in the competent and incompetent layers can therefore be calculated from

$$\frac{\gamma_{t1}}{\gamma_{t2}} = \frac{1}{(t_1/t_2 + 1) \sec^2\alpha - t_1/t_2} .   (22)$$

The model therefore predicts that for any fold with thickness ratio t_1/t_2, the ratio of the shear rates in the fold limbs is only a function of angle of dip (or, using (7), the amount of total shortening accomplished). Figure 19 illustrates these variations, and from these curves it is clear that the model implies that as the fold grows, the strain rates in the less competent material always become progressively large compared with those in the more competent material.

The shear stress that acts on the actual interface between any competent and incompetent layer must be the same in both materials, and if the materials possess linear stress–strain rate properties, then the shear strain rates in the two materials on either side of the interface will be from (16), in proportion to the reciprocals of their viscosity moduli:

$$\frac{\gamma_{t1}}{\gamma_{t2}} = \frac{\mu_2}{\mu_1} .   (23)$$

This quantity always has a fixed value, yet we have seen from (22) that the analysis of the geometric model predicts that the shear strain rates in the two layers must vary with angle of dip α. Two solutions to resolve this conflict are possible:

1. If the conditions expressed in (22) predominate and the geometry of the model is stable, then the strain rates inside each layer must vary, implying that the shear stress parallel to the layer contents varies across each layer.

2. If the conditions expressed in (23) predominate and the flow rates in each layer are constant and in proportion to their viscosities, the geometry of the model

must be modified. The shear strain rates must be retarded or accelerated compared with those predictions from the rates based on the geometric model. The consequence of this is that geometric modifications of the model must take place in the hinge zone of the structure, because Figure 4 shows that hinge shape and limb deformation are intimately related.

We can investigate the practical consequences of these changes by examining the progressive development of a chevron fold

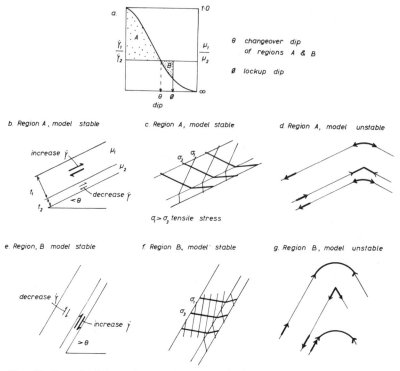

Figure 20. Geometric solutions to deviations of geometric model of simple shear from constraints of flow rate parameters μ_1 and μ_2. Diagram a shows appropriate shear rate curve from Figure 19; b, c, and d show changes in shear strain rate, average stress trajectories and displacements that must occur in field A; e, f, and g show changes in same three factors in field B.

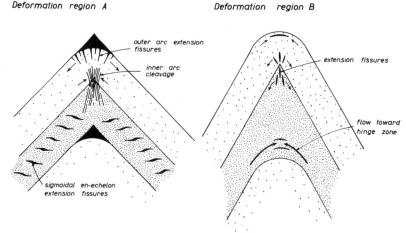

Figure 21. Potential development of small-scale structures in folds under conditions imposed in regions A and B of Figure 20.

Figure 22. Structures typical of deformation region A in hinge zone of anticlinal chevron fold. Outer arcs of competent layers show development of quartz-filled tensile fissures, whereas inner arcs show development of slaty cleavage. Culm series, Damehole Point, North Cornwall, England.

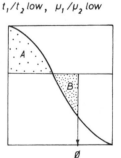

Figure 23. Structures typical of deformation region B. Note tensile fissures parallel to bedding in outer arc of competent sandstone, and late quartz extension fissures developed in cleaved shale on outside of thick competent layers. Zone of sigmoidal en echelon quartz fissures is related to limb thrust according to principles of Figure 8b. Dyer's Lookout, Hartland, North Cornwall, England.

t_1/t_2 high, μ_1/μ_2 high t_1/t_2 high, μ_1/μ_2 low t_1/t_2 low, μ_1/μ_2 high t_1/t_2 low, μ_1/μ_2 low

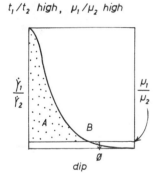
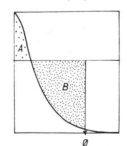
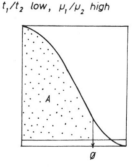

Figure 24. Variation of intensity of deformation structures of regions A and B, with variation of thickness t_1/t_2 and ductility ratio μ_1/μ_2.

in a multilayered sequence where $t_1/t_2 = 3/0$ (probably typically small), and the viscosity ratio μ_1/μ_2 is 3/1. Figure 20 illustrates the conflict between the results of (22) and (23). In this developing fold structure, the shear strain rate ratios $\dot\gamma_{t1}/\dot\gamma_{t2}$ on the fold limbs in region A are consistently higher for the model than for the conditions of constancy of viscous flow across each layer, whereas in region B the reverse is the case. In the example investigated here, the changeover position from region A to region B that marks an equality of the solutions of (22) and (23) occurs at an angle of dip θ of about 43° and a total shortening of about 25 percent.

REGION A, MODEL $\dot\gamma_{t1}/\dot\gamma_{t2} > \mu_2/\mu_1$

If the model geometry of Figure 4 is the predominating influence, the average shear stresses parallel to the bedding must rise in the more competent layer and fall in the less competent layer to enable the materials to flow at the rates required by the stability of the model (Fig. 20b). These changes in value of shear stresses can be accomplished in two ways: first, by increasing the stress deviator, and second, by changing the direction of the principal stresses so that the bedding surface in the more competent layer makes a higher angle with the principal stress axes and a lower angle in the less

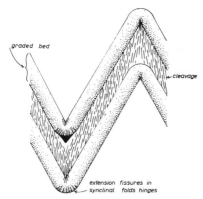

Figure 25. Appearance of small-scale structures developed in chevron folded graded beds.

competent material (see Fig. 20c). Changes in the values and orientations of the principal stresses might have considerable repercussions in fracture development, and in the competent layers, tensile practical formation might occur normal to the trajectory of greatest tensile stress (Fig. 20, c, f).

If the stresses do not change greatly to alter the flow rates in the fold limbs, the geometric restraints imposed by the model must be relaxed; the outer arcs of competent layers on the limb must be retarded compared with the inner arcs, and the reverse situation must occur in the incompetent layers (Fig. 20d). These displacements in the fold limbs must be accommodated by changes of shape in the hinge zones; the outer arc of a competent layer in the hinge must be extended parallel to the bedding surfaces, and the inner arc must be contracted. The effects in the incompetent layer are reversed. The small-scale structures likely to develop depend upon the rock ductility and fluid pressure; various possibilities are illustrated in Figures 21A and 22. The increased flow toward the hinge zone in the outer arc of the incompetent strata will inhibit any saddle reef dilation in this region but will assist such dilation spaces in the inner arcs of the incompetent layers.

REGION B, MODEL $\dot{\gamma}_{t1}/\dot{\gamma}_{t2} < \mu_2/\mu_1$

In this region, a completely different sequence of events takes place:

If the model geometry is retained, the shearing stresses parallel to the layering are modified in the opposite way to that in region A; they must be decreased in the competent layer and increased in the incompetent layer (Fig. 20e), which could again be accomplished by changes in the values of the principal stress differences or rotation of stress axes according to Figure 20f.

If the fold geometry is modified, the outer arcs of competent layers now must have an accelerated displacement rate toward the hinge, relative to those rates on the inner

Figure 26a. Small-scale structures typical of synclinal fold hinges in folded, graded graywacke, Culm Series, Milook, southwest England.

Figure 26b. Small-scale structures typical of anticlinal fold hinges in folded, graded graywacke, Culm Series, Milook, southwest England.

arc, and vice versa in the incompetent layer (Fig. 20g). In the competent layer, this leads to a contraction of the previously stretched layers in the outer arc and an expansion of the previously contracted layers in the inner arc. The overall strain history in different parts of the competent layers at the fold hinge is likely to be very complex, and small-scale structures are likely to develop that show these reversals in the directions of the maximum and minimum incremental strains. It is possible for late tensile fissures to develop in the inner arc of the competent layer and to be superposed on compressive features such as cleavage and conjugate thrust faults, often opening up these previously formed planar weaknesses in the rock (Figs. 21B, 23).

In the incompetent layers, the geometry of the structure is modified by slowing down the shear strain increments, an effect that leads to a preferential flow of incompetent material into the inner arc of the incompetent layer at the hinge and a retardation of the flow in the outer arc of the layer (Figs. 20g, 21b).

The proportions of these structural developments characteristic of the two regions A and B in any fold clearly depend upon the parameters of thickness ratio of competent and incompetent layers and the viscosity contrast. Generally, in situations of high viscosity contrast (Fig. 24, a, c), which is the natural environment for chevron fold development, the structural development characteristic of region A will be typical throughout most of the fold development, irrespective of the t_1/t_2 ratio. As the viscosity contrasts become less marked (Fig. 24, b, d), late-stage region B effects become more common and superposed on region A effects developed during early stages of folding. This effect is particularly dominant for folds having high t_1/t_2 ratios. Although these conclusions have been developed from a study assuming linear stress–strain rate relations of the materials, it should be clear that similar types of geometric constraints are likely to be imposed on the simple shear model of Figure 4 under conditions of nonlinear flow.

EFFECTS OF GRADED COMPETENT BEDS

As it is particularly common to find chevron folds developed in regularly bedded sequences of graded turbidite, it should be noted that the grading often exerts a strong control on the development of minor structures, and small-scale structure in the anticlinal hinge zones are often quite different from those in the synclinal hinge zones. In the folded Culm strata of southwest England, it is particularly common to find quartz-filled tension gashes perpendicular to the bedding surfaces in the outer arcs of synclines but not in the anticlines (Figs. 25, 26). In the anticlines, the internal deformation has been one of ductile extension with associated decrease in thickness in the outer arcs of the competent units.

SUMMARY

An analysis of a geometric model of chevron folds incorporating features observed in naturally formed chevron folds has shown that these folds possess special kinematic properties. This analysis has been used to show why in regularly bedded multilayer sequences, chevron folds develop in preference to other styles and why certain types of small-scale structures are localized in certain parts of the folds.

The chevron fold style is shown to be stable if the thickness of the competent layers is fairly constant. This conclusion is a strong one if the ratio of competent layer thickness to limb length is more than 1/10, and it can be relaxed if the ratio is lower than 1/10. It is possible to incorporate a single anomalously thick competent layer in the structure by the development of bulbous hinge forms, or hinge thrust modifications, whereas single anomalously thin competent layers may be incorporated if they are extendable by flow or can develop boudinage.

Hinge dilation is a potential feature of chevron folds; it is a function of amount of shortening and proportion of competent layer thickness to limb length, and maximum dilation always occurs during the latest stages of folding. The dilation zones may develop to form saddle reefs; other ways of adjusting the structure to potential dilation when saddle reef development does not take place are by the flow of less competent strata into the hinge zone and by hinge collapse.

Once formed, the fold development accelerates rapidly through a stage of open interlimb angles. Later the fold growth decelerates, and if the rocks are plastic, this leads to a state of locked closure with an interlimb angle of about 60°. In perfectly viscous material, the later stages of chevron fold development are not likely to be stable, and the folds will undergo geometric modification by hinge thickening, together with thinning of competent and incompetent layers.

Small-scale fractures, folds, and cleavage found at the hinge zones of chevron folds may show very complex geometric development because of the controls on the development of the hinge zone imposed by flow variations in the fold limbs. Under conditions of high viscosity contrast, the outer arcs of the competent layers are subjected to layer-parallel lengthening and the inner arcs to shortening. Under conditions of less marked viscous contrasts, an early strain history of this type is subjected to reversals, and during the later stages of folding, it is possible to produce extension features perpendicular to the bedding planes on the inner arcs of competent layers.

REFERENCES CITED

Biot, M., 1964, Theory of internal buckling of a confined multilayered structure: Geol. Soc. America Bull., v. 75, p. 563–568.

——1965, Theory of similar folding of the first and second kind: Geol. Soc. America Bull., v. 76, p. 251–258.

Chapple, W. M., 1968, A mathematical theory of finite amplitude rock folding: Geol. Soc. America Bull., v. 79, p. 47–68.

——1969, Fold shape and rheology: The folding of an isolated viscous-plastic layer: Tectonophysics, v. 7, p. 97–116.

Cobbold, P. R., Cosgrove, J. W., and Summers, J. M., 1971, Development of internal structures in deformed anisotropic rocks: Tectonophysics, v. 12, p. 23–53.

Currie, J. B., Patnode, H. W., and Trump, R. P., 1962, Development of folds in sedimentary strata: Geol. Soc. America Bull., v. 73, p. 655–674.

de Sitter, L. U., 1956, Structural geology: New York, McGraw-Hill Book Co., 552 p.

——1958, Boudins and parasitic folds in relation to cleavage and folding: Geologie en Mijnbouw, v. 20, p. 272–286.

Dieterich, J. H., 1970, Computer experiments on mechanics of finite amplitude folds: Canadian Jour. Earth Sci., v. 7, p. 467–476.

Dieterich, J. H., and Carter, N. L., 1969, Stress history of folding: Am. Jour. Sci., v. 267, p. 129–154.

Durney, D. W., and Ramsay, J. G., 1973, Incremental strains measured by syntectonic crystal growth, in de Jong, K. A., and Scholten, R., eds., Gravity and tectonics: New York, Wiley Interscience, p. 67–96.

Fleuty, M. J., 1964, The description of folds: Geologists' Assoc. Proc., v. 75, p. 461–492.

Ghosh, S. K., 1966, Experimental tests of buckling folds in relation to strain ellipsoid in simple shear deformation: Tectonophysics, v. 3, p. 169–185.

——1968, Experiments of buckling of multilayers which permit interlayer gliding: Tectonophysics, v. 6, p. 207–249.

Ramberg, H., 1959, Evolution of ptygmatic folding: Norges Geol. Tidsskr., v. 39, p. 99–151.

——1961, Contact strain and folding instability of a multilayered body under compression: Geol. Rundschau, v. 51, p. 405–439.

——1964, Selective buckling of composite layers with contrasted rheological properties: Tectonophysics, v. 1, p. 307–341.

Ramsay, J. G., 1962, The geometry and mechanism of formation of 'similar' type folds: Jour. Geology, v. 70, p. 309–327.

——1967, Folding and fracturing of rocks: New York, McGraw-Hill Book Co., 568 p.

Stephansson, O., and Berner, H., 1971, The finite element method in tectonic processes: Physics Earth and Planetary Interiors, v. 4, p. 301–321.

MANUSCRIPT RECEIVED BY THE SOCIETY NOVEMBER 30, 1972
REVISED MANUSCRIPT RECEIVED APRIL 22, 1974

From
Tectonics
by Jean Goguel

Kinematic Interpretation of Tectonic Deformations

Conservation of volume in deformation—Conservation of length of beds in parallel folding—Stripping and slipping of the cover—Deformation of the basement, disharmony—Intercutaneous scale—Movement of a fault, contraction or corresponding extension—Rhenish trench—Deformations not parallel to a given plane—Transverse faults—En echelon folds; folds with obliquely faulted side—Study of deformation in a horizontal plane, representation by a grid—Succession of several deformations; folded faults, successive faults, successive folds—Folding of nappes—Erosion between successive phases—Diverticulations.

WE HAVE just seen what gross uncertainties can be introduced if we apply geometric considerations to the study of the present form of mineral masses by prolonging the observations made on outcrops. It may seem that by not limiting ourselves to the present state, but attempting to follow the evolution of forms during the course of time, we greatly complicate the problem before us. In reality we introduce supplementary data that make it possible to state present forms with some precision.

Let us consider, for example, a folded sedimentary series; we know that before deformation, its beds were practically horizontal. Since mass is conserved during deformation, and since the density of deformed rocks differs very little from that of the original rocks (it may increase slightly, owing to a reduction in porosity), their volume is conserved quite completely. If the thickness of a bed is maintained, the same will be true of its area and probably of all the lengths measured on its surface, at least to the degree indicated above, that is, to the ratio of the change in density. The length along the

147

curved beds of a simple anticline, as seen in cross section, is greater than the straight-line distance between two points on opposite limbs; that is, the distance between two points situated on either side of an anticline is diminished during folding, which is accompanied by lateral contraction. According to what we have seen above, within a group of beds involved in parallel folding, the amount of contraction is the same for each bed. What appeared to us as a geometric consequence of the form of the folds is a necessity from the kinematic point of view. If a fold or a sheaf of folds is included between two undeformed regions in which the superposition on a given vertical of the different beds has not been disturbed, the lateral contraction corresponding to its formation, which is the distance by which the regions situated on either side have approached each other, evidently has a well-defined value. For all the beds that have undergone only a change of form (incurvation) without stretching in their plane, this lateral contraction is the difference between the length of a bed (measured on a transverse section) and the distance in a straight line of the extremities of the latter. It is to such beds that the term "competent" is applied.

For beds affected by parallel folding, this does not tell us much more than did the geometric analysis of Chap. 9. But with regard to other types of folding, kinematic considerations enable us to indicate the possible forms much more precisely.

As a first approximation (insufficient in many cases), we may assume that all movements took place parallel to a plane perpendicular to the axes of the folds. Then it is possible to make a complete study of the deformation by preparing cross sections.

If certain beds were stretched, or contrariwise increased in thickness by the lateral compression, evidently we cannot apply to them the calculation of the lateral contraction; but we know that (with an approximation related to the disappearance of the porosity) volume was conserved in the course of the deformation. This is not sufficient to determine the present form of the upper and lower surfaces of a stratigraphic unit, but it may suffice to remove an uncertainty. In the same way, if the anticline is broken by a longitudinal fault (for an oblique fault it is doubtful that the hypothesis of uniquely transverse movements can be applied), the equality of length of all the beds for which the thicknesses are conserved imposes on the tracing of the section a condition that permits excluding the possibility of certain conformations that might have been considered.

Beyond an incompetent bed which permits a disharmony, becoming deformed without the length and thickness remaining separately constant, but with conservation of the volume (area on the section), the amount of lateral contraction of competent beds will be the same as for the other competent beds, at least between two regions free of deformation (which may be difficult to prove).

Even in thin-bedded, incompetent formations, we can still apply the preceding considerations to the detailed contortions that affect it, but they are often difficult to completely unravel. We may attempt to study them statistically by evaluating the relation of lateral contraction (inverse of the thickening) to the detail of the folding. The study of the latter takes us back to the study of elementary deformation.

Let us apply the rule of the conservation of volume (or the area of a section) below a competent reference horizon of known form between two verticals situated on either side of a fold. The lateral contraction is the difference between the length of the folded competent bed and the straight-line distance between two points on opposite limbs of a fold. Let us calculate the depth of a bed such that, if there has been no change of form, the volume situated above this level may be the same before and after deformation. To do this, we divide the section of the intumescence (outlined by the reference horizon above its initial position) by the value of lateral contraction. This is an easy calculation to make with the aid of a planimeter and a curvimeter on a carefully drawn section after an accurate cartographic survey (Fig. 100).

It is impossible to affirm that the horizon whose depth is determined in this way is not deformed, but only that, if it was lifted above its initial position, it was also equally depressed below. Another horizon situated lower would have undergone essentially a subsidence. When trying to imagine what happens at more considerable depths we run into great difficulties. In numerous cases it is probable that the depth calculated as above corresponds actually to that of a horizon that has not been deformed and which consequently has not undergone the same contraction as had the overlying beds: the latter must have slipped on it, either on one side of the fold, or even on both sides of it, but for unequal distances. Kinematic analysis then shows us that the mechanics of folding of parallel beds can be applied only to a series of limited thickness, a "cover," to use an expression consecrated by use.

A whole series of considerations, certain of which will be found in the next chapter, show that the mechanics of slipping décollement of the cover have played an essential role in the Jura (as Buxtorf has shown), and in many other analogous regions, such as the sub-Alpine mountain chains. In order to effectively apply this method we must have a thorough knowledge of the form of the folds and in particular of the displacements of the faults that may be involved. It is not applicable to ruptured folds involving an overthrust fault of unknown amplitude, but it may be combined with the method whose principle was indicated in the preceding chapter and which consists of tracing the different stratigraphic units according to their known thicknesses. In this way, it is possible to determine the depth of the basement (socle) on which the "cover" has slipped as well as the stratigraphic composition of the latter.

In the Jura, in spite of the considerable variations of facies and of thickness of the Jurassic stages, the basement always corresponds to the same stage—the beds underlying the saliferous beds of the Trias. In the southern sub-Alpine mountain chains, an analogous identification can be made; in the

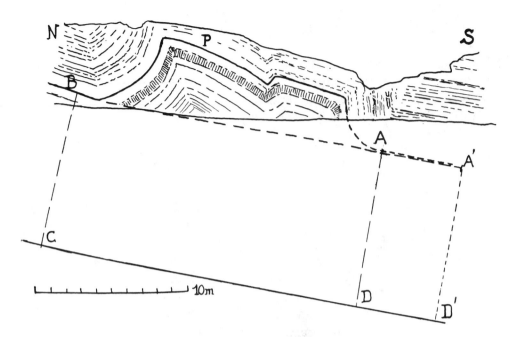

FIGURE 100. *Anticlinal fold and its kinematic interpretation. The section at the top of the figure, traced from a photograph, is visible on the left bank of the Calavon to the north of Céreste (Basses-Alpes). It affects well bedded Oligocene lacustrine limestones.*

The deformed zone is separated by a very clearcut limit from the blocks on both sides, which have remained unchanged; there is no progressive accordance, but an angle in the beds at the crossing of this limit. There is every reason to believe that outside these limits the beds remained rigid when they slipped in relation to each other in the deformed zone. At times, small hollow spaces are observed between them.

By prolonging one of the beds (traced in dashes), which may be very safely done from the attitude of the adjacent beds, we find that the unchanged blocks occupy positions that are the prolongations of each other (line BA). On the section, measurements show that the length of BPA surpasses the direct distance by 5.4 m, and that the area between PBA and the straight line BA is 51.4 m². Determining the quotient 51.4 × 5.4, we find that a horizon CD situated 9.90 m lower may have remained unchanged; we have APB = A'B and surface APBCD = surface A'BCD'.

By following the section more to the north we find, in fact, that about ten meters below the line APB, the calcareous sheaf rests on a marly mass; so it is probable that the flag limestone has slipped on the marls, undergoing a slight dislocation.

It is difficult to say whether it is a question of a gravity slip, aided by the slight dip, which continues very far toward the north, or whether the contraction of the calcareous bed does not result rather from the fact that the underlying beds are bent into a syncline with a very large radius of curvature.

The same method of interpretation may be applied to an anticline of large dimensions whose form will stand out in a detailed survey made on an accurate topographic base.

150

FIGURE 101. *Folds* (plis de couverture) *of the Alpes-Maritimes. The folds situated to the south of the Permian Dome of Barot indicate the slipping of the Mesozoic cover. The folds of the Muschelkalk to the north of the Barot Dome show, according to Bordet, that there has been a relative slip of the Jurassic cover. So it is probable that there has been an overall slip followed by the uplift of the Barot Dome.*

next chapter, we shall see what causes this localization. A confirmation of this interpretation results from the fact that older beds are never found in the heart of anticlines. But the height of the basement is not constant; undulations that are of great interest in understanding the localization of the folds of the cover may have formed prior to formation of the cover, but they also result from postdepositional movements. Thus, in the Permian Barot dome (Alpes-Maritimes) the basement has risen sufficiently to be exposed by erosion. It is thus seen that the basement forms a broad arched hump of an order of size entirely different from the folds of the surrounding cover and is itself not folded (Fig. 101).

The case of a horizon that favors slipping and of the unique disharmony that marks the base of the cover (at least tectonic, as from the stratigraphic point of view the basement may include sedimentary beds, such as the transgressive basal sandstones of the sedimentary series) is the most simple, but there may also be disharmony at other levels. It would be hard to explain why all the beds should preserve a constant thickness as they are deformed when their incurvation necessarily implies a change of form of the elements situated on the concave surface of a bed, on a convex surface, or on both.

Observation shows that certain beds (the marly and clayey beds in particular) that constitute thick, nonstratified masses undergo quite appreciable changes of thickness. For this reason, the calcareous beds that surround them do not have parallel forms (Fig. 102). That is the very definition of disharmony, but it may appear in different degrees. The folds of beds situated on both sides may correspond in general, but the minute folds of detail may not correspond; similarly, the forms of the principal folds may not be the same. For example, in the Massif des Bauges, Lugeon noticed that the anticlines were always irregular; they were sometimes split at the level of the Upper Jurassic, but were simple and regular at the top of the Neocomian (Urgonian). But, insofar as their position or general form is concerned, these anticlines correspond perfectly.

If the series that undergoes changes of thickness and presents disharmony

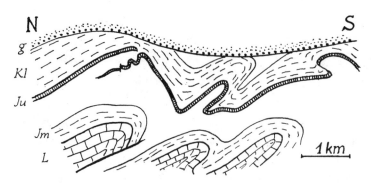

FIGURE 102. *Section of the folds between Chaudon-Norante and Barrême (Basses-Alpes). There is disharmony between the Lias and the Upper Jurassic, as well as between the Lias and the earlier terrains. The Oligocene (O) is discordant.*

is thicker, things may go farther. In the same Bauges massif (and also farther south in the Chartreuse—transverse valley of the Isère), the Middle Jurassic is never found in the heart of the anticlines; instead, only Oxfordian marls are found. It seems that the Bajocian or the Lias were not involved in the folding. This situation persists to the south of the Drôme, where the erosion never reaches, in the heart of the anticlines, a lower stage than the Callovian-Oxfordian marls (Fig. 103). The form of most of these anticlines, which are narrow and sharp, is such that, in a cross section, it is difficult to see how to include Dogger (Middle Jurassic) anticlines in the Drôme structures. Only the anticline situated to the north of the Montagne de Lure is wide enough to suggest that the Dogger should form an anticlinal structure below the level reached by erosion. Except at this location, there is a complete absence of correspondence between the folds of the Lias and the Dogger on the one hand, and those of the Tithonian and the Cretaceous on the other.

In terms of overall structure, the sedimentary units that show disharmony are those that are capable of undergoing a change of thickness in the course of deformation; this change of thickness is of course accompanied by an inverse change of length. As part of such a unit thickens, it undergoes lateral contraction in the same way as does the mass of the cover in the course of its folding, and if it is stratified, it may adapt itself equally by folds unrelated to

FIGURE 103. *Section in the south of the department of the Drôme. The anticline of the Buis (prolongation of the anticline to the north of Lure) is the only one whose form suggests that the Middle Jurassic should be included in its makeup.*

those of the adjacent units. These folds will of course be of very slight length and amplitude, since they are found only within the limits of a given formation.

We have seen above how to interpret minute folds, which are often quite visible in outcrop. They convey information not on the general structure, but on the local deformation of the mass, in much the same way that the deformation of fossils suggests the nature of elementary deformation in a sample of rock. Our efforts would be in vain if we were to seek a direct relation between the form of these minor folds and that of the major folds. In order to understand minor folds properly, it is necessary to consider them within a dynamic framework; the minute minor folds suggest the magnitude of the pressure to which the containing mass was subjected. We shall examine this aspect of the question in the next chapter.

Paul Fallot has noted a particularly interesting example of disharmony in the north of the Alpes-Maritimes for which he created the name of intercutaneous scale ("écaille intercutanée").[1] In a zone where the Mesozoic cover is sheared off at the level of the Upper Trias and has slipped in its whole mass, an overthrust, which in places attains an amplitude of 5 kilometers and which brings the Trias to repose on the Neocomian, does not continue above this horizon. It seems to fade out in the mass of the Aptian marls, and the Upper Cretaceous exhibits folds of only slight importance or local minor folds (Fig. 104). The attitude of the overthrust in the Jurassic indicates a contraction of

FIGURE 104. *The intercutaneous scale* (écaille intercutanée) *of La Roya.* [*After Paul Fallot.*]

the series reaching 5 kilometers. Since the movement certainly occurred follow deposition of the Annot sandstone (Oligocene), the Upper Cretaceous and the Nummulitic Paleogene evidently underwent an equivalent contraction, although it is not indicated by any important feature. Slips on the Aptian marls brought about the distribution of the contraction along quite a long distance; it must have taken place in the marls through a mass compression and in the heavily bedded Upper Cretaceous through minor folds.

The fact that a compression of such magnitude may escape a first examination, at least in certain beds, demonstrates the great care with which kinematic considerations must be put into operation, always paying strict attention to the lithological constitution. Here, the folds in the limestone of the Upper Jurassic indicate the deformations that the bed has undergone. It typifies the "competent" bed, whereas the Upper Cretaceous, and especially the Middle Cretaceous, behave "incompetently." The folds of the Basses-Alpes and of the Alpes-Maritimes furnish numerous examples of a disharmony in the same beds, which are, in general, less marked.

Paul Fallot pointed out later[2] that complications of detail described in different places in the Alps, which involve a division of the lower beds of an overthrust series into scales (*écailles*), might be considered as "*écailles intercutanées.*" The presence of such *écailles* may furnish proof of an overall slipping of the cover.

The application of kinematic considerations to faults, and especially to systems of faults, might also provide interesting results. We have already seen how the movement of a fault seems to behave and seems to result from a series of very rapid shocks of very limited magnitude that may continue over a long period. The geometric nature of the relative displacement is evident, since it results from slipping along a fault plane.

When a series of normal or reverse faults comes into play, presumably roughly parallel, the vertical throws may compensate one another, but the horizontal components are added together and cause an extension in the overall mass in the case of normal faults, a contraction if the faults are reverse. In the latter case, folds could have caused an analogous contraction; as a matter of fact, we often observe reverse faults closely associated with folds that may be broken and overthrust. On the other hand, normal faults produce an overall deformation of a sense contrary to that which would result from folds. Folds and normal faults are not produced concurrently. By virtue of the conservation of volume, the action of normal faults should, in general, cause a diminution of thickness together with an increase of breadth.

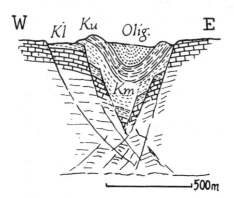

FIGURE 105

Transverse section of the Broves trench, near Comps (Var). An attempt has been made to suggest the stretching in depth of the faulted Jurassic to which the Lower Cretaceous remains connected. It is the Middle Cretaceous that admits a disharmony and the flexible subsidence of the Upper Cretaceous. Ku = Cretaceous; Km = Middle Cretaceous; Kl = Lower Cretaceous.

Among the deformations of the Mesozoic cover in the northern part of the department of the Var, which result from its slipping on an undeformed base, a certain number of complications of detail seem to indicate an extension that appears in the form of a series of relatively narrow north-south trenches, whose trend is perpendicular to the folds of the region and which lie between Jurassic plateaus. These trenches are bordered by faults that primarily affect the Jurassic and the Lower Cretaceous. The Upper Cretaceous, and the Eocene in places, outlines relatively deep synclines along their axes. It seems evident that these synclines do not indicate compression but that, on the contrary, the extension of the Jurassic may have determined a hollow, or at least a greatly thinned zone, above which the Upper Cretaceous subsided while being stretched (Fig. 105).

Along the edges of these trenches, the faults seem to be approximately vertical, but it is possible that their dip may change with depth, just as in the superficial slips of an argillaceous mass the surface along which movement takes place is, in general, a curve that becomes steeper upward and reaches the surface vertically (Fig. 51).

The absolute equality of height of the different horizons on both sides of each of the Comps trenches enables us to assert that they do not correspond to faults in the basement. This is an exceptional characteristic, inasmuch as normal faults generally seem to represent the surface repercussion of an analogous displacement affecting the whole of the crust. The Rhenish faults that delimit the Alsatian trench, whose throw probably reaches 4000 meters, are about 40 km from each other (Fig. 74). If we compare the subsidence, and the east-west extension, that results from their action, we find that conservation of volume would only be assured by relation to a surface situated at a depth of 20 km if these faults have a dip of 45°, or at a depth of 34 km if the dip is 60°. There is no reason to assume that a series of this thickness slips on an independent base. On the contrary, it is probable that the breaks affect the whole crust, even to a thickness of 30 to 50 km;

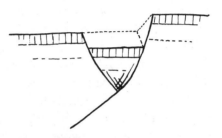

FIGURE 106

Top, Section through the village of Chateau-Chalon to the north of Lons-le-Saunier (Jura). The Middle Jurassic limestones form a wedge in the midst of the Lias (Lower Jurassic) marls. Bottom, Theoretical figure showing how, if a deep fault is tilted as it approaches the surface, conservation of volume requires the complementary subsidence of a wedge.

following the action of isostatic equilibrium, an uplift of the margins compensates in part for the subsidence of the basin (Chapter 14). This by no means indicates that such a fault traverses the deep terranes and their sedimentary cover in a straight line. In the Ledonian Jura, that is, along the margin of the Bresse, north-south belts, dropped between two normal faults, commonly separate plateaus of different levels (Fig. 106). Such a feature may be interpreted by assuming a deep fault of moderately steep dip (45°, to set a figure), and by assuming that in stratified sedimentary terranes, faults tend to approach the vertical; the conservation of volume may then be assured by making subsidence greater in the sedimentary terranes than in the underlying platform, at least along a certain extension.

Up to this point we have limited ourselves to the study of exactly transverse deformations which take place parallel to a plane and that can be studied in a plane section. This is never more than an approximation; many structures can be understood only if we take into account the fact that deformation took place in three-dimensional space, in which displacement could take place along in any direction. Such is the case for transverse faults that cut one or several anticlines, usually in a direction oblique to their axes. Generally the forms of folds on opposite sides of a fault are quite different; thus, we cannot assume that folding took place before faulting, since this would simply have displaced the two blocks relative to one another. When we compare the deformations undergone by the two margins, it becomes evident that the relative displacement varies in direction and amplitude from point to point. The fault has permitted the two margins to be folded differently. In particular, the anticlinal axes do not correspond. Let us assume, as is frequently the case (Jura, Grande Chartreuse, Fig. 107 and Fig. 108),[3] that the faults are oriented at 45° to the fold axes and that the displacement of the anticlinal axes is in the sense indicated by Fig. 109. We realize easily that the action of the fault is expressed by an elongation along the direction of the folds, whether the anticlinal axis was continuous at its origin, or, since the anticlines did not originate in a mutual prolongation, whether the mutual displacement in the region included between their axes is directed as indicated in the figure. We cannot

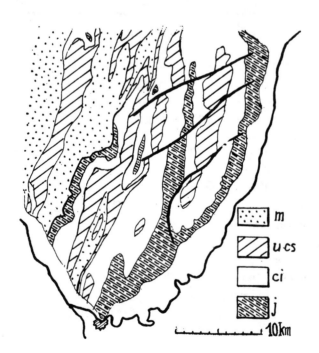

FIGURE 107

Geologic map of the Grande-Chartreuse Massif, showing the three transverse faults that cut the most internal folds. m, Miocene; u-cs, Urgonian and Upper Cretaceous; ci, Lower Cretaceous; j, Jurassic.

m

u-cs

ci

j

10km

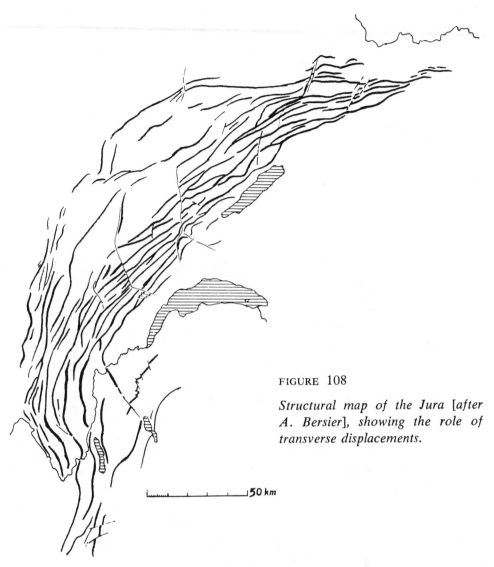

FIGURE 108

Structural map of the Jura [after A. Bersier], showing the role of transverse displacements.

⌐......ı......ı......ı......ı50 km

dissociate the ensemble of folding and transverse fault, which were produced together and whose kinematic result was a transverse contraction that took place at the same time as the elongation along the direction of the folds.

In the Jura, a whole series of displacements are arranged in the manner just indicated; the elongation along the axes of folds may be related to the

FIGURE 109

Diagram of a displacement, with indication of the relative offset of the two margins.

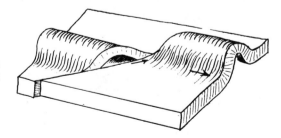

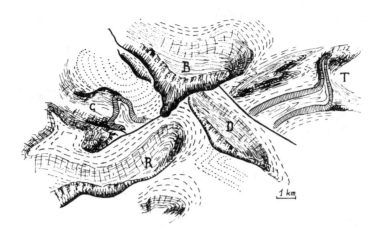

FIGURE 110. *Structural sketch of the environs of Castellane (Jurassic indicated by continuous heavy lines, Cretaceous by dashes, Tertiary by dots). The two anticlines of the Cadieres de Brandis (C) and of the Teillon (T), each of quite complex form with tilting to the north, are limited by two oblique transverse faults. In the wedge thus determined, the two anticlines of Rougon (R) and of Destourbes (D) have radiating strikes and seem to be related to the lateral compression undergone by the median wedge at the time of the advance of the anticlines C and T on its sides. In the north, the overthrust of the Braches (B) conceals the end of the median wedge.*

curve of the mountain chain, the arc evidently being longer than its chord. This tends to confirm the overall slipping of the folded cover and enables us to specify that it is the south-east side of the sheaf of folds that was displaced, the plateaus situated to the northwest and west having remained immobile.

The arrangement of transverse faults may be very complex; thus, near Castellane (Basses-Alpes), two transverse faults separate two strongly folded regions from a median compartment into which these folds are not prolonged (Fig. 110). The advance, on the sides of this median wedge-shaped compartment, of the zones that were folding brought about a contraction that was expressed by the appearance of radial folds.

During the formation of a regular cylindrical fold, the relative displacement of the regions situated on both sides is exactly normal to the elongation. A slight obliquity of this relative displacement may be expressed by different sorts of irregularities. Instead of a single fold, a series of folds may appear, each of which is normal to the relative displacement and relaying each other (folds "en echelon," Fig. 111).

To the north of the Buis (Drôme), a particular anticline exhibits one nor-

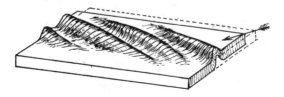

FIGURE 111

Diagram of folds en echelon with indication of the relative displacement.

mal side; the other is cut by a series of oblique faults, in all of which the displacement is horizontal and of the same sense. An experiment with a sheet of paper suffices to demonstrate that this arrangement expresses a relative displacement oblique to the general elongation (Fig. 112). The question that is posed in this case, as in the case of a sheaf of an echelon folds, is why anticlines normal to the relative displacement did not appear. This relative displacement no doubt had a different direction in an early phase; the broad fold that originated then continued to act later on at the time when the relative displacement of the two parts of the cover that it separated was taking place along a different direction.

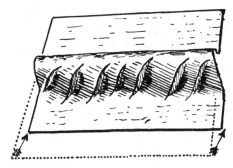

FIGURE 112

Relative displacement of the two sides of a fold when one of the sides is affected by oblique faults (experiment made with a sheet of paper).

When a fold is broken and an overthrust fault develops, there is no reason to believe that the relative displacement is exactly normal to the elongation. More rarely, a vertical fault extends along a straight anticline, following its axis or one of its sides; this exceptional arrangement may result from the fact that the relative displacement of the two sides of the anticline involved a longitudinal component or that the fault antedated the anticline, as we shall see later on (p. 000). The two cases may be difficult to distinguish.

In dislocations of detail there is a tendency to produce either faults or flexures. When a combination of the two is observed (Fig. 113) it will often be the indication of a longitudinal component of the displacement which cannot happen in a flexure.

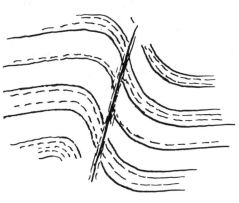

FIGURE 113

Combination of a flexure of beds with a fault whose displacement is probably longitudinal.

A careful study of the form of the tectonic features enables us to specify the direction of relative diplacements which have caused each of them. In the south of the Drôme, it is easy to recognize that a certain number of folds result from displacements oblique to their direction. We shall see later on how to interpret the latter as marking the location of earlier complications.

The analysis of such complex deformations, in which the direction of the relative displacements vary from point to point, is difficult to conduct. The magnitude of the relative displacement of numerous complications cannot be directly determined (such as faults with a horizontal displacement, or overthrusts).

The fact that the present form of reference horizon, which is assumed not to have undergone notable changes of thickness during the course of deformation, was derived through the sum of these displacements from a continuous flat form may enable us to remove these indeterminations. Let us suppose we cut up a map, following all the structural complications (transverse faults, anticlines, overthrusts), and separate the pieces from one another at a suitable distance and in the right direction, the divisions being separated by complications for which it has been possible to evaluate the relative displacement (which moreover may not be the same the whole length of the complication); it often happens that the data are sufficient to determine the relative positions of all the fragments into which the map has been cut. From these positions, it is possible to deduce an evaluation of certain relative displacements that remained undetermined.

To express the result of such a reconstruction of displacements in plan it is possible to imagine a horizontal projection of a kilometric grid, supposedly traced on the reference horizon before the tectonic deformation. The relative positions of grid fragments corresponding to the basins and synclinal depressions result immediately from the reconstruction of their mutual positions, as indicated above. The form of the folds enables us to draw the projection of the grid, supposedly folded in the same way as the beds to which it is connected (Figs. 114 and 115).

If we fold a sheet of paper so that it assumes the form of the known complications, we can see how the complications whose detail we have been unable to clearly define may appear. Such an experiment shows that this method can only be used in very simple cases. If a sheet of paper is very flexible, it is

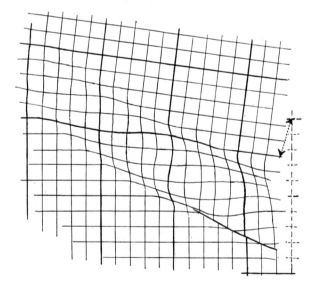

FIGURE 114

Representation by the projection of a grid supposedly traced on the reference horizon before deformation of an anticline attenuating toward the west and passing, in the east, to an overthrust toward the south.

FIGURE 115

Representation, as in Fig. 114, of an overthrust anticline limited by a transverse fault.

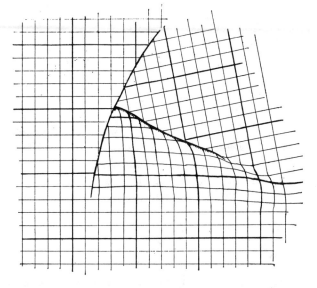

absolutely inextensible in its plane. Aside from the tears and breaks that might occur, its form is that of a developable surface, in the mathematical sense of the term. On the other hand, a sedimentary bed, even if it is relatively resistant and does not appreciably change in thickness, will assume forms much more varied than will a sheet of paper when subjected to stress; a better representation would be obtained with a thin sheet of plastiline or tin.

In order to study a somewhat complex deformation in which the form of the beds differs from a series of parallel cylindrical folds other than by the action of some faults, and to employ the method that has just been indicated, we have been led to consider only the deformation of a reference horizon. If we can transpose into three-dimensional space the consideration of the overall deformation of the series in the study of a plane section and not merely of a single reference horizon there is no doubt that very important precise data would result. However, the problem needs further investigation. As will be indicated in the next chapter, another method, which takes into consideration the forces that come into play, produces very important results.

So far we have attempted to follow the progress of a given deformation without being able to specify its rate. Kinematic considerations are of greatest importance when they enable us to reconstruct the spacing in time of different phenomena, such as distinct deformations by faults or folds, which are eventually separated by phases of erosion or sedimentation. In the latter case, the presence of dated sediments enables us to distinguish quite easily the successive phases of movement, according to whether they do or do not affect the deposits. We shall return in Chapter 17 to the determination of the age of movements by comparison with the age of sediments that are or are not affected.

Masses are frequently found of such form that their deformation cannot be

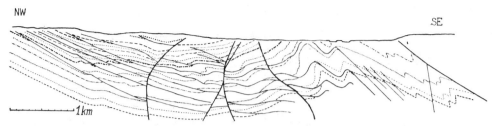

FIGURE 116. *Section of the Carboniferous basin of Liége. The veins of coal are indicated by different sets of dots and dashes, the faults by unbroken lines: distinguished are a first series of faults with very little dip that were folded at the same time as the beds and affected by later faults. [After Humblet, E., Revue Universelle des Mines, 8ᵉ série, t. 17, no. 12, Liége, 1941.]*

simply explained except by the succession of two phases of movements of quite different character. Thus in the Franco-Belgian Carboniferous basin, it has been possible to follow incontestable fault surfaces that are not flat but are curved approximately in the manner of the surrounding beds (Fig. 116). The interpretation is evident: in the first phase these faults acted along planes; in the second phase, the series was folded. The same explanation is valid for the celebrated Grenchenberg section in the Jura (Fig. 124).

Succession in time is no less clear when two faults cut across each other; we can readily distinguish the order of faulting, that is, determine which fault evidently came into play before the other (Fig. 117). Things are a little more difficult when the successive structural complications, instead of being a fault and a fold or two faults, are two folds. However there are numerous cases in which the genesis of quite complex forms can be explained by the succession of simple movements, comparable to those whose development we have just analysed.

Often we must take into consideration the erosion that has taken place between the phases of deformation. However, the necessity for this did not occur to geologists until relatively recently. It was seen that a convenient first approximation in tectonic reconstruction consisted of considering the folded mass as being undetermined toward the top. In order to take into account the effects of erosion, on the contrary, we must try to determine the position of the topographic surface at each of the phases of the deformation, which is

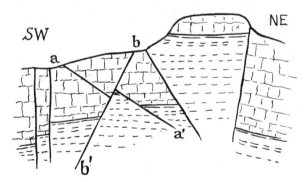

FIGURE 117

Faults affecting the base of the Urgonian limestone, Pas de l'Inferney (Vercors). It is evident that the fault aa' is anterior to the fault bb'. Note that the displacements are not necessarily in the plane of the section.

somewhat difficult to do. But the results obtained in this way are well worth the effort.

Although we cannot claim to have exhausted the variety of structures to which the superposition of different deformations may lead, we are going to try to review the results of the succession of two faults, of faults superposed on a fold, of folds of faulted regions, of two folds, of folds and nappes, of several nappes, and then we shall examine the effect of erosion during the course of the deformations.

We have seen, in Fig. 117, examples of the succession of two faults in which the order of faulting is evident. The succession of faulting has especially been studied in connection with mineralization, the nature of which almost always varies with time. In mining districts in which different sorts of veins exist, it is almost always possible to determine the order of mineralization by observing the intersections where one of the veins, the younger, displaces the vein-filling of the other (Fig. 118). However, the intersections are often marked by a particular mineralization, which is generally richer than the vein mineralization; for this reason, miners carefully seek the zones of intersection or crossings (*"filons croiseurs"*).

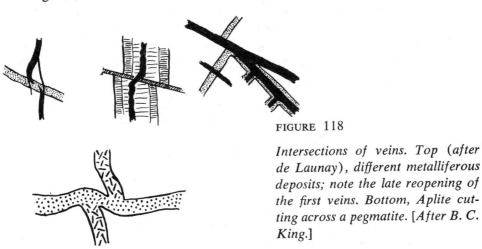

FIGURE 118

Intersections of veins. Top (after de Launay), different metalliferous deposits; note the late reopening of the first veins. Bottom, Aplite cutting across a pegmatite. [After B. C. King.]

The action of a fault affecting an earlier folded region is very easy to imagine. However, care must be taken not to confuse this case with that of a transverse fault contemporaneous with folding. In the first case, the two margins of the fault should have the same form, such that a simple reversal of the fault movement would be enough to bring the margins back into coincidence. On the other hand, if the forms of the two margins are different, the fault must have been active during folding, and it should be considered as a transverse fault, which does not exclude the possibility of its being active again later on. Transverse faults having the same age as the Alpine folds seem much more frequent than faults later than the folds.

If we consider more ancient folds, some of which were accompanied by a certain metamorphism and followed by erosion, the displacements resulting from recent movements, for example movements of Tertiary age in a Hercynian massif, are usually caused by faults. The existence of a sedimentary cover of intermediary age, which has been affected by faults but not by folds, often makes it possible to separate the two phases very clearly. This difference of behavior at the time of the two phases of deformation is essentially due to mechanical conditions; if certain beds of the block of old terranes are not indurated they may react other than by faulting at the time of more recent movements (see Chapter 12). Very frequently, a relation of position between recent faults and ancient folds may be due only to a difference in the mechanical resistance of the different strata.

The example shown in Fig. 116 corresponds to the subsequent folding of "flat faults," that is, of thrust faults. Glangeaud[4] has shown for the Jura the very varied modalities that may affect the folding of a region previously faulted. The subsidence movements of the Bresse and the Rhenish Trench are essentially of Oligocene age, and it is very probable, a priori, that they had numerous counterparts in the region now occupied by the Jura. Later, toward the end of the Miocene, the Hautes Chaines of the Jura were folded, primarily as a result of the slipping of the Mesozoic cover on the saliferous beds of the Trias. The existence of earlier faults disturbed the action of this phenomenon in different fashions. The faults that interrupted the mechanical continuity of the cover were transformed into transverse faults during subsequent folding, and their two margins folded in different ways. The folding by slipping of the cover assumes that the platform and the overlying saliferous beds were flat.

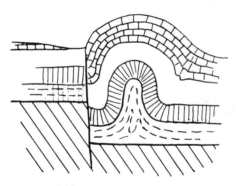

FIGURE 119

Localization of an anticline of the cover in contact with a fault affecting the platform, determining a sort of anchorage.

If they are broken through the action of faults the movement of the covering will be modified. The disharmonic action of plastic beds may lessen the effect on the cover of certain inequalities of the platform, but Glangeaud has described examples of folds adjacent to a vertical fault whose localization seems due to a sort of anchorage of the cover opposite a fault of the platform (Fig. 119).

But the effect of Oligocene faults becomes much more marked in the tabular Jura (Franche-Comté style described by Glangeaud, in contrast to the Helvetic style of the Hautes Chaines). The lines of structural complications that separate the plateaus often present a quite complex structure, which Glangeaud has shown to be the result of the crushing of pre-existing fault zones. We have seen how groups of such faults often delimit a collapsed trench, which compensates for the greater dip of the faults in the cover than in

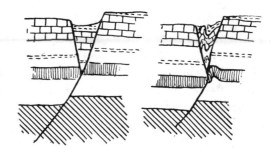

FIGURE 120

Diagram of a "pinch" resulting from the crushing of a collapsed belt between two faults.

the basement. The crushing of such collapsed or "pinched" belts, which may assume a folded character, produces varied complications (Fig. 120). The crushing of a sheaf of faults having the same sense may produce analogous complications (Fig. 121).

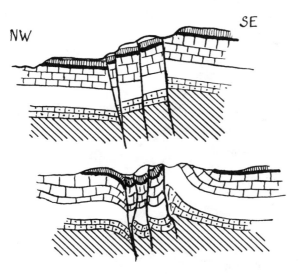

FIGURE 121

Structure of the Besançon sheaf [after Glangeaud]; posterior crushing of a faulted band.

If resistant beds on one fault margin face plastic beds of the opposite margin, certain complications may result. If a resistant bed is exposed as a result of erosion, the fault may be transformed into an overthrust fold as shown in Fig. 122 (fault fold of Glangeaud). To sum up, Glangeaud succeeded in explaining a number of characteristics of the Jura as being the result of the superposition of folds of Miocene age upon Oligocene faults.

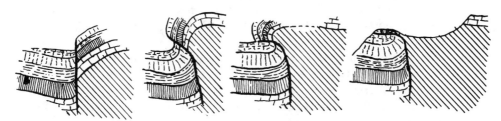

FIGURE 122. *Arguel fault-fold. (After Glangeaud.) Miocene pressure has twisted the plane of the Oligocene fault and determined a local overthrust.*

The superposition of two phases of folding rarely appears with simple characters. The profound reason for this will not appear clearly until the next chapter; we shall see that the cause of deformation by folding (from the mechanical point of view) lies in the stratified structure of the sedimentary series. After a first folding, the strata are greatly disturbed, thus the tendency of the series to fold a second time is greatly reduced. More than fifty years ago, Zurcher[5] tried to interpret the structure around Castellane by distinguishing a series of groups of parallel folds in each of which the folding was assumed to spread laterally from a central point. The manner in which the folds belonging to two distinct sheafs cut across each other would enable us to recognize the older, considered as having prevented the propagation of the most

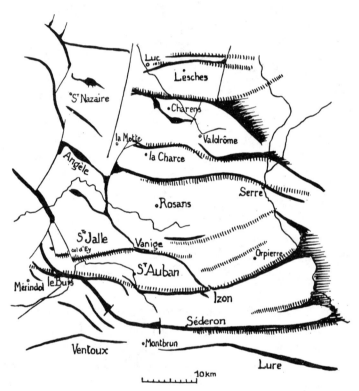

FIGURE 123. *Tectonic sketch of the Baronnies (south of the Drôme), indicating, in black, the post-Miocene anticlines and, by shaded belts, the earlier anticlines. The faults indicated by the fine lines are transverse faults. The uniquely post-Miocene folds run NW-SE, that is, they indicate a compression along a line running NE-SW. The existence of earlier folds of different direction (about E-W) has imposed this direction on a certain number of post-Miocene folds. This deviation is partially compensated, from the point of view of the overall relative displacement, by the action of transverse N-S faults, at times adjacent to an anticline. Note that, to the south of Valdrome, and to the south of Saint-Auban, the relay of the post-Miocene folds, which indicates the obliquity of the relative displacement in relation to the general direction E-W. (Cf. Fig. 111, Bull. Carte géol., no. 223, t. XLVI, Fig. 13, 1947.)*

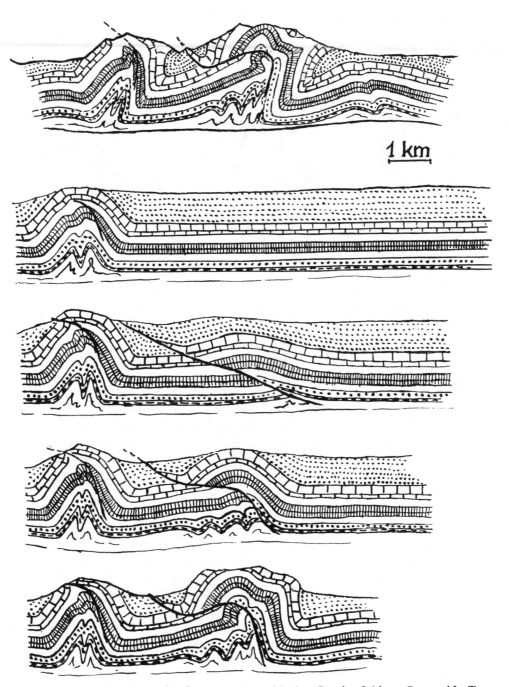

FIGURE 124. *The Grenchenberg section (Swiss Jura). [After Buxtorf.] Top, Present section from observations in the Granges-Moutier tunnel. Bottom, Successive stages of the deformation, showing (1) the action of the faults and (2) the folding which deformed them.*

recent fold. However ingenious this theory may be, it does not give very satisfactory results and has apparently been abandoned. But we must not eliminate the possibility that a similar method might give interesting results.

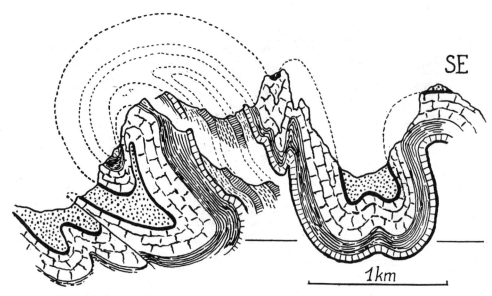

SE

1km

FIGURE 125. *Section of the Hundstein in the Säntis Massif. [After Albert Heim.]*
The heart of the anticlines is occupied by the Lower Cretaceous; the
Urgonian forms the principal calcareous mass; and the synclines are oc-
cupied by the Middle Cretaceous. The genesis of such a structure is better
understood if we accept that the fold of the anticlinal head (on the left)
was produced in a second phase, following the formation of a straight
fold. We might also ask if the movements of this second phase do not have
the character of a "collapse structure" (see p. 106), resulting from a
simple gravity subsidence, after erosion was already far advanced, without
any deep deformation. In this hypothesis the dotted lines traced above the
section would not at all indicate the form of the beds before erosion, but
a simple stratigraphic accordance.

Usually, however, the interaction between two successive folds takes place
in an entirely different manner. In a slightly folded region a deformation by
accentuation of old folds is quite easy, but the development of folds of different directions is very difficult. Consequently, there will be a tendency to
folding along old folds, both by accentuation of the latter and through the aid
of the thinning by erosion of the sedimentary series at the positions of the
anticlines.

But the general direction in which the cover tends to be compressed may
very well not be orthogonal to the old folds. Different complications may in

NW   SE

1km

FIGURE 126. *General character of the nappe of* schistes lustrés *(shading) in Haute-*
Tarentaise and Haute-Maurienne (Raguin); the mass was folded after
formation of the nappe. The crosses indicate Bonneval gneiss.

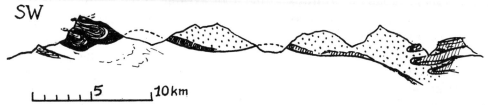

FIGURE 127. *Diagrammatic section of the Ubaye-Embrunsis nappe, to the south of the Durance. [After Schneegans.] The autochthon is left blank, the flysch of the nappe is indicated by dots, and the Mesozoic by shading. Right. Frontal folds of the Briançonnais constituting superposed scales (écailles). (See Fig. 94.)*

part make up for this obliquity, such as longitudinal slip along the ruptured anticlines, transverse faults, and so forth. In the southern part of the department of the Drôme, an admirable example of this type of complication exists,[6] which is indicated by the appearance of a grid of anticlines limiting a series of oval synclinal cuvettes (Fig. 123).

When the first phase of folding has progressed to the formation of nappes, whether by accentuation of folded beds, or by the action of flat faults, the overall mass resumes a roughly stratified structure and may easily fold again. The folded faults of the Carboniferous of Belgium (Fig. 116) constitute an example of this case, as does the famous Grenchenberg section, in which it is evident that displacement along a flat fault preceded overall folding[7] (Fig.

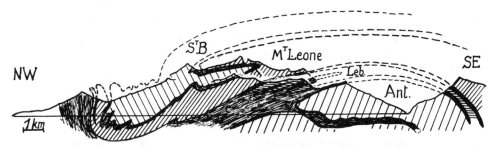

FIGURE 128. *The Simplon section. [After Argand.] The patterns represent the cores of the different nappes that contain, at least in places, a strip of Trias and are enveloped in a mass of schistes lustrés (Bundschiefer in German) (shaded). Note the folds affecting all the nappes of Mont Leone and the Saint Bernard, which are evidently posterior to their emplacement.*

124). There are also relatively superficial folds whose genesis is more easily understood if separated into several successive phases (Fig. 125).

In the preceding chapter, we noted the special nomenclature used for nappe folds, which, in general, result from later movements, the emplacement of the nappe hardly being possible except on an approximately plane surface, even a descending one (Fig. 126).

Inversely, the region in which nappes originate may happen to have been previously folded. As we shall soon see, certain folds may constitute the embryonic shape of later overthrusts. If the folds existing in the region where the overthrust takes place are eroded, they will simply produce a tectonic discordance (Fig. 99). It is possible for non-eroded folds to be torn away from their base and carried into the overthrust; this is how Gignoux and Schneegans interpret the presence of a great mass of Mesozoic with sub-Briançonnais facies that occurs in the frontal part of the Ubaye-Embrunais nappe (Morgan, Ancelle), which is essentially Tertiary and Cretaceous flysch (Fig. 127).

Isolated nappes are quite exceptional; usually we find a series of superposed nappes. Their mutual relations lead us to believe that they were formed successively, generally beginning with the highest. At the time of formation of the most recent nappes, the structure of already existing nappes could have been refolded, which could be the cause of very complex arrangements (see Fig. 128); we have seen (Fig. 97) an example of involution in which an upper nappe (here, ultra-Helvetic) is pinched under one of the Helvetic nappes. Such an arrangement is very easy to understand if we reconstruct the series of movements that produced the different nappes. On the other hand, it enables us to establish the order of succession of the latter. In this case, since the most recent nappes were formed under the earlier nappes, erosion could hardly have acted during the movements. It is different in the examples that we are now going to consider.

We have already pointed out the possible influence of erosion on the folds of a first phase. In Provence (Lutaud), a certain number of folds, perhaps but little accentuated, seem to have been formed as early as the Upper Cretaceous and subsequently eroded, while continental sedimentation was taking place in the adjacent synclines. At the time of the second tectonic phase (Lutetian), the resistant calcareous beds of the Jurassic, were eroded along the anticlinal axis, and one of the flanks of the fold slipped, aided by the plasticity of beds of the Trias and passed above the other flank, over-

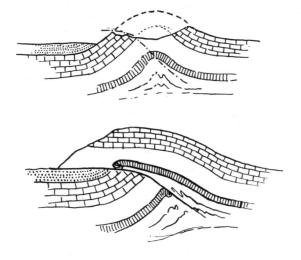

FIGURE 129

Diagram showing how erosion of an anticline (above) facilitates the rupture and the overthrust in a subsequent phase (case of Provence). [After Lutaud.]

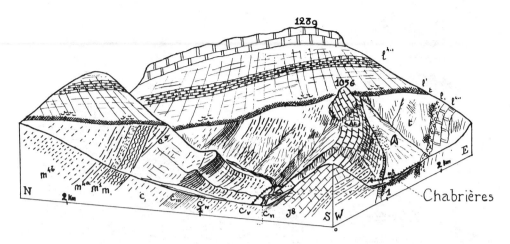

FIGURE 130. *Overthrust of the Cousson (Dourbes series), on the right bank of the Asse (Basses-Alpes). The overthrust series (Lias and Upper Trias) rests in discordance on the Jurassic, the Cretaceous, and the Tertiary, which outline an anticline; the Tertiary had been eroded before the overthrust, which took place on the surface of the ground. But the harder Jurassic determined a projection that was shoved forward (hill 1056). (Tectonic Description of the Border of the Alps from Bléone to the Var, Mém. Carte, 1937, Fig. 37.)*

thrusting upon the adjacent syncline; thus, erosion had prepared, even permitted, the rupture of the anticline (Fig. 129). But naturally, the degree of erosion must have varied from place to place, causing the lack of continuity of these Provencal overthrusts in the longitudinal sense.

We have already seen the role of erosion, which can prepare a tectonic discordance when an overthrust is extended to the very surface of the ground. Such is the case of the Cousson overthrust around Digne (Basses-Alpes), which rests on a leveled anticline in the valley of the Asse; however, a remnant, corresponding to the hard horizon in the Upper Jurassic, was shoved forward (Fig. 130).

Erosion causes even greater complications when it intervenes during the formation of nappes. Lugeon[8] has given a remarkable example of this in his description of the "diverticulations," certain complications of ultra-Helvetic nappes (Fig. 131).

Above the Helvetic nappe of the Wildhorn (shown cross-hatched in the figure), and folded with it, we find the three following nappes, from the bottom up: Plaine Morte (Cretaceous and Flysch), Mont Bonvin (Jurassic and Flysch) and Laubhorn (Trias and Lias). When the effect of the later folds is removed, these stages appear in normal succession in each one of these nappes. The hypothesis of diverticulations is based on the assumption that these are elements of a given series, which were emplaced by slipping one after the other, with the result that the oldest group of strata settled in

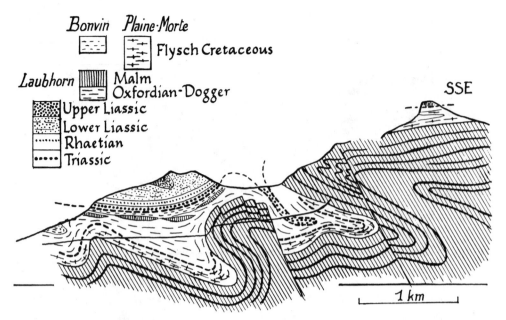

FIGURE 131. *Section of diverticulations of ultra-Helvetic nappes to the south of Lenk. [After Lugeon.]*

place above the others. The presence of flysch in several of these units probably resulted from the fact that their emplacement was contemporaneous with its deposition.

This diverticulation concept greatly simplifies the interpretation of a structure that would have been inextricable if it had been necessary to apply the hypothesis of distinct units. It may also shed some light on a number of other abnormal successions resulting from local slips. In this manner, Lugeon[9] proposed that the blade of granite caught in the flysch beneath the Morcles nappe was emplaced by a slide at the time of deposition of the flysch.

We shall see in Chapter 13 that emplacement by sliding under the influence of weight alone is not limited to units of small dimensions, but that it is the very essence of "tectonic flow" (*tectonique d'ecoulement*).

The question may be asked whether notions analogous to the diverticulation concept could not be applied on a much larger scale. In the Pre-Alps of the Chablais (compare Fig. 93), the Median nappe presents quite striking stratigraphic analogies with the Briançonnais strata, but it is overlain by the Breccia Nappe, whereas the Jurassic is composed almost entirely of sedimentary breccia. Analogous breccias are to be observed in Tarentaise in a more external zone (that is, more westerly) than that of the Briançonnais. Could the position of the Breccia and the Median nappes have been inverted relative to the position of the corresponding roots, comparable to the one produced in the diverticulations of the ultra-Helvetic units? If the block of Median slipped, under its own weight, to its present position (reckoning from Briançonnais units) and was subsequently uplifted, could the upwarping have progressed toward the west until it produced the slip of the Breccia Nappe, thus bending it back and bringing it to rest on the rear of the Median?

However rich kinematic analysis of the deformation and the distinction of successive phases may be, it could not suffice to explain such complex features; we have already alluded to the forces that have determined the emplacement of elementary nappes in the diverticulation. Dynamic analysis, which will be the subject of the next chapters, is necessary for a satisfactory understanding of tectonic deformation.

Notes and References

1. Fallot, P., *Les chevauchements intercutanés de Roya (Alpes-Maritime*s), Annales Hébert et Haug (livre jubilaire Ch. Jacob), vol. 7, pp. 161–169, 1949.

2. Fallot, P., *Remarques sur la tectonique de couverture dans les Alpes bergamasques et des Dolomites,* Bull. Soc. géol. France, 5ᵉ ser., vol. 20, pp. 183–195, 1950.

3. Goguel, J., *La rôle des failles de décrochement dans le massif de la Grande Chartreuse,* Bull. Soc. géol. France, 5ᵉ ser., vol. 18, pp. 227–235, 1948.

4. Glangeaud, L., *Le rôle des failles dans la structure du Jura externe,* Bull. Soc. Hist. Nat. Doubs, no. 51, pp. 17–38, 1944.

———, *Les nouvelles théories sur la formation du Jura,* ibid., Bull., no. 52, pp. 5–16, 1948.

———, *Les caractères structuraux du Jura,* Bull. Soc. géol. France, 5ᵉ ser., vol. 19, pp. 669–688, 1949.

See also Extraordinary Session of the Société Géologique de Belgique, Sept. 25, 1947 in Bull. Soc. géol. Belgique, vol. 73, pp. 53–150, 1949.

5. Zurcher, Ph., *Note sur la structure de la région de Castellane,* Bull. Serv. Carte géol. France, vol. 7, no. 48, pp. 299–335, 1895.

6. Goguel, J., *Recherches sur la tectonique des chaînes subalpines entre le Ventoux et le Vercors,* Bull. Serv. Carte géol. France, vol. 46, no. 223, pp. 533–578, 1947.

7. Buxtorf, A., *Grenchenbergtunnel, Prognosen und Befunde,* Naturforsch. Ges. Basel, vol. 27, pp. 185–254, 1915.

———, *Theoretische Profile zur Erklärung der Tektonik des Grenchenbergtunnels, entworfen December 1915,* ibid., Verhandl., vol. 27, plates x–xiii, 1915.

8. Lugeon, M., *Une nouvelle hypothèse tectonique: La diverticulation (note préliminaire),* Bull. Soc. Vaud. Sci. Nat., vol. 62, no. 260, pp. 301–303, 1943.

9. Lugeon, M., *Hommage à A. Buxtorf, et digression sur la nappe de Morcles,* Naturf. Ges. Bàsel, Verhandl., vol. 58, pp. 108–131, 1947.

Reprinted by permission of the Geological Society of America
from Rodger T. Faill, *Geological Society of America Bulletin*,
v. 84 (1973), p. 1289-1314.

RODGER T. FAILL   *Pennsylvania Geological Survey, Harrisburg, Pennsylvania 17120*

Kink-Band Folding, Valley and Ridge Province, Pennsylvania

ABSTRACT

It has become generally accepted that the Alleghanian deformational structures in the Valley and Ridge province of the central Appalachians were produced by northwestward movement of the Paleozoic rocks on décollements at depth. The folds in central Pennsylvania represent conversion of transport on the décollements into deformation of the overlying moving block. Abundant slickensides on bedding surfaces and constant bed-normal thicknesses indicate that they are flexural slip folds. But they are not concentric in profile. Rather, the folds possess planar limbs and narrow hinges. The presence of large and small kink bands, and the similarity of geometry (large zones of constant bed orientation, abrupt changes in bed attitude) and deformational mechanism of kink bands and these folds indicate a genetic relation. That is, kink bands can be combined to produce all the observed varieties of fold profiles. Thus, the folds in this province appear to be a widespread, large-scale kink-band deformation.

In cylindrical kink-band folds, bedding in both limbs has been rotated about the same axis. The presence of two differently oriented slickenside sets on single bedding surfaces indicates that in many of the folds, each limb has been rotated about a different axis. The resulting fold geometry is noncylindrical, giving rise to the doubly plunging, en echelon, and other complex folds that are prevalent in the province. These noncylindrical structures, reflecting a pervasive interfingering of kinematic domains (kink bands) of different orientations, constitute the regional arcuation of the Valley and Ridge folds in Pennsylvania.

INTRODUCTION

The structure of the Valley and Ridge province in central Pennsylvania (Fig. 1) consists predominantly of flexural slip folds with a distinctive cross-sectional profile—in the limbs, bedding maintains a constant orientation from hinge to hinge, and the change in bed orientation from one limb to the next occurs over a relatively short distance (Fig. 2); that is, the fold hinges are narrow with respect to the fold wave lengths. This cross-sectional profile occurs in virtually all the folds in the province, which range in size (wave length) from a few centimeters to 18 km (Fig. 3). These folds are flexural-slip folds because they exhibit the characteristic properties of uniform bed-normal thickness across the fold, and the presence of slickensides on the bedding surfaces indicates that slip between the beds was an important deformational mechanism.

In addition to these "angular" flexural slip folds, kink bands, ranging in width from a few centimeters to hundreds of meters (Fig. 4) can be observed in outcrop. The geometric and kinematic congruence between the folds and the kink bands indicates that they are genetically related, and Faill (1969a) has demonstrated how a suitable arrangement of kink bands can reproduce the observed fold profiles. The limitation of outcrop exposure precludes the delineation of kink bands of sufficient size to generate the largest folds in the province, but the geometrical similarity between small folds and the largest folds in the province suggests that the processes and mechanisms that gave rise to the largest folds were identical with those that produced the smaller, observable folds.

The largest folds in the province are more than 200 km long (Fig. 1). Smaller folds are arranged on the limbs and in the hinges of these major folds, imparting a complexity which justifies the terms anticlinoria and synclinoria. These complexities include such geometries as doubly plunging and en echelon folds, termination of folds, and abrupt changes in fold trend. The concept of kink-band folding readily explains all of the observed two- and three-dimensional variations in the folds.

The ensuing structural analysis is derived

Geological Society of America Bulletin, v. 84, p. 1289–1314, 31 figs., April 1973

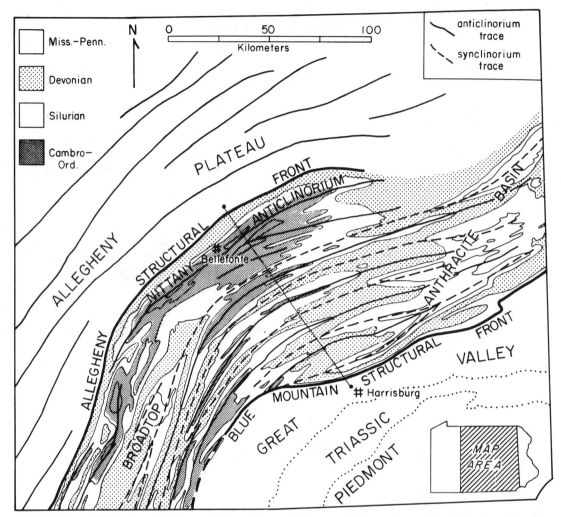

Figure 1. Generalized geologic map of the Valley and Ridge province in Pennsylvania, delineating the axial traces of the major anticlinoria and synclinoria in the province and in the adjacent Allegheny Plateau.

principally from mapping in the Millerstown 15′ quadrangle in the central part of the Valley and Ridge province (Faill and Wells, 1973; Faill, 1974). Reconnaissance and data from other published maps have shown that the structures in the Millerstown 15′ quadrangle are similar to those throughout the rest of the province.

TWO-DIMENSIONAL GEOMETRY OF KINK-BAND FOLDS

The basic structural element in kink-band folding is the kink band itself, a generally parallel-sided zone within which bedding has been rotated relative to the enveloping bedding outside of the kink band (Fig. 5). Although the profile geometry of a kink band is not identical to that of a Valley and Ridge fold, there are common elements. In the kink band, bedding is sharply bent at the kink

planes and the beds between the kink planes are planar. In the folds, the bedding in the limbs is planar and the beds are abruptly bent only at the hinge (Fig. 2). A fold geometry of this type is produced where two kink bands join or intersect (Fig. 6). The surface of junction between the two kink bands is the axial surface of the fold. The kink axis, the line about which the bedding is apparently bent[1],

---

[1] This definition differs from that of Faill (1969a) in order to be consistent with prior usage. The *kink axis*, the axis of apparent bed rotation, lies in the kink plane and parallels bedding within and outside of the kink band; the previously used term, flexure axis, is abandoned. The *rotation axis*, replacing the previous usage of kink axis, is the real axis of bed rotation. It lies in the kink plane, is normal to slickensides on adjacent bedding surfaces, and may or may not coincide with the kink axis. The kink axis is directly measurable, whereas the rotation axis orientation can only be inferred from the slickensides.

Figure 2. Characteristic flexural slip fold in the Valley and Ridge province, exhibiting planar bedding in the fold limbs and abrupt changes of bed attitude within narrow hinges. Wills Creek Formation (Silurian), U.S. Route 22, 10 km northwest of Huntingdon, Pennsylvania.

is analogous to the hinge line (fold axis) of the fold. In three-dimensional orientation, the kink bands and the folds in the province are geometrically and kinematically congruent (Fig. 7).

Both the kink axes and the fold axes plunge

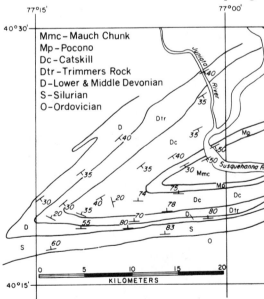

Figure 3. Geologic map of the Cove syncline in the southern part of the province near Harrisburg. The fairly straight map patterns and abrupt trend changes at the fold hinge indicate that this very large fold possesses the narrow hinge and planar bedding in the limbs that is characteristic of smaller folds. Map simplified from Dyson (1963, 1967).

gently to the northeast (and to a lesser extent to the southwest). The kink planes, and the slickensides on bedding in both the kink bands and the folds, are arranged along a great circle that is normal to the fold-axis (kink-axis) direction.

The wide range in size of folds and kink bands in the province has led Nickelsen (1963) to divide the folds into five orders: first-order folds include the largest in the province, those ranging in wavelength from 11 to 18 km; the fifth order includes folds of microscopic to hand specimen size; second-, third-, and fourth-order folds are of intermediate size. The cross-sectional geometry of planar limbs and narrow hinges is most clearly evident in fifth-, fourth-, and those third-order folds that are exposed in outcrop, but the similarity of geometry of second- and first-order folds can be inferred from their map pattern (Fig. 3). In exceptionally large exposures, the hinges of the first-order folds are approximately concentric. However, the radii of curvature in these fold hinges range from only 0.5 to 1.2 km; the wavelengths of the folds range up to 18 km, and the structural relief is as much as 4 km (Faill, 1974). Clearly, the radii of curvature in these first-order folds are much too small for them to be concentric folds. As a consequence, the methods of reconstructing fold cross sections as outlined by Busk (1929) cannot be used in a terrain possessing this type of fold geometry.

Figure 4. A 150-m-wide steeply north-dipping kink band in subhorizontal Catskill Formation (Upper Devonian) at the confluence of the east and west branches of the Susquehanna River, Northumberland, Pennsylvania.

The varieties of cross-sectional fold profiles reflect variations in the kink-band structures. For example, conjugate folds possess the same attributes as the simple folds with single axial surfaces—bedding in the limbs is planar, and the hinges are narrow with respect to the fold wave length. The significant difference is that conjugate folds have two sets of hinge lines rather than one, and thus, instead of only two limbs, they possess an additional central, relatively unrotated *interlimb* between the two axial surfaces (Fig. 6). The simple fold and the conjugate fold are actually two different parts of the same kink-band structure. Where two kink bands are joined, a simple fold occurs; where the two kink bands are separated (where the single axial surface of the simple fold bifurcates into two kink planes), a conjugate fold geometry exists.

Where two kink bands are of equal width, equal inclination to the enveloping bed attitude, and equal (and opposite) amounts of

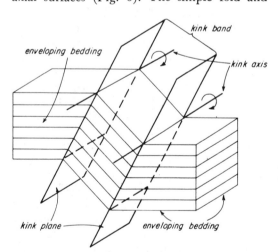

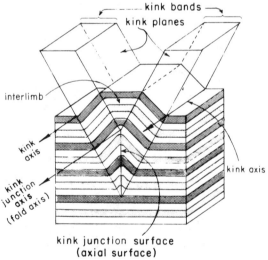

Figure 5. Idealized kink band, within which bedding has been rotated relative to bedding outside the kink band (the enveloping bedding). Generally parallel-sided, the boundaries of the kink band are the kink planes, within which all of the bending of the layers occurs. The axis around which bedding appears to have been rotated is the kink axis, which lies in the intersection of bedding within the kink band and the enveloping bedding.

Figure 6. Idealized kink-band fold, produced by the joining of two kink bands that are inclined toward each other and of opposite senses of rotation. Each kink band constitutes one limb of the fold, and thus the planar bedding characteristic of the kink bands is incorporated into each limb of the fold. In the upper portion of the structure, where the kink bands are separated, the geometry is that of a conjugate fold. The two limbs are separated by an unrotated domain, the *interlimb*.

bed rotation, the resulting fold is upright (the axial surface is perpendicular to the enveloping bedding) and symmetric (each limb is the mirror image of the other limb; Fig. 8A). Variations of width, inclination, and amounts of bed rotation between the two kink bands result in folds than are upright and asymmetric (Fig. 8B) or inclined and asymmetric (Fig. 8C).

The "space problem" in the cores of flexural slip folds has been discussed by a number of authors (deSitter, 1956, p. 198; Goguel, 1962, p. 122; Hills, 1963, p. 231; and Gwinn, 1964). It is reasoned that, because the decreasing radius of curvature toward the fold core results in a crowding of beds, some other mechanism, such as thrust faulting or disharmonic folding, must occur to relieve the congestion in the core. The implicit assumption is that all the layers in the fold must be shortened identically (Fig. 10A). The decrease in shortening possible by the flexural slip mechanism must therefore be compensated by an increase in displacement on a thrust fault, or by increasingly complex disharmonic folding. Because the beds are not folded below the fold, a strain discontinuity (detachment fault, or décollement) must exist between the folded and nonfolded layers (Fig. 10A). In contrast, no de-

tachment fault separates a kink-band fold from the nonfolded layers below (Faill, 1969b). The shortening of each layer (in a single limb) is $s = \ell (1 - \cos \delta)$, where ℓ is the layer length within the kink band, and δ is the amount of bed rotation (dip) relative to the enveloping bedding (Fig. 9B). Because the kink-band widths in both limbs decrease toward the fold core, the layer shortening progressively diminishes to zero.

Nonuniformity of flexural slip shortening in single kink-band folds does not mean that an entire stratified sequence must be nonuniformly shortened. Kink-band folds can be developed in a stratified block in such a way that local nonuniformities combine to produce an over-all uniform shortening.

In sequences with marked contrast in layer thickness (for example, Fig. 10), small kink-band folds may develop preferentially in the thinner bedded parts. In the anticline-syncline fold pair of Figure 2, a laminated bed occurs between two very thick beds, and has developed within it an extensive kink-band array, the geometry of which changes greatly across

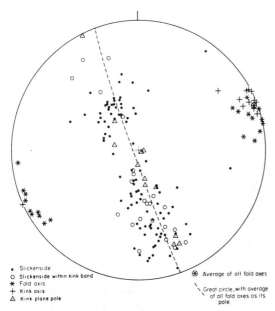

Figure 7. Fold axes and kink axes, and slickensides on bedding surfaces outside and within kink bands. Equal-area, lower hemisphere projections, as are all stereonets in this paper. Data from Millerstown quadrangle (Faill and Wells, 1973).

- Slickenside
○ Slickenside within kink band
✳ Fold axis
+ Kink axis
△ Kink plane pole

⊗ Average of all fold axes

╲ Great circle, with average of all fold axes as its pole

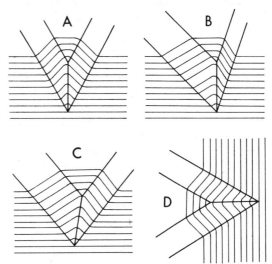

Figure 8. Variation of fold profiles as a function of variations in the kink bands. (A) The kink bands are of equal width, amount of rotation (though opposite), and inclination to the enveloping bedding The resulting fold is symmetric and upright. (B) The kink bands are of different widths, amounts of rotation, and inclination to the enveloping bedding: the resulting fold is asymmetric, though upright. (C) Similar to (B), except the resulting fold is both asymmetric and inclined. (D) Similar to (A), except the enveloping bedding is vertical instead of horizontal: the fold is thus symmetric and recumbent.

A

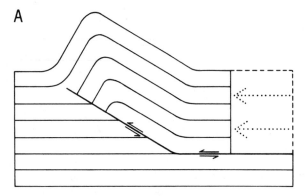

B

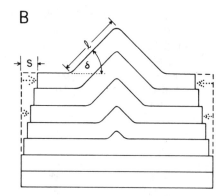

Figure 9. (A) Fold with thrust fault in core to relieve congestion of beds. The implicit assumption is that all the folded layers have been equally shortened, and thus there must be a bed-parallel detachment fault between the folded and nonfolded layers. (B) Kink-band fold. For a given dip of bedding (δ), the amount of shortening (s) of a layer is a function of the length of the layer (ℓ) in the fold. Because the layer length decreases toward the fold core, the amount of shortening similarly decreases, and thus there is no bed-parallel detachment fault between the folded and nonfolded layers.

the anticline-syncline pair. In the fold hinges the kink-band structure in the laminated layer is symmetric, whereas in the fold limbs it is asymmetric (Fig. 10). The counter-rotation kink bands (those in which the laminae rotation was in a sense opposite to that of the enveloping bedding of the fold) are narrow with a large amount of laminae rotation within them, whereas the synrotation kink bands (those in which the laminae rotation was in the same sense as that of the enveloping bedding) tend to be wider, with less rotation, and are more nearly normal to the bedding (Fig. 11). The asymmetry of these small kink-band structures in the laminated layer is a function of the amount of dip of the enveloping bedding— with increasing steepness, the divergence between the syn- and counterrotation kink bands increases with respect to inclination, width, and laminae rotation. The axial surfaces of the small kink-band folds in the laminated layer are inclined to the enveloping bedding and are subparallel with the axial surfaces of the larger folds. A similar geometry of axial surfaces of small folds parallel to those of enclosing larger folds has been observed in experimentally deformed foliated micaceous materials (Williams and Means, 1971).

The amount of shortening implied by this kink-band folding is maximum in the central part of the laminated layer and decreases to zero toward the boundaries with the massive beds, where the kink bands diminish and terminate. Either the layer shortening through such a sequence is not uniform, or some shortening mechanism other than kink-band folding must have been operative in the massive beds. On the other hand, the asymmetries of the kink-band structures suggest that they may not represent simple shortening; they may reflect a distributed slip between the two massive beds, a slip that would have been confined to a single common bedding surface had the laminated layer not been present.

The differences in widths between the syn- and counterrotation kink bands makes it unlikely that these structures were initially asymmetric throughout and subsequently internally rotated by simple shear parallel to the enveloping bedding. It is more probable that they developed in the later stages of the larger folding as a distributed slip between the more massive beds. Paterson and Weiss (1968), after experimentally deforming an inhomogeneously layered phyllite, found that small kink bands were developed in the more finely foliated parts after the larger kink bands had formed. Because the larger folds in the province are also kink-band structures, it is probable that the relations of subordinate fourth-, third-, and even second-order folds in the limbs and hinges of even larger folds are similar to those between the laminated layer and enclosing folds of Figure 2.

A continuously folded terrain can be created with a high density of kink bands. The wide range in size of kink bands in the Valley and Ridge province has given rise to the presence of smaller folds in the limbs and hinges of the larger folds. In that a complex arrangement of kink bands results in a complex pattern of folds, the fold geometry expressed by a given

Figure 10. Laminated shaly siltstone bed between very thick-bedded siltstone beds in the left limb of the syncline of Figure 2. In the laminated bed, an extensive, asymmetric kink-band array is developed, whereas similar kink bands are not present in the thick-bedded layers above and below. Wills Creek Formation (Silurian), U.S. Route 22, 10 km northwest of Huntingdon, Pennsylvania.

bed depends on the part of the kink-band structure in which a bed occurs. If the bed traverses a few large, simple kink bands, the resultant folds are large and simple, with considerable structural relief. If the bed traverses a greater number of smaller kink bands, the resultant folds are smaller and more complex, with less structural relief. Thus, in a single cross section, beds in one part of the stratigraphic section may exhibit a simple fold geometry, whereas beds higher or lower in the section may exhibit a quite different fold geometry. This can obtain purely as a function of position within the kink-band structure, and need not depend on the specific lithologies of the beds.

FAULTS

Faults in the Valley and Ridge province occur at all scales, ranging from small dislocations with displacements measured in centimeters to large, mappable offsets with displacements of hundreds and even thousands of meters. Virtually all the faults can be divided into two categories: wedge faults and cross faults. The wedge faults (Cloos, 1961, 1964) lie at a small angle to bedding; displacement on them resulted in lateral shortening and duplication of beds. The cross faults are subnormal to the tectonic trends of the folds. Both types of faults represent deformation that was subordinate to the folding, constituting local discontinuities in the folding process.

Wedge Faults

Wedge faults represent a transfer of bed-parallel slip from one bedding surface to another along a slip surface that lies at a small angle (10° to 30°) to bedding (Fig. 12). The consequent shortening results in a duplication of at

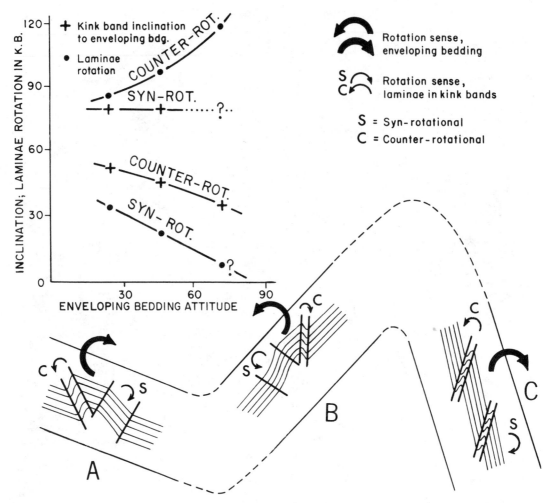

Figure 11. Angular relations of kink bands (in the laminated bed in the fold pair of Fig. 2) as a function of enveloping bedding attitude (dip of bedding in the limbs of the larger folds). The syncline is comprised of the left and middle limbs; the anticline, the middle and right limbs. Asymmetry of kink-band structures increases with increasing dip of the enveloping bedding. See text for discussion.

least part of the faulted beds. Mappable faults in the province appear to be wedge faults, because there is usually a repetition of stratigraphic section associated with these faults. Faults have been found at depth by drilling in the major anticlinoria where hundreds and even thousands of meters of stratigraphic repetition have been encountered (for example, Moebs and Hoy, 1959; Gwinn, 1970), suggesting that these faults are also of the wedge type. However, rather than being subordinate to the folding as are the faults at the surface, these faults at depth probably fulfill a more fundamental role in the deformation.

In surface exposures, the congruence of slickenside directions on the wedge faults (Fig. 13) and those on the bedding surfaces (Fig. 7) indicates that the wedge fault movements were kinematically related to the flexural slip movements. In many of the wedge faults, the intersection of the wedge surface with bedding is subparallel to the fold axis. Such surfaces represent the main slip surface of the wedge block. In other wedge faults, the intersection of the wedge surface with bedding lies at a large angle, even subperpendicular, to the fold axis, although the slickensides on the surfaces are normal to the fold axis. These wedge faults probably represent lateral terminations of the wedge blocks. That is, many of the wedge blocks appear to be lens shaped, thinning not only in the direction of transport, but also laterally in the direction of the fold axis (Fig. 14).

Wedge faults observed in outcrop occur predominantly in sequences of interbedded

Figure 12. Wedge faults in interbedded siltstone and shale sequence. Trimmers Rock Formation (Upper Devonian), U.S. Routes 22 and 322, 6.5 km north of Duncannon, Pennsylvania.

lithologies of contrasting mechanical properties. Commonly, wedge faults cross only beds of siltstone or sandstone that are surrounded by shale (for example, Fig. 12). Most of the large mappable faults similarly occur in sequences

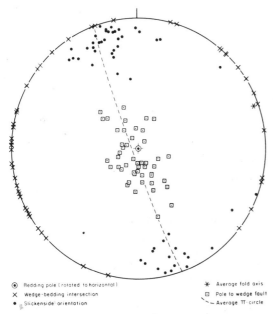

⊙ Bedding pole (rotated to horizontal)   ✳ Average fold axis
✕ Wedge-bedding intersection   ▫ Pole to wedge fault
● Slickenside orientation   ↘ Average π-circle

Figure 13. Poles to wedge faults, slickenside orientations on wedges, and the intersections of wedges with bedding. For the purpose of comparing the angular relations between the wedges and adjacent bedding, each set of data has been rotated about horizontal axes (the strike of bedding) such that the bedding pole is vertical for all 35 stations. Data from Millerstown quadrangle (Faill and Wells, 1973).

of interlayered contrasting lithologies. The proliferation of faults (relative to the rest of the province) in the Anthracite Region is undoubtedly attributable in part to the contrasting mechanical properties of the Pennsylvanian sequences of interbedded coal and sandstones.

Cross Faults

Cross faults, subperpendicular to the regional fold-axis direction, are less common than wedge faults. These cross faults are undoubtedly related to wedge faulting because in outcrop their slickensides are invariably at a small angle to the fault-bedding intersection (Fig. 15), congruent with those on adjacent wedge faults. Regionally, cross-fault slickensides possess the same distribution relative to the folds as those on wedges (Figs. 14, 15). They probably represent either vertical breaks between two parts of a wedge block, or lateral terminations of the block (Fig. 16). Cross faults have been observed only in outcrop: large cross faults have not been mapped in the province, with the possible exceptions of some cross faults in the southern Anthracite Region (Wood and others, 1969, p. 108–109).

KINK-BAND FOLDING AND DÉCOLLEMENTS AT DEPTH

Accumulated geophysical, geological, and drilling data indicate that the top of the crystalline basement beneath the folded Paleozoic rocks is relatively smooth, gently southeastward sloping, and does not reflect the structural relief exhibited in the Paleozoic rocks at the surface. This dichotomy between the top of basement and the overlying fold structures has led to the concept that the Alleghanian deformation in the Valley and Ridge province (and in the Allegheny Plateau and the Great Valley) occurred only in the Paleozoic rocks and was separated from the crystalline rocks below by widespread décollements (see Gwinn, 1970, for the most recent summary of the evidence). This change in structure with depth raises a question as to the relation of the kink-band folding observed at the surface to the inferred décollements at depth.

Geometrically, the first-order anticlinoria could be simple kink-band folds with no associated faulting or décollements, but this would indicate that the Paleozoic section was nonuniformly shortened, with the greatest short-

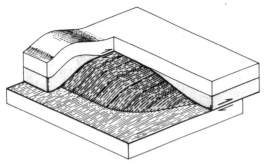

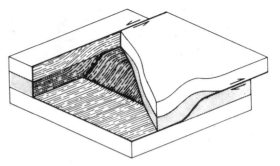

Figure 14. Idealized lens-shaped wedge block. Heavy lines represent wedge-fault surface; discontinuous lines on this surface represent slickensides which are subnormal to the fold-axis direction.

Figure 16. Idealized wedge block terminated laterally by a cross-fault. Heavy lines represent the wedge and cross-fault surfaces; discontinuous lines on these surfaces represent slickensides, which are subnormal to the fold-axis direction.

ening in the youngest, uppermost beds. In addition, the large stratigraphic separations in faults exposed at the surface, particularly in the Nittany anticlinorium (for example, Butts and others, 1939; Moebs and Hoy, 1959), and the frequent encountering of faults with large stratigraphic duplication by drilling in first-order anticlinoria, indicate that décollements and splays from décollements (wedges) did play an important role in the development of these largest of folds.

Many wedge faults are simple, originating in one bed surface and terminating in another bed surface above or below; however, a number

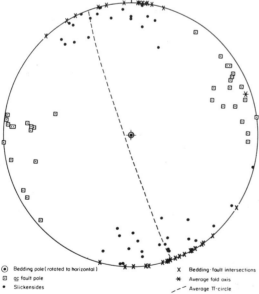

Bedding pole (rotated to horizontal)
qc fault pole
Slickensides

X   Bedding-fault intersections
*   Average fold axis
- - Average π-circle

Figure 15. Poles to cross faults, slickenside orientations, and intersection of cross faults with bedding. As in Figure 13, data have been rotated such that bedding poles are vertical for all 23 stations. Data from Millerstown quadrangle (Faill and Wells, 1973).

of wedge faults terminate in a fold (Fig. 17). Rather than a simple transference of displacement from one bedding surface to another, displacement of the overlying block is terminated in a zone at a large angle to bedding, representing a conversion of bed-parallel slip into deformation of the overlying block. The deformation process in this high-angle zone is folding, and the cross-sectional geometry is that of a kink band or kink-band fold. This latter type of structure may represent the process by which the largest folds in the province were generated above décollement surfaces.

Surface expression of the large wedge faults is most common in the Nittany anticlinorium (the northwesternmost anticlinorium of the province) and decrease in occurrence toward the southeast, suggesting a nonuniformity in the nature of the major structures across the province. However, the encountering of similar faults at depth under several other major anticlinoria indicates that perhaps all of the major anticlinoria overlie such faults. Furthermore, the enveloping surface (connecting the same stratigraphic unit on the crests of the anticlinoria) slopes to the southeast at a small angle (2° to 3°) parallel to that of the top of basement (Fig. 18). Thus, lower structural levels are exposed in the anticlinoria to the northwest than in those to the southeast, as reflected by the progressively older rocks exposed in the anticlinoria hinges from southeast to northwest. Apparently the amplitudes (amounts of structural relief) of the anticlinoria are relatively uniform across the province. The wave lengths of most of the anticlinoria are also quite uniform (12 to 18 km), not only among the several anticlinoria, but also along

Figure 17. Wedge fault terminating in a kink-band fold. The change from a fault to a fold represents a mechanical conversion of slip on the fault into deformation of the overlying beds. Bald Eagle Formation (Ordovician), U.S. Route 322, 1 km south of Reedsville, Pennsylvania.

each of their lengths from the Broad Top synclinorium to the eastern end of the Anthracite basin. Thus, the presence of faults and the uniformity of size among the anticlinoria indicate that the major structures are probably similar throughout the province (differing only in the arrangement of the subordinate structures), and that the absence of surface outcrops of faults in the southeast is only a function of the present attitude of the basin.

MECHANISMS OF KINK-BAND FOLDING

Two principal hypotheses of the mechanisms of kink banding have been developed out of recent experimental work with strongly foliated rocks: a "migration" hypothesis (Paterson and Weiss, 1966); and a "rotation" hypothesis (see Donath, 1968, for the current thought on the rotation hypothesis, and a comparison of criteria for each hypothesis). The migration hypothesis requires that the kink planes bisect the angle between bedding within the kink band and the enveloping bedding (approximately 120°), whereas the rotation hypothesis requires no such fixed relation. Kink bands in the Valley and Ridge province possess a wide range of angles of the kink planes to the enveloping bedding (45° to 90°), and the amount of bed rotation within the kink bands ranges from 10° to 120° (Faill, 1974). It appears, then, that the rotation hypothesis is more applicable to the kink-band deformation in this province.

A primary aspect of kink-band deformation (rotation hypothesis) is that the individual layers are essentially undeformed except at the kink planes. In this province, bed-normal thickness across fold limbs is relatively constant, and macroscopic deformational features within beds are not common. The isolated wedge faults in the fold limbs are viewed as local perturbations in the slip mechanism (a simple transfer of displacement from one bedding surface to another) rather than as a pervasive deformation of each bed. A poorly developed cleavage occurs in some shale and shaly limestone units but is believed to have been formed prior to folding and externally rotated with bedding (Nickelsen, 1963). The rock between the cleavage planes is essentially undeformed (Nickelsen, 1963), and the whole-rock strain associated with it is apparently small (Geiser, 1971). Preliminary measurements on deformed fossils throughout the province indicate a bulk strain in the rocks of up to 10 percent [similar to that found by Nickelsen (1966) in the Appalachian Plateau province to the northwest], but the derived strain ellipsoid orientations only partially reflect the fold geometry, and it is tentatively believed by the author to represent a prefolding deformation.

By far the greatest amount of intrabed deformation occurred in the fold hinges and kink planes, where the ratio of radius of curvature to bed thickness ranges from more than 1,000 (in the hinges of some first-order folds) to less than 2. In those folds with low ratios, the strain within each layer has been large (for example, see Donath and Parker, 1964, Fig. 5), but many of the beds exhibit little or no evi-

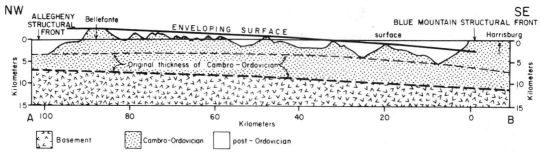

Figure 18. Simplified cross section of the Valley and Ridge province from near Bellefonte on the northwest to Harrisburg on the southeast (see Fig. 1 for location of section). Stippled area represents present distribution of Cambro-Ordovician units; dashed line indicates predeformation thickness of these rocks. The enveloping surface parallels the top of basement, indicating a constancy of amplitude of first-order anticlinoria across the basin.

dence of megascopic deformation (beyond simple bending), even in quartzite beds. It might be concluded that some flow process was operative (such as cataclasis in the coarser grained clastic rocks) or that the beds were not completely lithified at the time of deformation. Strain by twinning has been recorded in limestone beds but is not of sufficient magnitude to account for the amount of apparent bending (Groshong, 1971). Apparently, some other mechanisms, such as pressure solution along stylolitic surfaces, were operative (R. H. Groshong, 1971, oral commun.).

Although most beds cross fold hinges and kink planes with little appreciable change in bed-normal thickness, a number of beds have been tectonically thickened in fold hinges by flow (particularly in shaly beds interlayered with coarser grained beds). In contrast to these finer grained beds, some coarser grained beds have been thickened in fold hinges by local wedge faulting.

THREE-DIMENSIONAL GEOMETRY OF FOLDS

The apparent simplicity, great length, and constancy of cross section of the largest folds give the impression that the folds may be cylindrical (see Turner and Weiss, 1963, p. 107). Yet in detail, the folds exhibit considerable complexity: folds diminish in size and vanish in the limb of another fold; many abruptly change trend, plunge, and diminish

Figure 19. Noncylindrical fold termination. In a single bed, the geometry changes from a kink-band fold in the background to a simple monocline in the foreground. See text for discussion. Wills Creek Formation (Silurian), west bank of Juniata River, 10 km northwest of Huntingdon, Pennsylvania.

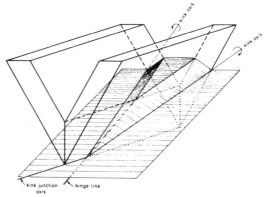

Figure 20. Noncylindrical fold termination resulting from a simple kink-band structure in which the kink-junction axis is inclined to the enveloping bedding. See text for discussion.

in size to the point of losing definition; some exhibit an en echelon pattern; others bifurcate. These complexities produce a pattern of inter-fingering folds not easily interpreted by a simple flexural slip fold model. On the other hand, a complicated fold terrain as this is well explained by use of the concept of kink-band folding.

The fold in Figure 19 encompasses all the salient aspects of the three-dimensional geom-etry of kink-band folds observed in the Valley and Ridge province. In the background, a simple fold exhibits the characteristic planar limbs and narrow hinge; toward the fore-ground, the fold amplitude diminishes to zero, and the fold is replaced by a simple monocline. This noncylindrical fold termination can be reproduced by the same kink-band structure as illustrated in Figure 6 with only one signifi-cant difference: the intersection of the kink bands or kink planes (the kink-junction axis), instead of lying in the enveloping bedding as in Figure 6, is inclined to the enveloping bed-ding (Fig. 20). [The presence of the monocline in the foreground of Figure 19 is not a sig-nificant feature—it can be reproduced merely by making the left kink band wider than the right one, as illustrated in Faill (1969a), Fig. 5.] The occurrence of these noncylindrical folds emphasizes the point that there are actually two structures present in these rocks. The first is the fundamental one, the kink-band structure, which generally is not observ-able in any completeness except in small, well-exposed structures (as in Figs. 10 and 19); the other is the more commonly observed fold that is the consequence of the kink-band deformation. Although two kink-band struc-tures may be identical (that is, the arrange-ment of kink bands is the same, as in Figs. 6 and 20), their orientation relative to the envel-oping bedding may differ, and it is this differ-ence in orientation that has the profound influence on the resultant fold geometry.

In kink-band folds where the kink-junction axis lies in the enveloping bedding (as in Fig. 6), the kink-junction axis, the kink axis of each kink band, and the fold axis (hinge line) are parallel (Fig. 21A). The cross-sectional profile of the resultant fold does not change along trend, and thus the fold is cylindrical. In contrast, the relations in noncylindrical folds are more complex. The four axes, instead of being parallel, are mutually divergent (Fig. 20; see also Fig. 21B, C, and D). The kink-

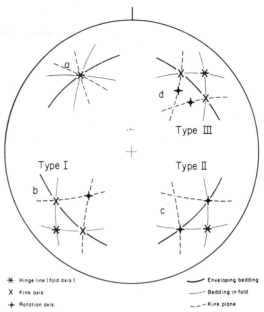

Symbol	Legend	Symbol	Legend
✳	Hinge line (fold axis)	—	Enveloping bedding
X	Kink axis	—	Bedding in fold
+	Rotation axis	- - -	Kink plane

Figure 21. Spatial arrangements of the rotation axis (or axes) relative to the enveloping bedding and the four measurable axes (hinge line, kink axes, and kink-junction axis). (A) Cylindrical kink-band fold. The intersection of the kink bands or planes (the kink-junction axis) lies in the enveloping bedding. A single rotation axis for both kink bands coincides with the kink-junction axis and the three other mea-surable axes. (B), (C), and (D) are noncylindrical kink-band folds. The intersection of the kink bands (the kink-junction axis) does not lie in the enveloping bedding, and the four measurable axes are mutually divergent. (B) Single rotation axis for both kink bands coinciding with the kink-junction axis (Type I). (C) Two rotation axes, one for each kink band, each coinciding with one kink axis (Type II). (D) Two rotation axes, one for each kink band, but neither coinciding with either of the kink axes, nor the kink-junction axis (Type III).

junction axis plunges (relative to the envelop-ing bedding) to the rear of the diagram. The trends of the two kink bands in the enveloping bedding are not parallel, and thus the two kink axes, both of which lie in the enveloping bed-ding, diverge from each other; and the hinge line of the fold plunges (relative to the envel-oping bedding) toward the foreground of the diagram.

The presence of four divergent axes in these noncylindrical folds raises a question as to around which axis the rotation of the rocks has occurred. In intracrystalline kink bands, the rotations occur around the kink axis which lies along the intersection of crystallographic planes

(Turner and others, 1954). With megascopic kink bands in foliated rocks, it is generally assumed that the rotations similarly occur around the kink axis (Paterson and Weiss, 1966; Borg and Handin, 1966). However, the *rotation* axis (the axis about which the real rotation of the rocks has occurred, as contrasted with the axis of apparent rotation, the kink axis) need not be constrained to the enveloping bedding, although it must lie within the kink plane. Ramsay (1962) has shown that a noncylindrical conjugate kink-band structure can develop when the rotation axis does not lie in the plane of foliation (enveloping bedding); that is, the rotation axis parallels the kink-junction axis, and not the kink axes (Fig. 21B). On the other hand, instead of a single rotation axis, there is the possibility of two rotation axes, each parallel to one of the two kink axes (Fig. 21C). Accepting that the rotation axes need not coincide with a kink axis, a third, more general possibility exists—the rotation axes coinciding with neither the kink axes nor the kink-junction axis, but rather lying at some other orientation within the kink plane (Fig. 21D). (The mechanical difficulty of rotating a layered sequence about an axis at a large angle to the bedding probably limits the angular divergence between the rotation axis and the enveloping bedding.) It should be emphasized that, in contrast to the kink-junction axis, the kink axes, and the hinge line (fold axis), each of which is measurable in outcrop, the rotation axis does not correspond to any physical feature in the rocks and thus cannot be directly observed or measured.

In these three types of noncylindrical fold structures, the four measurable axes are mutually divergent. But, as can be seen in Figure 21, the relation of the four axes to the enveloping bedding is identical. The kink-junction axis lies on one side of the enveloping bedding and the hinge line lies on the opposite side (in stereonet), and both kink axes lie in the enveloping bedding. Hence, the three types of noncylindrical structures are not distinguishable on the basis of relative orientation of these four measurable axes. However, even if the structures were distinguishable by these axes, all four axes are rarely, if ever, exposed. Although fold axes are commonly measurable, the kink planes of the kink bands comprising the fold are rarely observed, particularly in terrains as the Valley and Ridge province, in which the

fold density is high. The kink-junction axis is even less frequently exposed.

In view of the fact that the measurable axes are not useful for discrimination, it might be expected that the resultant fold geometry in three dimensions might differ among the three types. For example, it would be expected that rotation about a single axis inclined to the enveloping bedding (Type I) would produce a small circle distribution of the bedding poles. Similarly, rotation about two different rotation axes (Type II) would produce a bedding-pole distribution consisting of two great circles (Type II), or two small circles (Type III), one circle for each limb of the fold. However, examination of the development of a kink-band fold reveals that the movements are more complicated than a simple rotation about a single axis (or two axes).

At the initiation of the noncylindrical fold, three of the axes (the two kink axes and the hinge line) lie in the enveloping bedding; the fourth axis (the kink-junction axis) is inclined to the enveloping bedding (all four axes pass through or terminate at the fold vertex, Fig. 20). As the fold develops, the greater amplitude of the fold away from the vertex requires that the hinge line is progressively rotated away from the enveloping bedding (Fig. 22). The mechanical constraint of little or no internal deformation in the beds (such as thinning or thickening) requires that the angle between the hinge line and each of the kink axes must remain constant. Therefore, as the hinge line increases in inclination to the enveloping bedding, the two kink axes must shift toward each

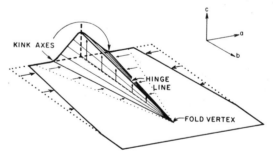

Figure 22. Kinematics of a noncylindrical kink-band fold. The dotted lines indicate the initial shape of the bed and locations of the kink axes; the light dashed line indicates the initial position of the hinge line. As the fold develops by a primary rotation about an axis (or axes) subparallel to *b*, the hinge line rotates away from the enveloping bedding (a rotation about *a*), and the kink axes rotate toward each other (rotations about *c*).

other (Fig. 22). This movement does not represent a shift of the kink axes through the rock—rather, it is a consequence of the rotation of the kink planes toward parallelism with the axial surface as the fold continues to develop. Thus, in addition to the primary kink rotation of the bedding approximately about the *b* direction, there are two additional rotations: the first is represented by the hinge line, a rotation about the *a* direction; the other is represented by the kink axes, rotations about the *c* direction (Fig. 22).

Primary rotations about an axis inclined to the enveloping bedding generally result in a bedding-pole distribution other than a great circle. But the mechanical constraint of no intrabed deformation (which adds the rotations about *a* and *c*) produces a great circle distribution (the pole of which is the hinge line) which shifts in orientation as the plunge of the hinge line increases (Fig. 23). Thus, this part of the fold (the hinge and adjacent limbs) is cylindri-

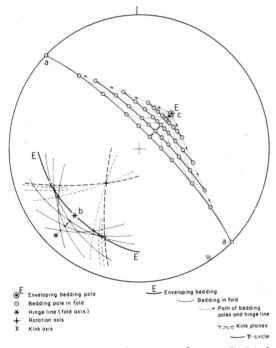

Figure 23. **Structural geometry of a noncylindrical fold. As the bedding dip increases in the fold limbs, the hinge line plunges more steeply (relative to the enveloping bedding), and thus the π-circle shifts away from the enveloping bedding pole. Concomitantly, the kink axes shift toward each other, resulting in a steepening of the kink planes. The π-circle remains a great circle distribution throughout fold development because the hinge and adjacent limbs are a cylindrical portion of the entire noncylindrical fold.**

⊙ᴱ Enveloping bedding pole
○ Bedding pole in fold
✳ Hinge line (fold axis)
+ Rotation axis
✕ Kink axis

——ᴱ Enveloping bedding
—— Bedding in fold
········ Path of bedding poles and hinge line
≈≈≈ Kink planes
——— π-circle

cal, although the entire structure is noncylindrical.

To distinguish the noncylindrical folds from the truly cylindrical fold, other criteria must be utilized. Slickensides are indicators of the local direction of movement on a surface: it is herein assumed that the slickensides on bedding, which were utilized as slip surfaces in the kink-band folding, are normal to the rotation axis of the kink band. Slickensides on beds of a single orientation are not sufficient to define the rotation axis—only slickensides measured on beds of two different orientations (such as beds inside and outside of a kink band, or in two limbs of a fold) can define the plane to which the pole is the rotation axis.

In a cylindrical fold, where a single rotation axis coincides with the kink-junction axis and lies in the enveloping bedding, the slickenside orientations lie on a great circle that coincides with the π-circle of the fold (Fig. 24A). In the Type I noncylindrical fold, the slickenside orientations are arranged on a great circle, the pole of which is the kink-junction axis: it differs from the π-circle of the fold which is normal to the hinge line (Fig. 24B). In the Type II structure, there are two rotation axes, each parallel to one of the kink axes (Fig. 24C). Neither of the slickenside great circles coincide with the π-circle of the fold. In the most general structure, Type III, there are also two distinct great circle distributions of slickenside orientations, but neither are normal to any of the measurable axes (Fig. 24D).

Regionally, the distribution of slickenside orientations along a great circle band that is subnormal to the regional fold-axis direction indicates that the rotation axes in general are subparallel to the fold axes (see Fig. 7). However, the dispersion of these orientations suggests that locally the rotation axes diverge from one another. Slickenside data from across a large fold are difficult to analyze because of possible contributions of slickensides associated with smaller structures within the larger fold. On the other hand, slickensides in a small fold are more likely to represent the kinematics of only that fold. In some small folds, a single set of slickensides falls on a great circle that coincides with the π-circle of the fold, indicating that the fold is cylindrical. In other folds, two divergent sets of slickensides appear on the same bedding surface (Fig. 25), and these two sets appear to have developed contemporaneously. The presence of two sets of

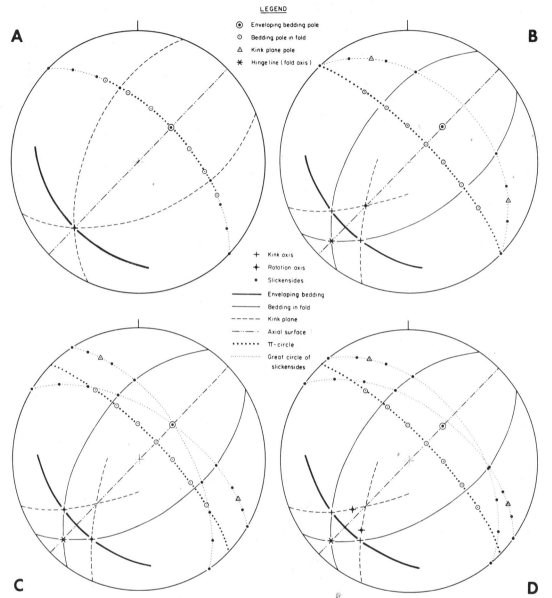

Figure 24. (A) Cylindrical fold. Because the rotation axis coincides with the other axes, the great circle distribution of slickensides coincides with the π-circle of the fold. (B), (C), and (D) are noncylindrical kink-band folds. (B) Type I. The single rotation axis coincides with the kink-junction axis, which is inclined to the enveloping bedding. Because of the consequent divergence between the hinge line and the rotation axis, the single great circle distribution of slickensides does not coincide with the π-circle of the fold. (C) Type II. Two rotation axes, one for each fold limb, and each of which coincides with one of the kink axes. There are two great circle distributions of slickensides, each normal to one of the kink axes, but neither corresponding to the π-circle of the fold. (D) Type III. Two rotation axes, neither corresponding to any of the four measurable axes. Neither great circle distribution of slickensides coincides with the π-circle of the fold, nor are either normal to any of the measurable axes.

slickensides indicates the presence of two rotation axes, which can be inferred to lie at the normals of the two great circle distributions of slickensides (Fig. 26).

Although the bed rotation in each limb is dominated by a single rotation axis, the presence of two sets of slickensides in a fold limb or hinge indicates that the movements associated with a given rotation axis extend beyond the boundaries of the kinematic domain dominated by that rotation axis. That is, although the bed rotation and final bed orientation in a limb are determined predominantly by the rotation axis associated with that limb, some of the movements reflect the kinematics of adjacent kinematic domains.

The occurrence of two sets of slickensides on single bedding surfaces is fairly common in the province, indicating that a fair proportion of these folds are comprised of two kinematic domains of divergent trends. Because of normal dispersion in the data, it is not possible to determine if these represent Type II or Type III noncylindrical folds. No exposures have yet been found which clearly indicate the presence of Type I noncylindrical folds.

ARCUATION OF THE VALLEY AND RIDGE PROVINCE

The arcuation of the Valley and Ridge province in Pennsylvania consists of a gradual change in fold trend from azimuth 025° at the southern border with Maryland to azimuth 070° (locally 080°) in the Anthracite region of eastern Pennsylvania. Although similar changes in trend occur elsewhere in the Appalachians, this one is distinctive because of the amount of arcuation and because it is exhibited in a more than 300-km-wide belt encompassing all the provinces of the Appalachian Mountain system. Significantly, this arcuate portion constitutes a boundary between the southern and northern Appalachians. To the south, the Valley and Ridge province extends as a relatively straight belt for 1,000 km from Maryland to Alabama. To the northeast, most of the major folds die out in eastern Pennsylvania.

The arcuation of the Appalachian system in Pennsylvania has been attributed (Drake and Woodward, 1963; Woodward, 1964) to a major dextral offset (the Cornwall-Kelvin displacement) between the northern and southern Appalachians that crosses New Jersey and southern Pennsylvania approximately along the N. 40° latitude with a displacement of 130 to 145 km. They suggest that movement on this displacement occurred intermittently throughout the Paleozoic and Mesozoic, bending an initially straight Appalachian geosyncline and Triassic basin into the present configuration, apparently by a rotation about a vertical axis. However, there is no surface expression of such a fault or fault zone anywhere across Pennsylvania. And it is unlikely that folds 200 km north of the fault zone would be rotated when those directly over the zone were not.

Furthermore, paleocurrent directions (McBride, 1962; Yeakel, 1962; Pelletier, 1958; and McIver, 1970) and paleomagnetic data (Roy and others, 1967; Knowles and Opdyke, 1968) do not reflect the structural arcuation, indi-

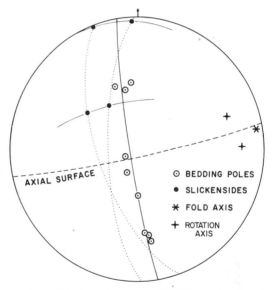

Figure 26. Data from a small fold in an abandoned quarry in Tonoloway Formation (Silurian) 0.8 km west-southwest of Millerstown, Pennsylvania. Two divergent sets of slickensides occur in the hinge and north limb, indicating the presence of two kinematic regimes during folding. The orientations of the rotation axes are inferred to be the normals of the great circles drawn through each set of slickensides. Because the kink planes were not exposed in the quarry, and thus the kink axes were not measurable, it is not possible to determine if this fold is a Type II or Type III structure.

Figure 25. Two sets of slickensides on a single bedding surface, indicating that two different kinematic regimes were operative during the development of the fold. Keyser Formation (Silurian-Devonian) in a quarry 0.8 km west of Mt. Pleasant Mills, Pennsylvania.

cating that the Valley and Ridge rocks have not been rotated in a horizontal plane about a vertical axis. These results do not, however, preclude the possibility that the depositional basin was initially linear and has been subsequently offset by the Cornwall-Kelvin displacement—but only by some mechanism other than rotation about a vertical axis (for example, a simple radial movement). In summary, it can be concluded that *the folds developed in their present orientation* and have not been subsequently rotated.

The change in trend of the fold structures (arcuation) in the Valley and Ridge province appears superficially to be smooth and continuous, but detailed examination of the folds reveals that they consist of two or more straight segments with small abrupt changes in trend between segments. Each fold consists of at least two kinematic domains (each domain comprising one fold limb) within each of which the bed orientation is constant. The boundary between contiguous domains along trend is marked by abrupt, though small, changes in the strike of bedding, or to a lesser extent the dip of bedding. Because the structures are well delineated by the topographic relief in this province, the boundaries of the domains can be quite easily delineated even on topographic maps.

Changes in trend do not occur for all the folds along single lines across the province. Although many folds parallel each other, there are numerous instances where a change in trend in one fold is not reflected in an adjacent fold—the corresponding change in the adjacent fold may occur many kilometers farther along trend. As a consequence, the province consists of a complex interfingering of kinematic trends. This interfingering tends to mask the small abrupt trend changes, and imparts to the province as a whole the superficial aspect of a smooth and continuous arcuation.

The geometrical similarity between the largest folds in the province and those observed in outcrop (Fig. 19) implies a common origin. The divergence or interfingering of kink bands that produces the small folds suggests that the divergent kinematic domains represent, in plan view, divergent kink bands that produced the large folds. The kink bands and the trend changes can be combined in numerous different manners, resulting in a rather wide variety of fold forms, of which five of the more common ones are illustrated.

1. Where the two domains in a fold segment are parallel, the bedding strike in each limb is parallel, the fold is cylindrical, and there is little or no plunge to the fold. The trend change may occur in both kinematic domains at the same place in a fold, whereby the fold abruptly changes trend with no change in the plunge of the fold. More commonly the trend change occurs in only one limb of the fold (Fig. 27). In such instances, the two domains comprising the fold are not parallel: the fold abruptly acquires a plunge, diminishes in size, and usually terminates within a relatively short distance along trend. In this fashion, every fold in the province terminates, at both ends, producing the observed doubly plunging fold form.

2. Few of the folds are greater than 100 kilometers in length—the continuation of the major, first-order anticlinoria and synclinoria for hundreds of kilometers is accomplished by the termination of one fold adjacent to the rising of another fold (Fig. 28). Essentially, two simple, noncylindrical anticlines plunge in opposite directions and terminate in the limb of the adjacent fold. With the appropriate changes in trend, one or both anticlines may persist as a subsidiary cylindrical fold within the limb of the adjacent major fold. This offsetting of second- and higher order folds within the first-order folds results in the en echelon fold pattern that is common in this province.

3. Rather than representing a change in fold trend, en echelon folding can occur in the hinge of a major fold which does not itself change trend (Fig. 29). The en echelon folds are confined to the hinge of the major fold, and are produced by the superposition of divergent kinematic domains on those of the major fold.

4. Trend changes are not restricted to fold hinges—folds also originate in the limbs of other folds without affecting the geometry or trend of that fold (Fig. 30). Originating as they do in the limbs of major folds, they are subsidiary to them. However, they can increase in size along trend and become major folds themselves, as with those south of the eastern end of the Nittany anticlinorium (see Fig. 2).

5. Some major folds bifurcate (particularly if they are conjugate), with one of the folds plunging and terminating, and the other persisting as a cylindrical fold for some distance along trend before terminating (Fig. 31). This represents a complex version of the simple termination.

The gross arcuation of the folds in the Valley

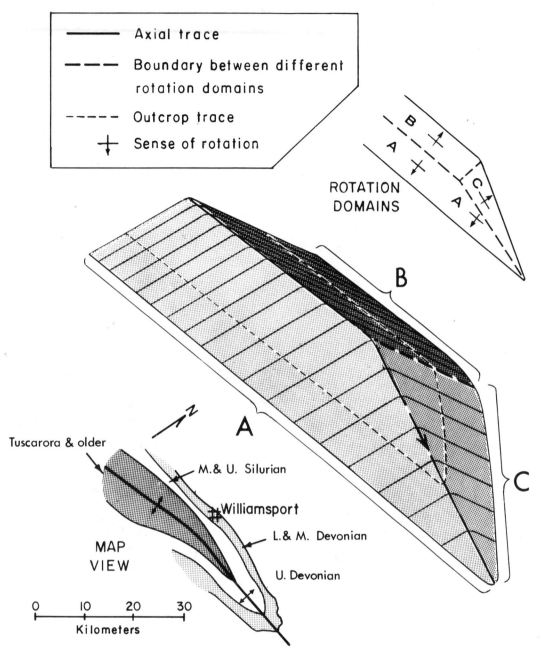

Figure 27. Simple termination of a fold by a trend change in only one limb of the fold. Kinematic domains A and B possess the same trend (although opposite senses of rotation), and where they are contiguous, the fold is cylindrical with no plunge. Domain C is a continuation of domain B, but with a different trend. Thus, where domains A and C are contiguous, the fold is noncylindrical, plunges, and terminates within a short distance. The geologic map is of the eastern end of the Nittany anticlinorium, near Williamsport, Pennsylvania (from Gray and Shepps, 1960).

and Ridge province suggests that the stress field responsible for the structures was radial in plan with a center of curvature to the southeast of the province. That slickensides are subnormal to the fold axes throughout the province supports this contention. It is simplest to assume that the radial stress field was smooth (no abrupt changes in orientation or magnitude) and continuous. In experimentally produced kink bands, the kink axis is normal to the maximum principal stress direction (Paterson and Weiss, 1966; Borg and Handin, 1966;

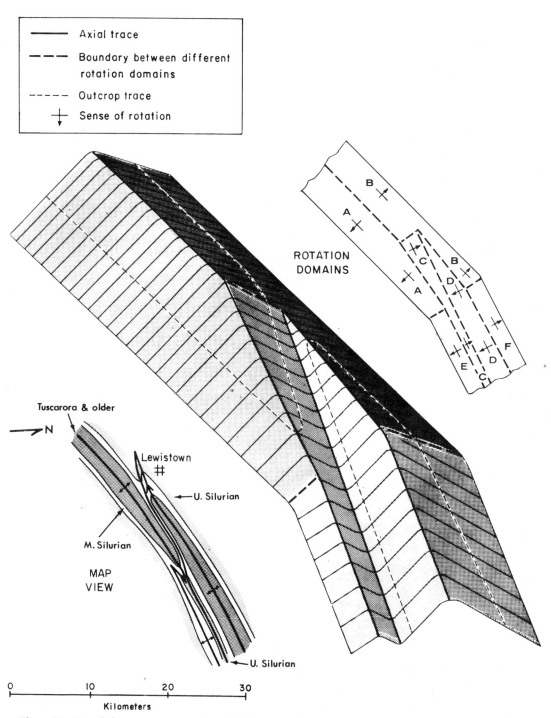

ROTATION
DOMAINS

Tuscarora & older

N

Lewistown
⌗

← U. Silurian

M. Silurian

MAP
VIEW

← U. Silurian

0   10   20   30
Kilometers

Figure 28. En echelon arrangement and termination of two adjacent major folds of different trends. Where contiguous domains possess the same trend (as in domains A-B, E-C, D-F, and C-D), the folds are cylindrical. Where the F trend changes to a B trend, the D-B fold plunges and terminates within the B domain of the A-B fold. Similarly, the change from B to C results in the plunging A-C fold. In contrast, though, the A domain changes to an E domain and thus, rather than terminating in the D domain, the fold becomes a cylindrical fold parallel and subsidiary to the major D-F fold. The geologic map is of the Shade Mountain anticlinorium, near Lewistown, Pennsylvania (from Gray and Shepps, 1960).

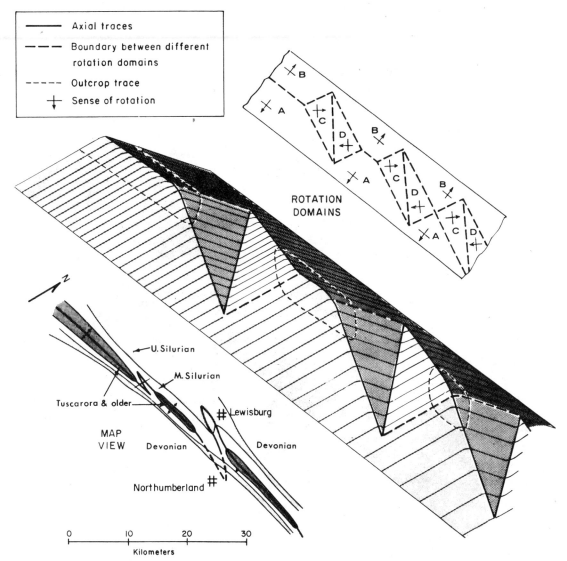

Figure 29. The dominant structure is the major fold comprised of kinematic domains A and B. Superimposed on this in the hinge are domains C and D, at some small angle to A and B (the angular relations are exaggerated here for illustrative purposes). This super-position results in short doubly plunging folds in the major fold hinge. The geologic map is of Jacks Mountain anticlinorium, near Northumberland, Pennsylvania (from Gray and Shepps, 1960).

Donath, 1968), and probably parallels the intermediate principal stress. The discordancies between the assumed smooth stress field and the observed discontinuous, interfingering kinematic domains may be attributable to mechanical constraints in the kink-banding process. The assumption is made (although there are no supportive or contradictory experimental data) that it requires less energy (and less deformation of the rock) to produce a rectilinear kink band (one in which the kink-axis orientation is constant along its length) than to produce one that is curved and conforming exactly to the curvature of the stress system. Yet, there is probably a limit to the angular divergence between the kink axis and the intermediate principal stress direction: where this limit is reached (5° to 10°?), the kink axis and the kink band abruptly change their orientation to be in closer accordance with the deforming stress system. Thus, the interfingering of domains, which constitutes the discontinuous curvature, probably represents a mechanical balance between a tendency for rectilinear

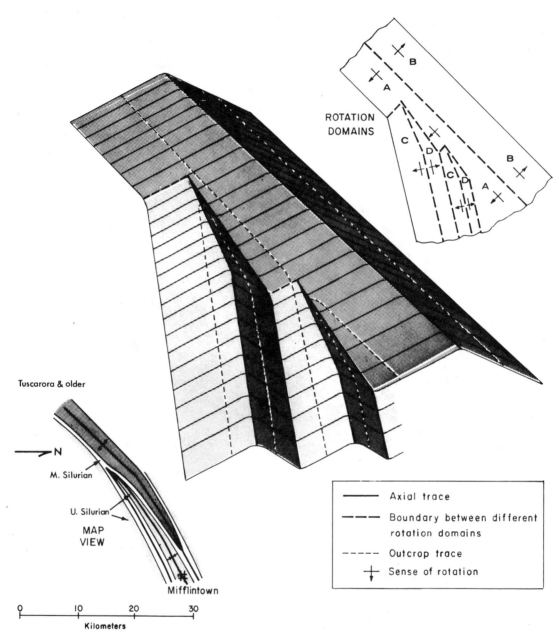

ROTATION DOMAINS

Tuscarora & older

M. Silurian

U. Silurian

MAP VIEW

Mifflintown

0   10   20   30
Kilometers

—————— Axial trace
— — — Boundary between different rotation domains
- - - - - Outcrop trace
+ Sense of rotation

Figure 30. Origination of folds in the limb of a major fold. The minor folds with trend C-D originate (or terminate) within the major A-B fold limb, and may or may nor persist for any great distance along their trend. Geologic map is of Shade Mountain anticlinorium, near Mifflintown, Pennsylvania (from Gray and Shepps, 1960).

kink-band deformation and a conformity to a uniformly curving stress system.

SUMMARY

The folds of the Valley and Ridge province in central Pennsylvania possess two features that distinguish them from concentric and similar folds: (1) the fold hinges are narrow in that the radii of curvature of the beds are small with respect to the fold wavelengths; and (2) bedding maintains a constant attitude in the fold limbs. They are flexural slip folds, in which the folding process has been slip on bedding surfaces (and subordinate wedges, or thrust faults) with little or no internal deformation of the beds. Yet they do not possess the cross-sectional profile of concentric folds. Although the similarity of bed form from one

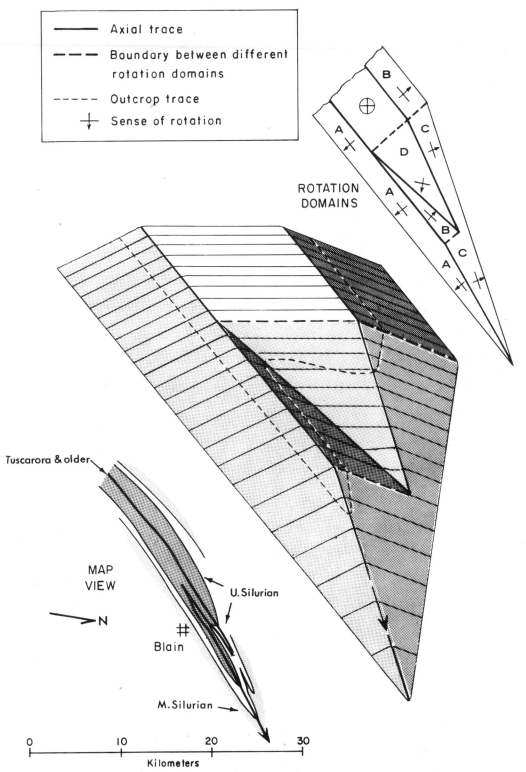

Figure 31. Complex version of a simple termination. In this structure, the major fold (which may or may not be conjugate) bifurcates where one of the major kinematic domains (for example, B) changes trend (into domain C). One of the bifurcation folds plunges abruptly, whereas the other persists along trend as a somewhat smaller cylindrical fold before it plunges and terminates as well. Geologic map is of Tuscarora anticlinorium, near Blain, Pennsylvania (from Gray and Shepps, 1960).

419

layer to the next suggests the geometry of similar folds, the constant bed-normal thickness across the folds indicates the bedding has not been attenuated in the limbs, nor appreciably thickened in the hinges. The profiles are also not sinusoidal as they tend to be in similar folds.

Rather, the Valley and Ridge folds are kink-band folds, consisting of two jointed kink bands, inclined toward each other and with opposite senses of rotation. This interpretation is supported by the presence of kink bands of greatly different sizes, the mechanical similarity between the kink-band process and the flexural slip process, and the common geometrical elements (zones of planar bedding, abrupt changes in bed attitude) between kink bands and the folds. More important, kink bands can be appropriately combined to reproduce all the variations of fold profile that are observed in the province.

As utilitarian as the kink-band concept is in describing the two-dimensional cross-section geometry of the folds, it is in three dimensions that the concept attains its full potential. The complexities in this province of doubly plunging folds, en echelon folds, fold terminations, and bifurcations defy adequate explanation by simple fold mechanics. On the other hand, in a kink-band fold, the two kink axes can, and do, diverge from each other, and the resulting folds are noncylindrical. These two concepts, the kink-band fold and the divergence of the kink axes, make possible the reproduction of all the three-dimensional variations of these noncylindrical folds. Because these noncylindrical folds are an integral part of the regional arcuation of the Valley and Ridge province in Pennsylvania, these two concepts enable a detailed delineation of the geometry and kinematics of this arcuation.

ACKNOWLEDGMENTS

The author expresses appreciation for the critical reviews by Fred A. Donath, Richard Groshong, and Richard P. Nickelsen which engendered significant improvements to the final manuscript.

REFERENCES CITED

Borg, I. Y., and Handin, J., 1966, Experimental deformation of crystalline rocks: Tectonophysics, v. 3, p. 251–367.

Busk, H. G., 1929, Earth flexures: London, Cambridge Univ. Press, 106 p.

Butts, C., Swartz, F. M., and Willard, B., 1939, Tyrone quadrangle: Pennsylvania Geol. Survey (4th ser.), Atlas 96, 118 p.

Cloos, E., 1961, Bedding slips, wedges and folding in layered sequences: Soc. Geol. Finlande Compte Rendus, v. 33, p. 106–122.

—— 1964, Wedging, bedding plane slips, and gravity tectonics in the Appalachians, in Lowry, W. D., ed., Tectonics of the southern Appalachians: Virginia Polytech. Inst. Dept. Geol. Sci. Mem. 1, p. 63–70.

de Sitter, L. U., 1956, Structural geology: New York, McGraw-Hill Book Co., 552 p.

Donath, F. A., 1968, The development of kink bands in brittle anisotropic rock, in Larsen, L. H., ed., Igneous and metamorphic geology: Geol. Soc. America Mem. 115, p. 453–493.

Donath, F. A., and Parker, R. B., 1964, Folds and folding: Geol. Soc. America Bull., v. 75, p. 45–62.

Drake, C. L., and Woodward, H. P., 1963, Appalachian curvature, wrench faulting, and offshore structures: New York Acad. Sci. Trans., v. 26, p. 48–63.

Dyson, J. L., 1963, Geology and mineral resources of the northern half of the New Bloomfield quadrangle: Pennsylvania Geol. Survey (4th ser.), Atlas 137ab, 63 p.

—— 1967, Geology and mineral resources of the southern half of the New Bloomfield quadrangle: Pennsylvania Geol. Survey (4th ser.), Atlas 137cd, 86 p.

Faill, R. T., 1969a, Kink band structures in the Valley and Ridge province, central Pennsylvania: Geol. Soc. America Bull., v. 80, p. 2539–2550.

—— 1969b, A solution of the "space problem" in the cores of flexural slip folds: Geol. Soc. America Spec. Paper 121, p. 347.

—— 1974, Structural analysis of the Valley and Ridge province, with particular emphasis on the Millerstown 15-minute quadrangle: Pennsylvania Geol. Survey, Gen. Geol. Rept. (4th ser.) (in press).

Faill, R. T., and Wells, R. B., 1973, Geology and mineral resources of the Millerstown 15-minute quadrangle: Pennsylvania Geol. Survey (4th ser.), Atlas 136 (in press).

Geiser, P. A., 1971, Deformation processes of some unmetamorphosed sedimentary rocks: EOS (Am. Geophys. Union Trans.), v. 52, p. 344.

Goguel, J., 1962, Tectonics: San Francisco, W. H. Freeman and Co., 384 p.

Gray, C., and Shepps, V. C., 1960, Geologic map of Pennsylvania: Pennsylvania Geol. Survey (4th ser.), Map 1.

Groshong, R. H., 1971, Strain in natural single-layer folds: EOS (Am. Geophys. Union Trans.), v. 52, p. 345.

Gwinn, V. E., 1964, Thin-skinned tectonics in the Plateau and northwestern Valley and Ridge provinces of the central Appalachians:

Geol. Soc. America Bull., v. 75, p. 863–900.

—— 1970, Kinematic patterns and estimates of lateral shortening, Valley and Ridge and Great Valley provinces, central Appalachians, south-central Pennsylvania, *in* Fisher, G. W., Pettijohn, F. J., Reed, J. C., Jr., and Weaver, K. N., eds., Studies of Appalachian geology: central and southern: New York, John Wiley and Sons, p. 127–146.

Hills, E. S., 1963, Elements of structural geology: New York, John Wiley and Sons, 483 p.

Knowles, R. R., and Opdyke, N. D., 1968, Paleomagnetic results from the Mauch Chunk Formation: a test of the origin of curvature in the folded Appalachians of Pennsylvania: Jour. Geophys. Research, v. 73, p. 6515–6526.

McBride, E. F., 1962, Flysch and associated beds of the Martinsburg formation (Ordovician), central Appalachians: Jour. Sed. Petrology, v. 32, p. 39–91.

McIver, N. L., 1970, Appalachian turbidites, *in* Fisher, G. W., Pettijohn, F. J., Reed, J. C., Jr., and Weaver, K. N., eds., Studies of Appalachian geology: central and southern: New York, John Wiley and Sons, p. 69–81.

Moebs, N. N., and Hoy, R. B., 1959, Thrust faulting in Sinking Valley, Blair and Huntingdon Counties, Pennsylvania: Geol. Soc. America Bull., v. 70, p. 1079–1088.

Nickelsen, R. P., 1963, Fold patterns and continuous deformation mechanisms of the central Pennsylvania Folded Appalachians, *in* Tectonics and Cambrian-Ordovician stratigraphy, central Appalachians of Pennsylvania: Pittsburgh Geol. Soc. Guidebook, p. 13–29.

—— 1966, Fossil distortion and penetrative rock deformation in the Appalachian Plateau, Pennsylvania: Jour. Geology, v. 74, p. 924–931.

Paterson, M. S., and Weiss, L. E., 1966, Experimental deformation and folding in phyllite:

Geol. Soc. America Bull., v. 77, p. 343–374.

—— 1968, Folding and boudinage of quartz-rich layers in experimentally deformed phyllite: Geol. Soc. America Bull., v. 79, p. 795–812.

Pelletier, B. R., 1958, Pocono paleocurrents in Pennsylvania and Maryland: Geol. Soc. America Bull., v. 69, p. 1033–1064.

Ramsay, J. G., 1962, The geometry of conjugate fold systems: Geol. Mag., v. 59, p. 516–526.

Roy, J. L., Opdyke, N. D., and Irving, E., 1967, Further paleomagnetic results from the Bloomsburg Formation: Jour. Geophys. Research, v. 72, p. 5075–5086.

Turner, F. J., and Weiss, L. E., 1963, Structural analysis of metamorphic tectonites: New York, McGraw-Hill Book Co., 545 p.

Turner, F. J., Griggs, D. T., and Heard, H., 1954, Experimental deformation of calcite crystals: Geol. Soc. America Bull., v. 65, p. 883–934.

Williams, P. F., and Means, W. D., 1971, Experimental folding of foliated materials: EOS (Am. Geophys. Union Trans.), v. 52, p. 345.

Wood, G. H., Jr., Trexler, J. P., and Kehn, T. M., 1969, Geology of the west-central part of the southern Anthracite Field and adjoining areas, Pennsylvania: U.S. Geol. Survey Prof. Paper 602, 150 p.

Woodward, H. P., 1964, Central Appalachian tectonics and the deep basin: Am. Assoc. Petroleum Geologists Bull., v. 48, p. 338–356.

Yeakel, L. S., Jr., 1962, Tuscarora, Juniata, and Bald Eagle paleocurrents and paleogeography in the central Appalachians: Geol. Soc. America Bull., v. 73, p. 1515–1540.

Manuscript Received by the Society March 29, 1972

Revised Manuscript Received September 20, 1972

[AMERICAN JOURNAL OF SCIENCE, VOL. 283, SEPTEMBER, 1983, P. 684-721]

GEOMETRY AND KINEMATICS OF
FAULT-BEND FOLDING

JOHN SUPPE

Department of Geological and Geophysical Sciences,
Princeton University, Princeton, New Jersey 08544

ABSTRACT. Since the initial study of folds in the hanging wall of the Pine Mountain thrust sheet in the southern Appalachians 50 yrs ago by J. L. Rich, it has become clear that many map-scale folds in sedimentary sequences are formed by the bending of fault blocks as they ride over non-planar fault surfaces. These structures, here called *fault-bend folds*, include "reverse-drag" or "rollovers" associated with normal faults that flatten with depth and the bending of thrust sheets as they ride over steps in decollement. This paper presents a number of geometric and kinematic properties of parallel fault-bend folds, the most important of which is a relationship between fault shape and fold shape for sharp bends in faults. These relationships are useful tools for developing internally consistent cross sections in areas of suspected fault-bend folding, particularly in fold-and-thrust belts.

INTRODUCTION

Many large-scale folds that have formed at shallow crustal levels, above the brittle-plastic transition, have origins that are intimately related to slip on adjacent faults. The important classes of fault-related folding include: (1) buckling caused by compression above a bedding-plane decollement, (2) fault-bend folding caused by bending of a fault-block as it rides over a non-planar fault surface, and (3) fault-propagation folding, caused by compression in front of a fault tip during fault propagation. Because of the shallow-level non-plastic nature of this folding, layer thickness is commonly preserved during deformation, that is, the folding is parallel. This report presents a number of useful geometric and kinematic properties of parallel fault-bend folding. A few examples of their application to real structures are included, many more are published elsewhere together with abbreviated fragments of this theory (Suppe, 1979, 1980a, b; Suppe and Namson, 1979; Namson, 1981).

FAULT-BEND FOLDING

If a fault surface is not planar there must be distortion within at least one of the fault blocks as they slip past one another. The distortion develops because the two blocks remain in tight contact along the fault surface during slip, the rocks not being strong enough to support large voids. If the rocks are layered they may fold in response to riding over a bend in a fault. We call this mechanism of folding *fault-bend folding*. This mechanism is well known in fold-and-thrust belts associated with steps in decollement, in so-called "reverse drag" associated with flattening normal faults (fig. 1), and in "flower structures" associated with bends in strike-slip faults. Fault-bend folding is closely related geometrically to

684

folding of preexisting faults and refraction of axial surfaces across angular unconformities (fig. 1).

This report presents an idealized two-dimensional geometric description of fault-bend folding, which has applications to all these phenomena (fig. 1). But the emphasis here will lie almost entirely with folds produced by thrust faults and their imbrications. Several sections on applications illustrate the quantitative use of the theory to decipher subsurface map-scale structure in fold-and-thrust belts.

FOLDING DUE TO A SIMPLE RAMP IN DECOLLEMENT HORIZON

Thrust faults do not run forever along a single bedding-plane decollement. The thrust normally steps up in the direction of slip to a higher decollement or to the land surface. As the thrust sheet rides over the bends in the fault it must fold. This fact was clearly perceived by Rich (1934), who applied the concept of fault-bend folding to the interpretation of folds in the Pine Mountain thrust sheet of the southern Appalachians (fig. 2). Rich realized that the Powell Valley anticline is the result of a ramp in decollement of the Pine Mountain thrust from the Lower Cambrian Rome Formation to the Devonian Chattanooga Shale. Analogous folding due to ramps in decollement, both across and along strike, is now widely recognized in many fold-and-thrust belts (for example, Rodgers, 1950; Douglas, 1950; Laubscher, 1965; Gwinn, 1970; Harris, 1970; Perry, Harris, and Harris, 1979; Roedder, Gilbert, and Witherspoon, 1978; Suppe, 1976, 1980a, b; Suppe and Namson, 1979).

The kinematics of fault-bend folding caused by a simple step in decollement along a thrust fault are illustrated in figure 3. Points X and Y,

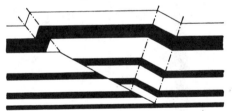 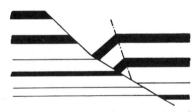

A. Step in decollement on thrust fault.   B. Reverse drag on flattening normal faults.

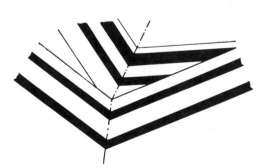 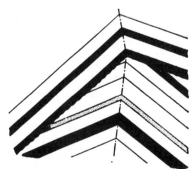

C. Folding of faults.   D. Folding of angular unconformities.

Fig. 1. Examples of some common types of fault-bend folds (A-C) and the geometrically related structure of a folded angular unconformity (D).

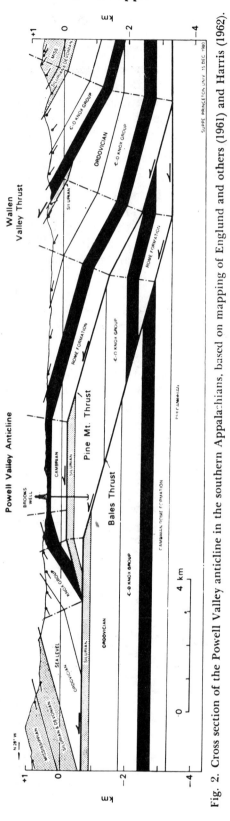

Fig. 2. Cross section of the Powell Valley anticline in the southern Appalachians, based on mapping of Englund and others (1961) and Harris (1962).

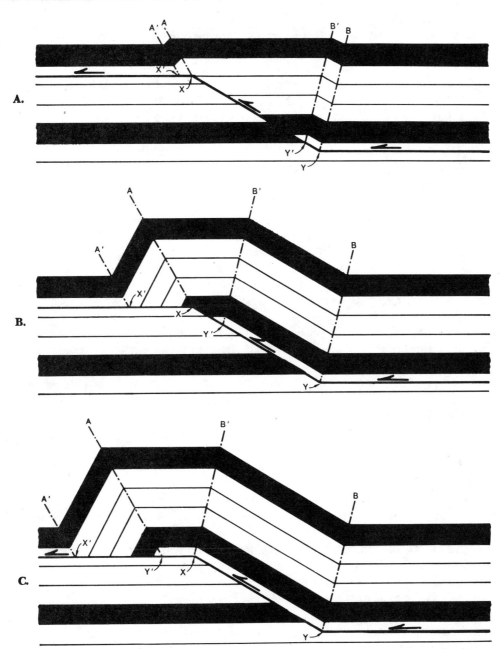

Fig. 3. Kinematic development of fault-bend folds in response to a simple step in decollement (after Suppe and Namson, 1979).

which are lines in 3-dimensions, bound the cross-cutting fault segment in the foot-wall block; points X' and Y' bound the cross-cutting segment of the hanging-wall block. Note that the folds are confined to the hanging-wall block. Axial surfaces A and B terminate along the fault at points (lines) X and Y in the foot-wall block; similarly axial surfaces A' and B' terminate along the fault at points (lines) X' and Y' in the hanging-wall block.

According to the simple geometry of figure 3, two kink bands, A-A' and B-B', form at the instant of initiation of slip. As slip continues, both kink bands grow in width, and the structural relief increases. Note that the slip is not constant along the fault but decreases about 60 percent on the left-hand side, because slip is taken up in kink band A-A'.

Axial surfaces A and B, associated with points X and Y in the foot-wall block of figure 3, remain fixed with respect to the foot-wall block; the beds of the hanging-wall block roll through these two axial surfaces as slip proceeds. In contrast, axial surfaces A' and B', which terminate at points X' and Y' in the hanging-wall block, are fixed in the hanging-wall beds and move with the thrust sheet.

The kinematics are a bit more complex, however, because when point Y' reaches point X, axial surface B' suddenly stops moving with the hanging wall and becomes fixed with respect to the foot wall at point X. At the same instant axial surface A is released from point X in the foot wall and begins to move with point Y' in the hanging wall. When Y' reaches point X the kink bands A-A' and B-B' cease to grow, although rocks still roll through axial surfaces B and B'.

Actual fault-bent folds may not develop precisely as shown diagrammatically in figure 3, because the kinematic details depend on the mechanical properties of the layers and the way in which the forces are applied. The drawings in figure 3 were constructed with the assumptions of (1) preservation of layer thickness, measured normal to bedding, (2) no net distortion where the layers are horizontal, and (3) conservation of bed length. Inclined layers have undergone only layer-parallel slip. We call this set of three assumptions *parallel behavior* in the following discussion.

Many fault-bend folds in unmetamorphosed sedimentary rocks are found to obey the three assumptions of parallel behavior; for example, map, well, and seismic data from the Pakuashan anticline in the Neogene basin of western Taiwan (fig. 4) can be fit to a cross section satisfying the assumptions. The cross section of the Pine Mountain thrust sheet (fig. 2) also obeys the assumptions of parallel behavior.

In most of the following discussion we assume parallel behavior but emphasize that many cases can be found for which the assumptions are invalid, for example in rocks that exhibit slaty cleavage. We are not developing a mechanical theory of fault-bend folding but rather a geometric and kinematic description of a specific material behavior known to be closely approximated in some fault-bend folds, such as the Pine Mountain thrust sheet (fig. 2). The usefulness of the theory comes from its combination of simplicity and rather wide applicability. Similar, but more complex, theories can be formulated to deal with modes of deformation other than parallel behavior. The mode of deformation must be determined by field and other observations (for example, Laubscher, 1975, 1976).

PARALLEL KINK FOLDING BY CHANGE IN DIP OF A FAULT

Introduction.—In many practical as well as scientific problems of subsurface exploration we wish to predict the complete shape of a system of folds and faults given presently available information, prior to continued exploration. It is a common situation for the structure to involve slip on non-planar faults or folding of preexisting faults. Both these situations are geometrically closely related to the structure shown diagrammatically in figure 3, although most situations appear to be more complex because of more bends in the faults or several branches to the faults. Even the classic Pine Mountain thrust (fig. 2) is more complex than the simple step in decollement of figure 3. In order to develop more precise predictions of subsurface structure, especially in fold-and-thrust belts, it would be of great practical help to have a relationship between

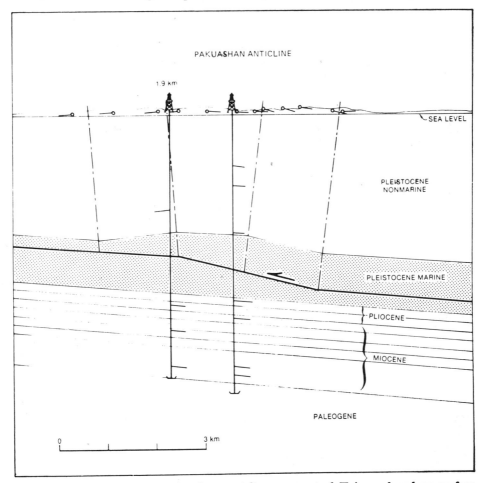

Fig. 4. Cross section of Pakuashan anticline, westcentral Taiwan, based on surface mapping, well data, and seismic data.

shapes of folds and the shapes of faults that are responsible for the folds through the mechanism of fault-bend folding.

In the following section we develop a simplified, yet widely applicable, two-dimensional geometric and kinematic theory of folding due to slip past a series of sharp bends in a fault (fig. 5). The theory is rather general. All shapes of sharp bends can be considered; curved bends can be treated as a series of sharp bends. Both convex and concave fault bends and associated anticlinal or synclinal folds can be treated. Even folding of faults, together with their adjacent beds, can be studied.

The primary geometric assumptions of the two-dimensional theory are sharp fault bends, conservation of area, and constant layer thickness normal to bedding (fig. 5), which imply conservation of bed length, deformation by layer-parallel slip, and angular kink (chevron) folds of infinite curvature and straight limbs. By these assumptions the axial surface bisects the angle between the two fold limbs ($\gamma_1 = \gamma_2$ in fig. 5); we call the angle γ the *axial angle*. Many actual fault-bend folds closely approximate these assumptions, as mentioned in the previous section, thus giving the theory some usefulness. More complex theories involving unequal axial angles or three-dimensional fault-bend folds may be developed with more effort. The next section presents the mathematical details. Applications are given in later sections.

Details of two-dimensional geometry.—We solve the geometric problem in two dimensions of what change in dip ϕ of a cross-cutting fault will produce a fold of a given axial angle ($\gamma_1 = \gamma_2$) (fig. 5). The initial angle between bedding and the fault, prior to slipping past the bend, is θ (fig. 5). It is assumed that the bedding thickness remains constant during folding, that the fold is angular, and that the deformation conserves area. It follows that the deformation is by slip parallel to bedding and that bed length is preserved during deformation. General shear within a fault-block is excluded under many circumstances by the assumptions, and for

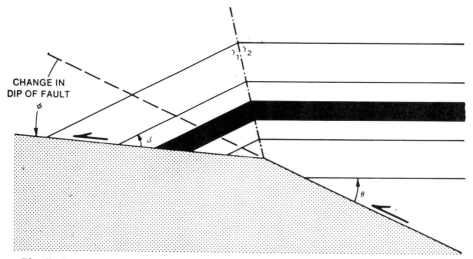

Fig. 5. Geometry of a generalized parallel-kink fault-bend fold (after Suppe, 1979).

the present we exclude all shear unrelated to riding over fault bends. Later we consider several important ancillary phenomena that involve additional layer-parallel shear.

The area of the deformed beds (triangle *a b d* in fig. 6) must equal original undeformed area (triangle *a b c*). Similarly, deformed and undeformed bed lengths must be equal; thus lines *b d* and *b c* are both of length *l* in figure 6. Given these two constraints we can solve for the change in dip of the fault ϕ in terms of θ and γ (figs. 5 and 6).

Line segment *b d* of length *l* is divided at point *e* into two segments *b e* and *e d* (fig. 6) where

$$b\,e = \frac{l \sin\theta}{\sin(2\gamma-\theta)} \tag{1}$$

by the law of sines. Similarly applying the law twice

$$e\,d = \frac{l \sin\gamma \, \sin\phi}{\sin(\phi+\gamma-\theta)\,\sin(2\gamma-\theta)} \tag{2}$$

Adding eqs (1) and (2) we have

$$b\,e + e\,d = l = \frac{l \sin\theta}{\sin(2\gamma-\theta)} + \frac{l \sin\gamma \, \sin\phi}{\sin(\phi+\gamma-\theta)\,\sin(2\gamma-\theta)} \tag{3}$$

Expanding this equation we get

$$1 = \frac{\sin\theta}{\sin(2\gamma-\theta)} + \frac{\sin\gamma \, \sin\phi}{\sin 2\gamma-\theta)\,[\sin\phi\cos(\gamma-\theta) + \cos\phi\sin(\gamma-\theta)]} \tag{4}$$

Multiplying by the denominator of the second term on the right and dividing by $\sin \phi$ we obtain

$$\sin(2\gamma-\theta)\,[\cos(\gamma-\theta) + \cot\phi\sin(\gamma-\theta)]$$
$$= [\cos(\gamma-\theta) + \cot\phi\sin(\gamma-\theta)]\sin\theta + \sin\gamma\,, \tag{5}$$

rearranging we find

$$-\cot\phi = \frac{\cos(\gamma-\theta)\,[\sin(2\gamma-\theta) - \sin\theta] - \sin\gamma}{\sin_{\cdot}\gamma-\theta)\,[\sin(2\gamma-\theta) - \sin\theta]}. \tag{6}$$

Finally we have

$$\phi = \tan^{-1}\left[\frac{-\sin(\gamma-\theta)\,[\sin(2\gamma-\theta) - \sin\theta]}{\cos(\gamma-\theta)\,[\sin(2\gamma-\theta) - \sin\theta] - \sin\gamma}\right] \tag{7}$$

The new angle β between the fault and bedding, after slip past the bend in the fault, is given by the equation

$$\beta = \theta - \phi + (180°-2\gamma) = \theta - \phi + \delta \tag{8}$$

where $\delta = (180°-2\gamma)$ is the change in dip across the axial surface. In some practical cases the angle β is more easily estimated than θ, ϕ, or even γ. In these cases eq (8) is a useful addition to eq (7).

We must establish certain sign conventions and ranges of angles. We define $90°{\geqslant}\theta{\geqslant}-90°$, $180°{\geqslant}\gamma{\geqslant}0°$, $90°{\geqslant}\phi{\geqslant}-90°$, $180°{\geqslant}\beta{\geqslant}-180°$, and

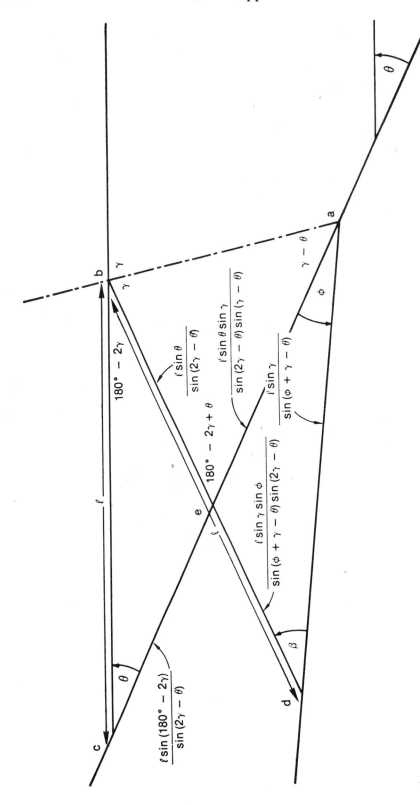

Fig. 6. Geometric relationships used in deriving the basic equations of parallel-kink fault-bend folding.

$180° \geqslant \delta \geqslant -180°$. Briefly, ϕ is measured from the projection of the first fault to the second fault and is positive for folds that are convex to the fault ("anticlines"). β and θ are positive in the same direction — clockwise or anticlockwise — as ϕ and are both measured from the fault to the same side of the bed (note $|\theta| \leqslant 90°$). δ is the same sign as ϕ if its direction is the same as ϕ (positive for anticlines). δ is measured from the projection of the bed in the same direction as θ.

The above sign conventions are unduly complex for most practical problems not involving slip over multiple positive and negative fault bends (+ and $-\phi$). The problem is simplified if we consider all folds convex toward the fault as "synclines" and all folds concave toward the fault as "anticlines"; we define $90° \geqslant \gamma \geqslant 0°$, $90° \geqslant \phi \geqslant 0°$, $90° \geqslant \theta \geqslant -90°$, $180° \geqslant \beta \geqslant -90°$, and $180° \geqslant \delta \geqslant 0°$. β and θ are measured from the fault to the same side of the bed (note $|\theta| \leqslant 90°$) and are positive for "anticlines" and negative for "synclines". Figure 7 presents graphs of eqs (7) and (8), using the simplified sign conventions. "Anticlines" are to the left, and "synclines" are to the right.

Equation of a simple step in decollement.—Folds that involve a simple step from one decollement to another (fig. 3) are sufficiently important that it is useful to derive the fault-bend folding relationship (eq 7) for this special case, namely

$$\phi = \theta \tag{9}$$

Combining with eq 4 we obtain

$$\sin (2\gamma - \theta) = \sin\theta + \frac{\sin\gamma\sin\theta}{[\sin\theta\cos(\gamma-\theta) + \cos\theta\sin(\gamma-\theta)]} \tag{10}$$

Simplifying using trigonometric identities we obtain

$$\tan\theta = \frac{\sin 2\gamma}{2\cos^2\gamma + 1} \tag{11}$$

and

$$\phi = \theta = \tan^{-1}\left[\frac{\sin 2\gamma}{1 + 2\cos^2\gamma}\right] \tag{12}$$

which is the relationship between cutoff angle θ or fault bend ϕ and fold shape γ for a simple step in decollement. A graph of eq (12) is given as part of figure 7.

Discussion of the fault-bend folding equations.—The above equations, especially 7, 8, and 12, have many practical applications in attempts to predict the subsurface geometry of structures. The assumptions are two dimensions, conservation of area and bed length, and no general affine shear. These imply slip parallel to bedding. If a structure cannot be successfully described by these equations then either it is not a fault-bend fold, the assumptions are not valid, or the structure is too complicated to solve given the limited data available. Thus the equations are useful whether or not the structure is actually of the parallel-kink fault-bend mechanism, because they are a standard against which the structure can be quantitatively compared.

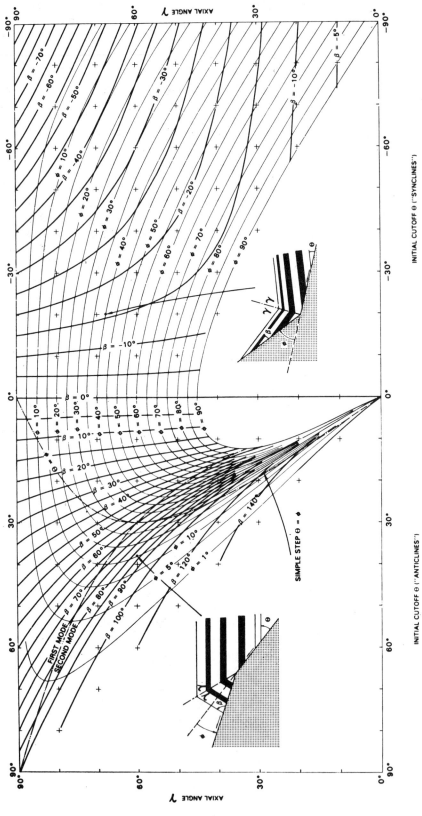

Fig. 7. Graph of relationships between fold shape γ, fault shape φ, and cutoff angles θ and β for parallel-kink fault-bend folding; based on eqs 7, 8, and 12.

A graph of eqs (7), (8), and (12) has been prepared (fig. 7) to allow quick visual analysis of the possible range of solutions to a given problem. For example, if an anticline has an observed axial angle γ of 83° then an associated change in fault dip ϕ cannot be greater than about 16° to 17°; furthermore ϕ of less than 10° is not possible for a large range of θ and β. If the bend is a simple step, then the cutoff angle $\theta = \phi$ is 14°. Thus the range of possible solutions can be quickly assessed. For this reason it is more efficient in most practical applications to use the graph than to use the equations directly.

The graph in figure 7 plots θ against γ showing lines of constant ϕ and β. A notable feature of the graph is that γ is a double-valued function of θ and ϕ for "anticlines"; thus, for a given fault bend ϕ and initial cutoff angle θ there are two possible shapes in the fold γ. For example, if $\theta = \phi = 20°$, then the axial angle γ can be 78° or 32°; the larger values of γ are called *first-mode folds,* and the smaller values are called *second-mode folds* (fig. 8). The boundary between the two modes on the graph is

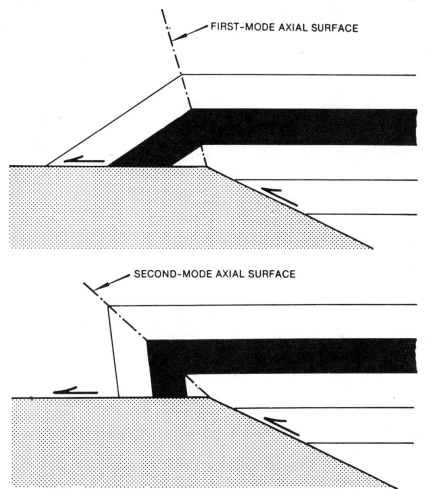

Fig. 8. The two modes of fault-bend folding for $\theta = \phi = 25°$ (after Suppe and Namson, 1979).

marked with a dashed line (fig. 7). It may be questioned whether or not both geometrically possible modes are mechanically possible; for example, the first mode might be favored by its smaller bending and bedding-plane slip. Nevertheless, there is some indication that second-mode folds might form; for example, Suppe and Namson (1979) present some subsurface interpretations involving second-mode simple-step folding. It is important to be cautious in interpreting folds with steep to overturned front limbs as second-mode fault-bend folds, because fault-propagation folds have a shape that is difficult to distinguish in the subsurface from second-mode fault-bend folds, as will be discussed elsewhere. Our present experience is that most fault-bend fold structures are first-mode folds or complex imbricated stacks of first-mode folds.

A second important aspect of the graph (fig. 7) is that for a given initial cutoff angle θ there exists a maximum angle of fault bend ϕ for which layer thickness is preserved in anticlinal folds because of the double-valued nature of eqs (7) and (12). For example in the case of a simple ramp in decollement the maximum angle of stepup ($\phi=\theta$) is 30° without thinning of beds (non-parallel folding). As another example, a reverse fault stepping up at 55° to bedding cannot flatten more than 4° without experiencing local layer thinning or secondary faulting. Therefore, information on layer thickness around suspected fault-bend folds offers a potential constraint on possible fault bends at depth. The range of possible fault bends for which layer thickness is preserved is even more restricted in cases of multiple imbrications, as is discussed later in this report.

Simple application of fault-bend fold equations.—The most straightforward applications of the fault-bend fold equations are in the solution of well-constrained problems or parts of problems involving simple bends in faults. For example, the fold shape (γ) (fig. 5) may be well known, and we may wish to compute the shape of a fault (ϕ) capable of producing the fold, given the orientation of the fault (θ or β) in one part of the structure. An actual example is given in the following paragraphs. The equations also may be used to solve more complex problems of imbricate structures as is explained near the end of this report.

Figure 9 presents as an example an incomplete cross section of the crest of the Hukou-Yangmei anticline in the fold-and-thrust belt of western Taiwan. The cross section is constrained by detailed surface mapping, two wells, and some seismic data. Well A is unusual because it encountered a double thickness of the distinctive Pliocene Chinshui Shale and normal thicknesses of formations below the Chinshui Shale, suggesting that the small fold on which well A sits does not extend below the Chinshui Shale. Let us make the hypothesis that the fold is a fault-bend fold associated with a fault that repeats the Chinshui Shale in well A; we now attempt to test this hypothesis.

This example from Taiwan is typical of simple applications of the fault-bend fold equations; we have a general structural hypothesis, and we wish to test it quantitatively. In order to do this we must guess a specific solution. Two guesses are shown in figure 10, both involving the

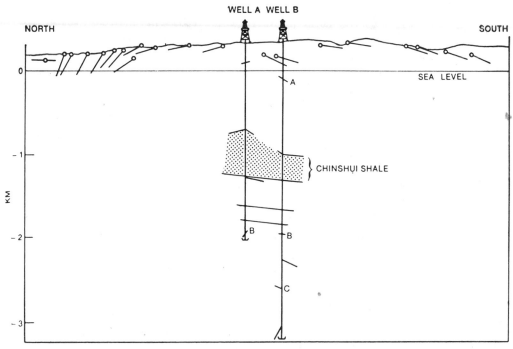

Fig. 9. Basic surface and subsurface data, Hukou-Yangmei anticline, northern Taiwan.

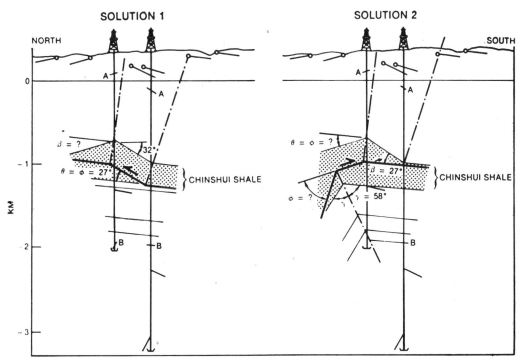

Fig. 10. Two potential solutions for the thickening of Chinshui Shale by fault-bend folding, Hukou-Yangmei anticline, northern Taiwan.

435

simple step of a thrust from one decollement to another in the Chinshui Shale. In solution 1 a thrust steps up to the north, whereas in solution 2 a thrust steps up to the south. The critical angular observations are that the dip at the base of the Chinshui Shale and below is 5°, whereas the minimum dip of the top of the Chinshui Shale, between the two wells, is 32° with similar but slightly lower dips observed at the surface off the line of section. Therefore we choose $32°-5° = 27°$ as $\theta = \phi$ in solution 1 and β in solution 2. Using figure 7 we determine $\beta = 34°$ for solution 1 with $34°-5° = 29°$ as the predicted surface dip. This surface dip is substantially greater than the observed surface dip of about 16° so we discard solution 1 as incorrect. Using figure 7 we determine $\phi = \theta = 22°$ for solution 2 and $22°—5° = 17°$ as the predicted surface dip, in good agreement with observation. We therefore consider solution 2 viable.

Further fault-bend computations are possible in this example. An anticline is present at depth, as shown by these and other wells. We now compute how the shallow fault in solution 2 will be folded by the deeper anticline ($\gamma = 58°$). The crosscutting fault block is on the footwall, convex toward the fault; therefore it corresponds to a "synclinal" geometry in figure 7, as may be seen by viewing figure 10 upside down. We observe $\theta = -22°$ and $\gamma = 58°$ for a "synclinal" geometry; therefore we determine from figure 7 that $\phi = 57°$ and $\beta = 15°$, which are in reasonable agreement with surface dips. A final version of the Hukou-Yangmei cross section (fig. 11), incorporates solution 2.

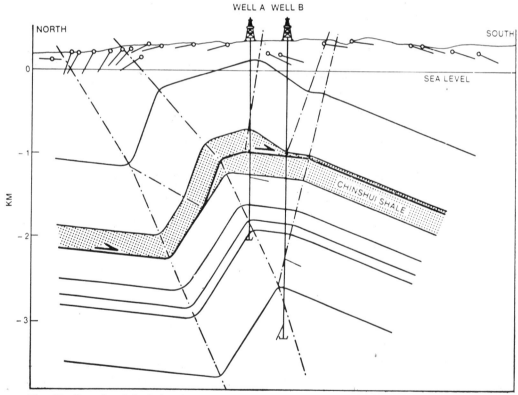

Fig. 11. Completed fault-bend folding interpretation of the shallow part of Hukou-Yangmei anticline, based on solution 2, figure 10.

ADDITIONAL ASPECTS OF PARALLEL-KINK FAULT-BEND FOLDS

In the present section we treat additional aspects of the geometric and kinematic description of parallel-kink fault-bend folds. These include (1) change in fault slip across a fault-bend fold, (2) multiple bends in a single fault, (3) shearing of fault-bend folds by layer-parallel slip in the thrust sheet, and (4) the branching of axial surfaces within the folded sheet. One further aspect, which is so important that it is treated in a separate section, is the geometric description of parallel-kink fault-bend folds involving multiple fault imbrications. This topic is discussed in the final section of this report, "Imbricate Fault-Bend Folding."

Change in fault slip across a fault-bend fold.—Fault slip is not conserved across a fault-bend fold if the fault cuts across bedding in the folded sheet. The slip may increase or decrease as a result of folding, although decreases are quantitatively more important. Continuing with the same assumptions as above, if $a\,c$ is the fault slip before the bend and $a\,d$ is the fault slip beyond the bend (fig. 6), then we may define a *ratio of slips R* as

$$R = \frac{a\,d}{a\,c} = \frac{\text{slip beyond bend}}{\text{slip before bend}} \tag{13}$$

By the law of sines we have

$$a\,c = \frac{l \sin (180°-\gamma)}{\sin (\gamma-\theta)} = \frac{l \sin \gamma}{\sin (\gamma-\theta)} \tag{14}$$

also

$$a\,d = \frac{l \sin \gamma}{\sin (\phi+\gamma-\theta)} \tag{15}$$

Combining (13), (14), and (15) we obtain an expression for the slip ratio

$$R = \frac{\sin (\gamma-\theta)}{\sin (\phi+\gamma-\theta)} \tag{16}$$

A graph of the ratio of slips R (eq 16) is given in figure 12. We first note that if the fault is not cross cutting ($\theta = 0°$), fault slip is preserved across the fault bend. If the fault is cross cutting ($\theta \neq 0°$) slip is increased in synclinal folding ($R > 1$) and decreased in anticlinal folding ($R < 1$). Note in figure 12 that decreases in slip are generally more substantial than increases in slip. Increases in slip are generally on the order of 10 to 20 percent or less ($R = 1.1$ to 1.2) for typical synclinal fault-bend folds, whereas decreases may reach 40 percent ($R = 0.6$) for first-mode simple-step anticlines and 60 percent ($R = 0.4$) for many second-mode simple-step anticlines. Therefore slip is not preserved across fault bends in thrust sheets that cross cut bedding. Fault slip is consumed or produced by folding within the thrust sheets.

Theory of multiple fault bends.—Many faults, especially thrust faults, have sufficiently large slip that the beds may have slipped past more than one bend in the fault. We now consider what effect the sequence of fault bends has on the final shape of the beds (γ, β). If the sequence of fault

John Suppe

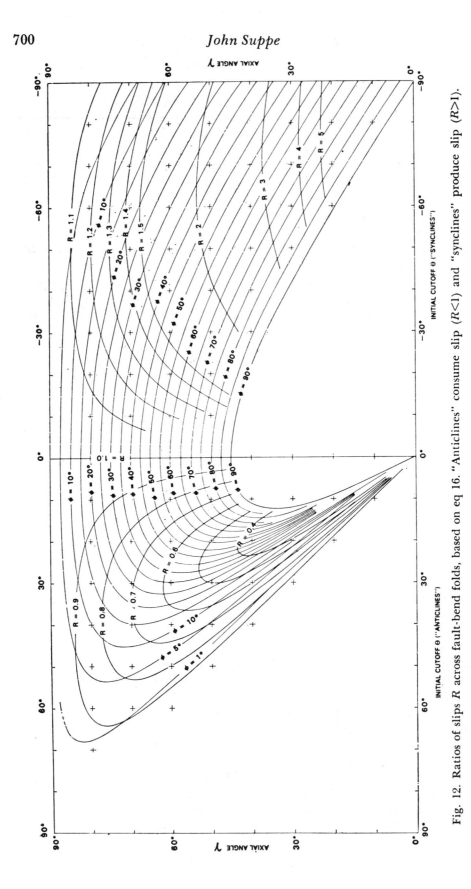

Fig. 12. Ratios of slips R across fault-bend folds, based on eq 16. "Anticlines" consume slip ($R<1$) and "synclines" produce slip ($R>1$).

bends is labeled $\phi_1, \phi_2 \ldots \phi_n$ then the *net bend* in the fault Φ between $i = 1$, and $i = n$ is

$$\Phi = \sum_{i=1}^{n} \phi_i. \tag{17}$$

For a sequence of bends, β_n of a previous bend is θ_{n+1} for the next bend (fig. 13); therefore from eqs (8) and (17) we have

$$\beta_1 = \theta_1 - \phi_1 + (180° - 2\gamma_1) = \theta_2$$
$$\beta_2 = [\theta_1 - \phi_1 + (180° - 2\gamma_1)] - \phi_2 + (180° - 2\gamma_2),$$

and

$$\beta_n = \theta_1 - \Phi + \sum_{i=1}^{n} (180° - 2\gamma_n) \tag{18}$$

The change in dip of the beds across the i-th fold γ_i is

$$\delta_i = (180° - 2\gamma_i). \tag{19}$$

Thus the *net change in dip* Δ is the last term of eq (18)

$$\Delta = \sum_{i=1}^{n} (180° - 2\gamma_i) = \sum_{i=1}^{n} \delta_i, \tag{20}$$

and eq (18) becomes

$$\beta_n = \theta_1 - \Phi + \Delta. \tag{21}$$

The net change in dip of beds, Δ, is obviously a function of the sequence of fault bends because it depends on the choice of mode of folding ($k = 1$ or 2) at each fault bend. Furthermore, even if we confine ourselves to a single mode, for example first mode, the final cutoff angle β_n and the net change in dip Δ both depend on the sequence of fault bends because they are not linearly related to θ and ϕ (eqs 7 and 8):

$$\beta_n (\phi_i, k_i) \quad i = 1 \rightarrow n, k_i = 1 \text{ or } 2$$

$$\Delta (\phi_i, k_i) \quad i = 1 \rightarrow n, k_i = 1 \text{ or } 2$$

The dependence of final cutoff angle on the sequence of fault bends is illustrated with two examples in figure 13, which also illustrate the effects of convex and concave ramps along thrust faults. In example 1 a net fault bend $\Phi = 30°$ with an initial cutoff of $30°$ is accomplished in one case (1A) with a single $30°$ bend ($\Phi = \phi_1 = \theta_1 = 30°$) producing a final cutoff $\beta = \Delta = 60°$, whereas in the other case (1B) the same net bend is accomplished by a series of three $10°$ bends ($\phi_1 = \phi_2 = \phi_3 = 10°$) producing a different final cutoff $\beta_3 = \Delta = 49°$. In example 2 a net fault bend $\Phi = 15°$ with an initial cutoff of $15°$ is accomplished in one case (2A) with a single $15°$ anticlinal bend ($\Phi = \phi_1 = \theta_1 = 15°$) producing a final cutoff $\beta = \Delta = 16°$, whereas in the other case (2B) the $15°$ net bend

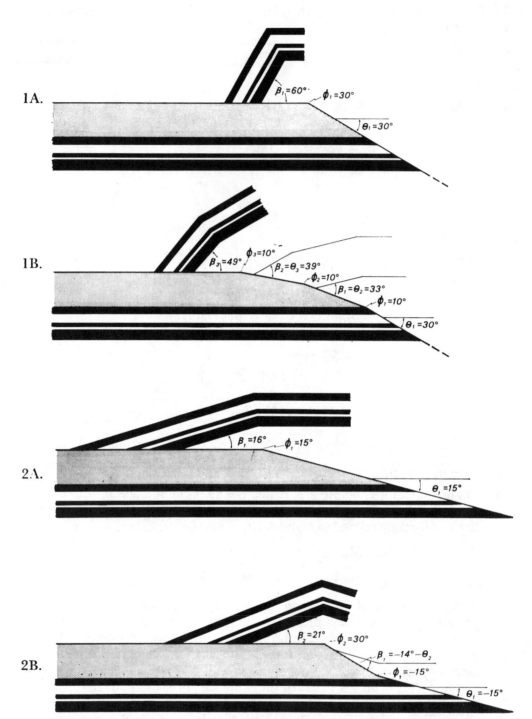

Fig. 13. Two illustrations of the dependence of final cutoff angle β_n on the sequence of fault bends (ϕ_1, ϕ_2, . . . ϕ_n) given the same net fault bend Φ and initial cutoff angle θ_1.

is accomplished by first a synclinal bend $\phi_1 = -15°$ and then an anticlinal bend $\phi_2 = +30°$ producing a final cutoff $\beta_2 = 21°$. These examples have no associated scale. In actual structures the fault-bend folds typically do not propagate a long distance into the overlying fault block if the distance between the bends is very small relative to the thickness of the fault block, instead the folds are dissipated in disharmonic, non-parallel behavior.

It should be noted that if we can measure the final cutoff angle β_n and the net change in dip Δ, which may be possible in practical problems, we can calculate $(\theta_1 - \Phi)$ using eq (21)

$$\beta_n\,(\phi_i, k_i) = \theta_1 - \Phi + \Delta(\phi_i, k_i) \tag{22}$$

because initial cutoff angle θ_1 and net fault bend Φ are independent of the sequence of bends or mode numbers.

Shearing of fault-bend folds.—Our theoretical development of the previous sections included the important constraint that the beds only undergo shear as they pass through a fault-bend fold; in particular the beds on the right side of the fold in figure 5 undergo no shear until they pass through the axial surface. We now relax this constraint to consider two important ways in which zones of layer-parallel shear within a thrust sheet or other fault block can deform parallel fault-bend folds: (1) shearing-out of flat fold crests and (2) general layer-parallel shear in parallel fault-bend folding.

1. *Shearing out of flat fold crests*: The ideal theoretical shape of a simple-step fold is shown in figure 3; however Suppe and Namson (1979) pointed out that several folds of western Taiwan exhibit an important deviation from this ideal shape. These modified structures exhibit a squeezing out of the flat crest of the ideal structure, as is shown in figure 14.

The squeezing out of the flat crest of the anticline is accomplished by simple shear within the thrust sheet. This shear is possible within the confines of the present theory when the line of the hanging-wall cutoff Y' is in contact with the line of the footwall cutoff X (figs. 3 and 14). Only at this stage in the deformation are axial surfaces A and B' in contact along X-Y' and able progressively to annihilate each other to form a new axial surface $(AB')^*$ by the mechanism shown in figure 14. The annihilation involves locking of the primary fault surface and slip in turn along progressively higher bedding surfaces, resulting in a layer-parallel shear of the thrust sheet above the lower decollement in the hanging wall. The active slip surface is always the bed in contact with the branch in axial surface (fig. 14). All slip is absorbed in the annihilation, therefore the thrust sheet is immobile along the upper decollement, beyond the anticlinal axial surface $(AB')^*$.

Applying the geometry of annihilation of the flat crest of a simple-step fold as shown in figure 15A we obtain an expression for the shortening or displacement d associated with annihilation along any given bed

$$d = c + b - a \tag{23}$$

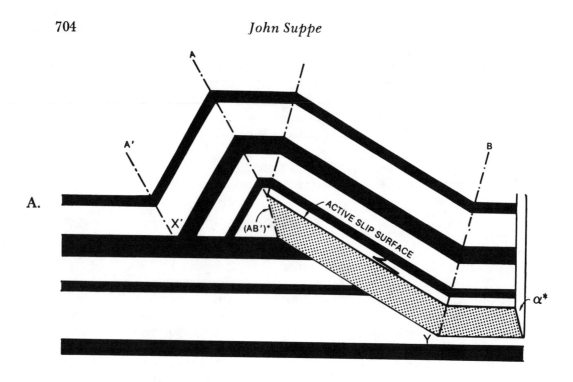

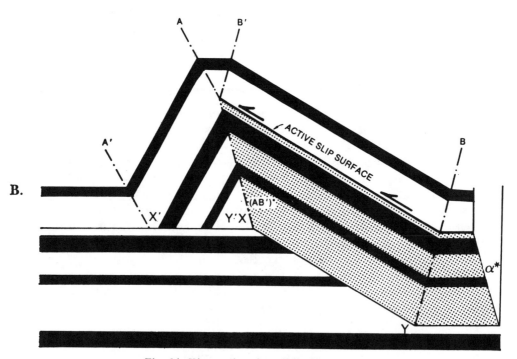

Fig. 14. Kinematics of annihilation of a flat fold crest.

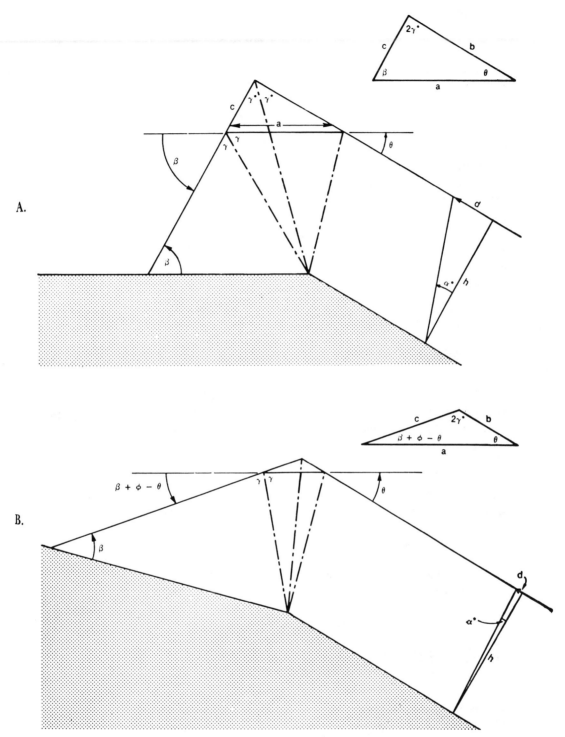

Fig. 15. Geometric elements for computing the shear associated with squeezing out of flat fold crests.

Applying the law of sines we obtain

$$d = a \left[\frac{\sin \beta + \sin \theta}{\sin 2 \gamma^*} - 1 \right] \tag{24}$$

where $\gamma^* = (180° - \beta - \phi)/2$ is the axial angle of the new axial surface $(AB')^*$. The width a of the annihilated flat top of the anticline is a function of stratigraphic height h above the lower decollement

$$a = h \left[\cot \gamma + \cot (90° - \theta/2) \right] \tag{25}$$

Combining eqs (24) and (25) we obtain the expression for the simple shear $S^* = d/h$ associated with the annihilation of the flat top of a simple-step fold (figs. 14 and 15A).

$$S^* = d/h = \left[\cot \gamma + \cot (90° - \theta/2) \right] \left[\frac{\sin \beta + \sin \theta}{\sin 2 \gamma^*} - 1 \right] \tag{26}$$

A more general form of the equation, not confined to simple-step folds (fig. 15B), is

$$S^* = d/h = \left[\cot \gamma + \cot (90° - \theta/2) \right] \left[\frac{\sin(\beta+\phi-\theta) + \sin\theta}{\sin 2 \gamma^*} - 1 \right] \tag{27}$$

The angular shear α^* is

$$\alpha^* = \cot^{-1} S^* \tag{28}$$

The angular shear α^* for all simple-step folds is shown in figure 16. For first-mode simple steps of less than 20° ($\phi = \theta$) the shear is negligible (less than 2°). The shear increases to 17.2° for a step of $30° = \phi = \theta$ and becomes progressively larger for second mode folds of smaller axial angle γ. Therefore substantial distortion of the hangingwall sheet is associated with annihilation. The distortion of faults farther back in the hanging wall sheet is, however, less severe because the faults will generally be at an angle to bedding of less than 30°. The dip of faults of the same angle ($\theta = \phi$) as the simple step is subject to a maximum flattening of 6.94° at $\theta = \phi = 23.5°$ (second mode). The flattening of faults in the hanging-wall sheet is less than 2° for most first-mode folds ($\theta = \phi \leqslant 27°$) and therefore needs only be considered in second-mode simple-step folding.

The shearing of a simple-step fault-bend fold with associated annihilation of axial surfaces, as shown in figure 24, is not inevitable. For example, the Pine Mountain thrust sheet (fig. 2) has not undergone this shear. Nevertheless, the shearing appears to be a widespread process (for example, Suppe, 1980b; Suppe and Namson, 1979), particularly in multiple imbrication structures (fig. 24), and may be a mechanism by which thrust sheets become locked. The shallow thrust in figure 11 is in locked position, for example. We recall that during annihilation, the primary fault is locked (fig. 14). If the stresses are high enough to release the fault, then the structure will be released from its associated ramp; at present we know no natural structures of this type. Apparently their formation is inhibited by resistance to the bending associated with axial surfaces that must form during their release. The propagation of a new fault or imbrication may require a lower stress. Locked thrust sheets with annihilated axial surfaces similar to figures 11 and 14 appear to be widespread.

2. *General layer-parallel shear in parallel fault-bend folding*: If we relax the constraint of no layer-parallel shear of the beds prior to passing through an axial surface, then a variety of new fold shapes is possible, of the sort shown in figure 17.

The old constraint of no layer-parallel shear was expressed as the requirement

$$b\,c = b\,d \tag{29}$$

in figure 6. We now allow layer-parallel shear of the sort shown in figure 17. The requirements for parallel folding then become $\gamma_1 = \gamma_2$ and

$$b\,c = b\,d + a\,b\,\sin\gamma\,\tan\alpha \tag{30}$$

where α is the angle of simple shear (fig. 17). The fault-bend fold equations may now be rederived including the last term in eq (30). In particular eq (7) becomes

$$\phi = \tan^{-1}\left[\frac{[\sin\theta\,\sin\gamma\,\tan\alpha - \sin(\gamma-\theta)]\,[\sin(2\gamma-\theta) - \sin\theta]}{\cos(\gamma-\theta)\,[\sin(2\gamma-\theta) - \sin\theta] - \sin\gamma}\right] \tag{31}$$

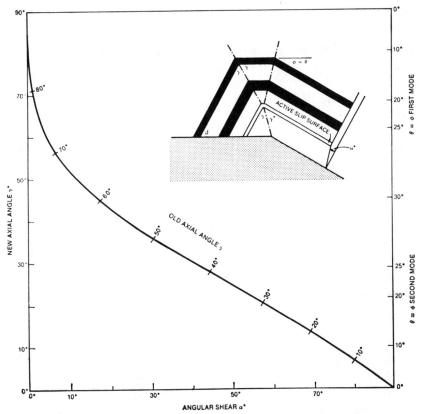

Fig. 16. Angular shear for squeezing out of flat fold crests in simple-step folds.

Applying the law of sines we obtain

$$d = a\left[\frac{\sin \beta + \sin \theta}{\sin 2\gamma^*} - 1\right] \qquad (24)$$

where $\gamma^* = (180° - \beta - \phi)/2$ is the axial angle of the new axial surface $(AB')^*$. The width a of the annihilated flat top of the anticline is a function of stratigraphic height h above the lower decollement

$$a = h\left[\cot \gamma + \cot (90° - \theta/2)\right] \qquad (25)$$

Combining eqs (24) and (25) we obtain the expression for the simple shear $S^* = d/h$ associated with the annihilation of the flat top of a simple-step fold (figs. 14 and 15A).

$$S^* = d/h = [\cot \gamma + \cot (90° - \theta/2)]\left[\frac{\sin \beta + \sin \theta}{\sin 2\gamma^*} - 1\right] \qquad (26)$$

A more general form of the equation, not confined to simple-step folds (fig. 15B), is

$$S^* = d/h = [\cot \gamma + \cot (90° - \theta/2)]\left[\frac{\sin(\beta+\phi-\theta) + \sin\theta}{\sin 2\gamma^*} - 1\right] \qquad (27)$$

The angular shear α^* is

$$\alpha^* = \cot^{-1}S^* \qquad (28)$$

The angular shear α^* for all simple-step folds is shown in figure 16. For first-mode simple steps of less than 20° ($\phi = \theta$) the shear is negligible (less than 2°). The shear increases to 17.2° for a step of $30° = \phi = \theta$ and becomes progressively larger for second mode folds of smaller axial angle γ. Therefore substantial distortion of the hangingwall sheet is associated with annihilation. The distortion of faults farther back in the hanging wall sheet is, however, less severe because the faults will generally be at an angle to bedding of less than 30°. The dip of faults of the same angle ($\theta = \phi$) as the simple step is subject to a maximum flattening of 6.94° at $\theta = \phi = 23.5°$ (second mode). The flattening of faults in the hanging-wall sheet is less than 2° for most first-mode folds ($\theta = \phi \leqslant 27°$) and therefore needs only be considered in second-mode simple-step folding.

The shearing of a simple-step fault-bend fold with associated annihilation of axial surfaces, as shown in figure 24, is not inevitable. For example, the Pine Mountain thrust sheet (fig. 2) has not undergone this shear. Nevertheless, the shearing appears to be a widespread process (for example, Suppe, 1980b; Suppe and Namson, 1979), particularly in multiple imbrication structures (fig. 24), and may be a mechanism by which thrust sheets become locked. The shallow thrust in figure 11 is in locked position, for example. We recall that during annihilation, the primary fault is locked (fig. 14). If the stresses are high enough to release the fault, then the structure will be released from its associated ramp; at present we know no natural structures of this type. Apparently their formation is inhibited by resistance to the bending associated with axial surfaces that must form during their release. The propagation of a new fault or imbrication may require a lower stress. Locked thrust sheets with annihilated axial surfaces similar to figures 11 and 14 appear to be widespread.

2. *General layer-parallel shear in parallel fault-bend folding:* If we relax the constraint of no layer-parallel shear of the beds prior to passing through an axial surface, then a variety of new fold shapes is possible, of the sort shown in figure 17.

The old constraint of no layer-parallel shear was expressed as the requirement

$$b\,c = b\,d \qquad (29)$$

in figure 6. We now allow layer-parallel shear of the sort shown in figure 17. The requirements for parallel folding then become $\gamma_1 = \gamma_2$ and

$$b\,c = b\,d + a\,b\,\sin\gamma\,\tan\alpha \qquad (30)$$

where α is the angle of simple shear (fig. 17). The fault-bend fold equations may now be rederived including the last term in eq (30). In particular eq (7) becomes

$$\phi = \tan^{-1}\left[\frac{[\sin\theta\,\sin\gamma\,\tan\alpha - \sin(\gamma-\theta)]\,[\sin(2\gamma-\theta) - \sin\theta]}{\cos(\gamma-\theta)\,[\sin(2\gamma-\theta) - \sin\theta] - \sin\gamma}\right] \qquad (31)$$

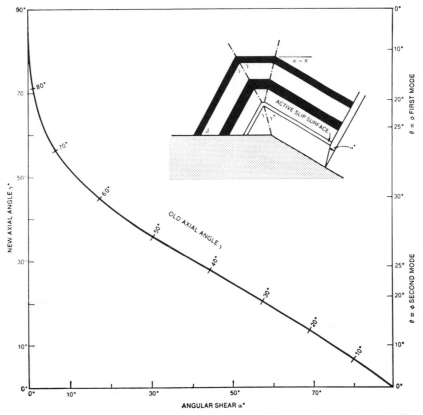

Fig. 16. Angular shear for squeezing out of flat fold crests in simple-step folds.

Eq (31) admits a wide variety of fold shapes that were impossible under the previous constraint of not allowing an arbitrary imposed shear α (fig. 17). Nevertheless our experience with actual structures whose shapes are well documented suggests that the imposed shear α is generally zero, but we know some exceptions to be discussed elsewhere. They involve reactivation of preexisting normal faults in thrust motion similar to the geometry of figure 17. Therefore they represent an exception to the rule that the highest cutoff angle $\theta = \phi$ for a simple-step parallel fault-bend fold is 30°.

Shear associated with branching axial surfaces.—Branching of axial surfaces is a widespread phenomenon in fold-and-thrust belts because of locking and shearing of thrust sheets as discussed above and because of interference of kink bands of nearby fault-bend folds. We have seen that branching of an axial surface is associated with a change in layer-parallel simple shear, as shown in figures 14 and 15. This property may be useful in predicting subsurface geology, because, in passing through a series of anticlines and synclines, shear may be conserved in a way somewhat analogous to conservation of bed length. If so, then branching in the anticlines, which is easily observed, must be balanced by equivalent branches of opposite effect in the synclines. The important point to note is that the opposite branches must occur at the same stratigraphic horizon in order to conserve layer-parallel shear. The elements of the theory are outlined below.

Any change in dip causes a shearing of the beds. If the unsheared state is horizontal, then the layer-parallel simple shear S is a simple function of dip angle δ. The geometry of the problem is shown in figure 18. By the Law of Sines

$$\frac{a}{\sin(\alpha - \delta/2)} = \frac{[a/\sin(\delta/2)]}{\sin(90 - \alpha)}$$

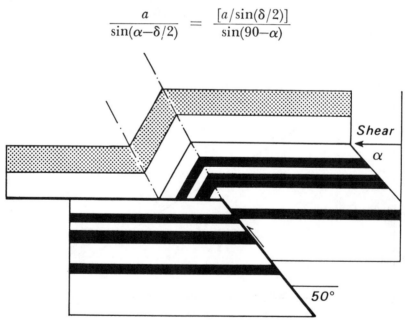

Fig. 17. An example of general layer-parallel simple shear in combination with fault-bend folding.

Rearranging we get

$$S = \tan \alpha = 2 \tan \delta/2 \qquad (32)$$

Thus, by knowing the dip we may immediately compute the shear.

Eq (32) is also the expression for the shear (change in shear) associated with the change in dip across an axial surface. We can add the shears associated with axial surfaces $i = 1, 2, \ldots n$ along a single layer

$$\sum_{i=1}^{n} S_i = S_1 + S_2 + \ldots S_n = 2[\tan \delta_1/2 + \tan \delta_2/2 + \ldots \tan \delta_n/2]$$

$$(33)$$

There is a change in shear across a node (fig. 19) associated with the merging or splitting of axial surfaces

$$\Delta S_{12} = S_1 + S_2 - S_{12} = 2[\tan \delta_1/2 + \tan \delta_2/2 - \tan \delta_{12}/2] \qquad (34)$$

ΣS can be defined for any layer and is constant for any stratigraphic interval that contains no nodes. From eqs (33) and (34) we then have for stratigraphic intervals X, Y, and Z

$$\Sigma S_x = \Sigma S_y + \Sigma \Delta S_{xy} = \Sigma S_z + \Sigma \Delta S_{xyz} \qquad (35)$$

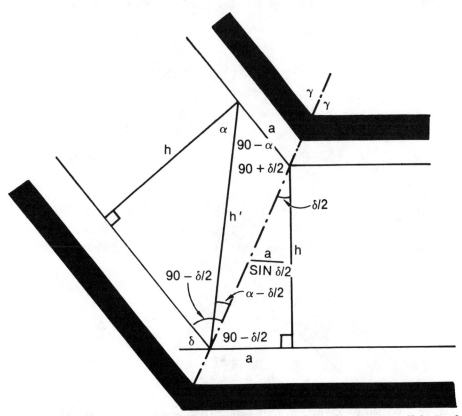

Fig. 18. Geometric elements used to determine the angular layer-parallel simple shear α associated with a change in dip δ (eq 32).

Therefore we can say that between pinning points of no shear or constant shear, the sum of the changes in shear along any bed containing nodes (nodel surface) is zero

$$\Sigma \Delta S = 0 \tag{36}$$

This fact reduces the problem of balancing a cross section across a "blind" syncline to the problem of balancing changes in shear along each nodal surface. The nodal surfaces can be discovered in the anticlines, and the amount of shear measured from the dips (eq 34). Changes in shear of the same magnitude but opposite sign must exist along the same bed in the intervening synclines.

IMBRICATE FAULT-BEND FOLDING

General experience in fold-and-thrust belts suggests that most fault-bend fold structures involve multiple imbrications. Even the type example of the Pine Mountain thrust sheet involves two imbrications (fig. 2). Therefore any useful geometric theory of fault-bend folding must be capable of dealing with imbrications in a straightforward fashion. In this section we consider imbricate fault-bend folding in which all imbrications are of the same vergence.

In general, to solve for an imbricated fault-bend fold structure such as figure 2, we need to know, or be able to compute, the undeformed cut-

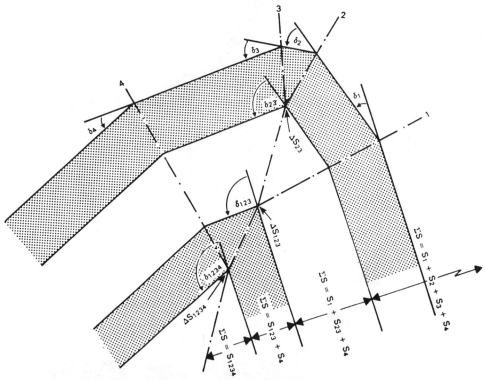

Fig. 19. Notation for changes in layer-parallel simple shear associated with branching axial surfaces. The change in shear ΔS associated with a branch is the difference in sums of shears, for example $\Delta S_{23} = (S_2 + S_3 - S_{23})$.

off angles $\phi_i = \theta_i'$ for each of the imbrications. During deformation, these cutoff angles may change because of shear associated with folding of the thrust sheets and annihilation of axial surfaces. Nevertheless, if we know the undeformed cutoff angles, here called the *fundamental-cutoff angles*, $°\theta_1, °\theta_2, \ldots °\theta_n$, we can solve all the angular relations of an imbricate structure as outlined below.

With each successive imbrication there is a quantum increase in forward and back dip. Therefore within an imbricate fault-bend fold, panels of rock bounded by axial surfaces may be classified according to the number of underlying imbrications (fig. 20). We will label the number of imbrications associated with a panel of rock using numerals 0, I, II, III . . . , with 0 indicating a panel of regional dip. To calculate quantum jumps in back dip and forward dip associated with successive imbrications we derive trigonometric relationships between the cutoff angles of imbrications ($°\theta_n$) and the dips of rock panels in the fault-bend folds.

Back dips associated with imbrication.—An imbrication of fundamental-cutoff angle $°\theta_1 = \delta$ produces a shear S_1 of angle α in the overlying thrust sheet as shown in figure 18 (eq 32)

$$S_1 = 2 \tan (°\theta_1/2) = \tan \alpha \tag{32}$$

If the overlying thrust sheet contains a crosscutting fault of fundamental-cutoff angle $°\theta_2$ prior to imbrication by fault 1, then considering the geometry shown in figure 21A, the new cutoff angle θ_2' is

$$\theta_2' = \tan^{-1}\left[\frac{1}{1/\tan °\theta_2 + 2 \tan (°\theta_1/2)}\right] \tag{37}$$

Similarly a third imbrication will have a crosscutting fault of new angle θ_3'

$$\theta_3' = \tan^{-1}\left[\frac{1}{1/\tan °\theta_3 + 2 \tan [(°\theta_1 + \theta_2')/2]}\right] \tag{38}$$

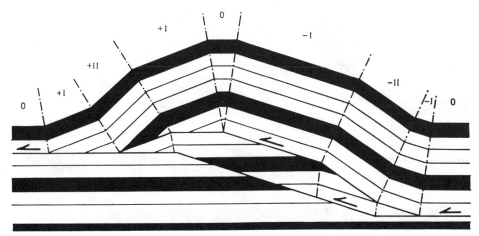

Fig. 20. Panels of forward and back dip in a simple-step fault-bend fold with two imbrications of $\theta = \phi = 18°$ (modfied from Suppe, 1980b).

and the *n*th imbrication will have a crosscutting fault of new angle θ'_n

$$\theta'_n = \tan^{-1}\left[\frac{1}{1/\tan {}^\circ\theta_n + 2\tan[{}^\circ\theta_1 + \theta'_2 + \ldots \theta'_{n-1})/2]}\right] \quad (39)$$

Therefore if we know the fundamental cutoff angles ${}^\circ\theta_1$, ${}^\circ\theta_2$, $\ldots$ ${}^\circ\theta_n$ we can calculate the back dips associated with each of the *n* imbrications. This is done for ${}^\circ\theta_1 = {}^\circ\theta_2 = {}^\circ\theta_7$ in table 1.

Forward dips associated with imbrication.—The key to computing the forward dip angle associated with imbrication is certain angular equivalences illustrated in figure 21B. The β-angle of a first or earlier imbrication (β_1, θ_1, ϕ_1) is the θ-angle when it is refolded by a second or later imbrication, for example

$$\beta_1 = \theta_{12} \quad (40)$$

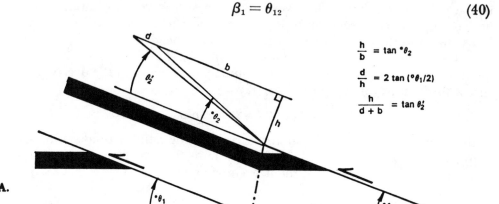

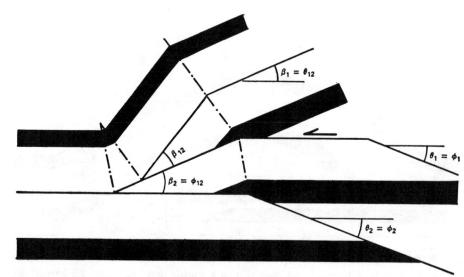

Fig. 21. Angular relationships used in computing the back dips (A) and foreward dips (B) caused by thrust imbrication.

TABLE 1

Forward and Back Dips assuming constant fundamental cutoff angle

Forward Dips							Fundamental Cutoff Angle° θ	Backdips						
VII	VI	V	IV	III	II	I		I	II	III	IV	V	VI	VII
61.6°	52.5°	43.0°	34.0°	25.2°	16.6°	8.2°	8°	8°	15.9°	23.4°	30.6°	37.3°	43.5°	49.3°
70.2°	59.2°	48.6°	38.3°	28.3°	18.6°	9.2°	9°	9°	17.8°	26.2°	34.0°	41.3°	47.9°	53.9°
80.6°	67.6°	55.2°	43.3°	31.9°	20.9°	10.3°	10°	10°	19.7°	28.9°	37.4°	45.1°	52.0°	58.2°
93.1°	77.3°	62.6°	48.8°	35.7°	23.3°	11.4°	11°	11°	21.6°	31.5°	40.6°	48.7°	55.9°	62.2°
109°	88.8°	71.0°	54.8°	39.8°	25.8°	12.6°	12°	12°	23.5°	34.1°	43.7°	52.1°	59.5°	65.9°
128°	102°	80.5°	61.5°	44.3°	28.5°	13.8°	13°	13°	25.4°	36.7°	46.7°	55.4°	62.9°	69.4°
160°	119°	91.3°	68.6°	48.9°	31.2°	15.0°	14°	14°	27.2°	39.1°	49.5°	58.4°	66.1°	72.5°
—*	146°	104°	76.3°	53.6°	33.9°	16.2°	15°	15°	29.1°	41.5°	52.3°	61.4°	69.0°	75.5°
	—*	124°	85.9°	59.0°	36.8°	17.4°	16°	16°	30.9°	43.9°	54.9°	64.1°	—*	
		—*	99.2°	65.6°	40.2°	18.8°	17°	17°	32.7°	46.2°	57.5°	—*		
			123°	73.1°	43.7°	20.2°	18°	18°	34.4°	48.4°	59.9°	—*		
			—*	82.2°	47.4°	21.6°	19°	19°	36.2°	50.6°	—*			
			—*	97.6°	52.0°	23.2°	20°	20°	37.9°	52.7°	—*			
				—*	57.0°	24.8°	21°	21°	39.6°	—*				
				—*	63.6°	26.6°	22°	22°	41.3°	—*				
					72.0°	28.4°	23°	23°	42.9°					
					—*	30.4°	24°	24°	—*					

* Thinning required in forward dips ($\theta_n = \phi_n > 30°$).

where the subscript 12 indicates the first imbrication folded by the second. Furthermore the β-angle of a second or later imbrication is the fault bend for the folding of a first or earlier imbrication, for example

$$\beta_2 = \phi_{12} \tag{41}$$

The angle of forward dip associated with each of n imbrications may be computed using similar angular relations to those in eqs (40) and (41) and the basic fault-bend fold equations (fig. 7), given the fundamental-cutoff angles $^\circ\theta_1, \,^\circ\theta_2, \ldots \,^\circ\theta_n$ associated with each of n simple-step imbrications. This is done for $^\circ\theta_1 = \,^\circ\theta_2 = \ldots \,^\circ\theta_i$ in table 1.

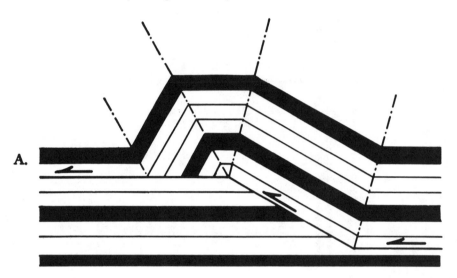

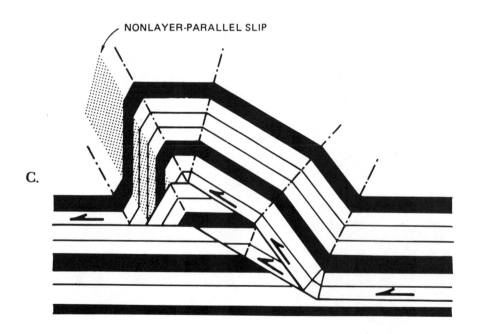

Constant-angle imbrications.—In order to calculate the forward and back dips associated with multiple imbrications we must know the fundamental-cutoff angle $°\theta$ for each fault. This procedure is not practical in solving most problems of subsurface geology, unless we have extensive drilling or excellent seismic-reflection profiling. In these problems we generally know the dips in certain regions and wish to solve for the fundamental-cutoff angles, number of imbrications associated with each dip panel (fig. 20), and ultimately the complete spatial arrangement of faults

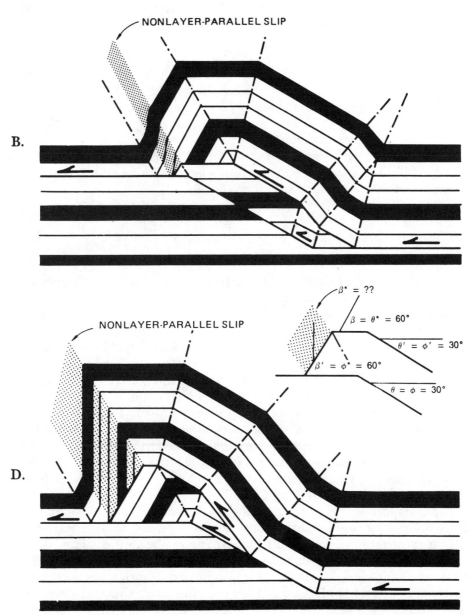

Fig. 22. Non-parallel folding caused by imbricate fault-bend folding with a high fundamental cutoff angle ($\theta = 30°$).

and folds at depth. In many cases we do not have sufficient information to solve the general inverse problem, but the problem may be considerably simplified and made soluble if we are willing to assume that each fault has the same fundamental-cutoff angle $°\theta$. This assumption may be mechanically reasonable in that the orientation of stresses relative to bedding may not change greatly between successive imbrications. Furthermore, applications suggest that some imbricate structures obey the constant-angle approximation rather closely.

The angular relations for forward dips are simplified by the constant-angle assumption. Eqs (40) and (41) for example, reduce to

$$\beta_1 = \beta_2 = \theta_{12} = \phi_{12} ; \tag{42}$$

therefore the refolding of an earlier imbrication by a later imbrication has the geometry of a simple-step in decollement ($\theta = \phi$) and is described by eq (12). The forward and back dips for constant fundamental cutoff angle up to seven imbrications are calculated using eqs (12), (39), and (42) and are presented in table 1.

It should be noted that the folding of an imbrication always involves a fault bend that is larger than the fundamental fault bend, because β is always larger than $\theta = \phi$ for anticlines (fig. 7).

$$\beta > \theta = \phi \qquad \text{for anticlines} , \tag{43}$$

and from eq (42) for constant-angle imbrications we have

$$\theta_1 = \phi_1 < \theta_2 = \phi_2 < \ldots \theta_n = \phi_n . \tag{44}$$

For imbrications at larger fundamental cutoff angles, refolding may be impossible with conservation of layer thickness because, as we see in figure 7, simple steps of $\theta = \phi$ greater than 30° cannot conserve layer thickness. Therefore, if two faults imbricate at $\phi_1 = \theta_1 = 30°$, as shown in figure 22, the refolding is impossible without change in layer thickness because $\beta_1 = \phi_2 = \theta_2 = 60°$. The maximum fundamental cutoff angle is $\phi = \theta = 23.79°$ for two imbrications conserving layer thickness. The maximum angle is about 20° for three imbrications, 18° for four imbrications, 16° for five imbrications, 15° for six imbrications, and 14° for seven imbrications (table 1). This result emphasizes the importance of observing layer thickness around imbricate fault-bend folds. If layer thickness is conserved then either the fundamental cutoff angle is less than about 20 degrees or some parts of the thrust sheets have undergone additional shear of the sort shown in figure 17.

It is important to realize that the surface shape of an anticline produced by two or more imbrications depends greatly on the amount of slip on each imbrication and on the spacing of the imbrications. A considerable variety of fold shapes can be produced by the mechanism of imbricate fault-bend folding. For example the imbricate structures in figures 23B, C, and D all have exactly the same fault slip and differ only in the distance between the two thrusts. In particular, compare examples B and C with example D; in B and C there is a steepening of forward dips by folding of the first imbrication whereas there is a flattening of forward dips in example D. We note that in D none of the dip angles differs from a single fault step. Only the short flat segment within the forward dips gives any real hint of imbricate structure at depth. Further slip on the faults of D will, however, produce the steeper dips characteristic of two imbrications.

When there are three or more imbrications substantially more complex geometric relationships exist between surface structure and underlying faults; in these cases stacked anticlines will occur. The complexity

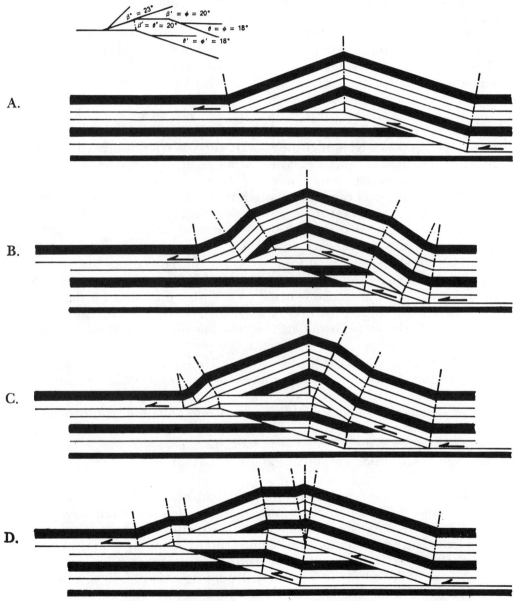

Fig. 23. Illustration of the effect of change in fault spacing on fold shape. Both imbrications have identical slip and cutoff angle. (after Suppe, 1980b).

of the problem of predicting subsurface structure is considerably reduced if we consider the angular relationships between the panels of forward and back dip before we attempt to solve the spatial aspects such as the locations of faults and folds at depth. We call the analysis of angular relations among dip data, *dip-spectral analysis*. This topic is treated in the following section.

Dip-spectral analysis of imbricate structures.—The quantum increase in forward and back dip of each successive imbrication is not a constant for a given fundamental-cutoff angle $°\theta$. The quantum change in forward dip becomes larger with each additional imbrication because β is always greater than θ (eq 43). In contrast the quantum change in back dip decreases with each additional imbrication because of shear within the thrust sheets (eqs 32 and 39). Therefore a distinctive and unique spectrum of forward and back dips is associated with each fundamental cutoff angle. For example three imbrications of $°\theta = 16°$ will produce forward dips of $17°$, $37°$, and $59°$ and back dips of $16°$, $31°$, and $44°$, whereas imbrications of $°\theta = 14°$ will produce forward dips of $15°$, $31°$, and $49°$ and back dips of $14°$, $27°$, and $39°$. Note that the forward-dip spectrum is a particularly sensitive indicator of fundamental cutoff angle. If dip data are available from surface mapping, seismic data, or dipmeter surveys, the observed and theoretical spectra can be compared. If the observed structure closely approximates the constant-angle assumption then we can estimate both the fundamental-cutoff angle and number of imbrications associated with each panel of forward and back dips.

The initial steps in producing structural interpretations are (1) assembly of data on a depth section, (2) dip-spectral analysis and assignment of regions of the cross section to the model dips predicted by the dip-spectral analysis, (3) construction of axial surfaces by bisecting the folds ($\gamma_1 = \gamma_2$). At this stage in the analysis any missing information may be noted and assumed or obtained by new measurements.

The next steps in producing structural solutions are substantially more difficult, because they involve guessing a solution and attempting to draw it. This step involves experience and intuition and is an integral part of all methods of subsurface structural interpretation. Nevertheless, it is in a sense more demanding using the present theory because bad guesses are quickly shown to be geometrically impossible. Two examples are given below, based on outcrop and well data from the Appalachians and Taiwan.

Pine Mountain Thrust (fig. 2).—The completed cross section of the Pine Mountain Thrust in the southern Appalachians was already presented in figure 2. We now outline how the structural interpretation was produced. The basic data include detailed surface mapping (England and others, 1961; Harris, 1962) and a well. Dip analysis suggests a fundamental cutoff angle of $°\theta = 15°$ with an essentially flat regional dip, in agreement with mapping to the west and nearby seismic data of Tegland (1978). The regions of the cross section were assigned to model forward and back dips based on the dip-spectral analysis. The inclinations of the axial surfaces

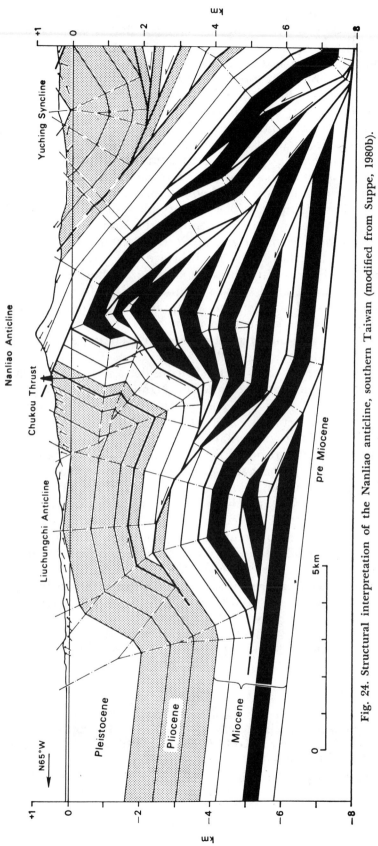

Fig. 24. Structural interpretation of the Nanliao anticline, southern Taiwan (modified from Suppe, 1980b).

459

are computed using the equal axial angle relationship ($\gamma_1 = \gamma_2$). The positions of some axial surfaces are much better located than others. These positions are adjusted by trial-and-error constrained by stratigraphic thickness and conservation of bed length. The Brooks well encountered the Pine Mountain Thrust below the Cambrian Rome Formation and entered flat-lying Silurian in the footwall. This footwall is elevated relative to the normal formation depths based on drilling and mapping to the west of the section; the normal depth of the Pine Mountain Thrust along the Silurian-Devonian decollement is shown on the left side of figure 2.

Next we must guess a qualitative structural solution. We choose a two imbrication structure based on the minimum of two imbrications deduced from the dip analysis. We choose a solution similar to the theoretical cross section in figure 20. The primary piece of information we do not directly have is the undeformed depth of the Rome décollement; we have two constraints: (1) the stratigraphic thicknesses observed and (2) the positions of axial surfaces. The axial surface east of the Wallen Valley thrust is particularly important, because, as seen in the theoretical cross section (fig. 20), it is the axial surface produced by the footwall cutoff of the Rome Formation along the Pine Mountain Thrust (see fig. 2). The cross section may now be constructed along the lines of figure 20. After some adjustment of positions of axial surfaces by trial and error we arrive at the final solution shown in figure 2.

Nanliao Anticline (fig. 24).—We now consider a much more complex and less well constrained cross section, the Nanliao anticline in western Taiwan (Suppe, 1980b). The basic constraints are the stratigraphic thicknesses, positions of axial surfaces, and surface and well dips. Based on the dip analysis we choose a fundamental-cutoff angle of 18° with a minimum of four imbrications. The regional dip of 6° is known from seismic and well data to the west. The next step in the interpretation was assignment of model dips and provisional location of axial surfaces. The only observed fault is the outcropping Chukou thrust which rides on a decollement in the upper plate. The deeper and higher decollement horizons used in the solution are approximately in positions of decollement in nearby structures. The final solution was then obtained by trial and error using the same methods as the Pine Mountain Thrust.

The proposed solution is obviously not well constrained in light of the uncertain positions of the decollement horizons and the large number of required imbrications. Nevertheless, it provides considerable insight into the structures that may be encountered in further exploration. In particular, the possible existence of structurally stacked reservoirs in the Nanliao anticline is identified by this analysis.

ACKNOWLEDGMENTS

I am indebted to many individuals for their important contributions to this study including J. Bialkowski, Stanley S. L. Chang, C. S. Ho, J. Namson, C. H. Tang, and Y. Wang. I thank D. Elliott, H. P. Laubscher, J. Rodgers, R. Stanley, and R. C. Vierbuchen for reviewing various versions of this manuscript. The subsurface interpretations are based largely on data generously provided by the Taiwan Petroleum Exploration Division, Chinese Petroleum Corporation. I am grateful to the Chinese Petroleum Corporation, Mining Research and Service Organization, and National Taiwan University for kind hospitality. This work was supported by N.S.F. grant EAR79-25446, "Geometric and Kinematic Properties of Map-scale Decollement Folds." I am grateful to the John Simon Guggenheim Memorial Foundation for gracious and unencumbered support in 1978-1979 during the initial development of these theories.

REFERENCES

Douglas, R. J. W., 1950, Callum Creek, Langford Creek, and Gap map areas, Alberta: Canada Geol. Survey, Mem. 255, 124 p.

Englund, K. J., Smith, H. L., Harris, L. D., and Stephens, J. G., 1961, Geology of the Ewing Quadrangle, Kentucky and Virginia (1:24,000): U.S. Geol. Survey Map GQ-172.

Gwinn, V. E., 1970, Kinematic patterns and estimates of lateral shortening, Valley and Ridge and Great Valley Provinces, Central Appalachians, South-Central Pennsylvania, *in* Fisher, G. W., Pettijohn, F. J., Reed, V. C., Jr., and Weaver, K. N., eds., Studies of Appalachian Geology: Central and Southern: New York, John Wiley & Sons, p. 127-146.

Harris, L. D., 1962, Geology of the Coleman Gap Quadrangle, Tennessee and Virginia (1:24,000): U.S. Geol. Survey Map GQ-188.

———— 1970, Details of thin-skinned tectonics in parts of Valley and Ridge and Cumberland Plateau provinces of the southern Appalachians, *in* Fisher, G. W., Pettijohn, F. J., Reed, V. C., Jr., and Weaver, K. N., eds., Studies of Appalachian Geology: Central and Southern: New York, John Wiley & Sons, p. 161-173.

Laubscher, H. P., 1965, Ein Kinematisches Modell der Jurafaltung: Ecologae Geol. Helvetiae, no. 58, p. 231-318.

———— 1975, Viscous components in Jura folding: Tectonophysics, v. 27, p. 239-254.

———— 1976, Geometrical adjustments during rotation of a Jura fold limb: Tectonophysics, v. 36, p. 347-365.

Namson, J., 1981, Detailed structural analysis of the western foothills belt in the Miaoli-Hsinchu area, Taiwan: I. southern part: Petroleum Geology of Taiwan, no. 18, p. 31-51.

Perry, W. J., Harris, A. G., and Harris, L. D., 1979, Conodont-based reinterpretation of Bain Dome—Structural reevaluation of Allegheny frontal zone: Am. Assoc. Petroleum Geologists Bull., v. 63, p. 647-675.

Rich, J. L., 1934, Mechanics of low-angle overthrust faulting as illustrated by Cumberland thrust block, Virginia, Kentucky, and Tennessee: Am. Assoc. Petroleum Geologists Bull., v. 18, p. 1584-1596.

Rodgers, John, 1950, Mechanics of Appalachian folding as illustrated by Sequatchie anticline, Tennessee and Alabama: Am. Assoc. Petroleum Geologists Bull., v. 34, p. 672-681.

Roedder, D., Gilbert, O. E., Jr., and Witherspoon, W. D., 1978, Evolution and macroscopic structure of valley and ridge thrust belt, Tennessee and Virginia: Knoxville, Univ. Tennessee Dept. Geol. Sci., Studies in Geology, no. 2, 25 p.

Suppe, John, 1976, Decollement folding in western Taiwan: Petroleum Geology Taiwan, no. 13, p. 25-35.

———— 1979, Fault bend folding (abstract): Geol. Soc. America Abs. with Programs, v. 11, p. 525.

———— 1980a, A retrodeformable cross section of northern Taiwan: Geol. Soc. China Proc., no. 23, p. 46-55.

———— 1980b, Imbricated structure of western foothills belt, south-central Taiwan: Petroleum Geology Taiwan, no. 17, p. 1-16.

———— 1981, Mechanics of mountain building and metamorphism in Taiwan: Geol. Soc. China, Mem. 4, p. 67-89.

Suppe, John, and Namson, J., 1979, Fault-bend origin of frontal folds of the western Taiwan fold-and-thrust belt: Petroleum Geology Taiwan, no. 16, p. 1-18.

Tegland, E. R., 1978, Seismic investigations of eastern Tennessee: Tennessee Div. Geology Bull., no. 78, 68 p.

STRUCTURAL TECHNIQUES

Reprinted by permission of the University of Chicago from J. Hoover Mackin, *Journal of Geology*, v. 58, no. 1 (1950), p. 55-72.

STUDIES FOR STUDENTS

THE DOWN-STRUCTURE METHOD OF VIEWING GEOLOGIC MAPS[1]

J. HOOVER MACKIN
University of Washington

ABSTRACT

Most contact lines on geologic maps are surface traces of inclined planes. If the map is viewed down the slope of these planes, it becomes, in effect, a section, and the map patterns are seen as structures. The "down-structure" method of looking *into* maps rather than *at* them can be introduced into the most elementary exercise in geologic map interpretation; it helps the student to grasp quickly the structural significance of patterns on maps made by others and increases his sensitivity in laying down lines on his own maps in the field. Used with proper caution and an understanding of its limitations, the method is applicable to a wide range of map patterns, from those of simple monoclinal structures to those of recumbent folds.

INTRODUCTION

1. Professor X is discussing Alpine structures with a small group of graduate students. After he has given a routine explanation of the areal geology of a nappe complex depicted on a wall map, the map is returned to its cabinet, and a sheet of sections is hung before the class. In answer to a question as to the factual basis for structural swirls high above and deep below the topographic profile, the professor tolerantly remarks that in dealing with some sections in the Alps it is necessary to distinguish between their author's facts and his fancies, and he draws a hearty laugh by pointing to a recumbent loop several thousand feet above the highest peaks and saying that he would like to put one question to the author: "Vas you dere, Charlie?" This is the climax of the lecture; the students file out, well satisfied that they are being trained by a scientist with both feet solidly on the ground.

2. One of the professor's young men, having completed his first job of mapping, prepares to walk over the area with an inspector. Forewarned that the inspector is a harsh critic, he plans a "Soviet" tour that threads its way along a route where he is confident of the geo-logic relations and gives a wide berth to four spots where stratigraphy and structure refuse to make sense and where he knows that he cannot defend his inferred contacts and faults. But, according to the young man's account of the affair, the inspector, on his arrival, spread the colored map on a table, twisted it about into odd positions as he scrutinized it from various angles, glanced briefly at several Soviet-type sections, asked a few searching questions, and finally put his finger neatly, in turn, on each of the four sore spots and said he would like to see each of them. The young man's interpretations in these spots prove to be wholly incorrect, and he later agrees with other young men who have had similar experiences with the same inspector that, in evaluating a geologic map and seeking out critical evidence in the field, the inspector either has X-ray eyes or is in league with the devil.

3. Another young geologist is sketching on a plane table after a round of shots. Working from the near side of the board for convenience, he transfers drainage, topography, outcrops, contacts of several types, and a minor cross fault from his notebook sketch to the plane-table sheet. Then he moves around the table to study the map from a different direction and vertical angle, is evi-

[1] Manuscript received September 26, 1949.

55

dently troubled by what he sees, trots back over part of his traverse to re-examine some of the features, and returns to erase the fault, modify an inferred contact, and change the shape of several outcrops. The revised map differs from that first drawn not only in pattern but also in geologic meaning—the first version was wrong and the second right.

These anecdotes relate to a method of geologic map interpretation that is a simple and effective aid in the understanding of maps made by others and— what is perhaps more important—in the drawing of geologic lines in the field. The writer was introduced to the method by E. B. Bailey in 1936, when he was privileged to watch Bailey apply it in grasping, literally at a glance, the structural significance of a swirling outcrop pattern in the Pennsylvania Piedmont that had been missed by two generations of competent geologists by whom the area had been mapped. Subsequent reading shows that the method has been in use in Europe for more than half a century, and there is no doubt that it is well known to many field geologists in this country. But it is rarely treated explicitly in geologic reports, probably (1) because the geologist is not inclined to turn aside from discussion of his area to explain a general method as such and (2) because no one area will provide the variety of examples needed for an exposition of the method. It came to the writer, then a graduate student, by word of mouth from Bailey to replace a clutter of rules-of-thumb, painted models, and block diagrams that had made up his formal education in the relationship between map pattern and structure. The method is outlined here in the hope that it will do the same for others. A secondary purpose is to clarify several cases (Bailey and Mackin, 1936; Mackin, 1944, pp. 16–17; 1947, pp. 13–14) in

which structural interpretations were arrived at via the down-structure method, without an adequate explanation of the reasoning involved.

Both the form and the substance of the text have benefited from constructive criticisms by H. L. James and A. E. Granger (U.S. Geological Survey), C. F. Park, Jr. (Stanford), and J. D. Barksdale and Peter Misch (University of Washington).

EXPLANATION OF THE METHOD

Nearly all geologic maps are framed by meridians and parallels, with printing and symbols so arranged that "north is up." This convention has many advantages for the cartographer; its great advantage to the user is that the fixed habit of viewing all maps "from the south" permits him to see the features of any individual geologic sheet in proper orientation with respect to a familiar regional geologic background. The geologic map of Colorado, viewed with "north up," reveals a striking assemblage of typical Rocky Mountain structural patterns; but the same map, hung upside down or with "east up," is a jumble of colors with as little meaning as a quotation out of context.

But most lines on geologic maps are surface traces of inclined planes, the orientation of which is wholly unrelated to the geographic framework of meridians and parallels; if the map is viewed down the slope of these geologic planes, the map patterns are seen as structures. Thus, if figure 1 is so oriented that the observer views it in the direction of dip or plunge[2] (i.e., from the east) and at an angle equal to the angle of dip or plunge,

[2] "Plunge" is used here for the angle measured in the vertical plane, in accordance with a recent report of a Map Symbol Committee of the U.S. Geological Survey (Goddard *et al.*).

the line of section S–S' becomes a hori-zontal surface, the portion of the map to-ward the observer is the structure below the surface along the line of section, and the portion of the map beyond the line is the structure removed by erosion above the surface. In practice the line S–S' may be a straightedge or slip of paper that can be shifted at will, so long as it is kept normal to the direction of dip or plunge of those structures on which at-tention is focused. Or one can, of course, dispense with the straightedge and see the structure in its entirety without ref-erence to any particular plane of section. Figure 1 illustrates four simple applica-tions of the method, each coupled with a more or less self-evident caution or quali-fication (discussed below) that must be kept in mind in its use.

A. The folds in the southern part of the map have been drawn with vertical axial planes, so that the surface traces of the axial planes correspond with the surface projections of the fold axes; complica-tions that arise when axial planes of plunging folds are inclined will be treated later. It will be noted that, in the down-plunge view, anticlines and synclines are simply *seen* as such; there is no need for recourse to the rules (oldest rock in cen-ter; or pattern convex in direction of plunge, etc., equals anticline) to distin-guish between them. The map pattern indicates that the folding is disharmonic, and it follows that the down-plunge view of the map provides a closer approxima-tion of subsurface structure than does any section based on downward exten-sion of bedding attitudes measured along the line of section. Along line S–S', for example, attitudes of bed *2* (solid black) in the limbs of synclines A and C are identical, but the trough form of bed *2* in syncline $\bar{A}$ is quite different from the trough form in syncline C. Minor plica-

tions in bed *3* in syncline C, observed at the surface near the line S–S', do not oc-cur in stratigraphic units beneath the

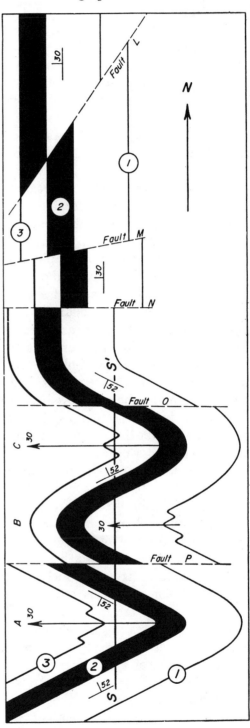

Fig. 1.—Diagrammatic map to illustrate several simple applications of the down-structure method.

surface, whereas observations along the line *S–S'* in anticline *B* fail to suggest the plications that do actually occur in bed *1* beneath the surface.

The use of the down-structure method in this case is obviously based on the assumption that the structural features in individual beds and corresponding parts of folds are continuous down the plunge. The extent to which this assumption is justified varies widely, of course, from district to district, depending on the rock types involved and local structural habits; the only generalization that can be made is that it is usually preferable to the alternative assumption that subsurface structures can be drawn in a structure section solely on the basis of graphical or mathematical manipulation of dip measurements recorded at the surface along the line of section. A corollary to this generalization is that, in an area of plunging folds, *strikes* measured at the surface up-plunge and down-plunge from a given line of section have quite as significant a bearing on the structure beneath the surface in the section as do *dips* measured along the line. To draw a structure section across plunging folds, it is usually necessary to map a zone, with as much attention to precision in strike measurements as in dip measurements.

B. A familiar elementary exercise in geologic map-reading requires that the student determine the relative movement on faults cutting homoclinal or folded strata on the basis of the outcrop pattern after erosion, with the assumption that the faults are vertical and the movement is dip-slip. Here again, as in interpretation of fold patterns, there is usually no need for a roundabout approach via rules-of-thumb; one simply *sees* the relative displacement of the blocks by looking into the map down the bedding planes (fig. 1).

Even if nothing is known or assumed as to the direction or amount of movement on the faults, the down-structure view provides at a glance a correct picture of the stratigraphic throw of any horizon within the moved blocks; on fault *N*, for example, the stratigraphic throw equals the thickness of bed *2*.

In the down-structure view, faults *M* and *L* may at first "look like" high- and low-angle thrusts, respectively. Both faults are actually vertical; they have been included in the diagram to emphasize this important point: that it is possible to look into a map along any sets of planes, but along only one set at one time. Attention is here focused on the bedding planes within the blocks, not on the fault planes between them; and it will be noted that the stratigraphic throw (on *L*, equal to the thickness of bed *2;* and on *M*, one-half the thickness of bed *2*) is readily seen, regardless of the relationship between the strike of the fault and the strike of the beds. Similarly, if one wishes to see the stratigraphic throw on faults *O* and *P*, one has only to twist the map so that the line of sight is down the dip of the bedding cut by these faults; the stratigraphic throw on *O* is one-half, and on *P* one and one-half, times the thickness of bed *2*.

C. The down-structure view of the map pattern of a plunging fold reveals, not the conventional vertical section, but a section that is normal to the inclined axis of the fold. The axis is, of course, by all odds the best line of reference as far as structural relations are concerned, and the "normal section" is unique in being the only type of section that shows the true form of the fold and the actual thickness of all stratigraphic units involved. Bed *2* in figure 1 is, for example, everywhere the same thickness; but this will be evident only if the pattern is viewed down the plunge of the fold (or, in the northern part of the map, down the dip). The point made here is simply that the down-structure view of a map pattern provides an undistorted normal section that differs from, and must not be confused with, the conventional vertical section, which is always distorted. Other differences between normal and vertical sections are treated later.

D. A third type of structure section is the horizontal section. The down-structure method of map interpretation is based on (1) the fact that each of these three types of sections (normal, vertical, and horizontal) can be derived from either of the other two if the angle of plunge is known and (2) the fact that a geologic map is a horizontal section except that the patterns are usually influenced in greater or less degree by topography. Figure 1, intended as a diagrammatic illustration of the method, is a horizontal section; it is, of course, necessary to take the influence of topography into account in applying the method to actual geologic maps.

The effect of topography on geologic map patterns is adequately treated in standard texts and needs no discussion here. For present purposes the best approach is via the fact that, on a relief model of a rugged area (or the area itself, seen from a plane), the outcrop pattern of a dipping bed is complex from every other point of view, but simple when the model is viewed down the dip. The eye quickly learns to accommodate the influence of topography on map patterns in the same way, namely, by seeing the contoured surface as a relief model. If a map does not carry contours, the down-structure method cannot be applied in matters of detail; the same limitation holds for all other methods of geologic map interpretation.

The examples that follow are intended to emphasize the critical difference between map-pattern trends and true structural trends in areas of plunging folds with inclined axial planes and to illustrate the wide range of use of the down-structure method by applying it in interpretation of map patterns of (1) a simple plunging monocline and (2) a recumbent fold. The first example is a special case, differing from the map-pattern–structure types of problem to which the method is usually applied.

USE OF THE GEOLOGIC MAP AS A RESTORED SECTION

Figure 2, A, is part of a large-scale outcrop map sketched from the "near side" of his plane table by the young geologist of anecdote 3 (actually, one of my associates on a U.S. Geological Survey war-minerals project). The dash symbol indicates a lithologic unit consisting chiefly of dark-gray shale and siltstone, well dated by fossils (hereafter called "gray shale"). The stipple is a fine-grained sandstone-siltstone, dominantly reddish, and not fossiliferous (hereafter called "red sandstone"). The conglomerate beds make good ledge outcrops; the gray shale and red sandstone are generally poorly exposed but can be distinguished on the basis of float where the dash and stipple symbols are shown. Blank areas are covered by a creep-wash mantle. All the beds dip steeply to the west on a surface that slopes gently westward; contours are not needed because topography does not significantly affect the map pattern.

On the map (and even more convincingly on the ground) the offset terminations of conglomerate outcrops A and B, both of which are underlain by gray shale and overlain locally by red sandstone, suggest faulting (fig. 2, A). Cemented breccia consisting of gray shale, observed in the float in the area between the terminations, tends to confirm the fault. An abnormal strike attitude in the same vicinity is readily taken to be a drag effect. Human nature being what it is, the south end of conglomerate D is regarded as another manifestation of the fault. On the basis of these several lines of evidence, the fault is sketched in with some confidence.

It so happens that the conglomerates contain pebbles of a distinctive felsite porphyry that occurs in place elsewhere in the district; the date of extrusion of the porphyry flows is a problem of considerable importance. The fact that conglomerate A appears to be underlain and overlain by fossiliferous gray shale is therefore especially significant; it indicates that the porphyry was available to streams during the time range of the gray-shale fossils. Tonguing of the gray shale into the conglomeratic red sandstone indicates, in addition, that there was no appreciable time interval between the deposition of these unlike, but intergrading, lithologic units; the fossils in the shale therefore serve to date the sandstone.

If the plane-table sketch in figure 2 were colored in as part of the final map on the basis of these interpretations, it would conflict with relationships observed in other parts of the district and would be a "sore spot" of the type noticed by the inspector in anecdote 2.

In the down-structure view of the sketch, from the east side of the plane table, outcrops A and B are seen not as offset segments of a conglomerate layer dipping about 40° westward on a map but as *flat* sheets of channel gravel at different levels in a section. The difference in level can, of course, be due to faulting of a once continuous bed, but it can also be (1) that the northern termina-

tion of the lower gravel sheet simply marks the northward limit of river-swinging at that level against a meander scarp cut in gray shale; (2) that the "gray shale" over this gravel sheet is fan wash shed from the cut bank and adjacent uplands after withdrawal of the river from the base of the meander scarp; (3) that these fan deposits, of local side-stream derivation, intertongue with, and are overlapped by, reddish sands and silts that are overbank deposits representing the suspended load of the through-flowing main stream that carried the gravel as a bed load; and (4) that the channel gravel sheet of outcrop B marks merely another position of the channel of the slowly aggrading main stream when, at a somewhat higher level, it planed away the upper part of a fan consisting of gray-shale debris and the higher relief features cut in (solid) gray shale from which the fan had been shed (fig. 2, B).

A second look at the field relations with these ideas in mind—and with a little digging where necessary—proves (1) that the critical "gray shale" above conglomerate A actually consists of lithic fragments of gray shale; (2) that the covered area over conglomerate B is red sandstone, not gray shale, as would be expected if there had been faulting; (3) that the supposed fault breccia rests on conglomerate A and is undoubtedly a talus breccia, etc. Revision of the sketch in accordance with these observations eliminates the fault and the associated unwarranted implication as to the date when the felsite porphyry became available for erosion, and it changes the supposed interlensing facies relationship between the gray and red sediments to an unconformity of unknown time value. The gray shale was lithified and dissected before the beginning of red siltstone sedimentation.

It may be noticed, incidentally, that

the revised map shows no local thickening of conglomerate *D*. This overturned esker (?) troubled the young geologist when he saw it in section, and, on more critical examination, it turned out to be a more or less continuous blanket of large conglomerate blocks creeping down the present westward-sloping surface over red sandstone. The change in shape of the conglomerate outcrop is inconsequential in so far as the geologic significance of the map is concerned. But if one subscribes to the philosophy of mapping that makes harmony of contact lines an end in itself (as set forth with contagious enthusiasm by Greenly and Williams [1930] in a chapter that ought to be required reading in every field methods course), then the revision of conglomerate *D* is perhaps no less "important" than the other changes.

The advantage of this special application of the down-structure method is that, by rotating the beds back to flatness, it helps to bring the geologic lines on the map to life as erosional or depositional surfaces, such as might be seen in a restored section or, more directly, in the walls of the Grand Canyon. The method thus tends to enable the student to carry over to his work on maps his "sense of rightness of things" as viewed in section; the case of the overturned esker is a very simple example. The use of the map as a section stimulates thinking as to the meaning of the lines as they are laid down in the field; a glimpse of an unconformity in section, for instance, will bring to mind a flood of possible analogies between that ancient landscape and landscapes now being modeled in similar rocks and structure by a variety of types of erosional processes under a variety of conditions of climate and relief, and these

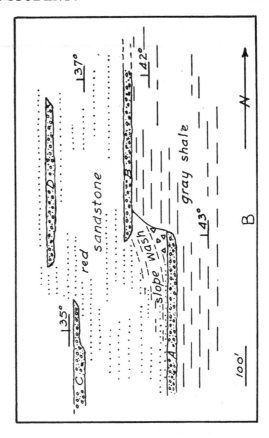

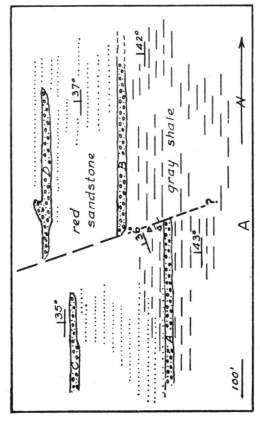

Fig. 2.—Plane-table sketch maps: *A*, as first drawn; *B*, as revised after a directed search for diagnostic evidence.

thoughts will almost invariably lead to a directed search for critical evidence that might be missed in routine "walking-out" of the contact.

It may be worth noting in this connection that the down-structure view of an ancient erosion surface may lead to consideration of a point that is likely to be neglected when the unconformity is merely a map line, namely, the extent to which relief features which developed during the erosion interval were modified or obliterated in the process of emplacement of the cover. The flat portions of the unconformity in figure 2 are *not* peneplanes mantled by basal conglomerate; they are lateral corrasion surfaces, cut and covered by the same process at the same time. As such, they belong in a category of unconformities in which the shape of the surface is determined *by the agency that deposited the cover*, and this "one-process" type of unconformity ought to be sharply separated from the "two-process" type, in which the surface formed by an erosional process during the hiatus is modified little or not at all by a later and distinctly different process of burial. If these two types of unconformities look the same on a detailed geologic map, there is usually something wrong with the mapping, and there is always something missing in the report.

AXIS AND AXIAL PLANE RELATIONSHIPS IN MAP PATTERNS OF PLUNGING FOLDS

GENERAL STATEMENT

The simple plunging folds in figure 1 are useful in introducing the down-structure method, but these folds, with vertical axial planes, represent a special case that is rare in nature. In the general case of the plunging fold with inclined axial plane, the surface trace of the axial plane diverges more or less markedly from the surface projection of the axis, the angle

of divergence depending on the plunge of the axis and the dip of the axial plane. The distinction between these two structural elements, stressed by Billings (1942) and others, needs special emphasis in this discussion of map patterns. The surface trace of the axial plane dominates pattern relationships of limbs and noses in plunging folds so completely that the eye tends to be deceived into regarding it as indicating the trend or "strike" of the fold. The "strike" of the fold axis, which is the true "strike" of the fold and by all odds its most significant directional element, may not appear at all in the map pattern. Figure 3 is intended to illustrate this relationship and to demonstrate that the down-structure view of the map pattern will provide a correct picture of the shape of a plunging fold *only if the line of sight parallels the axis*.[3]

The critical difference between the trends of (1) structures and (2) the map patterns produced by them under the influence of plunge has implications that extend far beyond the scope of this discussion of a method of map-reading. Two of the more important of the implications are outlined briefly here because the tendency of the down-structure method to lead the map-reader (or maker) to think along these lines is perhaps its chief long-term advantage.

STRUCTURAL TRENDS VERSUS PATTERN TRENDS

While the direction of the fold in figure 3, as a structure, is due north, essen-

[3] A map and vertical section used in another connection by Billings (1942, fig. 50, p. 63) is an excellent example of the same relationship. The map pattern suggests that the fold trends about N. 45° E.; the influence of the pattern on the eye is so strong that it is necessary to pencil in the surface projection of the axis (about N. 36° E., normal to line of section M–N) before the down-structure view of the map will match the section.

tially all inclined planar elements in it strike east of north. This rather anomalous situation results from the fact that the strike is (for practical purpose of measurement with spirit level and compass) necessarily referred to a horizontal plane and is therefore in some respects badly suited for discussion of structures of the type shown in figure 3, in which the strike of every inclined planar element is influenced by plunge. In dealing with structures of this type, horizontal

FIG. 3.—Diagrammatic map and section, showing plunging folds with inclined axial planes. North is up. The structural strike is north-south; the plunge is 20° due north. The line symbol used for two of the rock units parallels the strike of the axial planes of the folds.

The section is a normal section, not a vertical section. Angles that appear in the section are structural dip angles, not apparent dips derived from the "true" (conventional) dips shown by figures on the map. The levels *1–6*, from top to bottom in the right-hand margin of the section, correspond with east-west horizontal lines, *1–6*, from north to south in the margin of the map.

If one looks down the structure in the direction of the surface trace of the axial planes (N. 15° E.), the folds will appear to be upright, with notable thinning of all rock units on the steep eastern limbs of the anticlines. The folds are, in fact, overturned, and there is no thinning on the limbs. The structure will be seen correctly only if the map patterns are viewed down the plunge of the fold axes.

473

and vertical planes are, like the meridians and parallels on a map, simply a convenient reference framework for descriptive purposes. Any genetic treatment must, on the other hand, be based on a framework referred to the axis of the structure, because the whole dynamic history that is expressed by the shape of the fold, as well as all local differential movements recorded by minor structural features within it, are genetically associated with the axis. Two self-explanatory terms help to emphasize the important difference between structural and pattern relationships on maps and sections —"structural strike" and "structural dip."

"Structural strike" obviously refers to the trend of a structure, not to the direction of the intersection of an inclined plane with a horizontal plane. The "structural strike" of a fold is the same as the "direction of the surface projection of the axis" of the fold, but the expression is less cumbersome. "Structural strike" is preferable to "axial strike" because it carries no connotation of applying only to axial areas and is less likely to be confused with the strike of the axial plane, which is an altogether different thing, except in the special case when the axial plane is vertical. Structural strike is the direction that would be assumed, as conventional strike, by all planar elements in the fold if the axis were rotated up to horizontality—it is the strike of bedding, cleavage, etc., *relative to the axis of the structure.*

The chief value of the concept of structural strike is that it guarantees that the field geologist will make a sharp distinction between structural directions and map-pattern directions. One might say, with regard to a group of ledges at *A* in figure 3: "North-south lineation defines the structural strike; the plunge is 20°

due north. Under the influence of plunge the bedding strikes N. 34° E., and the axial plane cleavage strikes N. 15 E."
A notebook record of this type for many ledges in the map area will make it evident why *B*-lineation, which is a manifestation of structural strike, is usually the most constant of all directional elements measured in the field and will properly emphasize the distorting effect of plunge on pattern relations (strikes) of the planar elements. It will remind the student of the fact, often overlooked, that about three-quarters of the complexity of his geology is a complexity of pattern, not of structure.

As indicated earlier, vertical cross sections of plunging structures always show a distorted cross-sectional form, incorrect thicknesses for all the stratigraphic members involved, and apparent dip angles. These defects are, of course, due to the fact that a vertical section (like a map or horizontal section) is an unnatural slice through the plunging structure; the dimensions and dip angles that appear in a vertical section are referred to a plumb line in a vertical plane, neither of which is genetically related to the structure itself. The normal section, on the other hand, shows the actual cross-sectional form of the structure, the correct thickness of all the stratigraphic members, and what may be called the "structural dip" of all planar elements (see figs. 1, 3, and 4).

"Structural dip" is the dip that would be assumed by the planar elements if the axis of the plunging fold were rotated back to horizontality; it is the inclination of the planar elements *relative to the axis of the structure.* Structural dip is thus closely related to structural strike; but, whereas the direction of structural strike can and should be shown on geologic maps (see below), structural dip cannot be expressed by any simple map symbol because the angle lies in an inclined plane. The direction of structural dip may or not correspond with the direction of A-lineation as observed in the field, but the extended discussion needed even to state this interesting problem has no place here.

We are so accustomed to dealing with vertical sections that we are, like the "captives in the cave" in Plato's *Republic*, likely to confuse the distorted shadows of the structure shown on these sections with the structure itself. The difference between vertical sections and normal sections may be negligible if the plunge is low. But if the angle of plunge is high, all the relationships seen in a vertical section are so badly distorted as to have little meaning for purposes of structural analysis. The geologist who wishes to see his steeply plunging structures as they really are will turn to normal sections, by viewing (1) his map patterns or (2) his vertical sections down the plunge.

The emphasis in this paper is on the utility of the down-structure method as a quick visual method of grasping structural relationships. The *drawing* of normal sections on the basis of conventional data involves all the problems involved in drawing vertical sections and additional steps that are more or less self-evident. Structural-dip angles, for example, can be derived from conventional dip angles by any of the standard methods of engineering drawing (see Beckwith's [1947] handling of the analogous problem of preparing fault-plane sections). Already prepared vertical sections can be transformed into normal sections by a variety of methods, one of the easiest of which depends on the fact that, if any vertical dimension in a vertical section is multiplied by the cosine of the angle of plunge, the product is the cor-

responding dimension in the normal section.

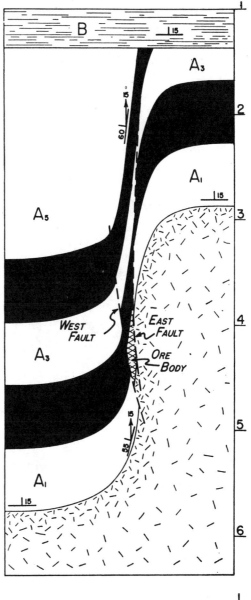

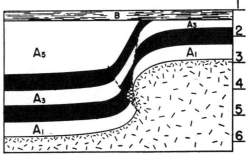

FIG. 4.—Diagrammatic map and normal section of an intrusive monocline.

NEED FOR SYMBOLS TO INDICATE DIRECTION OF STRUCTURAL STRIKE ON GEOLOGIC MAPS

The last implication of the axis–axial-plane relationship that merits mention here is a simple matter of procedure in the drafting of geologic maps. Surface traces of axial planes (sometimes mislabeled as "axes") are often drawn in and, because they help to distinguish anticlines and synclines at a glance, are of some use to the map-reader. But, as indicated earlier, the surface trace of the axial plane is clearly defined by the pattern relationships of the rock units involved in the folds—both strike and *dip* of the axial planes are at once evident if the maps in figures 3 and 4 are viewed down the plunge. Therefore, especially if they clutter a map crowded with other detail, symbols to show the surface traces of axial planes can be dispensed with.

The direction of structural strike, on the other hand, is the most significant directional element in the geology of a folded area. It is the "constant" direction from which all other elements of the map pattern diverge in various directions under the influence of plunge, and the structural significance of these zigzagging patterns can be understood only if the direction of structural strike is known. Because this critical direction cannot be closely inferred from the pattern, it ought to be shown on geologic maps, either by numerous *B*-lineation symbols or by lines to show surface projections of fold axes or by some other method appropriate to the scale of the map and the nature of the geology.

APPLICATION OF THE DOWN-STRUCTURE METHOD TO PLUNGING MONOCLINES

The following is a discussion of relationships along the margin of the intrusion in figure 4 as viewed in the usual manner, that is, with the line of sight normal to the page.

The intrusive border consists of three parts, differing in trend and structure, but grading into each other. In the northeastern and southwestern portions of the area the intrusive contact is concordant at the base of sedimentary sequence A; the sedimentary rocks strike east-west and dip gently northward off the intrusion. In the central portion of the area the same sedimentary rocks dip steeply to the west and are partly concordant and partly in fault contact with the intrusion. South from the faulted segment of the contact the dip of the sedimentary rocks and of the intrusive contact decreases as the strikes swing around to the west. The faults diverge to the north and pass into a monocline that trends slightly east of north. It is noteworthy that the east fault dips into the intrusion and is therefore a thrust at the intrusive contact, but reverses its dip and becomes a normal fault as it grades northward into the monocline.

The faults and the flexure were evidently produced by emplacement of the intrusion. The faults are generally poorly exposed, but drag effects indicate a considerable component of strike-slip movement suggesting that the east (intrusive) side moved northward. Sedimentary sequence B, along the northern margin of the map, rests on an unconformity that bevels the intrusive monocline, and is therefore clearly post-intrusive.

There are replacement ore bodies in limestone unit A 1 in the faulted segment of the contact. Mineralization is evidently limited to this segment and dies out northward into minor alteration effects in the breccia on the faults.

Having drawn this particular map from the section, the writer *knows* what the structure is. The first two paragraphs are not an analysis of structural relations along the intrusive contact (which they purport to be) but a description of outcrop patterns in which each statement emphasizes the incidental at the expense of the essential. Depending on wording, most of the statements range from quarter-truths to three-quarter truths. The interpretation in the third paragraph is entirely incorrect.

The major structure is a faulted intrusive monocline, plunging due north. It is not the intrusive contact but the *surface trace of the contact under the influence of plunge* that diverges from the northerly alignment of the monocline to form the northeastern and southwestern homoclinal segments of the outcrop pattern on the high and low sides of the monoclinal flexure. The structural strike is north-south; it is only the strata of the monoclinal limb that, under the influence of plunge, strike east of north. Decrease in the angle of dip south of the faulted segment is not due to a change in structure from north to south but to the fact that erosion truncates, at progressively lower levels from north to south, a structure that was originally identical in form throughout the map area. The faults do not pass northward into the flexure; they pass *upward* into the flexure, and they are continuous down the plunge beyond the north margin of the map. The change of the east fault from reverse to normal is merely a characteristic of its outcrop; the fault itself is a plane of such curvature that all straight lines within the plane plunge northward. The so-called "drag" that was taken to indicate strike-slip movement is the effect of plunge on the outcrop pattern of beds dragged by dip-slip movement. The faulting was not due to a northward bulging of the eastern part of the intrusion relative to the western part; the faults are breaks in a stretched flexure, along which the roof of the eastern part of the intrusion was raised higher than the roof of the western part. Alteration associated with mineralization does not die out *northward;* it dies out *upward* along the faults. Mineralization is not confined to the place where the faulted part of the contact happens to be exposed at the surface; the same favorable structural relationship continues

down the plunge beyond the north margin of the map.

It is often difficult or impossible to determine whether the plunge of a structure is genetically related to the structure or whether it was imposed on the structure by later and independent crustal movement. In this case the fact that the angle and direction of plunge is the same as the dip of sedimentary sequence *B* indicates that the intrusive monocline was originally horizontal and that the plunge was imposed by postintrusion northward tilting.

Figure 4 is, of course, greatly oversimplified. Complicating elements that might be expected in nature include (1) a variety of complexities in the pre-intrusion stratigraphy; (2) one or more generations of deformation structures predating the intrusion; (3) minor structures related to the emplacement of the intrusion, such as bulging along the monocline, cross faulting, and injection of apophyses; and (4) one or more generations of postintrusion structures. It is, in addition, virtually certain that a faulted monocline produced in this manner will actually change in form down the plunge and that its structural strike will be curving rather than rectilinear. Finally, the critical contacts may be poorly exposed and the outcrop patterns may be distorted by topography.

It is in enabling the geologist to look through this maze of extraneous detail, to see the plunging intrusive monocline, that the down-structure method is most helpful. Lineations, joint patterns, and other structural elements in the sedimentary rocks will be understandable only when viewed in their structural setting; they will not fit outcrop patterns influenced by plunge. Contrasted types of primary fracture-and-flow structures in the igneous rock along the east-west and north-south–trending segments of the intrusive border will fall into place under a roof restored on the basis of the down-structure view of the map pattern. The fracture-and-flow structures in the south-central part of the map area will be different from those anywhere else in the "interior" part of the intrusion and will be in good order or "meaningless," depending on whether the investigator realizes that the intrusive monocline did not terminate where the trace of the contact swings westward but was formerly continuous beyond the south margin of the map. And so forth. In general, without a grasp of the over-all picture, each day's recording of "all of the facts" will only "add to the confusion" and will be a dull business leading to a report of the type that introduced this section. When the picture is understood, the area will offer an exciting opportunity to study a variety of structural relationships in an intrusive monocline exposed from top to base as though seen in a vertical cliff a mile or so in height, except that, instead of being lowered down the cliff on ropes, the investigator merely walks southward to get deeper into the structure. And if the purpose of the study of the district is economic, the grasp of the picture provided by the down-structure view may mean the difference between success and failure of the economic venture.

APPLICATION OF THE METHOD TO COMPLEX STRUCTURES

Bedding attitudes measured at the surface in areas of upright folds and other simple structures can usually be extended downward in a structure section, with reasonable assurance, far enough to provide an adequate picture of subsurface relationships. But it is obvious that this approach will not serve in areas of *décollement* folding or in piled-up recumbent folds and thrust plates or, in general, wherever there are major tectonic discontinuities between structures exposed at the surface and those in depth. It is doubtless for this reason, and because necessity is the mother of invention, that the down-structure method was first developed and has been most used in districts of complex structure. Axial culminations are often said to be key areas in complexly deformed ranges. Perhaps it would be more accurate to say that the culminations are key-*holes;* the key that fits them is the method of looking down into depth along the plunge of structures revealed by map patterns.

Professor X in anecdote 1 missed this point himself and made it impossible for his students to understand the Alpine district under discussion in his class by exhibiting the sections apart from the map on which they were based.

The recumbent fold in the Pennsylvania Piedmont, mentioned earlier, was recognized as such by Bailey on the basis of down-structure study of the geologic map and therefore serves well as an example of the application of the method. The following axioms are much more readily understood against the background of the present paper than they were in the original very brief statement of our case:

1. *If the two boundaries of a formation, under the influence of plunge, follow each other with rough parallelism across the regional strike of fold axes, then the one boundary is (structurally) at the top, and the other is (structurally) at the bottom of the formation.*

In the southern part of figure 1 the regional strike of fold axes is east-west, and the stratigraphic units trend generally north-south under the influence of a westerly plunge. The western boundary of each unit is structurally at the top, and the eastern boundary is structurally at the bottom.

2. *If the two boundaries mentioned in axiom 1 represent the stratigraphic top and the stratigraphic bottom of the formation respectively, then the formation must be involved in a normal sequence of plunging folds.*

This is the case in figure 1, where the structural top and bottom of each unit correspond with its stratigraphic top and bottom.

3. *If the two boundaries mentioned in axiom 1 are stratigraphically equivalent, then the formation must be recumbently folded.*

Thus, in figure 5, the Baltimore gneiss–Setters quartzite contacts (1) near Chatham and (2) at Woodville, both convex to the southwest in pattern under the influence of plunge, are the same contact, and it follows that one of them must be upside-down.

4. *If the observer orients the map so as to look at it in the direction of plunge, he will see the formations disposed on the flat surface of the map in much the same attitude as they would present in a cross section, though, of course, with different proportions.*

In other words, if the observer looks down the plunge, he will see that it is the Woodville contact that is inverted and that the gneiss is the core of a recumbent fold that has its root somewhere off the map beyond *C*, and a down-bent tip at *E*.

A somewhat less mechanical approach to the problem is as follows:

The distinctive map patterns of broad foreland domes and basins, with interformational boundaries obviously influenced by topography, are indicative of low-dipping contacts. When these sprawling patterns (as in fig. 5) are encountered in belts of high-grade dynamic metamorphism, the student will do well to ask himself this simple question: "Do these low dips represent the sum total of the deformation that has occurred in these rocks?" If the ledges reveal intense stretching in small- and large-scale boudinage structures, minor isoclinal folds of types producible only by plastic flowage, or other evidences of great differential movement within and between the lithologic units, the answer to the question is clear. The sum total of deformation certainly must have been very much more than the gentle tilting indicated by the low-angle contacts; the contacts are subhorizontal because deep-seated orogenic flowage on a regional scale has developed folds with subhorizontal axial planes and limbs. Step 1 in this alternative approach is, then, a realization that large-scale inversions are not necessarily restricted to the mountain belts of Eurasia but may be suspected wherever metamorphic rocks show persistent low-angle foliation and contacts, dome-and-basin outcrop patterns superficially resembling those of forelands, or outcrop patterns that swirl across the structural strike.

In such patterns it is possible for the student to draw *one* cross section anywhere, and, if he pays attention only to dips of bedding and foliation and disregards map patterns and plunge, he can draw his one section without showing large-scale inversion, even though half the rocks are upside down. Thus, in figure 5, a section can be drawn along the line *A–A′* to show the structure as a simple overturned anticline with a core of Baltimore gneiss; or, a section can be drawn along the line *B–B′* showing two anticlines with gneiss cores separated by a syncline with the Wissahickon schist in the center. But *both* these sections cannot be drawn because they fail to comport with each other or with the map pattern and plunge.

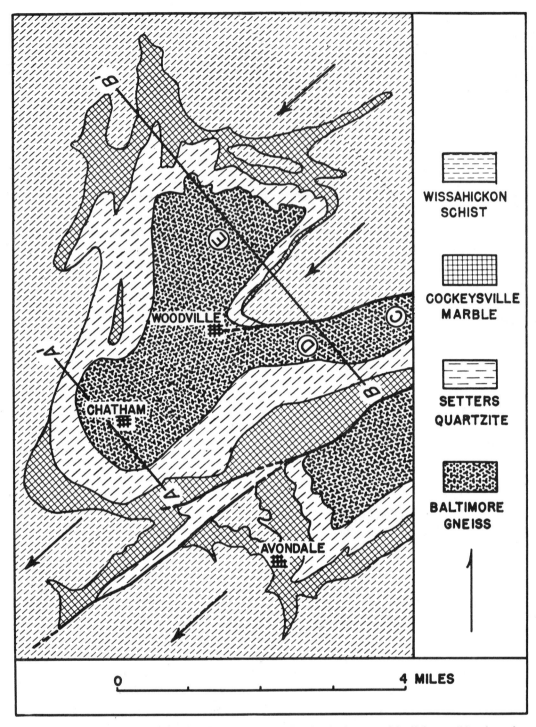

FIG. 5.—Geologic map of the Woodville area, Pennsylvania. Redrawn, with slight modification, from U.S. Geological Survey Folio 223.

The pair of sections do not match each other, because they place a major anticline (in the section along $A-A'$) directly on structural strike with a major syncline (in the section along $B-B'$). This is impossible because the same belt of Baltimore gneiss is involved in both structures, for this belt cannot be anticlinal on one boundary (the southwest) and synclinal on the other boundary (the northeast). The section along $B-B'$ would fail to match the pattern due to plunge, because the pattern requires that the supposed Wissahickon syncline northeast of Woodville plunge to the northeast, while all the types of B-lineation in this critical area actually plunge to the southwest. The area northeast of Woodville might be said to be "synclinal in stratigraphy" because the youngest rock unit is in the center; but the map pattern and lineation indicate that this area is definitely anticlinal in structure and that the younger Wissahickon is plunging to the southwest under the older gneiss. The Woodville structure must therefore be an arch in the inverted limb of a recumbent fold; the Chatham structure is the same arch in the upper limb.

As far as relationships along the line $B-B'$ are concerned, the gneiss at E might well be the core of an anticline. But when a map shows lineation on the nose plunging under a core that consists of the oldest of the rock units involved in the fold, and when the section shows the structure as a simple anticline, then there must be something wrong with the section or with the mapping. The termination of the gneiss belt at E is anticlinal in stratigraphy, but definitely synclinal in structure; it is the diving tip of the recumbent fold.

These conclusions are based on the general principle that, especially in folded metamorphics, linear elements are far more reliable as guides to structure than are planar elements and on the axiom that is the theme of this article, namely, that a section across a plunging structure *must* match the map not only along the line of section but also up- and down-plunge from the line of section.

The down-structure method is helpful in investigation of areas of the type shown in figure 5 because it guarantees that the field worker will grasp the concept of recumbent folding, as a working hypothesis, as soon as his B-lineation symbols are sufficiently numerous to guide his eye in viewing his map patterns. If the rocks are, in fact, recumbently folded and if the concept of recumbent folding is not part of the field worker's mental field equipment, then each day's observations will be at variance with what he thinks they should be on the basis of the previous evening's analysis of the mapping to date, and every tentative structure section will conflict with every other tentative section. A little of this type of frustration is perhaps good for the soul; too much of it is likely to drive a doctorate candidate into the life-insurance or brokerage business. The difference between heartbreaking bafflement at every turn and the happier situation in which each ledge fits its mates as snugly as pieces of a jigsaw puzzle may be only a simple twist of the wrist to orient the map so that the patterns are seen as structures. It is, of course, not the down-structure method but the snug fitting of the factual evidence on which the structural interpretations must be based.

Presentation of these interpretations poses a problem that is separate and distinct from the procedures of investigation that led to them. Any section through a complex area of the type in figure 5 must be extended to great depth if it is to show the inferred structure in its entirety. The use of dotted lines might properly express uncertainties as to details in the deeper parts of the section, but dotted lines are too weak to express the virtual certainty that the section is correct in depth in all essential relationships. It is, in other words, difficult to show in a section, clearly and honestly, all the nice variations between fact and inference, together with the grounds for the inferred parts. There is, moreover, always the danger that the section may be viewed apart from the map on which it is based, in which case it will look fantastic to any geologist accustomed to drawing sections on the basis of dip attitudes alone.

One possible answer to this problem of presentation is to avoid the use of vertical sections, which require long-distance extrapolation from surface observations, and, instead, to transform the map into a horizontal section at any convenient level, as midway between hilltops and valley floors. This involves a minimum of extrapolation and no theory at all. With the topographic distortion of the map patterns eliminated in this manner and with an abundance of symbols showing the direction and plunge of linear elements as well as dip and strike of planar elements, the revised map is, when viewed down the plunge, the most complete and honest manner of illustrating the structure because it is wholly factual. This method requires the co-operation of the reader to the extent of twisting the map, but this is no defect—map-twisting is good exercise for geologists.

REFERENCES CITED

BAILEY, E. B., and MACKIN, J. H. (1936) Recumbent folding in the Pennsylvania Piedmont—preliminary statement: Am. Jour. Sci., vol. 33, pp. 187–190.

BECKWITH, R. H. (1947) Fault problems in fault planes: Geol. Soc. America Bull. 58, pp. 79–108.

BILLINGS, M. P. (1942) Structural geology, New York, Prentice-Hall.

GODDARD, E. N., and OTHERS (no date) New list of map symbols. Obtainable free of charge from Map Editor, U.S. Geological Survey, Washington, D.C.

GREENLY, E., and WILLIAMS, H. (1930) Methods in geological surveying, London, Thomas Murby & Co.

MACKIN, J. H. (1944) Relation of geology to mineralization in the Morton cinnabar district, Washington: Washington State Division of Mines and Mining, Rept. Inv. no. 6, pp. 1–47.

——— (1947) Some structural features of the intrusions in the Iron Springs district, Utah: Guidebook to the Geology of Utah no. 2, pp. 1–62.

THE USE OF PLUNGE IN THE CONSTRUC-
TION OF CROSS-SECTIONS OF FOLDS[1]

C. H. STOCKWELL[2]

INTRODUCTION

This paper outlines a graphical method for the construction of sections of folded rock formations by projection of formational contacts down the plunge to the plane of the section[3]. Many plunge determinations are required, not only in the region of the axes but also on the flanks of the fold. They are given by the lines of intersection of planes represented by bedding strike and dip symbols on a map. The method is illustrated by the construction in plan and section of ideal synclinal folds of two types, one with the geometric characteristics of cylinders in which all the plunges parallel one another, and the other with the geometric properties of cones, in which the plunges diverge from a vertex. Domes, basins, and canoe-shaped folds are dealt with by dividing them into several parts and by treating each segment as if it were a portion of either a cylinder or a cone as the case may be.

In constructing the sections it is necessary to assume, unless there be evidence to the contrary, that the thickness of beds remains constant down the plunge, but any variation transverse to this direction, such as thinning on the limbs and thickening toward the axial plane, is brought out by the construction, as is also the curvature of the fold.

The plunges, as will be seen, define the fold in three dimensions and any section may then be drawn, or structure contours may be made on any particular horizon marker. In the case of cylindrical folds, sections normal to the plunge are preferred for purposes of study, for these give a view equivalent to that of a vertical section of a horizontal fold and accordingly reveal the true curvature and the actual thickness of the formations.

The advantages derived from viewing folds in the direction of the plunge have been pointed out recently by Mackin (1950) in a paper which emphasizes a means of obtaining, quickly, a visual picture of structures from that point of

---

[1]Published with the permission of the Director-General of Scientific Services, Department of Mines and Technical Surveys.

[2]Geologist, Geological Survey of Canada, Ottawa.

[3]The plunge of a line is the angle between the line and the horizontal as measured in the vertical plane. The direction of its horizontal projection is called the bearing of the line. An axial line or axis of a fold is a line along a surface of a particular bed where the degree of curvature, considered in three dimensions, is greatest. The axial plane or axial surface of a fold is the plane or surface that contains all the axial lines. A crest line is a line along the highest part of a bed and the crest plane or crest surface is a plane or surface containing all the crest lines. A trough line is a line along the lowest part of a bed and the trough plane or trough surface is a plane or surface containing all the trough lines. The above definitions follow rather closely those given by Billings (1942).

97

view. In the more geometric treatment of the subject, as presented in the following pages, many of the points brought out by Mackin are again mentioned and some additional features of interest are brought to light. The lines of intersections of planes, as represented by bedding strike and dip measurements, have been used for determining the plunge of the axes of folds by Hutton (1874), Harker (1884) and Eardley (1938) but, as far as known, have not been used previously to determine the three-dimensional shape of a fold.

The method can be applied only to plunging folds. Sections of horizontal folds may be made by the downward projection of dips, either by the method of concentric arcs (Busk, 1929) or by the method of evolute and involutes (Mertie, 1940, 1947, 1948). Both of these methods apply, in general, only to cylindrical folds of the parallel type, that is, to cylindrical folds in which the beds on the limbs are of the same thickness as those at the axial plane.

It is noted that the plunge method is independent of secondary features such as bedding-cleavage intersections (Leith, 1923) and lineations (Turner, 1948), but if these are dependent (Derry, 1939) on the folding they may be used to great advantage, especially in tight or isoclinal folds where many beds are so nearly parallel that they give poor intersections. This applies especially to cylindrical folds and, in this type, an essential agreement between the plunges determined by the two methods indicates that the secondary features are of the dependent type and that they may be used with confidence. Bedding-cleavage relationships in conical folds are not fully understood but it is obvious that plunges so determined on the limbs do not give the plunge of the axes even when the cleavage is dependent on the folding.

In applying the method to actual cases irregularities inherent in bedding planes of rock formations are smoothed out by taking averages.

The writer is indebted to officers of the Geological Survey, particularly to C. E. Cairnes, J. M. Harrison, and J. O. Kalliokoski for their interest and helpful suggestions. The mathematical parts of the paper were kindly checked by J. A. Rottenberg of the Dominion Observatory.

IDEAL CYLINDRICAL FOLDS

A cylindrical surface may be generated by a straight line moving parallel to itself so as always to touch any given curve. Such a surface can be formed by bending a sheet of paper, without buckling or stretching, into a trough or arch to represent a syncline or anticline. All straight lines drawn on the paper, and called generators, or elements, are parallel with one another and with the axis of the cylinder. All sections cut normal to the cylinder axis, called right sections, are alike in size and shape and may be the arcs of circles, or any other curve regular or irregular. When one cylinder lies within another, representing two surfaces of a bed or formation, the two cylindrical surfaces need not be the same shape, as seen in right section, but all the elements of both surfaces will be parallel with one another so long as the thickness of the bed or formation remains constant along the direction of the elements. The right section then reveals both the true thickness of the beds and any variations such as thinning on the limbs and thickening toward the axial line. If the formation is crenulated the gross thickness is revealed. The right section also shows the true curvature of the fold and on it may be accurately drawn the trace of the axial plane and of the crest or trough planes.

98

Each plane tangent to the surface of a simple or complex cylinder lies along an element and, since all the elements are parallel, the lines formed by the intersections of all such planes are also parallel. The converse is also generally true, namely, that if the lines of intersection of all tangent planes, as represented by many strike and dip symbols on a map, are parallel the fold is cylindrical[1]. The bearing and plunge of these lines give the bearing and plunge of the fold for they define the attitude of all the elements, including the axial line.

Moreover, each plane that cuts a cylindrical surface and lies parallel with each tangent plane cuts the surface along an element. All these planes, in turn, intersect one another in lines that likewise parallel the elements. In folded rock formations such planes may be represented by cleavage planes. Cleavage parallel with the axial plane, or axial plane cleavage, is a common type but other dependent cleavages may be symmetrically inclined with respect to it. For example, it is common to find that the angle of inclination changes as the cleavage passes from an incompetent bed to a competent one.

Accordingly, all the bedding and cleavage planes in ideal cylindrical folds are co-axial and, for convenience in visualizing the relationships, may be considered as being rolled around the elements, as a card may be rolled around a pencil.

SYMBOLS
To Accompany Figures 1, and 3-13

CONTACT, showing dip. . . . .

BEDDING, (horizontal, inclined, vertical, overturned) . . . .

CLEAVAGE, (horizontal, inclined, vertical) . . . .

ANTICLINE, trace and dip of axial plane (inclined, vertical); . . . .
    trace and dip of crest plane (inclined, vertical) . . . .

SYNCLINE, trace and dip of axial plane (inclined, vertical); . . . .
    trace and dip of trough plane (inclined, vertical) . . . .

MINOR FOLD (crest or trough). . . . .

DRAG-FOLD . . . .

LINEATION, bearing and plunge, horizontal, vertical. . . . .
May be used alone or in combination with
other symbols to indicate the intersection
of planes, elongate pebbles, slickensides,
axial line, crest line, trough line, etc.
(Only one type of arrow is needed in this paper
but in practice each type of lineation should
be indicated by a special type of arrow).

The elements lie in the direction of the b fabric axis which, in flexural folds, is the axis of rotation. Lineations may develop either along b or along a, which stands at right angles to b and lies either on the cleavage planes or along bedding planes as a result of differential slipping between beds during folding. Cleavages or lineations of the types described may begin to form before the folding is complete and may themselves later become folded without destroying the relationships outlined.

---

[1]Exceptions include a hemisphere in which all dips at the surface are vertical and also an isoclinal fold which has been re-folded.

99

Figure 2.

A duo-stereogram, or graph, for the determination of the line of intersection of two planes, and for the solution of other problems involving true and apparent dip.

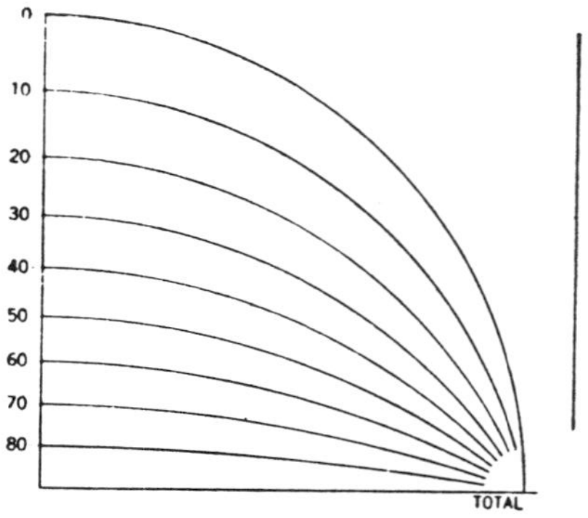

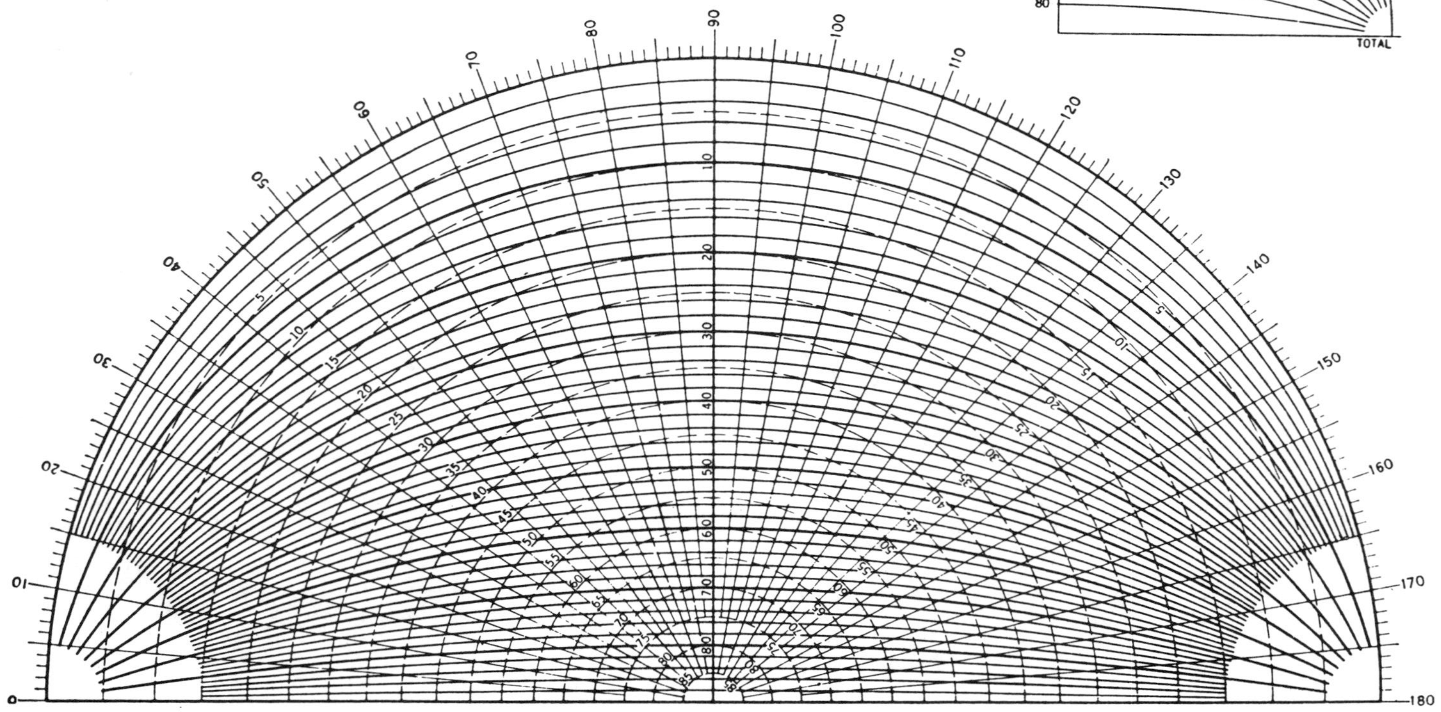

100

In the following diagrams of ideal folds the right section was drawn first and the plan was constructed from it. This is necessary in order to determine the plan of a fold of some assumed three-dimensional shape. However, it is more appropriate to describe the method of construction in reverse order for, in practice, the plan, or map, is first made and it is then required to draw the section. The method is illustrated in Fig. 1. The plan shows two horizon markers, or contacts between formations, and several localities where the attitude of beds has been recorded. The bearing and plunge of the lines of intersection of planes represented by strike and dip symbols, taken two at a time, may be determined by descriptive geometry, by means of a stereonet (Bucher, 1944) or, more quickly, by use of a modified stereonet which may be called a duo-stereogram (Fig. 2). This graph also gives a ready solution to other problems involving true and apparent dip, which frequently arise in subsequent constructions[1]. In Fig. 1 all these lines of intersection are found to parallel one another

---

[1]To use the graph the reader is obliged to make an exact tracing of the meridian lines of the lower half of the figure. The tracing is made on transparent material and is numbered as shown, on reduced scale and with some lines omitted, in the inset figure, and the word total is marked where indicated. The tracing, called graph A, is laid over the original, called graph B, so that the right angled corner of A coincides with the centre point on the left hand margin of B. If many readings are to be made it will be found convenient to insert a thumb tack from the underside at this point, so that the transparent overlay can be rotated about this point as centre.

To find the line of intersection of two planes the horizontal angle between the two strikes is measured with a protractor, care being taken always to measure the angle toward which both planes dip. To see this angle clearly it may be necessary to extend one or both strike lines beyond their point of intersection as shown in Fig. 1. This angle is called the total angle.

The procedure is best explained by using an example. Thus in Fig. 1 two bedding planes are shown, one dipping 39 degrees and the other dipping 70 degrees. The total horizontal angle measures 146 degrees. Rotate the transparent overlay until the edge marked total is opposite 146 on the margin of graph B. With this setting find by inspection the point where meridian line 70 on graph A crosses meridian line 39 on graph B. Hold a pencil at this point so that it will not be lost when the eye is moved and read the plunge (30 degrees) on the concentric, broken-line, circles and read the bearing (134 degrees) on the radial lines. With a protractor set off the bearing at 134 degrees from the strike of the bed which dips 39 degrees. It is essential that the greater of the two dips be read on the transparent overlay A and the lesser dip on B and that the bearing is set off from the strike of the bed of lesser dip.

When only one plane is being considered and it is required to find an apparent dip, graph B is used without the transparent overlay. Thus, given a true dip of 39 degrees and a horizontal angle of 134 degrees between the strike of the bed and the bearing of the apparent dip, the point is found on graph B where meridian 39 crosses radial line 134 and the apparent dip (30 degrees) is read on the concentric circles. Obviously, any one of the three variables, true dip, apparent dip, or horizontal angle, can be determined when the other two are known.

The writer uses a duo-stereogram similar in principle to the one illustrated but easier to read because of less confusion of lines. Both the underlying graph, which is printed on a card, and the transparent overlay contain only meridian lines. A third transparent overlay consists of a pivoting arm so graduated as to make the concentric circles unnecessary and the arm itself replaces the radial lines. It is, however, more difficult for the reader to make but would be supplied by the writer should there be sufficient demand.

101

(they plunge 30 degrees south). When a considerable number of such determinations have been made it may then be concluded that all the elements of the fold are parallel and that the fold is cylindrical, barring the exceptions noted which should normally be indicated by field work. If required, the dip of a bed at any point on the horizon markers can be determined since the strike, being a line tangent to the curve, and the apparent dip in the direction of the plunge are known.

The section plane, normal to the plunge, is drawn at AB in plan and N′M′ in longitudinal projection. The element LM penetrates the erosion surface at L and L′ and penetrates the section plane at M, M′ and M″, the latter point being on the required right section. It should be noted that in projecting this element as a straight line to depth M′ it is assumed that the thickness of the beds is constant down the plunge and the element, therefore, cannot be projected beyond the section plane which is presumed to be the limit of the cylindrical portion of a larger fold. Similarly, point P in plan is projected down its element to Q″ in the section. Many such points must be found in the section before a curve may be drawn through them and for this purpose it is much quicker to use proportional dividers set in the ratio L′N′ : N′M′. With

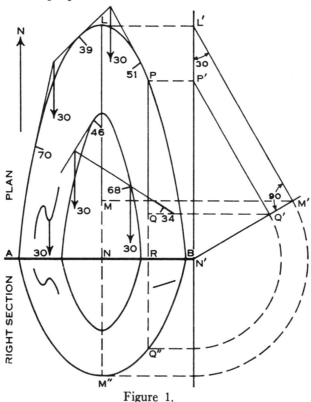

Figure 1.

Method of determining plunge and the construction of a right section of a cylindrical fold.

this ratio, distance LN is reduced to NM″, distance PR to RQ″ and so on until a sufficient number of points have been determined. The right section is an end view looking up the plunge, or equivalent to a vertical section of a horizontal fold. The axial plane is vertical, the formation maintains a uniform

102

thickness on the limbs and in the axial region alike, and the fold may be described as a cylindrical, upright syncline of the parallel type plunging 30 degrees south.

The figure shows a strike and dip symbol that is not on one of the horizon markers. The plane which this symbol represents is carried to the section by measuring proportional distances from each end of the symbol and the trace of the plane then appears in the section as the line shown. It crosses the general trend of the formations as if being part of a drag fold. In field work such cross strikes should be recorded especially on the limbs of folds, for there they give good intersections which could not be obtained otherwise, since most of the beds are here nearly parallel with one another. In measuring the attitude of contorted beds it is better to take a few strikes and dips accurately than to estimate an average for, although such features may have no significance in projecting trends, they are important in determining plunge. If no horizon

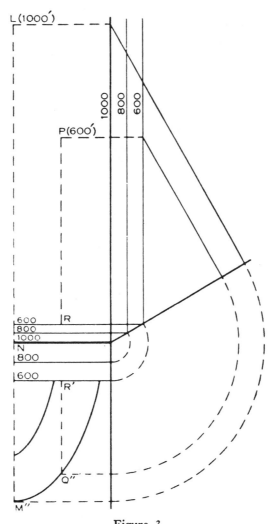

Figure 3.
Construction of Figure 1 adapted to contoured map.

103

markers or contacts are shown on the map, a large number of strike and dip symbols projected to the right section may give a good picture of the curvature and attitude of a fold. The horizontal trace of a drag fold is also shown, true in shape and orientation but exaggerated in size. It is evidently of the dependent type for its crest line plunges in accordance with the plunge of the main fold. In assymetrical folds the crest line of drag folds is more readily measured than the axial line and in case neither is well exposed, the plunge may be calculated from the strikes and dips on the limbs. When projected to the right section the drag fold appears in true perspective. Similarly the crest lines or trough lines of other dependent minor folds, rolls, or crenulations ideally plunge parallel with the plunge of the main fold.

In the construction of Fig. 1 it is assumed that the plan is a horizontal plane. Fig. 3 illustrates a simple means for allowing for differences in elevations of points on the mapped contacts. Contour intervals are drawn true to scale in the side view and are shown in plan and right section by the construction indicated. Thus, let point L be at an elevation of 1,000 feet, point P at 600 feet and assume a plunge of 30 degees south so that the proportional dividers are set with the same ratio as before. Distance LN becomes $\overline{NM}''$ and distance PR becomes $R'Q''$. The construction may also be made without proportional dividers by drawing the plunges in side view and carrying them to the section as shown. In rugged country this method might be found useful in projecting contacts, up or down the plunge, to slopes where no exposures are available.

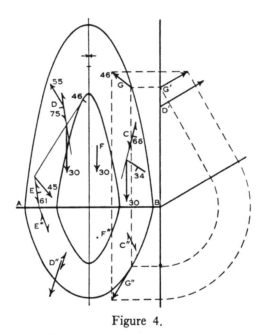

Figure 4.

Same as Figure 1, showing cleavage and lineations.

Cleavages and lineations are diagramatically illustrated in Fig. 4. This is the same fold as represented in Fig. 1 but redrawn to avoid confusion of symbols. In field work the bearing and plunge of the intersection of cleavage and bedding, or of two or more cleavages, may be measured directly, but if

104

490

exposures are not suitable for this they can be obtained just as well by measuring the strikes and dips of the two intersecting planes and calculating the plunge. Cleavage is shown on the plan by strike and dip symbols at C, D and E. The lines of intersection of cleavages C and D with bedding and with one another plunge 30 degrees south, in accordance with the plunge determined previously from the bedding intersections, and the cleavages are therefore dependent. Since these cleavages are co-axial with the fold the symbols may be projected to the section in the same way as was done for bedding planes. Cleavage at E is evidently independent for its line of intersection with bedding differs from that of the fold as a whole. Planes of independent cleavage, or of faults that do not follow an element of the fold, intersect the cylinder along curved lines. The trend of such a plane in right section is determined by finding the line of intersection between the plane and the right section and projecting this to the longitudinal projection plane and thence to the right section.

Lineation F along the *b* axis appears in the section as a point at F″. Other lineations appear as lines and it is desired to know their attitude with respect to *b*. Thus, a lineation at G plunges 46 degrees northwesterly. On calculating the depth of the head of the arrow and, on projection, the arrow is seen at G′ to stand normal to *b* and at G″ to lie on the bedding plane. Another lineation at D plunges 55 degrees northwesterly. By a similar method it is found, at D′, to stand normal to *b* and, at D″, to lie in a cleavage plane. The attitudes of

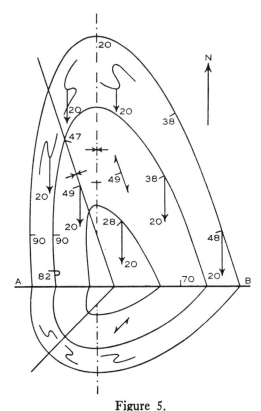

Figure 5.

A cylindrical, parallel, overturned syncline plunging 20 degree south.

105

491

various types of lineations with respect to the geometry of the fold are thus made apparent and serve as confirmatory evidence in fixing the plunge of the elements.

The method of constructing the right section of a cylindrical upright fold requires no modification for the inclined, overturned, or recumbent positions. A few diagrams of some of these assymetrical types, however, serve to emphasize some features of interest.

Figure 5, representing a cylindrical, parallel, overturned syncline plunging 20 degrees south, was constructed in the manner described above. The position of the axial plane was determined in the right section and was then projected to the plan. The noteworthy result of this operation is the lateral displacement of the surface trace of the axial plane. This is only an apparent displacement due to the manner in which the horizontal surface truncates the fold obliquely. The axial plane is seen to fall in proper relationship to drag folds, which are conventionally drawn, and to axial plane cleavage, both of which would have been interpreted incorrectly had the trace of the axial plane been drawn through the apparent nose of the fold as is commonly done. This apparent displacement of the axial plane occurs in all inclined and overturned folds, but is less conspicuous when the axial plane dips steeply or when the nose of the fold is very sharp. The dip of the axial plane is readily obtained from the plunge and the angle between the bearing of the plunge and the strike of the plane. The trough plane is also first drawn in right section. In parallel folds, such as here illustrated, the trough plane dips vertically, regardless of the inclination of the axial plane and it meets the axial plane along a line plunging parallel with the elements. At every point on this line the beds must form a sharp V, that is they must break, in order to maintain a constant thickness.

It should be noted also that the vertical beds strike parallel with the bearing of the plunge and that the trough beds dip down the plunge. These features may sometimes be found useful in making a preliminary estimate of the general attitude of a fold before any calculations are made.

Parallel and similar folds are special cases of a series of other types, all of which owe their characteristics to a systematic variation in the thickness of beds, and consequently in the curvature of one contact with respect to another. The patterns formed by the traces of the axial and trough planes for several of these types are illustrated in Fig. 6. In parallel folds, as already mentioned, the trough plane of a syncline and the crest plane of an anticline dip vertically. In similar folds, that is, in those in which a tracing of a right section of one curve can be superimposed exactly over any other, the trough plane parallels the axial plane. The thickness of beds, as measured in a direction parallel with the axial plane, is constant. In types intermediate between parallel and similar the trough plane lies in an intermediate position, as illustrated. In flexural folds in which competent beds alternate with incompetent ones the trough surface is vertical in the one bed and inclined in the other, although the trace of the axial plane may remain straight. As a consequence of attenuation in the axial region, parallel folds pass into supratenuous folds where the trough plane dips away from the axial plane in synclines and the crest plane dips toward it in anticlines. With thickening in the axial region beyond the requirements for similar folds the latter pass into a type that may be called acuminate folds because they

106

492

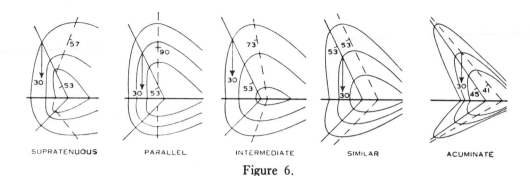

SUPRATENUOUS PARALLEL INTERMEDIATE SIMILAR ACUMINATE

Figure 6.

Trace of axial plane and trough plane in plan and right section of various types of cylindrical synclines.

taper to a point; in these the trough plane dips in the same general direction as the axial plane but the strikes converge toward the pointed end of the fold.

Figure 7 shows a series of synclines in various attitudes from an upright to a recumbent position, all plunging 30 degrees south. As the right sections show, the folds represent various stages in rotation about the axial lines directly toward the east. Allowing for an original tilt of 30 degrees to give the plunge, the resultant of the two rotations is indicated by the arrows below each of the diagrams. This ideal example suggests that a correlated and truer picture of movement (not external, casual forces) can be obtained in this way than by relating movements to the surface trace and dip of the axial plane, which gives a quite different conception, as indicated by the arrows above each of the diagrams.

IDEAL CONICAL FOLDS

A conical surface is a surface generated by a straight line that always passes through a fixed point and always touches any fixed curve. The fixed point is known as the vertex. In a trough or arch of the conical type the shape of the fold, as seen on all cross sections drawn parallel with one another, is the same but the size decreases at a uniform rate as the vertex is approached. Straight lines, or elements, drawn on the surface fan out from the vertex. The bearing and plunge are constant along any one element but they vary from one element to another and, consequently, except in upright folds, the axial line is not parallel with the crest or trough line. A plane held tangent to the surface of a cone of simple or complex curvature and rolled around it passes through the vertex in all positions and always lies along an element. The line of intersection of any two tangent planes, as represented by strike and dip symbols on a map, although not coinciding with any of the elements, likewise passes through the vertex. The position of the vertex is fixed in space by the point of intersection of two such lines. When one cone lies within another so as to represent the two surfaces of a bed or formation, each conical surface has its own vertex. There need be no crowding together of material as the vertex is approached for, as the radius of curvature decreases, the beds of a conical syncline pass on either side into a conical anticline of opposite plunge and of increasing radius.

The right section, following geometric usage, is one cut normal to the axis of the cone. This section does not have the advantage of a right section of a

107

493

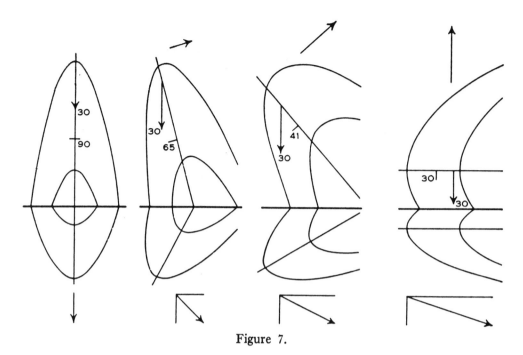

Figure 7.

A series of similar cylindrical synclines plunging 30 degrees and rotated on their axial lines through 0, 30, 60 and 90 degrees. Arrows below indicate resultant movement from the upright horizontal position. Arrows above represent the apparent rotation of the axial planes from the vertical and give a misleading notion of the movement.

cylinder for neither this, nor any other section that is not a curved surface, reveals the actual thickness of beds nor the true curvature. If, however, the curvature of the beds is circular the apparent thickness of one bed relative to another is shown and the curvature indicates whether the fold is of the parallel, similar, or some other type. In this case all the elements, as measured from the vertex to the plane of the section, are of equal length and the right section therefore may be drawn even when the position of the cone axis is unknown.

If the curvature is other than circular the right section gives a distorted view of the thickness and curvature. In this case the elements are of unequal length but a good overall picture may be obtained on a section that cuts three elements at points equidistant from the vertex, the one being the axial line and the other two being far out on the limbs and opposite one another. No one section, however, can give all the information that may be desired and supplementary sections or calculations are necessary to determine the true thickness of beds.

In the fold illustrated in Fig. 8 the section is a vertical one drawn normal to the bearing of the trough element. This section is chosen by way of illustration because the trough element can be determined at an early stage in the construction. The bearing and plunge of the two lines of intersection of three bedding planes have been determined at C and D. (In this illustration the plan is presumed to be a horizontal plane. Had there been differences in elevation between the points of observation of the attitudes of beds these planes would first be projected to a common datum level, because the position as well as the bearing of the plunge must be known). Both lines of intersection lie in

108

the bedding plane that occupies the intermediate position of the three chosen. Being in one plane, they must either parallel one another or meet at a point. If they parallel one another they represent elements of a cylinder. If not, as in this case, they represent elements of a cone that has a vertex at V. The height of the vertex above the ground or datum level is readily obtained because the distance VC or VD can be measured and the angle of plunge of each of these lines is known. If the lines of intersection of many other bedding planes on the same horizon marker likewise intersect in the same point, then the whole fold is a single cone with vertex at V. Having thus determined the position of the vertex the construction lines and the plunges, such as at C and D, which pass beneath the horizon marker, are of no further interest and may be erased.

It is now a simple, though tedious, matter to find the bearing and plunge of many elements on the horizon marker itself and to carry them down to the section. Since the plane of the section is to lie normal to the bearing, or horizontal projection, of the trough line, this element is found first. This, in the example given, is the element of steepest plunge and therefore nearest the vertex and is readily seen to penetrate the surface at E. The bearing of the element is VE. Project this to F, a point on the required section plane and draw the section line AB at right angles to it. Draw the longitudinal projection plane parallel with VE and plot the vertex V′ at the height previously determined. Project E to E′, and V′E′ to F′ on the vertical section plane. E′F′ is a side view of the trough element in true perspective and its plunge can be measured. (In projecting the element to depth F′ it is presumed that the conical characteristic of the fold is known to extend somewhat beyond the plane of the section). Project F′ to F″, a point on the required section. Similarly, any other element GH is found to penetrate the section at H″. By drawing many elements in this way the section of the horizon marker is finally made.

The plunge of the element GH does not appear directly in the side view but, if required, may be obtained from graph B (Fig. 2) for, considering the plane whose strike is GG′, the angle that G′H′ makes with the horizontal is the true dip and the angle G′GH is the horizontal angle between the strike of the plane and the bearing of the plunge. Also the dip of the bed at G may be determined since the plunge is known and the strike, being tangent to the horizon marker, can be measured. The vertical beds strike toward the vertex and, in contrast with cylindrical folds, do not trend at right angles to the strike of the trough beds.

The whole construction is repeated for each horizon marker or contact between formations because each surface has its own vertex. Thus, point W,W′ is the vertex for the inner horizon marker shown in the diagram. The trough plane, which contains the lowest points on each horizon, can now be drawn and its dip determined.

The axial plane, determined by points at which the curvature of the beds is sharpest, cannot yet be found for both the plan and the section give distorted views of the curvature. The axial plane may, however, be found on a stereographic projection. To accomplish this, place a piece of tracing paper over stereogram B (Fig. 2) and, imagining the vertex of the cone to be at the centre of the stereogram, plot the bearings and plunges of many elements to give the curve shown in Figure 9, where the bearings have been plotted with reference to the trough line which has been given an arbitrary bearing of 90 degrees.

109

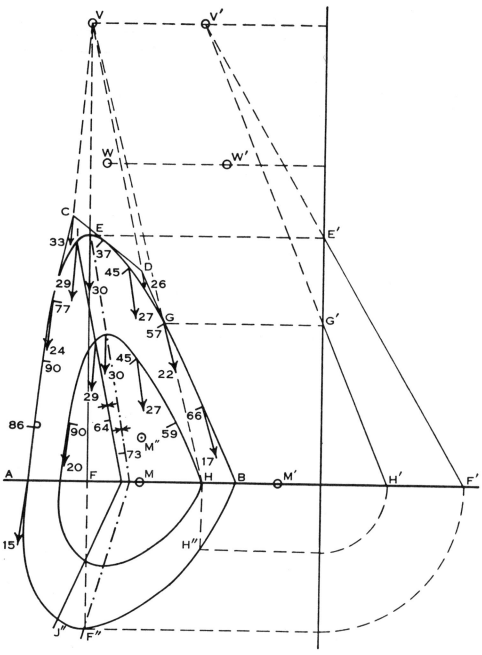

Figure 8.

A conical syncline, vertical section normal to bearing of trough line.

For example, the element at A, Figure 8, bears 82 degrees with reference to the trough element, and plunges 15 degrees. In Figure 9 this bearing is measured on the margin of graph B and the plunge of 15 degrees is found on the concentric circles, to give point A. Other points are found in similar manner and are joined to give the curve shown in the figure. This curve is a stereographic

110

496

projection of the intersection of the conical surface of the fold with the lower hemisphere of the stereogram. All elements are equal in length to the radius of the hemisphere and it gives, for any small area, the true curvature of the fold. The point of maximum curvature, or position of the axial line, is found at I. This line plunges 29 degrees, as read on the concentric circles, and its bearing is 86 degrees as read on the margin of the diagram. These data permit the location of the axial line at I in Figure 8. Similarly, the position of the axial line is found for the second horizon marker and the trace of the axial plane is accordingly determined in plan and can be projected to the section. Our main purpose, that of drawing a section of the fold, has been accomplished.

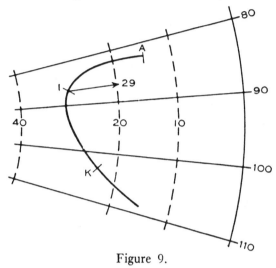

Figure 9.

Stereographic projection of the conical fold of Figure 8.

Other sections may be required to give additional information. For example, if a better picture of the curvature and relative thickness of formations is desired, a section is made to cut, at equal lengths, the axial line and two symmetrically placed elements on the limbs. The stereographic projection already made does not accomplish this purpose because the scale is distorted, but it serves to locate the bearing and plunge of two elements A and K (Fig. 9) on the limbs and approximately symmetrically placed with respect to the axial line I. These are transferred to the plan of the fold which is redrawn in Fig. 10. By a separate construction, which need not be shown, a plane which cuts the three elements equidistant from the vertex is found to strike in the direction AB and to dip 70 degrees north. On projecting the plan to this section the fold is seen to be intermediate between parallel and similar. In this view the vertices appear at V″ and W″. As might be expected by analogy with cylindrical folds (Fig. 6) the intermediate character is also indicated by the relative attitudes of the trough and axial planes. Similarly, a partial section normal to the axial lines gives the actual thickness of the formation along the axial plane, for this plane stands normal to the beds. The ratio between the actual thickness at this locality and the apparent thickness, in Fig. 10, can then be calculated. Since Figure 10 shows, approximately, the relative thickness of the beds, the above ratio may be applied to give, closely enough for most practical purposes the actual thickness anywhere on the limbs.

111

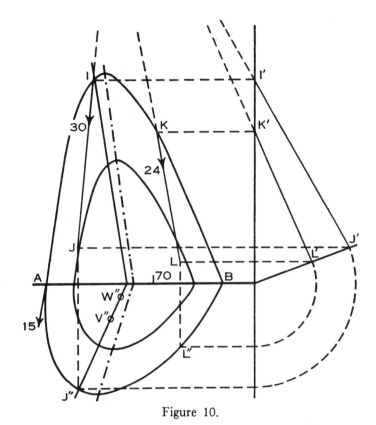

Figure 10.

The conical fold of Figure 8, section plane cuts elements A, I and K at points equidistant from the vertex.

For the purpose of obtaining an idea of the amount of tilt and rotation that the fold has undergone, as was done for cylindrical folds (Fig. 7) it seems best to refer the movement to the cone axes. One point of each of the two cones axes has already been determined at V and W (Fig. 8). Another point on each may be estimated at M″ (on the section plane, Fig. 8), which is a point on the axial plane where the beds on the limbs, or on a continuation of the limbs above the ground level, become parallel with the axial plane. It is necessary to estimate the curvature when reconstructing the limbs above the ground level. This point appears at M′ in side view and at M in plan. It is noticed that M lies on VW projected and that M′ lies on V′W′ projected, showing that the axes of the two cones coincide. The plunge of the cone axes may then be determined and the angle of rotation of the fold about this axis would show on a right section.

In conical folds in general the cone axes do not necessarily coincide, and it would be interesting to study various possibilities in folds of the parallel, similar, and other types with various degrees of curvature, circular and otherwise.

In conical folds each element is an axis of rotation about which the movement increases outwards from the vertex. The crests of drag folds ideally plunge away from the vertex. A drag fold or a minor crenulation is normally so

112

498

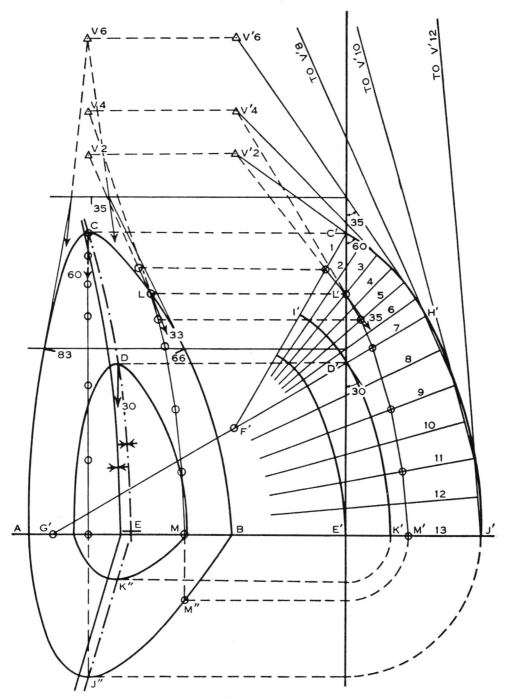

Figure 11.

Method of construction applied to an elongate basin.

113

small that all its elements are practically parallel with one another and, therefore, when the crest line cannot be measured, the line of intersection of bedding planes on the limbs serves to give the plunge of the crest. Slickensides that result from contemporaneous movement between beds lie in the bedding planes and, at any point, trend normal to an element.

The relationship of cleavage to bedding and to the elements is uncertain and is a matter that requires field and laboratory investigation. Three possibilities suggest themselves. The first is that the cleavage planes rotate, as it were, about each element. The second possibility, which is a special case of the first, is that the cleavage planes rotate about the cone axes. In both of these cases the line of intersection of cleavage with bedding coincides with an element. The third possibility is that the cleavage planes are of regional origin and rotate about lines parallel with the cone axes. In this case the lines of intersection of cleavage with bedding are curved and do not coincide with the elements, but are nevertheless dependent on the folding. In either hypothesis it is important to note that the intersection of bedding with cleavage on the limbs of a conical fold does not parallel the plunge of the axial line.

IDEAL FOLDS OF DOUBLE CURVATURE

Cylinders and cones are folds of single curvature and can be formed by folding a sheet of paper without buckling or stretching. Folds of double curvature cannot be reproduced in this way. In rock formations such folds are represented by domes, basins, and canoe-shaped structures. Such folds may be considered to be made up of successive parts of either cylinders or cones and an approximate construction may be made by methods already given. Thus, an elongate dome might be formed by bending a cylinder and it could be constructed by drawing a series of simple cylinders each tangent to the fold. Or, instead of the radius of curvature of a fold decreasing at a uniform rate, as it does in conical folds, it may change at a variable rate and the resulting surface of double curvature could be treated as a series of tangential cones, each with a different vertex. As some modifications of the methods previously described are required it seems advisable to give an ideal illustration.

Figure 11 is a hypothetical example of a canoe-shaped syncline, of which only the north half is shown. Its canoe-shaped character is evident in plan for the beds on the trough vary in dip (directly down the plunge) from 60 degrees on the outer horizon to 30 degrees on the inner horizon, and to horizontal at the central part of the fold. The longitudinal projection is made parallel with the bearing of the plunge of the trough line and on it the three dips are plotted at C', D', and E', and the two trough lines, corresponding to the two horizon markers are extended in depth by the method of concentric arcs. That is, draw radius C'F' normal to the 60 degree dip, draw radius D'F'G' normal to the 30 degree dip, and draw radius E'G' normal to the horizontal bed. With centre F' draw arcs C'H' and D'I' (the latter being above the ground level) and with centre G' draw arcs H'J' and D'K'. These arcs represent the trough lines of the two horizon markers, on the assumption that the thickness of formations remains unchanged down the plunge. Divide sector F'C'H' into an even number of equal parts by radii 2 to 6 and divide sector G'H'J' into an even number of equal parts by radii 8 to 12. These radii are side views of east-west striking planes that cut the fold into as many parts, successive pairs of which will be considered

114

500

as portions of cones (or cylinders) as, for example, cone 2 is bounded by planes 1 and 3, cone 4 by planes 3 and 5, etc. Greater accuracy will be obtained by cutting the fold into a larger number of parts than illustrated and it is especially desirable to make the subdivisions small near the nose of the fold.

The vertex for each cone cannot be determined as was done previously for a simple cone for no three surface strikes and dips are on the same cone, but two of these may be used and the third obtained at depth on the trough line. Proceed as follows, considering for the present, only the outer of the two horizon markers. For cone 2 draw, at right angles to radius 2, a chord between radii 1 and 3 and project it toward V'2, for cone 4 draw another chord between radii 3 and 5 and project it toward V'4 and so on. Each of these chords gives directly the dip of the trough bed, as, for example, the trough bed at the central part of cone 6 dips 35 degrees south. Its trace with the surface or other datum plane is carried over on the plan as shown. In the plan two other beds on the horizon marker and also in the central part of cone 6 dip 83 degrees and 66 degees, respectively, and their lines of intersection with the 35 degree dipping plane meet at V6 (and V'6), which is the vertex of cone 6. The positions of V2, V4, V8 and so on are found in similar fashion.

Any element on a boundary plane of one of the cones can now be extended both above and below the surface. Thus, in the longitudinal projection, the element L' that lies on the ground level at plane 3 is projected upward toward V'2 as far as plane 1, and downward from V'4 as far as plane 5. From plane 5 the element is continued downward from V'6 as far as 7, and thence from V'8 as far as 9 and so on to M' on plane 13. In plan the same element penetrates the surface at L on plane 3 and is extended above the ground level toward V2 as far as plane 1, and downward from V4 as far as plane 5, and so on to M on plane 13 (AB). In practice the one element in side view serves to give the position of two in plan, one on each side of the trough line. Similarly, other elements that lie on the bounding planes of other cones are constructed in side view and in plan. (If the segments into which the fold has been cut are small, any element within a particular cone can be projected in the same way without appreciable error). It is noted that the trough element, when plotted in this way, falls along the chords already drawn. Its position is accurately determined at the ends of each of the chords and, between them, falls on a smooth curve, circular in this case, passing through these points. Likewise smooth curves are passed through the determined points of the other elements. A tangent line at any point on the curved elements gives the bearing and plunge of the element at that point, such as L' where the plunge measures 35 degrees and is equivalent to the true plunge of 33 degees at L. The whole construction is repeated for the inner of the two horizon markers.

Having drawn the elements, the doubly curved surface of the fold is now defined at depth as well as for some distance above the ground level, and any section may now be drawn, such as a vertical section on AB, as illustrated. This particular section, being at a place where the fold is cylindrical, gives the true curvature and the true thickness of beds at this locality and, on it, are drawn the traces of the trough plane and of the axial plane. Axial lines are projected to the plan by reversal of the procedure by which the other elements were traced.

It is obvious that the method requires an accurate determination of the trough elements in the field and that it also requires that strikes and dips be

115

known at fixed points, such as L, even when they cannot be measured in the field. The attitude of beds at required points either on the trough or on the limbs is determined by interpolation on a smooth curve that is constructed by plotting, on squared paper, known strikes on one co-ordinate against corresponding known dips on the other. This method is recommended not only for the interpolation of strike and dip at any required point but also for correction of observed readings in order to average out irregularities.

It appears that, in general, the plunge of the curved elements, as found by the method outlined, should correspond with the plunge of drag folds. An exception, however, is found in a basin or dome that is a part of a sphere, for in this instance the construction leads to enveloping cylinders with elements that bear parallel with one another although having different plunges. This limiting form, however, has no specific trough nor is a systematic development of minor folds to be expected. The construction, nevertheless, gives the correct cross-section of such a fold.

EXAMPLES

In order to apply the above methods a large scale, detailed geological map is required. The field work should be done with the application of the method in mind. Field work of this type is necessary before the usefulness of the method can be appraised. Most published maps of plunging folds are on too small a scale and do not give sufficient information of the kind required. However, two examples have been selected for purpose of illustration.

The Sheila Lake fold (Fig. 12) is suitable for study for it has been mapped in considerable detail and the published map shows several formations and numerous strike and dip symbols, many more than have been reproduced in the figure. The bearing and plunge of lines of intersection of planes represented by strike and dip symbols are shown by the arrows. Each was obtained from a pair of bedding symbols and no one symbol was used twice. The pairs were chosen so as to be approximately on the same stratigraphic horizon and, at the same time, close to one another and diverging in strike sufficiently to give good intersections. This would have brought out any overall conical characteristc had there been one but, instead, the plunges, although irregular, show a pattern approaching that characteristic of cylindrical folds. The average plunge is 28 degrees, bearing north 86 degrees east. This average was used in constructing the right section on AB, which shows the fold to be a plunging recumbent syncline, tilted through 28 degrees and rotated from north to south through nearly 90 degrees. The trace of the axial surface can be drawn accurately in section and then projected to the plan. This surface dips, on the average, very nearly 28 degrees east.

The right section shows that the layer of hornblende-plagioclase gneiss is competent, as is also one of the layers of granitoid gneiss and these form approximately parallel types of folds. Other layers of granitoid gneiss and the beds of gneissic quartzite are incompetent and form folds approximating the similar type. The average is intermediate between parallel and similar and, consequently, the curvature broadens toward the north. Also it can be seen that the strata on the inverted limb are generally thinner than on the other limb. The trough surface, which is not shown, strikes east and dips vertically where it crosses the competent layer of hornblende-plagioclase gneiss and evidently

116

502

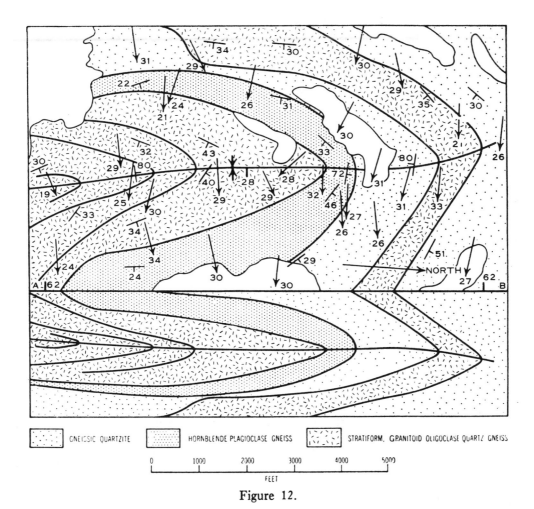

Figure 12.

Plan and right section of Sheila Lake fold, Sherritt Gordon mine area, Manitoba. (Plan after Geological Survey of Canada Map 44-4, by J. D. Bateman, 1943).

must change abruptly so as to lie nearly parallel with the axial surface where it crosses the incompetent formations, as dictated by the ideal examples given previously.

A note of caution should be added to the above interpretation of the structure because foliation everywhere parallels the bedding. This might be taken to suggest that the stratigraphic units have been first isoclinally folded and that such isoclinal folds were later refolded to form the recumbent structure. In this case it is difficult to know to what depth the formations may be carried down the plunge.

The map of the Oldham Gold District (Fig. 13) is considerably generalized from the published map, which shows several faults that are omitted from the figure and the formations have been reconstructed, as well as may be, to their pre-fault positions. The rocks consist of interbedded quartzite and slate and the best horizon markers, such as those in the figure, are quartz veins which generally follow the bedding planes and form saddle-reef structures for which

117

the district is well known. The figure serves as an example of a part of an elongate dome-like structure, which, when treated in segments by the method previously given, resolves into a succession of enveloping cones.

Plunges that have been calculated from bedding plane intersections are shown by the plain arrows while those that have been measured in the field on "barrels" are indicated by an arrow combined with the symbol for a minor fold. The agreement between the calculated and observed values is only fair. A better interpretation of the structure could have been obtained if a larger number of field observations on plunge had been given and if these had been used in conjunction with the calculated values to give average position of the vertices. Also the interpretation would be improved had topographic relief been known.

However, considering the structure as it has worked out, the vertical section on AB, which crosses the cylindrical, central part of the dome, shows a type of folding intermediate between parallel and similar, although with a considerable degree of thickening toward the axial plane. The construction permits section AB to be drawn only above the ground level. The vertical section on CD can be carried to considerable depth, where, it is interesting to note, the curvature reverses as if the fold here passes on either side into adjacent synclines. It is considered, however, that little reliance can be placed on this interpretation and that structures at such depth should be checked by working over a broader surface area.

DISCUSSION

MR. IRWIN: I found Dr. Stockwell's method extremely interesting. I have had occasion to work in an area for a couple of summers where the plunge of folds is rather important to get the picture of the pattern of folding. I cannot make any comments on applying Dr. Stockwell's method of using pairs of bedding planes. However, I might mention other means of getting the plunge of folds. One that I found very valuable was by the determination of the joints. The majority of the joints were perpendicular to the plunge of folds and a statistical plot of these joints gave an average plunge to the folding. Also the plotting of lineations particularly crumpling and the use of intersection of bedding and cleavage, which Dr. Stockwell mentioned, and also the use of the long dimension of pseudo-conglomerates. The pseudo-conglomerates were formed in beds of limestone and quartzite between beds of argillite. The argillite just flowed around broken pieces of the limestone and the quartzite and by getting the long dimension of that, one could actually find an amazing agreement of the lineations with the major folds. On occasion, actually, one could find the point, you might say the apex of the folds, because most of those folds showed up as as chevron type and the apex was fairly definite but it was very seldom observed in the field.

MR. FRANK EBBUTT: Mr. Chairman and gentlemen: Dr. Langford gave me the opportunity to review this paper a few days ago and I found it extremely interesting and I think the big point is that it relieves us of a great deal of calculating. I tried out this graphical arrangement and found that I was able to apply it quite rapidly and that gave me a lot of satisfaction. But I think the greatest satisfaction is the fact that a meeting like this is getting down to

118

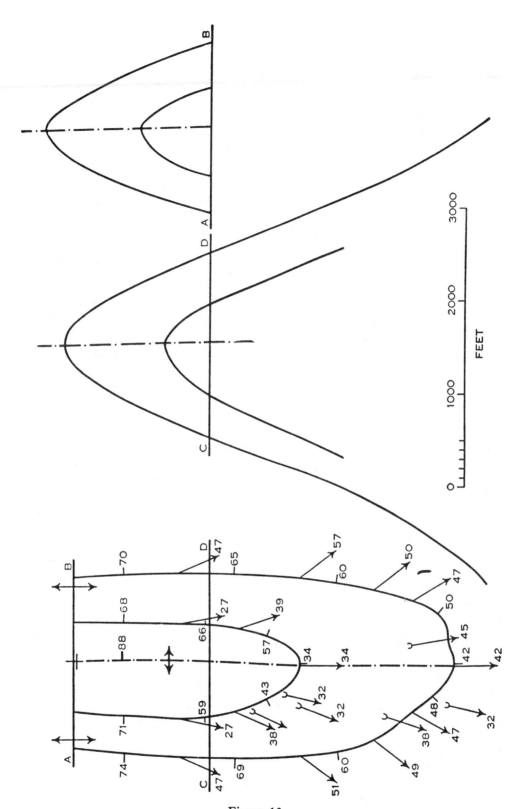

Figure 13.
Plan and vertical sections, Oldham Gold District, Nova Scotia (plan generalized from Geological Survey of Canada Map 642 by E. R. Faribault, 1898).

119

cases on structure. We don't have to look back so many years and we find that 10 or 20 years ago structure, in relation to ore deposition especially, was given a very small pumpkin indeed. But times are changing and here we have some of the good thinkers of the geological survey going to great lengths to develop such a method as this that will speed up the plotting of such things as we find in the field.

One criticism that I might make, not to be critical in any sense, but to perhaps try to add to the picture, is that very frequently in many sections of the country it is difficult to get sufficient outcrops to give you sufficient details perhaps to apply the method. The same thing is true in the position where you might like to have them in relation to the fold that you visualize so that certain parts of the picture perhaps are missing and when I think the important thing is a knowledge of the particular camp. If a man has had some experience locally and knows what happened to his last fold, then he is in a position, perhaps, to predict the funny things that may happen to the next fold and you have such things as dykes and sills and tongues and faults displacing folds and complicating the picture quite a bit.

I think Dr. Stockwell is to be congratulated on what he has presented here. I think it is a move in the right direction and I am sure a lot of you will be able to apply it quite successfully.

CHAIRMAN: Would anyone else like to add any comments to what Mr. Ebbutt has said?

MR. FERGUSON: I have just returned from Broken Hill, Australia, and the thing I would like to say is that over there we were very well off in making structural determinations of strike and plunge. Limitations were probably on the scale of the mapping rather than on the number of observations that one could take.

In spite of the fact that you could, nearly everywhere, accurately get the plunge, when we had that worked out we were convinced that the observed plunge was considerably more than the net plunge. That in effect there were plunge reversals which would give you, instead of getting a nice smooth line down the plunge, you would get a reversal which would, looking at it in the relation to dip, you don't know whether it was upright or overturned. Although hand in hand with that you could check in some ways on a few beds that you knew the thickness of. Very often in scammering there is no very sharp limit on the thickness of any individual bed. It was only within a certain range but you could see that the bed either had to be much thicker than you had always thought it was or it would have cleared the surface. Another thing you would see in the field was, besides lovely folds where you had no indication of that, the plunge line wasn't smooth. It is very interesting where you get a chain of basin structures. You get very steep dips generally but when you get into an area where they are just little closed basins, then, no one could really work out the answer but you would have to compromise between assuming that the net pitch was fairly flat or that adjacent folds of pitch were different.

REFERENCES

Billings, M. P. (1942): Structural Geology; Prentice-Hall, New York.
Bucher, Walter H. (1944): The Stereographic Projection, a Handy Tool for the Practical Geologist; Jour. Geol. Vol. 52, pp. 191-209.

120

Busk, H. G. (1929): Earth Flexures; Cambridge University Press.

Derry, D. R. (1939): Some Examples of Detailed Structure in Early Precambrian Rocks of Canada. Quart. Jour. Geol. Soc. London, Vol. XCV, pp. 109-134.

Eardley, A. T. (1938): Graphic Treatment of Folds in Three Dimensions; Bull. Amer. Assoc. Petrol. Geol., Vol. 22, pp. 483-489.

Harker, A. (1884): Graphical Methods in Field Geology; Geol. Mag., pp. 154-162.

Hutton, F. W. (1874): Suggestions for Geological Surveyors; Geol. Mag., p. 44.

Leith, C. K. (1923): Structural Geology; Henry Holt and Company, New York.

Mackin, J. Hoover (1950): The Down-structure Method of Viewing Geologic Maps; Jour. Geol., Vol. 58, pp. 55-72.

Mertie, J. B. Jr. (1940): Stratigraphic Measurements in Parallel Folds. Bull. Geol. Soc. Amer., Vol. 51, pp. 1107-1134.

———— (1947): Delineation of Parallel Folds and Measurement of Stratigraphic Dimensions; Bull. Geol. Soc. Amer., Vol. 58, pp. 779-802.

———— (1948): Application of Brianchon's Theorem to Construction of Geologic Profils; Bull. Geol. Soc. Amer., Vol. 59, pp. 767-786.

Turner, Francis, J. (1948): Mineralogical and Structural Evolution of the Metamorphic Rocks; Geol. Soc. Amer. Mem. 30.

121

Reprinted by permission of the Geologists' Association from
Proceedings of the Geologists' Association, v. 78, part 1
(1967), p. 179-209.

The Geometry of Cylindrical and Conical Folds

by GILBERT WILSON

CONTENTS

ABSTRACT: Cylindrical and conical folds can be distinguished from each other, not only by means of stereographic plots but also with reference to the forms and spacing of stratum contours and by the use of serial sections. Methods of constructing a profile of a cylindrical structure are described, and the difficulties liable to be encountered in the projection of a conical structure to depth are discussed. The need to distinguish the two styles of folding is emphasised.

1. INTRODUCTION

SINCE THE TIME, now thirty-five years ago, when I had the privilege and pleasure of working under, or should I write with Professor Hawkins at Reading, that branch of our science known as Structural Geology has expanded almost beyond recognition. New ways of looking at rocks and new techniques have been developed; sometimes these have quietly assumed their places in geological investigation, sometimes they have been heralded in with a fanfare of trumpets—though a study of early geological literature not uncommonly shows that the Old Men had recognised them before. They may not have worked with statistical methods backed by computers, but the structural significance of many observations in the field did not altogether escape them, even though there may have been a long hiatus before it was rediscovered. Lineation, for instance, was 'known to all geologists, but up till now it has not always been properly noticed and valued'—this was written by Neumann in 1839 (Cloos, 1946, 1).

Similarly, the principle of looking down the plunge of folded strata and mentally following the folds longitudinally was used by Swiss geologists for many years before Argand in 1911 illustrated its effectiveness in his work on the Pennine nappes. Pierre Termier had used the same conception when he recognised that the zone of *Schistes lustrés* of the Pennine Alps, on arrival at the Rhine Valley in eastern Switzerland 'hides itself in a

179

tunnel—a tunnel formed by a pack of nappes thrown over it. The zone of the *Schistes lustrés* continues its course along the interior of the tunnel, for presently it reappears . . . in the windows of the Lower Engadine and again in the Hohe Tauern. . . .' 'To the east of the Hohe Tauern, the zone of the *Schistes lustrés* plunges into another tunnel, to reappear no more' (Bailey, 1935, 135–6). Argand in 1911 described the axial plunge of the great Pennine nappes, and to illustrate his synthesis of the structure he published the well-known block diagram showing their continuation in depth. In this paper he recognised a continuous series of recumbent folds which rose and fell as they were followed longitudinally—'sea-serpent structure' to the irreverent—in a series of culminations and depressions from east of the Gotthard Pass to the Mediterranean. By noting the amount and direction of the plunges of the folds he was able to project the evidence shown by surface outcrops down (and up) the plunge, and so obtained a complete series of sections of these complex structures.

The same principle was fully recognised by such Masters as Lugeon, Heim and others, and was tacitly used in explaining and illustrating the structures of several Alpine regions. One has only to look at the block diagram of the High Calcareous Alps drawn by Arbenz to illustrate Lugeon's geology to realise the way in which major structures can be traced along their plunge direction (Collet, 1927, figs. 14 and 30; Bailey, 1935, fig. 22). Even better than consulting these references is to go to Nax on the south side of the Rhone Valley, above Sion, and looking north to see the real thing for oneself. Another natural demonstration can be seen looking southwards from the Rigi, across Lake Lucerne, Fig. 1.

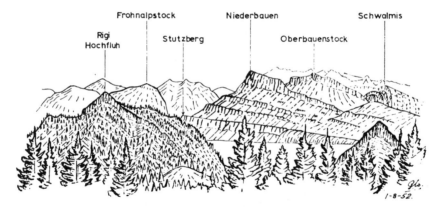

Fig. 1. Gently plunging fold structures in the High Calcareous Alps seen from Rigi Scheidegg; from a note-book sketch. The cliffed exposures of Cretaceous limestones on the Oberbauenstock and Niederbauen are connected by an overturned syncline; the hinge zone of a recumbent anticline plunges gently from Stutzberg to the right just above the level of the lake. The end-view of these folds is seen in Heim (1921, v. II, pl. XIX, section 5)

One important inference that can be derived from this visible evidence is that the structures are continuous, and if serial or *coulisse* sections are drawn normal to the fold trends, the hidden down-plunge structure can still be inserted below the rocks which actually crop out and mask it on the surface. Argand used to give his students a laboratory problem based on the principle that folds could be projected up or down their plunge without loss of fold-form.

In the mid-1920s, Eugène Wegmann, who had been Argand's assistant at Neuchâtel, went to Norway and thence to Finland. There he found, instead of the great natural sections such as he had been accustomed to see in the Alps and Norway, that the contorted Pre-Cambrian rocks cropped out on an ancient, ice-eroded peneplain with little topographic relief. In order to elucidate the complex structures present, he first measured and plotted the plunges of folds and linear structures, and then by modifying Swiss techniques to fit Finnish conditions, projected the surface structural evidence to depth. His methods of procedure and the results he was able to obtain are given in Wegmann (1928, 1929). By applying these methods he was able to construct cross-sections, etc., of greater clarity, accuracy and to greater depths than could have been obtained by simply plotting dip directions.

Wegmann's constructions and results were little known in this country until 1951, when Donald McIntyre, himself a student of Wegmann's, published a paper on the 'Tectonics of the Area between Grantown and Tomintoul (Mid-Strathspey)'. By applying Wegmann's methods he produced a structural profile or section normal to the regional plunge of the area, extending to a depth of six or seven miles, and based entirely on geometrical projection of the surface evidence. Though McIntyre was the first to demonstrate these methods of structural analysis in Britain, the same idea had been recognised previously: J. Hoover Mackin (1950) described how the late Sir E. B. Bailey introduced him to the method of reading a map by looking at it obliquely down the plunge in 1936, and Bailey's sections of the South-West Highlands clearly indicated that he was following Swiss techniques, even though he did not explain the methods by which he arrived at his conclusions.

The distances over which folds may continue longitudinally as single entities can be seen in those folds that have horizontal crests. The photograph with which G. M. Lees illustrated his Presidential Address to the Geological Society (1952, pl. I) shows an anticlinal ridge, admittedly slightly twisted, some fifty miles in length. Other Persian fold structures appear to be continuous over comparable distances. It requires little imagination to appreciate Termier's likening such folds to tunnels; and when one considers them in three dimensions one realises that they are not just simple two-dimensional arches, such as one sees on the page of a text-book, but like '*the wabe*' they go 'a long way before it, and a long way behind it'. Nevertheless, the fact that folds can be represented by single cross-sections implies that their structure or form will be the same in the next section; or as Clark & McIntyre (1951, 594) pointed out, 'many complex folds maintain remarkably constant profiles (sections at right-angles to the axis) even when the sections are spaced at distances many times greater than the amplitude'. Such folds, whose forms in serial cross-sections or profiles show little or no change, approximate to open cylinders, and are termed *cylindrical folds*. Once a fold or a group of folds has been recognised as being cylindrical, one can reasonably assume that the structure will continue more or less unchanged along its axial direction, whether this is horizontal or plunging; and, if the latter, the general form of the structure will still be present at depth even a long way from the area where it was seen to crop out on the surface.

But no fold continues for ever; at some point or other it loses amplitude, tapers out, and comes to an end. Examples occur at the ends of periclines, at the ends of *en échelon* folds (Campbell, 1958) or of pod-folds (Mendel-sohn, 1958–9; Wilson, 1961, fig. 39). These folds tapering along their lengths approximate in form to a longitudinal segment of a cone, and so are termed *conical folds*.

Naturally, if a cylindrical fold begins to change shape by contracting and developing into a conical fold, it cannot be projected in depth down the plunge just as if nothing had happened. It is, moreover, seldom easy for one to say while looking at a geological map whether the tapering end of a fold seen in plan is caused by a regular cylindrical fold plunging into the ground; or whether the termination of the outcrop has resulted from the fold narrowing and becoming conical in form. The cylindrical fold can be projected with considerable accuracy along its direction of plunge; but the projection of a conical fold will be uncertain. The evidence by which these two fold types can be recognised and distinguished, one from another, forms the main subject of this contribution.

2. CYLINDRICAL FOLDS

Most folds resulting from the action of uniformly directed tectonic forces have a dominantly *monoclinic symmetry* over much of their lengths. Thus, in any longitudinal unit of the fold, any plane drawn at right-angles to the fold-plunge acts as a plane of symmetry: the shape of the fold on one side of the plane is the mirror image of its shape on the other, irrespective of the complexity of the folding. Hence, the profiles of these folds carried along their lengths show relatively little change in the structure over considerable distances. Each bedding plane within the fold, no matter how contorted, can therefore be considered as an open *cylindrical structure* in the mathematical sense, formed by a straight line *generatrix* moving parallel to itself and to the axis of the cylinder (Fig. 2).

The generatrix, in the geological sense, was defined by Wegmann, 1929, as the nearest approach to the straight line which, moved parallel to itself, generates the fold and is termed the *fold-axis* or simply the *axis* (Clark & McIntyre, 1951; Wilson, 1961). In conformity with crystallographic terminology, this axis is known as the *b*-axis of the structure.

The *b*-axis of a cylindrical fold is thus declared by the direction of any straight line that can be drawn on a curved bedding plane within the fold in question. It is commonly represented in the field by some linear structure. The hinge-line of the fold, the fold crest, and the trough-lines of synclines are therefore all special cases of the axis; and in order to differentiate the general *b*-axis which has an orientation, but no fixed position in space, from the hinge-line around which the beds or some particular horizon have been rotated, the latter is referred to as the *B*-axis. Where the *b*-axes (of symmetry, or linear structures) = *B* (the fold hinges) so that throughout a given area the structures all have a uniform, parallel plunge, the regional structure is said to be *homoaxial*. The relationship can be likened to the courses of bricks in a series of parallel straight railway tunnels: they are

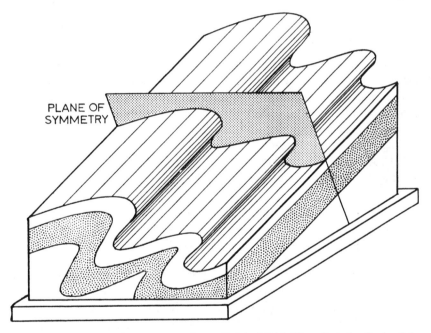

Fig. 2. The monoclinic symmetry of a cylindrical fold. Note that the form of the bedding planes can be represented by parallel straight lines which would also be parallel to the main cylindrical axis, and that the structure is symmetrical on either side of the (shaded) plane drawn normal to that axis

all parallel to the centre-lines and arches of the tunnels. If the group of tunnels were tipped up longitudinally at an angle, the brick courses would be similarly inclined and would declare the inclination of the tunnel axes, that is the *plunge* of the structure as a whole to the horizontal.

Because cylindrical folds can be considered as being made up of an infinite number of parallel straight lines drawn on bedding planes, their geometrical forms remain constant along their lengths—downwards into the crust, or upwards above the present erosion surface. They can therefore be projected in either direction by means of descriptive geometrical methods. Nevertheless, as nature rarely produces pure geometrical structures, the extrapolated results will not necessarily be accurate; but they will yield the closest approximations to the structure that can be obtained from the surface evidence.

Before any such extrapolation or projection is made, however, *it is essential that the structure in question be proved to be cylindrical*; unless this is done the results are very liable to be wrong and misleading. The geometry of a cylindrical structure is such that its traces on each of a series of parallel planes, no matter what their orientation, will have similar forms. Thus the cutting of a circular tube obliquely by a series of parallel saw-cuts will result in a series of ellipses each of which is identical with the others. A cylindrical structure can therefore be recognised by means of:

(*a*) *Parallel Cross-sections*. The structure is cylindrical if, on a series of equally spaced parallel cross-sections, it shows a constant fold-form; if the structure is plunging, and the sections are not normal to the trend of the plunge, the structure on each section will show a uniform vertical and lateral displacement relative to that on the next section. If equally spaced cross-sections cannot be drawn, the displacement of the structure between any two sections should vary in proportion to the spacing between them. Though it is desirable, it is not essential that the parallel cross-sections be at right-angles to the trend of the folding. The similarity of the two vertical end-sections of the block diagram, Fig. 2, can be taken as an example.

Sections drawn parallel to the plunge direction of a cylindrical structure will show the stratification as parallel straight lines running parallel to the plunge itself (Fig. 2). This is closely approached by the exposures on the hill-side below the Niederbauen in Fig. 1.

(*b*) *Similar Stratum Contours*. If a cylindrical structure is plunging, and is cut by a series of horizontal planes of known elevation above datum, the traces of the formation-boundaries on these planes will correspond to stratum contours drawn on those boundaries. The plunge of the structure will be declared by the direction and slope of the line (or lines) drawn through the hinge (or hinges) of the fold (or folds) seen at different elevations; its direction is also found to be parallel to that of the strike of vertical beds (Wegmann, 1929). In a cylindrical structure these plunge directions and slopes will be uniform throughout. If they are not, the structure is not cylindrical. The folding can also be recognised as being cylindrical if the spacing between successive contours on the same geological boundary, *measured parallel to the trend of the plunge of the structure*, is constant (Fig. 13, A).

(*c*) *Parallel Traces of Bedding Plane Intersections.* The dip and strike of a bedding plane, where observed at a point on an exposure, record the orientation of a plane tangent to the folded surface. If the fold is cylindrical, this tangent plane is in contact with the surface along a straight line; but, as mentioned above, all straight lines on a cylindrical fold are parallel and form *b*-axes. Hence, according to Stockwell (1950, 99), 'the lines formed by the intersections of all such [tangent] planes are also parallel. The converse is also generally true, namely, that if the lines of intersection of all tangent planes, as represented by many strike and dip symbols on a map, are parallel the fold is cylindrical. The bearing and plunge of these lines give the bearing and plunge of the fold. . . .' The lines of intersection can be obtained by drawing straight-line stratum contours for each dip and strike observation and then prolonging them until they intersect other similar stratum contours drawn from other points. The tangent planes represented by these contours drawn from any two points will meet each other in a single straight line trace. As Stockwell has pointed out, if the many traces so obtained are all parallel, the structure is cylindrical, and its plunge is declared by the orientation of the lines of intersection (Fig. 3).

Cleavage and schistosity planes can be used in addition to bedding tangent planes in determining the geometry of a fold. The line of intersection of bedding and cleavage surfaces—the trace of one on the other—declares the local plunge direction of the structure (Read & Watson, 1962, fig. 278 C, 462; Wilson, 1946, fig. 45, 270; 1961, fig. 21, 469). If these straight-line traces are constant in direction and plunge the structure can be considered cylindrical, and the angle of plunge can be estimated from that of the cleavage-bedding intersections: $b=B$.

(*d*) *Stereographic Projection.* The plotting of dip and strike data, together with those of cleavage and lineations on a stereogram, enables one to determine the plunge of the structure and whether it is cylindrical or not more accurately, and with less expenditure of time than the methods outlined in (*c*) above (Wegmann, 1929; Dahlstrom, 1954; Haman, 1961). The data can be plotted either as great circles which represent planes of bedding or cleavage, or as poles to such planes. The structure is cylindrical if the great circles all intersect in a common point or small area, which is referred to as the β-axis, and under ideal conditions $\beta=b=B$. The cylindrical form of the structure can also be accepted if the poles to the bedding

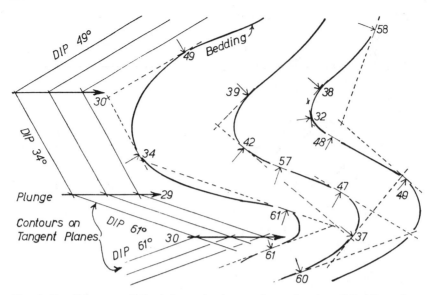

Fig. 3. Parallel traces derived from the intersections of pairs of bedding plane observations on a cylindrical fold structure

and other surfaces fall in a single great circle. These poles are known as π-poles, and the pole of the circles itself declares the direction and amount of the plunge of the structure. It is considered better and quicker to plot a π-pole diagram than a β-diagram under most circumstances (Ramsay, 1964). Both procedures are fully explained in Coles Phillips's book on stereographic projection (1954) and need not be elaborated here.

3. THE PROFILE AND ITS CONSTRUCTION

The *Profile* of a structure is a variety of cross-section drawn at right-angles to the local or regional plunge; the plane on which it is drawn corresponds to the plane of symmetry of a monoclinic structure, hence the construction of a profile is only valid if the structure in question is cylindrical. It is thus equivalent to the *orthogonal section* (Fisher, 1881) or *right section* (Stockwell, 1950) in Descriptive Geometry, and to the *normal cross-section* of Challinor (1945). Although Challinor carefully defined his term in his text, his use of the word 'normal' was unfortunate, because of the various meanings it has in English, ranging from 'perpendicular to something' to 'usual' or 'common or garden'—and the usual cross-section is the vertical one. Wegmann (1929) used the German expression '*Profil senkrecht zur Axialrichtung*' and later just *Profil* alone for the orthogonal section, in contradistinction to *Querprofil* (vertical cross-section) and *Längsprofil* (longitudinal section).[1] The term profile was first introduced into Britain by McIntyre (1951) who constructed one to illustrate the Grantown–Tomintoul structure; he commented on its construction, but did not explain the procedure. This, however, had already been done by Wegmann (1929), and again independently by Stockwell (1950). The principles upon which the construction is based have been further illustrated by Kalliokoski (1953).

The object of the profile is to give a true-to-scale picture of the geological structure in the third dimension, based on a geometrical projection of the data provided by the map. A vertical cross-section provides this only if the fold axes are horizontal; if the axes plunge the structure seen on a vertical section is distorted (Challinor, 1945). In a vertical section the horizontal scale remains correct, but the vertical scale becomes more and more exaggerated as the plunge of the structure increases. The profile presents the structure on the same scale as the map in both the horizontal and vertical directions. Provided the folding is cylindrical, it does not matter whether the structure is formed by parallel folds, shear folds or disharmonic folds; in fact one object of the profile is to show such distinctions. One must, however, assume that the strata shown maintain a constant thickness in the direction of the plunge, that is along the lengths of the folds.

The relationship of the profile to the mapped structure is illustrated in Fig. 4. Any point (*l*) on the map marks the point of emergence of one of the multitude of straight lines which together make up the cylindrical structure. The line is thus a *b*-axis, it is oriented parallel to the direction of folding *B*, and can therefore be prolonged upwards and downwards. If prolonged downwards it emerges from the front of the block at point *m*; if prolonged upwards it forms a perpendicular to the profile plane at point *n*. By drawing a number of similar lines all parallel to the plunge from different points on the mapped geological boundaries, a true orthogonal section or profile of the folded structure will appear on the profile plane as illustrated. It will be seen that though the horizontal length of the profile plane is the same as the width of the map or upper surface of the block, its height, *YP*, will vary in proportion to φ, the angle of plunge: $YP = XY . \sin\varphi$.

Once the amount and direction of the plunge over the area under consideration have been determined, one can construct a profile by different methods:

(*a*) *By Transferring the Mapped Boundaries to the Profile by Means of a Grid* (Fig. 5). This construction can only be used in areas where the

---

[1] In French, the term *coupe* is most commonly used for the vertical cross-section, but *profile* is given in Larousse Universel as having the same meaning. Bonte (1953) refers to the geological section as a whole as *la coupe*, but to the line representing the topographic surface as *la profile*.

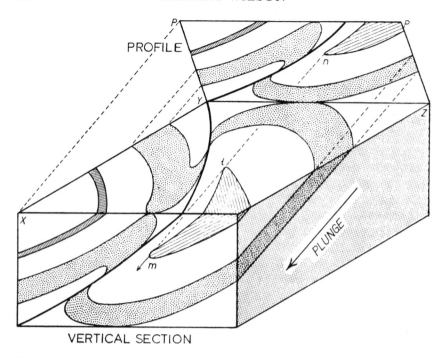

Fig. 4. Block diagram showing the relationship of the profile to the plunge of the structure seen on the map

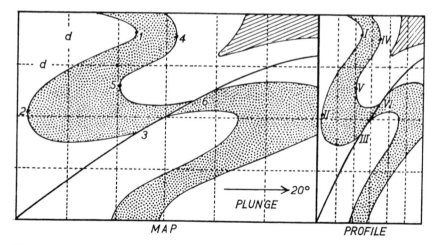

Fig. 5. The construction of a profile from a map of an area of flat ground by gridding the map and profile. The height of the rectangles drawn on the profile plane are equivalent to $d. \sin \varphi$, where φ is the angle of plunge.

ground is flat. The map is divided into squares of convenient size (*d*) by a grid, the lines of which are drawn parallel to and at right-angles to the trend of the plunge. The paper on which the profile is to be drawn is also marked with a grid in which the vertical lines are spaced at the same distances apart as the grid lines parallel to the direction of plunge on the map. The horizontal lines on the profile are drawn at intervals of $d . \sin\varphi$, where φ is the angle of plunge. The geological boundaries (points 1, 2, 3 . . .) are then transferred from the map to the profile (points I, II, III . . .) and are joined up by smooth curves. By placing the sheet of paper on which the profile is to be drawn at the down-plunge edge of the map, as shown in Fig. 5, the resulting profile is seen as it would appear looking down the plunge. This method has the advantage of requiring little draughting equipment.

(*b*) *By Geometrical Construction* (Wegmann, 1929; Stockwell, 1950). The ground surface is presumed to be flat (Fig. 6). Decide on the area that the profile is to cover, and along one edge draw *XY* parallel to the plunge direction, and *YZ* at right-angles to it. *YZ* now represents the line where in plan view the profile-plane meets the map surface. The side view or elevation of the structure will be drawn above the line *XY*.

Draw the line *YP*, which represents the edge view of the profile plane at an angle of $90°-\varphi$ to the line *XY*. We now have the plan and elevation of the map and profile plane as shown in Fig. 4, but in orthographic projection. In order to see the profile plane full face it must be swung back either into the plane of elevation, that is, as if it were hinged along the line *YP*, or it must be swung into the horizontal plane as if hinged along *YZ*—like closing a trap-door. This latter rotation is indicated by a curved arrow. The profile plane is now lying in the plane of the map and is outlined by a heavy line. This construction whereby a subsidiary plane is swung into one or other of the principal planes of projection is known as *rebatment*.

The problem now is to locate the position where any particular axes passing through points 1, 2, 3 . . . will appear on the profile plane. Firstly, one must draw the side view of each axis, projecting each point on to the line *XY* at points 1′, 2′, 3′. . . . As these points represent the positions, in side view, where the axes penetrate the map or ground surface, they can then be projected at an angle φ until they meet *YP* at points 1″, 2″, 3″ . . . which are now at their correct heights above the hinge line *YZ* on the profile plane. With centre *Y* swing these points down to the edge of the profile plane, i.e. the continuation of *XY*, as shown by the dotted arcs, and project them perpendicular to *XY* on to the plane itself. Each projection line is thus parallel to the hinge line *YZ* and at the requisite height up the profile plane. Secondly, one must fix each point at its correct distance on the profile from the edge of the map, i.e. from the line *XY*:

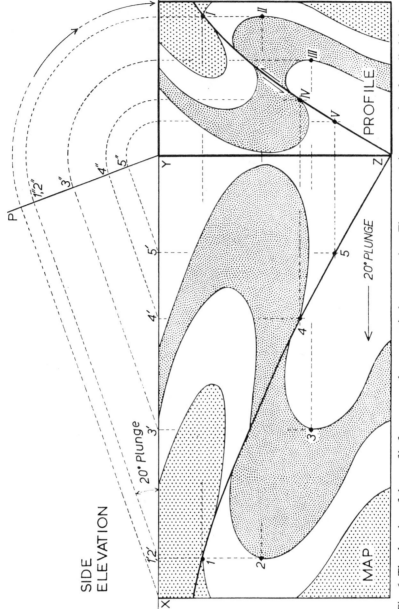

Fig. 6. The drawing of the profile from a map by geometrical construction. The structure is assumed to be cylindrical, and the ground surface flat. The plunge (φ) is taken as 20°

project the plan view of each axis on to the profile by drawing a line from each point 1, 2, 3, . . . parallel to XY until it cuts the corresponding line which had previously been projected up the plunge. The intersections I, II, III . . . give the correct positions on the profile of the axes that passed through the map at points 1, 2, 3 . . . and the profile can now be completed. The profile drawn in this way is seen looking *up* the plunge; to draw it looking down the plunge one would need to put the plane of side elevation below the map, and the profile at the down-plunge end of the map. There would also be added complications if topography were to be taken into account.

(*c*) *The Profile in an Area of Irregular Topography* (Fig. 7). Many areas underlain by folded rocks are, if not mountainous, at least part of a hilly region; it is therefore necessary in constructing a profile to allow for differences in elevation of the points where the fold axes emerge from the ground. Thus, in Fig. 7, A, an axis at sea-level (*p*) will also appear on the higher ground surface at point *q*, at an elevation of *h* units above datum. Its point of emergence will be shifted in plan in the up-plunge direction an amount *m*, where

$$m = h \cdot \cot\varphi, \text{ where } \varphi \text{ is the plunge.}$$

This can be allowed for by projecting point *q* not on to the XY line as previously done but to a point *q'* on the plane of side elevation at a height *h* above the line XY (Fig. 7, B). Once the point *q'* has been fixed the axis which it represents can be projected from it, up the plunge, as described in the previous section.

The easiest way to allow for elevation is to draw lines spaced as units of height parallel to, and above XY on the side elevation plane, Fig. 7, B, and Fig. 8. The vertical interval between the lines must be drawn on the same scale as the map and profile. Each point 1, 2, 3 . . . is then projected from the map or plan on to the line corresponding to its height above

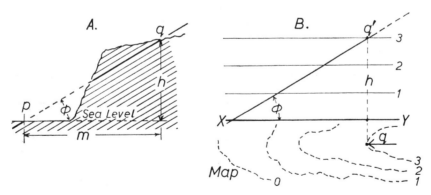

Fig. 7. The principle of allowing for elevation in the construction of the profile

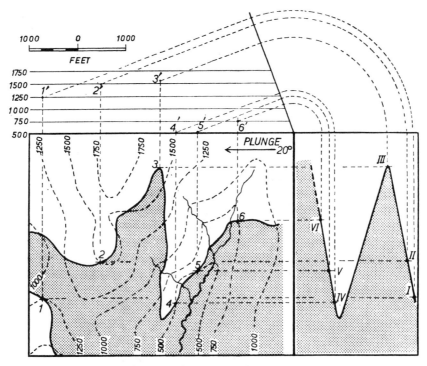

Fig. 8. The construction of a profile from a map of an area having irregular topography

datum on the side view, points 1′, 2′, 3′. . . . Intermediate elevations can be interpolated between the lines. The rest of the construction follows as described in sub-section (*b*) above. Since the construction is solely concerned with the geological structure, topography does not appear on the profile.

(*d*) *Profiles of Undulating Fold Structures* (Fig. 9). The axes of many folds show a reasonable constancy in direction or trend, but as the folds are followed along their lengths the angle of plunge may vary; excellent examples are shown in Wegmann, 1928, fig. 1. Regionally horizontal folds would thus tend to develop 'sea-serpent' structure, rising to culminations and dropping to depressions. The general effect is that envisaged by Termier when he likened the structure of the *Schistes lustrés* beneath the Eastern Alps to that of an undulating tunnel. Despite its rising and falling the main structure is still continuous and its trend is undeflected. If the main structure has a marked regional plunge in one direction the undulations would show simply as steepening and flattening in the degree of plunge along the folds.

Though such folds are not homo-axial, limited longitudinal sections of them may closely approximate to cylindrical structures, and from these

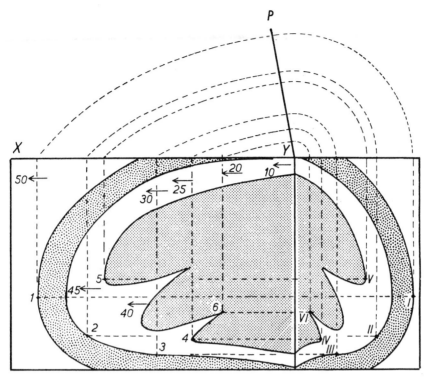

Fig. 9. The construction of a single profile of an undulating fold by means of concentric arcs. The arrows and figures show the direction and amounts of plunge

sections one may construct individual profiles: one, as it were, above the other. As these profiles will all be drawn on the same scale, and will be co-linear in the vertical plane containing the plunge, they can be combined; but any gaps between them will need to be filled in by interpolation.

An alternative method of construction is to consider the structure as a whole as a cylinder bent along horizontal axes perpendicular to the trend of the plunge. The various angles of plunge can then be set off along the line *XY*—allowance being made for elevation if necessary—and treated as if they were dips in a Busk construction (Busk, 1929). The plunges of the structure thus appear on the elevation plane as a series of parallel, undulating curves (Fig. 9), instead of straight lines as in Fig. 6. Individual axes can be assumed to conform with the curves drawn and can be projected up the plunge to the profile plane *YP* by the method of concentric arcs. It must, however, be emphasised that, because the structure is not homoaxial, the profile obtained in the way discussed above is liable to contain errors; but it will nevertheless give the closest approximation to the truth that can be obtained from the data available. It is obvious,

I hope, that folds which are turned back on themselves like those described by Dearman, Shiells & Larwood (1962) around Eyemouth cannot be analysed by these methods. Always one must remember that the greater changes in the plunge angles the greater the inaccuracy of the profile. Nor does the construction hold if the fold axes are bent or twisted laterally. Under those circumstances separate profiles of those sections of the folds which approximate to cylindrical structures seem to be the only solution. It may, depending on circumstances, be possible to combine these into a single composite profile; but here again errors are more than likely to occur.

4. FOLD ENDINGS

No fold, cylindrical or otherwise, can continue indefinitely—somewhere or other it will die out. Either it will lose amplitude until it merges into horizontal or uniformly dipping strata, or it will be pinched out in some way and so come to an end as an individual structural unit. Those folds which end by uniformly tapering to a point approximate to the general form of a longitudinal segment of a cone, and are referred to as *conical folds* (Stockwell, 1950).

A cone is a surface generated by a straight line, one end of which passes through a fixed point while the other is free to describe any given curve. A conical fold thus differs from a cylindrical one in that the generatrix of any surface within the structure does not move parallel to itself, but its successive positions all pass through a single point. This point is the *apex* or *vertex* of the cone formed by some particular surface or bedding plane. Beyond the apex the shape of the fold—if it were formed by a single surface—might be expected to be reversed, that is, an anticlinal cone should develop into a synclinal one, and vice versa. This, as will be considered later, does not occur in nature.

Folds which are diminishing in size or approximate to cones have no plane of symmetry perpendicular to the cone-axis, the fold hinge or any other element: they cannot be considered monoclinic structures. The fact that those with circular or elliptical right-sections have a longitudinal plane of symmetry does not imply (*pace* Haman, 1961, 32), that they have monoclinic structural symmetry. This lack of symmetry means that they cannot be projected to depth in the same way that one can project cylindrical structures. Conical folds, from their very shape, are obviously coming to an end within a relatively short distance, and what happens to the strata beyond the apex when one is dealing with solid rocks rather than surfaces is largely a matter of conjecture.

It is important to realise that if the fold approximates to a cone, the structure seen in cross-section near the apex should be identical in form to, but folded on much smaller radii of curvature than that of the same beds nearer the theoretical base of the cone; if this were not so, then the true conical form would not be present. Unfortunately the structural behaviour of strata at or near the ends of conical or dying folds seems to have received little or no attention from geologists. This may be because the critical part of the structure is rarely visible in surface exposures, or because nothing noteworthy has been observed. We can, however, consider theoretically what might happen within the area where folds end under different structural settings.

Probably the most simple setting is that exemplified in Argand's unconfined virgations (Argand, 1924, figs 1 and 2; Collet, 1927, fig. 5 (1 and 2)), in which the structure is tightly compressed in the central portion of the fold-bundle, but at the wings or ends the folds gradually fade away into horizontal or gently dipping strata. The suggestion is that the maximum tectonic forces were active in the central zone, that they diminished away from it, and that at the distal ends the forces had dropped completely. The ease with which such fold systems can be reproduced by pushing the centre of a table-cloth with one's hands, or as Tokuda (1926–7) did with wet rice-paper, and Lee (1929) did with wet tracing paper, suggests that many of these structures were formed at relatively shallow depths and are *plis de couverture* probably associated with *décollements*. The folds tend to flatten longitudinally as the stress per unit length diminishes, but their magnitudes and the relationship between their amplitudes and inflexion widths[2] will be largely controlled by the thicknesses of the competent strata forming the structure (Willis, 1923, 165–6). The evidence for this structural control has been demonstrated experimentally and mathematically by Biot (1961), Ramberg (1961) and Ramsay (1967), who have shown that, other things being equal, the curvature of a folded competent bed is a function of its thickness. Hence, for a given set of conditions, there is a maximum curvature or minimum radius of curvature beyond which a competent bed cannot be folded. The shape of a dying fold under these conditions will be dependent on the flexibility or otherwise of the strata involved, and is unlikely to have a true conical structure beyond the limiting point where the competent beds become the controlling factor in its development. It will certainly flatten as the forces available to lift the fold diminish, but its plan width will not necessarily taper in proportion.

The effect of limiting curvature on the form of a simple fold tapering into otherwise flat strata is shown in Fig. 10. The side elevation of the *Crestal Line* and a series of vertical sections *A*, *B*, *C* and *D* across the

[2] I have used *inflexion width* in preference to Bailey Willis's term *inflexion length* because we are considering distances across the fold. Inflexion length—the length of the chord across the arc of a fold between the points where the sense of curvature changes—though clearly defined, tends to suggest a longitudinal distance, culmination to culmination, rather than a measurement normal to the fold trend as Willis, 1923, 165, had in mind.

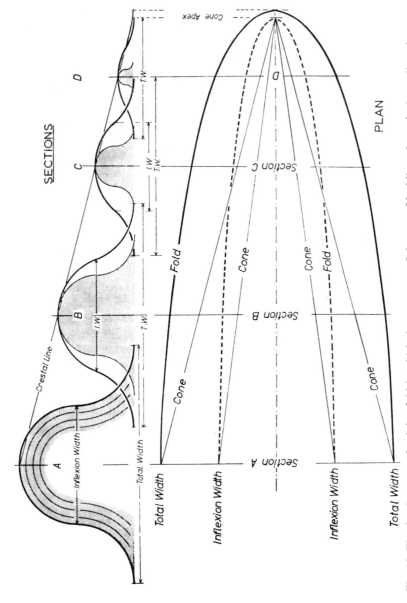

Fig. 10. The development of a dying fold in which the curvature of the arcs of bedding, shown by heavy lines, is constant throughout the structure, as if controlled by a thick group of competent strata. Sections of a true conical structure having the same crestal plunge are shaded. The bedding around the fold is assumed to be flat

structure are shown in the upper part of the diagram. The plan view of the fold is drawn in the lower part. The *Total Width* (*T.W.*) represents a single wave length of the fold, and the *Inflexion Width* (*I.W.*), here equivalent to half a wave-length, is also shown. The plan views of these elements as they would appear if the structure were truly conical are indicated by light full lines marked *Cone*, and the corresponding cross-sections have been stippled. It is assumed that the permitted limit of curvature of the strata was reached at Section *A*; the heavy lines in Sections *B*, *C* and *D* are all drawn with the same radii of curvature as in *A*, and the fold widths so obtained have been transferred to the plan view. The heavy full line in the plan outlines the total width of the fold thus derived, and the heavy, dashed line marks the plan of the inflexion width. The diagram brings out clearly the fact that a fold may grade into horizontal strata without necessarily having a conical form.

The stratum contours for the structure illustrated in Fig. 10 would be very similar to those that could be drawn on a cylindrical fold having the same plunge. This is because the curvature of the fold-arcs throughout remains constant, as seen on the different cross-sections. One could, however, tell from the map that the fold was losing height relative to the flat beds on either side, and the fact that there was no further folding beyond the apex would declare that the fold had ended.

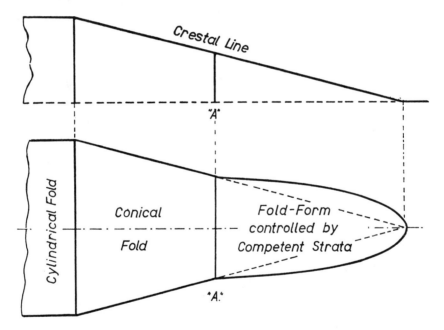

Fig. 11. Plan and elevation of a tapering fold in which control of the curvature by competent strata does not become effective till Section *A* is reached

A point which emerges from Fig. 10 is that folds containing competent beds which control the curvature will, over most of their lengths, tend to close in the plan view less rapidly than a true conical fold having the same plunge and basal section. If, beyond Section *A*, i.e. to its left, the folding continues to be controlled by the limit of curvature of the competent beds the fold as a whole will continue as a cylindrical structure of which Section *A* would be a representative cross-section. On the other hand, if the radius of curvature of the main cylindrical portion of the structure was greater than that of Section *A*, and the fold began to taper before the control by the competent beds became effective, the width of the fold in plan could contract in harmony with the plunge as a true cone until the dimensions of Section *A* were reached (Fig. 11). The fold limits in plan view under these circumstances will be represented by sigmoidally curved lines; and the rate of contraction will be greatest over the longitudinal section where the fold approximates to a cone.

It seems therefore unlikely, from the discussion above, that complete truly conical folds can develop fully in rocks which contain massive competent strata. With reduction in competency the limiting radius of curvature would be reduced, and the tapering of the fold would tend to approach more and more closely to that of a theoretical conical structure. One might therefore expect conical folds to be confined to structures or those parts of structures in which competent rock groups are absent or unimportant, or to those regions in the earth's crust where increase in plasticity or rock-flowage tends to render all rocks more or less equally incompetent.

5. CONICAL FOLDS

Despite the way in which the forms of tapering folds may be controlled or modified by their lithology, there is no doubt but that many are truly conical over considerable portions of their lengths. Three-dimensional sketches of these can be made by straight lines, which represent positions of the generatrix of surfaces within them, converging towards an apex (Fig. 12). In the field such lines may appear as linear structures, as shown by Sutton & Watson (1954, pl. III) between the north-west end of Loch Fannich and Loch a' Bhraoin (Fleuty, 1964, fig. 8a; and Stauffer, 1964, fig. 3). The poles of these lineations when plotted on a stereogram are not concentrated around a point maximum as for a cylindrical fold, but lie on a curved line. For a right circular cone, this line forms an arc of a small circle (Dahlstrom, 1954, fig. 9). The π-poles of bedding planes for a similar type of cone yield a girdle which also falls on a small circle lying at 90° to that formed by the lineations. The plots of poles of lineations and π-poles of bedding for any other variety of cone result in curves which are neither great nor small circles (Haman, 1961). This is because

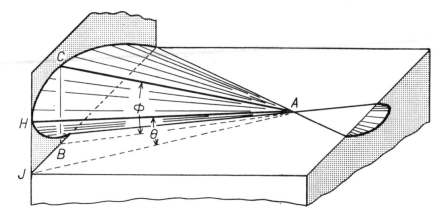

Fig. 12. An overturned conically folded surface represented by straight lines converging towards an apex. The horizontal trends of the crestal line (plunge φ) and hinge line (plunge θ) are also shown

any cone which is not a single right circular cone will be a composite structure made up—in a way analogous to a 'Busk construction'—of two, several or many circular cones. Each of these will have its own cone-axis, each of which will be differently oriented, but all will converge to a common cone-apex. The descriptive geometry of such a structure is complicated, because no single plane surface can be drawn at right-angles to these non-parallel cone-axes. Similarly, as Stockwell has pointed out (1950, 98–9), neither the right section of a cone, 'nor any other section that is not a curved surface reveals the actual thickness of beds nor the true curvature'. The nearest approximation to a true section of the strata forming a conical fold which has a circular or elliptical right section, is at right angles to the central axis of symmetry of the cone. This is the K-axis of Haman (1961) and the κ-axis of Tischer (1963).

Because conical structures have definite geometrical forms which allow them to be projected to depth, but only to a limited extent, it is essential that one should be able to recognise them when they do occur. In the field the orientation of linear structures may indicate their presence, but on a map structural symbols may be sparse or absent and the evidence that one has to look for is not so obvious. In particular, the differentiation of conical folds from cylindrical ones is the most important; I have therefore outlined below the evidence by which a true conical structure can be identified from a map, in the same order as that given for the recognition of cylindrical folds.

(a) *Parallel Serial Sections* drawn across the fold will all show the same structural style, but the magnitude of the structure will diminish in proportion as the cone-apex is approached, even though the apparent thicknesses of the beds remains constant, as shown by the shaded areas in Fig. 10.

The apparent dips of strata on longitudinal sections parallel to the cone-axis, but not passing through the apex, will appear as hyperbolic curves steepening in the direction of the cone-apex.

(*b*) *Stratum Contours*, whether the cone-axis is horizontal or plunging, will not show the same general accordance or similar shapes that one finds in cylindrical folds. Contours remote from the cone-apex will be open and gently curved, but as the apex is approached their curvature around the crestal zone of the fold increases until, at the apex itself they should theoretically meet at the angle formed by the two straight line asymptotes (Fig. 13, B).

(*c*) *Traces of Bedding Plane Intersections* in a conical fold have been shown by Stockwell (1950, fig. 8) to converge towards the cone-apex. They are not parallel, as in the case of cylindrical folds. Similarly, accordant linear structures *on any one particular horizon* will converge towards

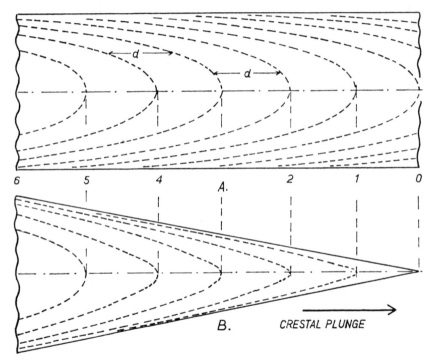

Fig. 13, A. Stratum contours drawn on a plunging semi-circular cylindrical surface, showing the equal spacing (*d*) parallel to the plunge between pairs of contours, and the similarity in form and curvature of all the contours. B. Stratum contours drawn on a semicircular conical surface. The contours are evenly spaced only along the crestal line, and their forms and curvatures vary as the cone-apex is approached

the cone-apex of that horizon. However, it should be noted that straight-line structures on different horizons may be parallel. The loci of such lines will be formed by surfaces which are normal to the bedding planes and pass through the cone-axis. In a non-circular cone, this locus will be a curved surface.

(d) *Stereographic Projection.* Dip and strike data plotted on a stereogram yield small circles for both π and β diagrams if one is dealing with circular cones, in contradistinction to the β point maximum and π pole great circle obtained from cylindrical folds (Dahlstrom, 1954; Evans, 1963). A cone of any other form will give an irregular curve on a stereographic plot; the extent to which this curve departs from a true small circle depends on the ellipticity and asymmetry of the fold (Haman, 1961; Tischer, 1963).

The plunge, both in amount and direction, of a conical fold varies over the whole structure. There is no definite axial direction, because the generatrix of a cone does not move parallel to itself. Its orientation is everywhere different. Consequently the fold plunge must be recorded with reference to some particular element. The most practicable element for reference purposes is probably the crestal line or the trough line of the fold, as these may be expected to be more or less co-linear with the crest or trough of the main cylindrical portion of the structure, and they measure the maximum angle of plunge under most circumstances. Alternatively one may refer to the hinge line, but it must be clearly stated which element is being used to declare the plunge. The divergence in plunge and trend between crestal and hinge lines in an irregular cone is illustrated in Fig. 12.

Conical folding is relatively easy to discuss and illustrate provided one is considering the folding of a single surface only; but beds or groups of strata are three-dimensional bodies with thicknesses that may be significant. To begin with, let us consider, as in Fig. 14, A, the folding of two parallel surfaces, one upper, one lower. The structure can be geometrically represented as two parallel cones, one within the other. On the left-hand side of the diagram the inner cone forms the core of the anticline, on the right-hand side it forms the envelope of the syncline; and because the annular separation between the two cones is constant, their apices do not coincide. These are lettered *Au* and *Al* (upper and lower respectively). This structure can easily be demonstrated by means of a wire model.

If, however, the separation between the two cones be solid, as it would be were it formed by a sheet of competent rock, the whole construction breaks down, and the anticlinal cone cannot be projected into the synclinal cone. The reason for this is shown in the three vertical sections drawn through *Au*, *Al* and an intermediate point *P* (Fig. 14, B, C, D). At *Au* and *Al* the full thickness of the bed between the upper and lower cone surfaces can be present as a competent but very tight fold—synclinal at one point, and anticlinal at the other. Between the two, however, an impossible geological structure develops: the upper surface of the bed appears as an anticline, the lower as a syncline, the one immediately above the other; and this will occur all along the locus of the apices, that is, along the axial line of the cone.

Away from the cone-axis, once the conical structure has been recognised, one can theoretically project the continuation of any anticlinally folded bed in depth as far as the cone-apex for its lower surface, *Al*. Synclinal beds can likewise be projected up the plunge as far as the apex of their upper surfaces, *Au*; but beyond these cone-apices the structural behaviour of any particular bed cannot be predicted graphically.

The extraordinarily tight folding that would be necessary to allow a conical fold to proceed to the limit of its apex rather suggests that this limit is seldom achieved by flexure folding, and it is here that information based on field observations is needed. That beds may be strongly plicated in one part of a fold relative to another has been recognised in disharmonic folding, especially where incompetent beds have been nipped into the core of a competent fold, as illustrated by Goguel (1952, fig. 61) and de Sitter (1964, figs. 121–2). In a cylindrical fold, one would expect this form

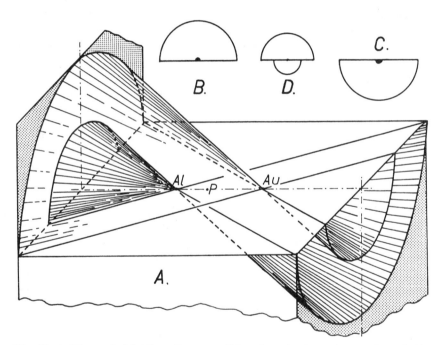

Fig. 14, A. The conical folding of two parallel surfaces having apices at *Au* and *Al*. B. Cross-section through *Au*. C. Cross-section through *Al*. D. Cross-section through an intermediate point *P*.

of disharmony to continue longitudinally in the same strata along the core; but in a conical structure, though it is likely to develop in certain beds near their cone-apices where the radius of curvature is small, it may equally well be expected to disappear as the beds in question are traced into regions where their radius of curvature is large. If this argument is correct—it awaits confirmation from the field—each bed as it approaches its cone-apex may be expected to be more strongly contorted, more fractured and perhaps more strongly cleaved than it was remote from the apex. Quite possibly this suggestion may have economic significance if one has to predict rock conditions in advance, as in engineering or mining geology.

The difficulty of directly observing the apical terminations of conical folds in normal field mapping has already been commented on. In mines, quarries or extensive civil engineering excavations, however, where the geologist is able to see and record the changing conditions while a heading or face is advancing, a three-dimensional picture of the structure can be obtained. Such observations were made by J. D. Campbell (1958) in Australia, and by F. Mendelsohn (1959) on the Roan Antelope Mine in Northern Rhodesia (Zambia). Campbell described the development of *en échelon* folds which as units build up more extensive fold systems, and Mendelsohn recognised the presence of pod-folds on the flanks of bigger structures. Both the individual *en échelon* and pod-fold units are periclinal folds, having elongated dome or basin shapes. They are more or less cylindrical in their central portions, but taper and disappear at either end where they apparently approximate to conical folds. Unfortunately neither author discusses the conditions observed at the ends of these folds, but the subject can hardly be included within the terms of reference of their respective papers. Nevertheless, the illustrations that they have presented show the folds converging conically to a point at which they disappear *into the limb of another fold*. Among Campbell's *en échelon* structures no anticlinal conical fold is shown running through the cone-apex to form a synclincal continuation and vice versa. As one fold begins to die out another starts to develop on one flank of it or the other, and the diagrams suggest that the crustal shortening across the whole series of such folds remained more or less constant. One can see the same thing happening in W. J. Mead's small-scale experiments in folding and in the photographs of Lee's structures: where one fold begins to taper and die another begins to grow alongside it (Mead, 1920, figs 11 and 12; Lee, 1929, plates XV, XVI, XVII). Each of these folds is a discrete unit on its own, and co-linear folds do not necessarily form a continuous undulating structure which is approximately cylindrical in form. Each separate fold begins to grow at some point; it continues until it reaches a maximum amplitude and then dies away again. Commonly each fold is staggered or *en échelon* in its relationship to the next; and, if we can consider the undulating continuous fold as forming a 'sea-serpent structure', a system of these discontinuous folds might be said to present a 'porpoise structure'.

An example of the way in which small periclinal folds may wedge out into the flanks of one another is shown in Plate 5. Here in a thinly laminated pelitic rock each dying fold merges gradually into its neighbour and disappears. There is no unseemly contortion in the cone apical region, and as far as can be judged the conical form persists to the end of the fold. I have also observed very similar phenomena in coarsely crinkled micaschists in the Moinian rocks of the Ross of Mull. It may be significant that where these types of structure have been noted the rocks containing them are to a greater or lesser extent metamorphosed. Either the relative competency of the strata has been reduced, or the material involved in the folding is made up of lamellae of schistosity rather than of beds of appreciable thickness: the modification of the conical structure as a result of lithological differences appears to have been wholly, or at least largely, eliminated. Hence tight folding at or near the cone-apex, possibly by shear-folding, seems reasonable.

An example of the behaviour of Pre-Cambrian strata near the apex of a conical structure could be seen in the Balaghat Manganese Mine in the Central Provinces of India (Fig. 15; Fermor, 1909, 714, *et seq.*). The ore-body, composed of bedded manganese oxides, extends for some 2500 yds. on a roughly north-east–south-west strike, and dips north-westerly. It gradually thins towards the north-east boundary of the property beyond which it can be traced no farther. The country rocks are quartzites and pelitic schists of the Chilpi Ghat series of the Dharwar system (Fermor, 1909, 281), and in them drag-folding and cleavage-bedding relationships indicate that the succession at the Mine is inverted.

At the time of my visit at New Year 1948, I had access to the detailed cross-sections of the ore-body, and from them the coulisse sections shown in Fig. 15, A, were prepared. The chainage distance in feet of each section from the north-east boundary is shown on the figure.

The ore-body dips evenly north-westwards from the boundary to chainage 600 ft., but by chainage 800 ft. it is strongly contorted, Fig. 15, B. At chainage 100 ft. it develops a knee-fold and a bulbous protuberance into the foot-wall, and thence south-westwards a recumbent fold closing to the north-west gradually develops and increases in amplitude to chainage 1800 ft. and beyond. Still farther to the south-west this fold diminishes and may die out completely, though diamond drilling indicates that the ore-body at depth maintains a steep dip to the north-west. The evidence seen nearly twenty years ago suggests that this growing fold to the south of chainage 1000 ft. is the beginning of what we would nowadays call a pod-fold, developed on the overturned limb of a much bigger fold-

Small-scale periclinal folds in low-grade metamorphosed shale. The dying conical folds can be seen to taper into and end on the flanks of growing folds. Half natural size. Torridonian, Rhinns of Islay. Scotland

[*To face p.* 204

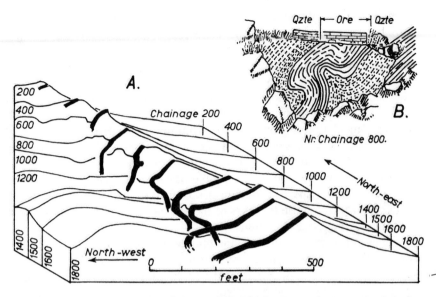

Fig. 15, A. Coulisse sections of the conical fold in the ore-body at the Balaghat Mine, C.P., India. *B*. Contortions in the Balaghat ore-body in the vicinity of the cone-apex

structure. From chainage 800–1000 ft. we have the contorted apical zone of a conical fold which increases steadily in amplitude to chainage 1500–1600 ft. The hinge has a uniform direction of trend, and an angle of plunge of between 15° to 20°. This uniformity changes around chainage 1600 ft. and the hinge can be seen to be side-stepped to the west; this occurs in an area where steeply plunging cross-folds were noted on the surface; to the south-west of this deflection the fold had the same general trend as that found elsewhere.

Though much of the termination of the conical fold at Balaghat had been mined out or was inaccessible because of caved workings, enough remained to show that here the apical zone was one of strong deformation which disappeared as the cone widened and the fold structure became more pronounced. Beyond the cone apex, i.e. to the north-east of chainage 600 ft., the evenly dipping ore-body gave no indication that such a fold as the one discussed was liable to develop.

6. CONCLUSIONS

Many fold structures maintain a remarkable uniformity for considerable longitudinal distances, but sooner or later the individual folds forming them come to an end and die out. Such structures over much of their lengths can be classified as being cylindrical; but before further analysis is made of them, it is essential that this be proved. If this type of structure can be demonstrated as being present, a single cross-section at right-angles to the axes of the folds will be more or less representative of the structure as a whole. Where the folds are horizontal, one or more vertical sections drawn in the usual manner will illustrate the structure; but where the folds plunge an orthogonal section or profile needs to be constructed. The cylindrical form can be recognised on geological maps by the shapes and spacing of stratum contours and from the similarity of serial sections, as well as from stereographic plots of structural data. Such methods are useful when analysing maps on which modern structural symbols are not shown. The construction of the profile has been described; and from it one can measure the true thicknesses or variations in thickness of the beds, and can recognise the style of the folding more accurately than one can by drawing vertical sections.

Some folds may undulate longitudinally—like a sea-serpent—and though they lose their true cylindrical form by doing so, they are nevertheless continuous and, provided the undulations are not too abrupt, a profile which shows their structure reasonably accurately can still be drawn. Profiles of folds which are bent sideways cannot be constructed by geometrical methods, though profiles of cylindrical portions between the points of deflection can be made.

The disappearance of folds seen on a map may be the result of such undulations, or it may be that the fold is dying out and ending. Those terminations which taper to a point are referred to as conical folds; and the distinction between them and cylindrical folds can be recognised from differences in the behaviour of the stratum contours as well as by stereo-graphic methods. Contours drawn on a cylindrically folded surface show uniform curvatures at different levels and are uniformly spaced when measured parallel to the direction of plunge. Contours on a conical surface are hyperbolic in form, their radii of curvature are not the same at different levels, and become less and less as the cone-apex is approached. In addition their separation parallel to the plunge is not constant. No cross-section showing the true thickness of beds and their curvature throughout a conical fold structure can be drawn, as was pointed out by Stockwell (1950).

The presence of competent strata in a tapering fold tends to modify the conical form of the structure, because the thickness of the beds will exercise a control on the curvature of the folding. Hence, the fold may flatten in amplitude without correspondingly tapering to a point in plan as might be expected. True conical folds, however, may develop under those conditions where the different competency of the strata is reduced or eliminated, that is, in rocks which are wholly incompetent, or were subjected to metamorphism whilst being folded.

That a conical antiform will pass through the cone-apex into a synform, or vice versa, appears to be very doubtful; because it is not a surface that is being folded, but beds of solid rock which have a definite thickness in the third dimension. In consequence, folds with conical terminations are not co-linear, but are either *en échelon*, each forming an individual unit—suggestive of a school of porpoises—or one fold grows out of the flanks of another as the latter begins to lose amplitude. In this way the total shortening across the whole structure remains roughly constant.

The structures developed at the apices of conical folds appear to have been little studied, but an example observed in the Balaghat Manganese Mine in India showed that the apical zone was one of strong contortion. This suggests that others may also form local zones of marked deformation, in which tight plication, fracturing or cleavage may be present. An appeal is made to those geologists who still do field mapping for further information on the endings of these somewhat elusive structures.

ACKNOWLEDGMENTS

Nothing has given my greater pleasure than preparing this contribution to Professor Hawkins's *Festschrift*, and I want to thank the organisers for including me among those invited to do likewise. While congratulating Professor Hawkins himself on this occasion, I would also like to express my gratitude to him for starting me on an academic career in his Department at Reading.

I also wish to acknowledge the debt I owe in this paper to the works of two of my friends, Dr. C. H. Stockwell and Professor Eugène Wegmann, who were both pioneers in the geometrical elucidation of fold structures. Some of the figures presented here have been developed from their original diagrams, and it has been difficult to elaborate their ideas further without unwittingly paraphrasing what they have already written. If my references to their publications are considered inadequate, I apologise. In conclusion, my thanks are extended to those of my colleagues who have read the text of this paper and have made helpful suggestions, and to Mr. J. A. Gee for the photograph reproduced in Plate I.

REFERENCES

ARGAND, E. 1911. Les nappes de recouvrement des Alpes pennines et leurs prolongements structuraux. *Mat. carte géol. de la Suisse*, N.S. 31e, Liv. Berne, p. 1–26.
———. 1924. La Tectonique de l'Asie. *Int. geol. Congr.*, 13, Brussels, 1922, **1**, 171–372.
BAILEY, E. B. 1935. *Tectonic Essays, Mainly Alpine*. London.
BIOT, M. A. 1961. Theory of Folding of Stratified Visco-Elastic Media and its Implications in Tectonics and Orogenesis. *Bull. geol. Soc. Am.*, **72**, 1595–620.
BONTE, A. 1953. *Introduction à la lecture des cartes géologiques*. Paris.
BUSK, H. G. 1929. *Earth Flexures*. Cambridge (Reprinted 1957, New York, U.S.A.).
CAMPBELL, J. D. 1958. En Echelon Folding. *Econ. Geol.*, **53**, 448–72.

CLARK, R. H. & D. B. McINTYRE. 1951. The Use of the terms Pitch and Plunge. *Am. J. Sci.*, **249**, 591–9.

CHALLINOR, J. 1945. The Primary and Secondary Elements of a Fold. *Proc. geol. Ass.*, **56**, 82–8.

CLOOS, E. 1946. Lineation. *Mem. geol. Soc. Am.*, No. 18.

COLLET, L. W. 1927. *The Structure of the Alps.* London.

DAHLSTROM, C. D. A. 1954. Statistical Analysis of Cylindrical Folds. *Trans. Can. Inst. Min. Metall.*, **57**, 140–5.

DEARMAN, W. R., K. A. G. SHIELLS., & G. P. LARWOOD. 1962. Refolded Folds in the Silurian Rocks of Eyemouth, Berwickshire. *Proc. Yorks. geol. Soc.*, **33**, 273–85.

EVANS, A. M. 1963. Conical Folding and Oblique Structures in Charnwood Forest, Leicestershire. *Proc. Yorks. geol. Soc.*, **34**, 67–80.

FERMOR, L. L. 1909. The Manganese Ore Deposits of India. *Mem. geol. Surv. India*, **37**, 1294.

FISHER, O. 1881. Oblique and Orthogonal Sections of a Folded Plane. *Geol. Mag.*, **18**, 20–3.

FLEUTY, M. J. 1964. The Description of Folds. *Proc. geol. Ass.*, **75**, 461–92.

GOGUEL, J. 1952. *Traité de Tectonique.* Paris.

HAMAN, P. J. 1961. Manual of Stereographic Projection. *West Canadian Research Publications*, Ser. 1, No. 1. Calgary, Alta.

HEIM, ALBERT. 1921. *Geologie der Schweiz.* **II**, pt. 1. Leipzig

KALLIOKOSKI, J. 1953. Interpretations of the Structural Geology of the Sherridon-Flin Flon Region, Manitoba. *Bull. geol. Surv. Canada*, **25**.

LEE, J. S. 1929. Some Characteristic Structural Types in Eastern Leipzig L. Asia and their Bearing upon the Problem of Continental Movements. *Geol. Mag.*, **66**, 358, 413, 457 and 501.

LEES, G. M. 1952. Foreland Folding. *Q. Jl geol. Soc. Lond.*, **108**, 1–34.

McINTYRE, D. B. 1951. Tectonics of the Area between Grantown and Tomintoul (Mid-Strathspey). *Q. Jl geol. Soc. Lond.*, **107**, 1–22.

MACKIN, J. H. 1950. The Down-structure Method of Viewing Geologic Maps. *J. Geol.*, **58**, 55–72.

MEAD, W. J. 1920. Notes on the Mechanics of Geologic Structures. *J. Geol.*, **28**, 505–23.

MENDELSOHN, F. 1959. The Structure of the Roan Antelope Deposit. *Trans. Instn Min. Metall.*, **68** (1958–9), 229–63.

PHILLIPS, F. C. 1954. *The Use of Stereographic Projection in Structural Geology.* London.

RAMBERG, H. 1961. Contact Strain and Folding Instability of a Multilayered Body under Compression. *Geol. Rdsch.*, **51**, 405–39.

RAMSAY, J. G. 1964. The Uses and Limitations of Beta-diagrams and Pi-diagrams in the Geometrical Analysis of Folds. *Q. Jl geol. Soc. Lond.*, **120**, 435–54.

————. 1967. *Folding and Fracturing of Rocks.* New York.

READ, H. H. & J. V. WATSON. 1962. *Introduction to Geology.* **I**. London.

SITTER, L. U. de. 1964. *Structural Geology.* (2nd Edit.). New York and London.

STAUFFER, M. R. 1964. The Geometry of Conical Folds. *N.Z. Jl Geol. Geophys.*, **7**, 340–7.

STOCKWELL, C. H. 1960. The Use of Plunge in the Construction of Cross-sections of Folds. *Proc. geol. Ass. Can.*, **3**, 97–121.

SUTTON, J. & J. V. WATSON. 1954. The Structure and Stratigraphical Succession of the Moines of Fannich Forest and Strath Bran, Ross-shire. *Q. Jl geol. Soc. Lond.*, **110**, 21–54.

TISCHER, K. 1963. Über κ-Achsen., *Geol. Rdsch.* **52** (for 1962), 426–47.

TOKUDA, S. 1926–7. On the Echelon Structure of the Japanese Archipelagoes. *Jap. J. Geol. Geogr.*, **5**, 41–76.

WEGMANN, C. E. 1928. Über die Tektonik der jüngeren Falten in Ostfinnland. *Fennia*, **50** (Sederholm Volume), No. 16, p. 1–22.

————. 1929. Beispiele tektonischer Analysen des Grundgebirges in Finnland. *Bull Commn. géol. Finl.*, No. **87**, 98–127.

WILLIS, Bailey. 1923. *Geologic Structures*. New York.

WILSON, Gilbert. 1946. The Relationship of Slaty Cleavage and Kindred Structures to Tectonics. *Proc. Geol. Ass.*, **57**, 263–302.

————. 1961. The Tectonic Significance of Small-scale Structures, and their Significance to the Geologist in the Field. *Annls Soc. géol. Belg.*, **84**, 423–548. Liège.

Gilbert Wilson
Department of Geology
Imperial College
London, S.W.7

Reprinted by permission of National Research Council Canada from *Canadian Journal of Earth Sciences*, v. 13, no. 1, p. 54–65.

Determining axes, axial planes, and sections of macroscopic folds using computer-based methods

H. A. K. CHARLESWORTH, C. W. LANGENBERG, AND J. RAMSDEN

Department of Geology, University of Alberta, Edmonton, Alberta T6G 2E1

Received 25 April 1975

Revision accepted for publication 20 August 1975

The fold-axis is the eigenvector associated with the smallest eigenvalue of a symmetrical 3×3 matrix of direction cosines of poles to the folded surface, only if the fold is cylindrical. Cylindricity can be tested using either a χ^2 or an F test. Sections showing the traces of macroscopic surfaces and of the axial plane may be constructed with the aid of computer plots that show the projection of each outcrop as well as the trace of the folded surface. The orientation of the axial plane can be calculated from the orientations of the fold-axis and the trace of the axial plane on a section normal to the fold-axis. These numerical procedures are illustrated by an analysis of four folds from the Rocky Mountains.

L'axe d'un pli correspond au vecteur propre associé à la valeur propre minimale d'une matrice carrée symétrique d'ordre 3 des cosinus directeurs des pôles de la surface plissée, seulement si le pli est cylindrique. La cylindricité peut être vérifiée au moyen d'un test F ou χ^2. Des coupes représentant les traces des surfaces macroscopiques ainsi que du plan axial peuvent être dessinées à l'aide des tracés furnis par ordinateur; elles montrent la projection de chaque affleurement ainsi que la trace de la surface plissée. L'attitude d'un plan axial peut être calculée à partir de l'orientation de l'axe du pli et de celle de la trace du plan axial sur une coupe perpendiculaire à l'axe du pli. Ces méthodes numériques sont illustrées par l'analyse de quatre plis dans les Montagnes Rocheuses.

Introduction

Most macroscopic folds are plane cylindrical or are divisible into plane cylindrical segments. The geometry of such folds or fold-segments is best described in terms of the orientation of the fold-axis and axial plane, and by various parameters that can be obtained from the profile. Graphical methods for determining axes and axial planes and for constructing sections and profiles are well known and used extensively. Not so well known and less widely used are the computer-based numerical procedures for performing these tasks. These are more accurate and cheaper to implement, and give results that are reproducible and commonly in a form suitable for statistical analysis.

This paper first discusses how to determine fold-axes. Many geologists still use a π diagram to do this, in spite of the inherent difficulty in escaping plotting errors and in estimating the best-fit great circle. Although the numerical method discussed below has been described previously, the available literature does not help the average geologist understand how and why it works. For this reason, and because the method is of major importance in all quantitative structural studies, a rigorous but simple explanation is given as an addendum.

Before calculating fold-axes and axial planes and plotting sections, the folds in question must be shown to be cylindrical. One advantage of the numerical method of determining fold-axes is that, unlike the graphical method, it provides the basis for statistical tests of whether or not a fold is cylindrical. These tests are described under the heading "Establishing Domains". After describing how to construct sections and determine axial planes, we illustrate all the above numerical procedures by analyzing four folds from the Canadian Rocky Mountains.

Computer programs to implement the procedures can be obtained from the senior author. The data-file used as input to these programs is readily created from field-data, and as far as computing costs are concerned, a complete analysis of a fold using data from 100 outcrops should cost considerably less than $5.00.

Fold Axes

To implement numerical procedures in structural geology, the orientation of a plane surface *s* should be specified in terms of the direction

Can. J. Earth Sci., **13**, 54–65 (1976)

cosines of its pole, *i.e.* the cosines of the angles between the pole and three mutually perpendicular reference-axes. If these axes are due north, due east, and straight down, the respective direction cosines $[l\ m\ n]$ of a pole of trend T and plunge P can readily be shown to be

[1] $\quad [l\ m\ n] = [\cos T \cos P \quad \sin T \cos P \quad \sin P]$

Let V_i, a vector with direction cosines $[l_i\ m_i\ n_i]$, be the ith of p unit vectors representing the poles to p measured s-surfaces, each of which is a small portion of a cylindrically folded surface. Let B, a vector with direction cosines $[L\ M\ N]$, represent the estimated fold-axis. In general, the angle Φ_i between B and V_i will not be exactly 90°. The deviation of Φ_i from 90° can be conveniently expressed by the cosine of Φ_i. By analogy with least-squares regression techniques, B may reasonably be chosen so as to minimize the sum of the squares of these deviations, *i.e.* so as to minimize S given by

[2] $\qquad S = \sum_{i=1}^{p} \cos^2 \Phi_i$

If V_i and B are the row matrices $[l_i\ m_i\ n_i]$ and $[L\ M\ N]$, respectively, $\cos \Phi_i$ is given by the matrix product[1]

[3] $\qquad \cos \Phi_i = V_i B^{\mathrm{T}}$

where B^{T} is the transpose of B. Using [2] and [3], the identity $V_i B^{\mathrm{T}} = B V_i^{\mathrm{T}}$, and the rules for matrix multiplication, we obtain

[4] $\quad S = \sum_{i=1}^{p} (V_i B^{\mathrm{T}})^2 = \sum_{i=1}^{p} B V_i^{\mathrm{T}} V_i B^{\mathrm{T}} = B T B^{\mathrm{T}}$

where

[5] $\quad T = \sum_{i=1}^{p} V_i^{\mathrm{T}} V_i$

$= \begin{bmatrix} \sum l_i^2 & \sum l_i m_i & \sum l_i n_i \\ \sum l_i m_i & \sum m_i^2 & \sum m_i n_i \\ \sum l_i n_i & \sum m_i n_i & \sum n_i^2 \end{bmatrix}$

Now it may be shown (see, *e.g.*, Watson 1965, Cruden 1968) that (1) the matrix T has three latent roots or eigenvalues; (2) each eigenvalue is associated with a three-element eigenvector; (3) S is minimized when B is the eigenvector associated with λ_3, the smallest eigenvalue of T; and (4) the minimum value of S is equal to λ_3. These

---

[1]Superscript ^{T} denotes a transposed matrix.

relationships are derived from first principles in the addendum to this article.

Establishing Domains

Before the eigenvector associated with the smallest eigenvalue λ_3 of the matrix T (λ_3-eigenvector) can be taken as the direction cosines of the fold-axis B, the fold from which the orientation data were obtained must be shown to be cylindrical. In the case of a correctly measured ideal cylindrical fold, the angle Φ between the λ_3-eigenvector and the measurement of an s-pole is invariably 90° and λ_3 is zero. For most natural folds, however, Φ is rarely 90° and λ_3 has a finite value that increases with p, the number of poles, and with the scatter of measured poles about the plane normal to the λ_3-eigenvector. How small must λ_3/p be for the folding to be cylindrical? In other words how much scatter is permitted in a cylindrical fold?

Cruden (1968) has pointed out that the departure of Φ from 90° may be separated into two components, namely those caused by (a) divergence of the measured from the true s-pole owing to errors associated with measurement and surface roughness, and (b) non-cylindricity of the fold. According to Watson (1965) the null hypothesis of coplanarity of s-poles (*i.e.* of cylindricity) is rejected with confidence $1 - \alpha$ if

[6] $\qquad K \lambda_3 > \chi^2{}_{p-2}(\alpha)$

where $\chi^2{}_{p-2}(\alpha)$ is the upper 100α percentage point of the χ^2 distribution with $p - 2$ degrees of freedom. The statistic K estimates measurement and roughness errors, and increases with decreasing error. As long as the features causing roughness are on a scale of an outcrop or less, this statistic can be determined by taking repeated measurements of the s-pole at each of q outcrops, and using the relationship

[7] $\qquad K = r \Big/ \sum_{i=1}^{q} r_i/k_i$

where r is the total number of measurements, r_i is the number of measurements at the ith of the q outcrops, and k_i is given by (Mardia 1972, p. 251)

[8] $\qquad k_i = (r_i - 2)/(r_i - R_i)$

where

[9] $\quad R_i^2 = \left(\sum_{i=1}^{r_i} l_i \right)^2 + \left(\sum_{i=1}^{r_i} m_i \right)^2 + \left(\sum_{i=1}^{r_i} n_i \right)^2$

Equation [7] is analogous to Watson's (1960) equation 2.8 except that k_i is given not by $r_i/(r_i - R_i)$ but by [8]. Each of the q outcrops should of course be small enough for the surface being studied to have escaped significant deformation by the fold under consideration. From [6] we can see that the larger the value of K (*i.e.*, the smaller the error) and the larger the value of λ_3 (*i.e.*, the larger the scatter of s-poles), the more likely the null hypothesis of coplanarity is to be rejected.

To apply the coplanarity χ^2 test, K is first determined from repeated measurements at several outcrops. Then measurements of the s-pole at p outcrops as randomly distributed as possible through the fold, one measurement per outcrop, are used to calculate λ_3. If the null hypothesis of coplanarity cannot be rejected, *i.e.* if $K \lambda_3 < \chi^2_{p-2}(\alpha)$, the observed scatter of s-poles about the λ_3-eigenvector expressed by λ_3/p can be attributed to errors associated with measurement and small-scale roughness features alone, so the fold may be regarded as cylindrical. If the null hypothesis is rejected, *i.e.* if $K \lambda_3 > \chi^2_{p-2}(\alpha)$, the fold may still be cylindrical if there are roughness features whose scale is larger than that of an outcrop. This could be the case, for example, where a macroscopic first-order fold under investigation is associated with several macroscopic second-order folds, or where there are macroscopic irregularities associated with lenticular layering. In such cases the value of k, and thus of K, determined from repeated measurements at an outcrop underestimates total roughness errors and is too high for use in [6].

The analysis of variance test described below, although not as rigorous as the coplanarity χ^2 test, offers one solution to the problem of how to test numerically the cylindricity of a fold with large-scale roughness features. We proceed by dividing the fold into two segments, by determining λ_3 and its associated eigenvector in each segment, and then by testing the null hypothesis that the measured s-poles in the two segments are clustered about planes with the same orientation. According to Watson (1965), this null hypothesis of coaxiality is rejected with confidence $1 - \alpha$ if

$$[10] \quad (p - 4)(\lambda_3 - \lambda_3 a - \lambda_3 b)/2(\lambda_3 a + \lambda_3 b)$$
$$> F_{2,p-4}(\alpha)$$

where λ_3, $\lambda_3 a$, and $\lambda_3 b$ are the smallest eigenvalues of the matrix T for the whole fold and for

segments a and b, respectively, and where $F_{2,p-4}(\alpha)$ is the upper 100α percentage point of the F distribution with 2 and $p - 4$ degrees of freedom.

Two conditions have to be met for the coaxiality F-test to be valid. First, the scatter of measured s-poles about the λ_3-eigenvector in one segment must be the same as that in the other. From Watson (1965), the null hypothesis of equal scatter is rejected with confidence $1 - \alpha$ where

$$[11] \quad (p_b - 2)\lambda_3 a/(p_a - 2)\lambda_3 b > F_{p_a-2,p_b-2}(\alpha/2)$$

or where

$$[12] \quad (p_b - 2)\lambda_3 a/(p_a - 2)\lambda_3 b < F_{p_a-2,p_b-2}(1 - (\alpha/2))$$

where p_a and p_b are the number of measurements in segments a and b. Secondly, the s-poles in each segment must be coplanar; *i.e.*, the segments must be cylindrical.

Let us now examine this second condition, first where the null hypothesis is rejected. Examination of [10] leads us to conclude in this case that λ_3 is significantly greater than $\lambda_3 a + \lambda_3 b$, *i.e.* that the scatter of measured s-poles in the whole fold is greater than in the segments. Here there is little doubt that the fold is non-cylindrical, a conclusion whose validity is unlikely to be affected by the possibility that the segments are also non-cylindrical. If, on the other hand, the null hypothesis cannot be rejected, the scatter in the whole fold is not significantly greater than in the segments. Now the scatter of measured s-poles in a segment of a fold can be expected to be significantly less than that in the whole fold if the fold is non-cylindrical, and to be essentially the same as that in the whole fold if the fold is cylindrical. We therefore conclude that if the null hypothesis cannot be rejected, both the whole fold and the segments may be regarded as cylindrical.

Sections

Knowing the coordinates of several points where a cylindrically folded surface has been precisely positioned, a section showing the configuration of this surface can be constructed by projecting these points parallel to fold-axis B onto the plane of section, and then by drawing a continuous line through the projected points. The projections can readily be carried out numerically by rotating axes as follows.

If the coordinates of a point, referred to one set of rectangular axes, comprise the row matrix $[x\ y\ z]$, and referred to another rectangular system having the same origin form the row matrix $[x'\ y'\ z']$, then

[13] $\qquad [x'\ y'\ z']^{\mathrm{T}} = [R][x\ y\ z]^{\mathrm{T}}$

where R is a 3×3 matrix whose first, second, and third rows are the direction cosines of the second set of axes referred to the original axes (see, *e.g.*, Ayres 1962, p. 88). Now the direction cosines of a unit-vector are also the coordinates of the terminus of the vector. Thus, if the direction cosines of a unit-vector, referred to the original set of axes, form the row matrix $[l\ m\ n]$, and referred to the second set make up $[l'\ m'\ n']$, then

[14] $\qquad [l'\ m'\ n']^{\mathrm{T}} = [R]\ [l\ m\ n]^{\mathrm{T}}$

Consider first the construction of a section (profile) normal to B. Let C be a $3 \times p$ matrix whose first, second, and third rows are the co-ordinates of the p points on the folded surface referred to north, east, and vertical axes respectively, *i.e.* the axes used to calculate direction cosines. Let X, Y, and Z be rectangular axes whose orientations are $[T+90\ \ 0]$, $[T\ \ P-90]$ and $[T\ P]$, respectively, where T and P are the trend and plunge of the downward-pointing normal to the plane of the section. Clearly, X and Y are suitable reference axes for the section (Fig. 1). If R is a 3×3 matrix whose first, second, and third rows are the direction cosines of X, Y, and Z, respectively, then from [13]

[15] $\qquad [NC] = [R][C]$

where NC is the $3 \times p$ coordinate matrix for the points referred to axes X, Y, and Z. Rows 1–2 of NC, which are the X and Y coordinates of the normal projections onto the plane of section of the points on the folded surface, are all that are needed to construct a profile normal to B.

Where a section oblique to B is required, the coordinates of points such as P′, the projection parallel to B of the point P onto the XY plane, must be calculated (Fig. 2). The direction cosines $[L'\ M'\ N']$ of B, and the coordinates $[X'\ Y'\ Z']$ of P referred to axes X, Y, Z, can be calculated from the original direction cosines and coordinates as described above. Since $[L'\ M'N']$ are also the coordinates of R, a point unit distance from O on the fold-axis OB, the coordinates of Q, the intersection of OB with the plane through

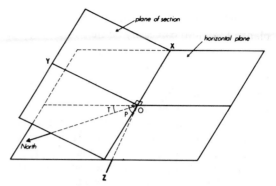

FIG. 1. Diagram illustrating the relationship between the plane of section, the trend and plunge of whose normal are T and P, and the mutually perpendicular X, Y, and Z axes whose orientations are $[T+90\ \ 0]$, $[T\ \ P-90]$, and $[T\ P]$, respectively.

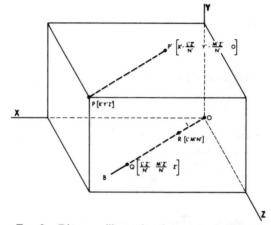

FIG. 2. Diagram illustrating how to calculate the coordinates of P′, the projection of P parallel to the fold-axis OB onto the plane of section whose normal is OZ and whose reference axes are OX and OY. $[X'\ Y'\ Z']$ are the coordinates of P and $[L'\ M'\ N']$ the direction cosines of OB referred to axes X, Y, Z.

P normal to OZ, are by proportion $[L'Z'/N'\ M'Z'/N'\ Z']$. The X and Y coordinates of P′ relative to O, since they are the same as those of P relative to Q, are therefore $[X' - L'Z'/N'\ Y' - M'Z'/N']$.

Where macroscopic surfaces cannot be accurately mapped, the above method has to be modified. Instead of plotting points representing the projections of localities where a macroscopic surface is exposed, each outcrop within the domain is projected onto the plane of section and represented by a short line parallel to the trace of *s*. The coordinates of the midpoints of each line are determined as described above. The pitch PT of the trace of s on the XY plane (the plane of section) can easily be shown to be given by

[16] $\tan PT = l'/m'$

where $[l'\ m'\ n']$ are the direction cosines of the s-pole at the outcrop, referred to axes X, Y, Z.

The above relationships can be programmed to instruct a computer to prepare a plot on which each outcrop is represented by a line parallel to the trace of s, beside which can be a single letter to identify the associated rock-unit. Two or more structures such as bedding and cleavage can be represented on the same plot. The trace of the topographic surface on the section can be represented by a series of crosses. To complete the section, the traces of the contacts between rock-units can be drawn parallel to the nearby traces of s.

Axial Planes

The axial plane of a plane cylindrical fold may be determined, knowing the strike of its trace on a horizontal surface and the orientation of the fold-axis B. However, as pointed out by Stauffer (1973), positioning the axial trace on a geological map is far from simple unless the axial plane is vertical. Thus, the axial plane should be constructed as the plane containing B and the axial trace on the profile. This trace can generally be approximated as the line connecting the points of greatest curvature on each macroscopic surface, and as the line that crosses the actual or inferred traces of s at right angles.

Once the pitch DE of the trace D of the axial plane on the profile has been measured, the trend T and plunge P of D may be calculated as follows (Fig. 3). The triangle DEF is a right-angled spherical triangle. Applying Napier's rule (see, e.g., Phillips 1972),

[17]
$$\tan EF = \sin p \cdot \tan DE$$
$$\sin DF = \cos p \cdot \sin DE$$

where p is the plunge of B. From Fig. 3

[18] $T = t + 90 + EF$ $P = DF$

where t is the trend of B.

The axial plane is parallel to B and D. If b and d are unit vectors specifying the direction cosines of B and D, respectively, the vector specifying the direction cosines of the normal to the axial

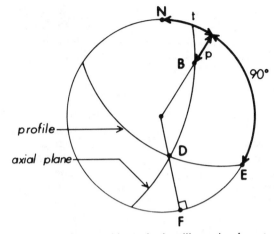

FIG. 3. Stereographic projection illustrating how to determine the orientation $[T\ P]$ of the trace D of the axial plane on the profile, given the pitch DE of D on the profile.

plane can be obtained by normalizing the outer product of b and d (see e.g. Shanks et al. 1965, p. 314). The trace of the axial plane on a geological map can be taken to be a line drawn between those outcrops whose projections delimit the axial trace on the profile.

Discussion

To apply the numerical procedures discussed above in an area of folded rocks, the area is divided, from general considerations, into the smallest possible number of 'domains' where folding appears cylindrical, and each 'domain' into two segments. A data-set is then compiled for each segment in which each outcrop or data-station is represented by (a) its coordinates referred to north, east, and vertical axes; (b) one or more measurements of the s-surface; and (c) a letter identifying the associated rock-unit.

Where applicable, a mean orientation (see e.g. Mardia 1972, p. 218) as well as a value for k (Equation [8]) are calculated for each outcrop, and a value of K is determined for each "domain" (Equation [7]). The eigenvalues and eigenvectors of the matrix T are then calculated using a single measurement of s from each outcrop, and the χ^2 coplanarity test (Equation [6]) is carried out. If the null hypothesis of coplanarity cannot be rejected with $\alpha = 0.05$, folding within the domain may be regarded as cylindrical.

If the null hypothesis of coplanarity is rejected, the eigenvalues and associated eigenvectors of the matrix T are calculated for each segment of the 'domain'. The F-tests for equality of scatter ([11] and [12]) and coaxiality (Equation [10]) are then performed. If the null hypotheses associated with these tests cannot be rejected with $\alpha = 0.05$, folding in the domain may be regarded as cylindrical. If on the other hand one of the null hypotheses is rejected, folding in the 'domain' can be assumed to be non-cylindrical, in which case each segment is examined in the way just outlined for 'domains'. As long as one has been careful in outlining the original 'domains', usually it will not be necessary to subdivide them any further before being able to demonstrate cylindricity. However, because data-sets are readily combined by the computer, but are difficult to split, to be on the safe side 'domains' can be divided into four parts, each initial segment consisting of two parts.

Once a domain within which folding is cylindrical has been established, the fold-axis is estimated by the λ_3-eigenvector of the matrix T. The map coordinates of the origin of the section as well as its orientation are then chosen. Using these values, the orientation of the fold-axis, the map coordinates of each outcrop, and the orientation of s at each outcrop, a computer plot is then obtained that shows the projection of each outcrop together with the trace of s. On this plot are drawn the traces of macroscopic surfaces, as well as the trace of the axial surface. Finally the orientation of the axial surface, if planar, is calculated, and its trace on the topographic surface drawn.

Examples

The following descriptions are by no means complete, and are included in this article only to illustrate the numerical procedures described above.

Folding Mountain Anticline

Part of the Folding Mountain anticline, situated in the Rocky Mountain Foothills of Alberta at latitude 53° 14'N and longitude 117° 46'W just south of Highway 16, was mapped using aerial photographs enlarged to a scale of 1:8000 (Fig. 4). The [x y z] coordinates of data-stations were obtained using a plane-table and alidade. Those data-stations that were not close to a surveyed station were transferred from the

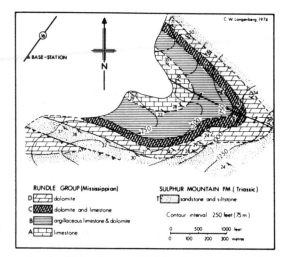

FIG. 4. Map of part of Folding Mountain anticline. The locations of some data-stations as well as the orientations of bedding are shown by the strike-and-dip symbols. Elevations are in feet relative to the elevation of the base-station, namely the foot of the northern post of the highway sign describing Folding Mountain.

aerial photograph to the map using a zoom transfer scope, and their elevations were determined by interpolation. Of the five stratigraphic units distinguished, the lower four are the divisions of the Mississippian Rundle Group established by Mountjoy (1960) and are equivalent to the Pekisko, Shunda, Turner Valley, and Mount Head Formations (Mountjoy 1962). The contact with the underlying poorly exposed Banff Formation could not be accurately mapped, whereas that with the fifth lithological unit, the Triassic Sulphur Mountain Formation, is relatively well exposed.

At all 44 stations, five measurements of bedding were taken and mean orientations and k values calculated (Equation [8]). The value of K for the whole area is 126 (Equation [7]). Using one original measurement of bedding per station, λ_3 of the matrix T (Equation [5]) was found to be 0.3018. Since $K \lambda_3 = 38$ and the value of $\chi^2_{(42)}(0.05)$ is 56, from [6] the null hypothesis of coplanarity cannot be rejected. This means that the scatter of bedding-poles normal to the λ_3-eigenvector expressed by $\lambda_3/p = 0.0069$ can be accounted for by measurement and small-scale roughness errors alone, so that folding in the whole area can be regarded as cylindrical, with the best estimate of the fold-axis (trend and plunge) being 301.6 0.6°, the λ_3-eigenvector of T compiled using all 44 mean orientations.

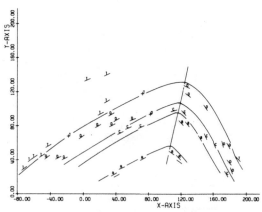

FIG. 5. Profile, looking down the fold-axis (301.6 0.6), of part of Folding Mountain anticline. For descriptions of units A, B, C, D, and T, see Fig. 4. The short lines are computer-plotted traces of bedding; the crosses mark the projections of localities where contacts were observed. The contacts between units have been drawn by hand, as has the trace of the axial plane. The origin of the profile is the projection of the base-station (Fig. 4). The X- and Y-values are in tens of feet.

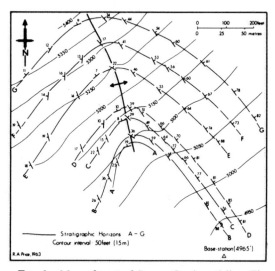

FIG. 6. Map of part of Squaw Creek anticline. The bedrock throughout the area belongs to the Mississippian Mount Head Formation, within which several stratigraphic horizons have been identified. The locations of some data-stations as well as the orientation of some bedding planes are shown by the strike-and-dip symbols. Elevations are in feet above sea level.

From an input of [x y z] coordinates, mean bedding orientations, and stratigraphic unit identifications, a computer plot was obtained that shows the projections of each station as well as the trace of bedding on a plane normal to B

(Fig. 5). Lines representing the traces of the contacts between stratigraphic units were drawn by hand on this plot. To complete the profile of the fold, the trace of the axial plane was also drawn by hand, and was found to have a pitch of 78°. Using the calculated orientation of this trace (Equation [18]) and that of B, the orientation (dip-direction and dip) of the axial plane was calculated to be 211.7 78.0°. The trace of the axial plane on the topographic surface (Fig. 4) was drawn between those data-stations whose projections delimit the axial trace on the profile.

Squaw Creek Anticline

The Squaw Creek anticline (Price 1965) is situated in the Taylor Range of the Canadian Rocky Mountains in southeastern British Columbia. Part of the anticline, 3/4 mi (1.25 km) east-northeast of Mt. Corrigan, at latitude 49° 24′N and 114° 39′W was plane-tabled by R. A. Price (Fig. 6). At this locality, the fold is entirely developed in carbonates belonging to the Mississippian Mount Head Formation.

The matrix T was compiled for the whole area, and λ_3 was found to be 0.2313 ($p = 69$). Series of repeated measurements of bedding at an outcrop from which to determine K, were not available. An overall value of 125 was, however, obtained from Mississippian carbonates at Folding Mountain. Using this value, $K \lambda_3 = 29$, and since $\chi^2_{(67)}(0.05) = 80$, the null hypothesis of coplanarity cannot be rejected. Thus the observed scatter of bedding-poles measured by $\lambda_3/p = 0.0034$ can be attributed to measurement and small-scale roughness errors alone, and the fold may be regarded as cylindrical.

Because of uncertainty over the value of K used in the above test, a coaxiality F-test was carried out. To do this, the area was divided into northwestern and southeastern segments, and the matrix T compiled for each segment. The λ_3 values of T for the two segments were found to be 0.1338 ($p = 34$) and 0.089 ($p = 35$), respectively. Since the value of the test statistic used in [11] and [12] is 1.55, and since $F_{32,33}(0.025) = 2.07$ and $F_{32,33}(0.975) = 0.48$, the null hypothesis of equal scatter cannot be rejected. Since the value of the test statistic used in [10] is 1.22, and $F_{2,65}(0.05) = 3.15$, the null hypothesis of coaxiality cannot be rejected. This confirms the conclusion reached above that the fold may be regarded as cylindrical. The best estimate of the fold-axis is 326.0 7.3°, the λ_3-

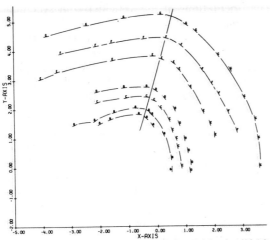

FIG. 7. Profile, looking down the fold-axis (326.0 7.3), of part of Squaw Creek anticline. The short lines are the computer-plotted traces of bedding. The traces of several stratigraphic horizons within the Mount Head Formation and of the axial plane were drawn by hand. The origin of the profile is the projection of the base-station (Fig. 6). The X- and Y-values are in hundreds of feet.

eigenvector of T compiled from all 69 measurements.

Using a set of $[x\ y\ z]$ coordinates and bedding orientations obtained from Price's contour map, a computer-plot was constructed that shows the projections of each data-station as well as the trace of bedding on a plane normal to B (Fig. 7). Lines representing the traces of six stratigraphic horizons were drawn by hand on this plot using the projections of each data-station, most of which are situated in the immediate vicinity of these bedding planes. The orientation of the axial plane, whose hand-drawn trace has a pitch of 73° on the profile, was calculated to be 238.2 73.1°. Using the relative positions of several stations and the axial trace on the profile, the trace of the axial plane on the topographic surface was then drawn (Fig. 6).

Jasper Folds

The Jasper folds are situated in the Main Ranges of the Rocky Mountains at latitude 52° 53'N and longitude 118° 5'W, at the north-western limits of the town of Jasper, Alberta, just behind the Ice Arena (Fig. 8). Two folds were mapped using a plane-table and alidade. Five lithological units, all within the Old Fort Point Formation of the Precambrian Miette Group (Charlesworth *et al.* 1967), were identified.

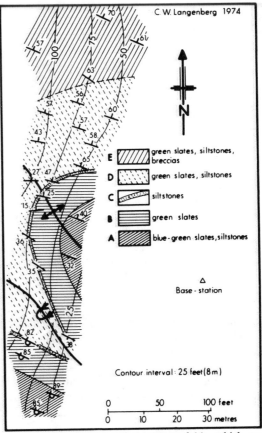

FIG. 8. Map of two of the Jasper folds, which are underlain throughout by the Precambrian Old Fort Point Formation. Elevations are in feet relative to the elevation of the base-station (southwest corner of the parking lot south of the Ice-Arena).

At all 70 data-stations, five measurements of bedding were taken and mean orientations and k values calculated. The value of K for the whole area is 87. Using one measurement of bedding per station, λ_3 of the matrix T was found to be 0.3605. Since $K\lambda_3 = 31$, and since $\chi^2_{(68)}(0.05) = 80$, the null hypothesis of coplanarity cannot be rejected. This means that the scatter of s-poles measured by $\lambda_3/p = 0.0051$ can be attributed to measurement and small-scale roughness errors alone, and that the folds may be regarded as cylindrical. The best estimate of the fold-axis, 292.5 5.7°, is the λ_3-eigenvector of the matrix T, compiled using a mean orientation rather than a single measurement from each outcrop.

From the usual input of $[x\ y\ z]$ coordinates, mean bedding and cleavage orientations, and identifications, a computer-plot was obtained

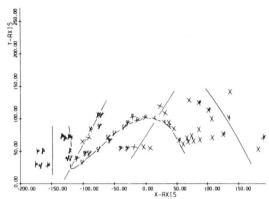

FIG. 9. Profile looking down the fold-axis (292.5 5.7) across two of the Jasper folds. For descriptions of units A–E, see Fig. 8. The short lines are computer-plotted traces of bedding and (in most cases) cleavage. The contacts between the units have been drawn by hand, as have the traces of the axial planes. The origin of the profile is the projection of the base-station (Fig. 4). The X- and Y-values are in tens of feet.

that shows the projection of most data-stations as well as the traces of bedding and (in most cases) cleavage onto a plane normal to B (Fig. 9). Lines representing the traces of the contacts between the various stratigraphic units and of the axial plane were drawn by hand on this plot.

The orientations of the anticlinal and synclinal axial planes, whose traces have pitches of 57° and 60° on the profile, were calculated to be 206.2 57.2° and 205.8 60.2°, respectively. The traces of the axial planes on the topographic surface were then drawn, using the relative positions of several stations and the traces of the axial planes on the profile.

Wynd Syncline

The Wynd syncline is situated in the Main Ranges of the Rocky Mountains at latitude 52° 52′N and longitude 118° 11′W, about 4 mi (6 km) west of Jasper, Alberta, just north of the CNR track and 1 mi (1.6 km) west of Wynd Siding (Fig. 10). It was mapped on an aerial photograph enlarged to a scale of 1:8000. Accurate horizontal and vertical control was obtained by carrying out a closed plane-table traverse and by using a Paulin altimeter and barograph and a zoom transfer scope. The eight arenaceous (largely sandstone) and argillaceous (largely slate) units established all belong to the lower member of the Wynd Formation of the Precambrian Miette Group (Charlesworth et al. 1967).

At each of the 78 data-stations, five measurements of bedding were taken and a mean orientation and k value calculated. The value of K for

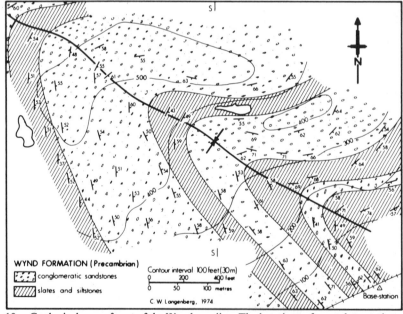

WYND FORMATION (Precambrian)

⬚ conglomeratic sandstones

▨ slates and siltstones

Contour interval 100 feet (30 m)

0   200   400 feet
0   50   100 metres

C. W. Langenberg, 1974

FIG. 10. Geological map of part of the Wynd syncline. The locations of most data-stations, as well as the orientations of bedding, are shown by the strike-and-dip symbols. Elevations are relative to the elevation of the base-station, namely the foot of the railroad sign "Wynd 1 mile". The line SS marks the position of the north–south vertical section (Fig. 13).

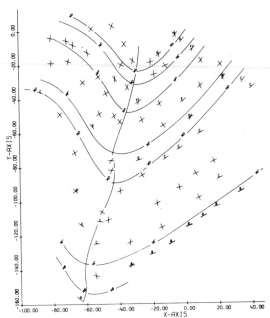

FIG. 11. Profile, looking down the fold-axis (111.9 48.8) of part of the Wynd syncline. Units A, C, E, and G are arenaceous units and B, D, F, and H are argillaceous units within the lower member of the Precambrian Wynd Formation. The short lines are computer-plotted traces of bedding and (in most cases) cleavage; the computer-plotted crosses mark the projections of localities where contacts were observed. The contacts between units have been drawn by hand, as has the trace of the axial surface. The origin of the profile is the projection of the base-station (Fig. 10). The X- and Y-values are in tens of feet.

From an input of $[x\ y\ z]$ coordinates, mean bedding and cleavage orientations, and rock-unit identifications, a computer-plot was obtained that shows the projections of each station onto a plane normal to B and, in most cases, the traces of bedding and cleavage (Fig. 11). The traces of the contacts were then drawn by hand on this plot. Difficulty was experienced in drawing the trace of the axial surface; the fold appears to be non-planar, mainly the result of the axial planes in the competent sandstones being offset from one another. Although no orientations were calculated, the relative positions of various stations and the trace of the axial plane on the profile were used to plot the axial trace on the topographic surface (Fig. 10). To illustrate the versatility of the plotting technique, computer plots were obtained that show the projections of each station onto horizontal (Fig. 12) and north–south vertical (Fig. 13) planes.

Acknowledgments

We wish to thank R. A. Price of Queen's University, Kingston, for allowing us to use his map of the Squaw Creek anticline. D. M. Cruden, D. H. Kelker, and E. A. Babcock of the University of Alberta read a preliminary draft of the manuscript and suggested several improvements. Mr. F. Dimitrov drafted the text-figures. The senior author gratefully acknowledges the

the whole area was found to be 76. Using only one original measurement of bedding per station, λ_3 of the matrix T for the whole area was found to be 1.2714. Therefore, since $K\ \lambda_3 = 97$ and $\chi^2_{(76)}(0.05) = 85$, the null hypothesis of coplanarity is rejected. The matrix T was then recompiled for the whole area and for northwestern and southeastern segments, using a mean orientation from each station. The values of λ_3 for the three matrices were found to be $0.7332\,(p = 78)$, $0.3036\,(p = 33)$ and $0.3915\,(p = 45)$, respectively. Using [11] and [12], the scatter of bedding-poles in one segment can be shown not to differ significantly from that in the other. Since $F_{2,74}(0.05) = 3.1$ and the value of the test statistic (Equation [10]) is 2.03, the null hypothesis of coaxiality cannot be rejected. Thus, the fold may be regarded as cylindrical. The fold-axis, taken to be the λ_3-eigenvector of the matrix T, calculated using the mean orientation at each station, has an orientation of 111.9 48.8°.

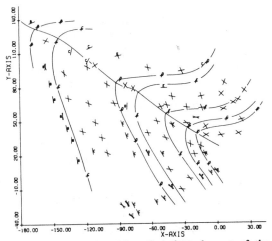

FIG. 12. A horizontal section through part of the Wynd syncline at the elevation of the base-station, which is also the origin for the section. The X- and Y-axes point east and north, respectively. The significance of the symbols and lines in this figure is the same as for Fig. 11. The X- and Y-values are in tens of feet.

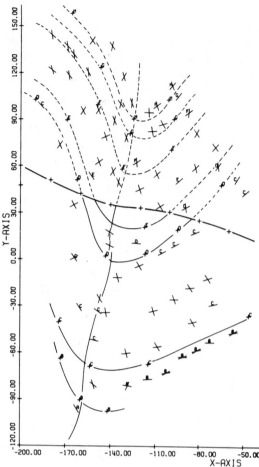

FIG. 13. A vertical, north–south section looking east through part of the Wynd syncline. The position of the section is given in Fig. 10 by the line SS and would appear on Fig. 12 as a north–south line 910 feet west of the origin. The significance of the symbols and lines in this figure is the same as for Fig. 11, except that the set of regularly spaced crosses marks the trace of the topographic surface on the plane of the section. The X- and Y-values are in tens of feet.

award of research grants from the National Research Council of Canada and the Geological Survey of Canada.

Addendum: Numerical Determination of Fold-Axes

From the discussion of fold-axes at the beginning of this paper, the direction cosines of the fold axis are those values of the row matrix $B = [L\ M\ N]$ that minimize S given by [4] and [5]. To minimize S, we must find the minimum value of the function

[19] $f = [X\ Y\ Z][T][X\ Y\ Z]^T = U$ (a constant)

where, since $L^2 + M^2 + N^2 = 1$, the variables $[X\ Y\ Z]$ are connected by the subsidiary condition

[20] $F = X^2 + Y^2 + Z^2 - 1 = 0$

Geometrically, the set of functions $f = U$ is represented by a set of similarly shaped and similarly orientated ellipsoids centered on the origin whose size increases with U (where U is zero the ellipsoid is a point at the origin), while the subsidiary condition $F = 0$ is represented by a sphere of unit radius also centered on the origin. The problem can thus be restated as follows: among the ellipsoids $f = U$, we must find the one with the smallest value of U that meets the condition $F = 0$. Clearly, the appropriate ellipsoid is the one whose long axis coincides with a diameter of the sphere, and the direction cosines of the fold-axis are those values of $[X\ Y\ Z]$ where the two surfaces touch. A two-dimensional geometric representation of the solution to this problem is shown in Fig. 14.

Now, if two surfaces touch, they have the same tangent. Thus, at the point of contact

[21] $\dfrac{\partial f}{\partial X} : \dfrac{\partial f}{\partial Y} : \dfrac{\partial f}{\partial Z} = \dfrac{\partial F}{\partial X} : \dfrac{\partial F}{\partial Y} : \dfrac{\partial F}{\partial Z}$

or, if we introduce the constant of proportionality λ

[22] $\dfrac{\partial f}{\partial X} - \lambda\dfrac{\partial F}{\partial X} = 0 \quad \dfrac{\partial f}{\partial Y} - \lambda\dfrac{\partial F}{\partial Y} = 0$

$\dfrac{\partial f}{\partial Z} - \lambda\dfrac{\partial F}{\partial Z} = 0$

Expanding [19], differentiating it and [20] with respect to X, Y, and Z, and substituting the differentials in [22] leads to the following three simultaneous equations

[23] $X\sum l_i^2 + Y\sum l_i m_i + Z\sum l_i n_i = \lambda X$
$X\sum l_i m_i + Y\sum m_i^2 + Z\sum m_i n_i = \lambda Y$
$X\sum l_i n_i + Y\sum m_i n_i + Z\sum n_i^2 = \lambda Z$

which can be expressed in matrix notation as

[24] $[T][X\ Y\ Z]^T = \lambda[X\ Y\ Z]^T$

where T is as specified in [5]. Rewriting [24], which is known as the characteristic equation in linear algebra,

[25] $[TP][X\ Y\ Z]^T = [0\ 0\ 0]$

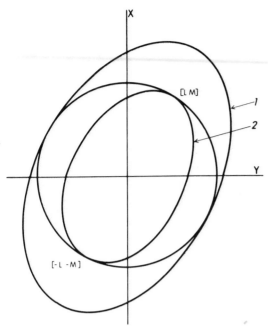

FIG. 14.   A two-dimensional geometric representation of the numerical determination of the fold-axis. Ellipses 1 and 2 represent two members of the family of ellipses

$$f = [X\ Y]\begin{bmatrix} \Sigma l_i^2 & \Sigma l_i m_i \\ \Sigma l_i m_i & \Sigma m_i^2 \end{bmatrix}[X\ Y]^T = U,$$

the two-dimensional equivalent of the family of ellipsoids represented by [19]. The circle $F = X^2 + Y^2 - 1 = 0$ is the two-dimensional equivalent of the sphere represented by [20]. To determine the two-dimensional fold-axis, among the ellipses $f = U$ we must find the one with the smallest value of U (*i.e.*, the smallest ellipse) that meets the condition $F = 0$. Clearly, the appropriate ellipse is ellipse 2, whose long axis coincides with a diameter of the circle, and the direction cosines $[L\ M]$ of the fold-axis are those values of $[X\ Y]$ where the two surfaces touch.

where

[26] $[TP] =$

$$\begin{bmatrix} \sum l_i^2 - \lambda & \sum l_i m_i & \sum l_i n_i \\ \sum l_i m_i & \sum m_i^2 - \lambda & \sum m_i n_i \\ \sum l_i n_i & \sum m_i n_i & \sum n_i^2 - \lambda \end{bmatrix}$$

Let us assume that there are solutions to [25] other than the trivial case where $[X\ Y\ Z] = [0\ 0\ 0]$. Applying Cramer's Rule to solve this set of simultaneous equations, because each element in the numerator contains a zero,

[27] $[X\ Y\ Z] = [0\ 0\ 0]/|TP|$

where $|TP|$ is the determinant of the matrix TP. If $[X\ Y\ Z]$ is not zero, it follows that

[28] $|TP| = 0$

Expanding [28] (see *e.g.* Davis 1973, pp. 146–151) yields a third-order equation in λ, whose three solutions $[\lambda_1\ \lambda_2\ \lambda_3]$, are known as the largest, intermediate, and least eigenvalues of the matrix T.

Premultiplying both sides of [24] by $[X\ Y\ Z]$ gives

[29] $[X\ Y\ Z] \cdot [T] \cdot [X\ Y\ Z]^T =$
$$[X\ Y\ Z] \cdot \lambda[X\ Y\ Z]^T$$

Therefore, from [19] and [29], since $[X\ Y\ Z] \cdot [X\ Y\ Z]^T = 1$,

[30] $U = \lambda$

Clearly, λ_3, substituted for U in [19], specifies that ellipsoid belonging to the set $f = U$, whose long axis coincides with a diameter of the sphere representing [20]. (Similarly, λ_2 and λ_1 specify those ellipsoids whose intermediate and short axes coincide with a diameter of this sphere.) Substituting $\lambda_3 = \lambda$ in [26] allows the matrix TP to be specified, and from it, using [25], the values of the direction cosines of the fold-axis can be determined.

AYRES, F., JR. 1962. Theory and problems of matrices. Schaum Publishing Co. New York.

CHARLESWORTH, H. A. K., WEINER, J. L., AKEHURST, A. J., BIELENSTEIN, H. U., EVANS, C. R., REMINGTON, D. B., STAUFFER, M.R., and STEINER, J. 1967. Precambrian geology of the Jasper region, Alberta. Res. Council Alberta, Bull. 23.

CRUDEN, D. M. 1968. Methods of calculating the axes of cylindrical folds: a review. Geol. Soc. Am. Bull. **79**, pp. 143–148.

DAVIS, J. C. 1973. Statistics and Data Analysis in Geology. Wiley, New York.

MARDIA, K. V. 1972. Statistics of Directional Data. Academic Press, London.

MOUNTJOY, E. W. 1960. Miette, Alberta. Geol. Surv. Can. Pap. 40-1959.

——— 1962. Mount Robson (southeast) map-area, Rocky Mountains of Alberta and British Columbia. Geol. Surv. Can., Pap. 61–31.

PHILLIPS, F. C. 1972. The Use of Stereographic Projection in Structural Geology. (3rd Ed.) Edward Arnold, London.

PRICE, R. A. 1965. Flathead map-area, British Columbia and Alberta. Geol. Surv. Can. Mem. 336.

SHANKS, M. E., BRUMFIEL, C. F., FLEENOR, C. R., and EICHOLZ, R. E. 1965. Pre-Calculus Mathematics. Addison-Wesley Publishing Co. Palo Alto.

STAUFFER, M. R. 1973. New method for mapping fold axial surfaces. Geol. Soc. Am. Bull. **84**, pp. 2307–2318.

WATSON, G. S. 1960. More significance tests on the sphere. Biometrika, **47**, pp. 87–91.

——— 1965. Equatorial distributions on a sphere. Biometrika, **52**, pp. 193–203.

Reprinted with permission from *Journal of Structural Geology*, v. 8, p. 897-909, N. J. White, J. A. Jackson, and D. P. McKenzie, The relationship between the geometry of normal faults and that of sedimentary layers in their hanging walls, Copyright 1986, Pergamon Press plc.

The relationship between the geometry of normal faults and that of the sedimentary layers in their hanging walls

N. J. White, J. A. Jackson and D. P. McKenzie

Bullard Laboratories, Madingley Road, Cambridge CB3 0EZ, U.K.

(*Received 7 August* 1985; *accepted in revised form* 30 *January* 1986)

Abstract—We derive an analytical expression that relates the shape of a fault in cross-section to the shape of the bedding horizons in its hanging wall block. The expression assumes that the hanging wall deforms by simple shear and that the footwall remains undeformed throughout. Although this paper concentrates on normal faults, the expression is equally valid and applicable to thrust faults. The direction of simple shear in the hanging wall block is arbitrary and has a dramatic effect on the predicted fault or bedding geometry. There is no reason to believe that the simple shear occurs on vertical planes, as is commonly assumed in graphical approaches to this problem, and ignoring the presence of inclined simple shear is likely to lead to considerable underestimates of the amount of extension across normal faults and in the amount of shortening across thrusts. Similar though more complicated expressions can be obtained when compaction within the hanging wall block is taken into account. For a planar normal fault such compaction may result in the development of a hanging wall syncline.

INTRODUCTION

IN REGIONS of extensional tectonics, a knowledge of the geometry and kinematics of large-scale faults is obviously of crucial importance, not only in evaluating particular commercial prospects, but also in understanding the nature and amount of extension involved. Although it is now clear that large-scale crustal and lithospheric stretching occurred during the formation of many continental sedimentary basins and margins, it is not always easy to reconcile estimates of the amount of stretching obtained from measurements of crustal thickness and subsidence with those obtained from the observed normal faulting (e.g. de Charpal *et al.* 1978, Le Pichon & Sibuet 1981, Wood & Barton 1983, Ziegler 1983). Much of this disagreement is probably attributable to a poor understanding of the geometry of the large normal faults that accommodate at least some, and perhaps most, of the extension at shallow crustal levels.

Recent reviews have tended to concentrate on observations of the faults themselves, using either outcrops and seismic reflection profiles (e.g. Wernicke & Burchfiel 1982, Anderson *et al.* 1983, Smith & Bruhn 1984) or seismological observations of earthquakes generated by active normal faults (e.g. Jackson & McKenzie 1983, Jackson 1986). This paper is concerned with a different approach to the same problem: what is the detailed relationship between the geometry of a normal fault and the geometry of the sediments in its hanging wall? The usual method of investigating their connection is graphical (Verrall 1981, Gibbs 1983, 1984) and assumes that the hanging wall is deformed by simple shear in vertical planes. We develop below a general analytic solution to the same problem. Our solution also assumes the deformation is by simple shear, but makes no assumption about the inclination of the shear planes to the vertical.

RELATIONS BETWEEN SEDIMENT AND FAULT GEOMETRIES

The problem

The problem is illustrated by Fig. 1, showing a listric normal fault that, for simplicity, becomes planar and horizontal at depth. (In general, faults need do neither of these things.) The geometry before movement is shown in Fig. 1(a). If movement now occurs such that all points in the hanging wall move a vector **h** relative to the footwall, the geometry would look like Fig. 1(b). In this

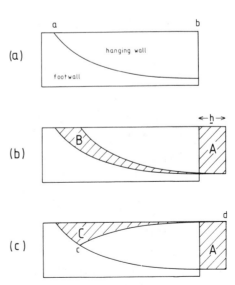

Fig. 1. Diagrams to illustrate the deformation of the hanging wall necessary to fill the potential void beneath it, if it moves a vector **h** relative to the footwall. After internal deformation of the hanging wall only points far from the fault outcrop have a displacement **h** relative to the footwall. Note that the footwall remains undeformed.

897

case **h** is horizontal with magnitude h. The cross-section has extended by area A, which, if there is no movement out of the plane of the section, is equal to area B. In reality no voids will occur and the hanging wall deforms, filling the gap beneath it, to leave a space above (Fig. 1c), such that area C = area A = area B. It is clear that the shape of the hanging wall surface (or 'rollover') in Fig. 1(c) is in some way related to the underlying fault geometry. Only at some remote point in the hanging wall, far from the surface outcrop of the fault, will the relative movement between hanging wall and footwall be represented by **h**. Closer to the fault outcrop the relative motion has been changed by the internal deformation of the hanging wall necessary to fill the (potential) void beneath it. A cross-section in which the areas A, B and C are equal is sometimes said to be 'balanced' (e.g. Gibbs 1983), though it is important to realize that this only refers to the preservation of hanging wall cross-sectional area during movement. A more powerful constraint is imposed by the use of this term in Dahlstrom's (1969) sense, in which bed length is preserved during movement. This sense is not applicable here, as the length ab in Fig. 1(a) is not equal to the length cd in Fig. 1(c).

The hanging wall cross-sectional area is preserved in Fig. 1, and it is possible to describe the change in its shape by simple shear in the plane of section. The graphical techniques used by Verrall (1981) and Gibbs (1983, 1984) to relate the fault and rollover geometries assume that the simple shear occurs on vertical planes. However, there is no reaon why this should be so, and indeed, antithetic faults observed within hanging wall blocks are generally not vertical. We will now develop more general analytical relations between fault and sediment geometries, that allow for non-vertical simple shear.

Note that the footwall in Fig. 1 remains undeformed throughout. This is an important assumption for both graphical and analytical methods, and will be discussed later.

The forward problem: from fault to sediment geometry

To begin with we will consider the movement in a coordinate frame attached to the footwall block. The planes in which simple shear occurs are parallel to the y' direction, which is not, in general, perpendicular to the Earth's surface (Fig. 2). In this frame, the velocity in the x' direction is a constant U_0 and that in the y' direction is $v = v(x')$. The shape of the bed is $B' = B'(x')$ and that of the fault is $F' = F'(x')$, where B' and F' are the y' coordinates of the bed and the fault.

Consider an element of bed ab whose length in the x' direction is $\delta x'$, which has been moved a small distance $U_0 \delta t$ in the x' direction and then deformed by simple shear parallel to the y' direction, such that point a moves to c and b moves to d. Let us suppose that before deformation the bed had a shape given by $R' = R'(x')$, with a dip of γ' at the point (x', y'). The coordinates of points a, b, c and d are

$$a = (x', y')$$
$$b = (x' + \delta x', y' + \delta x' \tan \gamma')$$
$$c = (x' + U_0 \delta t, y' + v(x') \, \delta t)$$
$$d = (x' + \delta x' + U_0 \delta t, y' + \delta x' \tan \gamma' + v(x' + \delta x') \, \delta t)$$

The new dip of element cd is ψ', given by

$$\delta x' \tan \psi' = Ed = v(x' + \delta x') \, \delta t - v(x') \, \delta t + \delta x' \tan \gamma'$$

$$\therefore \tan \psi' = \frac{dv}{dx'} \cdot \delta t + \tan \gamma'. \tag{1}$$

However, if no voids are to form, the hanging wall must remain in contact with the fault surface, and the velocity at x' must always be parallel to the fault, of dip $\theta'(x')$

$$\therefore \frac{v}{U_0} = \tan \theta'$$

and

$$\frac{dv}{d\theta'} = \frac{U_0}{\cos^2 \theta'}. \tag{2}$$

Combining (1) and (2) gives

$$\tan \psi' = \frac{U_0 \, \delta t}{\cos^2 \theta'} \frac{d\theta'}{dx'} + \tan \gamma'. \tag{3}$$

But $U_0 \, \delta t$ is the displacement in the x' direction and constant throughout the hanging wall. If $U_0 \, \delta t = h'$, then (3) may be rewritten

$$\tan \psi' - \tan \gamma' = h' \frac{d}{dx'} (\tan \theta') \tag{4}$$

or

$$\frac{d}{dx'} (B' - R') + h' \frac{d^2 F'}{dx'^2}. \tag{5}$$

Hence integration gives

$$B' = h' \frac{dF'}{dx'} + R' + C', \tag{6}$$

where C' is a constant. This expression is only valid if

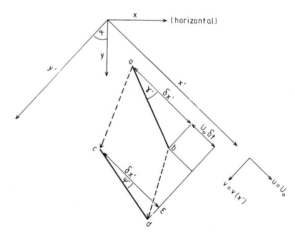

Fig. 2. Coordinate systems and geometrical relations used to derive the analytical expressions in the text.

dF'/dx' is continuous. In most cases the initial shape of the bed before deformation may simply be described by $R' = x' \tan \gamma_0' + D'$, where D' is a constant and γ_0' is the regional dip in the (x', y') frame. The value of C' depends on the origin chosen for the coordinate frame. If the origin is chosen such that $R' = R_0'$, $B' = B_0'$ and $dF'/dx' = \tan \theta_0'$ at $x' = 0$, then

$$C' = B_0' - R_0' - h' \tan \theta_0', \qquad (7)$$

when (6) becomes

$$B' - B_0' = h'\left(\frac{dF'}{dx'} - \tan \theta_0'\right) + R' - R_0'. \qquad (8)$$

These expressions are all in the (x', y') co-ordinate frame, in which simple shear is in the y' direction. But the (x', y') frame is rotated through an angle α with respect to the (x, y) frame, in which y is the downward pointing vertical (Fig. 2). Therefore the fault $F(x)$ and regional dip $R(x)$, defined in the (x, y) frame, must be rotated into the (x', y') frame using the relations

$$\begin{aligned} x' &= x \cos \alpha + y \sin \alpha \\ y' &= -x \sin \alpha + y \cos \alpha, \end{aligned} \qquad (9)$$

where y is F or R. The heave h', must also be calculated in the (x', y') frame from the fault displacement vector before eqn (6) can be applied. A simple differentiation will then yield the geometry of the bedding in the hanging wall, B', in the (x', y') frame, which can be returned to the (x, y) frame using the relations

$$\begin{aligned} x &= x' \cos \alpha - y' \sin \alpha \\ y &= x' \sin \alpha + y' \cos \alpha. \end{aligned} \qquad (10)$$

Equations (7) and (8) show that, given an observed fault geometry $F(x)$, the bedding in the hanging wall $B(x)$ may be determined if (i) the shape of the bed before movement, $R(x)$, (ii) the vector displacement on the fault, **h**, and (iii) the angle α between the downward vertical and the direction of simple shear in the hanging wall, are all specified. Note that the graphical constructions of Verrall (1981) and Gibbs (1983, 1984) assume that the simple shear is in vertical planes (i.e. $\alpha = 0°$). As will be shown later, this assumption greatly affects the predicted geometry of the hanging wall sediments.

The inverse problem: from sediment to fault geometry

From eqn (8)

$$\begin{aligned} F' = \frac{1}{h'} \int_0^{x'} &\{B' - B_0' - (R' - R_0') \\ &+ h' \tan \theta_0'\} \, dx', \end{aligned} \qquad (11)$$

where, once again, the boundary conditions B_0', θ_0' and h' must be known. Thus, given an observed bed geometry, the fault geometry may be calculated for various angles of simple shear, α. Because the inverse problem is an integration, it is more stable than the forward problem, which involves differentiation. This difference is fundamental and is not an artefact of the method used to solve the problem. Therefore the geometry of the beds in the hanging wall, determined by (8), will be strongly affected by small variations in the dip of the fault.

It is fortunate that the geologically important problem involves an integration, since numerical differentiation is not accurate, even when proper precautions are taken.

Assumptions

The main assumptions inherent in the derivation of equation (6) are:

(i) All displacements are small.
(ii) There is no movement out of the plane of section.
(iii) Deformation of the hanging wall is accomplished by simple shear. This is clear, since

$$\frac{\partial v}{\partial y'} = \frac{\partial u}{\partial x'} = \frac{\partial u}{\partial y'} = 0$$

and

$$\frac{\partial v}{\partial x'} = \frac{U_0}{h'} \cdot \frac{d}{dx'}(R' - B').$$

It is worth noting that simple shear on parallel planes is probably a reasonable assumption for the deformation in the hanging wall as it allows finite motion to occur on fault planes that do not intersect.

(iv) The footwall remains undeformed throughout.
(v) Sediment geometry has not been altered by compaction.

Of these, (iv) is probably the most important, and is least likely to be correct when applied to the deeper parts of faults that penetrate basement and are responsible for extension on a crustal scale. However, for growth faults of the type found in the Gulf of Mexico and Niger Delta, where extension of the basement does not take place, this assumption is more likely to be justified. Compaction may alter the geometry of beds within the hanging wall, particularly when syntectonic deposition occurs. A method which takes compaction effects into account is outlined in the Appendix.

On seismic reflection profiles the scales in the x and y directions are usually not equal. Provided that the exaggeration is constant (i.e. that the y scale is not a function of y), the expressions (6) and (7) will still be valid, and lead, of course, to corresponding exaggeration in the x and y scales of B and F. It is not therefore necessary to convert published seismic sections to true-scale sections; they can be digitized directly and the equations will then give the fault geometry at the same vertical exaggeration. Though this approach is clearly not accurate, it is often the only one possible (see below). However, unless **h** and α are accurately known, there is probably little purpose in making detailed corrections for velocity variations and compaction. If this is not done, it is important to remember that the value of α which should be used is the apparent, and not the true, dip of the planes of simple shear in the time section.

SYNTHETIC EXAMPLES

An analytic test

Fig. 3. The shape of a simple curved fault, given by $F = \tan^{-1} x$.

The use of the equations may be illustrated by a simple analytic example, with a fault shape described by $F = \tan^{-1} x$ (Fig. 3). Let us also assume that $R = 0$ and $\alpha = 0°$, in which case $x = x'$ and $y = y'$. Note that the dip of the fault is zero at $y = \pi/2$ and 45° at $y = 0$. From (6)

$$B = \frac{h}{1 + x^2} + C.$$

Since $B \to 0$ as $x \to \infty$, $C = 0$ and

$$B = \frac{h}{1 + x^2}. \tag{12}$$

Let us suppose the bed meets the fault at $h = x = \varepsilon$, where $\varepsilon \ll 1$. $(\varepsilon, \varepsilon)$ therefore satisfies both (12) and $y = \tan^{-1} x$.
Then

$$B = \frac{\varepsilon}{1 + x^2}. \tag{13}$$

The inverse problem can now be posed. Given equation (13) as the geometry of the bed, determine the fault geometry $F(x)$ if $\alpha = 0°$ and $R = 0$.
From (11)

$$F = \frac{1}{\varepsilon} \int_0^x \left(\frac{\varepsilon}{1 + x^2} + C \right) dx$$

$$= \tan^{-1} x + Cx + D. \tag{14}$$

Since the fault goes through the origin, $D = 0$. Clearly F is indeterminate unless dF/dx is given somewhere.

$$\frac{dF}{dx} = \frac{1}{1 + x^2} + C$$

If $dF/dx = 1$ at $x = 0$ (i.e. $\theta_0 = 45°$), then $C = 0$ and $F = \tan^{-1} x$, as it should.

The importance of inclined simple shear

The forward problem is illustrated in Fig. 4 using a fault with a dog-leg geometry, the two legs being joined by a circular arc. For simplicity $R = 0$. Two beds are drawn, both of which have the same infinitesimal displacement down the fault plane, and the same apparent horizontal heave on the fault, $h = \varepsilon$. However, in one case the hanging wall has been deformed by vertical simple shear ($\alpha = 0°$) and in the other case simple shear has occurred inclined at $\alpha = 45°$. The resulting shapes are very different, though as x becomes large, both return to the regional level of $R = 0$. Note that the displacement of the bed on the fault is the same in each

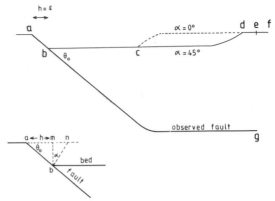

Fig. 4. Illustration of the forward problem using a dog-leg fault whose segments dipping at 45° and 0° are linked by the arc of a circle, so that dF/dx is continuous. The shape of the hanging wall surface is shown for simple shear at $\alpha = 0°$ (dashed) and $\alpha = 45°$ (solid line). The two surfaces are the same between b and c and d and f; although for $\alpha = 0°$ the final point on the fault, g, gives information on the bedding only as far as e. Equation (6) is only valid if ε is small, so the vertical scale of the hanging wall surface has been exaggerated arbitrarily to illustrate the difference between $\alpha = 0°$ and $\alpha = 45°$. The original level of the hanging wall surface is given by point a. The inset, bottom left, shows why, given a displacement ab on the fault with heave h, the horizontal movement of the rigid part of the hanging wall (i.e. overall mass transport or extension of the hanging wall block) depends on α. By referring to Fig. 1, it is clear that, for vertical simple shear point b has apparently come from m, and the horizontal extension is am = h. If the simple shear is inclined at α, point b has apparently come from n, and the horizontal extension is an = $h (1 + \tan \theta_0 \tan \alpha)$.

case, and that cross-sectional area is conserved. Why then is the area of the depression in each hanging wall different? The reason is that, in the case where $\alpha = 0°$, the horizontal displacement of the rigid part of the hanging wall is h, but where $\alpha \neq 0°$, the rigid part of the hanging wall is displaced $h(1 + \tan \theta_0 \tan \alpha)$. In this case, where $\theta_0 = 45°$ and $\alpha = 45°$, the inclined shear example represents an additional extra extension of 100%. If $\theta_0 = 60°$ and $\alpha = 60°$ this would rise to 300%. This example illustrates one of the most important results of this study: that the dip of the fault and the apparent displacement of a bed on it are not sufficient to work out the amount of extension, if that bed has also deformed in the hanging wall. The inclination of the simple shear in the hanging wall (perhaps given by the dip of minor antithetic faults) is also needed.

The inverse problem is illustrated in Fig. 5 using a bed whose shape is given by $B = h/(1 + x^2)$. Given $R = 0$,

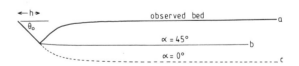

Fig. 5. Illustration of the inverse problem. Two different fault geometries are predicted from the same observed bed shape and apparent offset, h. The dashed line is the fault predicted for $\alpha = 0°$ and the solid line the fault predicted when $\alpha = 45°$. Note that when $\alpha = 45°$ the observed bed contains no information on the fault geometry beyond point b, whereas when $\alpha = 0°$ the fault may be extended as far as point c (vertically below a).

$\theta_0 = 45°$ and h, two different fault geometries are shown: one for $\alpha = 0°$ and the other for $\alpha = 45°$. These examples clearly illustrate the dramatic effect of inclined simple shear in the hanging wall block, and the danger of assuming $\alpha = 0°$.

Real examples and applications

The method described here should really be used only on seismic sections that are depth migrated and not affected by compaction. Unfortunately few such sections are available in the literature, and there are even fewer in which footwall and hanging wall stratigraphies are also shown, so that the heave, h, may be estimated. Nonetheless three examples that do not meet all these ideal requirements will now be briefly discussed. Differences between the results obtained below and those obtained using equations which deal with compaction (see Appendix) do not significantly alter our conclusions.

Predicting fault shapes or simple shear in the hanging wall

The first example is taken from Gans *et al.* (1985). Figure 6(a) shows a line drawing from a migrated seismic line across the Spring Valley in east-central Nevada. The alluvial fan and playa lake deposits that underlie Spring Valley define a wedge of west-dipping to sub-horizontal reflectors truncated on the west by a large normal fault (the Schell Creek Fault), which outcrops at the base of the Schell Creek Range. The seismic line runs approximately perpendicular to the strike of the Schell Creek Fault and motion is thought to be almost entirely in the plane of section.

Near the base of the layered sedimentary wedge, a prominent band of reflectors, labelled event E, may be traced. This event was identified as a disconformity between lacustrine sediments and underlying volcanic rocks by drill logs from the Yelland well (SP 1630). The dip of the Schell Creek Fault near the surface can be estimated as about 45° from the truncation of the layered reflectors in the hanging wall. This value is typical of other such faults in the Basin and Range Province (Smith & Bruhn 1984). Since the infilling sediments are predominantly lacustrine, we can assume that the regional dip at time of deposition was approximately horizontal, i.e. that $R = 0$. Can we now use the shape of horizon E to estimate the shape of the Schell Creek Fault at depth?

Figure 6(b) shows the shape of horizon E and the shape of the Schell Creek Fault beneath it, calculated using eqn (11), for two cases: one in which simple shear is vertical ($\alpha = 0°$) and the other in which it is inclined at 45° towards the west ($\alpha = 45°$). A minimum heave, h, is estimated from the truncated relectors as shown, $\theta_0 = 45°$ and $R = 0$. The two fault shapes are clearly very different. Interestingly, the shape calculated using $\alpha = 45°$ coincides at depth with reflectors labelled L in Fig. 6(a), which Gans *et al.* (1985) tentatively suggest may represent a deep part of the Schell Creek Fault. If

this identification is correct it implies that $\alpha = 45°$ and the hanging wall is pervasively sheared by small faults dipping 45° towards the west. It is worth noting that Gans *et al.* identify a few such faults in their line drawing (Fig. 6a).

Although Fig. 6(a) is migrated, the y axis shows two way travel time rather than depth. The vertical exaggeration is unlikely to be uniform with depth, and, until the section is depth corrected, no firm conclusions can be drawn from this experiment. Nonetheless this example illustrates one use of the method: the main unknowns are the shape of the fault and the direction of simple shear, α. A priori knowledge of one of these (from reflections off the fault plane or observations of antithetic faulting in the hanging wall) could be used to predict the other.

Testing structural models

The second example is taken from Wernicke & Burchfiel (1982). Figure 7(a) shows an interpreted seismic section across a normal fault. Wernicke & Burchfiel (1982) also show the same section uninterpreted (their fig. 15), which is remarkable for the clarity of reflections from the fault plane and from the beds within the hanging wall. Note that their interpretation includes numerous sub-parallel small faults in the hanging wall, thus also establishing a likely direcion of simple shear ($\alpha = -15°$). Thus F, B and α are all indicated. Is this interpretation self-consistent?

No stratigraphy was given with this example, and therefore the most difficult parameter to estimate is the heave, h. In the absence of any other information we assume that the top marked reflector, A, is offset from the surface, and that the heave is h, as marked in Fig. 7(b). Since the surface is not horizontal we assume an initial dip of $\gamma_0 = 4°$. θ_0 is estimated as 55°. Three calculated fault shapes are shown in Fig. 7(b), with values of α of -15, 45 and 90°. Clearly, that of 45° agrees best with the observed fault shape, and the value of $-15°$ implied by Wernicke & Burchfiel's interpretation of minor faulting in the hanging wall leads to a poor prediction of the fault shape.

On the face of it a better interpretation would include minor faulting dipping at 45° towards the main fault plane. This is compatible with the uninterpreted observed seismogram, which simply shows severe internal deformation of this part of the hanging wall. In reality of course, with no stratigraphic control on the heave and no corrections for compaction (which may be important, given the obvious growth across the fault and the high deposition rates implied by the non-horizontal surface) or non-uniform vertical exaggeration, this example serves mainly to illustrate the use of the method in testing structural models.

Use of redundant data

It is apparent from the first two examples that the main obstacle to determining the shape of the fault plane

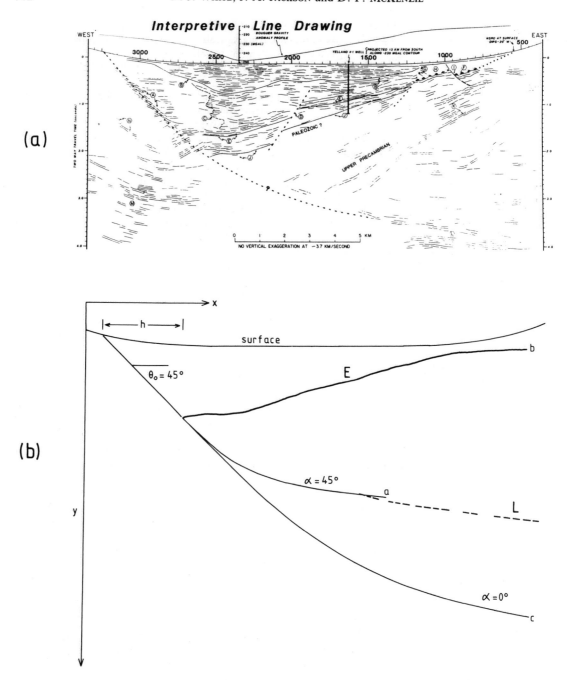

Fig. 6. (a) Line drawing across Spring Valley, east-central Nevada, taken from Gans *et al.* (1985). (b) Shapes predicted for the Schell Creek Fault from the geometry of reflector E shown as lines a ($\alpha = 45°$) and c ($\alpha = 0°$). Note that when $\alpha = 45°$ the easternmost point of reflector E contains no information on the fault geometry beyond point a. The position of reflectors L in part (a) are shown by dashed lines in part (b). Both figures have the same horizontal and vertical scales, with no substantial vertical exaggeration in the top part of the section. The values of θ_0 and α are thus approximately true. Reproduced by permission.

(even if depth-corrected, decompacted sections are available) is the unknown direction of simple shear, α. With one observation (B) and two unknowns (F and α) the problem is indeterminate. What progress may be made if the geometry of more than one bed is observable?

Simple shear in the hanging wall is likely to be accom-

modated by pervasive sub-parallel small faults. Numerous observations suggest that it is easier to continue using an existing fault than to create a new fault, even if the applied stress changes slightly. For this reason the anisotropy or 'grain' imparted to the hanging wall by small faults taking up an early episode of simple shear may well control the direction of simple shear during the

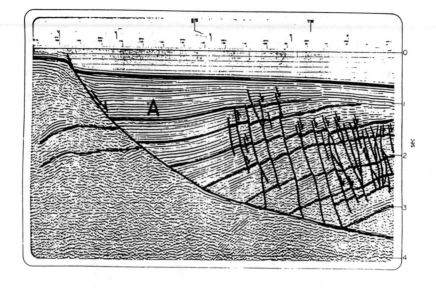

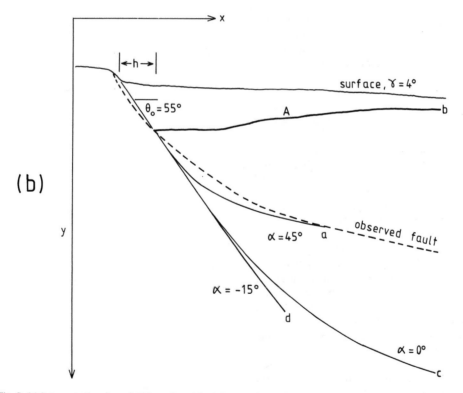

Fig. 7. (a) Interpreted section taken from Wernicke & Burchfiel (1982). (b) Fault shapes predicted from the geometry of bed A using three values of α: -15, 0 and 45°. The vertical and horizontal scales in (a) and (b) are equal. The observed fault is shown by a dashed line in (b). Note that the extent of bed A provides no information on the fault beyond point a when $\alpha = 45°$. For the case of $\alpha = -15°$ the fault extends belows point d, but has been prematurely truncated in this picture. No horizontal scale was given in the original picture of Wernicke & Burchfiel, so vertical exaggeration is uncertain. Reproduced by permission.

subsequent deformation of later, overlying horizons. If this happens, the small faults in the hanging wall will be subparallel at all stratigraphic levels. Since the deformation of the hanging wall is assumed to be by simple shear only, the planes of simple shear are not rotated in the (x', y') frame of the footwall during deformation.

Hence, α should remain constant throughout the deformation of the hanging wall. Since the fault geometry also remains unchanged, observations of two bed geometries should be sufficient to determine the two unknowns F and α. With three or more bed geometries known, the assumption of constant α can be tested, and a formal inversion scheme applied to find the 'best' fault geometry and direction of α which will allow simultaneous fitting of all observed beds. Such a scheme requires both the initial shape, R, of each bed and the displacement on the fault to be known, as there is no reason why they should be the same for different beds.

This approach is illustrated using a section from Bruce (1973; his fig. 7) taken from the Texas coastal area (Fig. 8a). The vertical exaggeration on this section is about 2:1. Is it possible to find a single fault geometry and angle of simple shear, α, that can account for the geometry of both beds A and B? Figures 8(b) and (c) show that the fault geometry suggested by Bruce in Fig. 8(a) can be predicted from both beds, using a common value for α of 26° (true dip 45°), in reasonable agreement with the small hanging wall faults in Fig. 8(a), with $\gamma_0 = 6°$ (true dip 3°) for bed A and $\gamma_0 = 17°$ (true dip 8°) for bed B. These differing values of γ_0 imply that R is not the same for both beds, though their actual values are not dips because of the vertical exaggeration. Note how a value of $\alpha = 0°$ predicts a fault geometry completely different from that suggested by Bruce (Fig. 8d). Once again, since no allowance has been made for variable vertical exaggeration, migration or compaction, this example should be regarded simply as illustrative of the method.

EXTRAPOLATION TO CRUSTAL SCALES

The method of relating fault and sediment geometries described here relies on the footwall remaining undeformed throughout. These ideas are likely to work reasonably well for faults which redistribute the sedimentary cover in a basin, rather than contribute to overall crustal extension. On a crustal scale, where faults penetrate the deep basement and do lead to crustal extension, there are likely to be two difficulties.

The first is that the deeper part of the footwall is likely to experience some form of distributed deformation. This is particularly probable below the maximum depth at which earthquakes nucleate (usually 6–15 km on continents), where, although 'faults' are thought to exist and are seen on some deep reflection profiles, how much motion is concentrated on them and how much is distributed in the blocks either side is unknown. This is discussed further by Jackson (1986). In spite of the uncertainty surrounding the nature of faults in the lower crust, large faults in the upper crust, above the nucleation depths of earthquakes, probably do represent concentrated simple shear with relatively little internal deformation of the footwall. The justification for this statement comes mainly from seismological observations: while aftershocks of major normal faulting earthquakes are common in footwall blocks, their cumulative seismic moment is usually insignificant compared to that of the mainshock. Thus the methods described in this paper should work in the upper parts of large crustal-scale normal faults.

In practice a second difficulty arises: that of large-scale

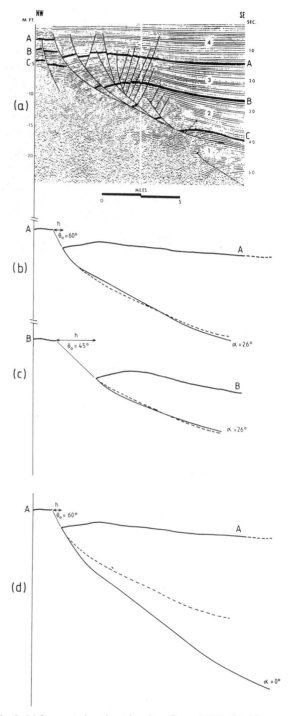

Fig. 8. (a) Interpreted section taken from Bruce (1973). (b)–(d) are drawn with the same vertical and horizontal scales as (a), and show the fault geometries predicted from the shapes of beds A and B using $\alpha = 26°$ (b and c) and $\alpha = 0°$ (d). The fault drawn by Bruce in (a) is shown dashed in (b)–(d). The vertical scale is exaggerated by about a factor of two. Reproduced by permission.

rotation of crustal blocks about a horizontal axis. This is a necessary consequence of trying to stretch the crust (pure shear) by movement on faults (simple shear) and is the justification behind 'domino-style' models of crus-

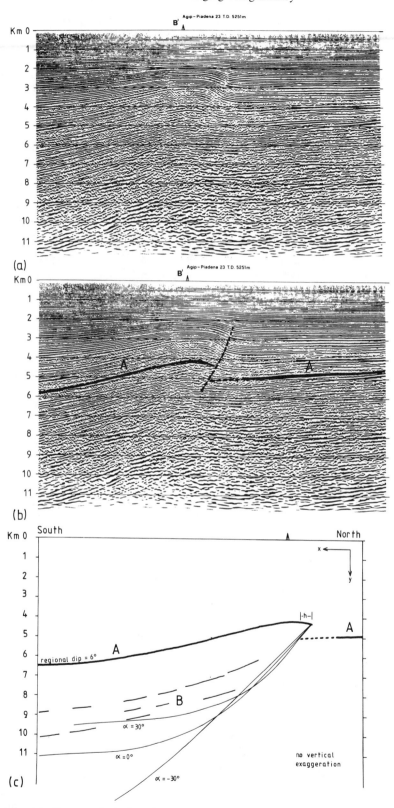

Fig. 9. (a) Migrated and depth corrected seismic section across the Po Plain, Italy, from Peri (1983). The horizontal and vertical scales are equal. (b) and (c) are drawn to the same scale as (a). (b) shows the interpreted position of reflector A, identified as near the top of the Miocene. It is offset by a small thrust fault. (c) Predicted geometry of the thrust with depth calculated from the shape of reflector A, with $\theta_0 = 40°$ and $\gamma_0 = 6°$. Some deeper reflectors, labelled B, are also shown. Three different predicted fault geometries are shown, corresponding to values of α of $-30°$, $0°$ and $+30°$. Reproduced by permission.

tal extension (e.g. Ransome *et al.* 1910, Morton & Black 1975). The rate of this rotation can be calculated if the rate of horizontal extension and the fault block geometry are known (see e.g. Le Pichon & Sibuet 1981; Wernicke & Burchfiel 1982), or can be estimated from stratigraphic arguments. Such rotation means that the footwall itself rotates, and allowance must be made for this when applying the methods used in previous sections.

A NOTE ON THRUST FAULTS

Equations (6) and (11) are equally valid if h is negative and shortening across a thrust fault occurs. The forward problem, of calculating the bed shape given the fault, can be illustrated using the geometry in Fig. 5. Given a bed whose initial shape is $R = h/(1 + x^2)$ and a fault $F = \tan^{-1} x$, what will be the shape of the bed after a small reverse displacement of heave $-h$ on the fault, assuming vertical simple shear ($\alpha = 0°$)?

From (6):

$$B = -h \frac{d}{dx} (\tan^{-1} x) + \frac{h}{1 + x^2} + C$$

or

$$B = \text{constant}$$

as it should.

A real example is provided by Peri (1983) in the Po Plain, N. Italy (his fig. 3.4.1–18). Figures 9(a) and (b) show a migrated, depth-corrected seismic section across a thrust that offsets a horizon labelled A, identified from a borehole as near the top of the Miocene sedimentary section. Figure 9(c), which extends a little further south of Figs. 9(a) and (b), shows the geometry of reflector A and also that of some deeper subparallel reflectors, labelled B. The question may be asked: at what depth (if any) does the thrust fault become parallel to the sedimentary layering?

The answer, of course, depends on the inclination of the simple shear that has led to the formation of the gentle fold in the hanging wall. In Fig. 9(c) three different predicted fault shapes are shown, corresponding to values of α of $-30°$ (pervasive imbricate thrusting in the hanging wall), $0°$ (vertical simple shear) and $+30°$ (pervasive back-thrusting in the hanging wall). In this case no reflectors appear to be continuous across the base of the section, even at the deepest levels, so that perhaps $\alpha < 0°$ (imbricate thrusting) is the most likely. In practice, it is harder to estimate h and θ_0 for thrusts as, unlike in the case of normal faults, abrupt truncations of sedimentary horizons are rarely seen. Figure 9 should therefore be considered only as illustrative of the technique as applied to thrusts.

DISCUSSION

The examples shown here demonstrate that fault shapes can be predicted with some confidence if the

direction of simple shear in the hanging wall is known. The assumption that the hanging wall deforms by simple shear is, of course, central to the derivation of the equations we use. This assumption may not be unrealistic, as it is equivalent to implying that the hanging wall deforms by motion on numerous parallel small faults that do not intersect (of the type illustrated in Fig. 7a). All the examples shown here suggest that this simple shear did not occur on vertical planes, but on planes whose true dip is inclined towards the fault at about 45°. This is consistent with the observation that antithetic faults in hanging walls are rarely vertical, even on the scale of faults that penetrate the entire brittle upper crust and generate earthquakes (see Jackson 1986). A dip of 45° is in the middle of the range of dips observed for seismically active normal faults worldwide, and in the absence of any other information it is probably sensible to assume that $\alpha = 45°$ (true dip) rather than 0° [the value used by Verrall (1981) and Gibbs (1983, 1984)]. Such a difference in the value of α leads not only to a great difference in predicted fault shapes (see Figs. 6–8) but also to a substantial difference in the estimated horizontal extension in the hanging wall. Assuming the faults eventually become horizontal at depth, then simple shear inclined at 45° leads to extensions of $2h$ for horizon E in Fig. 6 and $4h$ for horizon A in Fig. 8: increases of 100 and 300% above that estimated from offset of the bed on the fault alone. In the case of thrust faults, ignoring the presence of inclined simple shear will lead to an underestimate of the amount of shortening in the hanging wall.

Compaction may be an additional complication particularly for normal faults. Its principal effects are discussed in the Appendix. Allowing for compaction will not alter the arguments presented in the main body of this paper concerning the importance of inclined simple shear. However one important observation, summarized in Fig. 10(b), is that differential compaction can lead to a pronounced downwarping of sediments in the hanging wall. This may result in the formation of a 'hanging wall syncline', giving the appearance of 'normal drag' with a long wavelength.

If such hanging wall synclines are used to calculate fault geometry without allowing for compaction, the predicted fault will have a convex-upwards shape (Fig. 11), whereas in fact the syncline is more likely to be due to differential compaction of hanging wall sediments above a planar fault.

CONCLUSIONS

Provided displacements are small and the hanging wall deforms by simple shear (thus preserving cross-sectional area), an analytical expression exists that relates the shape of a normal fault in cross-section to the shape of the bedding horizons in its hanging wall block. The expression also assumes that the footwall remains undeformed throughout. The expression may be integrated provided the boundary conditions are specified.

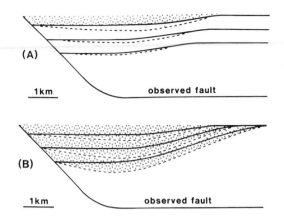

Fig. 10. Bed geometries have been calculated for a given fault with compaction effects taken into account. Solid lines show beds without compaction while dashed lines show beds with compaction. (A) Beds represented by solid lines were deposited, and thus partially compacted, prior to faulting. Motion on the fault causes a depression to form in the hanging wall. This fills with sediment (dotted) causing the original beds to compact further. Note that the deepest bed compacts less than the shallowest one since it was already partially compacted. $\alpha = 45°$, $\phi_0 = 0.6$, $\lambda = 2$ km. (B) All of the beds shown were deposited during faulting. After each increment of slip, the depression formed fills with sediment causing deeper beds, which were originally at the surface, to compact. Note that in (B) the effect of compaction increases with depth whereas in (A) it decreases with depth.

Perhaps the most important result is that the direction of simple shear within the hanging wall block has a very strong influence on the shape of the bedding horizons within it. Graphical techniques relating fault and sediment geometries have been described by Verrall (1981) and Gibbs (1983, 1984), which assume that the simple shear occurs by movement in vertical planes. This condition is clearly a special case, and the examples described here, as well as the observation that minor hanging wall faults are not always vertical, suggests it is not, in general, valid. A substantially different geometry is obtained for the fault if the simple shear planes in the hanging wall are inclined to the vertical, and estimates of the amount of extension in the hanging wall may change by a factor of two or more. Indeed, this study demonstrates that the amount of horizontal extension across a normal fault cannot be estimated simply from the appar-

ent offset of a bed on the fault, if the bed has been deformed in the hanging wall: the dip of the simple shear planes in the hanging wall must also be known.

Although this study has concentrated on normal faulting, the method is equally applicable to thrust and reverse faults, as demonstrated by Fig. 9.

Acknowledgements—We thank Peter Verrall for first drawing our attention to this problem at a lecture he gave for the Joint Association for Petroleum Exploration Courses (JAPEC) in London, and D. I. Rainey for many stimulating discussions. Comments made at an early stage by Graham Yielding were very helpful. This work was supported by NERC. N.J.W. gratefully acknowledges a British Council F.C.O. studentship and generous support from Merlin Profilers Ltd. Cambridge University Department of Earth Sciences contribution no. 712.

REFERENCES

Anderson, R. E., Zoback, M. L. & Thompson, G. A. 1983. Implications of selected subsurface data on the structural form and evolution of some basins in the northern Basin and Range Province, Nevada and Utah. *Bull. geol. Soc. Am.* **94**, 1055–1972.

Bruce, C. H. 1973. Pressured shale and related deformation: Mechanism for development of regional contemporaneous faults. *Bull. Am. Ass. Petrol. Geol.* **57**, 878–886.

De Charpal, O., Guennoc, P., Montadert, L. & Roberts, D. G. 1978. Rifting, crustal attenuation and subsidence in the Bay of Biscay. *Nature (Lond.)* **275**, 706–711.

Dahlstrom, C. D. A. 1969. Balanced cross-sections. *Can. J. Earth Sci.* **6**, 743–757.

Gans, P. B., Miller, E. L., McCarthy, J. & Ouldcott, M. L. 1985. Tertiary extensional faulting and evolving ductile–brittle transition zones in the northern Snake Range and vicinity: new insights from seismic data. *Geology* **13**, 189–193.

Gibbs, A. D. 1983. Balanced cross-section construction from seismic sections in areas of extensional tectonics. *J. Struct. Geol.* **5**, 153–160.

Gibbs, A. D. 1984. Structural evolution of extensional basin margins. *J. geol. Soc. Lond.* **141**, 609–620.

Hamblin, W. K. 1965. Origin of "reverse drag" on the downthrown side of normal faults. *Bull. geol. Soc. Am.* **76**, 1145–1164.

Hobbs, B. E., Means, W. D. & Williams, P. F. 1976. *An Outline of Structural Geology.* John Wiley, New York.

Jackson, J. A. 1986. Active normal faulting and crustal extension. In: *Continental Extension Tectonics. Geol. Soc. Lond, Spec. Publs* in press.

Jackson, J. A. & McKenzie, D. P. 1983. The geometrical evolution of normal fault systems. *J. Struct. Geol.* **5**, 471–482.

Le Pichon, X. & Sibuet, J. C. 1981. Passive margins: a model of formation. *J. geophys. Res.* **86**, 3708–3721.

Magara, K. 1978. *Compaction and Fluid Migration, Devs. petrol. Sci.* **9**, Elsevier, New York.

McKenzie, D. P. 1984. The generation and compaction of partial melts. *J. Petrology* **25**, 713–765.

Morton, W. H. & Black, R. 1975. Crustal attenuation in Afar. In: *Afar Depression of Ethiopia* (edited by: Pilger, A. & Rosler, A.) Interunion commission on Geodynamics, Sci. Rep. No. 14. E. Schweizerbart'sche Verlagsbuchhandlung, Stuttgart, 55–65.

Ransome, F. L., Emmons, W. H. & Garrey, G. H. 1910. Geology and ore deposits of the Bullfrog district, Nevada. *Bull. U.S. Geol. Surv.* **407**, 130 pp.

Peri, M. 1983. Three seismic profiles through the Po plain. In: *Seismic Expression of Structural Styles* (Edited by: Bally, A. W.), *Bull. Am. Ass. Petrol. Geol*, Studies in Geology Series No. 15, **3**, 3.4.1–8-19.

Sclater, J. G. & Christie, P. A. F. 1980. Continental stretching: an explanation of the post mid-Cretaceous subsidence of the Central North Sea basin. *J. geophys. Res.* **85**, 3711–3739.

Smith, R. B. & Bruhn, R. L. 1984. Intraplate extensional tectonics of the eastern Basin-Range: inferences on structural style from seismic reflection data, regional tectonics and thermal–mechanical models of brittle–ductile deformation. *J. geophys. Res.* **89**, 5733–5672.

Steckler, M. S. & Watts, A. B. 1978. Subsidence of the Atlantic margin of New York. *Earth Planet. Sci. Lett.* **41**, 1–13.

Verrall, P. 1981. Structural interpretation with application to North Sea problems. Course notes No. 3, Joint Ass. for Petroleum Exploration Courses (UK).

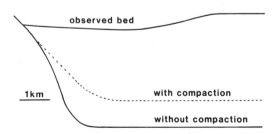

Fig. 11. Illustration of the inverse problem where the fault geometry is calculated given a compacted bed geometry. Solid line shows the fault calculated when compaction is neglected. Note the convex-upwards bulge close to where the bed meets the fault in this case. Dashed line shows the fault geometry obtained with compaction taken into account. Here the upper part of the fault is planar. Parameters as in Fig. 10.

Wernicke, B. & Burchfiel, B. C. 1982. Modes of extensional tectonics. *J. Struct. Geol.* **4**, 105–115.

Wood, R. & Barton, P. 1983. Crustal thinning and subsidence in the North Sea. *Nature (Lond.)* **302**, 134–136.

Ziegler, P. 1983. Crustal thinning and subsidence in the North Sea. *Nature (Lond.)* **304**, 561.

APPENDIX

Unconsolidated sediments generally contain considerable amounts of water (often at least 50% by volume). During burial, this water is lost and the sediment compacts. Such a process will obviously change the shape of a sedimentary horizon within the hanging wall of a normal fault.

The purpose of this appendix is to investigate the importance of this process using expressions which are approximately correct and which should be adequate to illustrate the effects one should expect. Most authors assume that the strain produced by compaction is uniaxial, the axis of shortening being vertical. This assumption is reasonable provided lateral variations in facies and thickness of the sedimentary layers can be neglected. However, it is unlikely to be an accurate description of the strain field in regions undergoing tectonic deformation during sedimentation. The equations governing the behaviour of such a system have recently been developed (McKenzie 1984). Unfortunately they are not easily solved. The principal difficulty is that the compaction rate of the matrix is governed by the pressure of the interstitial water, which is in turn controlled by the water flow within the whole region. Hence local changes in porosity are governed by the behaviour of the whole system. Under these conditions, the porosity cannot be obtained from depth of burial alone.

No attempt has been made here to solve this more general problem. Instead, we have simply modified the standard expressions relating porosity to depth of burial (Magara 1978, Steckler & Watts 1978, Sclater & Christie 1980) so that some of the geometric consequences of compaction can be explored. The resulting eqn (A15) only provides an approximate solution to the full problem and, for this reason, should be applied with care.

Equations developed earlier assume that the hanging wall deforms by simple shear alone. If the strain rate distribution is more complicated, these equations do not hold. This situation is avoided by constraining the strain field produced by compaction to be uniaxial with the axis of shortening parallel to the shear direction within the hanging wall. Under these conditions, the deformation caused by compaction can be treated separately to that caused by movement on the fault. Hence the problem can be solved. It is important to stress that the above constraint has been arbitrarily imposed so that a solution may be obtained with ease. Nevertheless it is unlikely to be any worse than assuming that compaction involves uniaxial shortening which is purely vertical.

The necessary expressions may now be derived. Given that compaction occurs by uniaxial shortening parallel to y', the resultant velocity field is calculated. This is required to be parallel to the fault at every point so that voids do not form at depth. The final result is a first-order differential equation which can be solved by iteration.

The porosity at any depth, d, below a pre-faulting surface of dip γ is

$$\phi = \phi_0 \exp\left\{-\frac{d}{\lambda}\right\} \tag{A1}$$

where

$$d = y - x \tan \gamma, \tag{A2}$$

ϕ_0 is the initial porosity and λ is a constant governing the change of porosity with depth. As before, the (x', y') co-ordinate frame is rotated through an angle, α, with respect to the (x, y) frame. Therefore the porosity at any point in the (x', y') frame is

$$\phi' = \phi_0' \exp\left\{-\frac{y'}{\lambda'}\right\}, \tag{A3}$$

where

$$\phi_0' = \phi_0 \exp\left\{-\frac{x' \sin(\alpha - \gamma)}{\lambda \cos \gamma}\right\} \tag{A4}$$

and

$$\lambda' = \frac{\lambda \cos \gamma}{\cos(\alpha - \gamma)}. \tag{A5}$$

As before, the fault, $F'(x')$, is considered fixed in the (x', y') co-ordinate frame. We determine the movement of the bed, given by $B'(x_0', t)$ where x_0' is the initial value of x' for some point on the bed. A Lagrangian reference frame x_0' is then used to follow the movement of a point on the bed. By differentiating with x_0' fixed, the velocities in the x' and y' directions are

$$U_0 = \left(\frac{\partial x'}{\partial t}\right)_{x_0'}, \qquad v = \left(\frac{\partial B'}{\partial t}\right)_{x_0'}, \tag{A6}$$

respectively. Neither $F'(x')$ nor $\phi_0'(x')$ are functions of t in the (x', y') frame. Differentiation of $F'(x')$ thus gives

$$dF' = \frac{dF'}{dx'}\, dx'.$$

Therefore

$$\left(\frac{\partial F'}{\partial t}\right)_{x_0'} = U_0 \frac{dF'}{dx'} \tag{A7}$$

similarly

$$\left(\frac{\partial \phi_0'}{\partial t}\right)_{x_0'} = U_0 \frac{d\phi_0'}{dx'}, \tag{A8}$$

where U_0 is the x' component of the velocity (constant within the hanging wall).

Both (A7) and (A8) take account of the effect of compaction on the velocity field.

The volume of solid material, V_s, in a vertical section between the bed and the fault is

$$V_s = \int_{B'}^{F'} (1 - \phi')\, dy'. \tag{A9}$$

Substitution of (A3) into (A9), followed by integration yields

$$V_s = F' - B' + \phi_0' \lambda' \left\{\exp\left(-\frac{F'}{\lambda'}\right) - \exp\left(-\frac{B'}{\lambda'}\right)\right\} \tag{A10}$$

Since compaction is uniaxial in the y' direction, V_s must remain constant in a frame fixed to the hanging wall. Therefore

$$\left(\frac{\partial V_s}{\partial t}\right)_{x_0'} = 0. \tag{A11}$$

Note that the condition

$$\frac{\partial V_s}{\partial t} = 0 \tag{A12}$$

is not satisfied because in a frame fixed to (x', y'), and thus to the footwall, the volume of sediment between the bed and the fault at any given value of x', must change as the hanging wall is displaced. In an extreme case, when the hanging wall moves far enough, the point where the bed meets the fault passes the chosen value of x' and there is no material left between the fault and the bed. Clearly conservation of sediment volume only occurs in a frame fixed to the hanging wall (x_0'). Therefore (A11) is the correct condition to impose. Differentiation of (A10) gives

$$\left(\frac{\partial F'}{\partial t}\right)_{x_0'} - \left(\frac{\partial B'}{\partial t}\right)_{x_0'} - \phi_0'\left\{\left(\frac{\partial F'}{\partial t}\right)_{x_0'}\exp\left(-\frac{F'}{\lambda'}\right) - \left(\frac{\partial B'}{\partial t}\right)_{x_0'}\exp\left(-\frac{B'}{\lambda'}\right)\right\}$$
$$+ \left(\frac{\partial \phi_0'}{\partial t}\right)_{x_0'}\lambda'\left\{\exp\left(-\frac{F'}{\lambda'}\right) - \exp\left(-\frac{B'}{\lambda'}\right)\right\} = 0. \tag{A13}$$

As before, movement should be parallel to the fault. This gives, from (6),

$$v\, \delta t = B' - R' - C', \qquad U_0\, \delta t = h'. \tag{A14}$$

Substitution of (A4), (A6), (A7), (A8) and (A14) into (A13) then yields

$$\frac{dF'}{dx'} = \frac{(B' - R' - C')\{1 - \phi_0' \exp(-B'/\lambda')\}}{h'\{1 - \phi_0' \exp(-F'/\lambda')\}}$$
$$- \frac{\phi_0' \tan(\alpha - \gamma)\{\exp(-B'/\lambda') - \exp(-F'/\lambda')\}}{\{1 - \phi_0' \exp(-F'/\lambda')\}}. \tag{A15}$$

This first-order differential equation can be solved by iteration either for F' when B' is given, or for B' when F' is given. When $\phi_0' = 0$, (A15) reduces to a previously derived expression. This provides an initial solution to (A15). Only three or four iterations are then required to find the correct solution since convergence is rapid.

Examples and implications

Figure 10 illustrates the effect of compaction on two different depositional situations. In Fig. 10(a) the beds were deposited, and hence partially compacted, prior to the onset of faulting. As a result of faulting, beds close to the fault move to deeper levels than those further away in the hanging wall and are buried by young sediments deposited in the depression adjacent to the fault. Therefore, beds close to the fault compact more than those further away in the hanging wall block. The effect on bed geometry is greater for shallow beds since deeper beds lost much of their porosity prior to faulting.

Figure 10(b) shows the effects of compaction on beds that were all deposited during faulting. In this case compaction leads to a pronounced downwarping of beds adjacent to the fault resulting in the creation of a 'hanging wall syncline'. This is similar in shape to what is often described in the lierature as 'normal drag' (Hamblin 1965, Hobbs *et al.* 1976), except that it is on a longer wavelength.

It is important to note that if such synclines are used to infer fault geometry without allowing for compaction, a convex-upwards fault is predicted (Fig. 11). In fact, it is more likely that hanging wall synclines arise due to differential compaction above an initially planar fault. The presence of features similar to those illustrated in Figs. 10 and 11 is probably a good indication that the effects of compaction are significant and should be allowed for.

BULLETIN OF THE AMERICAN ASSOCIATION OF PETROLEUM GEOLOGISTS
VOL. 38, NO. 5 (MAY, 1954), PP. 854-877, 23 FIGS.

SUBSURFACE INTERPRETATION OF INTERSECTING FAULTS AND THEIR EFFECTS UPON STRATIGRAPHIC HORIZONS[1]

GEORGE DICKINSON[2]
Houston, Texas

ABSTRACT

The effects of intersecting faults in the subsurface are most difficult to comprehend. A graphical treatment for the study of hypothetical combinations of faults is developed whereby it is hoped more realistic subsurface interpretations may be deduced for complexly faulted structures.

INTRODUCTION

Faulting is the most frequent complication encountered in subsurface geological studies. A thorough understanding of it is therefore of utmost importance to the geologist engaged in the exploration for, and the exploitation of, oil and gas. The effects of intersecting faults in subsurface geology are most difficult to comprehend. In complexly faulted structures there are rarely adequate data available from wells, even after development drilling is complete, from which a correct interpretation of the fault pattern may be derived. Long experience frequently plays an important part in the interpretation of such structures, and it has been far from easy to pass on this intangible knowledge to young geologists.

The writer has been teaching subsurface geological methods to groups of graduate geologists as part of their initial training period in the petroleum industry and during the last 5 years has gradually devised special methods to facilitate a better understanding of intersecting faults by his students. The study of certain hypothetical combinations of intersecting faults and bedding planes soon appeared to offer the best method of approach to this problem. There is usually much difficulty in visualizing three dimensions from two-dimensional drawings; however, prior study of simple block models illustrating similar conditions was found to solve this problem.

The purpose of this paper is to explain the methods used for illustrating intersecting faults and their effects on stratigraphic horizons, and to point out some of the unexpected results obtained in these hypothetical combinations. It is hoped thereby to enable more realistic subsurface interpretations to be deduced for complexly faulted structures.

METHODS OF STUDY

INTERSECTING FAULTS ON EROSION SURFACE OR UNCONFORMITY

Fault-contour maps are the most satisfactory method of showing the attitude of fault planes. Such a map of two intersecting faults is shown in Figure 1-c.

[1] Manuscript received, December 3, 1952; revised manuscript, February 1, 1954. Published by permission of the Shell Oil Company.

[2] Chief production geologist, Technical Services Division, Shell Oil Company. The writer expresses appreciation to the management of the Shell Oil Company for permission to publish this paper. Thanks are also due to L. D. Hillyer and C. F. Martin for their valuable assistance in the original work which made this paper possible.

854

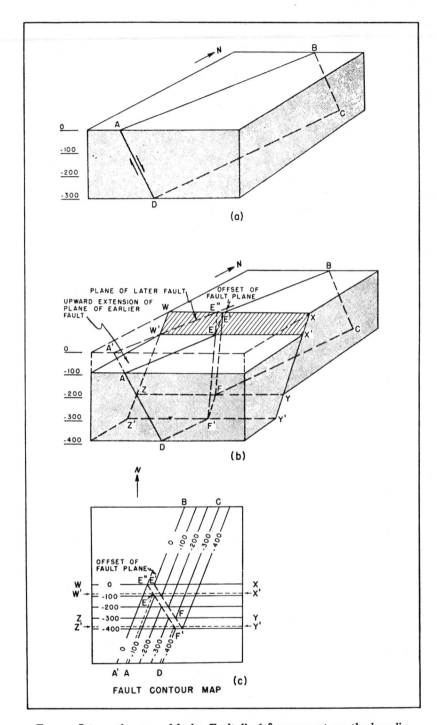

FIG. 1.—Intersecting normal faults. Fault dip 60°, movement exactly downdip.

The conditions shown in this map are illustrated in the block diagram above it. The original fault plane ABCD (Fig. 1-a) has been cut by a later fault plane WXYZ, along which movement was exactly down its dip (Fig. 1-b). If the downthrown part of the earlier fault plane AEF'D is extended upward to a level equivalent to the upper surface of the upthrown block, that is to A'E," it is clear that the fault line E'B will be offset from the fault line A'E" in a direction down the dip of the earlier fault.

The horizontal projections in Figure 2 illustrate conditions where various combinations of faults, intersecting at different angles, meet the surface or an unconformity. The intersection of the intersected fault and the surface is invariably offset in the direction of its dip in the block upthrown by the intersecting fault, irrespective of the angle of intersection of the faults providing the direction of movement of the intersecting fault is dip slip.

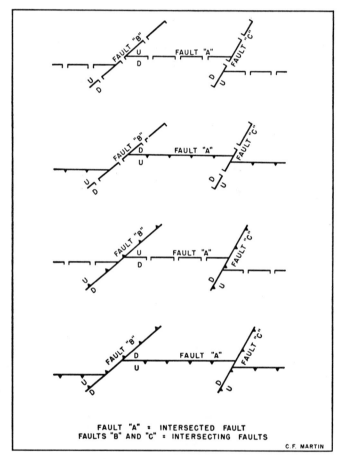

FIG. 2.—Intersecting normal and reverse faults. Direction of offset of intersected fault on erosion surface or at surface as seen on map or in horizontal projection. Dip-slip movement. Hachures on normal faults and arrow points on reverse faults point down slope of fault planes.

INTERSECTING FAULTS IN HORIZONTAL BEDS

The effects of intersecting faults in the subsurface are much more complicated than is apparent at the surface. Two intersecting faults cut the beds into four blocks and result in a vertical displacement which may be different for each block. It is evident, therefore, that the intersection of the faults and any stratigraphic horizon may be at four different subsurface elevations (Fig. 3). The diagrams are similar to those in Figure 1 and have a similar lettering system in order to facilitate comparison. The original block consists of six horizontal beds

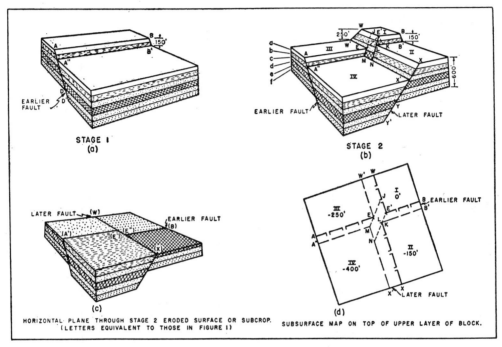

FIG. 3.—Intersecting normal faults in horizontal strata. Fault dip 60°, movement exactly downdip.

each 100 feet thick which have been cut by the fault plane ABCD and displaced 150 feet, as shown in stage 1. The faulted block of stage 1 has been cut by a later fault WXYZ, having a throw[3] of 250 feet to give the conditions shown in stage 2. The movement of both faults is assumed to be directly downdip so that point E′ in the upthrown block is equivalent to point E in the downthrown. The intersections of the earlier fault on the top of the upper layer of the blocks, AE and E′B, are in one and the same straight line despite the fact that the plane of the earlier fault is offset by the later one as shown in Figure 3-c. The upthrown intersections of the later fault and the same horizon are WE′ and LX, and it is obvi-

[3] The term "throw" in this paper represents the amount of section missing or repeated as is normally determined in approximately vertical oil wells. If the beds dip appreciably it is not the same as the vertical throw as defined by Billings in his textbook, *Structural Geology*, but is convenient to use since it represents a measured interval determined by correlation of well logs.

ous from Figure 3-b that these lines are offset by an amount equal to LK, even though the plane of the second fault goes straight through without interruption. A subsurface structure-contour map of the top of the block of formations is shown in Figure 3-d.[4] Thus, in this case, the offsetting of the intersections of the faults and the horizon in the subsurface map is the opposite of that shown for the same faults in a surface geological map or a subcrop map of an erosional unconformity of the type shown in Figure 3-c.

The conditions resulting from the intersection of reverse and combinations of reverse and normal faults in horizontal beds are shown by similar block diagrams and geologic maps in Figures 4, 5, and 6.

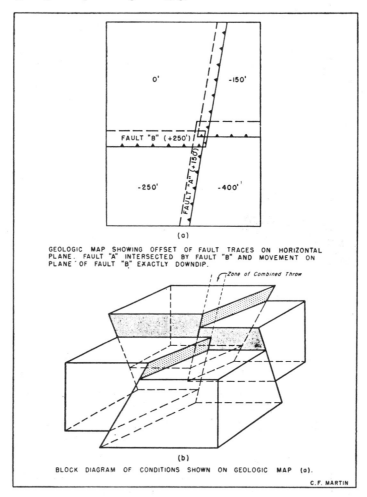

(a)

GEOLOGIC MAP SHOWING OFFSET OF FAULT TRACES ON HORIZONTAL PLANE. FAULT "A" INTERSECTED BY FAULT "B" AND MOVEMENT ON PLANE OF FAULT "B" EXACTLY DOWNDIP.

(b)

BLOCK DIAGRAM OF CONDITIONS SHOWN ON GEOLOGIC MAP (a).

C.F. MARTIN

FIG. 4.—Intersecting reverse faults in horizontal strata. Fault dip 60°, movement exactly dip slip.

[4] It is not usual to show contour lines of faults on the subsurface structure maps used in the petroleum industry. These contours if required are drawn on a separate transparent overlay so that they may be used in conjunction with structure maps of different horizons.

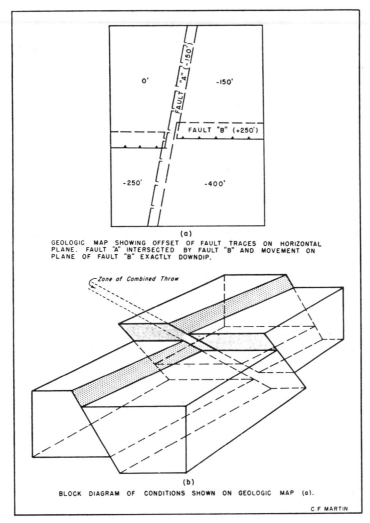

(a)

GEOLOGIC MAP SHOWING OFFSET OF FAULT TRACES ON HORIZONTAL PLANE. FAULT "A" INTERSECTED BY FAULT "B" AND MOVEMENT ON PLANE OF FAULT "B" EXACTLY DOWNDIP.

(b)

BLOCK DIAGRAM OF CONDITIONS SHOWN ON GEOLOGIC MAP (a).

C.F. MARTIN

FIG. 5.—Normal fault intersected by reverse fault in horizontal strata. Fault dip 60° movement exactly dip slip.

ZONE OF COMBINED THROW

The term "zone of combined throw" had been applied to the segment of the intersecting fault which lies between the offset limbs of the intersected fault. In this zone a well will encounter only one fault, for example, in the rhomboid JE′NM of Figure 3-d or in its upward or downward extension. The throw of this single fault is a function of many variables, including the dips of the formation, and the throws and relative movements of the individual faults. In the special case shown in Figure 3-d it is equal to the sum of the throws of the intersecting faults. Thus at point N, a well would pass from the top of bed (a) to the top of bed (e); that is, beds a, b, c, and d, or 400 feet of section, would be missing. The

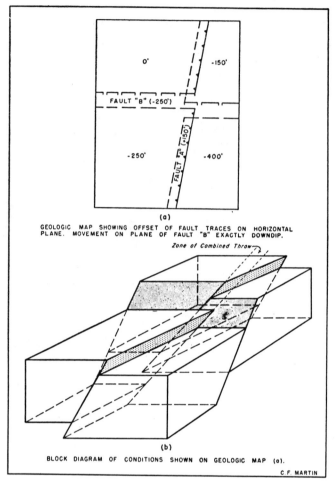

0'   -150'

FAULT "B" (-250')

-250'   -400'

(a)
GEOLOGIC MAP SHOWING OFFSET OF FAULT TRACES ON HORIZONTAL
PLANE. MOVEMENT ON PLANE OF FAULT "B" EXACTLY DOWNDIP.

Zone of Combined Throw

(b)
BLOCK DIAGRAM OF CONDITIONS SHOWN ON GEOLOGIC MAP (a).

C.F. MARTIN

Fig. 6.—Reverse fault intersected by normal fault in horizontal strata. Fault dip 60°,
movement exactly dip slip.

schematic cross sections in Figure 7, showing the displacements of intersecting normal, reverse, and combinations of normal and reverse faults, with horizontal beds, give an indication of the complexity of structural interpretation when a zone of combined throw has been drilled.

The effects of intersecting normal and reverse faults are illustrated in more detail in Figures 8–11. The upper diagrams are contour maps depicting the interruption of a dipping horizon "O" by various combinations of intersecting normal and reverse faults.[5] The cross sections in the lower diagrams of each figure have been drawn to show the structural conditions more clearly and horizon "P" has

[5] It can readily be demonstrated when the formation strike is the same on both sides of a fault that the throw of the fault is represented by the numerical difference between the contours in the line of strike and not at right angles across the fault.

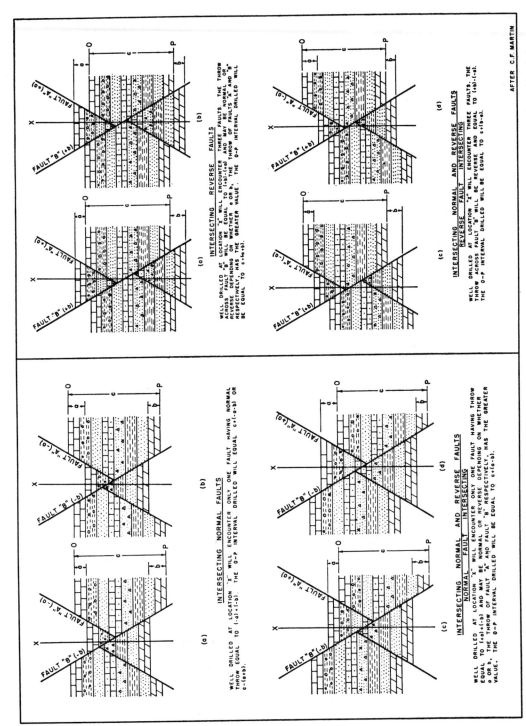

FIG. 7.—Effects of relative throws of intersected and intersecting faults on apparent displacement of strata across zone of combined throw.

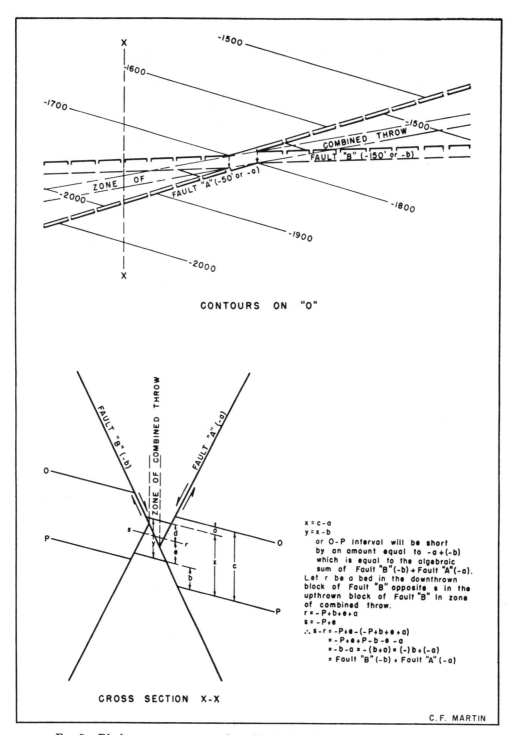

FIG. 8.—Displacement across zone of combined throw for intersecting normal faults.

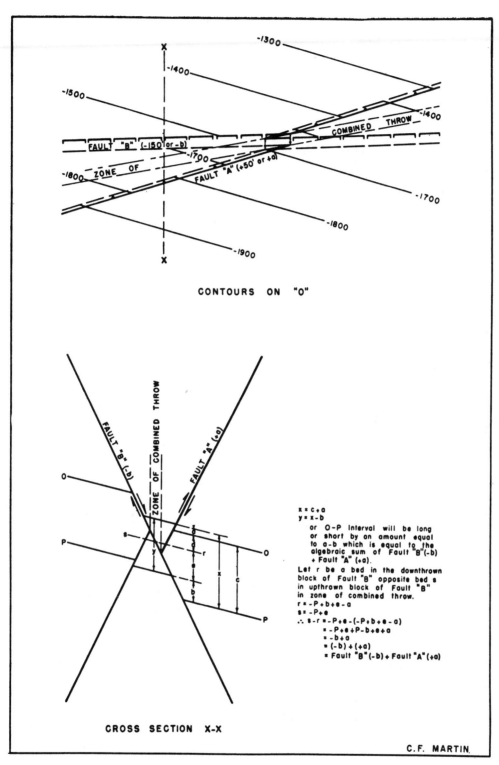

FIG. 9.—Displacement across zone of combined throw for reverse fault
intersected by normal fault.

577

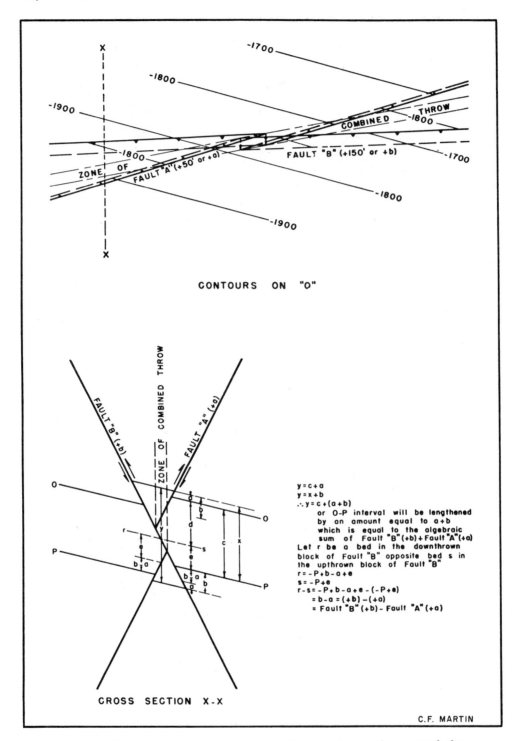

Fig. 10.—Displacement across zone of combined throw for intersecting reverse faults.

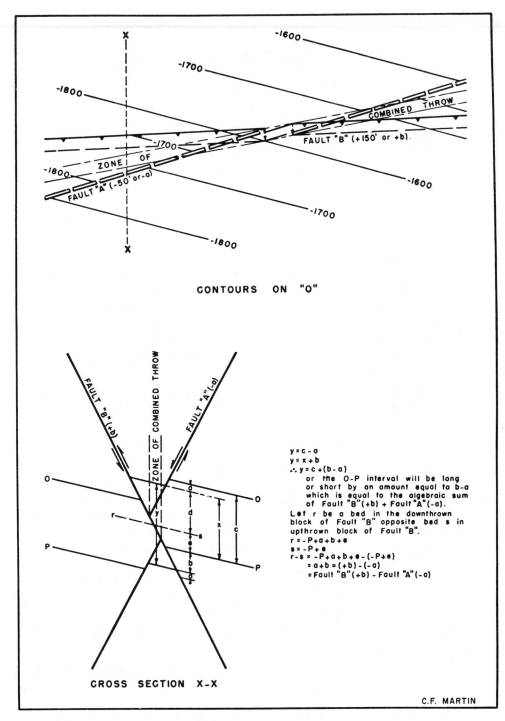

CONTOURS ON "O"

CROSS SECTION X-X

$y = c - a$
$y = x + b$
$\therefore\ y = c + (b - a)$
 or the O-P interval will be long
 or short by an amount equal to b-a
 which is equal to the algebraic sum
 of Fault "B"(+b) + Fault "A"(-a).
Let r be a bed in the downthrown
block of Fault "B" opposite bed s in
upthrown block of Fault "B".
$r = -P + a + b + e$
$s = -P + e$
$r - s = -P + a + b + e - (-P + e)$
 $= a + b = (+b) - (-a)$
 $=$ Fault "B"(+b) $-$ Fault "A"(-a)

C.F. MARTIN

FIG. 11.—Displacement across zone of combined throw for normal fault
intersected by reverse fault.

been drawn at a distance "c" below "O" for convenience in determining the effects of the intersecting faults. The cross sections show that the "O-P" interval in the zone of combined throw is invariably equal to the normal "O-P" interval plus the algebraic sum of the fault throws.[6] These figures should be used in conjunction with the block diagrams in Figures 3, 4, 5, and 6 for studying the conditions prevailing in the zone of combined throw.

When the intersecting fault is normal (Figs. 8 and 9), the cross sections show that the throw of the intersecting fault in the zone of combined throw is equal to the algebraic sum of the throws of the intersecting and intersected faults, that is, it equals (−b) plus (±a). It is evident that, when both faults are normal, the throw across the intersecting fault in the zone of combined throw will also be normal. However, when the intersected fault is reverse the throw across the zone of combined throw will be normal only when the intersecting fault has the greater throw and will be reverse when the intersected reverse fault has the greater throw.

The conditions prevailing when the intersecting fault is reverse are shown in Figures 10 and 11. In these cases the throw across the intersecting fault in the zone of combined throw is equal to the algebraic difference between the throws of the intersecting and intersected faults, that is, it equals (+b) minus (±a). It is evident, therefore, that the apparent throw across the zone of combined throw will be normal when the intersected fault is reverse and has a throw greater than that of the intersecting fault. When the intersected fault is reverse and has a throw smaller than that of the intersecting fault and when it is normal, the throw across the zone of combined throw is always reverse for a reverse intersecting fault.

GRAPHICAL CONSTRUCTION OF EFFECTS OF INTERSECTING NORMAL FAULTS AND DIPPING STRATA

A graphical construction for illustrating all variations of dip of strata and faults and of angle of intersection of faults is simple but necessitates much care in order to avoid errors. The basic steps are shown in the series of diagrams in Figures 12–16. The solid lines represent contours on the intersecting fault B. The broken lines in Figure 12 are contours of the intersected fault A. The dotted lines in Figure 13 are contours on a stratigraphic horizon X in the structurally highest block, and their intersections with the appropriate fault contours give the upthrown fault intersections of faults A and B in block I. To simplify the construction, it is assumed that faults A and B have throws of 200 feet and 400 feet, respectively. The contours on horizon X and three arms of the intersecting fault intersections shown in Figure 14 can be constructed by normal methods. The position of the intersections of the intersected fault A on horizon X on the

---

[6] The throw of a normal fault is negative (section cut out) and that of a reverse fault is positive (section repeated).

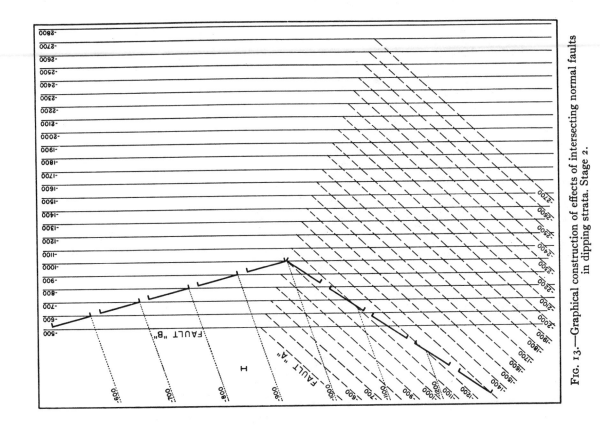

FIG. 13.—Graphical construction of effects of intersecting normal faults
in dipping strata. Stage 2.

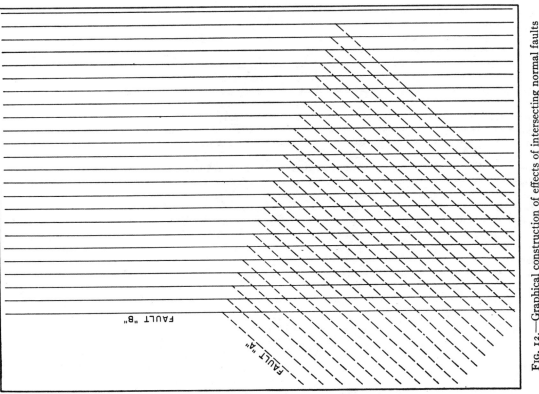

FIG. 12.—Graphical construction of effects of intersecting normal faults
in dipping strata. Stage 1.

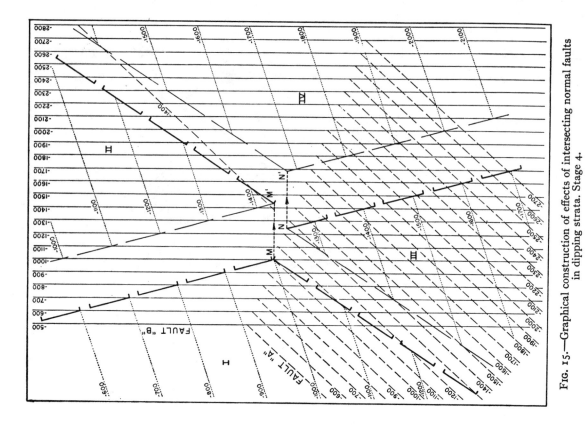

Fig. 15.—Graphical construction of effects of intersecting normal faults in dipping strata. Stage 4.

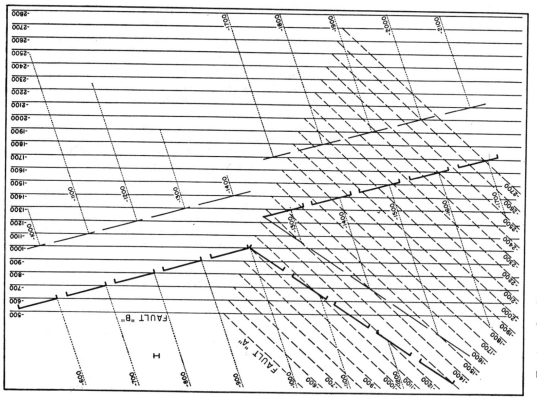

Fig. 14.—Graphical construction of effects of intersecting normal faults in dipping strata. Stage 3.

downthrown side of fault B must now be determined (Fig. 15). Assuming that the movement on fault B is directly down the dip, then point M must move along a line perpendicular to the fault contours until this line meets the downthrown intersection of fault B and horizon X at point M'. Similarly, point N must move to point N'. It is axiomatic that the desired intersections of fault A and the horizon must meet the intersections of fault B at M' and N'. Since fault A is assumed to be straight, its intersections with horizon X on the downthrown side of fault B will be parallel with its intersections on the upthrown side of fault B and can be drawn from M' and N' as shown in Figure 15. These lines, therefore, represent the intersections of fault A and horizon X on the upthrown and downthrown sides of the fault in blocks II and IV. The points, where the minus 1,400-feet contours on horizon X meet the intersections of the fault and the horizon, must lie on the minus 1,400-feet contour on fault A (Fig. 15). Since the fault is straight and has constant dip, the contours will be equally spaced and parallel through the other intersection points, as shown in Figure 16. These contours must stop, of course, at the equivalent contours on fault B and so delineate the zone of combined throw. The photographs in Figure 17a are oblique and vertical views of a block model of conditions similar to those shown graphically in Figure 16.

EFFECTS OF VARIOUS HYPOTHETICAL COMBINATIONS OF INTERSECTING NORMAL FAULTS AND DIPPING STRATA

Dip-slip movement.—Similar diagrams for different angles of intersection between the faults show that the intersection of the intersecting fault and the horizon is invariably offset in the direction of its dip on the downthrown side of the intersected fault, and that when the fault throws and the fault and horizon dips are unchanged, the amount of offset remains constant for all angles of fault intersection. On the other hand, the apparent vertical displacement of the intersected fault plane and the offset of its intersections with the contoured horizon vary as the angle between the faults changes. If the intersected fault A is rotated to reduce the obtuse angle between the faults bounding block I, there is an intermediate position where the intersection of fault A and horizon X is perpendicular to the strike of the intersecting fault B, as shown in Figure 18. The intersection of the intersected fault and the horizon is then in line with the direction of movement on the dip-slip intersecting fault and, therefore, it will not be offset on either side of the intersecting fault, and the apparent vertical displacement of the intersected fault will be equal to that of the contoured horizon. Where the angle between the faults in block I is less than that required for no offset of the intersection of the intersected fault and the horizon, for example, the acute angle shown in Figure 19, the direction of offset of the intersections of the intersected fault and the horizon is reversed; that is, it is offset in the direction of its dip on the upthrown side of the intersecting fault instead of on the downthrown side as in the case of the obtuse angle shown in Figure 16. Moreover, the

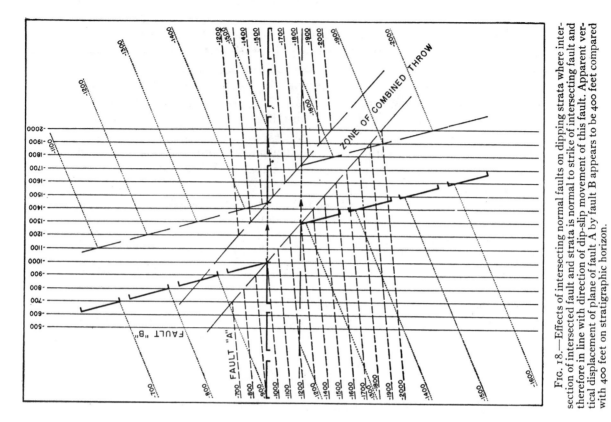

FIG. 18.—Effects of intersecting normal faults on dipping strata where intersection of intersected fault and strata is normal to strike of intersecting fault and therefore in line with direction of dip-slip movement of this fault. Apparent vertical displacement of plane of fault A by fault B appears to be 400 feet compared with 400 feet on stratigraphic horizon.

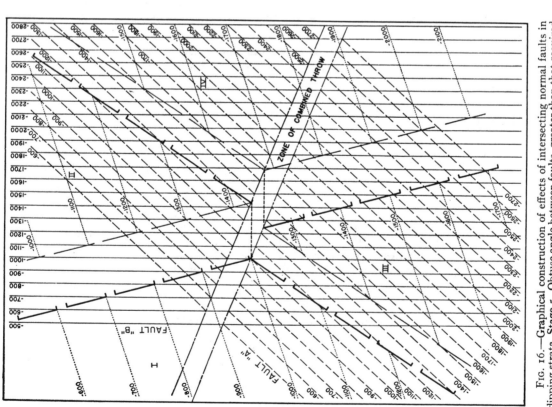

FIG. 16.—Graphical construction of effects of intersecting normal faults in dipping strata. **Stage 5.** Obtuse angle between faults, greater than that required for conditions shown in Figure 18. Dip-slip fault movement. Apparent vertical displacement of plane of fault A by fault B appears to be 120 feet compared with 400 feet on stratigraphic horizon.

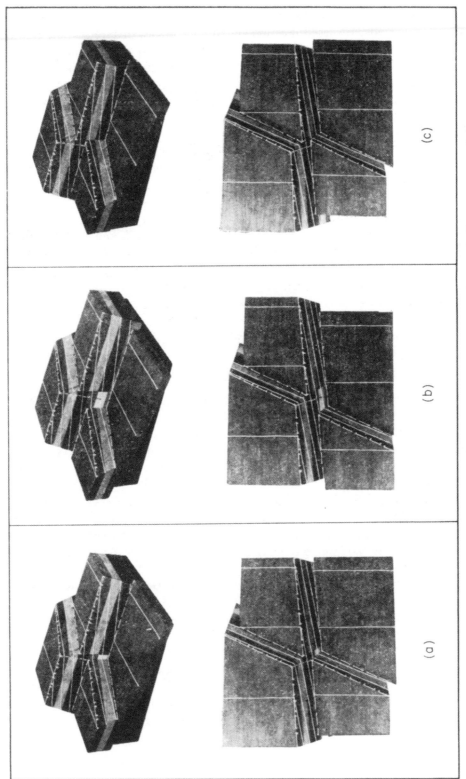

FIG. 17.—Photographs of block models illustrating conditions similar to those shown in Figures 16, 20, and 21.

fault plane is displaced vertically by an amount greater than the contoured horizon compared with less than this amount for angles of intersection greater than that shown in Figure 18. For example, it is evident that the minus 1,000-feet contour on fault A in the block upthrown by fault B is on strike with the minus 1,120-feet contour on fault A in the downthrown block in Figure 16 and with the minus 1,660-feet contour on fault A in Figure 19, representing vertical displacements of 120 feet and 660 feet, respectively, compared with 400 feet displacement for horizon X in both cases.

Oblique-slip movement.—In the case of oblique-slip faults, the intersections of the intersected fault and the horizon are not offset wherever the movement along the intersecting fault is in a direction parallel with it, as shown in Figures 20 and 17b.

The zone of combined throw occurs in most combinations and is used as a clue to the fault pattern prevailing. However, this zone will vanish if the direction of movement along the intersecting fault should be oblique directly along the line of intersection of the faults, as shown in Figure 21, and the photographs of a block model showing similar conditions, as shown in Figure 17c. The same may also happen for true dip-slip faults, if the line of intersection of the fault planes should be perpendicular to the strike of the intersecting fault. In such cases, the intersected fault plane appears to be continuous, and there is no apparent displacement of it by the intersecting fault.

IMPORTANT FAULT RELATIONS RECOGNIZED

The effects of various hypothetical combinations of intersecting normal and reverse faults based on the foregoing discussion are summarized graphically in Figure 22. Diagrams for normal faults intersected by reverse and for reverse faults intersected by normal ones show similar effects but are not illustrated here. A study of these diagrams leads to the recognition of the following important fault relations.

The *intersections of the intersected fault* and the horizon are always offset horizontally and vertically, except in two special cases where the offset is entirely vertical. This occurs where either the beds are horizontal and the faults intersect at right angles or the intersections of the intersected fault and the horizon are perpendicular to the strike of the intersecting fault.

The apparent vertical displacement of the intersected fault plane is different from the apparent vertical displacement of the horizon, except in the special case where the intersections of the intersected fault and the horizon are not offset.

The direction of offset of the intersections of the intersected fault and the horizon and the apparent vertical displacement of the intersected fault plane is complicated by the angle between the intersecting faults and the direction of movement along the fault planes, so that no apparent simple relationship exists.

The *intersections of the intersecting fault* and the horizon are invariably offset

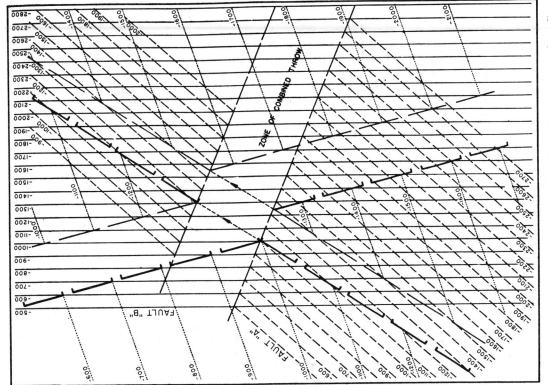

FIG. 20.—Effect of intersecting normal faults on dipping strata where direction of movement on intersecting fault is parallel with intersection of intersected fault and strata. Apparent vertical displacement of plane of fault A by fault B appears to be 400 feet compared with 400 feet on stratigraphic horizon.

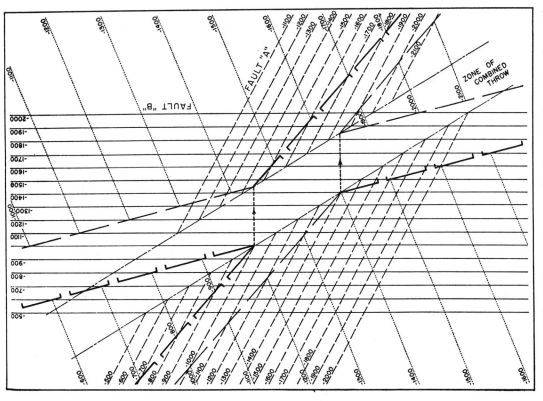

FIG. 19.—Effect of intersecting normal faults on dipping strata where angle between faults is less than that required for conditions shown in Figure 18. Apparent vertical displacement of plane of fault A by fault B appears to be 660 feet compared with 400 feet on stratigraphic horizon.

587

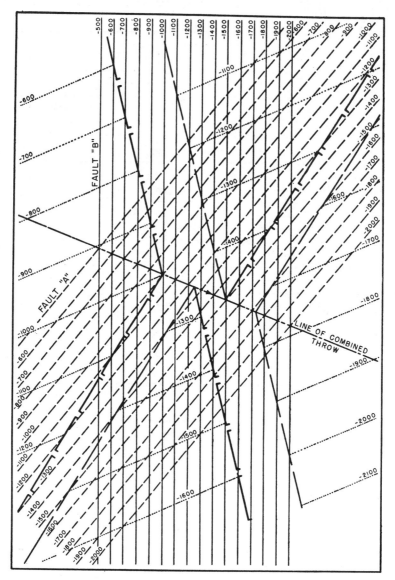

FIG. 21.—Effect of intersecting normal faults on dipping strata where direction of movement is along line of intersection of faults. Apparent vertical displacement of plane of fault A by fault B appears to be o feet compared with 400 feet on stratigraphic horizon.

and the amount of offset is constant regardless of whether the faults are normal or reverse, and of the angle of intersection between the faults as long as the dips of the faults and the horizon, the strike of the horizon and the fault throws remain constant. This offset is in the direction of the fault dip in the block downthrown by the intersected fault.

The apparent displacement of an intersecting fault in the zone of combined

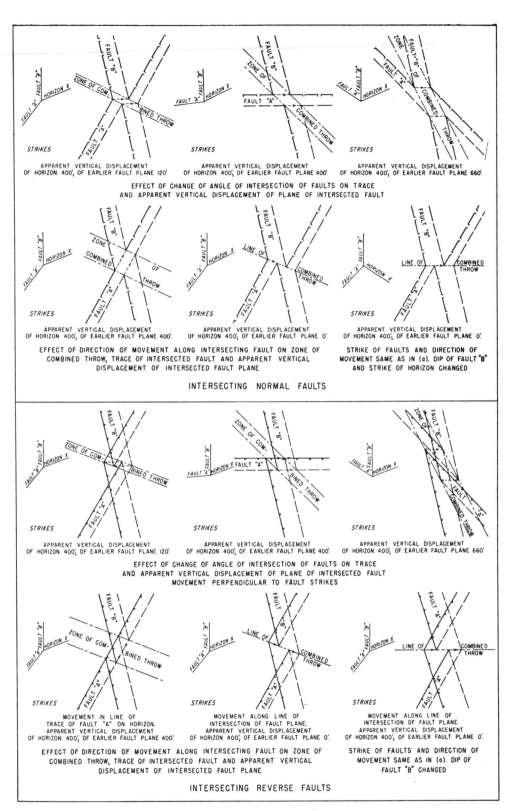

FIG. 22.—Summary of effects of intersecting faults. Fault B with throw of 400 feet intersects fault A which has throw of 200 feet.

throw may be in a direction opposite to the true displacement; for example, when a reverse fault intersects a normal fault of greater throw.

The zone of combined throw becomes a line where the movement down the intersecting fault plane exactly coincides with the line of intersection of the two faults. In this case there is no apparent vertical displacement of the intersected fault plane, and the contours on this plane cross the intersecting fault plane without interruption.

These conclusions are of the utmost importance in the structural interpretation of complexly faulted structures such as salt domes. Further complications may be introduced by the lack of adequate subsurface control, the presence of unconformities, and the possibility of the presence of transcurrent faulting.

Figure 23 is typical of the more complexly faulted type of deep-seated salt dome. Subsurface information indicated fault 3 to be a through fault so that in order to test the upthrown block on the east at the optimum structural posi-

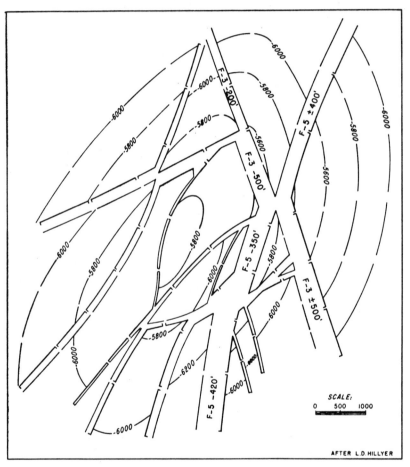

FIG. 23.—Structure-contour map of deep-seated salt dome in Texas showing intersecting faults.

tion it was necessary to know the probable location of the intersected fault 5. The position shown was deduced by the use of the methods discussed in this paper.

Geologic data obtained from the development of the pre-Permian formations in West Texas and deeper drilling in the Pennsylvanian of southern Oklahoma have shown the presence of both normal and reverse faulting closely associated within the same structure. Application of the results of this study of intersecting faults has aided the understanding of some of these structural complexities.

It is hoped the foregoing discussion of the effects of various hypothetical combinations of intersecting faults will facilitate a more accurate structural interpretation of complexly faulted structures and thereby lead to the discovery of additional hydrocarbon accumulations.

Reprinted by permission of the Geological Society of America
from C. A. Bengtson, *Geology*, v. 8, no. 12 (1980), p. 599-
602.

Structural uses of tangent diagrams

C. A. Bengtson
Chevron U.S.A. Inc.
San Francisco, California 94119

ABSTRACT

Tangent diagrams are polar coordinate graphs on which the attitude of planes and lines is represented by the end point of vectors, proportional in length to the tangent of the angle of dip. They provide convenient and easily visualized vectorial solutions for such problems as finding apparent dip from true dip, true dip from two apparent dips, and the line of intersection of two planes. In addition, they have proved to be especially useful for orienting cylindrical and conical folds by graphic analysis of dip data and distinguishing cylindrical folds from the two possible kinds of conical folds. Dip measurements at random locations on cylindrical folds define straight-line "statistical" patterns on tangent diagrams. Dip data for conical folds, however, define two kinds of curved lines corresponding, respectively, to the two possible kinds of conical folds. Lines concave toward the center identify conical anticlines that narrow up-plunge (or synclines that narrow downplunge) and lines concave away from the center define conical anticlines that narrow downplunge (or conical synclines that narrow up-plunge). Although the first kind of conical fold is more common than the second, only the second kind has been treated in the literature.

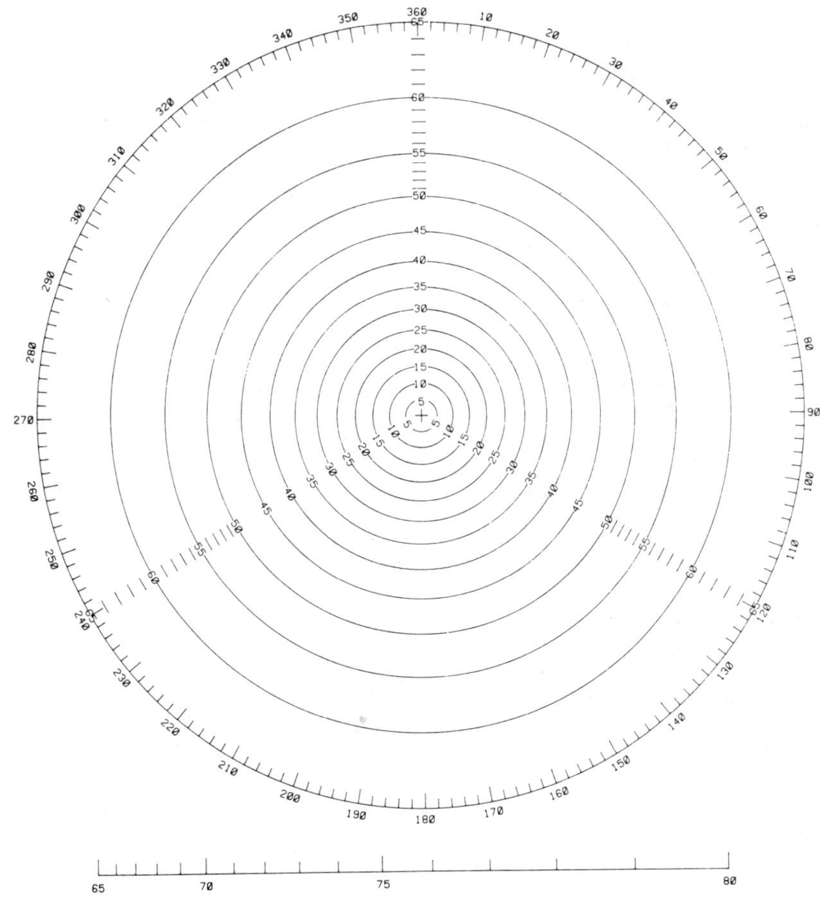

VECTOR OPERATIONS ON TANGENT DIAGRAMS

Tangent diagrams, such as the example shown in Figure 1, are special polar coordinate graphs that provide convenient graphic solutions for many problems of structural geology. Direction of dip is read at the circumference, and angle of dip is read from the concentric circles. Notice, however, that the radius of each circle is proportional to the tangent of the angle of dip. High dips, therefore, plot proportionally farther from the center than low dips. Figure 1 accommodates dips from 0° to 65°, and the auxiliary scale at the bottom extends the range to 80°, beyond which the exaggeration of the dip scale becomes excessive. The distinctive feature of this method of display is that planes can be represented by true vectors. Although tangent diagrams are more easily applied than stereonets to many problems, their structural uses are apparently not mentioned in the literature.

Figure 1. Polar tangent diagram for dips from 0° to 65°. Auxiliary scale extends range to 80°.

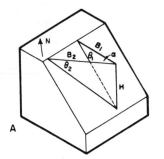

$$\tan \theta_1 = H/B_1 \ , \ \tan \theta_2 = H/B_2, \quad B_1 = B_2 \cos \alpha$$
$$\tan \theta_2 = \tan \theta_1 \cos \alpha$$

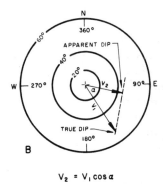

$$V_2 = V_1 \cos \alpha$$

Figure 2. (A) Block diagram showing trigonometric relation between apparent and true dip. (B) Tangent method of finding apparent dip from true dip.

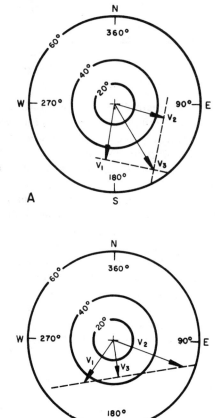

Figure 3. (A) Finding true dip from two apparent dips. (B) Finding line of intersection of two planes.

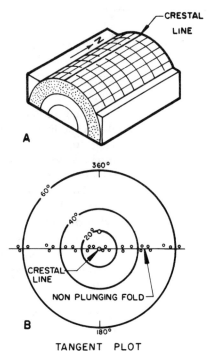

Figure 4. (A) Block diagram of nonplunging cylindrical fold, showing generating lines. (B) Tangent plot of dip data.

Figure 2A, a block diagram of a sloping plane, illustrates the basic principle of the tangent diagram. Line B_1 is a horizontal line in the direction of true dip, and B_2 is another horizontal line making an angle α with B_1; θ_1 is the angle of true dip, and θ_2 is the angle of apparent dip in the direction of line B_2. The trigonometric relations entered on this drawing demonstrate that the tangent of apparent dip in any direction is equal to the tangent of the true dip times the cosine of the angle between the directions of true dip and apparent dip. The tangent of dip, therefore, is a true vector that obeys the cosine law of vector addition and resolution.

Finding Apparent Dip from True Dip

The problem of finding apparent dip from true dip can be resolved vectorially on a tangent diagram, as shown in Figure 2B.

1. Plot V_1, the true dip, as a vector from the origin with length proportional to the tangent of the angle of dip.

2. Draw a line in the direction of the apparent dip.

3. From the terminus of V_1, draw a line perpendicular to the direction of apparent dip.

4. Read the apparent dip, V_2, from the intersection of the two lines.

Finding True Dip from Two Apparent Dips

Figure 3A shows how the tangent diagram is used to find true dip from two apparent dips.

1. Plot V_1 and V_2, the two apparent dips.

2. Draw perpendicular lines through their end points.

3. Read the true dip, V_3, from the intersection of the perpendicular lines.

Finding the Line of Intersection of Two Planes

If two planes intersect, they have equal apparent dips in the vertical plane containing their line of intersection. Figure 3B shows how this principle is used to find the line of intersection of two planes.

1. Plot V_1 and V_2, the true dip vectors for the two planes.

2. Connect the end points of the two vectors with a straight line.

3. Draw V_3, the perpendicular from the origin to the straight line. This vector gives the bearing and plunge of the line of intersection of the two planes.

Figure 3B provides insight for an important statistical application of tangent diagrams: The lines of intersection of planes tangent to the bedding on the same or opposite flanks of an ideal cylindrical fold are parallel to the crestal line. Dip measurements obtained at random locations on such a structure will fall on a straight line when plotted on tangent diagrams, as exemplified by the dashed line in Figure 3B.

CYLINDRICAL AND CONICAL FOLDS

It has been recognized for a long time that the geometry of most folds is well approximated by either cylindrical or conical bulk curvature (Dahlstrom, 1954; Stockwell, 1950; Tischer, 1962; Wilson, 1967). Dip data from random locations

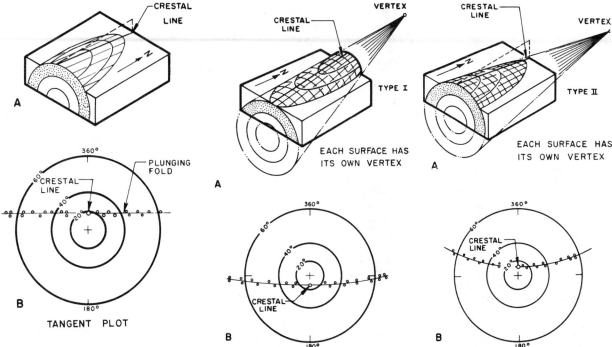

Figure 5. (A) Block diagram showing plunging cylindrical fold. (B) Tangent plot of dip data.

Figure 6. (A) Block diagram of type I conical fold; anticlinal vertex up-plunge. (B) Tangent plot of dip data.

Figure 7. Block diagram of type II conical fold; anticlinal vertex downplunge. (B) Tangent plot of dip data.

on such structures define smooth "statistical" curves when plotted on tangent diagrams, provided the data are accumulated from one plunge only. Similar patterns are also developed from drag zones of dip-slip faults.

A cylindrical fold is defined as a structure whose bedding surfaces can be represented by a moving straight line that remained parallel to a given fixed line called the *b* axis. A conical fold is defined as a structure whose bedding surfaces can be represented by a moving straight line that passed through a fixed point called the vertex. Each bedding surface has a different vertex, but the crestal line of each bedding surface remains parallel to a fixed line that could be called a local *b* axis. Cylindrical folds, or approximately cylindrical folds, are more common than conical folds, but both occur in nature.

Nonplunging and Plunging Cylindrical Folds

Figure 4A shows a nonplunging cylindrical fold. The straight lines on this figure show various positions of the generating line and also delineate the lines of intersection of neighboring dip-strike planes. Except for scatter, all dips are either due east or due west and therefore define a "statistical" straight line passing

through the center of the tangent diagram of Figure 4B. Notice that the azimuth of the crestal line, which is known to be north-south, is perpendicular to the line defined by the data points.

If the structure of Figure 4A is tilted to the north, it becomes a plunging cylindrical fold, such as shown in Figure 5A. The dip data, which show variable strike and dip, again define a straight line on the tangent diagram (Fig. 5B) but one that does not pass through the center. This line is parallel to the nonplunging data line of the previous example and therefore demonstrates that each dip on a cylindrical fold comprises a constant component in the direction of the crestal line (equal to the angle of plunge) and a variable component in the perpendicular direction ranging from zero at the crest to a maximum on the flanks. The perpendicular vector from the center to the data line therefore establishes both the direction and amount of plunge of the crestal line. Cylindrical fold data are almost always easier to interpret on tangent plots than on stereonet pole plots—especially when the data are restricted to one flank, a situation that commonly arises when orienting folds from dipmeter data.

Type I and Type II Conical Folds

Figure 6A is a block diagram of a conical anticline with its vertex up-plunge. The corresponding conical syncline would have its vertex downplunge. Such folds, which will be called type I conical folds, are much more common than the type II folds described in the next section. The straight lines, as before, represent instantaneous positions of the generating line. Dip data accumulated on conical folds, as might be expected, define curved rather than straight lines when plotted on tangent diagrams. The data curve for an ideal type I conical fold is a hyperbola (Adams, 1919; Deetz and Adams, 1945) that opens to the center, such as shown in Figure 6B. The shortest line from the center to the curve gives the bearing and plunge of the crestal line.

If the conical fold of Figure 6A is tilted to the north, the angle of plunge will diminish and finally reverse, and the vertex will now be downplunge rather than up-plunge (Fig. 7A). Anticlines with vertices downplunge and synclines with vertices up-plunge will be called type II conical folds. The dip data of a type II conical fold also define a hyperbola when plotted on a tangent diagram, but one that opens away from the origin, as shown in Figure 7B. The curvature of the data line is the opposite of that for a type I conical fold, but the bearing and plunge of the crestal line again are given by the shortest line from the center to the data line. Conical fold data are usually much easier to interpret on tangent diagrams than on stereonet pole plots. This is especially true when dealing with data from type II conical folds.

Lines that connect points of equal strike (and dip) converge up-plunge on contour maps of type I conical anticlines and downplunge on contour maps of type II conical anticlines. Because the latter behavior is seldom seen, it would appear that type II conical folds are quite rare, a conclusion that runs counter to the limited literature on conical folding (Rech, 1977; Stauffer, 1964; Stockwell, 1950; Wilson, 1967), which implies that all conical anticlines are type II conical folds.

PROJECTING STRUCTURAL CONTROL ONTO CROSS SECTIONS OF CYLINDRICAL AND CONICAL FOLDS

The procedure for projecting structural control onto cross sections of cylindrical folds is (1) find the crestal line from the tangent plot; (2) project the control parallel to this axis. The comparable procedure for conical folds is (1) project data in crestal positions parallel to the crestal line; (2) project data in flank positions parallel to appropriate "local projection axes." To find a local projection axis, draw a line tangent to the data curve of the tangent plot. The perpendicular line from the center to this line is the required local projection axis.

CONCLUSIONS

Solutions to many structural problems are more easily accomplished on tangent diagrams than on stereonet pole plots because (1) there is no need to convert planes to poles and poles back to planes; (2) there is no need to rotate tangent diagrams, because operations that require tracing a great circle on stereonets require only straight-line constructions on tangent diagrams; (3) dip data from ideal cylindrical folds define straight-line patterns on tangent diagrams rather than great circle patterns, as on stereonets; and (4) dip data from conical folds, especially type II conical folds, define more easily recognized patterns on tangent diagrams than on stereonets.

REFERENCES CITED

Adams, O. S., 1919, General theory of polyconic projections: U.S. Coast and Geodetic Survey Special Publication 57.

Dahlstrom, C.D.A., 1954, Statistical analysis of cylindrical folds: Canadian Institute of Mining and Metallurgy Transactions, v. 57, p. 140–145.

Deetz, Ch. H., and Adams, O. S., 1945, Elements of map projection with applications to map and chart construction [fifth edition]: U.S. Coast and Geodetic Survey Special Publication 68.

Rech, W., 1977, Zur Geometrie der geologischen Falten: Geologische Rundschau, v. 66, p. 352–373.

Stauffer, M. R., 1964, The geometry of conical folds: New Zealand Journal of Geology and Geophysics, v. 7, p. 340–347.

Stockwell, C. H., 1950, The use of the plunge in the construction of cross-sections of folds: Geological Association of Canada Proceedings, v. 3, p. 97–121.

Tischer, G., 1962, Über X-Achsen: Geologische Rundschau, v. 52, p. 426–447.

Wilson, G., 1967, The geometry of cylindrical and conical folds: Geological Association [London] Proceedings, v. 78, p. 179–210.

ACKNOWLEDGMENTS

Reviewed by John M. Crowell, C. Dahlstrom, Nicholas Christie-Blick, and Paul Karl Link. Their suggestions for improvement are greatly appreciated.

MANUSCRIPT RECEIVED MAY 22, 1980
MANUSCRIPT ACCEPTED SEPT. 22, 1980

The American Association of Petroleum Geologists Bulletin,
v. 65, no. 2 (February 1981), p. 312-332.

Statistical Curvature Analysis Techniques for Structural Interpretation of Dipmeter Data[1]

C. A. BENGTSON[2]

ABSTRACT

Variations of dip from point to point in the subsurface and variations in the angle of dip as a function of azimuth of dip are related to the bulk curvature of the structural setting. The bulk curvature and related transverse and longitudinal structural directions of any well's setting can usually be determined by statistical curvature analysis of the well's dipmeter data. The bulk curvature can be assigned to one of four general categories: planar, singly curved, plunge reversal, or domal bulk curvature. Structural bulk curvature and related transverse and longitudinal structural directions provide the means for extracting "true structural dips" from erratic or ambiguous dipmeter data. They also provide the means for calculating the bearing and plunge of crestal and trough lines of folds—finding the strike and dip of crestal, axial, and inflection planes of folds—and the strike and direction of dip of dip-slip faults. These capabilities can usually be used to draw partial maps and cross sections centered at the well. A field example shows how the three-dimensional geometry and productive limits of the doubly plunging Railroad Gap field in California could have been predicted from the dipmeter data of the discovery well.

INTRODUCTION

This paper describes statistical curvature analysis techniques, designated by the acronym SCAT, that have proved to be useful for structural interpretation of dipmeter data. SCAT is based on four unfamiliar, but empirically well-verified geometric concepts: structural bulk curvature, transverse and longitudinal structural directions, structural surfaces and related dip-profile special points, and dip isogons.

Most readers are aware that much useful information concerning structural (as well as stratigraphic) interpretation of dipmeter data is available in Schlumberger's manual of dipmeter interpretation (Anonymous, 1970). The most obvious difference between SCAT and the Schlumberger approach is that SCAT uses five machine-plotted data displays, whereas the Schlumberger system relies on a single all-purpose display. The Schlumberger manual tacitly assumes that the interpreter is seeing data pertaining to a transverse cross section. SCAT actually resolves the data into mutually perpendicular transverse and longitudinal (or

T-and L-direction) components and plots these components as a function of in-hole depth. For the time being, the T-direction is defined as the direction of cross section (through the well) that shows the greatest structural change, and the L-direction as the direction of cross section that shows the least structural change. The T- and L-directions are specified in advance of machine plotting, using the best estimates available. Often, the preliminary estimates will be sufficiently close to the true situation to let them stand. One purpose of SCAT is to provide a method for calculating the T- and L-directions directly from dipmeter data. If these calculations show that the preliminary estimates were too far in error, these displays should be replotted in the correct directions. Additional plots, showing the component of dip in any other direction of interest, can also be programmed.

Resolving dipmeter data into transverse and longitudinal components provides an important statistical advantage because the trend lines to be fitted to T- and L-direction dip-component versus depth plots must conform to various empirically verified constraints, the most important of which are:

1. Longitudinal dip-component versus depth plots show lower and less variable dip than dip-component plots for any other direction. The average L-direction component of dip is zero for planar and non-plunging fold settings and equal to the angle of plunge for plunging fold settings. On plunge reversal settings the average L-direction component of dip shows a reversal of dip (and hence plunge) with depth. The only exceptions occur in wells cut by cross faults. However, longitudinal dip-component plots may show considerable scatter in zones of steep dip.

2. Transverse dip-component plots show steeper or more variable dip than dip-component plots for any other direction. The patterns may be simple or complex. However, they are usually simpler and never more complex than patterns on Schlumberger-type angle-of-dip versus depth plots. For example, angle-of-dip versus depth plots show complex patterns when a well crosses a crestal plane[3], but transverse dip-component plots, in contrast, show smooth trend lines that cross the zero dip-component axis. Because angle of dip is neither positive nor negative, there is no chance on an angle-of-dip versus depth plot for negative scatter to cancel positive scatter in a flat dip situation. Therefore, a zone

[1]Manuscript received, March 3, 1980; accepted, May 30, 1980.
[2]Chevron U.S.A., Inc., San Francisco, California 94119.
The writer appreciates the encouragement of many colleagues, especially C. Dahlstrom, P. Verrall, and D. L. Zieglar. Published with permission of Chevron U.S.A. Inc.

[3]In the discussion to follow, the term "plane" used in expressions such as "crestal plane," "axial plane," etc, refers to a surface that may be either planar or curved.

of zero dip is falsely perceived as a zone of 5 to 10° average dip on an angle-of-dip versus depth plot. On a dip-component versus depth plot, however, half of the points will fall to the right of the zero dip axis and half to the left, correctly indicating zero average dip.

3. The shape of the "statistical trend line" on a transverse dip-component versus depth plot has a one-to-one relation with the details of bedding curvature on a transverse cross section. A trend line conforming to uniform dip indicates planar curvature. A smoothly curved trend line with no bends or reversals indicates uniform or smoothly varying curvature, whereas a trend line with bends or reversals will show one or more of eight mathematically definable patterns or special points. Six of these points serve to locate and identify structural surfaces (axial planes, kink planes, inflection planes, secondary inflection planes, minimum-curvature planes, and zero strain boundaries) that intersect the well, and two serve to locate dip-slip faults, distinguishing faults that dip to right from faults that dip to the left.

Finally, it should be stressed that SCAT has the capacity to find the bearing and plunge of crestal and trough lines of folds, the strike and dip of crestal, axial, and inflection planes of folds, and the strike and direction of dip of dip-slip faults. In addition to transverse and longitudinal dip-component versus depth plots, SCAT also employs azimuth-of-dip versus depth plots, angle-of-dip versus depth lots, and dip versus azimuth plots, for a total of five basic SCAT plots. Interpretation of these plots is accomplished by fitting internally consistent hand-drawn "statistical trend lines" to all five data displays. This procedure is illustrated by a series of idealized models of increasing complexity and summarized by a field example that shows how the SCAT plots of the discovery well of the Railroad Gap field in California could have been used to draw reliable cross sections and maps of this doubly plunging structure in advance of follow-up drilling.

END-MEMBER CATEGORIES OF STRUCTURAL BULK CURVATURE AND RELATED DIP VERSUS AZIMUTH PATTERNS

The term "structural bulk curvature" implies that each bed of a structure shares certain curvature properties or regularities with other beds on the same structure. These regularities can be classified according to the manner in which angle of dip varies as function of azimuth of dip, regardless of location within the structure. Three different methods are available for determining this relation. The first and most familiar method is to enter the basic data on stereonet pole plots. Another method is to plot the data on polar tangent diagrams. However, the most advantageous method for dipmeter interpretation (based on literally hundreds of examples from many different oil provinces) is to plot the data on "Mercator-type" displays.

Figure 1 illustrates the proposition that (for dipmeter interpretation) any structural setting can be assigned to one of seven end-member categories of structural bulk curvature. Each category (except the first) is characterized by a specific kind of contour-map pattern, and each (except the first) is characterized by mutually perpendicular transverse and longitudinal structural directions that can be determined either from contour-map analysis or from dip-data analysis.

The concept of limited varieties of structural bulk curvature holds true regardless of whether the beds are planar, folded, or faulted. The reasoning is: a given set of beds is either planar or curved; if planar, the beds are either horizontal (Fig. 1A) or dipping at a low (Fig. 1B) or moderate to high angle (Fig. 1C). The zero dip setting shows no structural change in any direction and hence has no T- or L-directions. In the low and higher homoclinal dip settings (Fig. 1B and 1C) the T-direction parallels the dip and the L-direction parallels the strike. If the beds are curved they are either singly or doubly curved. If singly curved,[4] their crestal or trough lines are either horizontal (Fig. 1D) or plunging at a constant angle (Fig. 1E). In either situation, the T-direction is perpendicular to the crestal or trough lines and the L-direction is parallel. If the beds are doubly curved, their structure contours are either elliptical or circular in plan. If elliptical (Fig. 1F), their geometry can be approximated by two singly curved plunges joined by a non-plunging central sector, in which case the T-direction parallels the short dimension and the L-direction parallels the long dimension. If the structure contours are circular (Fig. 1D'), the transverse directions will converge radially toward the center, and the longitudinal directions will be disposed circumferentially around the center.

One method of testing the concept of limited varieties of geologic bulk curvature is to prepare nonpolar "Mercator-type" displays of dip data (Fig. 2) that show angle of dip plotted as a function of azimuth of dip, regardless of depth or location on the structure. Dip is measured along the vertical axis and azimuth on the horizontal axis. North azimuth appears at the center and south azimuth appears at the right and left margins (because this display is really a rolled out cylinder). When such plots of dipmeter data are made, only one of the six patterns or, more usually, part of one of the six patterns shown in Figure 2 appears, regardless of whether the structure is planar, folded, or faulted, and regardless of whether the well is straight or deviated.[5] If a well passes through an unconformity or certain kinds of faults, different patterns result above and below the structural discontinuity. The arrows labeled T and L on the dip versus azimuth plots identify the transverse and longtitudinal directions of the setting penetrated by the well, and they also correspond with the T and L arrows appearing on the block diagrams of Figure 1. Because

---

[4]In this discussion, we tacitly assume that singly curved beds display cyclindrical curvature. Conical curvature also occurs in nature (Stockwell, 1950; Dahlstrom, 1954; Ramsay, 1967) but usually only in special circumstances. Distinguishing cylindrical curvature from the two possible kinds of conical curvature is readily accomplished by use of polar-tangent diagrams, a separate topic beyond the scope of this paper (Bengtson, 1980).

[5]Under certain circumstances, the pattern obtained from a deviated well on a circular dome may violate this generalization.

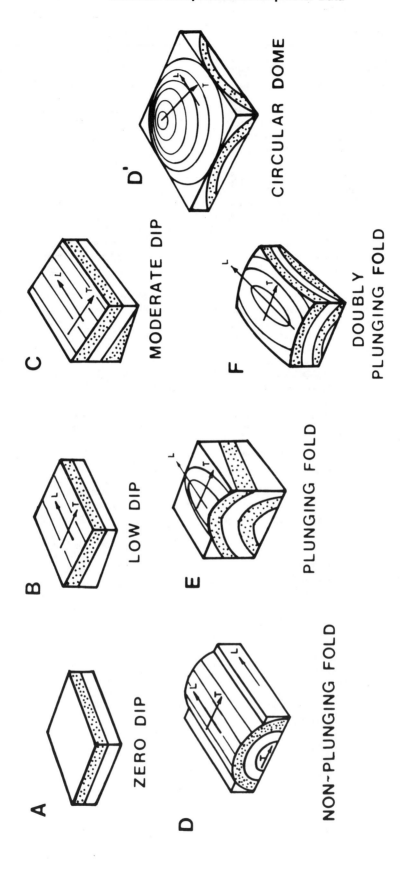

FIG. 1—Idealized block diagrams illustrating basic categories of geologic bulk curvature. Arrows designate mutually perpendicular T- and L-directions.

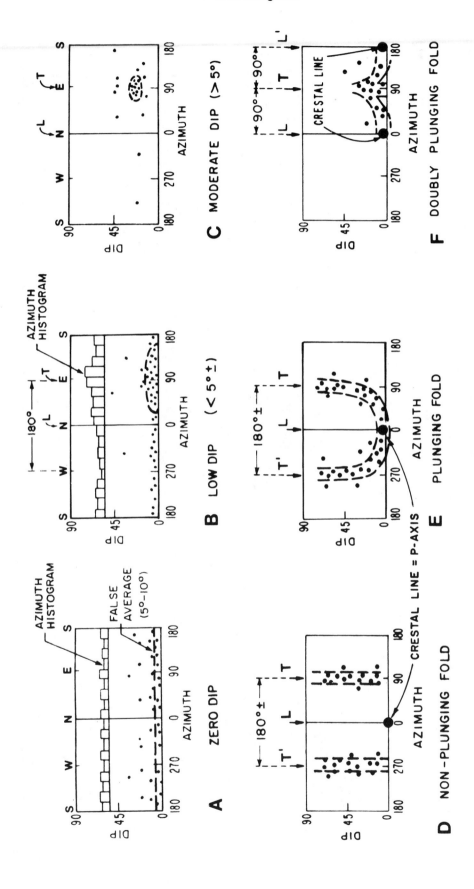

FIG. 2—Six basic dip versus azimuth patterns corresponding to bulk-curvature categories of Figure 1.

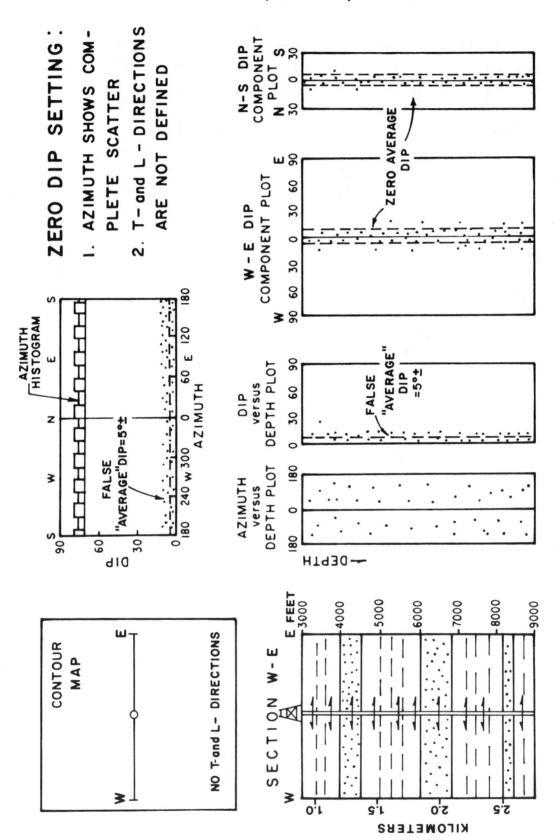

FIG. 3—SCAT plots, contour map, and W-E cross section for zero homoclinal dip setting. Dip versus azimuth plot and azimuth versus depth plot show display-related scatter.

the data on these displays are plotted independently of location, the data for neighboring wells on the same structure can be combined in a single display or, alternatively, can be combined with surface data or seismic cross-spread data. Stratigraphic convergence may distort the patterns in Figure 2 but, in folded and faulted settings, no distortion occurs if the convergence is confined to the transverse structural direction.

DIPMETER SCATTER

The scatter shown on the dip versus azimuth plots of Figure 2 and the SCAT plots for the idealized structures of Figures 3 through 14 is typical of well-bedded deposits and is due to second-order stratigraphic and structural effects and to instrumental error. Similar scatter is observed when surface dip data are displayed in a comparable fashion. More severe scatter is often observed in real situations, such as the stratigraphic scatter present in the Point of Rocks sandstone in Figure 15. Stratigraphic causes of scatter include cross-bedding, slumping, channeling, and the like; structural causes include drag folding and minor faulting. Instrumental causes include poor development of resistivity curves, as in massive carbonate or volcanic rocks, and poor hole condition which then is usually related to some geologic cause such as the presence of sloughing or heaving shale, conglomeratic zones, or zones of fracturing near faults. One of the main advantages of SCAT is that its system of multiple data displays very often provides a means for "seeing through" this scatter, thereby making it possible to obtain "true structural dips" from otherwise erratic data.

PLANAR DIP SETTINGS

Figure 3 shows the SCAT plots for the simplest of all structural settings—zero planar dip. This is the only setting that has no T- or L-directions. The upper display on the right of the contour map is the dip versus azimuth plot and, because zero dip has no azimuth, the data points are randomly scattered at the bottom of the display. The same random pattern is present on the azimuth histogram at the top of the display. The first graph on the right of the cross section is a plot of azimuth of dip as function of in-hole depth. This plot again shows random scatter. The next panel on the right is a graph of angle of dip versus depth. Because of scatter, this plot incorrectly indicates an average dip of about 5°. This is a correct arithmetic average but an incorrect vector average, because the vector average of 5° of dip in every direction is not 5°, but 0°. The third panel on the right, which gives the component of dip in a W-E direction, correctly shows zero average dip, as does the N-S dip-component plot on the far right.

Tilting the zero dip setting of Figure 3 slightly to the east results in the low dip setting shown on Figure 4. The dip versus azimuth plot again shows much scatter, but the azimuth histogram shows a preponderance of easterly dips and a deficiency of westerly dips. The azimuth versus depth plot also shows considerable scatter, as

would be expected. The angle-of-dip versus depth plot shows dips whose average is somewhat too high, but the W-E dip-component plot, which is the transverse dip-component plot, shows low homoclinal dip, and the longitudinal dip-component plot on the far right shows zero average dip, as expected. If the low dip setting is again tilted to the east, as in Figure 5, the various SCAT plot patterns come into relatively sharp focus requiring no further explanation.

NON-PLUNGING FOLD SETTING

Our next example, the non-plunging fold of Figure 6, is characterized by constant strike but variable dip. Referring to the azimuth versus depth plot, the direction of dip is uniformly to the west down to the crestal plane, where an abrupt change to constant east direction occurs. Similar bands of constant west and east azimuth in a "railroad track" pattern are seen on the dip versus azimuth plot at the top of the figure.

The transverse dip-component plot shows west dip in the upper part of the hole that decreases steadily with depth, and reverses to east dip at CP, the crestal plane. Below the crestal plane the east dip increases at a progressively higher rate, until a point of maximum rate of change of dip (corresponding with a point of maximum bed curvature) is reached at AP, the axial plane. The dip continues to increase below AP, but at a progressively slower rate, until point IP, an inflection plane, is reached. At IP the change of dip reverses from increase with depth to decrease with depth concurrently with a change of curvature from anticlinal curvature above to synclinal curvature below. Three structural surfaces, corresponding to special points on the transverse dip-component versus depth "statistical trend line," have been pinpointed: (1) the crestal plane, whose intersection with the well bore is marked by a point of change of dip from west dip to east dip; (2) the axial plane, whose intersection with the well bore is marked by a point of maximum rate of change of dip; and (3) the inflection plane, whose intersection with the well bore is marked by a point of local maximum dip and zero rate of change of dip.

If the cross section of Figure 6 is turned upside down, the inflection plane is still an inflection plane, with anticlinal curvature above and synclinal curvature below; the axial plane is still an axial plane, but is now a synclinal axial plane, rather than an anticlinal axial plane; and the crestal plane is now a trough plane. The transverse dip-component plot is still valid, provided the labels for direction of W and E component of dip are reversed. It should be mentioned for completeness that four additional kinds of structural surfaces, not illustrated in Figure 6, are present in nature: kink planes or surfaces that separate zones of different semiplanar dip; minimum curvature planes located between opposing axial planes on box folds and marked by points of minimum rate of change of dip; secondary inflection planes, present only on disharmonic structures and marked by points of local minimum dip and zero rate of change of dip; and zero-strain boundaries or surfaces

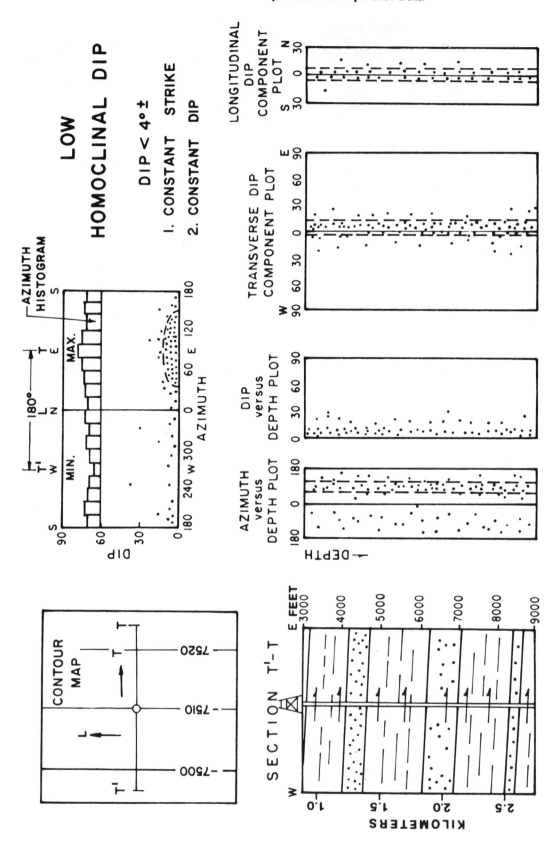

FIG. 4—SCAT plots, contour map, and transverse cross section for low homoclinal dip setting.

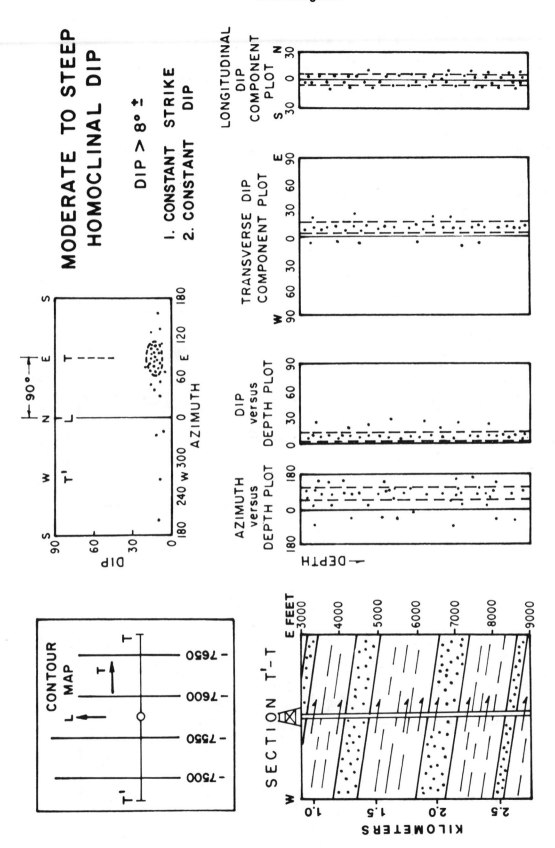

FIG. 5—SCAT plots, contour map, and transverse cross section for moderate to steep homoclinal dip setting. This setting has no display-related scatter.

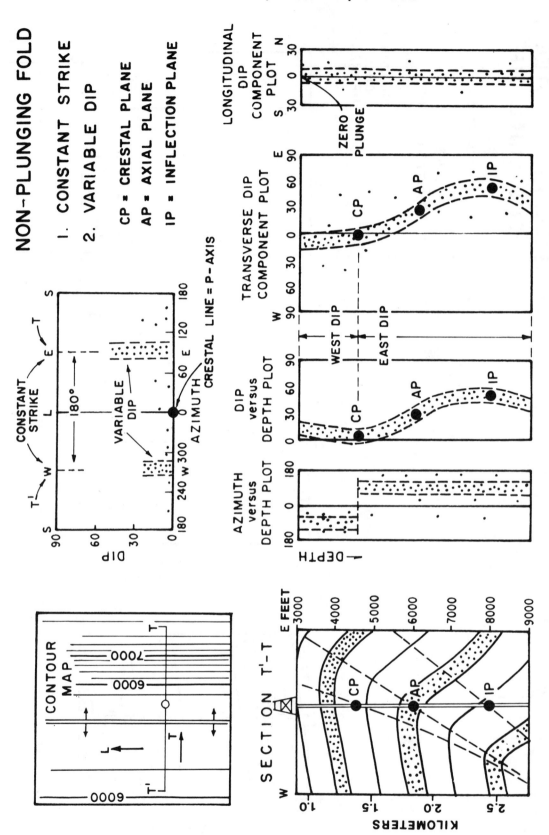

FIG. 6—SCAT plots, contour map, and transverse cross section for non-plunging-fold setting.

that separate zones of planar bedding from zones of curved bedding.

The non-plunging attitude of this fold is confirmed by the longitudinal dip-component plot, which shows zero component of longitudinal (or N-S) dip at all depths.

PLUNGING FOLDS AND PLUNGING PROJECTION AXES

If the non-plunging fold of Figure 6 is tilted to the north, as in Figure 7, we see a setting characterized by variable dip and variable strike, but a constant component of longitudinal or north dip equal to the tilt of the previously horizontal crestal line. Referring to the dip versus azimuth plot, we see that the "railroad track" pattern of the non-plunging fold has changed to a kind of "horseshoe" pattern. The low point on this pattern gives the azimuth and plunge of the crestal line. The point where the horseshoe pattern, if extended, would intersect the 90° dip axis is the T-direction. The azimuth versus depth plot, instead of shifting abruptly from west dip to east dip at the crestal plane, as in the previous non-plunging example, changes gradually from NW to NE dip. Because rapid change of azimuth occurs only with low dip, the maximum rate of change of azimuth occurs at the crestal plane, where the dip is lower than anywhere else and equal to the plunge of the crestal line. The point where the crestal plane cuts the well is identified by: (1) a point of maximum rate of change of azimuth on the azimuth versus depth plot; (2) a point of minimum angle of dip on the angle-of-dip versus depth plot; (3) a point of minimum dip and maximum rate of change of azimuth on the dip versus azimuth plot, with azimuth and dip values corresponding, respectively, with those of (1) and (2); (4) a change from dip to the left to dip to the right on the transverse dip-component plot; and finally (5) a constant value of L-direction component of dip equal to the lowest values present on the angle-of-dip versus depth plot and the dip versus azimuth plot. Also, the point where the inflection plane intersects the well is identified by: (1) a point of maximum swing to the east on the azimuth versus depth plot; (2) a point of maximum angle of dip on the angle-of-dip versus depth plot; and (3) a point of maximum east component of dip on the transverse dip-component plot.

The preceding discussion demonstrates that the correct way to project data into cross sections on plunging folds is: (1) find the L-direction, (2) find the average component of dip in the L-direction, and (3) project the data in the plane of the L-direction and raise or lower the elevation of the points in question in accordance with the average L-direction dip component. This common projection direction, or axis, is identified by the label "P-axis" on Figures 2, 6, 7, and 9 through 12.

PLUNGE-REVERSAL SETTING DETECTED BY DIPMETER DATA ALONE

Figure 8 illustrates the capability of SCAT to identify plunge-reversal settings. The contour map shows a well drilled just north of a north-dipping plunge-reversal zone marked by a dotted line. North of this line the contours are similar to those of the uniformly plunging structure of Figure 7; south of the line the contours indicate south plunge. The transverse dip-component plot and angle-of-dip versus depth plot for this well are essentially similar to those of the well in Figure 7, but the dip versus azimuth plot shows a partial horseshoe pattern, indicative of north plunge, and a second partial horseshoe pattern, indicative of south plunge, joined to a vertical limb indicative of zero plunge. Such a pattern could be called a "Christmas tree" pattern. The azimuth versus depth plot likewise shows a pattern that is different from its counterpart in Figure 7 but the difference is not so striking. The key to these differences is in the longitudinal dip-component plot, which shows uniform north plunge reversing to uniform south plunge at about 5,600 ft (1,707 m).

"Christmas tree" patterns are also present in structural saddles and synclinal plunge-reversal settings. The more customary stereonet pole plots used for orienting cylindrical folds (Stockwell, 1950; Dahlstrom, 1954) are unsuited for orienting plunge-reversal settings because they fail to generate easily recognized plunge-reversal patterns.

IDENTIFICATION AND ORIENTATION OF DIP-SLIP FAULTS

The cross section of Figure 9 shows a well on an east-dipping homocline cut by a N-S striking normal fault whose drag steepens the regional dip, and the cross section of Figure 10 shows a well on a similar homocline cut by a N-S striking normal fault whose drag flattens and reverses the regional dip. The SCAT plots for the two wells show patterns reminiscent of those for non-plunging folds, but the transverse dip-component plots show distinctive cusp patterns not seen before. The transverse dip-component plot of Figure 9 shows a cusp pattern (superimposed on the regional dip pattern) that points to the right, that is, in the direction of the dip of the fault, and Figure 10 shows a similar cusp pattern that points to the left, again in the direction of dip of the fault.

Geometric constructions demonstrate that the curvature of beds in a drag zone increases as the point of the cusp (i.e., the fault) is approached from either side and that the curvature reverses from anticlinal to synclinal (or vice versa) at the point of the cusp. Fault drag can be regarded as a special kind of cylindrical folding that conforms to the following empirically verified rules:

1. Because of drag, the upthrown side of any fault is characterized by anticlinal curvature and the downthrown side by synclinal curvature, regardless of whether the fault is a normal or a thrust fault.

2. The transverse dip-component plot for a well cut by a dip-slip fault displays a cusp pattern that points in the direction of dip of the fault if the fault is a normal fault, and in the opposite direction if the fault is a thrust fault. However, there is no way to distinguish normal from thrust faults using dipmeter data alone.

3. The one exception to rules (1) and (2) occurs with

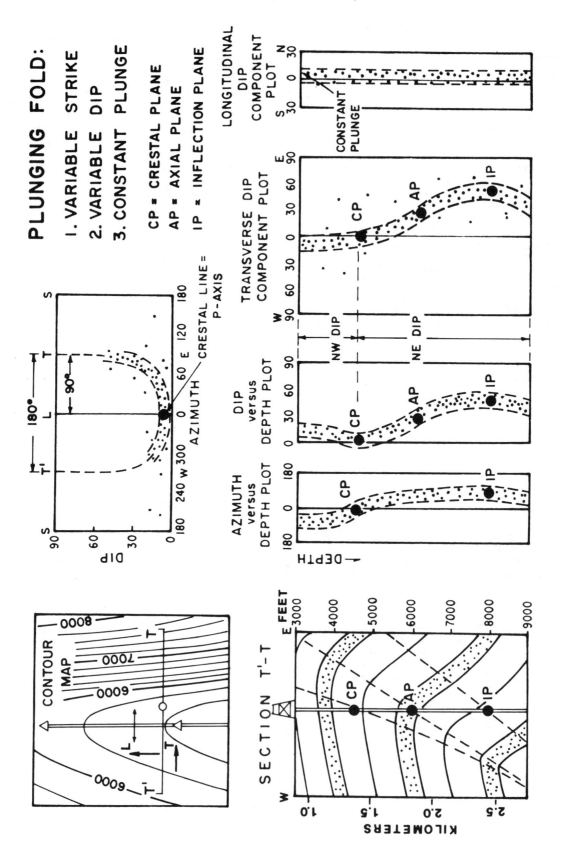

FIG. 7—SCAT plots, contour map, and transverse cross section for plunging-fold setting produced by uniform northward tilting of non-plunging setting of Figure 6.

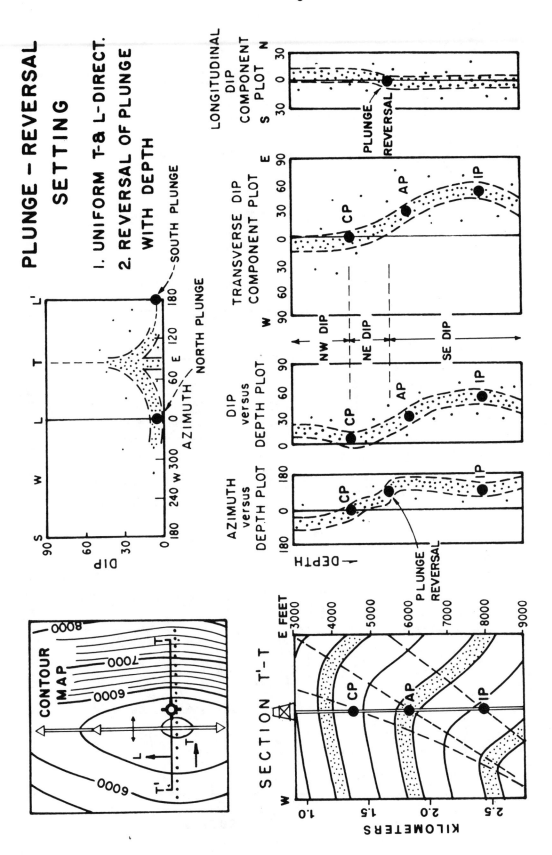

FIG. 8—SCAT plots, contour map, and transverse cross section for plunge-reversal setting.

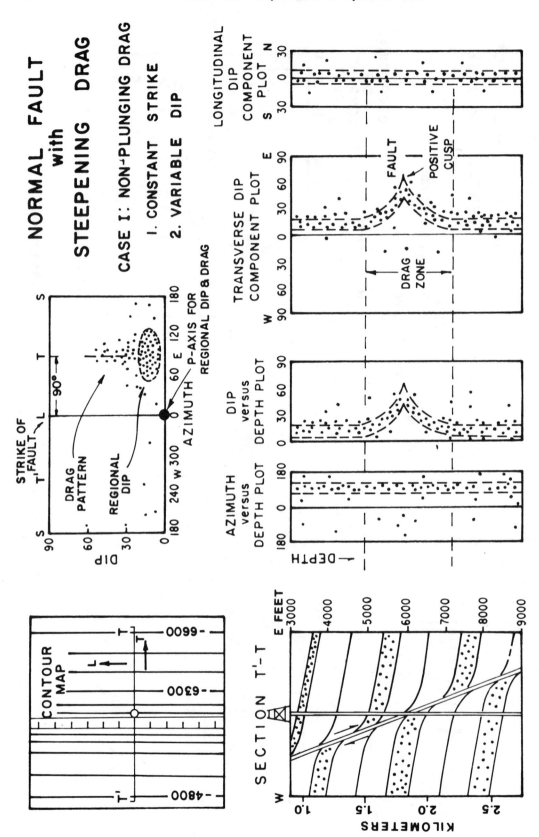

FIG. 9—SCAT plots, contour map, and transverse section for normal fault whose drag steepens regional dip. Strike of fault is same as strike of regional dip.

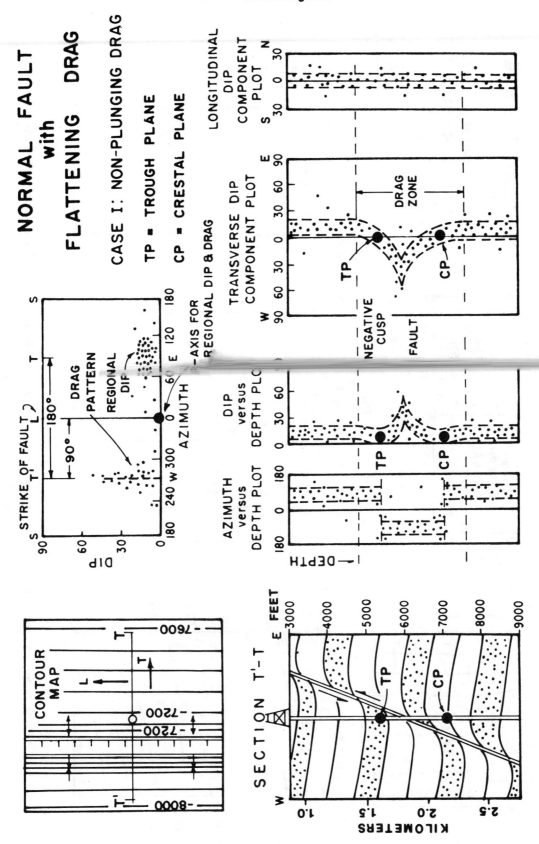

FIG. 10—SCAT plots, contour map, and transverse section for normal fault whose drag flattens and reverses regional dip. Strike of fault is same as strike of regional dip.

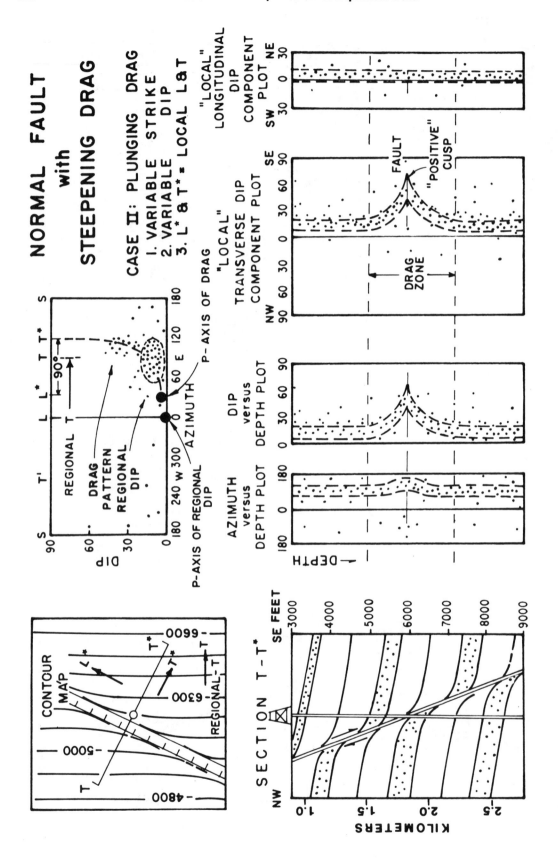

FIG. 11—SCAT plots, contour map, and transverse section for normal fault whose drag steepens regional dip. Strike of fault is inclined to strike of regional dip.

listric normal faults (Fig. 13). Transverse sections of such faults show anticlinal curvature on the downthrown block and transverse dip-component plots show a pseudo-half-cusp pattern pointed opposite to the dip of the fault.

4. Drag zones, like folds, are characterized by mutually perpendicular T- and L-directions. The T- and L-directions of drag zones, however, may differ from those of their regional structural settings. Faulted settings, therefore, may have local T- and L-directions in the vicinity of faults and regional or background directions elsewhere.

5. The maximum effect of drag, as reflected by build-up of cusp patterns, is on local T-direction plots and the minimum effect (usually zero) is on local L-direction plots.

Inspection of the SCAT plots of Figure 9 and 10 reveals that, in both examples, the axes for projecting the regional dip and the axes for projecting the faults and their associated drag zones are horizontal and oriented N-S. Regarding the orientation of the faults shown in Figures 9 and 10, the following is known from dipmeter data: (1) The orientation of a horizontal line contained in the fault plane (i.e., its strike), (2) the direction (but not amount) of dip, assuming that we already know the fault is a normal fault from other information.

The maps and cross sections of Figure 11 show an east-dipping homocline cut by a normal fault that steepens the regional dip, as in Figure 9; Figure 12 shows an east-dipping homocline cut by a normal fault that flattens and reverses the regional dip, as in Figure 10. The transverse dip-component plots of Figures 11 and 12 are nearly identical to their counterparts in Figures 9 and 10, but the other SCAT plots show plunging-fold patterns superimposed on regional dip, rather than the non-plunging patterns of Figures 9 and 10. The reason is that the strike of the faults in Figures 11 and 12 is inclined to the strike of regional dip, whereas the strike of the faults in Figures 9 and 10 is parallel with the strike of the regional dip. The dip versus azimuth plots of Figures 11 and 12 show partial horseshoe patterns corresponding with those for plunging folds. The extrapolated low point of the dip versus azimuth plot of Figure 11 and the actual low point on the dip versus azimuth plot of Figure 12 give the azimuth and plunge of the projection axes of the faults and also establish the local T- and L-directions for the region affected by fault drag. All effects of faulting are removed in the local L-direction dip-component plots which show no disturbance of the component of regional dip in the direction parallel with the L-direction of the drag zone. The local L-direction has the same azimuth as the apparent strike of the fault (i.e., its trace on a structure contour map). The cross sections of Figures 9 and 11 show no structural surfaces, except the drag zone (or zero strain) boundaries and the faults. The cross sections of Figure 10 and 12, likewise, show no structural surfaces except the drag-zone boundaries, the faults, and the crestal and trough planes.

From dipmeter data regarding the orientation of the faults of Figures 11 and 12, we know (1) the orientation of a line contained in the fault plane whose azimuth is the same as that of the local L-direction and whose plunge is equal to the average local L-direction component of dip; this line is the projection axis for the fault and its associated drag zone; and (2) the direction (but not amount) of dip in a section parallel with the local T-direction (assuming that we already know from other information that the fault is a normal fault).

Thrust faults, like normal faults, are characterized on transverse dip-component plots by cusp patterns that either steepen or flatten the dip that would be observed if the fault were not present. However, the cusp patterns for thrust faults point in the direction opposite to the dip of the fault. The T- and L-directions of thrust faults and thrust fault drag are usually the same as the T- and L-directions of the setting in which they occur and, for this reason, the dip data for the drag zone fall into the same dip versus azimuth and L-direction dip-component patterns as the data for the rest of the setting. Tear faults and cross faults, in contrast, generate dip versus azimuth and L-direction dip-component patterns that conflict with the patterns for the setting as a whole.

To conclude our survey of fault analysis, consider the listric normal fault shown in Figure 13 and the listric thrust fault shown in Figure 14 and recall that movement on curved faults creates a space problem resolved by rotation of beds in the hanging wall. In listric normal faults (Fig. 13) the hanging-wall rotation overwhelms the predisplacement drag resulting in a so-called rollover anticline with associated "reverse" drag. Rollover anticlines, in contrast to compressive or diapric anticlines, have no true axial planes because the maximum rate of change of dip occurs at the listric fault. The transverse dip component plot for the listric thrust fault of Figure 14 shows a cusp pattern that develops very steep left dip with a continuation representing overturned dips appearing in the right panel.

FIELD EXAMPLE OF DIPMETER PREDICTION

Figure 15 shows the SCAT plots for the discovery well of the Railroad Gap field in Kern County, California. The diagram at the upper right of Figure 15 is the dip versus azimuth plot for all depths from 4,300 to 13,000 ft (1,311 to 3,963 m). The middle diagram is for depths from 4,300 to 9,800 ft (1,311 to 2,579 m) and the bottom diagram is for depths from 9,800 to 13,000 ft (2,579 to 3,963 m). These plots show that the well sampled all four flanks of a doubly plunging fold whose crestal line strikes 300 to 120° with 5° of NW plunge and 7° of SE plunge.

Two crestal planes, one axial plane, two inflection planes, and three southwest-dipping thrust faults are identified and marked by appropriate symbols on the transverse dip-component plot of Figure 15 (third plot from the left). These special points serve to anchor the "statistical trend lines" that provide the basis for establishing the "true structural dip" along the well bore.

Guidelines for extrapolating the traces of structural

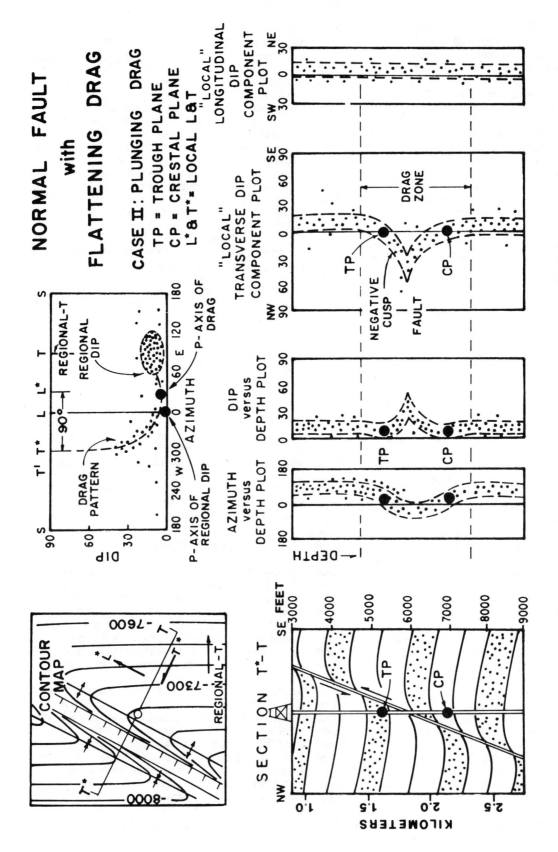

FIG. 12—SCAT plot, contour map, and transverse section for normal fault whose drag flattens and reverses regional dip. Strike of fault is inclined to strike of regional dip.

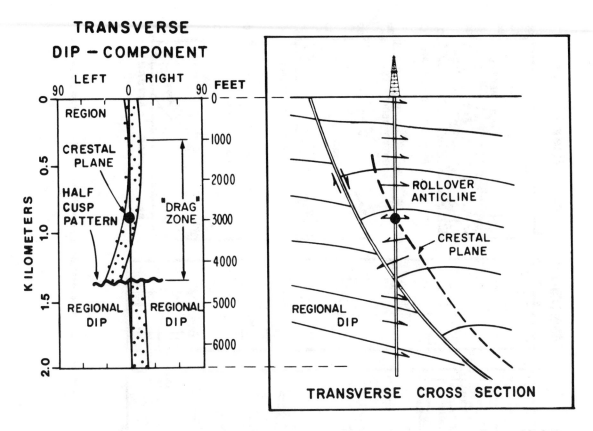

FIG. 13—Transverse dip-component plot and transverse cross section for listric (concave upward) normal fault.

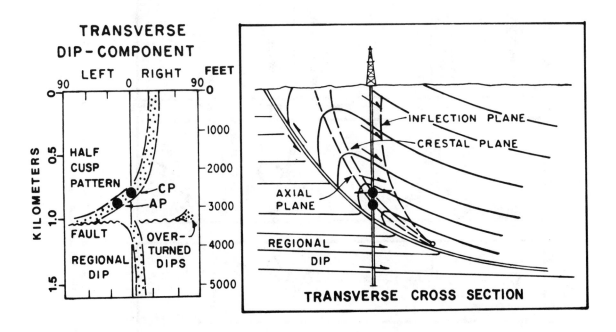

FIG. 14—Transverse dip-component plot and transverse cross section for listric (concave upward) thrust fault.

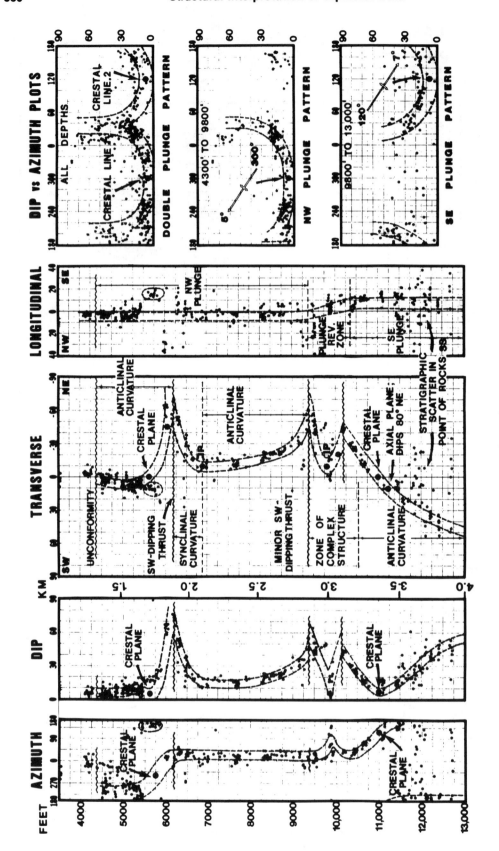

FIG. 15—SCAT plots for discovery well of Railroad Gap field in California. Patterns of transverse dip-component plot are complex and contrast with simple pattern of longitudinal dip-component plot.

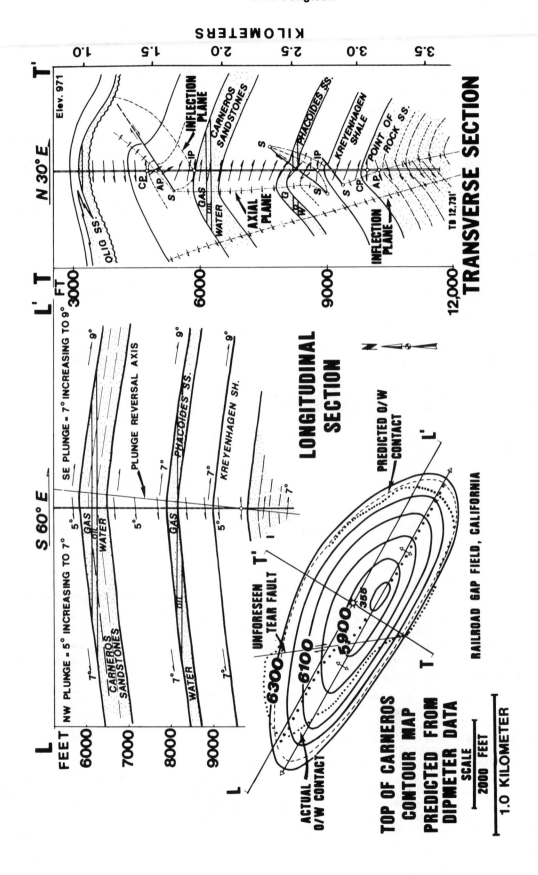

FIG. 16—Predicted transverse and longitudinal cross sections and structure contour map derived from SCAT plots of Figure 15.

surfaces on transverse sections can be formulated by comparing the geometry of real folds with the geometry of ideal folds discussed in structural textbooks. First, we note that inflection planes and crestal planes as well as dip isogons (i.e., lines that connect points of equal dip on transverse cross sections; Ramsay, 1967) are perpendicular to the bedding on idealized concentric folds but parallel with the axial plane on idealized similar folds. On most real folds, their orientation is intermediate to these two end-member conditions. Axial planes, however, are (ideally) perpendicular to the bedding regardless of whether the structure is concentric or similar and regardless of whether the axial plane is planar or concave upward or downward. These principles provide guidance for drawing the various structural surfaces (including the fault planes) present on the transverse section. Additional references on using dip data to draw cross sections are in papers by Ickes (1925), Busk (1929), Coates (1945), Mertie (1947, 1948, 1949), and Gill (1953). None of these papers, however, treats the important topic of structural surfaces and related dip-profile special points.

Using the direct dip control along the well bore and the secondary dip control afforded by the extrapolated structural surfaces, we derive the predicted transverse cross section present in the right panel of Figure 16, which is much like the true section revealed by later drilling. The points marked "S" on the transverse section are singular points. Such points, which occur when two or more structural surfaces (including faults) meet and terminate, are always present on disharmonic structures, such as the present example.

The longitudinal dip-component plot of Figure 15 is characterized by a reversal of plunge at about 9,800 ft (2,579 m). A plunge-reversal plane is entered at this depth on the longitudinal section of Figure 16 and is extrapolated upward and downward with an estimated 80° dip. On the basis of relations observed on nearby structures, the beds are given a slight curvature such that the 5° NW dip increases to 7° and the 7° SE dip increases to 9° at the ends of the section. The transverse and longitudinal sections provide secondarily derived control for the structure contour map of Figure 16. The predicted crestal line is shown by a solid line and the actual crestal line is shown by a dotted line. The predicted and actual oil-water contacts are shown by dashed lines. The chief cause of disagreement between the predicted and actual map is the presence of an unforeseen right-lateral tear fault that offsets the crestal line and structure contours.

SUMMARY AND CONCLUSIONS

This paper describes statistical curvature analysis techniques for structural interpretation of dipmeter data in accordance with the following five-step program. (If the well passes through an unconformity, or certain kinds of faults, the portions of the well above and below the structural discontinuity should be treated separately.)

1. Determine which of six possible complete or partial statistical patterns appears on the dip versus azimuth plot, thereby establishing the structural bulk curvature and T- and L-directions of the well's structural

setting.

2. Prepare transverse and longitudinal dip-component versus depth plots. The longitudinal dip-component plot should show either zero dip or low dip that varies little or not at all, except in plunge-reversal settings (or settings affected by cross faults). The transverse dip-component plot should show steeper or more variable dip than plots in any other direction. The "statistical trend line" to be fitted to this display will be a straight line, a continuously curved line, or a line that displays one or more of eight possible kinds of special points that serve to identify and to locate structural surfaces and dip-slip faults.

3. Make sure that the "statistical trend lines" fitted to the dip versus azimuth plot, the transverse and longitudinal dip-component versus depth plots, and the angle-of-dip and azimuth-of-dip versus depth plots are mutually consistent and structurally reasonable.

4. Plot the well course on a transverse section and enter "true structural dips" (derived from the transverse dip-component plot's "statistical trend line") at appropriate locations along the well bore. Enter any structural surfaces or faults that have been identified and extrapolate them for reasonable distances beyond the well bore. Do the same for selected dip isogons. This procedure establishes secondary control for drawing a partial transverse cross section centered at the well.

5. Project the cross-section data longitudinally "up plunge" and "down plunge" away from the well, using the average L-direction component of dip to raise or lower the indicated structural elevations. This procedure establishes secondary control for drawing a partial contour map centered at the well.

In rare conditions, as in the field example, it will be possible to reconstruct the geometry of an entire structure using the data of a single well. More often it will be possible to reconstruct the geometry of part of a structure, and to derive more structural information from a dipmeter survey than just the dip along the well bore.

REFERENCES CITED

Anonymous, 1970, Fundamentals of dipmeter interpretation: New York, Schlumberger Ltd.

Bengston, C. A., 1980, Structural uses of tangent diagrams: Geology, v. 8, p. 599-602.

Busk, H. G., 1929, Earth flexures: Cambridge, England, Cambridge University Press.

Coates, J. S., 1945, The construction of geological sections: Geol., Min. & Met. Soc., India, Quart. Jour., v. 17, p. 1-11.

Dahlstrom, C. D. A., 1954, Statistical analyses of cylindrical folds: Canadian Inst. Mining and Metallurgy Trans., v. 57, p. 140-145.

Gill, W. D., 1953, Construction of geological sections of folds with steep-limb attenuation: AAPG Bull., v. 37, p. 2389-2406.

Ickes, E. L., 1925, The determination of formation thicknesses by the method of graphical integration: AAPG Bull., v. 9, p. 451-463.

Mertie, J. B., 1947, Delineation of parallel folds and measurements of stratigraphic dimensions: Geol. Soc. America Bull., v. 58, p. 770-802.

_____ 1948, Application of Brianchon's theorem to construction of geologic profiles: Geol. Soc. America Bull., v. 59, p. 767-786.

_____ 1949, Stratigraphic measurements in parallel folds: Geol. Soc. America Bull., v. 51, p. 1107-1134.

Ramsay, J. G., 1967, Folding and fracturing of rocks: New York, McGraw Hill, 568 p.

Stockwell, C. H., 1950, The use of the plunge in the construction of cross-sections of folds: Geol. Assoc. Canada Proc., v. 3, p. 97-121.

American Association of Petroleum Geologists Memoir 32,
The Deliberate Search for the Subtle Trap, edited by Michel T.
Halbouty, copyright 1982, p. 31-45.

Structural and Stratigraphic Uses of Dip Profiles in Petroleum Exploration

C. A. Bengtson
Chevron, U.S.A.
San Francisco, California

Dip profiles are graphs that show apparent dip as a function of distance along selected horizontal, vertical, or inclined lines on cross sections. Such profiles can be derived from dip-related data of any kind—surface data, subsurface data from maps and cross sections, seismic data, or dipmeter data, singly or combined. Their advantages for petroleum exploration are that (1) they organize dip-related structural and stratigraphic control into a single numerical package that can be evaluated on a "statistical" basis; (2) they often provide subtle clues to structural or stratigraphic conditions that are either missing, or not readily discernible, from other modes of data display; and, (3) they often provide surprisingly effective procedures for predicting deep structure, including drastic change of shape with depth, from shallow data.

The geometric basis for these capabilities is that the slope of a dip profile is directly related to the curvature of the bedding, as seen on a (transverse) cross section. Bedding curvature, however, is a conservative property of subsurface space and is characterized by various regularities that limit the range of permissible shapes that a dip profile may display. A dip profile accordingly, may show either zero slope, progressively increasing slope, progressively decreasing slope, or slope that is marked by the appearance of one or more of 10 possible "special points."

Dip profiles are graphs that show apparent dip as a function of distance along selected horizontal, vertical, or inclined lines on cross sections. Dip profiles can be generated from structural controls of any kind including surface dips, migrated seismic dips, dipmeter dips, or dips derived from cross sections or contour maps, or any combination of the above. Dip profiles have many structural and stratigraphic uses, in petroleum exploration, in particular, they provide the following.

1. Objective procedures for integrating different kinds of dip-related data into a numerical package that can be interpreted on a "statistical basis." This capability is especially advantageous for resolving conflicts of ambiguous or erratic data.

2. Objective methods for extrapolating and interpolating structural control on cross sections that are especially effective for predicting deep structure from shallow data.

3. Subtle clues to structural or stratigraphic conditions that are not readily apparent from other methods of data display.

RELATION OF DIP PROFILE SLOPE TO BEDDING CURVATURE OBSERVED ON CROSS SECTIONS

The slope of a dip profile bears a direct relation to the curvature of the beds along the line of traverse. Unless otherwise specified, the dip profiles figured in this paper represent "smoothed" data and the cross sections represent transverse cross sections (i.e., cross sections perpendicular to the structural grain, as defined by Bengtson, 1981). The upper center drawing of Figure 1 is a cross section of beds that dip uniformly to the left. The drawing at the right is a dip profile for the horizontal line on the section and the drawing at the left is a similar dip profile for the vertical line. The slope of the data line (with reference to the zero dip axis) is zero on both profiles. These relations demonstrate that a dip pro-

Read before the Association June 2, 1981, at the AAPG Annual Convention. Accepted for publication July 1, 1981.

31

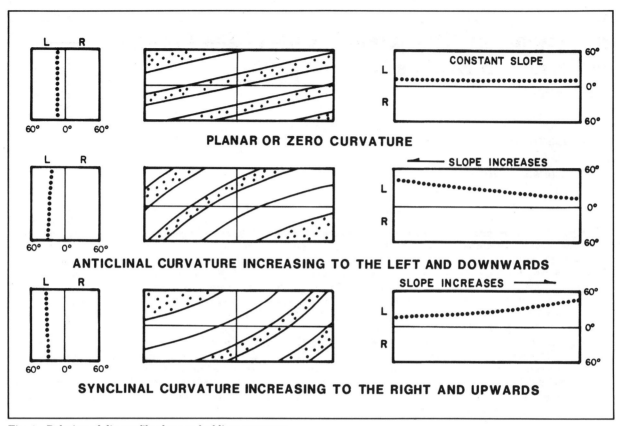

Fig. 1—Relation of dip profile slope to bedding curvature.

file of zero slope corresponds to bedding with zero curvature.

The next series of drawings represents a cross section and horizontal and vertical dip profiles for a group of anticlinally curved beds with curvature that increases progressively to the left (and downward). The slope of the horizontal dip profile increases toward the left, reflecting the increase of bedding curvature in that direction and the slope of the vertical dip profile increases downward, reflecting the increase of bedding curvature downward. The bottom drawings depict a cross section and horizontal and vertical dip profiles for a group of synclinally curved beds with curvature that increases to the right and upward. The increase of curvature is reflected by an increase of slope of the dip profiles in the direction of increasing curvature. The foregoing considerations demonstrate that the curvature of a group of beds is related to the change of dip per unit distance along a dip profile; the greater the rate of change (or the steeper the slope of the dip profile), the greater the curvature of the beds. The rare exception occurs in a dip profile that is fortuitously aligned along a dip isogon (i.e., a line that connects points of equal dip on a cross section). No points on the dip profiles of

Figure 1 could be singled out as having a special property not shared by neighboring points—in other words, all points might be called "ordinary points," in contrast to the "special points" discussed in the next section.

RELATION OF DIP PROFILE SPECIAL POINTS TO CROSS-SECTION CURVATURE ANOMALIES

Figure 2A is a transverse cross section of a simple anticline that maintains its overall shape with depth. Figure 2B shows horizontal and vertical dip profiles for line A-A' and the vertical well, respectively. These profiles, in contrast to the profiles of Figure 1, are characterized by the presence of special points, or points where the slope of the dip profile exhibits a property not shared by points on either side. For example, the points marked 1P α on the horizontal and vertical dip profiles are points of zero slope that are also points of maximum left dip (points 1 and 2) or maximum right dip (point 5). Such points always fall on the trace of an inflection plane,[1] or locus of points on a cross section where the curvature of the beds changes from anticlinal curvature on one side to synclinal curvature on the other.

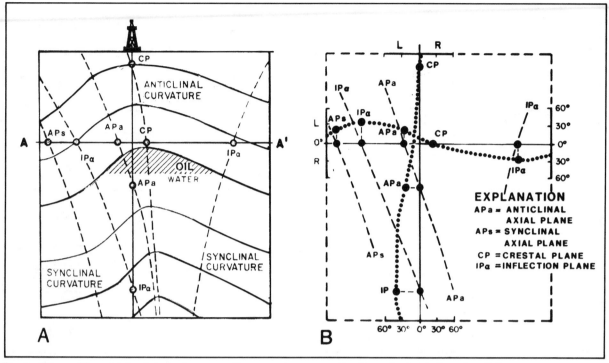

Fig. 2—Dip profile special points and cross-section curvature anomalies for a simple conformal anticline.

The points marked APa on the horizontal and vertical dip profiles correspond to points of maximum dip-profile slope. As such, they identify points of maximum change of dip per unit distance along the respective dip profiles and hence serve to locate points of maximum anticlinal curvature. Such points, by definition, fall on the trace of an axial plane; in this instance, the anticlinal axial plane. In a similar fashion, a synclinal axial plane is identified at the point marked APs on the horizontal dip profile.

The locus of points on a cross section where the bedding has a special curvature property not shared by points on either side will hereafter be called a cross-section curvature anomaly. As explained in the next section, nine kinds of curvature anomalies, each characterized by a distinctive kind of dip profile special point, exist in nature. The points labeled CP on the horizontal and vertical dip profiles define the trace of the crestal plane. It should be noted that although crestal (and through) planes are structural anomalies, they are not curvature anomalies, except by chance.

Most of the geometry of Figure 2A could be reproduced in Figure 2B simply by projecting the inflection and axial planes along their indicated trajec-

tories and assuming that both the dip and the curvature property remains the same at each point along a particular trajectory.

CATALOG OF DIP PROFILE SPECIAL POINTS

Figure 3 presents a summary review of the 10 possible kinds of special points observed on horizontal and vertical dip profiles. Nine of these special points correspond, respectively, to one of nine possible kinds of cross-section curvature anomalies, whereas the tenth—namely, a break in trend, as illustrated by Figure 3J—marks an abrupt increase or decrease of dip as observed across an angular unconformity or certain kinds of faults. All cross-section curvature anomalies, except an IP γ-type inflection plane (Fig. 3F), are seen on both vertical and horizontal dip profiles. An IP γ-type special point, however, is seen only on vertical or highly inclined dip profiles on disharmonic structures, such as the example shown in Figure 5.

Figure 3A, the first example, is a point of initial slope that marks the boundary between planar beds to the left and curved beds to the right. A point of maximum slope, as in Figure 3B, corresponds to an ordinary axial plane, such as shown in Figure 2. A point of minimum slope, as in Figure 3C, corresponds to symmetry plane of minimum curvature between opposing axial planes, as on a flat-topped anticline,

---

[1] The term "plane" as used in expressions such as "axial plane," and "fault plane," is intended to mean surfaces that are either plane or curved and the term "trace of" will usually be omitted when referring to such features.

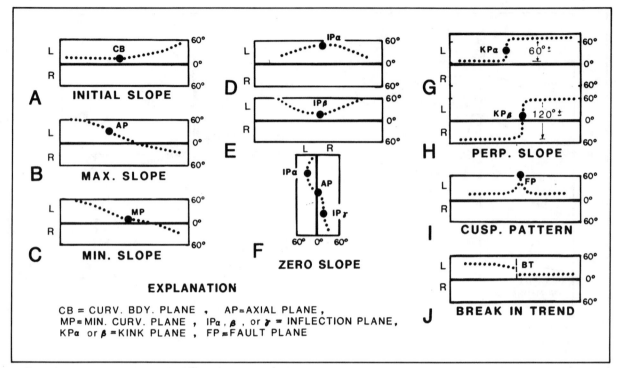

Fig. 3—Ten kinds of special points observed on dip profiles. First nine (A to I) are related to cross-section curvature anomalies. The tenth (J) is related to either an angular unconformity or certain kinds of faults.

such as shown in Figure 6, or a flat-bottomed syncline.

Figure 3D, as already seen in Figure 2, shows a point of zero dip profile slope at a point of maximum local dip that corresponds to an ordinary, or IP α - type inflection plane. Figure 3E shows a point of zero dip profile slope at a point of minimum local dip that corresponds to an IP β -type inflection plane, and Figure 3F shows a point of zero slope on a vertical dip profile, labeled IP γ, that is neither a point of maximum or minimum dip. The presence of IP β and IP γ points, if of structural origin, are always indicative of disharmonic structure. The special points of Figures 3D and 3E are seen on both horizontal and vertical dip profiles, but the point labeled IP γ on Figure 3F is seen only on vertical or steeply inclined dip profiles.

Figures 3G and 3H show points of very steep or nearly vertical slope corresponding to zones of very abrupt change of dip. Such points identify kink planes. A kink plane is a species of axial plane that bisects the angle of intersection of two bands of different semiplanar dip, thereby producing a narrow zone of extremely sharp bedding curvature. The bands of differing semiplanar dip usually differ by either 60° ± or 120° ±.

Figure 3I shows a cusp pattern characterized by dip that increases at an ever-increasing rate as the point of the cusp is approached from either side. If the cusp pointed downward instead of upward, it would correspond to dip that decreased at an increasing rate and eventually reversed. Cusp patterns, as explained later, serve to locate dip-slip faults and determine their direction of dip.

DIP PROFILE SPECIAL POINTS RELATED TO LISTRIC NORMAL FAULTING

Figure 4 is a transverse cross section showing a listric normal fault and related rollover anticline. The upper horizontal dip profile is for line A-A' on the cross section, the bottom horizontal dip profile is for line B-B', and the vertical dip profile is for the well. The points labeled CB on all three dip profiles correspond to points of initial dip profile slope (cf. Fig. 3A). Such points define the trace of a "curvature boundary plane" that separates planar beds (showing regional dip) from the curved beds comprising the rollover anticline. The curvature of the beds in the rollover anticline increases progressively to the right and reaches a maximum at the listric fault. The lower horizontal dip profile shows an abrupt return to regional dip to the right of the listric fault, and the vertical dip profile shows a similar return to regional dip below the fault. The distinctive pattern labeled FP on these profiles could be called a "half-

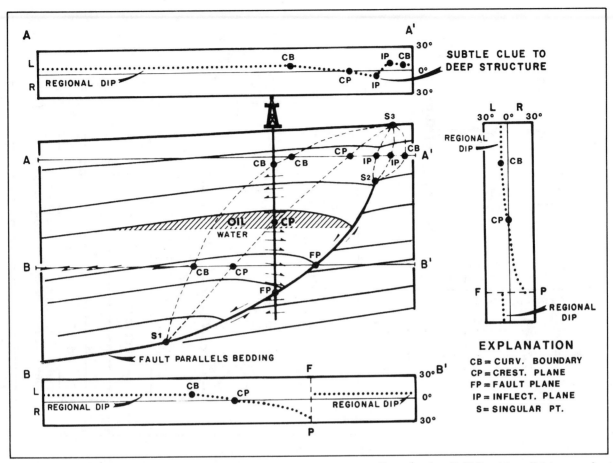

Fig. 4—Dip profile special points and related cross-section curvature anomalies related to Gulf Coast-type listric normal faulting.

cusp pattern." Referring to Figure 3 it is seen to be a combination of the patterns of Figure 3I and 3J. Such patterns are indicative of listric faulting, either normal or thrust. Alternative explanations (if faulting is ruled out) include proximity to a flat-bottomed bar or reef, or a channel cut-and-fill configuration.

Because of downward flattening, the listric fault becomes a bedding-plane fault when traced downward and to the left. At or beyond this point (depending on the fault displacement), the beds above and below the fault are parallel. This point is labeled S_1 on Figure 4. CB, the curvature boundary plane, must therefore terminate by contacting the fault at point S_1. Because of compaction and drape, departure from regional dip persists upward from point S_2 to point S_3, above which regional dip prevails. A curvature boundary plane that extends from S_2 joins the larger left-flank curvature boundary plane at point S_3. Inflection planes extending from S_1 to S_3 bound a narrow zone of synclinal cur-

vature that terminates upward and downward at S_3 and S_2, respectively. The crestal plane of the rollover anticline terminates downward at S_1, and upward at S_3. The rollover anticline of Figure 4, unlike the compressive anticline of Figure 3, has no axial plane because the curvature increases progressively from the left-flank curvature boundary plane to the listric fault. As a general rule, axial planes are observed only on structures of compressional or diapiric origin and are absent on structures of extensional origin, such as the Figure 4 example.

The points marked S_1, S_2, and S_3 on Figure 4 are called "singular points," the name given to points where one or more curvature anomalies join. Such points are always found on disharmonic structures, such as the examples shown in Figures 4, 5, and 6. An important rule concerning cross-section curvature anomalies and singular points can now be formulated. A cross-section curvature anomaly, once established, will either (a) extend across the entire cross section, as in Figure 2; (b) terminate at a fault

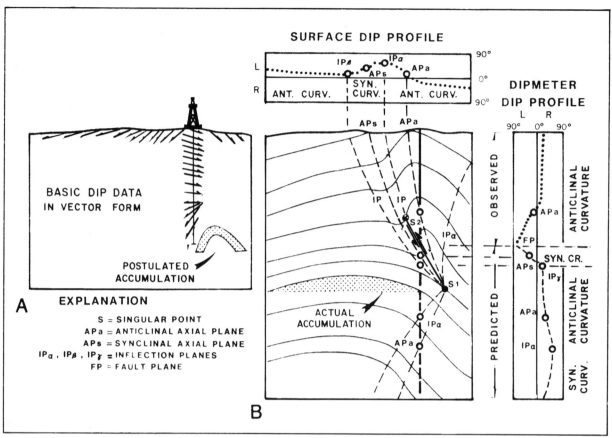

Fig. 5—Predicting disharmonic structure from subtle dip profile clues.

or unconformity; or (c) join one or more other curvature anomalies at a singular point, as in Figure 4.

In Figure 4, the configuration of either the vertical dip profile or horizontal dip profile B-B′ provides clear indications of the true structural situation. In the absence of deeper data, the pattern displayed on dip profile A-A′ suggests, but does not prove, the possibility of listric faulting. If similar patterns from nearby sections define a consistent trend when plotted on a constant level map, the strength of this subtle clue is reinforced. In this connection, it should be noted that entering dip profiles from seismic lines on constant time (or depth) maps affords a supplementary method of mapping seismic structure that quickly reveals true structural conditions and does not have to be redone when new data are acquired.

PREDICTING DEEP STRUCTURE FROM SHALLOW DATA

A band of shallow dip data displayed in vector form appears at the top of the cross section of Figure 5A (these data could be either surface dips or shallow seismic dips). A similar band of vectorial dipmeter data is displayed along the well bore. Visual inspection of these data suggests a possible accumulation to the right of the well. The same data are displayed in dip profile form above and to the right of cross-section B. The surface dip profile from left to right shows a point of zero slope and minimum left dip labeled IP β (cf. Fig. 3E), a point of maximum slope to the left labeled APs (cf. Fig. 3C), a point of zero slope and maximum left dip labeled IPa (cf. Fig. 3D), and a point of maximum slope to the right labeled APa (cf. Fig. 3B). The presence of IP β, an indicator of an inflection plane located at a point of minimum rather than maximum dip, tells us immediately that we are dealing with disharmonic structure. The only special point seen on the vertical profile is APa, an anticlinal axial plane. This is the same axial plane seen on the surface dip profile, so a corresponding axial plane can be drawn on cross-section B and projected downward and to the right beyond the well. The deepest part of the vertical dip profile suggests a buildup toward a cusp pattern. Such a pattern is entered on the predicted portion of the dip profile. Because this pattern corresponds to a thrust fault dipping to the right, as indicted in Figure 7, a thrust

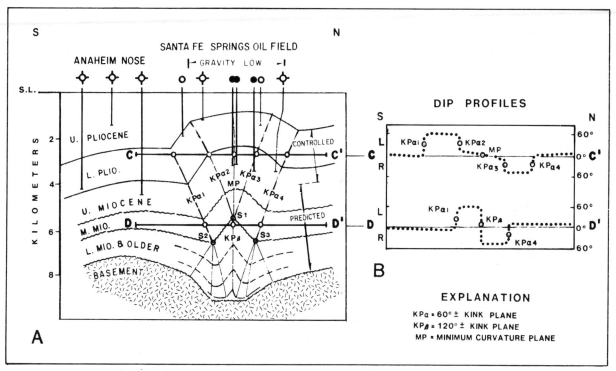

Fig. 6—Predicting drastic change of shape at depth from shallow dip profile evidence.

fault is entered on the cross section and extended upward to join the inflection plane corresponding to IP α at singular point S_2. This thrust fault is then extended downward to intersect APa at singular point S_1. The question concerning how to project the cross-section curvature anomalies corresponding to points APs and IP β on the surface dip profile remains. APs, the synclinal axial plane, and inflection plane IP β are projected downward and to the right to join APa and the thrust fault at point S_1.

At points where projected curvature anomalies intersect the downward projection of the well bore, points of secondary "control" on the extrapolated portion of the vertical dip profile are defined. By elimination, it is seen that the inflection plane represented by an IP β special point on the horizontal dip profile must correspond to an IP γ -type special point on the vertical dip profile. (As previously noted, such special points are observed only on vertical or highly inclined profiles.)

A postulated anticlinal axial plane and an IPα -type inflection plane are projected downward and to the left from point S_1. The intersections of these curvature anomalies with the projected well bore provide two additional "control points" for the predicted vertical dip profile. To complete the cross section, another IPα -type inflection plane is projected upward and to the right from point S_1.

The above operations provide the basis for reconstructing the cross-sectional geometry shown in Figure 4B. The steps involved can be summarized as follows.

1. First, convert any data bearing on structural configuration to dip profile form.

2. After "smoothing" the data, determine which, if any, of the 10 special points shown in Figure 3 occur on the dip profile.

3. Project the cross-section curvature anomalies corresponding to the special points (determined by step 2) into areas of no data. (More rigorous procedures for this step will be developed in later sections.)

4. By trial and error, bring the analysis into agreement with any other information bearing on the problem, such as regional stratigraphic knowledge and known or inferred tectonic style.

DRASTIC CHANGE OF SHAPE PREDICTED BY DIP PROFILE ANALYSIS

Figure 6A is a cross section of the Santa Fe Springs oil field in the Los Angeles basin. The interpretation for the portion of this cross section that is designated as "controlled" is based on abundant well data. (No usable seismic data are obtained in this area below the Pliocene-Miocene contact.) A curious

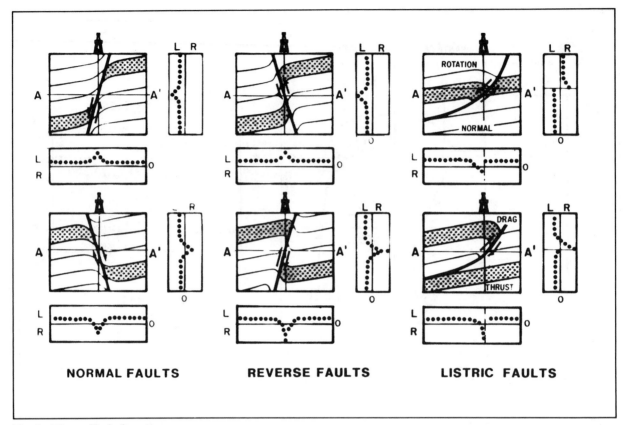

NORMAL FAULTS   **REVERSE FAULTS**   **LISTRIC FAULTS**

Fig. 7—Dip profile fault patterns.

feature of the Santa Fe Springs anticline is that although it is a large high-amplitude feature, its location is marked by a residual gravity low.

Figure 6B shows dip profiles for lines C-C' and D-D' of Figure 6A. The dip profile for line C-C' as previously indicated, is based on excellent subsurface control, but the dip profile for line D-D' is based entirely on projected data because it lies below the depth of direct control. From left to right, dip profile C-C' shows a point of maximum (nearly vertical) slope labeled KPα_1 and a second point of nearly vertical slope labeled KPα_2. These points define the boundaries of a quasi-kink band that constitutes the south flank of the Santa Fe Springs structure. The difference between the steep flank dips and the dips to the left and the right is nearly 60°. These points, therefore, correspond to KPα-type kink planes (Fig. 3G). Similar KPα-type kink planes on the north flank of the Santa Fe Springs structure are labeled KPα_3 and KPα_4. A point labeled MP, located approximately midway between the flanks, marks a point of minimum dip profile slope and, accordingly, defines the trace of a minimum curvature plane (Fig. 3-C). KPα_2, MP, and KPα_3 are projected downward

and intersect at S$_1$, a singular point. From the geometry of the cross section, it is obvious that a KPβ-type kink plane characterized by a dip difference of 120°± must project downward from this point. Kink planes KPα_1 and KPα_4, if projected downward, would intersect KPβ near the estimated top of the basement. Such an arrangement, however, would result in excess shortening of the middle Miocene and younger beds with respect to the lower Miocene and older beds, a condition that violates the principle of balanced sections (Dahlstrom, 1969). S$_2$ and S$_3$ are the lowest points along the trajectories of KPα_1 and KPα_4, respectively, above which proper bed length balance prevails.

Geometric reasoning, therefore, requires that extra curvature anomalies fan out from these points in such a fashion as to balance the shortening below with the shortening above. When this is done, it is seen that the basement surface underneath the Santa Fe Springs anticline has the shape of a syncline, rather than an anticline, thus explaining the gravity minimum. Notice that the horizontal shortening across the basement syncline and that of the upper level anticline are approximately balanced.

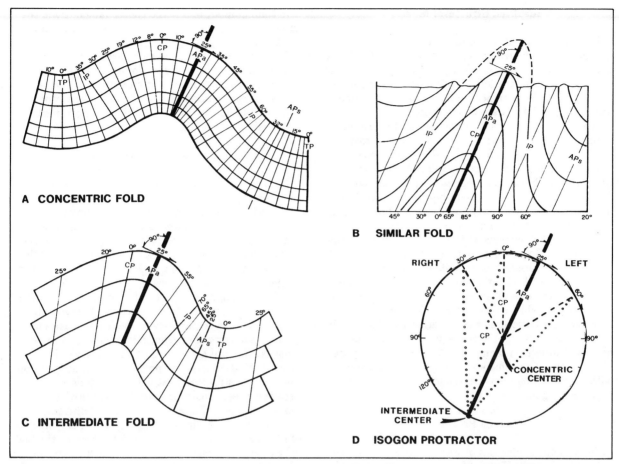

Fig. 8—Isogon patterns of idealized compressive folds: A, concentric fold (isogons are perpendicular to the bedding). B, similar fold (isogons are parallel to the axial plane). C, intermediate fold (isogons are intermediate between A and B).

DIP PROFILE FAULT PATTERNS

Dip-slip fault planes can be considered as inflection planes across which differential displacement has occurred. Because most dip-slip faulting is preceded by the formation of drag zones, such faults are bounded by zones of very abrupt change of dip that show distinctive cusplike patterns on dip profiles, as illustrated by the examples in Figure 7. The rate of change of dip corresponding to cusp patterns is too great to be explained by any folding mechanism. Faulting is the only structural process that can explain these patterns. (Although kink planes also show very abrupt change of dip, they are characterized by steplike, rather than cusplike patterns.) Because drag is a local disturbance imposed on the main trend, it may either steepen, flatten, or reverse the background dip.

Figure 7 shows that fault drag, with one exception, is characterized by anticlinal curvature on the upthrown block and synclinal curvature on the

downthrown block. The exception occurs with listric normal faults, whose "reverse drag" is caused by rotation of the hanging wall toward the foot wall. Figure 7 also illustrates an important rule concerning dip-profile fault interpretation: The cusp points in the direction of dip of the fault if the fault is a normal fault, and in the reverse direction if the fault is a reverse fault. The one exception, again, occurs with listric normal faults.

SUMMARY OF ISOGON ANALYSIS

Isogons (Ramsay, 1967) are lines that connect points of equal dip on those portions of a cross section where the bedding is curved. The trace of a crestal plane, or isogon of zero dip, is a familiar example. If an anticline as a whole is tilted 10° to the right, its crestal plane becomes an isogon of 10° right dip and the previous isogon of 10° left dip becomes the new crestal plane.

Idealized Structural Models

Figure 8 shows three idealized fold models characterized by straight isogons. The anticlinal and synclinal axial planes of all three structures have the same dip—65° to the left. The isogons of the concentric fold (Fig. 8A) are perpendicular to the bedding and will therefore intersect when projected downward on the anticlinal portion, or upward on the synclinal portion. The isogons of the similar fold (Fig. 8B), in contrast, are parallel with the axial planes (APa and APs), and therefore have no upward or downward limit. The isogons of the intermediate fold (Fig. 8C) are intermediate between those of 8A and 8B but have upward and downward limits—like a concentric fold. Such a structure, if examined in detail, would show isogons that have a fine-scale zigzag pattern. Each isogon is perpendicular to the bedding in the competent beds, but refracts and becomes parallel with the axial plane in the incompetent beds. The overall effect on cross sections of normal scale is a smooth trajectory of intermediate inclination. On structures with straight isoclines, such as those of Figure 8, the curvature anomalies (i.e., axial and inflection planes) are also straight and are also isogons. The thickness of any bed on a structure with straight isogons is constant when measured along an isogon.

Isogon Protractor

The isogon regularities of Figure 8 are summarized in graphic form in Figure 8D, an isogon protractor. To construct such a protractor, overlay a transparent circular protractor on a cross section and draw a line through the center parallel with the axial plane. Solid line APa of Figure 8D is such a line.

When this protractor is overlain on the concentric fold (Fig. 8A), the isogon for any point where the bedding is tangent to the circumference coincides with a radius of the protractor. The dashed line on Figure 8D are examples. Notice in particular that the crestal plane, CP, is vertical. If the protractor is overlain on the similar fold (Fig. 8B), it is seen that all isogons are parallel with the axial plane, as expected. When the same operation is performed on the anticlinal portion of the intermediate fold, it is seen that the intermediate isogons (represented by dotted lines on Fig. 8D) converge at the point where diameter APa intersects the circumference below the concentric center. The isogons for the synclinal portion converge at the point where diameter APa intersects the circumference above the concentric center. Notice that the crestal plane, CP, is inclined in contrast to the concentric crestal plane, which is vertical. If the center of isogon convergence falls inside the circumference, the structure is more concen-

tric than similar, and if the center falls outside the circumference, the structure is more similar than concentric.

With a protractor, such as shown in Figure 8D, and a single horizontal dip profile, it is possible to reconstruct the geometry of an entire transverse cross section in accordance with either the concentric, the similar, or the intermediate isogon mode, or any desired combination.

Idealized Real Structures

Figure 9 shows five structures that are obviously more analogous to real structures than the highly idealized models of Figure 8. Neither the curvature anomalies nor the isogons are straight, and the curvature anomalies are not isogons, as in the idealized models. The isogons steepen or flatten with depth and reverse their dips on crossing an inflection plane. Under these circumstances it is not possible to develop an isogon dip protractor for a cross section as a whole. However, from relations observed on portions of a cross section where the isogons are known from actual control or from relations observed on cross sections of structures of similar style, it is possible to derive isogon protractors for selected portions of the structure that will aid in constructing a network of secondary or predicted dip profiles. The cross sections of Figures 2, 5, 8, and 9 illustrate an important rule of dip-profile analysis. In the absence of stratigraphic convergence, the tangent to the bedding along the trace of an axial plane (i.e., the locus of points of maximum slope on dip profiles) is ideally perpendicular to the trace of the axial plane. This rule is equally valid for both curved and straight axial planes. A similar rule applies to minimum curvature planes.

Isogon "Rules"

Although the subject of isogon analysis is almost as large as the subject of descriptive structural geology, much of its practical value can be expressed by various "rules of thumb." A summary of the rules discussed and implied thus far is set forth below:

1. Isogons tend to be straight or smoothly curved, except in zones of disharmonic structure, stratigraphic convergence, or structural "growth."

2. Isoclines show a zone of misfit at angular unconformities.

3. Irregularities of isogon behavior, not explainable by structural causes, are due to stratigraphic effects.

4. Isogons disappear only at faults, unconformities, or singular points.

5. Isogons converge toward areas of greater bed

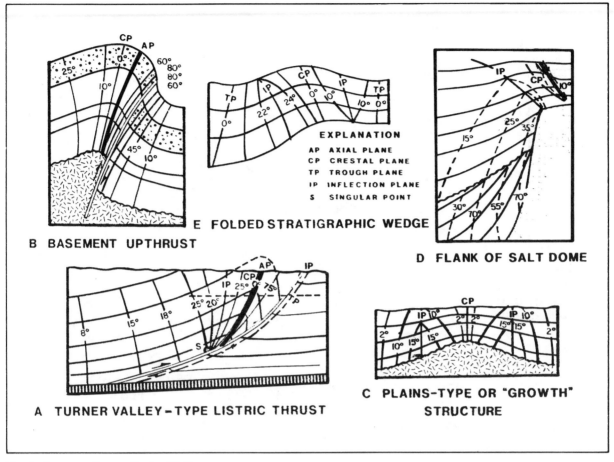

Fig. 9—Isogon patterns of actual folds.

curvature.

6. Isogon spacing on unfaulted structures is closest at ordinary axial planes.

7. Isogons near dip-slip faults are closely spaced and subparallel to the fault because of drag.

8. Isogon behavior tends to be more "similar" in zones of closely spaced isogons and more "concentric" in zones of widely spaced isogons.

9. Isogons (and opposing flank inflection planes) converge downward on compressional anticlines and upward on compressional synclines (except on ideal chevron or similar folds, where they remain parallel with the axial plane).

10. Isogons (and opposing flank inflection planes) diverge downward on extensional (or horst-type) anticlines (related to normal faulting) and upward on extensional (or graben-type) synclines.

11. If the dip on a flank of a structure increases with depth, the isogons form an anticlinal pattern roughly bisected by the inflection plane. If the dip decreases with depth, the isogons form a synclinal pattern roughly bisected by the inflection plane.

EXPANDING STRUCTURAL CONTROL BY CURVATURE ANOMALY AND ISOGON PROJECTION

Figure 10B depicts a well that had good shows at an apparently low structural position. No structural information concerning the possible size and location of the indicated discovery is available, other than the well's dipmeter survey. Figure 10D is a polar tangent or "true vector" plot of the dipmeter data (Bengtson, 1980) that shows an east-west-trending straight line pattern displaced 6° northward from the center of the display. This configuration indicates that each dip on this structure (disregarding scatter) has a constant north component of 6° (equal to the angle of plunge of the associated anticline) and a variable east-west (or transverse) dip component. Figure 10A is a vertical dip profile for a transverse (or east-west) cross section through the well, and Figure 10E is a similar dip profile for a longitudinal (or north-south) section through the well. The transverse dip profile identifies a left-flank anticlinal

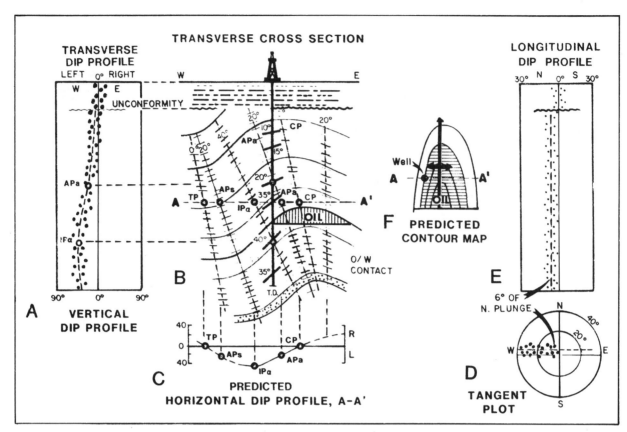

Fig. 10—Three-dimensional structural prediction from limited dip control.

axial plane, APa, and a left-flank inflection plane, IPa. Figure 10E (disregarding scatter) shows a constant component of 6° dip to the north, thus verifying the conclusions derived from the polar tangent plot.

The dip at point APa on the transverse dip profile is 20° to the left. The axial plane, therefore, should ideally dip 70° to the right, and is so drawn on cross-section B. The next step is to construct a synthetic horizontal dip profile for a conveniently located line such as A-A'. We now make the reasonable assumption that the isogons conform to the intermediate style of folding and project all available dip control into profile A-A' (Fig. 10C) in accordance with this assumption. The resulting curve, shown as a solid line on Figure 10C, has numerous points of "known" dip and two points of "known" slope. This curve can be extrapolated for a reasonable distance to the right and left with rather good certainty. The dashed portion of profile A-A' reflects such an extrapolation. Additional isogons, based on the extrapolated portion of profile A-A', can be entered on the cross section. Six isogons are shown as dashed lines with

light dip bars. Secondary dip control provided by the light bars is combined with primary dip control provided by the heavy bars along the well bore to establish the cross-section geometry developed in Figure 10B. Various analytic techniques for semimechanical cross-section interpretation are found in papers by Busk (1929), Coates (1945), Gill (1953), Ickes (1925), and Mertie (1947, 1948, 1949). Although the methods of these papers can be considered as special applications of the isogon approach, none of them deal specifically with the important topics of dip-profile special points and related cross-section curvature anomalies.

Because the dip on any longitudinal section, either through the well, or through any other vertical line on the structure, is 6° north (as determined from the polar tangent plot) it is a simple matter to draw a predicted contour map of the top of the oil zone, as shown in Figure 10F. These procedures, repeated for nearby wells, would yield a tightly controlled and highly reliable three-dimensional structural interpretation.

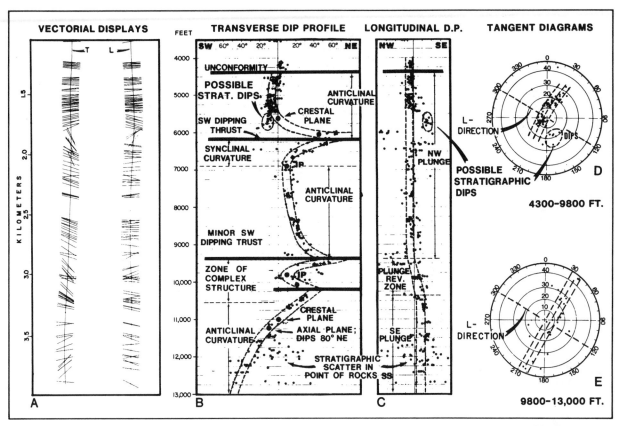

Fig. 11—Transverse and longitudinal dip profiles for the discovery well of the Railroad Gap oil field, Kern County, California.

DIP PROFILE ANALYSIS
OF THE RAILROAD GAP OIL FIELD

In concluding the structural portion of this paper, consider the two left-hand panels of Figure 11, which are vectorial displays of dipmeter components in the direction of transverse and longitudinal cross sections through the discovery well of the Railroad Gap oil field in Kern County, California. Although the data of these displays have been "smoothed" by low-order vectorial averaging, the erratic appearance of these displays defies interpretation by conventional methods. The circular displays at the right of Figure 11 are tangent diagrams. The first is for depths from 4,300 to 9,800 ft (1,311 to 2,579 m) and the second is for depths from 9,800 to 13,000 ft (2,579 to 3,963 m). These displays indicate that above 9,800 ft (2,579 m) this structure plunges about 6°N 60°W and that below 9,800 ft (2,579 m) it plunges about 8°S 60°E. The data on these displays were also filtered by low-order vectorial averaging. The two central displays are dip profiles for transverse and longitudinal cross sections through the well. These displays are not vec-

torially smoothed and show all of the calculated dipmeter picks. Disregarding scatter, the longitudinal profile shows uniform northwest dip (6° ±) to about 9,800 ft (2,579 m) where a reversal to uniform southeast dip (8° ±) occurs. The transverse dip profile, in contrast, presents a very complicated appearance. In particular, we note the presence of three southwest-dipping thrusts, two β-type inflection planes, two crestal planes, one axial plane, and three singular points. Applying the principles previously discussed, the transverse cross section of Figure 12 is derived. This section is much like the true section developed by later drilling. (An early, nearly identical version of this section was drawn soon after the dipmeter data became available.) The longitudinal section of Figure 12, whose construction is self-explanatory, together with the transverse section, provide secondarily derived control for drawing a structure contour map of the top of the Carneros oil zone. The predicted and actual oil-water contact agree quite well. The main disagreement is due to the effects of an unforeseen right-lateral tear fault.

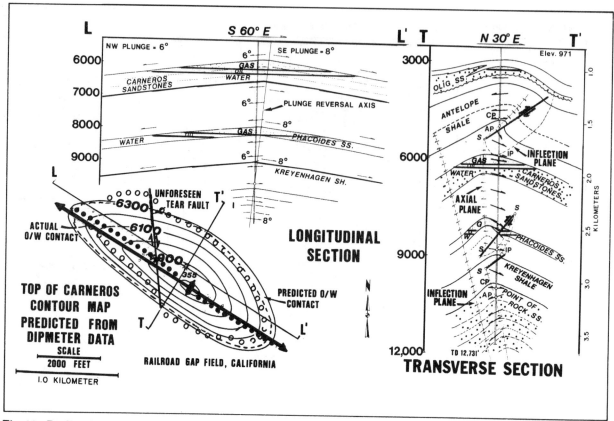

Fig. 12—Predicted transverse and longitudinal structure sections and contour map derived from the dip profiles of Figure 11.

STRATIGRAPHIC USES OF DIP PROFILES

The first step in applying dip-profile analysis to stratigraphic interpretation is to determine the structural background. In spite of claims to the contrary, there is no way that dip data can be used for stratigraphic purposes without first resolving the structural setting on which anomalous dips of possible stratigraphic origin have been imposed. The basic rule is to identify isogon irregularities and dip-profile special points that cannot be readily attributed to structural causes. For example, the half-cusp pattern that characterizes listric faults could also be due to compaction over a sandbar or a reef. Channel cut and fill provides another explanation (Gilreath and Maricelli, 1964). Other effects, such as slumping or sliding during deposition, can also be visualized. In any event, the basic principle is that a cross-section curvature anomaly of stratigraphic origin persists for only a limited distance on a cross section before disappearing at a singular point or an unconformity.

A group of dips that clearly does not conform to the overall trend of either the transverse or longitudinal dip profiles of Figure 11 occurs in the interval between 5,500 and 5,999 ft (2,020 and 2,167 m). These dips are enclosed in a dashed band on these displays, and also on the upper tangent plot to the right. If valid, they call for either a stratigraphic or structural explanation. One possibility, as explained by Gilreath and Maricelli (1964) is that these dips relate to the north side of a channel trending ENE-WSW. Inspection of the E-log, however, indicates that the entire interval is comprised of Antelope shale and we conclude, therefore, that these dips, if valid, are of tectonic origin and related to an ENE-WSW-trending tear or cross fault, probably confined to the Antelope shale.

SUMMARY AND CONCLUSIONS

The dip profile approach provides a seminumerical system of structural and stratigraphic analysis that results (within limits) in a "statistically most probable" interpretation of a given set of data—regardless of the subjective notions of the interpreter. These methods are applicable to any kind of dip-related data; surface data, subsurface data from maps or cross sections, seismic data, or dipmeter data—singly or combined. Dip-profile analysis, which is based on geometric principles alone and is therefore independent of theories of origin, has the following advantages for petroleum exploration.

1. It reveals clues to structural or stratigraphic relations that are often difficult or impossible to detect from other methods of data display and analysis.

2. It provides objective methods for extrapolating and interpolating dip control in three dimensions, and is especially effective for predicting deep structure—including drastic change of shape with depth—from shallow data.

3. It is especially effective for resolving erratic or ambiguous data and for extracting a maximum amount of information from a minimum amount of data.

REFERENCES CITED

Bengtson, C. A., 1980, Structural uses of tangent diagrams: Geology, v. 8, p. 599-602.

_____ 1981, Statistical curvature analysis techniques for structural interpretation of dipmeter data: AAPG Bull., v. 65, p. 312-332.

Busk, H. G., 1929, Earth flexures: Cambridge, England, Cambridge Univ. Press, 106 p.

Coates, J. S., 1945, The construction of geological sections: India Geol., Mineralog. Meteorolog. Soc. Quart. Jour., v. 17, p. 1-11.

Dahlstrom, C. D. A., 1969, Balanced cross sections: Canadian Jour. Earth Sci., v. 6, p. 743-757.

Gill, W. D., 1953, Construction of geological sections of folds with steep-limb attenuation: AAPG Bull., v. 37, p. 2389-2406.

Gilreath, J. A., and J. J. Maricelli, 1964, Detailed stratigraphic control through dip calculations: AAPG Bull., v. 48, p. 1902-1910.

Ickes, E. L., 1925, The determination of formation thickness by the method of graphical integration: AAPG Bull., v. 9, p. 451-463.

Mertie, J. B., 1947, Delineation of parallel folds and measurements of stratigraphic dimensions: Geol. Soc. America Bull., v. 58, p. 770-802.

_____ 1948, Application of Brianchon's theorem to construction of geologic profiles: Geol. Soc. America Bull., v. 59, p. 767-786.

_____ 1949, Stratigraphic measurements in parallel folds: Geol. Soc. America Bull., v. 51, p. 1107-1134.

Ramsay, J. G., 1967, Folding and fracturing of rocks: New York, McGraw Hill, 586 p.

ACKNOWLEDGMENTS

The writer appreciates the encouragement of many Chevron colleagues, especially C. Dahlstrom and P. Verrall.

Reprinted by permission of National Research Council Canada from *Canadian Journal of Earth Sciences*, v. 6, p. 743-757.

Balanced cross sections

C. D. A. DAHLSTROM

Chevron Standard Limited, Calgary, Alberta

Received March 26, 1969
Accepted for publication April 30, 1969

Post-depositional concentric deformation produces no significant change in rock volume. Since bed thickness remains constant in concentric deformation, the surface area of a bed and its length in a cross-sectional plane must also remain constant. Under these conditions, a simple test of the geometric validity of a cross section is to measure bed lengths at several horizons between reference lines located on the axial planes of major synclines or other areas of no interbed slip. These bed lengths must be consistent unless a discontinuity, like a décollement, intervenes. Consistency of bed length also requires consistency of shortening, whether by folding and (or) faulting, within one cross section and between adjacent cross sections.

The number of possible cross-sectional explanations of a set of data is reduced by the fact that, in a specific geological environment, there is only a limited suite of structures which can exist. This imposes a set of local "ground rules" on interpretation. When these local restrictions are coupled with the geometric restrictions which follow from the law of conservation of volume, it is often possible to produce structural cross sections that have a better-than-normal chance of being right.

The concept of consistency of shortening can be extrapolated to a mountain belt as a whole, thereby indicating the necessity for some kind of transfer mechanism wherein waning faults or folds are compensated by waxing en echelon features. These concepts are illustrated diagrammatically and by examples from the Alberta Foothills.

Introduction

The purpose of this paper is to describe a method of checking cross sections for geometric acceptability. In the elementary schools of a bygone era children were taught to prove their answers to arithmetic problems by reversing the process: subtraction was checked by addition and division by multiplication. To apply this principle to a geological cross section, one would flatten out the deformed beds and return them to their depositional position. If this restoration could be done, one would conclude that the cross section was geometrically possible (although not necessarily true) but, if the beds could not be restored, one would conclude that the cross section was geometrically impossible. Construction of restored cross sections is tedious so they are seldom used to check structural interpretation. However, there are shortcuts which make the principle easy to use.

The method to be discussed is now being used consciously by a few geologists and unconsciously by many more to check their cross sections. The essence of the method has been discussed by Hunt (1957), mentioned by Goguel (1952), and illustrated by the published works of Carey (1962) and Bally *et al.* (1966), where it is obvious that the method is being consciously applied. However, the rules have not been set forth and discussed so that many

geologists have not been exposed to the ideas. This paper is intended to remedy this oversight because the writer believes that a "balanced" cross section is a better cross section.

Source of Restrictions in Cross Section Construction

Relatively few geologists still demand for themselves the license of the artist, that is, the right to put on their cross section any interpretation which their imaginations can conceive. Aside from the data itself, most geologists recognize that other restrictions impose boundary conditions within which the imagination must be confined. These restrictions derive from two sources, (1) generalizations derived from observed facts, and (2) geometric principles.

This paper discusses one of the geometric principles and shows how specific rules of interpretation can be derived from that principle. The derivation involves simplifying assumptions which are valid only for specific structural environments. Therefore the same geometric principle will give rise to interpretive rules which vary from one structural environment to another. The writer will not attempt to develop all the possible variants. Discussion is restricted to one simple geological environment, the marginal part of an orogenic belt, and to one example of that environment, the

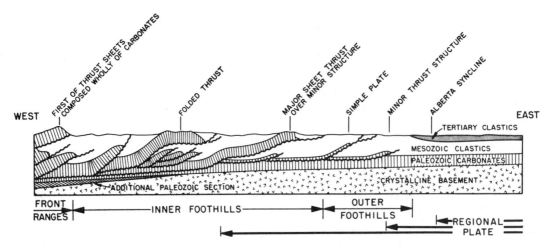

Fig. 1. Schematic cross section of Alberta Foothills (after R. E. Daniel, Chevron Standard). Note that the regional plate extends progressively farther west as older rocks are considered until, at Precambrian level, it extends completely across the section.

Alberta Foothills. To avoid misunderstanding, let it be emphasized that the interpretive rules derived in this paper are literally applicable only to the Alberta Foothills. With minor variants, they can be used in other marginal belts, but they cannot be applied to another environment, such as a salt dome province. However, the principle and method can be used to develop rules that do apply to those other environments.

Generalizations of Observed Facts

As a consequence of intensive government and industry search for oil and gas, the Alberta Foothills is well mapped (Bally *et al.* 1966, and the authors in their bibliography) and sufficient data are available for sound generalizations as to the pattern of structural behavior within this structural province. These generalizations establish the local ground rules for structural interpretation. For example, one of the local ground rules is that folds are ordinarily concentric, rarely chevron but never similar. Recognition of such local interpretive rules is an acknowledgment of the existence of "familial associations" of structures. According to this idea a specific geological environment contains a cratonic family of structures. The "foothills family of structures" comprise:

1. concentric folds,
2. decollement,
3. thrusts (usually low angle and often folded)
4. tear faults, and
5. late normal faults.

The first restriction then in constructing Foothills structure sections is that one is limited to some variant of these basic structural forms.

The area being used as an example is shown in diagrammatic cross section in Fig. 1. Of the many generalizations from observed data which can be made about the Alberta Foothills belt, the following are pertinent to the subsequent discussion.

1. The Precambrian basement extends, unbroken, beneath the Foothills structures.
2. The rocks east of the Foothills are essentially undeformed.
3. The Foothills structures were formed in the latter stages of the Laramide orogeny long after most of the rocks now preserved were deposited.
4. There is no thinning or thickening of beds by "flow" and therefore folding is concentric.

The through-going basement and the lack of deformation east of the Foothills edge limit the number of solutions which can be postulated for particular geometric problems. In an earlier and simpler day, it was the custom of geologists to draw shallow Foothills cross sections wherein geometric problems were solved by faults which dribbled off the bottom of the cross section. A large low-angle thrust, which would be perfectly reasonable in Foothills cross sections, would be as incongruous in a cross section through the plains of Saskatchewan as an elephant on the tundra, because thrusts are not part of the

Seismic and drilling data (Bally *et al.* 1966) have established the basement configuration so that now cross sections like Fig. 1 are closed systems except at the western end. Now the sophisticated geologist draws deep Foothills cross sections with a through-going basement, wherein geometric problems are solved by faults which are gathered into flat sole faults and hustled westward through the only available exit.

The Laramide orogeny moved progressively from west to east with the Foothills being deformed in the Paleocene Epoch. With the probable exception of the youngest beds in the Alberta syncline, all of the Paleozoic and Mesozoic rocks had been deposited long before the deformation. During the course of their depositional history in the miogeosyncline, these Paleozoic and Mesozoic rocks had indeed been subject to epeirogenic fluctuations and consequent erosion but they had not been orogenically deformed prior to Paleocene time. This single, post-depositional deformation of the Foothills is an important point because it simplifies structural interpretation by eliminating from consideration:

1. pronounced angular unconformities,
2. structures growing during deposition, or
3. substantial amounts of compaction during deformation.

The foregoing terse comments touch only upon those topics which are necessary background for the subsequent geometric discussion. No attempt has been made to summarize Foothills geology.

Conservation of Volume

The law of elementary physics (which Einstein amended) that matter can neither be created nor destroyed is paraphrased in geology as the "law of conservation of volume" (Goguel 1952, p. 147). A rock consists of mineral particles and voids filled with fluids. In the initial stages of clastic deposition, the proportion of voids to mineral grains is large and density is low, but this stage is brief because the compaction rate is very high during the first few hundred feet of burial. Thereafter density increases only slowly with depth. It is worth noting here for future reference that the volume change due to load compaction does reduce the

thickness of a bed but it does not alter the areal extent. Volume reduction on account of deformation is thought to be negligible, particularly in the earlier stages of tectonism represented by the marginal part of the Cordillera, because no significant deformation-dependent decrease of porosity (or increase of density) is recognized. For practical purposes, one may assume that the law of volume conservation applies and that rock volume does not change during Foothills type of deformation.

Volume is three dimensional which makes the law of conservation of volume awkward to apply. However, in specific instances it can be made simpler. To start the simplification process one can use the three mutually perpendicular tectonic axes as the three dimensions: "a" is the direction of tectonic transport of rock during deformation, "b" is the direction of the fold axes, and "c" is the third axis which is perpendicular to the "ab" plane.

In the Foothills, the fold axes and the fault strikes are parallel, the folds are concentric, and the faults are dip-slip thrusts. Therefore the "action" takes place in the ac plane, which is a vertical transverse cross section, and virtually nothing happens in the b direction. By ignoring changes in the b direction as insignificant, the law of conservation of volume becomes a two-dimensional statement that the cross-sectional area of a bed does not change during deformation. Cross-sectional areas are manageable quantities so the law has been applied in this form to check cross sections (Hunt 1957) or to calculate depth to detachments (Bucher 1933).

Even further simplification is possible when folding is concentric because bed thickness does not change during deformation. This has been demonstrated in the Foothills, directly by detailed studies of the anatomy of folds (Price 1964), and indirectly by many stratigraphic studies, which do not detect any tectonic thinning of units whether they be on gentle upright limbs, on vertical limbs, on overturned limbs of folds, or on far-travelled thrust sheets. Because the cross-sectional area of a bed is a function of bed length and bed thickness and because the thickness remains constant, it is possible to eliminate one more dimension and to state the law of conservation of volume in terms of length alone: *In concentric regimes the cross-*

sectional length of a bed remains constant during deformation.

Consistency within Cross Sections

In simple deposition the initial areal extent (or length in cross section) of any bed is the same as that of the beds above and below it. Since bed area (or length in cross section) is not appreciably altered by either compaction or by post-depositional concentric deformation, it follows that bed lengths in a cross section must be consistent with one another (Figs. 2a and 2b), unless a discontinuity such as a sole fault or décollement intervenes (Fig. 2c) between the longer and shorter beds.

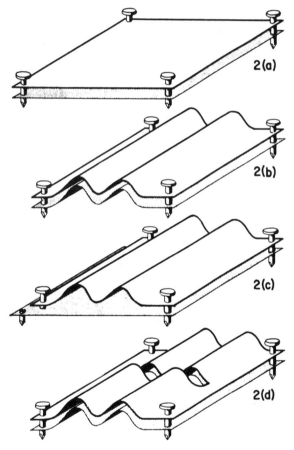

FIG. 2. Consistency of bed length.

Using the idea of consistent bed lengths to check a cross section is very simple. The first step is to establish a pair of reference lines at either end of the section in areas of no interbed slip. These reference lines should be the axial planes of major anticlines or synclines or other

planes of no slip such as a plane perpendicular to the regional dip of the undistorted "plains" section at the eastern end of a regional Foothills cross section. The second step is to measure the bed lengths of selected horizons between these reference lines. They should all be the same so that bed lengths "balance". If the cross-sectional lengths do not balance then the cross section must show a valid explanation of why they do not. In regional Foothills cross sections for instance, there is never a balance between the length of the Mesozoic and Paleozoic beds and the length of basement. The "cover" beds are always too long, which is explained in cross section (Fig. 1) by a sole fault along which the upper beds have been moved (shoved? glided?) into the cross section from the west. This is not simply a way of sloughing one's problems into an adjacent area, because the necessary implication of such a section is that the sole fault continues to the west until the bed length anomaly is resolved by shortening of the basement (compressional hypothesis), or extension of the Mesozoic and Paleozoic "cover" rocks (glide hypothesis). One should note that checking Foothills cross sections for balance by measuring bad lengths is purely a geometric test which is quite independent of the genesis of the structures.

The fundamental rule, that the cross-sectional length of a bed remains constant during concentric deformation, has led to the idea that bed lengths in a cross section ought to be consistent with one another. If this is true, then the displacement on thrust faults ought to be consistent as well. In Fig. 3 the displacement at B must be the same as at A unless it is postulated that elastic, plastic, or compactional deformation of the rock takes place between A' and B'. None of these things happen in a concentric regime.

Despite this conclusion that thrust displacement ought to be consistent, there are many instances where the displacement can be observed to change along the fault plane. There are only two basic ways in which this paradox (Fig. 4a) can be resolved:

1. by interchanging fold shortening and fault displacement (Fig. 4b), or
2. by imbrication (Fig. 4c).

Thrust faulting and folding are both mechanisms for making a packet of rock shorter and

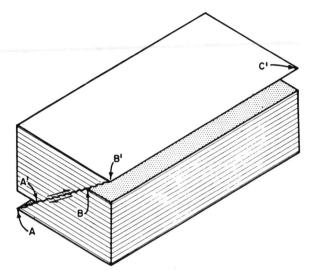

FIG. 3. Consistency of thrust displacement.

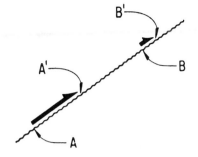

a. APPARENT ANOMALY
IN AMOUNT OF FAULT DISPLACEMENT

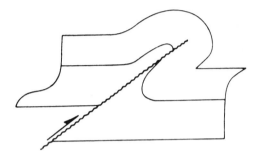

b. ACCOMMODATION BY FOLDING

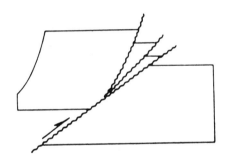

c. ACCOMMODATION BY
UPWARD IMBRICATION

FIG. 4. Changes in thrust displacement. See text for explanation of curved ends on some fault blocks.

thicker than it was originally, so that one could expect the two mechanisms to be interchangeable. Imbrication is the distribution of displacement from one large fault to several minor ones. Figure 4c shows the common imbricate pattern, but imbricates can develop in the footwall rather than the hanging wall of the main thrust, and they can be in the hanging wall but dipping in the opposite direction to the principal fault ("back thrusts").

In Fig. 4 the thrust faults were arbitrarily and diagrammatically represented as simple planar features. With planar faults and constant bed length, a substantial amount of interbed slippage is required, which would alter the originally vertical ends of the blocks to the curved shapes shown. From these block diagrams it is fairly evident that:

1. faults with changing displacement are apt to be curved in cross section rather than planar (can be confirmed by observation);

2. interbed slippage is a necessary part of the thrust faulting process just as it is in concentric folding;

3. interbed slippage can contribute to the change of displacement along a fault plane and, in extreme instances, could become a species of imbrication.

The Turner Valley cross section (Fig. 5) is based on good well, seismic, and surface geological control. It shows a remarkable change from more than 2 miles (3.2 km) of fault displacement at depth to virtually none at surface and the accommodation of this change by fold-

ing. This is an excellent example of a balanced transition from shortening by faulting to shortening by folding. In checking this cross section for consistency, one would probably use the vertical ends of the cross section as reference lines even though strictly speaking, neither is a proper reference line. The eastern end is in the undisturbed plains section, but the reference line should be perpendicular to the regional dip rather than vertical. The western end of the section does not extend as far as the synclinal axial plane, which would be the proper place for the second reference line. However, at this end of the

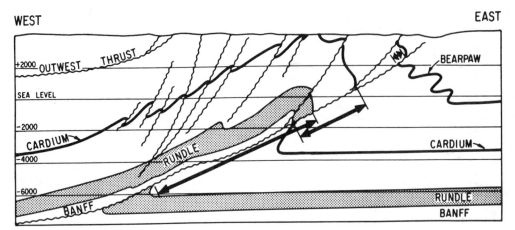

FIG. 5. Changes in thrust displacement in the Turner Valley structure (after Gallup 1951, with the permission of the Amer. Assoc. Petrol. Geol.).

section, the Rundle member has "returned to regional", which provides for practical purposes an acceptable place to draw a reference line, although it too should be perpendicular to the regional dip. (Please observe the behavior of the top of the Paleozoic rocks in the central part of Fig. 1 for an appreciation of the significance of a horizon's return to its regional elevation.)

Examples of the Use of Balanced Cross Sections

Interpreting a cross section through the culmination of the Panther River anticline posed an interesting problem when a well drilled on the crest of the structure penetrated a thrust of Cambrian over Jurassic rocks, a stratigraphic throw of some 8000 ft (2440 m). This was particularly perplexing because at the surface, the west-dipping fault on the east flank of the structure thrust Jurassic over Lower Cretaceous rocks, a throw of only a few hundred feet. Some well control, surface geology (Fig. 6), and seismic data were available to provide the critical data shown in Fig. 7. Data in the line of section could be interpreted in two basically different ways (Fig. 8). The interpretation with one major thrust appeared most likely to be correct because it correlated two major thrusts with stratigraphic throws of the same order of magnitude. The interpretation with two major thrusts seemed unreasonable because it linked a major fault in the subsurface with a minor fault at the surface (one order of magnitude difference in stratigraphic throws). Which alternative one selected had considerable economic significance, because it affected where and at

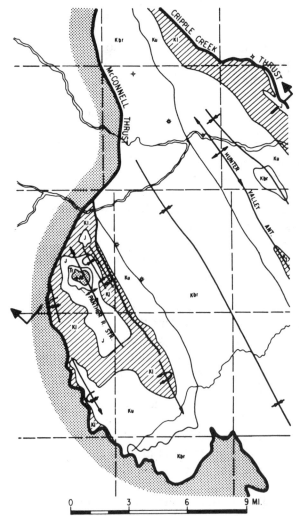

FIG. 6. General geology of the Panther River area.

what depth one could expect to find closure in the potentially hydrocarbon-bearing objective horizon at the top of the Paleozoic section.

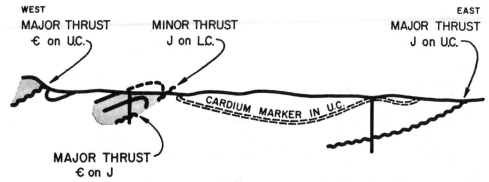

WEST
MAJOR THRUST
Є on U.C.

MINOR THRUST
J on L.C.

EAST
MAJOR THRUST
J on U.C.

CARDIUM MARKER IN U.C.

MAJOR THRUST
Є on J

FIG. 7. Basic data available for the original interpretation of the Panther River cross section.

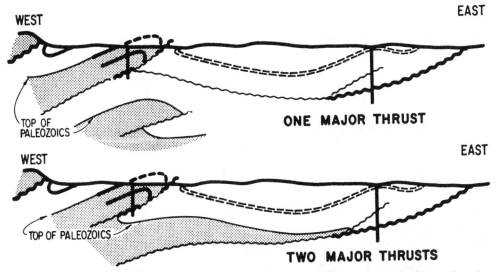

WEST   EAST

TOP OF
PALEOZOICS

ONE MAJOR THRUST

WEST   EAST

TOP OF PALEOZOICS

TWO MAJOR THRUSTS

FIG. 8. Two alternate cross sections which would satisfy the data on the original line of section.

However, the geological map provided some data off the line of section, which had not been used. R. E. Daniel of Chevron Standard statistically analyzed the fold (Dahlstrom 1954), found it to be cylindrical, and calculated plunge values. The data on the geological map were projected up plunge (Stockwell 1960) into the plane of a vertical cross section passing through the culmination to provide the information shown above ground level at the west end of the section in Fig. 9. On this cross section it is apparent that the axial planes of the anticline and syncline are converging in depth, which is confirmed by the reduced length of the steep fore limbs in successively lower stratigraphic horizons. Having the fold disappear with depth would produce a geometrically unacceptable discordance in bed lengths unless the fold disappearing downward could be compensated by the fault dying out upward. This would be a repetition of the Turner Valley situation in

Fig. 5. Once this fundamental point was grasped, it was apparent that the superficially unlikely looking cross section in Fig. 8 with the two major thrusts was the correct one. The cross section which R. E. Daniel constructed according to this concept was subsequently confirmed by the drilling of a well on the deep structure immediately to the east, and by the deepening of the original well.

Hypotheses of regional significance can develop from balancing cross sections. Figure 10 shows two alternative interpretations of seismic, well, and surface data. The data in the two sections are the same, but in Section A, the simply migrated seismic section, the vertical dimension is essentially two-way transit time, whereas B is the normal natural scale structural cross section. The Mesozoic and Paleozoic sections are both cut by thrusts, but even a casual inspection shows that the shortening in the Mesozoic section is far in excess of that in

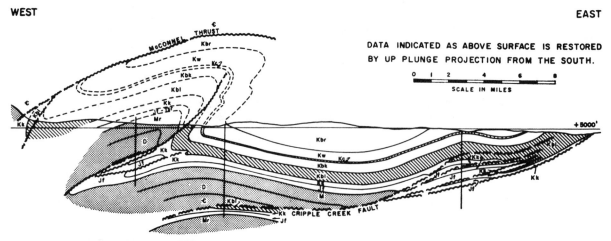

FIG. 9. Panther River cross section as subsequently shown by drilling (after R. E. Daniel, Chevron Standard).

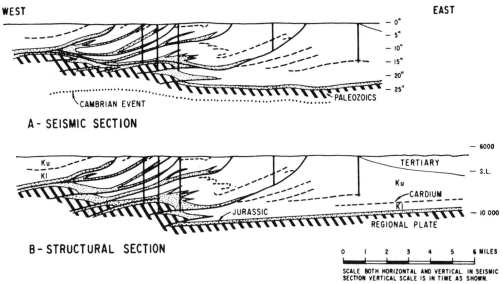

FIG. 10. Two alternate interpretations of surface, well and seismic data in a Waterton area cross section. Minor fault traces have been omitted but their existence is shown by marker bed offsets. Note the inconsistencies of fault displacement in Section B which make this alternate untenable.

the Paleozoic. This difference in shortening precludes the kind of interpretation that was attempted in Section B, because any simple fault that one tries to draw will have far more displacement at Mesozoic level than at Paleozoic level. To avoid an unexplained inconsistency it is necessary to postulate a décollement separating the Mesozoic rocks from the Paleozoic rocks. This décollement would be in the basal Jurassic beds just above the top of the Paleozoic section. Since this horizon is itself now deformed by both folding and faulting, it becomes necessary to postulate a two-stage sequence of deformation. According to this hypothesis the Mesozoic rocks were first de-

formed by folding and thrusting above a sole fault (décollement) in the basal Jurassic beds, and the deformation in the Paleozoic rocks is a subsequent stage of the tectonic cycle. Balancing this cross section shows that the western (upper) thrust system was formed before the eastern (lower) one, and demonstrates (Bally et al. 1966, Fig. 6) that, in the regional sense, deformation advanced from west to east.

Consistency between Cross Sections

The foregoing discussion has been concerned with a one-dimensional check of the internal consistency of an individual cross section by simple measurement and comparison of bed lengths. During the stage of reducing the law of conservation of volume from three dimensions to one, there was a two-dimensional stage wherein one could state that deformation changes the form but does not alter the surface area of an individual bedding plane. This stage is represented in Fig. 2d, where it is evident that no abrupt change can occur in fold form and (or) location unless there is a discontinuity. Similarly, in the block diagram of Fig. 3, the fault displacement between B′ and C′ cannot change abruptly unless a discontinuity intervenes. Such transverse discontinuities (tear faults) do occur in a few places in the Foothills, where they may affect either or both thrusts and folds. From such considerations of the two-dimensional version of the law of conservation of volume one can derive the rule that: *In adjacent cross sections the amount of "shortening" at a specific horizon between comparable reference lines must be nearly the same unless there is a tear fault between them.*

In this context, "shortening" is the difference between actual bed length and the horizontal distance it now occupies. This statement does not deny the possibility of a gradual change in the shortening along an individual structure nor along a mountain belt as a whole, but it does deny that these changes can occur abruptly without tear faulting.

The reader can convince himself of the reasonableness of this rule by taking a long thin strip of light paper and attempting to produce a long parallel system of folds and faults. Paper is a suitable medium for this kind of demonstration because its lack of plasticity obliges one to maintain a constant bed area. The Rocky Mountain belt is 1000 miles (1610 km) long from the Idaho batholith to the Liard River. The deformed width of say 100 miles (160 km) is, perhaps, half of the width of the rocks as originally deposited. These figures provide some approximate dimensions for the paper experiment.

Having performed the experiment actually or mentally, the reader may now have reason to suspect that there should be comparable amounts of shortening in adjacent cross sections. Figure 11 shows the north end of Turner Valley, where there is an abrupt change in structure from the simple faulted anticline of Section B–B to the sheaf of imbricates represented by Section C–C. Despite this change in structural form, the shortening at the top of the Paleozoic level in both Sections B–B and C–C in the published sections is in phenomenally good agreement at a figure of 15 000 ft (457.2 m). These two sections are consistent. Consider the third section D–D. In Section D–D the well control does not establish how far to the left (west) the regional plate extends under the Turner Valley sole fault. The natural tendency is to put the footwall "cut-off" of the top of the Paleozoic rocks at X. If this is done, the shortening in Section D–D becomes approximately 6000 ft (1830 m), some 9000 ft (2740 m) less than in Section C–C. A 60% reduction in shortening from C–C to D–D would be quite inconsistent, and prompts an interpretation where the top of the Paleozoic beds in the regional plate extends back to Y. This would also produce better internal consistency within section D–D by making the fault displacement on the Home Sand (R–S) equivalent to that on the top of Paleozoic section (Y–Z).

The two southern sections have provided a clear example of consistency of shortening in adjacent sections despite rather drastic changes in structural form, and the third section shows how the rules of consistency can be used to choose between alternate interpretations in the absence of definitive well control.

Transfer Zones

Previously it was stated that shortening on the local scale in individual structures and in the regional scale in a whole mountain belt could change, but that the change would be gradual. Using the same word "gradual" for change at both scales is not really appropriate because the rate of change is substantially greater for individual structures than it is for the mountain belt as a whole. The Lewis Thrust, for instance (Dahlstrom *et al.* 1962), has a minimum of 23 miles (37 km) of thrust displacement at the United States border and 135 airline miles (217.2 km) to the north, the fault displacement is zero. Over the same distance the overall shortening in the mountain belt as a whole may have diminished, but certainly not by 23 miles (37 km). Since the whole does not change as rapidly as its component parts, it

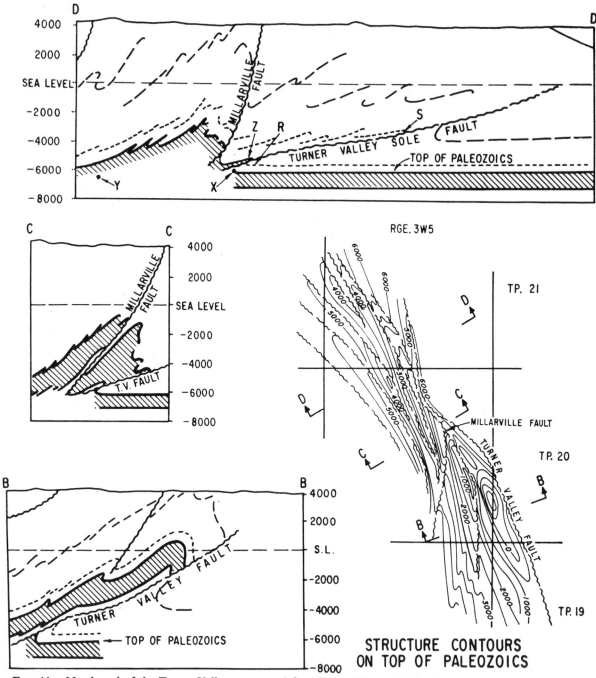

FIG. 11.   North end of the Turner Valley structure (after W. B. Gallup, 1951, with the permission of the Amer. Assoc. Petrol. Geol.). Note that the cross-sectional trace of minor faults has been omitted, but their presence is indicated by offsets in the marker horizons. See text for discussion.

follows that there must be some sort of compensating mechanism at work whereby displacement is "transferred" from one structure to another. Such mechanisms have been observed in the *ac* cross-sectional plane (Fig. 5), so it is not unreasonable to expect that comparable phenomenon would function in the horizontal *ab* plane (as in Fig. 11).

The compensatory mechanism for thrusts is a kind of lap joint wherein the fault whose displacement is diminishing is replaced by an echelon fault whose displacement is increasing. Clearly such a "transfer zone" could not exist unless all of the faults involved in the transfer zone are rooted in a common sole fault. Figure 12 shows a relatively simple transfer zone consisting of three faults. In each of the five cross sections the shortening is exactly the same, although at one end virtually all of the displacement is on fault C and, at the other, on Fault A. In natural examples, the pattern is often complicated by folds and folded thrust imbricates as shown by the two mapped examples in Fig. 13.

through-going basement. Consequently, in the gross view, one would expect some transfer of displacement between the six thrust zones.

It is not proposed to discuss en echelon folding, although it is a part of the Foothills movement pattern (Fitzgerald 1968) wherein the transfer mechanism operates for folds as it does for faults. En echelon folds are tied to

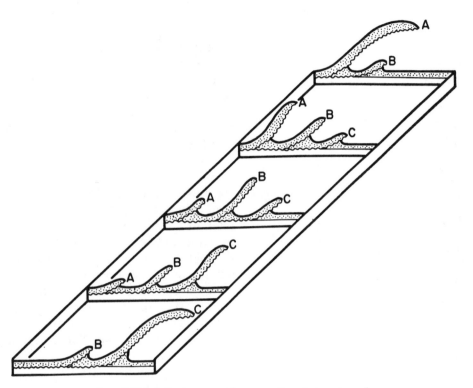

FIG. 12. Shift of displacement between thrusts in a transfer zone.

Recognition of transfer zones enables correlations to be made between thrust faults. Although one fault terminates, its place is taken by another and the zone of thrusting persists. On this basis, a frontal zone and five principal zones of thrusting can be identified within the Foothills and Front Ranges over a distance of some 400 miles (643.7 km) (Fig. 14). At the north end correlation fails because the surficial deformation was primarily folding, and faults are not prominent. The persistence and parallelism of these zones of thrusting over substantial distances lends credence to the suggestion that overall shortening is reasonably consistent along the mountain trend. In one respect, designating six zones of thrusting may be rather arbitrary because at depth all of these zones will join to a common sole fault above the

one another by a sole fault (décollement) and maintain consistent shortening by replacing a dying structure with an en echelon growing equivalent (Wilson 1967).

Palinspastic Maps

The ultimate check of cross sections in deformed terrane is whether or not the process can be reversed and the beds put back into their depositional position without introducing inexplicable bed length anomalies. Since the construction of restored cross sections is a tedious process, the discussion has been concerned with shortcuts which would be adequate for checking the geometric acceptability of individual cross sections and suites of cross sections. However, when stratigraphic studies are being done in deformed terrane, the shortcuts are

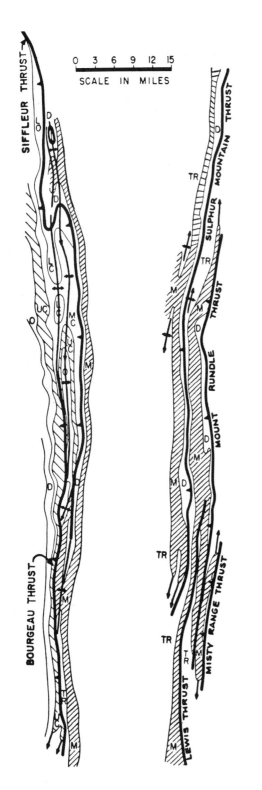

FIG. 13. Two transfer zones in the Front Ranges of Alberta. The abbreviations designate formational ages (i.e., M = Mississippian).

inadequate, and it becomes necessary to make a restored palinspastic map (Dennison and Woodward 1963), which shows beds in their depositional position.

When stratigraphic data obtained by study of deformed terrane are recorded and interpreted on an ordinary map, it is possible to make serious errors. In Fig. 15a, seven pieces of surface and subsurface stratigraphic data are shown in cross-sectional view. On the normal geographic presentation this data would appear in the sequence 1, 4, 5, 6, 2, 3, and 7. Obviously, what is needed for stratigraphic work is a restored map where the data are returned to their original depositional sequence of 1, 2, 3, 4, 5, 6, and 7. The normal presentation can also produce misleading stratigraphic trends (Figs. 15b and 15c).

To avoid these pitfalls, one begins a palinspastic map by constructing a suite of cross sections at regular intervals across the study area. These cross sections must be checked for internal consistency and for consistency between contiguous cross sections according to the methods previously discussed. When a consistent suite of sections is available, then shortening is determined for each fault plate and a palinspastic map constructed wherein individual thrust sheets are unfolded and pulled back to their original locations. Some geographic reference points must be maintained on the palinspastic map so that stratigraphic data can be plotted and the interpretations applied to present-day land positions.

The commonest error in palinspastic map construction is using improper values for shortening in the reconstruction. In Fig. 15d, two horizons A and B are shown where virtually all of the shortening in A takes place at position S, while B is shortened at R. In calculating shortening for palinspastic map construction, it would be very easy to add the shortening at R and the shortening at S together, to arrive at an answer that was twice the proper value. The best method to determine shortening is to do all the measurement on one horizon. If one is forced to change horizons, this can only be done in an area where there is no discontinuity between the reference horizons. In Fig. 15d, one could change from horizon A to horizon B at positions V or T, but certainly not at position U.

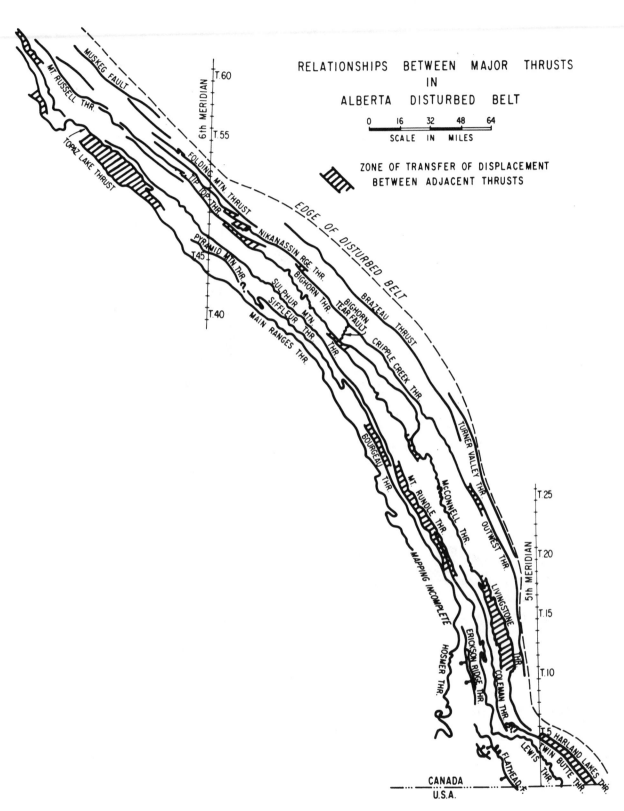

RELATIONSHIPS BETWEEN MAJOR THRUSTS
IN
ALBERTA DISTURBED BELT

SCALE IN MILES

ZONE OF TRANSFER OF DISPLACEMENT
BETWEEN ADJACENT THRUSTS

FIG. 14. Use of transfer zones in correlating zones of thrusting.

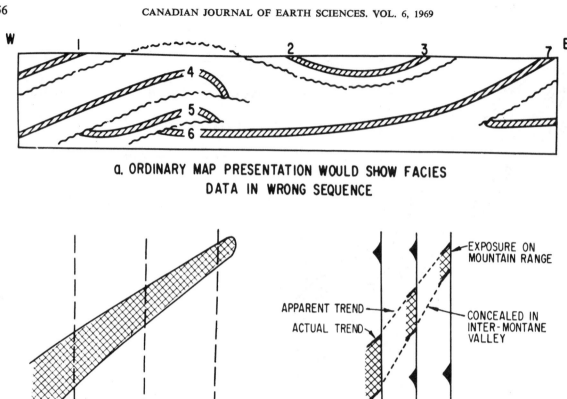

a. ORDINARY MAP PRESENTATION WOULD SHOW FACIES
DATA IN WRONG SEQUENCE

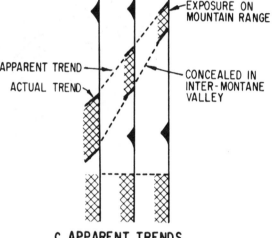

b. TRENDS BEFORE THRUSTING

c. APPARENT TRENDS
AFTER THRUSTING

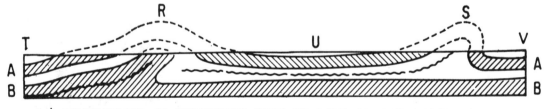

d. CALCULATIONS OF SHORTENING MUST BE MADE ON ONE HORIZON ONLY

FIG. 15.   Elements of palinspastic map construction. See text for discussion.

Conclusions

The quality of structural cross sections can be improved by testing them for geometric validity. In a concentric regime where the structure is post-depositional there is no significant change of rock volume during deformation. Since bed thickness remains constant, it follows that the surface area of any bedding plane remains constant during deformation. Because the area of successive beds in an undeformed depositional sequence is constant, it is necessary that bed lengths in individual cross sections be consistent and that adjacent cross sections have consistent amounts of shortening. Apparent inconsistencies in or between cross sections develop in consequence of discontinuities such as sole thrusts, décollement, or tear faults. Actual inconsistencies are ordinarily due to conceptual, observational, or drafting errors, which might pass undetected were there no tests for geometric validity.

In some lines of geological endeavor, cross sections are significant only as diagrams that are used to convey concepts. In geology applied to oil and mining exploration or to engineering projects, cross sections are used to convey predictions as to rock behavior and they must be conceptually and, more important, geometrically correct. In these areas it is important to the geologist and to his client that there be some way of checking cross-section interpretations prior to drilling. It should be emphasized that a cross section which passes the geometric tests is not necessarily correct, because completely ridiculous cross sections can be drawn which abide by the law of conservation of volume. However, if a cross section passes the geometric tests, it could be correct, and if it has been drawn with due regard for the "local ground rules" it probably is correct. On the other hand, a cross section that does not pass the geometric tests could not possibly be correct.

The rules developed in this paper pertain to a simple, concentrically deformed packet of "layer cake" geology in the Alberta Foothills. The rules themselves cannot be transported to a more complicated area, but the basic concept and method can be used to generate sets of rules which apply to other environments. Thus it is possible to devise amended rules, which apply to concentric extensional structures and to diapiric structures that grow during deposition or to "similar fold" regimes. Each of these areas requires a different set of rules, which must be derived from the basic geometric principles and observational restrictions. The interpretational rules for similar folding were discussed by Carey (1962). In other areas the geologist will have to develop his own rules, an endeavor that he should find interesting and rewarding.

Acknowledgments

The author is indebted to Chevron Standard for permission to publish this paper. It grew from his attempts to provide a rational explanation of normal interpretational practice to structural geologists, new to the ways of the Foothills geology.

BALLY, A. W., GORDY, P. L., and STEWART, G. A. 1966. Structure, seismic data, and orogenic evolution of southern Canadian rocky mountains. Bull. Can. Petrol. Geol., **14**, pp. 337–381.

BUCHER, W. H. 1933. The deformation of the Earth's crust. Princeton Univ. Press. Princeton, N.J. 518 pp.

CAREY, S. W. 1962. Folding. J. Alta. Soc. Petrol. Geol., **10**, pp. 95–144.

DAHLSTROM, C. D. A. 1954. Statistical analysis of cylindrical folds. Trans. Can. Inst. Mining Met., **57**, pp. 140–145.

DAHLSTROM, C. D. A., DANIEL, R. E., and HENDERSON, G. G. L. 1962. The Lewis thrust at Fording mountain. J. Alta. Soc. Petrol. Geol., **10**, pp. 373–395.

DENNISON, J. M. and WOODWARD, H. P. 1963. Palinspastic maps of central Appalachians. Bull. Amer. Assoc. Petrol. Geol., **47**, pp. 666–680.

FITZGERALD, E. L. 1968. Structure of British Columbia foothills, Canada. Bull. Amer. Assoc. Petrol. Geol., **52**, pp. 641–664.

GALLUP, W. B. 1951. Geology of Turner Valley oil and gas field, Alberta, Canada. Bull. Amer. Assoc. Petrol. Geol., **35**, pp. 797–821.

GOGUEL, J. 1952. Tectonics (1962 translation). Freeman and Company, San Francisco, 384 pp.

HUNT, C. W. 1957. Planimetric equation. J. Alta. Petrol. Geol., **5**, pp. 259–264.

PRICE, R. A. 1964. Flexural slip folds in the Rocky mountains, southern Alberta and British Columbia. Seminars on tectonics—IV, Dept. Geol. Sci., Queen's Univ., Kingston, Ont. (Also as Geol. Surv. Can., Reprint 78, 16 pp.)

STOCKWELL, C. H. 1960. The use of plunge in the construction of cross sections of folds. Proc. Geol. Assoc. Can., **3**, pp. 97–121.

WILSON, G. 1967. The geometry of cylindrical and conical folds. Proc. Geol. Assoc. London, **78**, pp. 179–210.

The construction of balanced cross-sections

D. ELLIOTT†

(*Received and accepted* 17 *December* 1982)

A GEOLOGICAL cross-section should integrate all known borehole, geophysical and surface geological data and, wherever possible, information off the line of section should be projected down-plunge on to the section. The line of section should be chosen to lie parallel to the slip or movement direction so that the section can be balanced. An oblique section through cylindrical fold structures can still be balanced because during deformation the amount of material leaving the cross-section equals the amount entering the section. However, sections through non-cylindrical fold structures can be balanced only in the slip direction. Sections which cross oblique or lateral ramps cannot, in general, be balanced.

One guideline which a geologist should apply during the section construction is structural style. That is, the structures drawn on the section are those that can be seen in the area in cliffs, road cuts, mountain sides, etc. The use of these structures leads to an *admissible* cross-section.

Additionally, a restored as well as a deformed-state cross-section should be constructed at the same time. Usually, plane strain is assumed to exist in the plane of the section. If a section can be restored to an unstrained state it is a *viable* cross-section.

By definition, a *balanced* cross-section is both *viable* and *admissible*.

There are four levels of confidence in geological cross-sections.

(1) *An unbalanced section.* This represents a preliminary investigation of the section showing conjectural structures.

(2) *An unrestorable cross-section.* This can arise from an unfortunate choice of line of section. For instance a section which crosses oblique thrust ramps cannot be restored to the undeformed state. Such a section could still be valid but the interpretation cannot be helped by balancing.

(3) *A restorable and admissible cross-section.* This section contains tectonic structures which satisfy clearly-stated rules (e.g. thrusts always cut up-section in the direction of transport, extension faults always cut down, etc.). Ideally a geologist should begin at this level of confidence by *simultaneously* constructing deformed and restored sections.

(4) *A valid balanced cross-section.* A balanced section is not a unique solution. But if the section integrates various sources of data in a quantitative manner and additional work is carried out at the surface, in down-plunge projection, in wells, using seismic, gravity and aeromagnetic data, a section may be found which is sufficiently restrictive with little room for alternative interpretation.

---

† This posthumous contribution by David Elliott was written by J. R. Hossack from his notes which were taken during a conversation between them.

101

Reproduced by permission of the Geological Society from The use of balanced cross sections in the calculation of orogenic contraction: a review, by John R. Hossack in *Journal of the Geological Society*, v. 136 (1979), p. 705-711.

J. geol. Soc. London, Vol. **136**, pp. 705–711, 5 figs. Printed in Northern Ireland.

The use of balanced cross-sections in the calculation of orogenic contraction: A review

John R. Hossack

SUMMARY: Balanced section calculations assume that the section has been deformed by a plane strain. If the section is underlain by a décollement, it can be restored to its pre-deformational length by dividing the area of the section by its original stratigraphic thickness. The thickness can be measured in undisturbed foreland beds or reconstructed by unstraining deformed beds. The restoration of the balanced section does not depend on any particular mechanism of deformation or folding, but in the special case of flexural slip folding the section can be unstrained by straightening out the sinuous bed lengths of the folds. However, in nature the area of the section may have decreased by 15–45%, but the assumption of plane strain always leads to a minimum estimate of shortening. Balanced section calculations suggest that the margins of orogenic belts have contracted by 35–54%.

Balanced geological cross-sections are drawn with the assumption that the area of section has not changed during deformation. It is assumed that an original sequence of flat-lying beds has been folded and faulted to form the present geological cross-section. If the section contains flexural slip folds with their axes normal to the section, the beds will suffer no shortening or elongation along the axes. Hence, the deformation in the plane of the section will be a plane strain. The area of the section before strain 'balances' the area after strain. However, it is not always necessary to assume that the folds in the plane of the section are flexural slip folds. Balanced section calculations can be made on geological sections simply by assuming that in the plane of the section there has been no change in area (Chamberlin 1910; Goguel 1962).

Many of the ideas which have been incorporated into balanced section construction have been developed by geologists working in the oil and gas fields of Alberta (Douglas 1950; Hunt 1957; Carey 1962; Bally *et al.* 1966) prior to the formal description by Dahlstrom (1969*a*). In the Alberta sections, most of the folds appear to have been formed by flexural slip and the thickness of the beds measured normal to the bedding remains constant. If the area of the cross-section remains constant, the bed lengths measured around the folds at different structural levels in the section must remain constant (Dahlstrom, *op. cit.*). The sinuous bed lengths around the folds should always be checked when a geological cross-section is constructed and if the bed lengths in any section do not 'balance' the section cannot be correct (Dahlstrom, *op. cit.*). The bed lengths should also be balanced across normal, reverse, and thrust faults. Dahlstrom (*op. cit.*) has shown that, at one level in the section, orogenic contraction can be accommodated by thrust faults, and at another level by folds. But the bed lengths in each of the levels must balance, and if they do not there must be a convincing structural explanation. Hence, the

term balanced cross-section can be used in two ways. It can either be applied to sections where it is assumed that the areas of the original and the strained sections are the same, or, in the case of flexural slip folding, where the bed lengths in the original and the strained sections are the same.

Balanced sections have one very useful property; they can be unstrained to restore the beds of the section back to their depositional position (Bally *et al.* 1966; Dahlstrom 1969*a*). Therefore, these sections can be used to carry out finite strain analyses on large sections of the Earth's crust. So far, the technique has only been used on gently folded rocks (Chamberlin 1919) or in geological cross-sections along the margins of orogenic belts (Chamberlin 1910; Bucher 1933; Laubscher 1962; Dennison & Woodward 1963; Bally *et al.* 1966; Dahlstrom 1970; Gwinn 1970).

Balanced section calculations

Balanced section calculations were first used to estimate the depth to the décollement underlying concentric folds (Chamberlin 1910, 1919; Bucher 1933; Goguel 1962; Laubscher 1962; Dahlstrom 1969*b*). Consider a series of beds of thickness t_0 in their depositional position which initially overlies a plane destined to become a thrust or décollement surface (Fig. 1). The initial section has an arbitrary length l_0 but it is important in section balancing that the ends of the section are fixed at points where there is no interbed slip. These points can be chosen as the planes which will become axial surfaces of folds or planes normal to the regional dip in the undisturbed beds of the foreland (Dahlstrom 1969*a*). The bed AB of length l_0 (Fig. 1) is transformed by folding into the strained state $A'B'$ with axial surfaces $A'D$ and $B'C'$ at each end of the section. The final length of the section after folding is AO or l_1. During the folding,

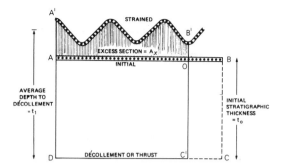

FIG. 1. Chamberlin's (1910, 1919) equal area calculation. The bed AB, originally at height BC above the décollement surface, is folded into a new position $A'B'$, $AB = l_0$, $AO = l_1$, $\cdot OB =$ shortening. During plane strain, $A'B'C'D = ABCD$, therefore excess section $A_x = OBCC'$. Initial depth to décollement $t_0 = A_x/OB$, or shortening $OB = A_x/t_0$. Within the area $A'B'C'D$ the rocks may be deformed by any style of folding or faulting.

the reference bed AB is uplifted to its strained position and the area A_x between the folded and the initial positions is described as the excess section (Gwinn 1970). The area of A_x can be measured with a planimeter or by Simpson's rule. The initial stratigraphic thickness t_0 is transformed to an average structural thickness t_1 (Fig. 1). If the deformation is a plane strain, the area of section $ABCD$ before strain will equal the area $A'B'C'D$ after strain. Because the area $AOC'D$ is common to both sections, A_x, the excess section, is equal to the area $OBCC'$. The latter is the product of the shortening and the initial depth to décollement (Fig. 1). If the folds have flexural slip geometry, the shortening can be estimated by unfolding the sinuous bed length around the folds. Hence the initial depth to décollement is A_x divided by the shortening (Chamberlin 1910, 1919; Goguel 1962; Dahlstrom 1969b). The initial height of the reference bed AB is usually estimated from the height of the same bed in the undisturbed beds of the foreland, although this assumes that there are no thickness or facies changes between the folded and undisturbed zones.

Chamberlin (1910) used a balanced section calculation to estimate that the depth to décollement in the Pennsylvania Appalachians lay between 9 and 52 km. He later estimated that the depth to décollement beneath the folds in the Colorado Rockies lay between 21 and 172 km (Chamberlin 1919). This latter result suggests that the surface folds die out in the Earth's mantle, clearly an absurd result. There are several possible sources of error in such a calculation. Firstly, the estimate of the initial height of the reference bed may be in error because of thickness or facies changes towards the undisturbed belt. Another possible error

could arise from the assumption of flexural slip folding. The folds could have similar geometry in which case the unfolded bed length will not give an accurate estimate of total shortening. However, recent applications of the technique by Laubscher (1962) in the Jura and by Dahlstrom (1969b) in Alberta have given satisfactory estimates of depth to décollement.

Another source of error was described by Bucher (1933). Chamberlin (1910, 1919) divided his geological sections into small segments and carried out depth to décollement calculations in each segment. Bucher (*op. cit.*) suggested that a more accurate estimate could be made if the calculation was carried out over the whole area of the section. He re-estimated that the depth to décollement in the Appalachians was 21 km and, in addition, that the depth to décollement in the Jura was 852 m. It is not clear why Bucher's technique should give a better estimate. However, it is likely that the division of the section into segments may produce serious over- and under-estimates of the depth to décollement which cancel out over the whole section.

The logic of the Chamberlin technique can be reversed to calculate the amount of shortening if the depth to décollement is known (Dennison & Woodward 1963; Gwinn 1970). This is potentially more useful because the depth to décollement may be known from borehole data or from seismic sections (Bally *et al.* 1966; Dahlstrom 1970). Also, it is not necessary to assume flexural slip folding in the plane of the section. The beds beneath $A'B'$ (Fig. 1) can be deformed internally by concentric or similar folds, faults, imbricate thrusts, or solution transfer mechanisms. All that is necessary to calculate the shortening is to assume plane strain. The calculations have not been described in detail by Dennison & Woodward (1963) or by Gwinn (1970) but the method has been described by Elliott (1977). The original length of the deformed geological cross-section l_0 is given by

$$l_0 = (A_x/t_0) + l_1 \text{ (Elliott 1977, after Gwinn 1970).}$$

Gwinn (1970) used this calculation to estimate that the Valley and Ridge Province of the Appalachians had undergone 43% orogenic contraction (where the contraction is measured by the conventional engineering strain $\varepsilon = (l_1 - l_0)/l_0$).

The original length of a geological cross-section can also be estimated from the total area of the section (Dennison & Woodward 1963; Kiefer & Dennison 1972). Imagine a section of layered rocks of original length l_0 and original stratigraphic thickness t_0 (Fig. 2). The section becomes shortened by folding and thrusting to a final length l_1 and an average structural thickness of t_1. Irrespective of the style of folding, if the deformation is a plane strain,

$$\text{area of section} = l_0 t_0 = l_1 t_1$$
$$\therefore \quad l_0 = l_1 t_1/t_0$$

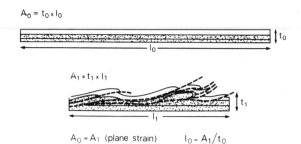

$A_0 = t_0 \times l_0$

$A_1 = t_1 \times l_1$

$A_0 = A_1$ (plane strain) $l_0 = A_1/t_0$

FIG. 2. Balanced section calculation using the total area of the section. A section of crust of original length and thickness l_0 and t_0 is contracted to l_1 with an average structural thickness of t_1. During plane strain $A_0 = A_1$; therefore $l_0 = A_1/t_0$. A_1 is measured more accurately with a planimeter.

This, of course, is simply a rearrangement of the Elliott equation above. The area of the excess section or the geological cross-section can be measured more accurately with a planimeter. Using the latter equation, Kiefer & Dennison (*op. cit.*) estimated that the Devonian of Alabama and Georgia had an average orogenic contraction of 35%.

I have applied this calculation to the rocks of the Etnedal nappes of the southern Norwegian Caledonides (Fig. 3). The original thickness of the nappe rocks (t_0) cannot be measured directly because the rocks of equivalent age in the foreland are much thinner and have a completely different stratigraphy. However, t_0 can be estimated by unstraining the distorted stratigraphic section using finite strain data (Cloos 1947; Borradaile & Johnson 1973; Tobisch *et al.* 1977; Hossack 1978). Calculations by these authors indicate that stratigraphic sections are frequently either halved or doubled in thickness during deformation. The stratigraphy of the Etnedal nappes has been measured carefully by R. P. Nickelsen (*pers. comm.*). The Eocambrian to Middle Ordovician sequence has a deformed thickness of 332 m averaged over several sections (Fig. 3). The rocks contain a slaty cleavage which is nearly parallel to bedding and deformed pebbles in conglomerates suggest that there was 20–25% ductile flattening normal to the cleavage (C. Peach, *pers. comm.*). The maximum 25% flattening value has been chosen because it leads to a minimum result in subsequent calculations. If all the beds of the section have undergone the same amount of flattening then the removal of the ductile strain gives an estimated original stratigraphic thickness (t_0) of 443 m (Fig. 3). The flattening strain has been measured in the most competent rocks of the section and hence this t_0 value must at least be a minimum estimate. The present cross-sectional area of the nappes is $4.15 \times 10^7 \, \mathrm{m}^2$. Dividing this area by the estimated t_0 suggests that the original length of this section between Mellane and the thrust front (Fig. 3) was 94 km. The

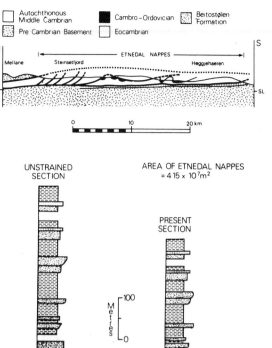

FIG. 3. Geological cross-section, with $\times 2$ vertical exaggeration, through the Etnedal nappes between Mellane and the thrust front to the S. Beneath is the stratigraphy of one of the nappes in the Steinsetfjord district averaged over several localities (R. P. Nickelsen, *pers. comm.*). Removal of 25% cleavage-forming strain from the present day thickness of 332 m restores the beds to a pre-deformational thickness (t_0) of 443 m. The area of the Etnedal nappes from Steinsetfjord southwards to the thrust front is estimated to be $4.15 \times 10^7 \, \mathrm{m}^2$.

present length of the section is 54 km, giving a contraction of 43%. It is fortunate that this calculation is insensitive to errors in the estimate of the original stratigraphic thickness (t_0). Dennison & Woodward (1963) showed in their calculations that variations of up to 365 m (1000 ft) in t_0 affected the calculated shortening by only 2–3%. Unfortunately, it is not possible to estimate from their figures what percentage 365 m represents in the total stratigraphic section. However, I have calculated from the Etnedal data that a 50% underestimate in the thickness t_0 would only reduce the original length of the nappes to 83 km.

The most sophisticated restorations of geological cross-sections have been made by geologists in the Alberta fold and thrust belt (Douglas 1950; Bally *et al.* 1966; Dahlstrom 1969*a*, 1970). In addition to

6

sinuous bed length and area balancing calculations, they have also had to take into account the stratigraphic separation along numerous major and imbricate fault zones of the region. Thus, the total shortening is the sum of both the folding and thrusting, and the orogenic contraction in Alberta is estimated to have been 54% (Price & Mountjoy 1970). In addition, the restored sections of Bally *et al.* (*op. cit.*) indicate the area loss by erosion from the frontal toes and risers of the nappes.

Reduction of area in balanced cross-sections

Geiser (1978) criticized the assumption of plane strain in balanced section calculations. For instance, Ramsay & Wood (1973) suggested that slates may undergo a volume decrease of 10–20% during finite strain. Many of the rocks in the marginal parts of orogenic belts show evidence of solution transfer (von Plessman 1964) and this could also lead to a volume loss during finite strain. Therefore it is pertinent to consider the effect of volume loss on balanced section calculations. The processes by which volume may be lost are compaction accompanying lithification, tectonic compaction, and pressure solution. In addition, the area of the section can decrease because of elongation along the orogenic strike.

Compaction during lithification

During burial, wet sediments may lose up to 50% of their volume because of the removal of pore water (Wood 1974). This type of compaction has uniaxial symmetry and merely reduces the stratigraphic thickness normal to the bedding. Oertel (1970) estimated that lapilli tuffs in the Lake District had been compacted by 52% before they suffered tectonic finite strain. Sanderson (1976) estimated that undeformed chalk may undergo 32–44% compaction during lithification. However, lithification volume loss should not affect balanced section calculations. These calculations normally compare the area of a sedimentary sequence in the cross-section with the area of the same *compacted* beds in the undisturbed foreland. Therefore, the lithification compaction strain has already been accounted for and does not enter the calculation.

Tectonic compaction

There is some evidence that previously lithified sediments may be compacted further during deformation. Wood (1974) described an increase in density as mudstones are deformed to form slates. The density may increase from 2.5 up to 2.7–2.85 g/cm^3, equivalent to a volume reduction of 10%. In contrast, Siddans (1977) found that there was no density increase between undeformed and deformed sediments along a

section in the French Alps. Hence, the little evidence that exists in the literature suggests that this kind of volume loss is less than 10%. It is likely that this loss will have biaxial or triaxial symmetry, but in order to simplify subsequent calculations (Fig. 5) the simple case of isotropic volume loss has been assumed. A 10% volume loss produces a 6.7% area reduction in a cross-section (Fig. 5).

Pressure solution

Von Plessman (1964) described the effects of tectonic pressure solution in cleaved rocks. Contractions of 20–30% are usual at right angles to the pressure solution cleavage and may go up to 50% (von Plessman, *op. cit.*; Alvarez *et al.* 1978). If the dissolved material is completely removed from the rock there will be a corresponding decrease in volume. However, much of the quartz and calcite which is dissolved away may be deposited locally within the rock in areas of lower pressure (Durney & Ramsay 1973; Elliott 1973; Alvarez *et al.* 1978). Veins and fibres of quartz and calcite are common, but not always present, in rocks showing pressure solution. It will be assumed in subsequent calculations that volume losses of 20–30% normal to the cleavage are possible by pressure solution and that this volume loss is likely to have uniaxial symmetry.

Elongations along orogenic strike

Most orogenic strains which have been measured lie within the flattening field of a Flinn plot (Ramsay & Wood 1973). An apparent flattening strain can be produced by the superposition of a volume loss on a plane strain (Ramsay & Wood, *op. cit.*). However, many of the natural strains would require volume losses of up to 60% to account for their symmetry by this mechanism. Therefore, most slates would appear to have undergone real flattening which involves extension along the Y finite strain axis.

Slaty cleavage is normally parallel or sub-parallel to the axial surfaces of folds. In the margins of orogenic belts, such as the Appalachians and Scandinavia, the axial surfaces and main cleavage are parallel to the strike of the orogenic belt (Cloos 1947). Geological cross-sections are usually drawn normal to the orogenic strike in the XZ finite strain plane (Cloos, *op. cit.*) with the Y axis normal to the plane of section. If Y suffers an elongation during the tectonic strain, there must be an area decrease in the XZ plane, even during constant volume deformation because

$$(1+\varepsilon_1)(1+\varepsilon_2)(1+\varepsilon_3) = 1+\Delta$$

where Δ = the volume strain $\delta V/V$.

By assuming no volume strain, it is possible to calculate the average strike elongation of an area in an orogenic belt by integrating the measured finite strains along the *Y* finite strain trajectories (Hossack 1978). The integrated elongation along the Caledonian strike in the Bygdin area, Norway was found to be between 11 and 15%. I have used Cloos's data (1947) to determine the strain integration around the South Mountain Fold (Fig. 4) and here the strike elongation

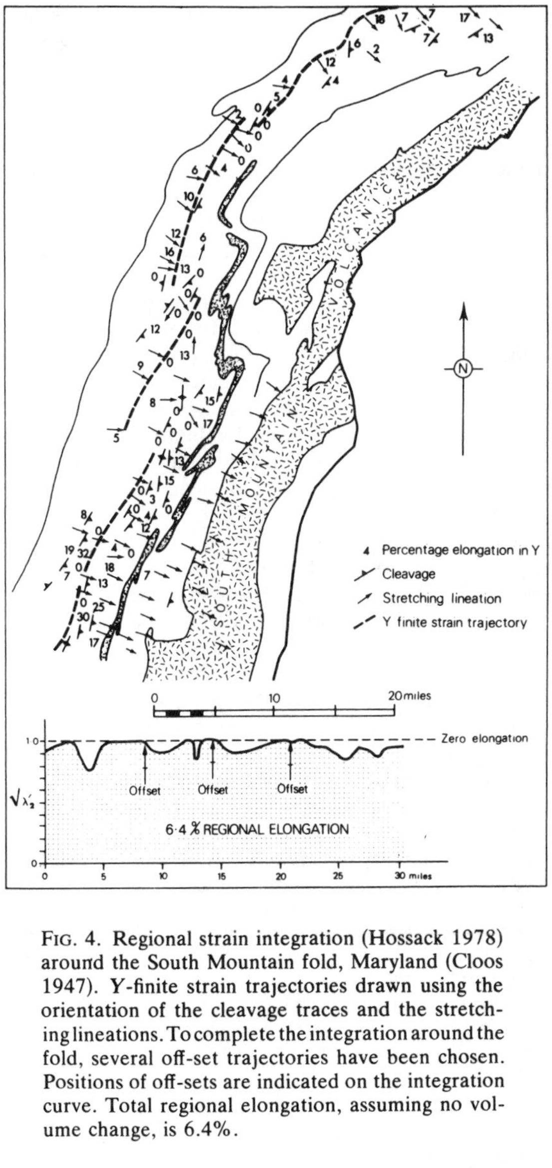

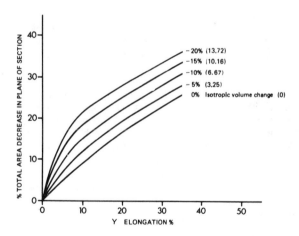

FIG. 4. Regional strain integration (Hossack 1978) around the South Mountain fold, Maryland (Cloos 1947). *Y*-finite strain trajectories drawn using the orientation of the cleavage traces and the stretching lineations. To complete the integration around the fold, several off-set trajectories have been chosen. Positions of off-sets are indicated on the integration curve. Total regional elongation, assuming no volume change, is 6.4%.

is 6.4%. This probably represents a real strike elongation, because Nickelsen (1966) described *Lingula* shells from the Allegheny foreland plateau which have extended by fracturing in the *Y* direction by an average of 4.4%. Borradaile (1979) suggested that the average strike elongation in the Dalradian of Islay was

8%. Hence, the average strike elongation from localities in the Scandinavian and Scottish Caledonides and the Appalachians suggests that strike elongations of 10% might be typical. This will produce an area change of 9% in an *XZ* balanced section where there is no volume change (Fig. 5). Strike elongations of up to 30% can produce area reductions of just over 20% in a section at right angles to the strike. Note, however, that if the strike direction of an orogenic belt suffers a shortening, there will be an area increase in the balanced section.

Errors in balanced section calculations

Strike elongation and tectonic compaction may each produce area reductions in geological cross-sections. Within the margins of orogenic belts each may typically have values of around 10%. Both kinds of area reduction are likely to have biaxial symmetry. Unfortunately there are an infinite number of ways by which two strain tensors can be combined. To simplify calculations, I have assumed that both kinds of area decrease have isotropic symmetry and that reductions in length are the same in all directions in the section. Fig. 5 has been prepared by combining varying values of strike elongation and tectonic compaction. Strike elongation combined with 10% tectonic compaction produces a total area decrease of 15% in a geological

FIG. 5. Total area changes in geological cross-sections with strike (*Y*) elongation combined with isotropic tectonic compaction. Percentage isotropic volume decrease indicated against each curve. Figures in brackets represent the corresponding area decrease in the section brought about by the volume decrease.

section at right angles to the strike. This area reduction will affect the estimate of original length (l_0) and thickness (t_0) of the undistorted section. By assuming plane strain with no volume reduction, the balanced area calculation produces original lengths and thicknesses that are 15% too low. However, the technique at least provides a minimum result. It is likely that the original stratigraphic thickness of the Etnedal nappes was closer to 510 m, if there had been 15% area reduction, rather than the 443 m calculated with the plane strain assumption. Similarly the original length of the section could have been closer to 108 km rather than the calculated 94 km. The former figure gives an orogenic contraction of 50% compared to the 43% calculated by assuming plane strain.

The effects of pressure solution are much more difficult to estimate. There is not much evidence in the literature describing what happens to material that is dissolved away from the solution seams. Some authors (Durney & Ramsay 1973; Elliott 1973, Alvarez *et al.* 1978) assumed that the material is precipitated locally in areas of lower pressure and that the total volume of rock is preserved. With no volume change, the pressure solution would produce an additional plane strain of 20 or 30% which could be added by tensor multiplication (Elliott 1972) to the ductile strain used to restore beds back to their pre-deformational thickness (Fig. 3). However, if the dissolved material is completely removed out of the plane of the section, there will be an area decrease of 20–30%. This decrease does not have isotropic symmetry but occurs in one direction normal to the solution seams. Many orogenic belts seem to fall into 2 possible geometric models. Firstly, many belts have crudely horizontal beds with near-vertical slaty cleavage (e.g. the N Wales slate belt, Wood 1973). In this case, if there is 20–30% pressure solution without precipitation, the calculated original length of the belt (l_0) will be 20–30% too small.

The Etnedal nappes seem to fall into the second geometric model where there are crudely horizontal beds with a near-horizontal slaty cleavage. Any loss of material in a vertical direction will obviously affect the original stratigraphic thickness (t_0), which may be 20–30% too small. Hence, allowing for possible unidirectional shortening as a result of pressure solution, the original beds of the Etnedal nappes may have been up to 610–660 m thick rather than the 510 m calculated

by allowing for isotropic area changes. Once again, the assumption of plane strain with no area change provides a minimum estimate.

There is an obvious need for mass transfer estimates to be carried out on rocks containing evidence of pressure solution to find out how much material actually leaves the rock, how much remains behind to be precipitated, and how far the removed material can be transported. When such estimates are available, more refined balanced section calculations will be possible.

Discussion

Balanced section calculations are an important tool for geologists because it is possible to estimate in a simple manner the orogenic contraction across large areas of the Earth's crust. The most simple calculation assumes that the area of a geological cross-section is the same before and after strain. Hence, the original length of the section is given by the area of the section divided by the original stratigraphic thickness of the beds in the section. Fortunately, the calculation is insensitive to errors or variations in the assumed original stratigraphic thickness. In spite of probable area decreases in the plane of the section during tectonic strain, balanced section calculations always lead to a minimum estimate of the amount of orogenic contraction. Balanced section calculations have been carried out in the marginal parts of several orogenic belts. The marginal thrust belt of the Appalachians has contracted by 35–43% (Dennison & Woodward 1963; Gwinn 1970; Kiefer & Dennison 1972), the Alberta thrust belt by 54% (Price & Mountjoy 1970) and the Etnedal nappes of S Norway by 43% (this paper). It is unlikely that balanced section calculations can be used in the internal parts of orogenic belts, but here the Eulerian destraining technique of Schwerdtner (1977) and Borradaile (1979) can be used to remove tectonic finite strain. It may eventually be possible to unstrain complete orogenic belts by using balanced section calculations and the Schwerdtner technique.

ACKNOWLEDGMENTS. I would like to thank Ray Skelhorn, Bob Standley, Mark Cooper and Mike Garton for criticizing an early draft of the manuscript; Nick Nickelsen for donating his stratigraphic sections from the upper Etnedal nappe; Alan Sutton for drawing the figures; and Dave Elliott for several discussions on balanced sections.

References

ALVAREZ, W., ENGELDER, T. & GEISER, P. A. 1978. Classification of solution cleavage in pelagic limestones. *Geology*, **6**, 263–66.

BALLY, A. W., GORDY, P. L. & STEWART, G. A. 1966. Structure, seismic data, and orogenic evolution of southern Canadian Rocky Mountains. *Bull. Can. Pet. Geol.* **14**, 337–81.

BORRADAILE, G. J. 1979. Strain study of the Caledonides in the Islay region, SW Scotland: implications for strain histories and deformation mechanisms in greenschists. *J. geol. Soc. London*, **136**, 78–88.

—— & JOHNSON, H. D. 1973. Finite strain estimates from the Dalradian Dolomitic Formations, Islay, Argyll, Scotland. *Tectonophysics*, **18**, 249–59.

BUCHER, W. H. 1933. *The Deformation of the Earth's Crust.* Princeton Univ. Press. 518pp.

CHAMBERLIN, R. T. 1910. The Appalachian folds of Central Pennsylvania. *J. Geol. Chicago,* **18,** 228–51.

—— 1919. The building of the Colorado Rockies. *J. Geol. Chicago,* **27,** 225–51.

CAREY, W. S. 1962. Folding. *J. Alberta Soc. Pet. Geol.* **10,** 95–144.

CLOOS, E. 1947. Oolite deformation in the South Mountain fold, Maryland. *Bull. geol. Soc. Am.* **58,** 843–918.

DAHLSTROM, C. D. A. 1969*a.* Balanced cross sections. *Can. J. Earth Sci.* **6,** 743–57.

—— 1969*b.* The upper detachment in concentric folding. *Bull. Can. Pet. Geol.* **17,** 326–46.

—— 1970. Structural geology in the eastern margin of the Canadian Rocky Mountains. *Bull. Can. Pet. Geol.* **18,** 332–406.

DENNISON, J. M. & WOODWARD, H. P. 1963. Palinspastic maps of central Appalachians. *Bull. Am. Assoc. Petrol. Geol.* **47,** 666–80.

DOUGLAS, R. J. W. 1950. Callum Creek, Langford Creek, and Gap map areas, Alberta. *Geol. Surv. Can. Mem.* **255,** 124pp.

DURNEY, D. W. & RAMSAY, J. G. 1973. Incremental strains measured by syntectonic crystal growths. *In:* DE JONG, K. A. & SCHOLTEN, R. (eds). *Gravity and Tectonics.* Wiley, New York, pp. 67–96.

ELLIOTT, D. 1972. Deformation paths in structural geology. *Bull. geol. Soc. Am.* **83,** 2621–38.

—— 1973. Diffusion flow laws in metamorphic rocks. *Bull. geol. Soc. Am.* **84,** 2645–64.

—— 1977. Some aspects of the geometry and mechanics of thrust belts. Part 1. *8th Annual Seminar Can. Soc. Petrol. Geol. Univ. Calgary.*

GEISER, P. 1978. Gravity tectonic removal of cover of Blue Ridge anticlinorium to form Valley and Ridge Province: Discussion. *Bull. geol. Soc. Am.* **89,** 1429–30.

GOGUEL, J. 1962. *Tectonics.* Freeman, San Francisco, 348pp.

GWINN, V. E. 1970. Kinematic patterns and estimates of lateral shortening, Valley and Ridge and Great Valley Provinces, Central Appalachians, South-Central Pennsylvania. *In* FISHER, G. W. *et al.* (eds). *Studies of Appalachian Geology: Central and Southern.* Wiley, New York, pp 127–46.

HOSSACK, J. R. 1978. The correction of stratigraphic sections for tectonic finite strain in the Bygdin area, Norway. *J. geol. Soc. London,* **135,** 229–41.

HUNT, C. W. 1957. Planimetric equation. *J. Alberta Pet. Geol.* **56,** 259–41.

KIEFER, J. D. & DENNISON, J. M. 1972. Palinspastic map of Devonian strata of Alberta and Northwest Georgia. *Bull. Am. Assoc. Petrol. Geol.* **56,** 161–66.

LAUBSCHER, H. P. 1962. Die Zwiephasenhypothese der Jurafaltung. *Eclog. geol. Helv.* **55,** 1–22.

NICKELSEN, R. P. 1966. Fossil distortion and penetrative rock deformation in the Appalachian Plateau, Pennsylvania. *J. Geol. Chicago,* **74,** 924–31.

OERTEL, G. 1970. Deformation of a slaty lapillar tuff in the English Lake District. *Bull. geol. Soc. Am.* **78,** 1173–87.

PLESSMAN, W. von, 1964. Gesteinlösung, ein Hauptfaktor beim Schieferungsprozess. *Geol. Mitt. Aachen.* **4,** 69–82.

PRICE, R. A. & MOUNTJOY, E. W. 1970. Geologic structure of the Canadian Rocky Mountains between Bow and Athabasca Rivers—A progress report. *Spec. Pap. Geol. Assoc. Can.* **6,** 7–25.

RAMSAY, J. G. & WOOD, D. S. 1973. The geometric effect of volume change during deformation process. *Tectonophysics,* **16,** 263–77.

SANDERSON, D. J. 1976. The determination of compaction strains using quasi-cylindrical objects. *Tectonophysics,* **30,** T25–32.

SCHWERDTNER, W. M. 1977. Geometric interpretation of regional strain analyses. *Tectonophysics,* **39,** 515–31.

SIDDANS, A. W. B. 1977. The development of slaty cleavage in a part of the French alps. *Tectonophysics,* **39,** 533–57.

TOBISCH, O. T., FISKE, R. S., SACKS, S., & TANIGUCHI, D. 1977. Strain in metamorphosed volcaniclastic rocks and its bearing on the evolution of orogenic belts. *Bull. geol. Soc. Am.* **88,** 23–40.

WOOD, D. S. 1973. Patterns and magnitudes of natural strain in rocks. *Phil. Trans. R. Soc. London,* **A274,** 373–82.

—— 1974. Current views of the development of slaty cleavage. *Ann. Rev. Earth Sci.* **2,** 369–401.

Received 15 March 1979; revised typescript received 5 June 1979.

JOHN R. HOSSACK, Geology Department, City of London Polytechnic, Walburgh House, Bigland Street, London E1 2NG.

The American Association of Petroleum Geologists Bulletin
V. 72, No. 1 (January 1988), P. 73-90, 22 Figs., 1 Table

Balanced Section in Thrust Belts
Part 1: Construction[1]

DECLAN G. DE PAOR[2]

ABSTRACT

Balanced geological cross sections are an important aid to understanding thrust-belt structures and to estimating their hydrocarbon-bearing potential. Two approaches to section balancing have been taken in the past: construction of retrodeformable sections from raw data, and modification of previous interpretations after an evaluation procedure. These approaches may be unified by use of a Langrangian grid. Projection of datum locations, dips, and stratigraphic thicknesses onto the section plane is performed directly by calculation, or graphically, using plunge lines. Spline interpolation and parallel-fold modeling are the simplest ways of filling gaps in the sectional data, but isogon interpolation yields rheologically more realistic results. Faults are interpolated using cutoff geometries as constraints. Fault tips are located from distance-displacement plots, whereas the depth to detachment is obtained by a modified Chamberlin construction. Interpolations made on the section are directed back to the geologic map via plunge lines. Although these techniques of section construction do not guarantee a balance, especially when complexities such as growth structures, diapirs, or strike-slip faults are present, they eliminate many potential errors.

INTRODUCTION

Two basic approaches exist for the construction of balanced structural cross sections (Figure 1). The first employs simple kink-style models of fault-related folding (Coates, 1945; Gill, 1953; Coward and Kim, 1981; Sanderson, 1982; Suppe, 1983; Jamison, 1987), and assumes either vertical simple shear over ramps (Jones and Linsser, 1986) or layer-parallel shear (Kligfield et al, 1986; Groshong and Usdansky, 1986). Such models are objective in that the same interpretation can be produced regardless of the user. Though suited to specific structural settings, they are limited in scope because primary bed forms, rock properties, and deformation mechanisms are not considered. The second approach is the evaluation of pre-drawn sections using rules or guidelines

regarding material conservation (Chamberlin, 1912; Dahlstrom, 1969; Hossack, 1979), fault-related fold geometry (Suppe, 1983; Medweleff and Suppe, 1986; Jamison, 1987), map-section compatibility (Hossack, 1983; Diegel, 1986), and bedding cutoff geometry (Crane, 1987; De Paor, 1987). This approach is subjective in that the user's concept of a structural style is a key element; however, the procedure objectively eliminates geometrical or mechanical impossibilities. Taken together, these rules and guidelines place strict constraints on the interpretative parts of sections, while accommodating the departures from ideal geometry that invariably occur in nature. A full account of how to construct and evaluate a balanced cross section using the above cited rules and guidelines has not been published.

This paper is the first in a series dealing with section balancing, and it addresses the first problem encountered, i.e., the construction of sections and maps from raw data. In subsequent papers we will discuss the evaluation of predrawn sections (De Paor and Bradley, in press), and practical applications to the Sverdrup basin margin in central Ellesmere Island.

TERMINOLOGY

Although the principles of section balancing apply to all tectonic regimes, most applications concern thrust belts; therefore, a brief introduction to thrust terminology is appropriate (Figure 2). A thrust is here defined as a dip-slip fault that contracted horizontal distances at the time of movement (division into low-angle thrusts and high-angle reverse faults is not justified because dips vary rapidly in time and space). The material above and below the thrust surface at the time of movement constitutes the hanging wall and footwall, respectively. Footwalls are commonly assumed to be fixed and hanging walls to be mobile; however, both sides of a fault surface may be mobile and capable of deforming while displacement proceeds. The portion of the hanging wall that undergoes finite displacement related to the footwall is termed a thrust sheet if it covers a wide region, or a thrust slice if it forms a narrow strip on the map. A thrust sheet or slice may be bounded by a combination of faults, in-situ rocks, and present or synorogenic erosion surfaces; if all boundaries are faults, the term "horse" is employed. An imbricate zone is an en echelon array of thrust slices; a duplex is an array of horses with bounding roof and floor thrusts. (Some duplexes crop out, but most are artifacts of the balancing procedure and must be treated with skepticism.) Thrusts terminate longitudinally (along the slip direction) and laterally (perpendicular to slip) at traces, tip lines, branch lines, cutoff lines, or shear zones.

[1]Manuscript received, September 12, 1986; accepted, September 10, 1987.
[2]Department of Earth and Planetary Sciences, Johns Hopkins University, Baltimore MD 21218. The evaluation of previously constructed balanced sections is dealt with in De Paor and Bradley (in press).
This research was funded by grant PRF 17526-G-2 from the American Chemical Society Petroleum Research Fund, with additional support from Texaco Inc., the Atlantic Richfield Foundation, Elf Aquitaine (S.N.E.A.(P.)), and Unocal.

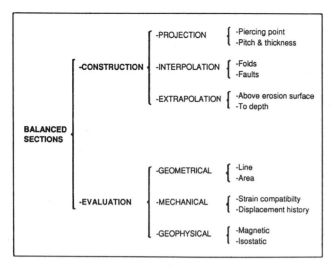

Figure 1—Elements of section-balancing procedure.

A fault may form a trace on the present topographic surface or it may be truncated under an unconformity. At tip lines, displacement drops gradually to zero, whereas at branch lines, faults splay or merge with a sudden change in net slip. Cutoff lines are caused by offset on later faults, and shear zones at the terminations of discrete faults serve to distribute displacement over an expanse of rock, enabling movements on individual planes to rapidly become infinitesimal.

At the boundaries of a thrust sheet or slice, beds may be cut off by faults or connected to in-situ beds across the axial plane of a fold. Bounding faults may parallel bed contacts forming flats, or they may cut across bedding in

ramps. Regional flats define a decollement or detachment surface. Thrust movements opposite the movement of the regional detachment are termed back thrusts. Wedges are thrust slices bounded above and below by faults of opposite displacement, or vergence. An imbricated wedge structure is termed a passive roof duplex.

Folding is a necessary strain accommodation of thrust displacement. Borrowing standard terms from continuum mechanics, one may identify material folds, which move with their thrust slice so that a constant set of material points occupy the axial planes, and spatial folds, which are pinned to bends in the fault trajectory and through whose axial planes material migrates (Suppe, 1983). Folds may result from initial fault propagation, but they will commonly be modified during subsequent displacement (Jamison, 1987). When one limb of a fold is truncated by a fault, the term "snakehead fold" may be employed.

On a regional scale, thrusts are generally grouped in belts or systems that occur in a variety of tectonic settings: compressional, transpressional, and wrench. Typically, a foreland thrust belt forms by lateral compression and/or gravitational instability of sedimentary basin strata (cover) and underlying sialic crust (basement). It is bounded to the rear by a magmatic arc or allochthonous terrane (hinterland) and to the front by an authochthonous continental interior (foreland). Topographical expressions of foreland thrust belts (foothills) are generally dwarfed by the high peaks of the allochthonous hinterland where ductile deformation that accompanies metamorphism permits recumbent folds to grow into enormous thrust nappes.

In every foreland thrust belt is a frontal thrust with a virtually undeformed footwall, either emergent at the

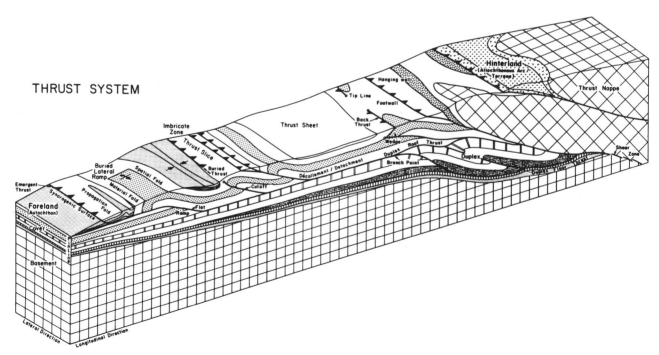

Figure 2—Illustration of features associated with terminology as defined in text. For simplicity, early normal faults have been omitted.

synorogenic surface or buried. A vertical section through the footwall constitutes a pin line relative to which displacement can be measured. Only those beds that are present in the pin, and their lateral correlatives, may be included in the balancing process. The hinterland end of a cross section is bounded by a loose line, a spatial line that is not inexorably linked to particular material points. Sections are balanced from the pin to the last fault trace only. Maintaining a vertical material line at the hinterland end of the section during restoration may lead to errors.

PRELIMINARY CONSIDERATIONS

Thrust belts commonly are shortened by 50 to 100%. Therefore, when preparing a section for balancing, two sheets of transparent drafting paper are needed; one sheet should be the length of the exposed section in the deformed state and the other sheet should be twice as long. Draft sections should extend from the highest present or synorogenic elevation to the deepest penetrating shear zone (commonly at the brittle-ductile transition); yet the scale should permit resolution of rheologically significant stratigraphic packages. Because the vertical and horizontal scales should always be equal, the drafting sheets may need to be very large. Ideally, several alternative interpretations should be compiled simultaneously. One begins by setting up horizontal X' and vertical Z' reference axes on the shorter drafting strip for the deformed state (Figure 3a). If the undeformed foreland strata dip moderately, or if the earth's curvature is significant, then a monoclinic or curvilinear reference frame may be necessary. The present topographic trace is added to the diagram; the outcrop, borehole, or seismic data that lie on the line of section are inscribed indelibly. Next, a horizontal axis X'_o representing pre-orogenic sea level is drawn on the longer drafting strip along with details of the frontal footwall. In the simplest cases, the latter comprises parallel layers joining contacts in the pin line to cutoff points on the footwall of the first fault (Figure 3b). Generally, these layers will have a gentle dip owing to lithostatic flexure under the thrust sheet (Price, 1973; Elliott, 1977; Mugnier and Vialon, 1986). As a rule, their depth below sea level on the longer strip should be reduced by four times the present-day relief to allow for isostatic off-loading during restoration. Assuming no major thickness variations, the footwall contacts may be extended across the entire width of the drafting strip to form a restoration template upon which initial bed segments and fault trajectories will be marked (more complicated non-layer-cake templates are commonly needed, however—see De Paor and Bradley, in press). It is important to fill in details on deformed and restored sections simultaneously, to avoid accumulating errors toward the hinterland.

METHODS OF SECTION CONSTRUCTION AND EVALUATION

Figure 1 outlines the steps in the construction and evaluation of a geologic cross section. To construct a cross section, one must collect all available data (well logs, seis-

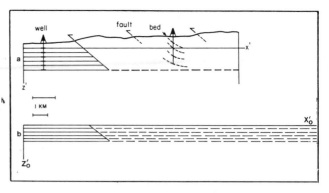

Figure 3—Preliminary data marked indelibly on deformed section (a) and restoration template (b).

mic profiles, and outcrop or dipmeter dips) from the region around the line of section and project them into the section plane. Contacts must then be interpolated between scattered data points and extrapolated up and down section. In thrust belts, faults and tilted strata generally maintain constant orientation over significant distances, so simple extrapolation along strike to meet the chosen section plane works adequately. However, where folds and listric faults are even moderately inclined, large errors ensue unless data are transferred along projection axes parallel to the structural plunge. The theory of plunge projection is well documented in the geological literature (Fisher, 1881; Lugeon, 1901; Argand, 1911; Ickes, 1925; Wegmann, 1929; Eardley, 1938; Mackin, 1950; Stockwell, 1950; McIntyre and Weiss, 1956; Ambrose and Carswell, 1962, Christensen, 1963; Lisle, 1980; Hoffman et al, 1987; King, 1987), although few practical applications on the lines of Kilby and Charlesworth (1980), Diegel (1986), Hoffman et al (in press), or King (in press) have been published. Traditionally, plunge projection has been accomplished by graphical methods (e.g., Ragan, 1985, p. 343). Charlesworth et al (1976), Langenberg et al (1977), and Diegel (1986) described how to program a computer to construct profiles, using rotation matrices to transform locational and directional data. However, their mathematical methods are restricted to exactly cylindrical forms, whereas most geologic structures are somewhat conical, especially at their lateral terminations. The simple equations and graphical techniques presented in this paper are relatively easy to use and are applicable to conical structures.

CHOICE OF SECTION PLANE AND PROJECTION AXIS

Before projecting a geologic surface (bed or fault) from a datum location onto the section plane, one must choose an ideal geometry to which the surface approximately conforms. If the surface is assumed to be planar, three point locations suffice to fix its spatial position (Rowe's classical "three point problem"), or a single clinometer or dipmeter reading may be used to construct a set of strike lines (e.g., Ragan, 1985). If the surface is assumed to be cylindrical, it may be modeled using structure contours (Ragan, 1985), or by the more elegant tech-

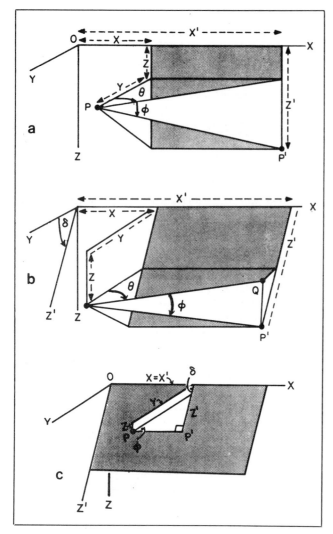

Figure 4—Calculation of piercing point P ′ = (X ′, Y ′, Z ′), corresponding to datum location P = (X, Y, Z). The symbol ϕ indicates the plunge of the projection axis, and θ is its trend relative to Y. (a) Vertical section plane; (b) section plane dipping at δ; (c) profile plane.

nique of "plunge line" construction described below. A conical surface is specified by radiating plunge lines; more complex form requires curved plunge lines with irregularly spaced spot heights.

The simplest projection employs a vertical section plane, but this may not be the most suitable section to balance if, for example, the entire region is tilted. Three choices are considered here: a vertical plane, a general or oblique section, and a "profile" plane drawn perpendicular to the projection axis. In each case, four steps must be followed: (1) calculation of positions where projection lines pierce the section plane, (2) calculation of pitches of structural traces at the piercing points, (3) calculation of stratigraphic thicknesses, and (4) graphical representation of numerical data on the section.

Step 1: Piercing Points of Projection Lines

When a datum location lies off the line of section, it must be projected into the section plane using the plunge

and trend of the chosen projection axis. For a vertical section plane (Figure 4a), if P = (X, Y, Z) is the datum location, its image P ′ = (X ′, Y ′, Z ′) is given by

$$X' = X + Y \tan \theta \qquad (1)$$
$$Y' = 0, \qquad (2)$$

and

$$Z' = Z + Y \sec \theta \tan \phi, \qquad (3)$$

where ϕ is plunge and θ is the trend of the projection axis relative to the normal to the section plane (see Table 1 for notation).

Equations 1, 2, and 3 are valid only for a vertical section. If the section plane dips at δ and contains the reference axis OX as zero-level strike line, then the piercing point for a datum location is calculated by first determining the horizontal length of the projection axis PQ in Figure 4b:

$$X' = X + PQ \sin \theta \qquad (4)$$

and

$$Z' = (Z + PQ \tan \phi)/\tan \delta \qquad (5)$$

where

$$PQ = (Y \tan \delta - Z)/(\tan \phi + \tan \delta \cos \theta). \qquad (6)$$

If we choose a profile plane at right angles to the projection axis,

$$X' = X, \qquad (7)$$

and

$$Z' = Z \cos \phi + Y \sin \phi. \qquad (8)$$

Such a section gives the "downplunge" view obtained by looking from a far distance along the projection axis. From this viewpoint, the "twist" angle defined below is particularly useful (De Paor, 1983; De Paor and Krome, in press).

Step 2: Pitch Calculation

Over a sufficiently small area, any structural surface may be approximated by its tangent plane. When such a plane is extrapolated to intersect a section, the line of intersection, or trace, is generally oblique to the true dip of the structure, so its pitch must be calculated. The simplest expression is for a vertical section plane,

$$m' = m \cos \theta, \qquad (9)$$

where m is the structure's slope (tangent of dip), m ′ is its apparent slope (tangent of pitch) parallel to the section plane, and θ is the angle subtended between the line of section and the true dip direction (Figure 5a). Equation (9) is a simplification of the apparent dip formula first derived circa 1911 by O. Fisher (see, e.g., Ragan, 1985, equation 1.4). To avoid possible confusion in the cases of non-vertical sections that follow, the term "apparent dip" is replaced by the equivalent "pitch."

For a non-vertical section plane (Figure 5b, c) with a dip of δ, we proceed in two steps. First, choose a unit length along the dip direction of the section (Figure 5b)

Table 1. Notation Used in Text

A	Area
C	Contour interval
A1, A2	Axes of a periclinal fold
A_{opt}, A_{av}	Optimal, average axes
F1, F2, F3	Faults
FWC	Footwall cutoff
HWC	Hanging-wall cutoff
K, S, S′	Spot-height spacings
L	Length of strike line
I, I′	Ramp lengths
M	Marker horizon
m, m′	Slope, apparent slope
MA	Material anticline
O	Origin, center of curvature
OA, OB, OC	Bed lengths
P = (X, Y, Z)	Point in space
P′ = (X′, Z′)	Projection of P
P, P′, P1, P2	Plunge lines
R	Rotation matrix
R	Radius of reference sphere
r	Radial polar coordinate
S1, S2, S3	Slip vectors
SA	Spatial anticline
SS	Spatial syncline
T	Depth to detachment
To	Fault tip
t, t′, t_A, t_B	Bed thicknesses
t_a	Apparent thickness
X′, Y′, Z′	Cartesian coordinate frame
$X′_o, Y′_o, Z′_o$	Pre-deformation coordinates
δ, δ′	Dip, apparent dip
φ	Plunge
ψ, ψ′	Pitch, twist
θ	Trend, cutoff angle
π_b, π_f	Pole to bed, fault
σ	Strike

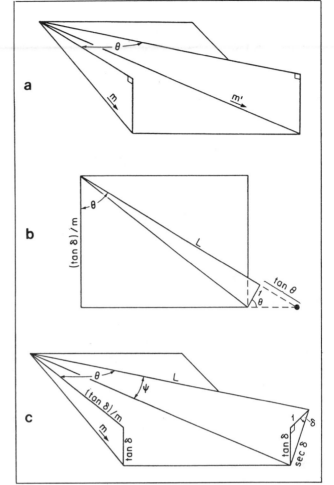

Figure 5—Illustration for calculation of apparent dip (pitch).

such that in the three-dimensional view (Figure 5c) the length of the dip line is sec δ and let L be the length of the section's strike line. Then,

$$L = \sec θ (\tan δ)/m - \tan θ, \tag{10}$$

where θ is the trend of the projection axis relative to the section plane normal. The desired pitch ψ of the structural trace is given by

$$\text{Tan } ψ = (\sec δ)/L. \tag{11}$$

For the special case of a profile plane, the twist angle of De Paor (1983) is most easily calculated or measured. Traditionally, linear structures in a plane are described using pitches to define the lines, and strikes and dips to specify the planes. These are analogous to a set of Euler angles. Trend and plunge are used independently to orient linear structures; however, they are but two of another triad, the third of which is termed "twist" (De Paor, 1983; the synonym "angle of departure" appears in my notes, but I know no source). Twist is a useful measure of a planar structure such as a fold axial plane, with an associated linear structure such as the fold axis; it is

defined as the angle measured about the linear structure (positive clockwise looking downplunge) from a vertical plane to the planar structure in question (Figure 6a). Thus, twist is the supplement of pitch in a profile section, though not in an oblique one. The convention for measuring twist on an equal angle stereonet is illustrated in Figure 6b. A line OT is drawn along the trend of the axis A, and AP is drawn tangential to the great circle representing the planar structure. The angle TAP is the twist of the plane about the axis A, and is measured with a protractor centered on A or by counting degrees along the great circle of pole A. Because of the angle-conserving property of stereographic projection, if the overlay is placed on a profile section with OT aligned along the Z′ coordinate axis, AP is automatically oriented parallel to the plane's trace in the section.

Step 3: Thickness Calculation

Given the true thickness t of a stratum in three dimensions, its apparent thickness in the plane of section is given by a formula from Coates (1945),

$$t_a = t(1 - \sin^2 δ \sin^2 θ)^{-1/2}, \tag{12}$$

665

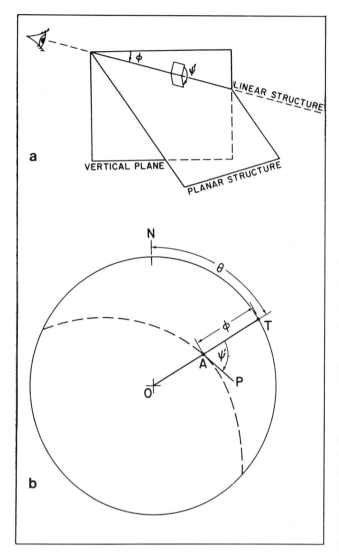

Figure 6—Illustration for definition of twist angle ψ' of a planar structure about a contained line of plunge ϕ and trend θ.

where δ is the true dip of the bed, and θ is the obliquity of the section line to the dip direction. The formula works only for straight bed segments. More commonly, it is necessary to project contacts into the plane of section independently; then the problem becomes one of determining the thickness of strata between two non-parallel pitch readings. If several intervening pitches ψ are available between locations labeled X'_1 and X'_2, then Ickes's (1925) integration technique may be employed, assuming parallel fold geometry,

$$ t = \int_{x'_1}^{x'_2} \sin \psi \, dx', \qquad (13) $$

with a correction factor $\Delta Z'/\cos \psi$ for an elevation difference $\Delta Z'$. For two pitch readings only, Hewitt (1920) proposed a construction (Figure 15a) that assumes concentric circular fold geometry. The thickness t is determined by drawing concentric circular arcs tangential to the given pitch lines (see discussion below).

Step 4: Representation

Whether one chooses a vertical, general, or profile section plane, it is necessary to represent the trace of a structural surface on the section plane. The simplest representation employs a short line segment drawn tangential to the trace curve and representing the intersection of the section plane and the structure's three-dimensional tangent plant extrapolated from the source location. Let such a segment have a half-length Δ chosen arbitrarily by the user to suit the size of section under construction. A line should be drawn connecting the point $(X' - \Delta \cos \psi, Z' - \Delta \sin \psi)$ to the point $(X' + \Delta \cos \psi, Z' + \Delta \sin \psi)$. In addition, it is advisable to label the upper side of the tangent line to correspond with the upper side of the source tangent plane, when such directional information is known, to avoid errors in interpretation of primary structures that have undergone inversion.

Matrix rotation.—The equations presented above are the simplest approach when direct calculation of section data is desired. For computer implementation, however, a matrix method is preferable. The position of a point P = (X, Y, Z) in the northeast-down reference frame is transformed to P' = (X', Y', Z') in an inclined frame using the matrix operation

$$ \mathbf{P}' = \mathbf{RP}, \qquad (14) $$

where $\mathbf{R}$ is a matrix of direction cosines of the inclined reference axes relative to the geographic axes, and (X', Z') are the piercing point coordinates in the section plane perpendicular to Y'. Full details of the computations involved are given in Charlesworth et al (1976), Langenberg et al (1977), and Flinn (1978), and are not repeated here. A PDP Pascal program to perform this projection, written by graduate student F. Diegel, and translated to VAX Pascal by graduate student R. Reed is available from the writer on request. Figure 7 shows program output from Diegel's field area (see Diegel, 1986).

CONICAL STRUCTURES AND OPTIMAL PROJECTION AXES

In reality, beds rarely form perfectly cylindrical folds; therefore, projection axes should not all be parallel, but should converge or diverge. Previous workers have employed an average fold axis as the projection axis, but this means projecting data along a line that is not actually parallel to the surface at a non-average location (Figure 8). To overcome this problem, the projection axis must be chosen from the lines that actually lie in the tangent plane to the folded surface. The chosen line should be the one that is closest to (i.e., subtends the smallest angle with) the average fold axis A_{av}. The average axis is determined by stereonet techniques (the π and β diagrams of Ramsay, 1967) or by calculating the eigenvectors of the Scheidegger orientation tensor, a topic beyond the scope of this paper. Having determined the average axis, consider the pole to bedding, π_b, at an arbitrary datum locality. The following double-cross product yields a vector A_{opt} of

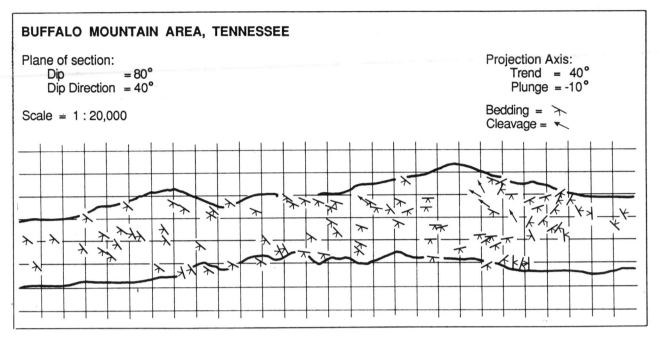

BUFFALO MOUNTAIN AREA, TENNESSEE

Plane of section:
    Dip   = 80°
    Dip Direction = 40°

Scale ≈ 1 : 20,000

Projection Axis:
    Trend = 40°
    Plunge = -10°

Bedding =
Cleavage =

Figure 7—Output from computerized downplunge projection by F. Diegel.

irrelevant magnitude, oriented in the desired optimal projection direction,

$$A_{opt} = \pi_b \times A_{av} \times \pi_b. \qquad (15)$$

This numerical solution is suited to microcomputer programing; an alternative approach is to use standard stereonet techniques to find the great circle that contains the projection axis and is perpendicular to the tangent plane's great circle. The intersection of these two planes defines the optimal projection axis. Remember that this axis is valid only for one bedding surface orientation, and must be recalculated whenever dips and strikes differ appreciably. The user must watch for the possibility of crossed projection lines leading to reflections of relative structural positions when structures depart markedly from the ideal cylindrical form. Such crossovers indicate that data are being projected too far from the source; the solution is either to choose a closer line of section, or to omit the data location farthest from the section plane.

PLUNGE LINES

The equations presented above permit the calculation of tangents to traces of structural surfaces on a section plane of vertical, oblique, or profile orientation. However, in real applications, complications requiring a graphical solution generally arise. For example, lateral structures may intervene locally between datum locations and section planes, or the line of projection may need to be curved to follow a periclinal structure. I here introduce a novel approach based on the "plunge line" concept of De Paor (1981), as illustrated in Figure 9. A plunge line is a line drawn on a map in the trend direction of a linear structure and marked with spot heights at equal intervals of altitude. Thus, straight plunge lines with evenly spaced spot heights represent straight linear

structures, in the way that strike lines represent planar surfaces; bent plunge lines with uneven spot-height spacing represent space curves in the way that structure contours represent curved surfaces in space. Indeed, the intersection locus of two strike-line or structure-contour sets define a plunge line (Figure 9b).

Figure 10 illustrates the use of plunge lines to track the underground extent of a cylindrical structure that is offset by movement on a lateral fault. The anticlinal fold is shown at the topographic surface (equally, it might be detected in a seismic section). Bedding strikes and dips are measured at two surface exposures, or inferred from serial seismic sections. From these, the fold-axis plunge and trend are calculated using standard stereonet meth-

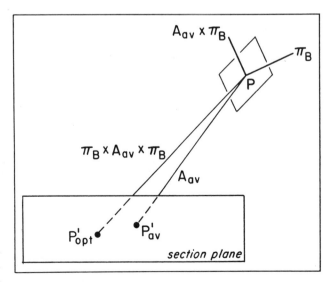

Figure 8—Illustration showing choices of optimal projection axis (see text).

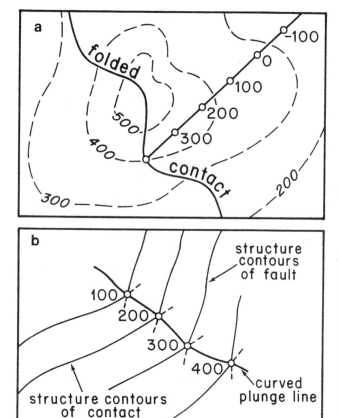

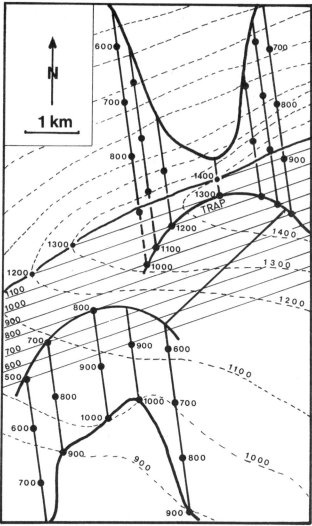

Figure 9—(a) Plunge line constructed in trend direction of fold and marked with spot heights at regular intervals of plunge. (b) Curved irregular plunge line defined by intersecting structure contours.

Figure 10—Use of plunge lines to project fold contact across lateral fault (see text). Potential hydrocarbon trap is labelled as "trap."

ods. Lines are drawn in the axial trend direction, beginning at each contour intersection on the topographic or seismic trace and extending downplunge indefinitely. Spot heights are marked at a horizontal spacing of

$$S = C \cotan \phi, \qquad (16)$$

where C is the contour interval. Where necessary, a plunge line may be interpolated between contour intersection points. In this case, the first spot height is located a distance

$$S' = C' \cotan \phi \qquad (17)$$

from the outcrop or seismic trace, where C' is the fraction of a contour interval represented by the interpolation. After the first spot height has been located at an interval of S', subsequent points are equally spaced at intervals of S. Interpolation is generally necessary to find the points where plunge lines pierce the lateral fault plane shown in Figure 10. The geologist must compare spot-height and strike-line values until the last spot height that is higher than the fault plane is located. Then interpolation of fractional contour intervals permit one to find the piercing point by successive overestimation and underes-

timation. The hanging-wall trace of the structure in Figure 10 is revealed by joining piercing points. Applying the fault's net slip vector to each point on the trace yields the footwall equivalent. To continue to follow the fold structure, a new set of plunge lines that begin on the fault's strike lines is chosen (thus obviating the need to calculate fractional spot-height spacings in the footwall). At any stage, spot heights of equal altitude may be connected to reveal the structure's contours. For example, it may be desirable to contour the potential hydrocarbon trap on the footwall side of the lateral fault in Figure 10. Alternatively, the plunge lines may be extended to intersect another seismic section, a chosen line of cross section, or (as in Figure 10) an unmapped topographic surface.

A more complicated situation is illustrated in Figure 11a. The plunge and trend of an upright fold axis are determined from limb dips and strikes at two well sites, but farther east a third well reveals changes in both dip and strike of one limb. Assuming a vertical axial plane, we can assign this change in limb orientation to a pericli-

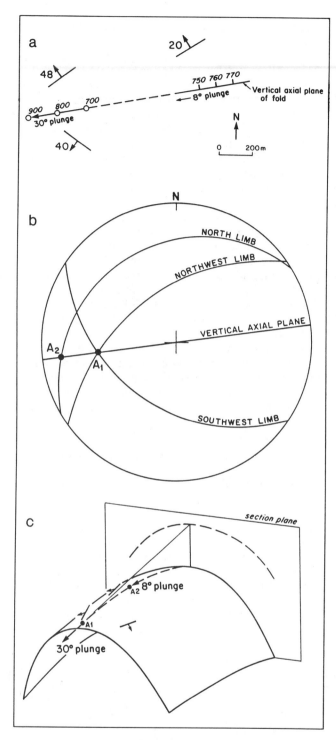

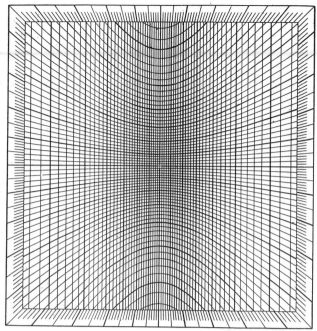

Figure 12—Gnomonic orientation net.

Figure 11—(a) Field data from periclinal fold. Circles mark steep plunge; ticks mark gentler plunge. (b) Stereonet construction for data in (a). A1 is steep plunging axis; A2 is gentle one. (c) Perspective view of structure. Dashed bedding trace is plunge projection based on constant steep axis.

tion at a greater depth than would have resulted from simply projecting along the original hinge direction (Figure 11c).

GNOMONIC PROJECTION

Two separate approaches are apparent for geometrical analysis of planar and linear features in structural geology: (1) the orientation-net technique in which directional data are analyzed independent of location, and (2) the strike-line or stratum-contour technique (here extended to include plunge lines) in which both directional and positional data are incorporated. Orientation nets in common use include the equiangular, equiareal, and orthographic types. However, a fourth type of projection also exists whose structural ramifications have not been fully exploited. It is termed the gnomonic projection, and it was introduced to geology by Hilton (1904). Gnomonic projection (Figure 12) transforms a line of plunge ϕ and trend θ into a point with polar coordinates (r, θ), where

$$r = R \cot \phi, \qquad (18)$$

and R is the radius of the reference sphere. Bengston's (1980) "tangent diagrams" are an example of gnomonic projection. Cruden (1981) has pointed out one disadvantage. Shallow plunges yield excessively large values of r; thus, this net has no primitive circle, but extends to infinity in all directions. Depending on the shallowness of dip or plunge in a particular problem, the net may be photo-enlarged or reduced so that all necessary angles fit on a finite sheet of paper. The great advantage of gnomonic projection for structural studies is that great circles are straight lines parallel to the strike lines of a plane and are separated from the net's center point by a distance equal

nal fold structure. Figure 11b illustrates the stereonet construction to determine the new fold-axis plunge, which is shallower than the original. In this way, it is possible to find the culmination of the periclinal structure (a potential trap), and to project the fold onto a cross sec-

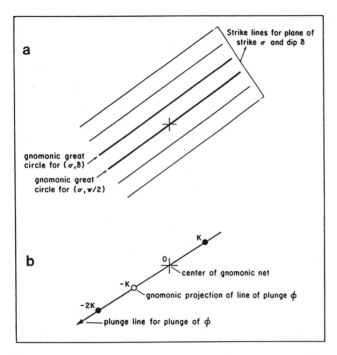

Figure 13—(a) Illustration of correspondence between strike lines and gnomonic great circles. (b) Point (open circle) on gnomonic net (center +) and plunge line (spot heights indicated by closed circles) represent the same line in space. Note equality of circle spacings.

to the strike line spacing on a map of contour interval R (Figure 13a). Similarly, points on the gnomonic net representing plunging lines are separated from the net's center point by a distance equal to the corresponding spot-height spacing of the plunge line (Figure 13b). The use of these properties is illustrated as follows.

Projection of Data Across a Pivotal Fault

Figure 14a illustrates a problem in which a plunging linear structure is deflected by a pivotal fault. Assuming a pivot of ψ degrees about the origin, the trend and spot-height spacing of the deflected plunge line may be calcu-

lated as follows. First, rotate the location of the hanging wall piercing point through an angle ψ about the pivot point to yield the footwall piercing point (Figure 14a, b). Then, determine the trend and spot-height spacing of the footwall plunge line using the gnomonic net construction of Figure 14c. Represent the fault plane on an overlay by its great circle F and pole π_f (these are located exactly as on a standard stereonet). The hanging-wall plunge line is represented by the point P. By rotating the net under the overlay, cause π_f and P to lie on a common great circle; then, measure the angle α they subtend. Lable the great circle's intersection with the fault plane P_1, and set off the pivot angle ψ from P_1 along F to yield a second point P_2. Rotating the net under the overlay, make a great circle pass through π_f and P_2, and locate the footwall plunge line P ' at an angle α from π_f. The plunge and trend of P_2 may be read off the net in standard fashion, or the plunge-line spacing may be transferred to a map of compatible scale factor R. Unless the footwall piercing point happens to lie on one of the fault plane's strike lines, a fractional spacing must be calculated for the first footwall spot height.

The use of gnomonic projection is no more tedious than standard stereographic projection, yet the advantages in this case are significant. Even small pivotal deflections may result in large errors of location or orientation when data are transferred significant distances along plunge.

INTERPOLATION IN SECTION PLANE

Having defined control points on the cross section by projection of map, well, seismic, or other data along plunge lines into the section plane, and having traced the tangents to surfaces at the piercing points of their projection lines one is invariably left with an incomplete section. The next step in section construction is to interpret the structure between control points. Again, a geometric model of the three-dimensional structure must be chosen, but it is important not to adhere to a model at the expense of observed data. For example, straight contacts may be interpolated between data points by least-squares

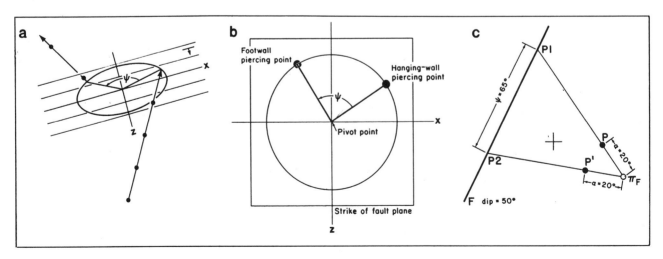

Figure 14—Projection across pivotal fault (see text).

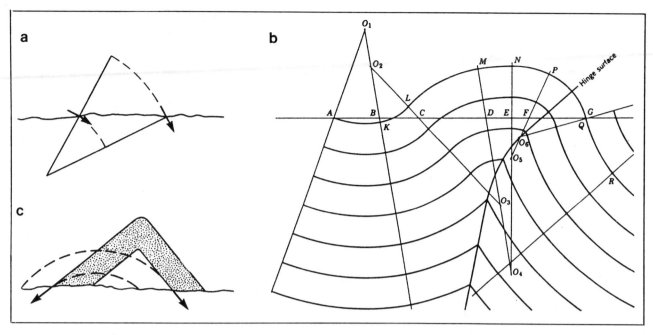

Figure 15—(a) Hewitt's construction, on which Busk interpolation (b) is based. (c) Error in construction is revealed when stratigraphic sequence is considered. Solid lines indicate actual structure; dashed lines indicate Hewitt-Busk construction.

analysis; however, given the unlikelihood of such a simple structure in nature, it is far better to spline contacts. A spline is a curve that passes through all of the data points on which it is based. Most computing centers have programs available for interpolating a cubic spline through a set of data points.

The procedure yields a smooth curve (i.e., each pair of neighboring cubic-curve segments share a common tangent at a data point) that passes through all projected data points, and it may be the most suitable form of interpolation where only the locations of contacts are known. However, where tangents to the projected structural surfaces have been plotted, their pitches may be in conflict with the interpolated spline. In such circumstances, it is preferable to use either parallel fold or Ramsay fold interpolation. Before such methods may be used, it is necessary to determine stratigraphic thicknesses in the plane of section.

FOLD INTERPOLATION

Parallel Folds

The best known method of fold interpolation is due to Busk (1929). It is based on Hewitt's (1920) technique for determining the thickness of strata between adjacent nonparallel dips by drawing normals to the dip lines and constructing circular arcs about their points of intersection (Figure 15a). In Busk's construction, circular arcs are interpolated between each pair of neighboring dip (or pitch) readings to form smooth splinelike curves (Figure 15b). As observed by Higgins (1962), this construction yields an apparently satisfactory interpolation when the stratigraphic sequence is ignored. However, the interpretation changes markedly whenever data locations are added or deleted, and the method breaks down when the

stratigraphic sequence is considered. The flaw in the Busk construction arises from Hewitt's technique of thickness calculation on which it is based. Figure 15c illustrates the problem. Two dips of known stratigraphic levels are recorded at the topographic surface. When dips alone are considered, Hewitt's construction yields a totally false (sequentially inverted) picture of the structure. Because of this fundamental flaw in Hewitt's construction, the Busk technique is invalid and should not be used. Busk was aware of the problem presented by stratigraphic identity, but seemed to understate its ramifications. For points in the same bed, his solution, modified by Higgins (1962), is to find two circular arcs that are tangential to the dip data points and that intersection in a point of common tangency. This solution works for pairs of points, but it is tedious and it breaks down when several stratigraphic levels are considered in unison (as in Reches et al, 1981).

A second method of parallel fold interpolation is the kink solution of Coates (1945) and Gill (1953), recently popularized by the retrodeformable sections of Suppe (1983, 1985) and Kligfield et al (1986). The method involves extrapolation of dips or pitches in straight lines (Figure 16a) to form "dip domains" separated by kink fold axial planes [the "boundary rays" of Coates (1945)]. When stratigraphic positions of the exposed bed contacts are unknown, the interpolation is underconstrained. Assuming constant limb thicknesses, the kink axial plane may be oriented to bisect the interlimb angle, but the plane's location is free to lie anywhere within the exposure gap, with consequent implications for the thickness of strata between the exposed layers. If the relative ages of the latter are known, this freedom is somewhat constrained. (It is naive to locate a fold axial plane midway between exposed limb dips without considering

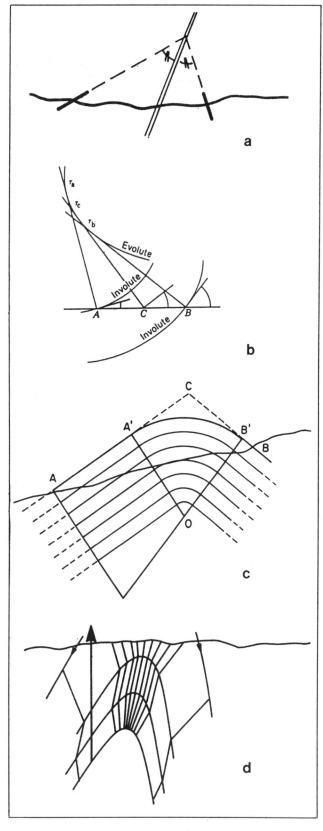

the relative ages of the limb strata.) Where kink folds merge toward the core of a box fold, the middle limb is eliminated and a new axial plane bisects the outer limbs' dihedral angle.

Though suited to certain lithologies, kink constructions sometimes yield excessively angular results and should be treated as an extreme case opposite to the excessively round constructions of Busk and Higgins. Mertie (1922, 1940, 1944, 1947, 1948) devised an interpolation technique using evolute and involute curves (Figure 16b). These curves provide parallel folds of continuously varying curvature, representing the smoothest possible interpolation between the extreme kinks and circular arcs noted above. However, the involution method is subject to the same problems of stratigraphic identity as is the Busk construction. Use of french curves or cartographer's splines to approximate involute curves is justified in view of the imperfect geometric forms that occur in natural settings.

The best method of parallel fold interpolation combines dip domains and circular arcs as illustrated in Figure 16c. Two outcrops of the same bed are labeled A and B; their dip lines are extended to intersect at C. Any fraction CA ' of line CA is chosen, provided an equal length CB ' can be set off along CB. Normals to A ' and B ' are constructed to intersect at O. From the construction, OA ' = OB ', and a circular arc may be drawn tangential to CA at A ' and to CB at B '. Concentric arcs are constructed between the radii OA ' and OB ', and these arcs are continued into the homoclinal dip domains beyond the two radii. Because the length CA ' was chosen as an arbitrary fraction of CA, the interpolated fold may be assigned any desired degree of angularity. However, if a third exposure of a different stratigraphic level exists, then this degree of freedom is eliminated, and the angularity must be adjusted until the third dip is correctly located in the structure. This method is much simpler than the involution technique and yields satisfactory results for normal bed thicknesses. For a multilayer stratigraphy, however, the parallel fold model is generally inappropriate; antiformal structures cannot be extended to depth as they pass through curvature centers and into cusp geometries, and synformal structures grow to large amplitudes unless unexposed box folds or other accommodation structures are invented. A more satisfactory solution is to employ the dip (pitch) isogon technique described below.

Isogons

Dip isogons (lines joining points of equal dip) were described by Agterberg (1961), Elliott (1965), and Ramsay (1967), who used them as a basis of fold classification. Their use in section construction was suggested by Bengston (1982) and developed further by Ramsay and Huber (1987). The procedure requires surface and/or well control and some estimate of the rheologies of strata. First, an isogon convergence factor is assigned to the strata exposed at the surface or starting point of a well log. Then the subsurface is interpolated by maintaining constant dips (or pitches) along the isogons (Figure 16d).

Figure 16—**(a) Kink interpolation for two dip readings on same bed. If beds were of unknown relative age, the kink axis location could lie anywhere in the exposure gap. (b) Mertie's involute construction. (c) Modified kink/Busk interpolation. (d) Isogon interpolation (see text for explanation).**

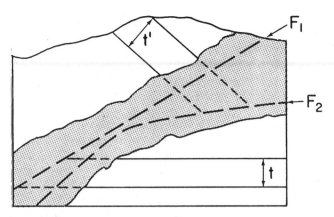

Figure 17—Fault interpolation. Dotted lines mark boundary of disturbed zone. Straight fault F1 leads to incompatible hanging-wall and footwall ramps. Bent fault F2 is preferred.

After the first contact is passed, a new convergence factor is chosen with regard for the rheological nature of the underlying bed. In this way, layers of alternating parallel and similar geometry may be continued to depth (see Ickes, 1923). The isogon method alone takes account of rheological properties of strata and permits the user to gradually increase or decrease the amplitudes of buried structures regardless of whether they are synformal or antiformal. No other method can produce interpolations of similar realism (the Fourier method of Hudleston, 1973, has potential in this regard, but has not been investigated). Kinks, circular arcs, involutes, and general parallel folds represent the special case where dip isogons are everywhere normal to bedding.

When constructing a cross section, in addition to interpolating fold shapes, it is important to identify types of folds whenever possible. For example, buckling and bending produce distinct characteristic strain patterns. Spatially fixed folds through which material has migrated may be identifiable by the presence of a deformation fabric on one limb only. In the absence of such evidence, geometric factors such as fold asymmetry and limb thickness changes may indicate the history of deformation (e.g., Jamison, 1987).

FAULT INTERPOLATION

Methods of fold interpolation assume structural continuity and, therefore, break down whenever a fault is encountered. The interpolation of fault traces between partially constrained folded thrust slices is generally the most difficult step in section construction. Sometimes faults are exposed at the topographic surface; however, more commonly their presence is indicated on maps, well logs, or seismic records by a disturbed zone lacking coherent structure. The principal controlling factor in fault interpolation is the relationship between the lengths and orientations of ramps and flats on opposite walls of the fault. In Figure 17, the outcrop of a fault is obscured by an exposure gap separating segments of a recognizable marker bed whose thickness is t in the footwall and t′ in the hanging wall. In the absence of evidence for internal strain (t′ = t), the footwall and hanging-wall values of

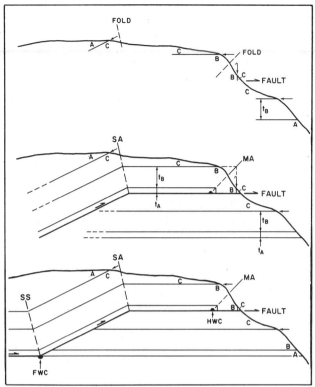

Figure 18—Fault interpolation based on field example (see text for details).

the ramp lengths and cutoff angles should be equal (I′ = I, θ′ = θ; Crane, 1987). Attempts to draw a straight fault F1 through the exposure gap do not obey Crane's rule, but a bent trajectory F2 yields an acceptable approximation. The bent fault also explains the tilted orientations of strata in the hanging wall. Trajectory F2 is not the only possible solution. If internal strain is evident, then any interpolation obeys De Paor's (1987) equation,

$$I'/I = (t'/t)(\sin\theta/\sin\theta'),\qquad(19)$$

yields a balanced section. When adding strata above and below the marker bed in Figure 17, net slip will change as material passes over the fault bend (see graphs in Suppe, 1983).

A more adventurous interpolation is illustrated in Figure 18, which is simplified from the writer's mapping of the Vesle Fiord thrust, central Ellesmere Island. The thrust's trace at the topographic surface is constrained by the outcrops shown. In the hanging wall, a fold MA is identified as a material anticline because of an axial plane breccia that is not present on the limbs, implying that no passage of material occurred through hinges of layers. Therefore, the fault is extended until it truncates the fold and the lowest bed in the core is identified at point HWC. The map reveals a second fold SA whose axial plane is not exposed. Interpreting this as a spatial anticline, a bend in the fault trajectory is inferred, creating a ramp in the footwall. The ramp is continued until a point FWC where the same stratigraphic level as HWC is encoun-

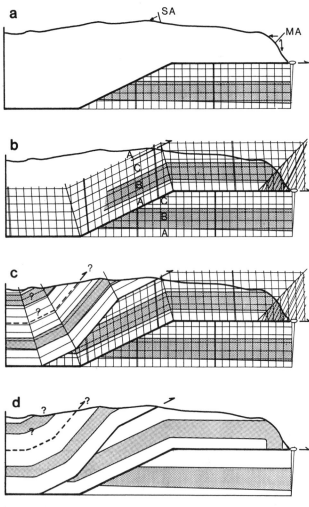

Figure 19—Illustration showing use of Lagrangian grid to permit calculation of fault's initial trajectory while constructing its deformed state trajectory. (a) Square grid marked on undeformed footwall. MA = material anticline, SA = spatial anticline. (b) Grid is extended through hanging wall of first fault, maintaining bed lengths and stratigraphic thicknesses A, B, C. (c) Extrapolation of second fault so that dip measured relative to deformed grid equals the dip of first ramp. (Slope = 1:2.) Note kink in second fault at base of first ramp. (d) Final section after erasing grid lines. Note that line and area balanced without iterative restoration.

tered; then a spatial syncline is inferred to return the fault to a flat-on-flat geometry. The fault's net displacement is calculated from the arc length FWC to HWC.

This example illustrates the power of Coates's (1945) kink interpolation method when combined with fault cutoff rules (Crane, 1987). However, it is important to remember that many alternative interpretations are equally valid. For example, the hanging-wall faults in Figure 18 may be replaced by low-angle ramps, or the kink geometry may be more rounded.

STRAIN GRID INTERPOLATION

One of the disadvantages of manual section balancing is that deformed sections must be drawn, evaluated, and redrawn iteratively. This tedious and time-consuming

procedure may be avoided by use of the Lagrangian grid, or "strain grid" method of De Paor and Bradley (1986). First, the undeformed footwall of the frontal thrust is constructed and marked with a square grid (Figure 19a). Then, the deformed equivalent of an initial square grid is superimposed upon the antiformal structure in the first thrust slice (Figure 19b). The grid may be constrained either by the choice of a folding model such as bed length and area conserving flexural slip, or neutral surface buckling, or by actual strain data. In the former example, conservation of bed-length results in grid nodes that are spaced at equal intervals along the traces of beds (fractional intervals may be interpolated at fault cutoff points). When constructing the second fault trace, the predeformational fault trajectory can be calculated using the coordinates of the deformed trace measured relative to the Lagrangian grid. Thus, when constructing a fault in the deformed state, it is clear that the initial dip was reasonable (Figure 19d).

EXTRAPOLATION ABOVE TOPOGRAPHIC SURFACE

For proper section evaluation, one must consider the eroded portions of the thrust systems. When deciding how far to continue a thrust sheet above the current erosion surface, isostatic conditions must be carefully considered. Metamorphic grades in the preserved portions of footwalls may indicate a maximum overburden; suitable indicators include fluid inclusions, illite crystallinity states, vitrinite reflectances, conodont color indices, and zeolites, as well as conventional mineral paragenesis in higher grade rocks. Stratigraphic separation diagrams (Elliott, 1977) and distance-displacement curves (Chapman and Williams, 1984; Morley, 1986) may give a qualitative indication of the distance to a fault tip (Figure 20). Examination of the fault's map trace may also offer clues to constrain its possible rate of slip dissipation. However, quantitative application of these measures assumes (1) no jumps in displacement at fault bends or branch points and (2) a linear relationship between distance and displacement.

Because thrust belts commonly form at shallow depths, the possibility of synorogenic emergence must be considered. Evidence for such emergence will be found in the stratigraphic record (Vann et al, 1986; De Paor and Anastasio, 1987). In areas of emergence, the restored section must be modified to make room in the thrust sheets for the molasse now deposited in the foredeep. Because the upper layers of a thrust sheet are eroded first, the allowance for emergence in balanced sections will steepen the inferred trajectories of faults in the initial state (Figure 21).

EXTRAPOLATION TO DEPTH

Chamberlin (1910) first applied a construction for the determination of depth to detachment in fold and thrust belts. His method, sometimes termed "excess area" or "excess section" calculation, is illustrated in Figure 22a. It is assumed that an initially rectangular stratigraphic package was subject to a shortening displacement S by means of movement on a detachment at a depth T below

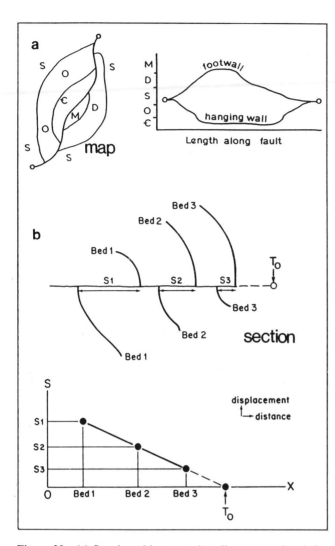

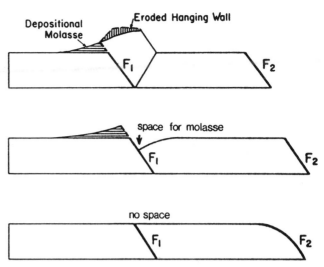

Figure 21—Restoration of emergent thrust fault F1 steepens inferred initial dip of hindward fault F2.

Figure 20—(a) Stratigraphic separation diagram used to infer rate of slip dissipation on fault. (b) Distance vs. displacement plot used to extrapolate fault to its tip point T_o. S1, S2, and S3 are slip vectors of beds 1, 2, and 3, respectively.

a chosen marker horizon M. It is also assumed that the displaced area A = St is conserved in the section plane; by implication, no material moves into or out of the plane, and volume is conserved. The displaced area must then be present above the marker horizon's undeformed trace. By measuring this area, using a digitizer, and by measuring S, using the line balance technique (see De Paor and Bradley, in press), we may estimate the depth to detachment, t = A/S. A useful property of the Chamberlin construction is that the calculated depth to detachment is entirely independent of the thrust's trajectory where it rises from the detachment to the surface. However, there is one drawback: it is implicitly assumed that the loose line at the hinterland end of the section remains vertical throughout strain, an unlikely scenario. The error involved in ignoring the possible angular shear of the hinterland loose line depends on the magnitude of the area swept out by that line during shear relative to the total area of the thrust sheet. If necessary, the error can be reduced using the strain grid construction described

above to determine the loose line's deformed shape. The excess area is balanced against the area bounded by that sheared line (Figure 22b).

PROJECTION FROM SECTION ONTO MAP

Having constructed a section whose prototype was derived from projected data, interpretations made on the section should then be relayed back to the source seismic section or map surface. Furthermore, it should be clearly understood that modification of source data to fit a particular structural model is not condoned; only the interpolated contacts between data points should be altered. For example, seismic stratigraphers may extend strong bedding reflectors to meet a fuzzy fault zone without an intervening rollover fold. They may feel that such a stylistic addition to the visible structure is unjustified. If section interpolation techniques described above reveal the existence of a rollover, then this structure should be incorporated in the seismic interpretation also. Projection from structural section to neighboring seismic section is a trivial case of the more general problem of projection onto a topographic surface, and will not be treated further.

As a second example, despite their undergraduate training, mappers frequently extend dipping contacts across valleys without the correct upstream or downstream "V." When projected onto a cross section, such errors appear as spikes in otherwise smooth contacts. If spikes are removed during section balancing, it is logical to transmit the modifications back to the source map. The method proposed here is suited to either manual or automated implementation based on a combination of the plunge-line concept and the projection technique of Turner and Weiss (1963, p. 164; Ragan, 1985, p. 351). First, project the section onto a horizontal datum as illustrated in Figure 23a, using equations 7 and 8 for profile plunge projection derived previously. Then translate this horizontal "section" along the structural trend (Figure 23b) until it lies at the level of a particular contour on the

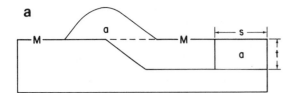

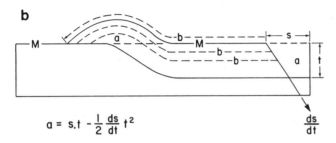

$$a = s.t - \frac{1}{2}\frac{ds}{dt}t^2$$

Figure 22—Illustrations of extrapolation to depth. (a) Chamberlin's (1921) construction; (b) Modified method (see text).

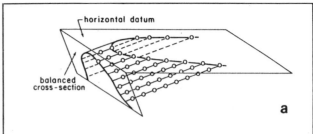

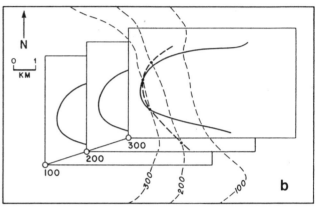

Figure 23—Construction of balanced map by projection of section data up plunge lines to intersect various contour levels (see text for explanation).

map. Record the points of intersection of this contour plane with geologic contacts on the horizontal section, and translate the section farther along the direction of the plunge lines to the next contour level. Repetition of this procedure for each contour of the topography yields a map projection of the interpretive section.

DISCUSSION AND CONCLUSIONS

During the past decade, the need to balance cross sections has become accepted in the geologic literature, but no single publication has suggested how to construct such a section from scratch. Instead, techniques and tests are scattered among various sources, and the skill of section construction has been transmitted mainly by example.

Projection of geologic map, well log, and seismic data along plunging axes onto a structural section is a well-established theoretical concept, but practical applications are not common, probably because precise details of the projection procedure are not widely known. This paper presents three approaches suited to different circumstances: tractible equations in terms of outcrop location, section orientation, and projection axis orientation; matrix rotation suited to computer solution; and the graphical technique of plunge-line construction. Using any of these techniques, it should not be difficult to project a wide variety of data types from spatial locations onto a section plane. When stratigraphic location, contact pitch, and bed thickness data have been accumulated in the section plane, fold and fault interpolation procedures are used to fill gaps in the data. Folds may be modeled as parallel structures comprising circular arcs, straight-limbed kinks, or intermediate shape; however, the Ramsay isogon interpolation technique yields rheologically more realistic results. Limb orientations and relative thicknesses may provide valuable clues to the origin and strain history of fault-related folds (Jamison, 1987). Fault interpolation is generally least well constrained.

The geometry of bedding cutoff points on either side of a fault must be consistent with the amount of resulting internal strain, subject to the complicating effects of growth sedimentation and fault reversal.

When known thicknesses of formations are added beneath the surface exposures of thrust sheets, the cover-basement contact is commonly found to occur at the same level in each sheet. This occurrence is good evidence for the existence of a detachment. The depth to detachment may be checked by the excess-area method as modified herein to consider penetrative angular shear. Where preservation of the detachment requires invention of buried duplexes, the alternative of basement involvement should be considered. Offsets in the cover-basement contact are easily missed in (or migrated out of) seismic sections, with profound consequences for total displacement calculations. Finally, the interpretations incorporated in a section must be relayed back to the source map to ensure mutual compatibility between the two views of the three-dimensional structure. Complex sections containing growth sediments, diapiric or piercement structures, and facies changes are best constructed by iterative evaluation and modification, as described in De Paor and Bradley (in press).

REFERENCES CITED

Agterberg, F. P., 1961, Tectonics of the chrystalline basement of the Dolomites in North Italy: Geologica Ultraiectina, no. 8, p. 1-232.
Ambrose, J. W., and H. T. Carswell, 1962, Right sectional block diagrams and random sections with the aid of a stereonet: Geological Association of Canada Proceedings, v. 13, p. 119-128.

Argand, E., 1911, Les nappes de recouvrement des Alpes Pennines et leurs prolongement structuraux: Beitrage zur geologischen Karte der Schweiz, new series, part 31, p. 1-26.

Bengston, C. A., 1980, Structural uses of tangent diagrams: Geology, no. 8, p. 599-602.

——— 1982, Structural and stratigraphic uses of dip profiles in petroleum exploration, in M. T. Halbouty, ed., The deliberate search for the subtle trap: AAPG Memoir 32, p. 31-45.

Busk, H. G., 1929, Earth flexures: Cambridge, U.K., Cambridge University Press, 106 p.

Chamberlin, R. T., 1910, The Appalachian folds of central Pennsylvania: Journal of Geology, v. 27, p. 228-251.

Chapman, T. J., and G. D. Williams, 1984, Displacement-distance methods in the analysis of fold-thrust structures and linked-fault systems: Journal of the Geological Society of London, v. 141, p. 121-128.

Charlesworth, H. A. K., C. W. Langenberg, and J. Ramsden, 1976, Determining axes, axial planes, and sections of macroscopic folds using computer-based methods: Canadian Journal of Earth Science, v. 13, p. 54-65.

Christensen, M. N., 1963, Structural analysis of Hoosac Nappe in northwestern Massachusetts: American Journal of Science, v. 261, p. 97-107.

Coates, J. S., 1945, The construction of geologic sections: Quarterly Journal of the Geological, Mining and Metallurgical Society of India, v. 17, p. 1-11.

Coward, M. P., and J. H. Kim, 1981, Strain within thrust sheets, in K. R. McClay and N. J. Price, eds., Thrust and nappe tectonics: Geological Society of London Special Publication 9, p. 275-292.

Crane, R. C., 1987, Use of fault cut-offs and bed travel distance in balanced cross-sections: Journal of Structural Geology, v. 9, p. 243-247.

Cruden, D. M., 1981, Comment and reply on 'Structural uses of tangent diagrams': Geology, v. 9, p. 242-243.

Dahlstrom, C. D. A., 1969, Balanced cross sections: Canadian Journal of Earth Science, v. 6, p. 743-757.

De Paor, D. G., 1981, Geological strain analysis: PhD treatise, National University of Ireland, Dublin, 104 p.

——— 1983, Orthographic analysis of geological structures. Part 1: deformation theory: Journal of Structural Geology, v. 5, p. 255-277.

——— 1987, Stretch in shear zones; implications for section balancing: Journal of Structural Geology, v. 9, p. 893-895.

——— and D. Anastasio, 1987, The Spanish External Sierra: a case study in the advance and retreat of mountains: National Geographic Research, v. 3, p. 199-209.

——— and D. C. Bradley, 1986, Algorithm for section balancing using strain compatibility criteria (abs.): AAPG Bulletin, v. 70, p. 580.

——— in press, The theory and practice of balanced section construction. Part 2: computerized line and area balance: Geobyte.

——— and J. Krome, in press, Strain analysis of the Loudoun Formation, west Maryland, using data restored from boulder to outcrop orientation: GSA Bulletin.

Diegel, F. A., 1986, Topological constraints on imbricate thrust networks, examples from the Mountain City window, Tennessee, U.S.A.: Journal of Structural Geology, v. 8, p. 269-281.

Eardley, A. T., 1938, Graphic treatment of folds in three dimensions: AAPG Bulletin, v. 22, p. 483-489.

Elliott, D., 1965, The quantitative mapping of directional minor structures: Journal of Geology, v. 73, p. 865-880.

——— 1977, Some aspects of the geometry and mechanics of thrust belts. Part 1. Eighth Annual Canadian Society of Petroleum Geologists Seminar, 95 p.

Fisher, O., 1881, Oblique and orthogonal sections of a folded plane: Geological Magazine, new series, v. 8, p. 20-23, 237.

Flinn, D., 1978, Construction and computation of three-dimensional progressive deformations: Journal of the Geological Society of London, v. 135, p. 291-305.

Gill, W. D., 1953, Construction of geological sections of folds with steep-limb attenuation: AAPG Bulletin, v. 37, p. 2389-2406.

Groshong, R. H., Jr., and S. I. Usdansky, 1986, Deformation in thrust-ramp anticlines and duplexes: implications for geometry and porosity (abs). AAPG Bulletin, v. 70, p. 59.

Hewitt, D. F., 1920, Measurement of folded beds: Economic Geology, v. 15, p. 367-385.

Higgins, C. G., 1962, Reconstruction of flexure folds by concentric-arc method: AAPG Bulletin, v. 46, p. 1737-1739.

Hilton, H., 1904, The gnomonic net: Mineralogical Magazine, v. 14, p. 18-22 and Plate 2.

Hoffman, P. F., R. Tirrul, J. E. King, M. R. St-Onge, and S. B. Lucas, in press, Axial projections and modes of crustal thickening, eastern Wopmay orogen, northwest Canadian shields, in S. P. Clark, Jr., ed., Processes in continental lithospheric deformation: GSA.

Hossack, J. R., 1979, The use of balanced cross-sections in the calculation of orogenic contraction: a review: Journal of the Geological Society of London, v. 136, p. 705-711.

——— 1983, A cross-section through the Scandinavian Caledonides constructed with the aid of branch-line maps: Journal of Structural Geology, v. 5, p. 103-112.

Hudleston, P. J., 1973, Fold morphology and some implications of theories of fold development: Tectonophysics, v. 16, p. 1-46.

Ickes, E. L., 1923, Similar, parallel and neutral surface types of folding: Economic Geology, v. 18, p. 575-591.

——— 1925, The determination of formation thickness by the method of graphical integration: AAPG Bulletin, v. 9, p. 451-463.

Jamison, W. R., 1987, Geometric analysis of fold development in overthrust terranes: Journal of Structural Geology, v. 9, p. 207-220.

Jones, P. B., and H. Linnser, 1986, Computer synthesis of balanced structural cross sections by forward modeling (abs.): AAPG Bulletin, v. 70, p. 605.

Kilby, W. E., and H. A. K. Charlesworth, 1980, Computerized down-plunge projection and the analysis of low-angle thrust-faults in the Rocky Mountain Foothills of Alberta, Canada: Tectonophysics, v. 66, p. 287-299.

King, J. E., in press, Thirty kilometers of obliquely exposed structural relief in th early Proterozoic Wopmay orogen, Canada: a composite down-plunge cross section of a metamorphic-internal zone: Tectonics.

Kligfield, R., P. Geiser, and J. Geiser, 1986, Construction of geologic cross-sections using microcomputer systems: Geobyte, v. 1, p. 60-66.

Langenberg, C. W., H. E. Rondeel, and H. A. K. Charlesworth, 1977, A structural study in the Belgian Ardennes with sections constructed using computer-based methods: Geologie en Mijnbouw, v. 56, p. 145-154.

Lisle, R. J., 1980, A simplified work scheme for using block diagrams with the orthographic net: Journal of Geological Education, v. 29, p. 81-83.

Lugeon, M., 1901, Les grande nappes de recouvrement des Alpes du Chablais et de la Suisse: Societe geologique de France Bulletin, 4th series, v. 1, p. 723-825.

Mackin, J. H., 1950, The down-structure method of viewing geologic maps: Journal of Geology, v. 58, p. 55-72.

McIntyre, D. B., and L. E. Weiss, 1956, Construction of block diagrams to scale in orthographic projection: Geologists' Association of London Proceedings, v. 67, p. 142-155.

Medweleff, D. A., and J. Suppe, 1986, Kinematics, timing, and rates of folding and faulting from syntectonic sediment geometry (abs.): EOS, v. 67, p. 1223.

Mertie, J. B., Jr., 1922, Graphic and mechanical computation of thickness of strata and distance to a stratum: USGS Professional Paper 129, p. 39-52.

——— 1940, Stratigraphic measurements in parallel folds: GSA Bulletin, v. 51, p. 1107-1134.

——— 1944, Calculation of stratigraphic thickness in parallel folds: AAPG Bulletin, v. 28, p. 1376-1386.

——— 1947, Delineation of parallel folds and measurements of stratigraphic dimensions: GSA Bulletin, v. 58, p. 779-802.

——— 1948, Application of Brainchon's theorem to construction of geologic profiles: GSA Bulletin, v. 59, p. 767-786.

Morley, C. R., 1986, A classification of thrust fronts: AAPG Bulletin, v. 70, p. 12-25.

Mugnier, J. L., and P. Vialon, 1986, Deformation and displacement of the Jura cover on its basement: Journal of Structural Geology, v. 8, p. 341-360.

Price, R. A., 1973, Large scale gravitational flow of supra-crustal rocks, southern Canadian Rockies, in K. A. De Jong and R. Scholten, eds., Gravity and tectonics: New York, Wiley, p. 491-502.

Ragan, D. M., 1985, Structural geology, an introduction to geometrical techniques, 3rd ed.: New York, Wiley, 393 p.

Ramsay, J. G., 1967, Folding and fracturing of rocks: New York, McGraw-Hill, 568 p.

——— and M. I. Huber, 1987, The techniques of modern structural geology, vol. 2: folds and fractures: New York, Academic Press, 700 p.

Reches, Z., D. F. Hoexter, and F. Hirsch, 1981, The structure of a monocline in the Syrian Arc System, Middle East—surface and subsurface analysis: Journal of Petroleum Geology, v. 3, p. 413-425.

Sanderson, D. J., 1982, Models of strain variation in nappe and thrust sheets: a review: Tectonophysics, v. 88, p. 201-233.

Stockwell, C. H., 1950, The use of plunge in the construction of cross-sections of folds: Proceedings of the Geological Association of Canada, v. 3, p. 97-121.

Suppe, J., 1983, Geometry and kinematics of fault-bend folding: American Journal of Science, v. 283, p. 684-721.

———— 1985, Principles of structural geology: Englewood Cliffs, New Jersey, Prentice-Hall, 537 p.

Turner, F. J., and L. E. Weiss, 1963, Structural analysis of metamorphic tectonites: New York, McGraw-Hill, 545 p.

Vann, I. R., R. H. Graham, and A. B. Hayward, 1986, The structure of mountain fronts: Journal of Structural Geology, v. 8, p. 215-228.

Wegmann, C. E., 1929, Beispiele tektonischer Analysen des Grundgebirges in Finland: Commission Geologique de Finlande Bulletin, v. 87, p. 98-127.

Reprinted with permission from *Journal of Structural Geology*, v. 5, p. 153-160, A. D. Gibbs, Balanced cross-sections construction from seismic sections in areas of extensional tectonics, Copyright 1983, Pergamon Press plc.

Balanced cross-section construction from seismic sections in areas of extensional tectonics

A. D. GIBBS

Britoil, 150 St. Vincent Street, Glasgow GZ2 5LJ, U.K.

(*Received 5 July* 1982; *accepted in revised form 20 January* 1983)

Abstract—The application of the technique of balanced section construction, initially developed for areas of compressional folding and faulting, is reviewed with reference to extensional tectonics. A number of examples are discussed where these techniques have been successfully applied in the North Sea. The interpretation of geoseismic sections is considered to be greatly assisted by careful application of geometrical balance and a consideration of strain even in areas of low crustal extensions. The nature of seismic sections, however, places limitations on the validity of balancing which must be borne in mind with such interpretations and wherever possible the balancing of a geoseismic section should be confirmed by complete depth conversion. The rapid testing of the integrity of the geoseismic section by attempting to balance the section at the interpretational phase can eliminate many problems as well as allowing the fullest use to be made of the geophysical information.

INTRODUCTION

THE CONSTRUCTION of accurate geological cross-sections is of the greatest importance to all branches of geology but is of paramount importance when large commercial investments are at stake. For this reason balanced section techniques have evoked considerable interest, in particular in areas of overthrusting (Elliott 1977). In extensional regimes, however, much less work has been done, partly because finite extensions are commonly fairly small and the tectonic style may be deceptively simple. This paper considers some of the problems of applying balanced section techniques to areas of extensional tectonics with reference to problems encountered in the North Sea.

While virtually all the geometrical constructions used in balanced sections have been worked out for regions of contractional tectonics particularly since the synthesis of Dahlstrom (1969a,b, 1970) there are a number of problems which are special to extensional regimes, where major errors can be made. In addition, the primary data is likely to consist entirely of seismic reflection lines with or without well control. In contrast, in many of the areas of contractional tectonics, where section balancing techniques have been successfully applied, field mapping can be used and supplemented by seismic and well data. Field observation allows accurate estimates of ductile strain to be made, and for this to be accounted for in the construction of any palinspastic sections. Removal of ductile strains (e.g. Hossack 1979, Schwerdtner 1977, Borradaile 1979, Elliott & Johnson 1980) in addition to rotational and translational strains may be possible. Strain heterogeneity may also be estimated and its possible effects on balance calculations deduced. This is rarely possible in areas of extensional tectonics such as in the North Sea. The availability of closely spaced seismic lines, coupled with well control, however, can give three-dimensional and stratigraphic control rarely possible in areas of orogenic contraction.

THE BALANCED SECTION

The basic approaches to section balance assume plane strain, or conservation of cross-sectional area. Chamberlin's (1910, 1919) calculation of equal areas for a section deformed above a décollement or detachment surface can be applied to extension as well as contraction. Figure 1 shows this where the area A before deformation equals the area B after deformation. The area C is common and hence the equation which expresses the relationship between the undeformed length, the deformed length of

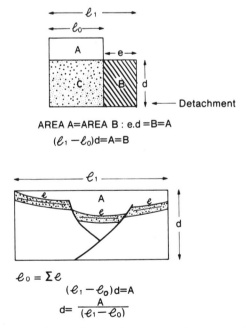

$$AREA\ A = AREA\ B : e.d = B = A$$
$$(\ell_1 - \ell_0)d = A = B$$

$$\ell_0 = \Sigma\ell$$
$$(\ell_1 - \ell_0)d = A$$
$$d = \frac{A}{(\ell_1 - \ell_0)}$$

Fig. 1. Area balance for extension. Above: l_0 is original length of section which is compared with length in deformed state and area. Below: the regional projected horizontal, to calculate depth to detachment. See text for details.

153

section and the depth to the décollement surface, d, is given by: $l_0 = l_1 + A/d$ (Elliott 1977; after Gwinn 1970) where A is the cross-sectional area. This can be expressed in terms of average stratigraphic thickness, d, above the décollement surface: $l_0 = l_1 d_1/d_0$, or in terms of the extension, e, measured as conventional engineering strain ($e = (l_1 - l_0)/l_0$) where the equation is written as $d_1 = A/e$. The extension factor, β (McKenzie 1978a,b), may be substituted for unit length l_0: $\beta = (1 + e)$; $\beta = l_1/l_0$.

These equations are seen to be identical to those for orogenic contraction (Hossack 1979) with the exception of the change in sign convention for e and are likewise independent of the style of deformation. The geometry represented on any cross-section should satisfy these equations and depending on the data available balanced lengths and/or areas are used as appropriate.

The sections may not appear to satisfy these conditions for a variety of reasons. Firstly, the section may not be normal to the tectonic strike, that is, it may not be parallel to the extension direction. In this case the data can be projected on to such a section or allowance can be made for the angular deviation of the section and the extension direction.

$$l_\lambda = (l_c^2 - (S \cdot \sin \alpha)^2)^{1/2},$$

where l_λ is length in the section normal to tectonic strike, and is observed in a section at α to this dip section and l_c is length of observed segment over horizontal distance S. The dip of l_λ, θ_λ is given by

$$\theta_\lambda = \sin^{-1} (S \cdot \sin \alpha/l_\lambda).$$

Secondly, there may be extension along the tectonic strike when the relationship: $(l + e_1)(l + e_2)(l + e_3) = l + \Delta$, where Δ is the volume strain, $\delta V/V$ will be satisfied. To allow for this case Hossack (1978) used the method of integrating the measured finite strains along the intermediate finite strain trajectories to calculate the average strike elongation assuming constant volume. Hossack's approach holds equally for extensional deformations.

The assumption of plane strain in balanced cross-sections has been criticized (e.g. Geiser 1978). The large volume changes which can accompany diagenesis (Sanderson 1976) and tectonic deformation (Wood 1974, Ramsay & Wood 1973) can lead to volume changes possibly in excess of 40%. Mass transfer by pressure solution may also be present and again large volume changes may occur (Plessman 1964, Durney & Ramsay 1973, Elliott 1973).

COMPACTION

Both compaction changes and pressure solution changes in volumes of this order of magnitude are recorded from the North Sea area. For contractional tectonics, Hossack (1979) discussed how these volume changes may be discounted, or have net effects which

are small in relation to other strains. For example, compaction usually occurs before deformation and, therefore, the tectonically undeformed compacted state is compared with the deformed state and the compaction is the same in each case. It may also be a valid assumption for pressure solution that the volume loss has uniaxial symmetry or that the material is redeposited locally.

In areas of extensional tectonics such as the North Sea many of these assumptions do not apply. Compaction is probably the most important element of the finite strain with the exception of brittle faulting. As the compaction may be both synchronous with, and post-date the tectonic deformation it is essential to account for this in any balance calculation, and subsequent palinspastic reconstruction. Estimates of compaction can, in some cases, be made from direct measurement of deformed objects such as reduction spots and burrows recovered during coring using a variety of techniques such as those outlined by Ramsay (1967). Unfortunately such measurements only apply to discrete intervals and often to only one lithology. Nevertheless they do allow a realistic estimate of average compaction strain to be used. An alternative approach is to use compaction curves for depth of burial (Steckler & Watts 1978, Sclater & Christie 1980) and to expand the section progressively during back stripping (see Wood 1981). However, with rapid burial or erosion the present depth may not reflect the actual compaction and a more direct method is to estimate compaction from the sonic log (Magara 1976a,b, 1978). In the case of the example illustrated in Fig. 2 compaction estimates using the direct measurement of reduction spots in cores and Magara's technique in a number of wells on the structure both gave comparable values of about 30% for volumetric strain. The application of this volume compaction to the section improves the reconstruction in the vicinity of areas of complex faulting (cf. Figs. 2b & c) and allows the very low observed dips and 'horse tailing' of the main fault (MF) in the shale sections overlying the reservoir to be attributed to progressive compaction (Fig. 2). 'Horse tailing' of normal faults is frequently seen on seismic interpretations and is analogous to splaying in strike-slip faulting. The analysis discussed here suggests a progressive 'younging' of the splay faults away from the footwall during syntectonic compaction. In addition it is necessary to postulate movement on the secondary fault (SF) to down-fault the crest of the horst after the formation of the main horst edge by movement on the main fault. Bedding-plane slip or flow in the overlying shale section must also be accommodated at this time. In this case the constraints of balancing the section provide an insight into a common and fundamental process of horst and graben formation with the active deformation moving outwards from the graben axis with time. The progressive retreat of horst lip away from the graben axis results in dip changes of the earlier-formed faults and may lead to faults generated as normal faults being rotated into a reverse geometry (Fig. 3) (see Profett 1977, Wernicke & Burchfiel 1982).

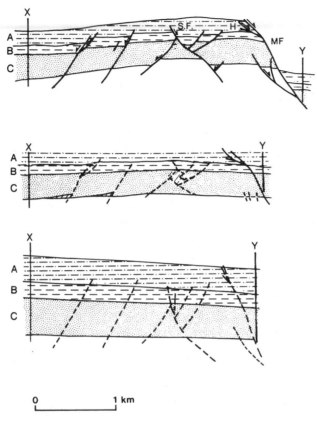

0   1 km

Fig. 2. Sections across a horst block showing respectively: (top) geological section compiled from seismic data; (middle) first attempt at reconstruction by 'jigsaw' technique, note area deficit shown by an oblique lined ornament; and (bottom) final reconstruction after decompaction and allowance for differential compaction.

LISTRIC FAULTS AND ROTATION

A further problem which is not normally encountered in contractional tectonics is that of growth faulting and syntectonic erosion and deposition. Growth faulting (Rider 1978, Crans *et al.* 1980, Crans & Mandl 1980) on a low-angle listric gravity slide may usually be distinguished on geoseismic sections from syn-sedimentary fault activity on deep-seated structures by the presence of roll-over and by use of a depth to detachment calculation. A case of shallow detachment which is controlled by basement faults is illustrated in Fig. 4 (after Heybroek 1975). Detachment is on Zechstein evaporites and comprises listric faulting and halokinesis towards the deeper parts of the basin. Such thin-skinned extensional tectonics may be much more prevalent than is realized even outside areas of Zechstein salt movement and can be recognized as both local and regional events. The distinction, however, on seismic sections between growth faults *sensu stricto* and synsedimentary sliding (growth faults *sensu lato*), which both obey the same geometric rules, is important in determining the effect on sedimentation. It may be advantageous to establish the timing of the syn-sedimentary fault activity by plotting a growth curve (Fig. 5) and discontinuities in the curve can be used to identify periods of erosion and reactivation on the fault.

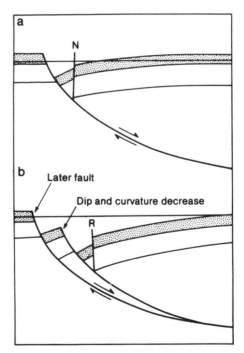

Fig. 3. Progressive migration and 'piggy-back' movement of listric normal fault resulting in (a) rotation of an early normal fault (N) on a roll-over and (b) an apparent reverse geometry (R) and flattening of the earlier synthetic fault.

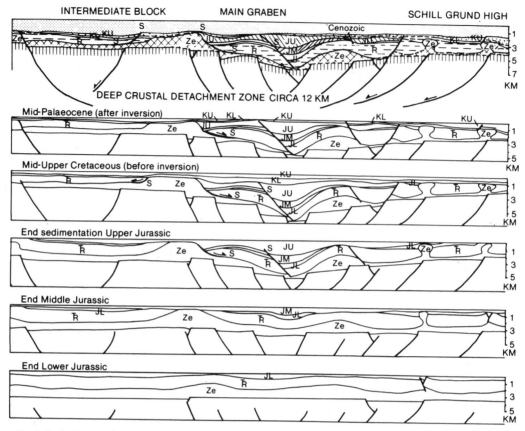

Fig. 4. Regional section balanced and reconstructed (after Heybroek 1975) by progressive back stripping of the sediments, removal of halokinetic effects and balancing the basement faulting. Note the decoupling zone in the Zechstein (Ze) combining low-angle listric normal faults (S) and halokineses towards the deeper parts of the basin. Also note the control of the higher level faulting by the deeper basement structures with drape and secondary faulting above the Zechstein.

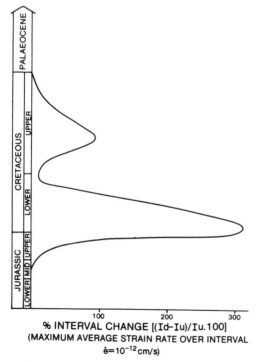

Fig. 5. Growth curve plotted for a fault from the Central graben showing two periods of fault activity and rapid growth. The average slip rate for the main phase is approximately 10^{-12} cm s^{-1}.

In order to calculate the true rate of faulting it is necessary to correct these curves for compaction on the foot and hanging-wall and then average tectonic strain rates ($\dot{e}$) may be calculated.

Well control may be necessary to confirm that the sequence on the upthrown footwall is condensed rather than eroded, and this is necessary before the section can be reconstructed. Partial reconstructions may be required for each formation and it may be impossible to balance the section in a single stage. Care should be taken that thickened sequences on the downthrown side of faults can be accounted for by the geological model if it exceeds that estimated from erosion of upthrown blocks within the immediate vicinity.

Dip changes seen across faults on geoseismic lines demonstrate that most faults are listric (Price 1977). Roll-over geometries can be used to construct the change of curvature of the listric fault (Fig. 6) and rely upon the necessity to avoid creating 'gaps' in the section during deformation. Errors in the use of this technique arise from differential compaction, but these should be small. An alternative approach is to use the rotation ϕ (in radians) given by the dip change across the fault and chord length, c, defined by the fault cut to calculate the rotation pole and radius of curvature, r:

$$r = \phi \cdot c/2\pi.$$

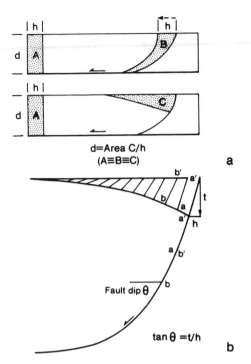

Fig. 6. (a) Roll-over geometry showing accommodation of area B for fault with heave, h. Areas A, B, and C are equal. (b) Fault plane constructed from shape of roll-over. Dip of fault plane segments a–a′, b–b′, etc. are given by tan θ = t/h and then projected to link up as fault plane.

In general, however, graphical methods are preferable as they will accurately generate faults with discontinuous rates of curvature. On seismic sections, however, antithetic and synthetic accommodation faults may be missed which can have a large effect on the constructed shape of the main listric fault. The requirement to maintain cross-sectional continuity has the important effect of predicting that a variety of syn- and antithetic accommodation faults are necessary in the absence of 'ductile' roll over. Figure 7 shows a variety of these

geometries which can accompany deep-seated listric faults bounding major horst blocks. Crestal terraces, grabens and platforms of the sort predicted geometrically are common features of many North Sea Oil Fields (e.g. Blair 1975, Hallett 1980). The possibility that a small number of faults in these secondary structures could be reverse while the main deformation is extensional is of particular interest. As yet no undoubted structures of this type have been proved in the North Sea although a small number of cases where this is a possible model have arisen, where the well trajectory and fault dip do not allow a distinction to be made between a very steep normal fault and a reverse structure (Fig. 8). Reverse faults may be present both as direct result of this genetic process (Fig. 7) and as a result of later rotation (Fig. 3).

OBLIQUE SECTIONS

In areas such as the North Sea, the main source of structural data is derived from reflection seismic lines which are interpreted with reference to available well control. For this reason, even in well-known areas, the exact stratigraphic significance of a seismic event may not be known. Detailed stratigraphic knowledge may be confined to a narrow band in the section which corresponds to the reservoir horizon. A further problem is that while every effort is normally made in designing the seismic survey to shoot the lines parallel to the structural dip, that is in the profile, or principal plane of the extension this may not always be possible, nor desirable. Where more than one structural element is present, or the geology is poorly known the seismic lines may lie oblique to the plane of extension.

Where the seismic line is oblique it may be possible and desirable to project whatever data is present on to the profile plane before attempting to balance the section. This, however, implies the use of a model, or knowledge of the deformation style and introduces yet another level of interpretation into the data. It may also

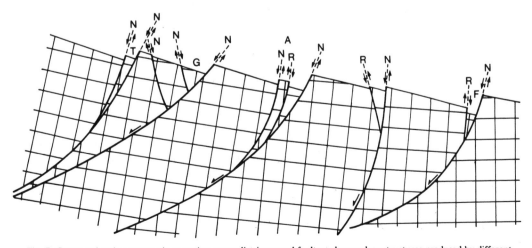

Fig. 7. Cartoon showing progressive rotation across listric normal faults and secondary structures produced by different combinations of syn- and antithetic accommodation faults. Note the crestal terrace (T), graben (G) horst (A) and horst-foot graben (F) as well as the possibility of a number of reverse faults (R).

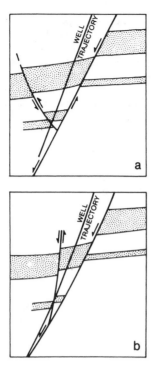

Fig. 8. Cartoon showing interpretation problems where data from an inclined well indicate a repeated section. (a) Interpretation with reverse fault and (b) with normal fault.

be, and frequently is, the case that the existing seismic line has been acquired to utilize well control and to help elucidate the structure between the wells. For these reasons it is often desirable to attempt to balance the oblique section rather than to extrapolate on to another section which may be of less value.

Balancing oblique sections is inherently difficult as area may not be maintained in the section. Two approaches are presently used to resolve this problem. The simplest technique is where the deformation has an 'orthorhombic' symmetry and the plane of section contains one of the deformation symmetry axes. This is frequently a valid assumption for crustal-parallel deformations of the type and scale of those investigated in the North Sea. Bed length measurements can be recalculated using the simple trigonometric relationship discussed above. Such a technique for oblique sections is described by Cooper (1983). Corrections for two way time (T) and velocity (v_i) over each interval must be made and a simplified form of the expression for l_λ becomes:

$$l_\lambda = (l_0 \sin \theta_0 + \left(\frac{T \cdot v_i}{2}\right)^2 - (l_0 \sin \alpha \sin \theta_0)^2)^{1/2},$$

where θ_0 is observed dip and segment length $S = l_0 \sin \theta_0$. Bed lengths are then summed over the appropriate segments and used in bed length and depth to detachment calculations. In many cases a rough check of lengths and areas using this sort of simplified equation to take velocity effects into account is sufficient to give a feel for the errors introduced in balancing time rather than depth sections. Where the section does not contain one of the deformation axes, or the deformation lacks orthorhombic symmetry a number of successive trigonometric operations can be made as outlined by Cooper (1983). As the strain tensor is non-commutative this approach is theoretically unsound and can lead to substantial errors. More sophisticated geometric manipulations may be attempted, but in general involve assumptions which cannot be justified in terms of known errors and uncertainty in the data.

The problem of adequately defining the strain tensor can be avoided largely by an empirical approach where a number of non-parallel seismic lines are available. Provided that the line spacing is less than the spacing of the structures an integrated area balance can be made. Accurately depth-converted sections are again desirable but over a small seismic interval may not be necessary in terms of other inherent errors. The technique hinges on the assumption that volume is conserved during deformation, and while volume may be apparently lost in one section it should be gained in another adjacent section. Careful planimetering or otherwise estimating volume on successive lines enables an areal volume balance to be established. This approach has the added advantage of giving valuable clues to the geometry of the finite strain ellipsoid. It should be possible to establish, for example, the plane of no finite longitudinal strain. Where parallel seismic lines cross an area of suspected strike-slip faults, comparison of areal balance on successive lines can assist in establishing displacement on the wrench fault system. This approach has been used to test the hypothesis of post-Jurassic strike-slip faulting in part of the Inner Moray Firth, and along with other data showed that such a model was not necessary to account for observed fault displacements and sediment distributions. Mapped fault patterns with apparent Riedel shear geometry could, in this case, be attributed to the later normal faults inheriting an earlier basement trend.

GEOSEISMIC SECTIONS

Seismic reflection data rarely allow fault geometries to be observed directly and geoseismic sections inevitably consist of interpretation and models of differing confidence. For example, the throw of the fault on the section plane may be known with some certainty and after drilling of an adjacent well the seismic prognoses may be confirmed. The dip of the fault plane and possible existence of antithetic accommodation faults on the other hand may not be supported by any information other than the structural model. The only reliable interpretation tool in such cases is that of geometrical balance.

As many of the North Sea reservoirs are highly faulted with, in some cases, pressure sealed fault compartments, the ability to test the geometrical integrity of a fault pattern mapped in section and plan is of paramount importance. Even if it remains impossible to achieve a uniquely satisfactory balance, the knowledge gained by interpretative attempts to balance the model can be important. For example, secondary horsts and crestal grabens on the larger structure may consist of numerous small faults with displacements less than the resolution of the seismic technique. The ability to define the extent of such areas where an unresolved space problem exists may assist in siting wells.

Balanced sections and reconstructions are ideally carried out on geological sections with equal vertical and horizontal scales. In some cases, as discussed above, it may be desirable to attempt to balance geoseismic sections where the vertical scale is given in two-way time. It is probable that the results of such an exercise will inevitably be inferior to a correctly balanced geological section where all of the geological information is observationally correct or the interpretation has a high degree of confidence. Geoseismic sections, however, invariably contain geometric information, in particular of deep events below the target formation which may be absent from the interpreted section and which give some qualitative control of the geometry. In addition the iterative application of balance techniques, particularly area, bed-length and fault trajectories derived from roll-overs on the hanging walls of listric faults are an essential part of the primary interpretation of the seismic data. Waiting until the geophysicist has depth converted his maps and derived true-scale geological sections from these may be too late. An interpretation which does not work geometrically may have already been built into the model. Attempts to balance sections should be made concurrently with the geophysical interpretation in order to maximize the evolving structural model.

EXTENSION CALCULATIONS

In extensional regimes it is also important to attempt to calculate depth to detachment for the basement faulting. Application of crustal models, such as those of McKenzie (1978a) and Le Pichon & Sibuet (1981), require that a brittle–ductile transition occurs at a depth of the order of 10–15 km. Where subsidence and extension values are required, regionally balanced sections are critical in providing a structural check on extension values calculated from the subsidence and thermal decay equations used in the McKenzie (1978a) model (e.g. Jarvis & Maclean 1980). These in turn may be important in maturation and subsidence history studies (e.g. Wood 1981). On a more fundamental level balancing regional lines for successive time intervals provides a direct test of the rapid stretching followed by thermal subsidence type of model. In many areas it is becoming clear that such a simple model, while adequate as a first order approximation, cannot be used in detail and that some cyclic stretch model may be more appropriate.

CONCLUSIONS

There are considerable benefits to be gained from improving structural interpretation in areas of exten-

sional tectonics by the use of balanced section techniques. In particular, an improved understanding of the structural pattern and tectonic evolution of such areas can result if such techniques are integrated into seismic interpretation at an early stage. While seismic data rarely permit a unique interpretation of the structure, balanced geoseismic sections should be constructed iteratively in order to derive a geoseismic model which makes geometrical sense as well as obeying stratigraphic constraints. Care should be taken in producing models which are as simple as possible while honouring the data. The interpretational errors inherent in seismic data will infrequently allow ultra sophisticated analysis.

Acknowledgements—I wish to thank the *British National Oil Corporation* (now Britoil) for permission to publish this work and I acknowledge the help of many people within BNOC who contributed to my understanding of the problems.

REFERENCES

Blair, D. G. 1975. Structural styles in North Sea Oil and Gas Fields. In: *Petroleum and the Continental Shelf of North-West Europe Volume 1* (edited by Woodland, A. W.). Institute of Petroleum, London, 327–335.
Borradaile, G. J. 1979. Strain study of the Caledonides in the Islay region SW Scotland: implications for strain histories and deformation mechanisms in greenschists. *J. geol. Soc. Lond.* **136**, 78–88.
Chamberlin, R. T. 1910. The Appalachian folds of Central Pennsylvania. *J. Geol.* **18**, 225–251.
Cooper, M. 1983. The calculation of bulk strain in oblique and inclined balanced sections. *J. Struct. Geol.* **5**, 161–165.
Crans, W. & Mandl, G. 1980. On the theory of growth faulting, Pt II (a), Genesis of the 'Unit'. *J. petrol. Geol.* **3**, 209–236.
Crans, W., Mandl, G. & Harembourne, J. 1980. On the theory of growth faulting; a geomechanical delta model based on gravity sliding. *J. Petrol. Geol.* **2**, 265–307.
Dahlstrom, C. D. A. 1969a. Balanced cross sections. *Can. J. Earth Sci.* **6**, 743–757.
Dahlstrom, C. D. A. 1969b. The upper detachment in concentric folding. *Bull. Can. Petrol. Geol.* **17**, 326–346.
Dahlstrom, C. D. A. 1970. Structural geology in the eastern margin of the Canadian Rocky Mountains. *Bull. Can. Petrol. Geol.* **18**, 322–406.
Durney, D. W. & Ramsay, J. G. 1973. Incremental strains measured by syntectonic crystal growths. In: *Gravity and Tectonics* (edited by De Jong, K. A. & Scholten, R.). Wiley, New York, 67–96.
Elliott, D. 1973. Diffusion flow laws in metamorphic rocks. *Bull. geol. Soc. Am.* **84**, 2645–2664.
Elliott, D. 1976. The motion of thrust sheets. *J. geophys. Res.* **81**, 949–963.
Elliott, D. 1977. Some aspects of the geometry and mechanics of thrust belts. Part 1. 8th Annual Seminar Can. Soc Petrol. Geol. Univ. Calgary.
Elliott, D. & Johnson, M. R. W. 1980. Structural evolution in the northern part of the Moine thrust belt, N.W. Scotland. *Trans. R. Soc. Edinb., Earth Sci.* **71**, 69–96.
Geiser, P. 1978. Gravity tectonic removal of cover of Blue Ridge anticlinorium to form Valley and Ridge Province: Discussion. *Bull. geol. Soc. Am.* **89**, 1429–1430.
Gwinn, V. E. 1970. Kinematic patterns and estimates of lateral shortening, Valley and Ridge and Great Valley Provinces. Central Appalachians, South-Central Pennsylvania. In: *Studies of Appalachian Geology—Central and Southern* (edited by Fisher, G. W. *et al.*). Wiley, New York, 127–146.
Hallett, D. 1981. Refinement of the geological model of the Thistle Field. In: *Petroleum Geology of the Continental Shelf of North-West Europe* (edited by Illing, L. V. & Hobson, G. D.). Institute of Petroleum, London, 315–325.
Heybroek, P. 1975. On the structure of the Dutch part of Central North Sea Graben. In: *Petroleum and the Continental Shelf of North-West Europe, Volume 1* (edited by Woodland, A. W.). Institute of Petroleum, London, 339–349.

Hossack, J. R. 1978. The correction of stratigraphic sections for tectonic finite strain in the Bygdin area, Norway. *J. geol. Soc. Lond.* **135**, 229–241.

Hossack, J. R. 1979. The use of balanced cross-sections in the calculation of orogenic contraction: a review. *J. geol. Soc. Lond.* **136**, 705–711.

Jarvis, G. T. & McKenzie, D. P. 1980. Sedimentary basin formation with finite extension rates. *Earth Planet. Sci. Lett.* **48**, 42–52.

McKenzie, D. P. 1978a. Active tectonics of the Alpine–Himalayan belt: the Aegean Sea and surrounding regions. *Geophys. J. R. astr. Soc.* **55**, 217–254.

McKenzie, D. P. 1978b. Some remarks on the development of sedimentary basins. *Earth Planet. Sci. Lett.* **40**, 25–32.

Magara, K. 1976a. Water expulsion from clastic sediments during compaction, directions and volumes. *Bull. Am. Ass. Petrol. Geol.* **60**, 543–553.

Magara, K. 1976b. Thickness of removed sedimentary rocks, paleopore pressure, and paleotemperature, southwestern part of western Canadian Basin. *Bull. Am. Ass. Petrol. Geol.* **60**, 554–565.

Magara, K. 1978. Compaction and fluid migration. *Devl Petrol. Sci.* **9**.

Le Pichon, X. & Sibuet, J. C. 1981. Passive margins: a model of formation. *J. geophys. Res.* **86**, 3708–3720.

Plessman, W. 1964. Gesteinlosung, ein Hauptfaktor beim Schieferungsprozess. *Geol. Mitt. Aachen* **4**, 69–82.

Price, N. J. 1977. Aspects of gravity tectonics and the development of listric faults. *J. geol. Soc. Lond.* **133**, 311–327.

Proffett, J. M. 1977. Cenozoic geology of the Yerington district, Nevada, and implications for the nature and origin of Basin and Range Faultings. *Bull. geol. Soc. Am.* **88**, 247–266.

Ramsay, J. G. 1967. *Folding and Fracturing of Rocks.* McGraw-Hill, New York.

Ramsay, J. G. & Wood, D. 1973. The geometric effects of volume change during deformation processes. *Tectonophysics* **16**, 263–277.

Rider, M. H. 1978. Growth faults in the Carboniferous of Western Ireland. *Bull. Am. Ass. Petrol. Geol.*, **62**, 2191–2213.

Sclater, J. G. & Christie, P. A. F. 1980. Continental stretching; an explanation of the post mid-Cretaceous subsidence of the Central North Sea Basin. *J. geophys. Res.* **85**, 3711–3739.

Schwerdtner, W. M. 1977. Geometric interpretation of regional strain analysis. *Tectonophysics* **39**, 515–531.

Steckler, M. S. & Watts, A. B. 1978. Subsidence of the Atlantic Margin of New York. *Earth Planet. Sci. Lett.* **41**, 1–13.

Wernicke, B. & Burchfiel, B. C. 1982. Modes of extensional tectonics. *J. Struct. Geol.* **4**, 105–115.

Wood, D. S. 1973. Patterns and magnitudes of natural strain in rocks. *Phil. Trans. R. Soc.* **A274**, 373–382.

Wood, D. S. 1974. Current views of the development of slaty cleavage. *A. Rev. Earth Planet. Sci.* **2**, 369–401.

Wood, R. J. 1981. The subsidence history of Conoco well 15/30-1, Central North Sea. *Earth Planet. Sci. Lett.* **54**, 306–316.

Reprinted by permission of Elsevier Science Publishers from
Tectonophysics, v. 79, no. 3/4 (1981), p. T43-T52.

A strain reversal method for estimating extension from fragmented rigid inclusions

C.C. FERGUSON

Department of Geology, University of Nottingham, Nottingham (England)

(Received June 3, 1981; accepted for publication July 23, 1981)

ABSTRACT

Ferguson, C.C., 1981. A strain reversal method for estimating extension from fragmented rigid inclusions. Tectonophysics, 79: T43—T52.

In the standard method of strain analysis using fragmented rigid inclusions (such as "stretched" belemnites) extension is estimated by comparing the initial length of the inclusion with its final (stretched) length. This approach fails to recognise that not all of the total extension is recorded by separation of the fragments, even when cross-fractures are present at the onset of deformation. This paper presents an alternative approach in which the extensional strain is reversed, the inter-fragment gaps being incrementally closed until the inclusion is restored to its original state. This allows an estimate of the "unrecorded" extension (accommodated by inhomogeneous matrix flow near the inclusion) as well as that recorded by fragment separation.

Model experiments, in which the fragmented inclusion is represented by wooden blocks set in a ductile matrix, allow the strain reversal method to be compared with the standard method. The mean percentage relative error is much smaller in the former (10.2%) than in the latter (37.2%). The main disadvantage of the strain reversal method is that it requires the collection of more data (fragment lengths and gap lengths in correct sequence), which must be processed by computer (a short BASIC program is provided). Although there is no one-to-one correspondence between strain estimates derived from the two methods, an approximate numerical transformation from one to the other may be possible. One such transformation (based on a study of natural boudinage) has been applied to the experimental data and reduces the mean percentage relative error in the standard method estimates to 11.5%.

INTRODUCTION

Naturally deformed rocks occasionally contain rigid inclusions which have been "pulled apart" during flow of the surrounding matrix. The fragmentation of belemnite guards is a common example which has often been used for strain estimation (Daubrée, 1876; Heim, 1919; Badoux, 1963; Ramsay, 1967). Disrupted crinoid stems and some forms of boudinage present a similar type of strain analysis problem. Belemnites are particularly valuable because they closely approximate ideal rigid-brittle inclusions. The standard method for estimating extensional strain (Ramsay, 1967 p. 248) involves measuring the initial length of the belemnite (the sum of the lengths of the separate fragments), the final extended length (sum of fragment lengths and inter-fragment gap lengths), and then calculating the extension as:

$$e = \frac{\text{final length} - \text{initial length}}{\text{intial length}} \tag{1}$$

In two recent studies (Hossain, 1979; Beach, 1979) large numbers of individual belemnite strain gauges are used in order to estimate the principal axes of the strain ellipse. Hossain uses the relationship:

$$\lambda' = \lambda_1' \cos^2\theta + \lambda_2' \sin^2\theta \tag{2}$$

in which λ_1', λ_2' are the principal reciprocal quadratic elongations, and λ' is the reciprocal quadratic elongation estimated from a "stretched" belemnite oriented at an angle θ to the stretching lineation (direction of maximum principal extension). Dividing both sides of eq. 2 by $\cos^2\theta$ yields the equation of a straight line referred to axes $\lambda'/\cos^2\theta$ (ordinate) and $\tan^2\theta$ (abscissa). Hossain suggests that if values from many belemnite strain gauges are plotted against these axes and a linear regression performed, the slope of the best-fit straight line will give an estimate of λ_2' while the intercept (on the ordinate) will give an estimate of λ_1'. Hossain's application of this method using 196 belemnites from Leytron, Switzerland reveals considerable scatter on the $\lambda'/\cos^2\theta$ vs. $\tan^2\theta$ plot. This is to be expected because the amount of apparent extension recorded by a belemnite depends on how much of the deformation had elapsed before cross-fracturing took place. As Beach (1979) points out, some belemnites lying parallel to strong stretching lineations may show no extension while others in the same outcrop show much extension. Therefore, the extension indicated by an individual belemnite will usually (and, arguably, will always) underestimate the true extension. Hossain's linear regression method fails to take this into account; belemnites showing the largest apparent extensions are "numerically cancelled" by those showing the least so that the final "best-fit" line will systematically (and often substantially) underestimate the principal strains.

In contrast Beach (1979), using samples of up to 100 belemnites per locality, plots elongation $(1 + e)$ against orientation and then draws an envelope around the field of points. This envelope is considered to define the variation in true elongation with orientation and explicitly recognises that most belemnites record an elongation less than the envelope value. This method does assume that "at least one belemnite recording the maximum elongation in the rock is included in the sample" (Beach, 1979 p. 131). Here lies the problem; even if initial cross-fractures are present, calculating the extension using eq. 1 assumes that the elongation is entirely accommodated by separation of fragments. This assumption is challenged in the next section and a new line strain gauge is proposed.

Consider a single elongate rigid inclusion, of length L, enclosed in a matrix deforming by ductile flow with uniform extension (in the far field) parallel to the inclusion length. Remote from the inclusion a passive marker line (also of length L) parallel to the extension direction will record the maximum extension exactly. Close to the inclusion, however, the flow field will be disturbed; that part of the rock occupied by the inclusion will record no extension while the matrix nearby will deform inhomogeneously. Although the local flow field will be complex, overall the matrix close to the inclusion must suffer extra deformation so that the remote flow field can be uniform. Now consider the same inclusion with a single cross-fracture. Will the far field extension now be recorded by separation of the two parts of the inclusion? Clearly it will not, for the flow field will still be distorted close to the two parts of the inclusion. Some of the bulk extension will be accommodated by separation and some (as before) by extra matrix deformation. As more fractures appear more of the far field extension will be recorded by fragment separation and less by inhomogeneous matrix flow. But even if many cross-fractures existed before the onset of matrix flow, some of the far field extension will still be accommodated by inhomogeneous matrix flow and therefore will not be recorded by fragment separation. Use of eq. 1 will always underestimate the true extension (that is, the extension that the matrix would have experienced had the inclusion not been present), and will often substantially underestimate it.

The detailed behaviour of a ductile rock containing a rigid-brittle inclusion will depend on the geometry of the remote flow field, the inclusion shape and orientation, the operating flow and fracture mechanisms, the solubilities and diffusivities of mobile components etc. A full analysis would require the integrated solution of many difficult problems. Some of the underlying problems have been tackled, albeit in a highly idealised form, in the mechanics literature (e.g. Tillett, 1970; Piggott, 1970; Kelly and Street, 1972; Goddard, 1976), and some of the geological aspects have received preliminary attention (Ghosh and Ramberg, 1976; Lloyd and Ferguson, 1981). But a rigorous analysis remains way beyond current capability. Therefore, the approach discussed in this paper makes no claims to be rigorous, but it does recognise that only some of the true extension will be recorded by fragment separation. The basic approach is illustrated in Fig. 1 in which it is assumed that the inclusion is rigid-brittle and that the flow field remote from it involves uniform extension parallel to the inclusion length. Until the appearance of the first cross-fracture (Fig. 1a) the inclusion preserves no record of the far field extension. Once fragmentation of the inclusion begins (Fig. 1b) we assume that the extension δl experienced by the passive line segment BC (of initial length l) will be recorded by separation, δl, between the two parts of the inclusion. The extension experienced by the passive line segments AB and CD will be accommodated near the inclusion by heterogeneous

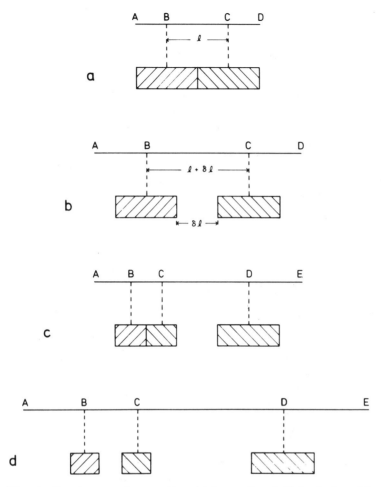

Fig. 1. Fracture and separation history of a rigid inclusion (shaded) compared with the extensional history of a passive marker line. See text for discussion.

flow of the matrix, and will not be recorded by separation. On the appearance of a second fracture (Fig. 1c) extension is recorded by separation in the two subsystems BC and CD (Fig. 1d) but not in the subsystems AB and DE.

In practice we are faced with analysing the final product of such a process. The method outlined above is therefore used in reverse. To do this the fragmented inclusion ($n + 1$ fragments) is partitioned into n subsystems of length L_i ($i = 1$, n), see Fig. 2a. We then find the subsystem requiring the least amount of shortening, parallel to the inclusion length, in order to close the inter-fragment gap (of course this need not be the subsystem with the smallest gap). This is achieved iteratively using increments of 0.5% shortening. Having identified this subsystem (L_2 in Fig. 2a) and the number K, of increments needed to close it, K increments of shortening are then applied to each subsystem in order to close the gap in L_2 and reduce the gap length in all the others. The total system is then relabelled (Fig. 2b) and the pro-

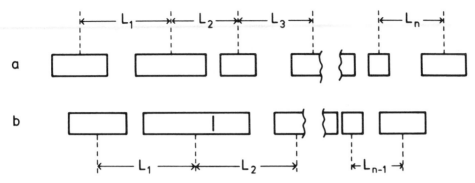

Fig. 2a. Initial subsystem definition in a train of rigid inclusion fragments, as used in the strain reversal method. b. Relabelling of system following closure of the last formed inter-fragment gap (i.e. that requiring the least number of shortening increments to close it).

cedure repeated n times until all gaps are closed and the system is restored to its original state. A listing of the BASIC program written to perform these operations is given as an Appendix.

An important feature of this analysis is that two inclusions having the same initial and final lengths need not have suffered the same number of increments of extension. In Fig. 3 for example one fragmented inclusion (A)

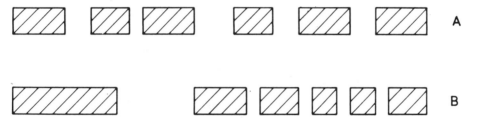

Fig. 3. Two inclusion fragment trains having the same initial length (sum of fragment lengths) and final length. Using the strain reversal method train B records more far-field extension than train A (see text).

represents 56.2% extension while the other (B) represents 67.2%. Both have the same initial and final length and therefore both represent the same apparent extension (45.5%) using eq. 1. The difference between A and B arises because, during an appreciable part of the total deformation, example B contained only two fragments separated by what is now the largest gap. During this time a relatively large proportion of the total extension is accommodated by inhomogeneous matrix strain and therefore is not recorded by separation. In example A the successive cross-fractures were separated by shorter time intervals so that the apparent extension calculated using eq. 1 is closer to the "true" extension. The strain reversal method always yields a greater extension than eq. 1 because it estimates both the recorded (by separation) and unrecorded (inhomogeneous matrix flow) components of the total extension. Needless to say, the methods used to define the subsystems (simple bisection) and to partition the two extension components (end-of-line subsystems record no extension, within-line subsystems record full extension) are obvious oversimplifications. But, as we shall see in the next section, the method does yield substantially better strain estimates than the standard method.

COMPARISON OF METHODS

Although I have applied the strain reversal method to numerous sets of fragmented inclusions (mainly rectangular boudin trains), none of these was associated with independent strain markers from which extension in the remote field could be estimated. Therefore in order to evaluate the method I have used the results of five published model experiments (rectangular wooden blocks set in a ductile matrix) together with four similar experiments of my own. The published experiments are in Ramsay (1967; figs. 3.46 and 3.47) and Ghosh and Ramberg (1976; figs. 29, 31 and 33). Ramsay's model apparatus, and my own, imposed a far field irrotational strain (maximum extension parallel to inclusion length); Ghosh and Ramberg's apparatus imposed a simple shear in which the rigid blocks (already partly separated) were initially aligned normal to the shear direction (two experiments) or at $60°$ to it (one experiment). For each experiment the fragment lengths and gap lengths were measured in correct sequence, and extension was estimated using both the strain reversal method and the standard (eq. 1) method. The results are plotted in Fig. 4 which shows that the extension, e, derived from eq. 1 considerably underestimates the true extension (relative errors range from 13.1% to 71.9% with a mean value of 37.2%). In contrast the extension, e^*, derived from the strain reversal method provides a reasonably good estimate (relative errors range from 0.8% to 26.4% with a mean value of 10.2%). The performance of both methods tails off at large extensions; unfortunately I cannot provide more data in this region because my apparatus fails to produce a homogeneous far field with strain ratios greater than about 3.0 (corresponding to 73% extension). In the region with true extension less than 70%, the maximum relative error in the e estimates is 45.7% (mean 29.2%) while in the e^* estimates it is 11.1% (mean 6.7%).

On the basis of this limited experimental data I conclude that the strain reversal method provides much better strain estimates than the standard method. Its main disadvantage is that it requires the collection of more data — fragment lengths and gap lengths in correct sequence — which then need to be processed by computer. In view of the fairly systematic differences between the two types of estimate it is worthwhile considering whether standard method estimates (which can be obtained fairly quickly) can be numerically transformed to become closer to the true values. The basis for this suggestion is my study of thirty-four natural boudin trains (each containing between five and seven boudins) in which extension has been

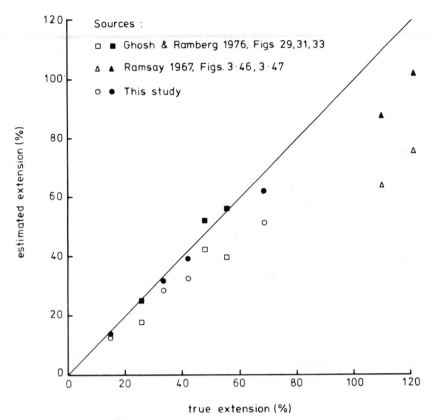

Fig. 4. Plot of true extension vs. estimated extension for nine model experiments (rectangular wooden blocks set in a ductile matrix). Open symbols: extension estimates using standard method (eq. 1). Filled symbols: extension estimates using strain reversal method. Ghosh and Ramberg experiments (squares) were in simple shear; all others were irrotational.

estimated using both methods. Linear regression analysis suggests that the "optimum" transformation for this data is:

$$e^* = 1.564 \, (e - 3.22)$$

Applying this transformation to the e data plotted on Fig. 4 (open symbols) produces extension estimates with relative errors in the range 3.2% to 21.7% (mean 11.3%), a performance which compares fairly well with the direct e^* estimates. Of course, this sort of transformation should be treated with caution especially as there is no one-to-one relationship between e and e^*. But, in the absence of "well calibrated" field examples, a careful model experiment programme could provide one or more numerical transformations which would allow the main advantage of the standard method (ease and speed of data collection and processing) to be retained.

CONCLUSIONS

The standard method (eq. 1) for estimating extension from a fragmented elongate rigid inclusion fails to recognise that, even when cross-fractures are present at the onset of ductile deformation in the matrix, not all the far field extension will be recorded by fragment separation. Extension estimates derived by incrementally reversing the separation history of the fragments (though in an oversimplified manner) are consistently larger than those derived from eq. 1. Model experiments indicate that the mean relative error in the strain reversal method is substantially less than in the standard method. This study suggests that a much more extensive set of model experiments would be worthwhile in order to evaluate the robustness of the strain reversal method. At the moment little is known in detail about how the performance of the method will be affected by variations in:

(a) The nature of the remote flow field.
(b) The amount of extension.
(c) Aspect ratio of fragments (i.e. cross-fracture spacing).
(d) History of cross-fracturing with respect to the matrix strain history.
(e) Detailed inclusion shape.
(f) Rheological properties of the matrix.

ACKNOWLEDGEMENT

This work developed from a study of boudinage undertaken in collaboration with Geoffrey E. Lloyd. I am grateful to him for providing an excellent set of field measurements of rectangular boudin trains.

APPENDIX

```
20 REM  ****************************************************************
21 REM     PROGRAM TO CALCULATE EXTENSION OF DISRUPTED RIGID
22 REM     INCLUSION BY INCREMENTALLY RESTORING SEPARATED
23 REM     FRAGMENTS TO THEIR ORIGINAL POSITION. DATA SHOULD
24 REM     BE TYPED IN, STARTING AT LINE 400, AS FOLLOWS :
25 REM        (1) N GAP LENGTHS IN CORRECT SEQUENCE
26 REM        (2) N+1 FRAGMENT LENGTHS IN CORRECT SEQUENCE
27 REM
28 REM     AUTHOR: C C FERGUSON, DEPT OF GEOLOGY, UNIV NOTTINGHAM
29 REM  ****************************************************************
40 DIM G(50),B(50),L(50)
50 PRINT "TYPE NO. OF GAPS "; \ INPUT N
60 PRINT "TYPE INCREMENTAL SHORTENING "; \ INPUT S
65 PRINT "TYPE LENGTH OF 'CLOSED' GAP "; \ INPUT T
70 L(0)=1 \ G(0)=10000 \ B(0)=0
74 REM  ********************************************************
75 REM     READ IN GAP LENGTHS AND FRAGMENT LENGTHS ;
76 REM              DEFINE SUBSYSTEM LENGTHS
77 REM  ********************************************************
80 FOR I=1 TO N \ READ G(I) \ NEXT I
82 FOR I=1 TO (N+1) \ READ B(I) \ B(I)=B(I)/2 \ NEXT I
```

```
83  FOR I=1 TO N \ L(I)=B(I)+B(I)=B(I+1) \ NEXT I
84  G4=T \ W=-1
85  REM  ***************************************************
86  REM     IS SMALLEST GAP LENGTH < T? IF SO, GOTO 200
87  REM               FOR SYSTEM RELABELLING
88  REM  ***************************************************
89  FOR I=1 TO N
90  IF G(I)<G4 THEN 92
91  GO TO 93
92  G4=G(I) \ W=I
93  NEXT I
94  IF W>0 THEN 200
95  K=100000
96  K=10000
97  REM    ***************************************************
98  REM     FIND THE SUBSYSTEM WHICH REQUIRES THE SMALLEST
99  REM     NUMBER, K, OF INCREMENTS IN ORDER TO REDUCE ITS
100 REM     GAP LENGTH TO < K. LABEL THIS SUBSYSTEM W.
101 REM  ***************************************************
104 FOR I=1 TO N
106 L=L(I) \ C=0
108 FOR J=1 TO 2000
110 IF L<(B(I)+B(I+1)+T) THEN 150
115 L=L-L*S \ C=C+1
130 IF L<(B(I)+B(I+1)+T) THEN 150
135 NEXT J
139 REM  ***************************************************
140 REM   IF LENGTH OF CLOSED GAP < 0, SET P=1 IN ORDER TO
141 REM    HALVE SHORTENING INCREMENT ON NEXT RUN THROUGH
142 REM  ***************************************************
150 IF L<(B(I)+B(I+1)) THEN LET P=1
152 IF C<K THEN 160
155 GO TO 170
160 K=C \ W=I \ L3=L
170 NEXT I
179 REM  ***************************************************
180 REM   REDUCE LENGTH OF ALL GAPS BY SHORTENING ALL SUBSYSTEMS
181 REM   BY K INCREMENTS. SUBSYSTEM L(0) RECORDS THE SHORTENING
182 REM            OF A PASSIVE MARKER LINE OF UNIT LENGTH.
183 REM  ***************************************************
192 FOR I=0 TO N
194 FOR J=1 TO K \ L(I)=L(I)-L(I)*S \ NEXT J
196 G(I)=L(I)-B(I)-B(I+1)
198 NEXT I
200 K=0
202 IF P>0 THEN S=S/2
204 IF N=1 THEN 310
205 IF P>0 THEN LET S=S/2
207 REM  ***************************************
208 REM   ELIMINATE GAP W AND RELABEL SUBSYSTEMS
209 REM  ***************************************
210 IF N=1 THEN 310
212 B(W)=B(W)+B(W+1) \ IF W=N THEN 220
216 G(W)=G(W+1) \ IF W=1 THEN 224
```

```
220 L(W-1)=B(W-1)+B(W-1)+B(W)
222 IF W=N THEN 230
224 L(W)=B(W)+B(W)+B(W+2)
230 FOR I=1 TO N
240 IF I>W THEN 244
242 GO TO 280
244 G(I)=G(I+1) \ B(I)=B(I+1) \ L(I)=L(I+1)
280 NEXT I
285 N=N-1 \ P=0
289 REM ********************************************
290 REM   RETURN TO LINE 84 AND REPEAT PROCEDURE
291 REM ********************************************
300 GO TO 84
310 PRINT \ PRINT "EXTENSION ESTIMATE IS"100/L(0)-100"PER CENT"
350 STOP \ END
400 DATA 10.5,11.4,11.2
401 DATA 10.4,11.7,10.5,11
```

READY

REFERENCES

Badoux, H., 1963. Les bélemnites tronconées de Leytron (Valais). Bull. Lab. Geol. Minéral. Géophys. Mus. Géol., Univ. Lausanne, 138: 1—7.

Beach, A., 1979. The analysis of deformed belemnites. J. Struct. Geol., 1: 127—135.

Daubrée, M., 1876. Expériences sur la schistosité des roches et sur les déformations des fossils etc. Bull. Serv. Carte Géol. Fr., 3e Sér. t.iv.

Ghosh, S.K. and Ramberg, H., 1976. Reorientation of inclusions by combination of pure shear and simple shear. Tectonophysics, 34: 1—70.

Goddard, J.D., 1976. The stress field of slender particles oriented by a non-Newtonian extensional flow. J. Fluid Mech., 78: 177—206.

Heim, A., 1919. Untersuchungen über den Mechanismus der Gebirgsbildung. In: Geologie der Schweiz. Tauchwitz, Leipzig.

Hossain, K.M., 1979. Determination of strain from stretched belemnites. Tectonophysics, 60: 279—288.

Kelly, A. and Street, K.N., 1972. Creep of discontinuous fibre composites. Proc. R. Soc. London, Ser. A, 328: 267—293.

Lloyd, G.E. and Ferguson, C.C., 1981. Boudinage structure: some new interpretations based on elastic-plastic finite-element simulations. J. Struct. Geol., 3: 117—128.

Piggott, M.R., 1970. Theoretical estimation of fracture toughness of fibrous composites. J. Mater. Sci., 5: 669—675.

Ramsay, J.G., 1967. Folding and Fracturing of Rocks. McGraw-Hill, New York, N.Y., 568 pp.

Tillett, J.P.K., 1970. Axial and transverse Stokes flow past slender axisymmetric bodies. J. Fluid Mech., 44: 401—417.

BULLETIN OF THE AMERICAN ASSOCIATION OF PETROLEUM GEOLOGISTS
VOL. 42, NO. 3 (MARCH, 1958), PP. 561-587, 7 FIGS.

CHIEF TOOL OF THE PETROLEUM EXPLORATION GEOLOGIST: THE SUBSURFACE STRUCTURAL MAP[1]

LOUIE SEBRING, Jr.[2]
Lafayette, Louisiana

ABSTRACT

This paper, prepared primarily for the inexperienced geologist, describes the subsurface structural map with stratigraphic additions and modifications, its preparation, and its uses in the search for petroleum. An accurate, relatively uncluttered base map with the wells spotted correctly on it, is a prerequisite. Well logs provide the bulk of the basic data required, and the most useful of these logs is the electric log. Electric-log correlation is based on electrical characteristics of the datum and its position in a sequence of correlative events. Correlation difficulty is a function of lateral stratigraphic variation and of structural complication.

The mapping datum should reflect the structure of prospective producing zones, should occur over a wide area, should be encountered by most wells, and the geologist must be able to recognize it and correlate it.

Mapping should begin in areas of greatest control, and should reveal producing characteristics and trapping elements of the fields already productive. The most reliable geological interpretations are based on production information, as well as on the correlation of well logs. When production information is sketchy and the geology complicated, mapping on multiple horizons is a useful technique to determine the correct geological solution.

The contour interval selected for regional mapping will be based on studies of fields that will reveal closure necessary for accumulation.

In exploratory mapping, the geologist will direct his primary search for traps already known in the area, but he should not overlook the possibility that traps not previously recognized may occur in the area studied.

The occurrence of oil or gas shows in a non-productive well is direct evidence of a trap, and their presence is an important aid in interpretation.

Fault-throw computations can give additional control on the opposite side of the fault plane, but as they are subject to many errors, they should be used with caution. Determination of the dip and strike of faults is best determined by contouring on the fault plane.

The reflection seismograph is the geologist's most important source of additional information, but its limitations should be recognized.

The modification of the basic subsurface structural map by the addition of boundary lines of permeability of the various prospective zones, will make it useful in exploration for combination structural and stratigraphic traps.

Current maintenance of exploratory maps, made necessary by intense competition, will also enable the geologist to evaluate best his previous mapping efforts and those of his predecessors.

The exploration geologist's proper attitude is described as reasonable optimism. This leads to the conclusion that exploration for large reserves in the older producing districts should be concentrated in areas where they are known to exist. In areas previously explored unsuccessfully, new basic approaches and techniques would seem most likely of success.

The geologist must actively "sell" his maps and its prospects to his management, since all of his previous efforts are useless unless his work is used in acquiring land and drilling wells.

Barring the discovery of a direct method of finding oil, the various subsurface methods based on well control will eventually tend to supplant all other methods in the search for petroleum.

INTRODUCTION

This paper was prepared mainly for the inexperienced geologist who is just beginning his exploratory work. The ideas expressed are based on ten years of

[1] Manuscript received, June 17, 1957. Presented before the Gulf Coast Association of Geological Societies at New Orleans, Louisiana, November 8, 1957. Published by permission of Champlin Oil & Refining Co.

[2] Geologist, Champlin Oil & Refining Co. Many of the ideas expressed in this paper are not originally the writer's, but were conveyed to him by geologists employed by Champlin Oil & Refining Co. and The Standard Oil Company of Texas. The assistance of Bernard Bonnecaze in drafting the figures and Mrs. Leo Franques in typing the manuscript is gratefully acknowledged.

exploratory work in Southwest Texas, South Louisiana, and the Denver-Jules-burg basin. The writer believes that the principles of exploratory mapping that apply in these areas will also apply in other areas where sandstone production predominates.

Subsurface geology may be defined as the study of the rocks and rock formations below the surface of the earth. A subsurface map is constructed from data supplied by wells penetrating these formations and from geophysical surveys that measure significant physical properties of the subsurface rocks.

Wells are now being drilled at a constantly increasing rate. More than 58,000 wells were completed in the United States in 1956. These wells provide an increasing wealth of subsurface information.

Many types and kinds of maps may be prepared from the available data. These include the subsurface structural map, the isopachous map, the various types of facies maps, paleogeologic maps, geophysical maps and many others. The maps most commonly used by the exploration geologist in his search for petroleum are the subsurface structural map, with some basic stratigraphic additions and modifications, and the various geophysical maps, most of which are also structural maps prepared from geophysical data. This discussion is limited to the subsurface structural map with stratigraphic additions and modifications, its preparation, and its uses in the search for petroleum.

The subsurface structural map presents in plan view a three-dimensional picture of the structure of a recognizable datum occurring mainly below the surface of the earth. The structure of this datum is represented by contours passing through points of equal elevation. These points of equal elevation are related to their vertical distance either above or below an arbitrary datum. The elevation on the datum bed is the algebraic difference between the surface elevation of the mapping datum, above or below the reference datum. In most parts of the country where subsurface mapping is carried on in petroleum exploration, the subsurface structural map can also be called a subsea map, since the reference datum is commonly sea-level, and the mapping datum is generally further below sea-level than the surface elevation of the well is above it.

BASE MAP

To prepare a useful subsurface map the geologist must have an accurate base map with his control points accurately spotted on it. These control points are obtained from logs of the wells that have penetrated to the mapping datum. The importance of a correctly spotted map can not be overemphasized. Huge amounts of capital have been spent buying leases and drilling wells on prospects based on incorrectly spotted wells, and undoubtedly much more money will be expended in the future because of similar errors. It is important for the best possible interpretation that the wells not only have the correct location in relation to the tract boundaries upon which they are located, but also they should be correctly related to wells on different tracts. In other words, the tracts themselves should be correctly dimensioned, shaped, and located. Misspotted or misshapen tracts can cause substantial errors in mapping.

Accuracy of the base map and its well locations is certainly the most important quality desired. However, a base map reasonably free from the clutter of land data is also desirable. The control points (wells) should be easily seen and not lost in a maze of lots and small tracts. The scale of the map used should be large enough so that the necessary detail of a complicated structure can be shown. It should be small enough so that sufficient area is covered to show the relation between several areas of present production and of possible future production. It follows, therefore, that the largest-scale maps will be used when the area mapped is complicated, highly developed, and/or mapped in extreme detail. It is desirable, when possible, to use the same base map for both leasing and mapping operations. Where a highly complicated ownership situation makes this impossible, a base with most of the ownership details removed is useful. Land ownership bases and regional map bases prepared by the same map company are desirable since ordinarily the relation of tracts between the bases is similar. The most accurate base maps are prepared from aerial maps and aerial surveys. Although maps prepared in this manner generally show tract relations correctly, the well locations are not nearly so reliable. This is because only a part of the wells drilled before the aerial survey can be located on the photographs. Locations of wells not pin-pointed on the photographs and wells drilled after the photographing must be spotted from other information. These locations that are spotted later are only as good as that information and as good as the draftsman who spots them.

WELL INFORMATION

Most of the information used by the geologist in preparing a subsurface map is obtained from wells drilled in the area to be mapped. This information is taken from logs of the wells, core descriptions, sample descriptions, completion information, *et cetera*. The more of this information that the geologist has at his disposal in readily accessible form, the more accurate will be his subsurface map.

WELL LOGS

Information taken from logs of the wells drilled in the area to be mapped provides most of the basic data required by the subsurface geologist. A well log may be defined as a record of the rocks and rock formations penetrated by the well or a record of certain physical properties of these rocks. The most common types of logs used are electric logs, sample logs, radioactivity logs, mud logs, core logs, temperature logs, drilling-time logs, drillers logs, caliper logs, *et cetera*. The most useful log for the exploration geologist is the electric log. Practically all wells now drilled in the search for petroleum are being electrically logged. No one appreciates this fact more than the geologist who has attempted to use drillers logs and sample logs for correlations where the stratigraphic section consists mostly of shale and sand. The electric log is now the basic correlation tool of the subsurface geologist. Most log libraries of companies engaged in the search for petroleum are made up mainly or wholly of electric logs. Indeed, many geologists in the Gulf Coast have never prepared a sample log.

Electric-log correlation.—Many volumes have been written about the electric log and its interpretation. It is beyond the scope of this paper and the ability of its writer to improve upon them. The discussion is limited to their uses in correlation. The elevation of the selected subsurface datum is determined by a sometimes not so simple comparison with another well log in which the datum has already been selected. This correlation is based on the characteristics of the datum and its position in a sequence of correlative events. The correlation may be relatively simple, as in the Denver-Julesburg basin, where wells tens of miles apart can be easily correlated; or very difficult, as in some parts of South Louisiana or South Texas, where wells only hundreds of feet apart can be correlated only with great difficulty. In these difficult areas micropaleontology is a great help—in some areas a necessity—in making the correlation. Difficulty of correlation is caused by two factors; first, rapid lateral stratigraphic variation and second, complicated structure. In areas where stratigraphic variations are small, and the structure is simple, the correlations and the mapping progress rapidly. An example of this is the "D" and "J" Cretaceous sand mapping in the Denver-Julesburg basin. In areas where stratigraphic variations are small, but the structure somewhat complicated (as in the Deep Wilcox trend of South Texas), or where the structure is simple, but the stratigraphic variations are great (as in some parts of the Yegua-Jackson trend of South Texas), the correlations and the mapping proceed somewhat slowly. When both structure and stratigraphy are complicated, progress is slow indeed. This condition is exemplified by the Miocene of South Louisiana and the Frio of the Rio Grande Valley of South Texas. The same geologist can map a county in the Denver-Julesburg basin faster than he can map some of the reasonably complicated fields in South Texas or South Louisiana.

The geologist who is familiar with correlation by electric log in areas of predominantly sand and shale deposition knows that the shales are more useful for minute correlations than are the sands. The electrical curves, particularly the so-called normal resistivity curve, exhibit what the geophysicist calls "character" in the shales. Most modern electric logs have an additional curve not present on the older logs, called the "amplified normal." This curve is just what it says it is, a horizontal amplification of the normal curve. In this manner, the minute variations in the normal resistivity curve are amplified, and correlation by "character" is greatly simplified.

Logging errors.—The elevation of a datum selected from any type of well log is subject to an error if the hole logged was not vertical. A hole that deviates greatly from the vertical will indicate abnormal thickness of formations and resultant greater distance from the surface to the datum point. A correction for this deviation must be made in order to determine the true elevation of the datum.

An error somewhat peculiar to the electric log is occasionally caused by improper splicing of different runs or parts of the log. This results in what might be called a "blueprint fault." A splice which omits a part of the log results in an apparent normal fault. A splice which repeats a part of the log results in an apparent reverse fault. These "blueprint faults" usually appear where an electric-log run is spliced onto the preceding run, or at a 100-foot or 50-foot depth-marker where the splice is often made.

Occasionally an apparent abnormal thickening or thinning of formations appearing on the electric log is due to differential stretching or shrinking of the material on which the log is printed. If the geologist encounters a log showing unusual thickening or thinning, he should check its vertical scale either by comparison with another log or by comparison with a ruler. In the ordinary log, this variation from true vertical scale is generally so gradual that it will not vitally affect the preparation of a subsurface map. In the preparation of a detailed structural cross section, however, it is necessary that this deviation be recognized and compensated at regular depth intervals. If this is not done the sum of the gradual scale errors compounded over the length of the log will result in a considerable error at the bottom of the log. Modern techniques of printing the log on a single strip of paper combined with modern printing materials have reduced blueprint errors to a minimum in recent years.

INDIVIDUAL WELL DATA

The geologist needs only a spotted map and some correlative well logs in order to prepare a subsurface structural map. However, the quality and reliability of this map can be considerably improved if the geologist has additional information at his disposal. This will generally consist of well information on the individual wells, geophysical maps of the area or parts of the area, and detailed field maps of fields in, or adjoining the area.

The importance of reliable well data on the individual wells in easily useable form can not be overemphasized. A reported show of oil or gas is generally direct evidence of a trap. These shows are as useful in the interpretation of a subsurface map as are the datum elevations themselves. Although reported shows are the most important information available from the individual well records, almost any information is useful. The record and description of cores and side-wall cores and the intervals and depths cored are especially useful, since they will eliminate from consideration zones that appear prospective on the electric log. The exact depths cored and a description of the results should be available to the geologist since he certainly can not condemn a zone as non-productive that has been cored without recovery. If a likely appearing zone has been side-wall cored without recovery, the condition of the bullets, whether broken, pulled off, or empty, will give the geologist valuable clues about characteristics of the formations cored. The records of drill-stem tests, potential tests, accumulated production, core-analysis data, and reserve estimates are particularly useful in evaluating a field, a prospect, an area, or a trend. In the absence of better data, casing-seat depths and mud weights can give clues to hazardous drilling conditions. Work-over dates will indicate the richness and life of individual reservoirs. Even the spud date and completion date can indicate the drilling difficulties encountered.

The large scouting staffs maintained by the major companies and the success of the many commercial oil-field reporting services indicate the importance of well information to these companies.

SELECTION OF MAPPING DATUM

The process of selecting a mapping datum may require much preliminary work, such as construction and correlation of regional cross sections and the determination and cataloging of the main producing zones and their characteristics, or it may consist simply of asking an experienced geologist what horizon he uses as a mapping datum. In either case several requirements should be met by the mapping datum selected. The map made on the datum should reflect the structure of the prospective zone or zones. It should occur over a fairly widespread area so that wells and fields in the area can be related. It should be shallow enough so that sufficient wells encounter the datum to supply enough control to construct an adequate map. And finally, the datum must be recognizable and the geologist must be able to correlate it.

Structure reflection.—The requirement that the map should reflect the structure of the prospective zone or zones is most important. For example, a map using the base of the "*Marginulina* lime" (Fig. 1) as a datum bears no recognizable resemblance to a map on an underlying producing zone (Fig. 2) at the Rayne field, Acadia Parish, Louisiana. The shallow map shows only regional dip. The difference is obvious between this shallow map and the underlying deep map with its large fault and the complete reversal and steepening of the rate of dip. One must conclude that a regional map on the base of the "*Marginulina* lime," which is the most commonly used mapping horizon in this area, would not be useful in the search for this deep accumulation. Its only use would be negative in nature, showing that in this trend a shallow anomaly is not a necessary indication for the presence of a deep structure.

Widespread occurrence.—The second requirement of widespread occurrence is not so important. Ideally, the entire mapped area can be related best by mapping on a single horizon. This is generally impossible except in stratigraphically simple areas of relatively flat rates of dip. An example of this is the Denver-Julesburg basin where several horizons can be correlated practically over the entire basin area. In the Gulf Coast, the more steeply dipping beds, abrupt lateral stratigraphic changes along strike, and the characteristic of progressively younger beds becoming the chief producing beds as the mapping proceeds from the inland areas toward the coast, generally prevents the geologist from mapping on a single horizon very far in either a dip or a strike direction. The geologist must remember that the mapping datum must reflect the structure of the main producing beds

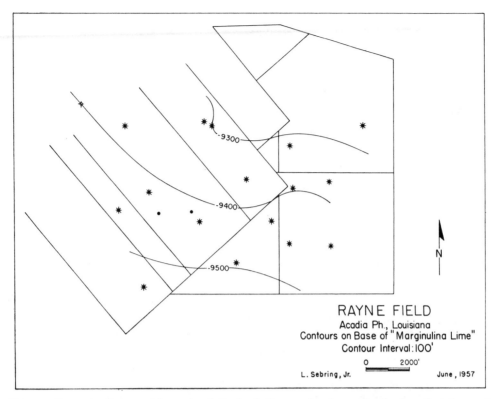

Fig. 1.—Example of commonly used, relatively shallow, regional mapping horizon that does not give any indication of underlying deep accumulation. Depths in feet. Scale in feet.

and he should not hesitate to jump datum points when this condition is no longer met. Adjacent areas mapped on different datum points should be related by an overlap area where both datum points are selected and posted to the map.

Sufficient well control.—The third condition to be met by the mapping datum is that enough wells should encounter the datum to supply sufficient control for the construction of an adequate map. It can be argued that at the time of the discovery of the Rayne field there was insufficient deep well control to map an anomaly in this area. In this case, the geologist should recognize that the subsurface structural mapping method can not be applied to the search for that type of structure at that time, and other methods must be used. The geologist should review his shallow mapping over such an area carefully in order to see if there are any clues, whether structural or stratigraphic, to indicate the presence of the deep accumulation.

Some textbooks describe rather elaborate methods of preparing a deep map from shallow control. These methods use estimated or calculated thicknesses of beds between the desired deep map and the shallow datum. These thicknesses are presumably based on control from a few deep wells that penetrate both datums so that rates of convergence can be computed or an isopachous map constructed. Unfortunately, this geologist has never been fortunate enough to work

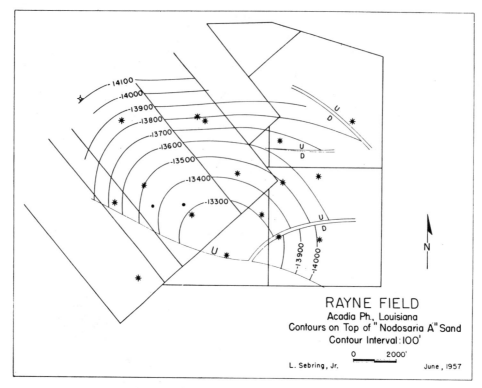

FIG. 2.—Map on deep producing zone bearing no recognizable resemblance to map on overlying regional mapping zone. Depths in feet. Scale in feet.

in an area where the rate of convergence can be determined with sufficient accuracy to compute a deeper subsurface map. Use of this method appears to impose a doubtful and possibly highly inaccurate thickness factor on a shallow subsurface map of substantially greater validity. The writer prefers to use the more reliable shallow map until sufficient deep control is obtained for the construction of a deeper map. In the case of the isopachous map, it is felt that if the data are sufficient to construct such a map, the data are sufficient to construct the deep structural map directly. This does not mean that the isopachous map is not a very useful tool in exploration, but that its use in the construction of a deep structural map would be questionable.

When a well does not reach the mapping horizon, its datum may be estimated, based on a shallow correlation. The geologist should remember that actually he is only projecting a shallow correlation and that unless the section is of a relatively uniform thickness, his projection may be subject to a large error. Estimating the datum by correlation with several deep wells will indicate the uniformity of the section and the validity of the estimated depth.

Recognition and correlation.—Finally, the geologist must be able to recognize and correlate the datum. He *must* be able to correlate it. The more easily recognizable the datum, the faster the mapping will progress. Correlations may be so easy that the geologist can select his datum and compute its elevation without even laying the logs beside another; or they may be so difficult that days may be spent in correlating a single log with the other logs in the area. There is no room for error in this matter; a miscorrelation is worse than none at all. All of the subsurface mapping and interpretation to follow is based on the correctness of the elevation of the datum and its location in respect to other control points.

Of course, the ideal mapping horizon would be a major producing zone in the area studied. It would be of widespread occurrence, a major drilling objective, easily recognized and correlated, and would reflect the structure of the other objective zones.

MAPPING TECHNIQUE

FIELD MAPPING

When the mapping datum has been selected, the geologist is ready to begin his mapping. He must remember his basic rules of contouring. These are that the contour line represents a line drawn along the intersection of a horizontal plane of a specified elevation with the datum bed, and that each contour line must pass between points of greater and less elevation than that of the contour. Also a contour can not cross over itself or another contour or branch, except it may cross over in the case of reverse faults or overturned anticlines.

The geologist should commence mapping in the areas of greatest control and work toward areas of less control. In other words, he should first map the fields or obtain detailed maps of these fields. These field maps should reveal the structure and producing characteristics of oil and gas fields already discovered in the area. They should show whether the trap which resulted in the known petroleum accumulation is of structural or stratigraphic origin or is a combination of the two. The development geologist must map all of the producing sands in the field, determine their thickness, their probable productive extent, and as much about their producing characteristics as he possibly can. He must do all this in order to help determine how to produce the field most economically to obtain the maximum oil and gas. The exploration geologist must map the field in sufficient detail to determine the trapping element or elements, and to determine if these elements can be extended beyond the limits of known production. The logical projection of these trapping elements can result in the discovery of additional reserves of oil or gas, either in an extension of the producing area, or in the discovery of a separate and distinct new field area where the entrapment conditions are similar.

If the structure and stratigraphy of the known producing area are complicated, the exploration geologist's study of the area may be long and tedious. Generally, the exploration geologist does not have at his disposal information as detailed or as reliable as does the development geologist. This is because the development geologist is usually charged only with studies of fields in which his company has production or contemplates acquisition of such production. Consequently his records of wells in these fields are likely to be excellent. The most reliable interpretation of any field is based on voluminous production information about fluid contacts, bottom-hole pressures, fluid characteristics, *et cetera*, as well as on the correlated well logs. The exploration geologist must study all of the producing fields regardless of ownership. His information on fields in which his company has no production will be gleaned from his electric logs, his probably sketchy core descriptions, a probably incomplete record of drill-stem tests made prior to completion, and the initial potential test. If the area is complicated, it is sometimes necessary for the exploration geologist to prepare maps on several horizons in order to determine the most nearly correct solution on any datum. An interpretation that satisfies all of the structural and productive requirements on numerous horizons is most probably a unique solution for the area studied. An example is a structural map on top of the "R" sand (Fig. 3), the chief producing zone in East White Lake field, Vermilion Parish, Louisiana. Although the production information available to the writer was meager, this interpretation is felt to be reasonably correct, as maps on seven horizons were prepared in arriving at the solution. Usually the exploration geologist is not required to prepare so many maps of a small area to fulfill his purposes. If his company has production in the area and a development department, he can generally obtain maps from them based on much more information than he has readily available. The geologist is often able to obtain more or less detailed field maps from sources outside his

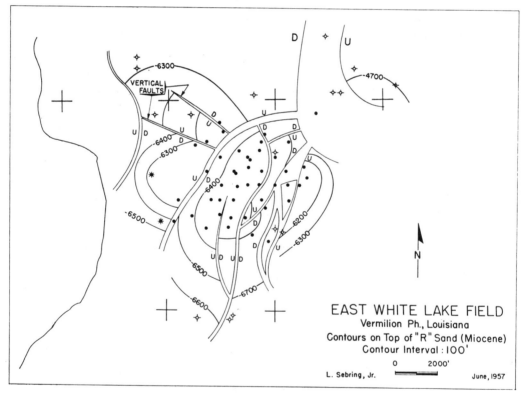

Fig. 3.—Example of complicated structure, solved by mapping on several zones. Depths in feet. Scale in feet.

company. However, he should use these with great care. He must carefully evaluate the quality of these acquired maps. In complicated areas, acquiring maps from three different sources may result in three different interpretations so different as to appear to be similar only in well names, locations, and title block. Under such conditions, the solution is for the geologist to prepare his own interpretation. If the geologist is not pressed for time, this is generally best, for in this manner he can attain the best understanding of the stratigraphy and structure of the area he is mapping. Areas of accumulation are anomalous and clues to their presence, either structural or stratigraphic, are best obtained by intense study of these already productive areas in order to determine what these clues may be. Presence of similar clues in areas not yet productive are signposts to new discoveries of similar type. Before the geologist makes his decision whether to map the fields himself or to obtain these field studies from other sources, he should remember that by acquiring these field maps from other sources he is probably eliminating them as a source of prospective production for his company. Although many geologists will be kind enough to supply their fellows with a field map, they will rarely supply him with such a map unless they have previously culled it for prospective areas of production. In most, if not all, of the older producing districts, the largest part of additional reserves found each year comes from extensions and new producing segments in and adjacent to the known producing areas, rather than from the discovery of entirely new and separate traps. Any exploration technique that ignores the additional possibilities in the fields themselves has questionable validity. This does not mean the exploration geologist should spurn completely the field maps prepared by others, because if he decides not to use them, they are still a valuable source of information to him, supplying him with alternate interpretations or the confirmation of his own. In extremely complicated fields they can show the geologist where he should commence his mapping by indicating which well is most likely to have a normal section. For example, in the East White Lake field, several of the rather shallow wells, 7,000–8,000 feet deep, are distinctly faulted in seven and eight places. A well log on one of these wells is certainly not the one to use to commence mapping.

EXPLORATION MAPPING

After the geologist has made a study of the producing fields in the area, he is ready to extend his study to adjacent areas of less control.

Contour interval.—The field studies will aid in selecting the contour interval to be used in regional mapping. If studies reveal that a closure of 25 feet will effectively trap hydrocarbons in commercial accumulations, obviously a regional map with a 200-foot contour interval would pass over many of these smaller anomalies with no hint of their presence. On the other hand, if the field studies reveal that several hundred feet of closure are required for production, a contour interval of 25 feet would just as obviously impose upon the geologist needless detailing in the preparation and maintenance of the map. The use of a contouring interval of less than 20 feet in regional mapping would not generally be recommended since a drill hole with only minor deviations from the vertical might show elevation variations that approach this figure. Contour intervals of less than 20 feet can reflect anomalies that are the results of errors in measurement and in exact correlation rather than actual subsurface structural variations.

Similar traps.—In exploratory mapping the geologist will first look for traps similar to those already producing in the area, and for the clues that might reveal their presence. The geologist should be cautioned, however, that although these familiar trap types should get his first attention, he should not overlook the possibility of a new type of trap that has not been known to produce in the area studied. This is particularly true in relatively new areas of production. A fairly recent example can be found in the Denver-Julesburg basin. The first oil found in this basin was in a closed anticlinal structure at the Gurley field in Nebraska. Later it was recognized that most of the accumulation in the basin and all of the best fields were actually caused by updip pinch-outs of permeability, either as a permeability re-entrant updip across essentially homoclinal dip, or as a permeability pinch-out across a nose. At first, probably because oil and gas were first found there, salt domes were thought to be the only structural type capable of producing petroleum in the Gulf Coast. In Southwest Texas, accumulation in fault traps was first discovered trapped against the upthrown side of up-to-the-coast faults. Early fault exploration sought this type of faulting to the exclusion of all others. Later it was discovered that oil could accumulate in anticlinal closures on the downthrown side to down-to-the-coast faults. Still later, and in relatively recent years, accumulation against the upthrown side of down-to-the-coast faults has been recognized and sought. The search for stratigraphic accumulation in South Louisiana is in its infancy and probably progress in this direction will be slow, although the geologist should not overlook the possibility in his search for oil. Companies will be understandably reluctant to drill 10,000-foot to 15,000-foot tests in search of an updip permeability pinch-out unless there is much improvement in exploration methods for this type of trap.

Rate of dip.—The rate of dip that the geologist will show will be determined in large part by the rate of dip in the areas of greatest control. The geologist should remember, however, that the rate of dip will be flatter over the highs and over the lows and will be steeper on the flanks of the structures. His contours will be parallel or nearly so, one with another, and any deviation from this or reversal or flattening of the dip indicates an anomalous structural condition that may be the clue to a structural trap containing hydrocarbons.

Oil and gas shows.—In areas of minimum control, one of the best clues to interpretation is an oil or gas show in a non-productive well. This is a direct indication of a trap and its importance can not be overemphasized. The trap that the show indicates is not necessarily of commercial value, but the geologist should study his evidence carefully before discarding the reported show as of no importance. In areas of little control, the geologist will base his interpretation of the structure in the vicinity of the show on predetermined producing structures in the trend. An example is shown in Figures 4 and 5. Geological conditions shown in plan view in the lettered diagrams of Figure 4 are shown in cross-section view, similarly lettered, in Figure 5. In Figure 4a are four non-faulted wells, A, B, C, and D. In the absence of any information on the normal rate of dip in the area, the geologist would contour this area as a simple structural nose with its axis through wells B and D. A cross-section view of this same condition is shown in Figure 5a. Suppose, however, that there is reported a show of "live" oil in well D in the top of the sand which is being used as a mapping datum. As a further condition, suppose that the mapping horizon is the top of a well developed permeable sand in all of the wells. Another rule must now be followed in contouring the area. Well D must now be separated in some manner from the other wells since it is structurally lower and contains a show, which they do not. Multiple interpretative possibilities now become apparent. The geologist will probably choose to keep the dip axis of this structure in about the same place since he now has two indications that the dip axis is as he has previously pictured it, that is, the structural contour passing through wells A, B, and C and the show in well D. A simple interpretation that could be applicable in almost any area of low dip and reasonably competent beds is found in Figures 4b and 5b. Here a high area is centered north of well D and separated from the other wells by a structural saddle. The high could just as well be south of well D, but a saddle must separate it from the

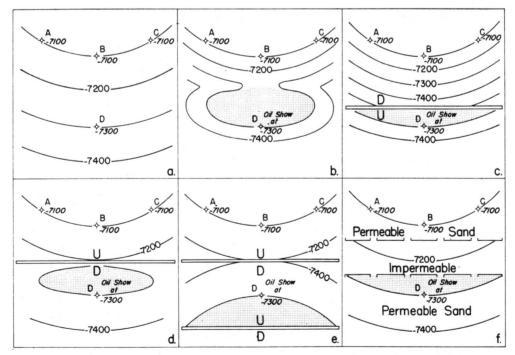

Fig. 4.—Examples of use of oil or gas shows in interpretation—map view. Prospective area, dotted pattern. Depths in feet.

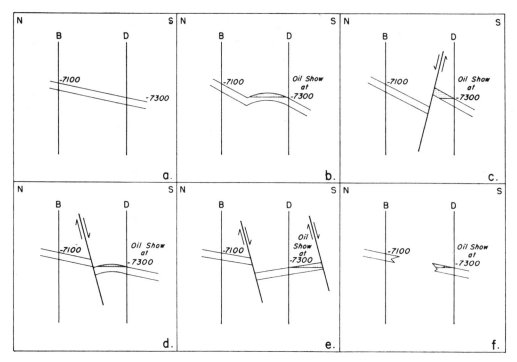

Fig. 5.—Examples of use of oil or gas shows in interpretation—cross-section view.
Prospective area, dotted pattern. Depths in feet.

wells on the north. In Figures 4c and 5c, the upper wells are separated from well D by an up-to-the-south fault so placed that it would not cut any logged part of the wells. Such an interpretation might be favored in any trend where this type of trap is common, like the Edwards, "Pettus," or shallow Wilcox trends of Southwest Texas. In Figures 4d and 5d, well D is separated from the upper wells by a down-to-the-south fault with an indicated high area north of well D. If the log of well D did not go much below the datum, and if it were lengthening considerably over the other wells north of D, the geologist would show his high as south of the well and the down-to-the-south fault just north of the well D. This interpretation would be a favorite almost anywhere in the Gulf Coast where this type of trap is probably the most common producing feature. A somewhat more complicated interpretation would show down-to-the-south strike faults both north and south of the well D, with the well being on the north flank of the indicated stucture. This interpretation is shown in Figures 4e and 5e, and is a very common producing feature in the deep Wilcox trend of Southwest Texas and in some areas of the Frio and Miocene trends throughout the Gulf Coast. Still another possible interpretation is shown in Figures 4f and 5f. In these figures, the show in well D is caused by an updip permeability barrier which traverses the nose. Then since it was stated earlier that the mapping datum was a sand containing a show only in well D, but developed and permeable in all of the wells, it is necessary to draw a downdip permeability barrier south of the three structurally higher wells. This type of trap, caused by repeating bands of permeability approximately paralleling the structural strike, is found in some places in the Yegua-Jackson and Frio trends of Southwest Texas. It is the most common type of trap in the Denver-Julesburg basin, but there the figure would be turned 90° to the right since the regional dip in the main producing area of the basin is toward the west.

There are of course myriad other ways in which these data could be interpreted; however, these seem to be the simplest and most common. If there is a known producing field along strike with these wells, the geologist would probably extend the trapping element of the known field across this area in order to explain the show at D, unless there was good reason not to do so.

Absence of shows.—The absence of reported shows in the wildcat wells mapped is not so important as their presence. A prospective area should not be condemned because wells on the flanks of the prospective area have no reported shows. The flank wells may have actually had oil or gas shows that were not recorded. This may be because the well was not cored sufficiently, or because the well-sitting geologist was inexperienced, or because the type of drilling fluid obscured the show, or most commonly, the show was not recorded on the well data that the geologist has available. This is another example of the extreme importance of good well data. One geologist may map an area and indicate a highly questionable anomaly. Another may map the same area in much the same manner, but with much greater faith in the anomaly because he has well information that a well or wells on the flank of the indicated anomaly had shows.

In extremely permeable productive zones, like the Miocene of Louisiana, wells only a few hundred feet from production may penetrate the producing zone below the oil-water or gas-water contact and have no show. This phenomena is explained by the extreme permeability of the sands. In other areas, such as in parts of eastern Nueces County, some sands contain oil shows at their top, regardless of their proximity to a producing structure. This may be due to small irregularities in the top of the sand that traps minute amounts of oil during migration.

Computing datum across fault.—In areas where faults are present and effective at the mapping level, additional control may be obtained by computing the datum elevation on the opposite side of the fault. This is commonly done in the case of the normal fault by subtracting algebraically the amount of vertical displacement (throw) of the fault from the upthrown datum elevation in order to compute the elevation of the datum on the downthrown side of the fault. If the well is faulted by a normal fault below the datum, an upthrown elevation can be computed by adding algebraically the throw to the downthrown datum. Commonly, the only direct evidence of faulting found in subsurface maps made with electric logs, is an omission of section in a well cut by normal fault and a repetition of section in a well cut by a reverse fault. The amount of throw of the normal fault is assumed equivalent to the thickness of the section missing.

A datum computed in this manner is only as reliable as the throw computation. And since this computation is subject to many errors, the computed datum generally should be used cautiously in the preparation of the map. This is due to several causes. If faulting went on contemporaneously with deposition, the value of the throw of the fault will generally vary with each normal well with which the faulted well is compared. Usually the well is not faulted exactly at the datum, and it is a rare fault indeed that has the same amount of displacement at all levels. Rigorous use of computed throws in figuring datum elevations across faults can result in structural interpretations that are misleading, no matter how carefully the geologist made the throw computation. An example is shown in Figure 6, where well A is faulted at −6,900 feet. Suppose that by comparison with well B the fault throw is computed to be 200 feet. The downthrown point for well A would be −7,330 feet and the geologist would indicate a possible high area east of A on the downthrown side of the fault (Fig. 6a). By comparison with well C suppose that the fault throw is determined to be 300 feet. In that case, the downthrown point would be −7,430 feet and essentially regional dip would be indi-

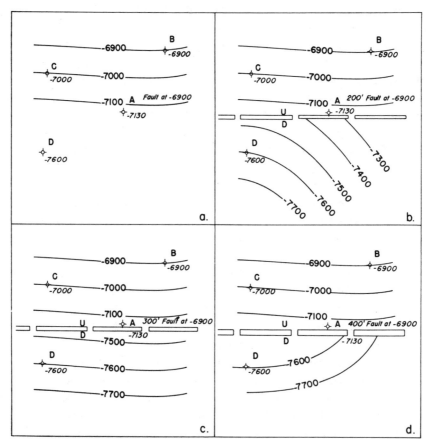

Fig. 6.—Example of effect of computed value of fault throw on interpretation. Depths in feet.

cated (Fig. 6c). Finally, suppose by comparison with well D the throw is found to be 400 feet. In this case, a high would be indicated west of well A on the downthrown side of the fault (Fig. 6d). In this quite common case where the value of the computed throw is so dependent on the normal well used in the computation, other clues such as shows, evidence of regional nosing, and geophysical anomalies would certainly be more reliable than a datum computed across a fault. If the throw is computed with the aid of a normal well along the stratigraphic strike with the faulted well, the value of the throw is much more likely to be accurate than when the normal wells are considerably updip or downdip from the faulted well. Another problem that arises with this type of datum computation is caused by the well being cut by more than one fault. Are the faults dipping the same directions such that their throws should be added in order to compute the datum across them? Or is one or more of the faults a fault dipping against the others? In this case, some of the faulting may act as compensating faulting so that its throw should be subtracted from the aggregate. This problem seems insoluble without adequate well control, and if the geologist has that much control, he usually does not need the additional information gained from the computed fault datum.

In a great many areas in the Gulf Coast, a large amount of reverse dip or turnover into the downthrown side of regional down-to-the-coast faults is a common characteristic. Almost everywhere, in this case, a computed downthrown point will be much higher than the true value of the datum. This is because a well drilled adjacent to the downthrown side of the fault will penetrate the steeply dipping beds at an angle, and each formation will show an indicated thickness greater than if the well were drilled normal to the bed surfaces. When deposition and faulting are contemporaneous, the upthrown section of the faulted well will be more similar to normal wells on the upthrown side of the fault. Also, the downthrown section of a faulted well will be more similar to normal wells on the downthrown side of the fault. Therefore, the geologist who uses upthrown normal wells in computing the upthrown datum, and downthrown normal wells in computing the downthrown datum, will get the best results with this technique.

Careful study of faulting in the areas of maximum control in the mapping area will indicate the reliability of the computed or restored datum. If the values are about the same regardless of the well used in the computation; if the fault throw is fairly constant in wells that are cut both shallow and deep; if the section is very similar in both upthrown and downthrown normal wells; and if the computed values in the areas of dense control fit into the well controlled structural picture; then, they may be considered fairly reliable in the wildcat areas. If these conditions are not met, the geologist is probably better off using the computed throw value with considerable latitude in his interpretation. He should never, of course, have his upthrown side lower than the downthrown side, and he should always have a finite difference in elevation across the fault. The fact that his fault indicates a throw of 200 feet at one point and 300 feet at a point a mile farther along its strike, should not be the cause of much concern. If he studies his fields carefully, he will probably find examples showing just this. He should remember that a fault as well as a fold is a simple result of a force or forces; that when he combines the two features into a faulted and folded, contoured, structural map his end result must explain the combined geological result of two structural features. In other words, if the geologist must traverse an area between two wells with 1,000 feet elevation difference, he will do it with structurally contoured dip, or faults, or both, as he may deem most reasonable. His only definite restriction is that whatever his interpretation, the final result must show by a combination of faulting and contours the correct difference in elevation. Field studies of most faulted fields will show that faults are commonly sinuous, have variable amounts of throw along their strike and along their dip, and show variations in degree of dip as well. When faults die out, they do so into a fold that may be minute or large according to the competency of the beds. The forces that form a fault must first fold the beds until they break. If the beds are incompetent, only a very small force is necessary to cause them to fault; if they are very competent, a very large force may be necessary.

In the special case when the datum is faulted out, it is necessary for the geologist to compute an upthrown and a downthrown datum. Fortunately, in this case, his computation should be fairly reliable. This is because the fault occurred exactly at the datum, and the throw computation is necessarily made at exactly the desired point.

Fault location.—In regional mapping, the position of the fault with respect to the well that is faulted should be shown as accurately as possible. This position is a function of the position of the fault in the well, of the direction of strike, and of the degree of dip of the fault plane. Generally, in areas away from the fields, it is difficult to find the necessary three faulted wells in a suitable geographical location so that the geologist can determine dip and strike by the three-point method. Fault studies in the fields will best determine reasonably accurate dip and strike of the faults. If there are more than three faulted wells in any area studied, the most accurate determination of dip and strike can be made by the construction of a map contoured on the fault plane. Multiple faulting in the wells may make it impossible to determine the dip and strike of the faults due to difficulty in relating the individual faults in the various wells.

Ordinarily, the scale of the map used in regional mapping is so small that it is not necessary for the geologist to show fault gap (zone where the datum surface is missing due to normal faulting). In the case of very large faults or large-scale maps, or both, it should be shown. The width of the gap shown will not only be a function of the throw of the fault but its dip as well. The larger the throw and the lower the angle of dip of the fault, the larger the fault gap.

When tracing tensional faults from areas of good control into areas of less control, the geologist will generally show them striking approximately parallel with the strike of the beds. They will commonly be downthrown in the direction of dip of the beds. This is a simple application of the theory that the faulting is a result of forces applied to the beds. The more anomalous dip faulting will be found in areas of great structural hiatus, such as around salt domes or in areas where there is a sharp change in the regional strike. The common graben structural form in salt-dome areas is a well recognized phenomenon in the Gulf Coast. The up-to-the-coast strike fault, common in some areas, can be regarded as essentially a compensating fault in many places developing in lieu of the sharp reverse dip into the more common down-to-the-coast fault.

ADDITIONAL INFORMATION

The subsurface geologist has now used most of the direct information available in interpreting the structure in areas of considerable control. He should now turn to any additional information, no matter how obscure, in his interpretation of the areas of less control. This information may range from detailed seismic maps, surface maps, core-drill maps, and shallow subsurface maps to rumors of another company's seismic high. This information will be of varying degrees of reliability, but it should all be carefully weighed before it is used or discarded in preparing the final interpretation.

A. I. Levorsen, in his book, *Geology of Petroleum*, has stated: "The petroleum geologist is in many ways like a detective—he is forever following up and evaluating clues that might lead to the discovery of a pool of oil or gas." It follows that the geologist who has the most clues or who uses the most clues will be the most successful in his search for petroleum.

Here again, a comparison of the available additional data in an area of considerable well control is probably the most useful way in which to determine their value. If the surface structural map, core-drill map, or shallow subsurface map indicates an anomaly over a producing area, then similar anomalies in wildcat territory may also indicate areas of future production. Gravity and magnetometer maps may be used in the same way. This type of evaluation of the additional information at the geologist's disposal will also help determine what type of additional exploration methods will be desirable in the future evaluation of some of his more nebulous prospects.

REFLECTION SEISMOGRAPH MAPS

Usually the most reliable deep additional information that the geologist can get without the drilling of wells is obtained from the reflection seismograph. This tool gives him a final map which attempts to show the subsurface structural conditions of the area explored at the desired depth in exactly the same manner as does the subsurface geological map. The subsurface geologist's attitude toward the seismograph map may range from sublime confidence in the map to complete ridicule of it. He may assiduously trace all of the seismic contours onto his map, or he may ignore them altogether. A compromise between the two extremes is usually the best solution. Although the seismograph is generally the best deep tool available to the subsurface geologist short of a wildcat well, it does have its limitations. These limitations and sources of error should be recognized by the geologist. It is generally stated that the seismograph's limit of error is approximately 50 feet. In areas of good records, constant weathering, and good velocity control, it may be considerably less than this. In other more difficult shooting areas, it may be considerably more than this. Probably the seismograph is most accurate in determining the direction of dip. It is also very good in determining the amount of dip. Estimations of actual depth to any horizon will vary in reliability with the available velocity control and the number of well ties. Excluding velocity and weathering problems, the seismologist generally has the most trouble with fault interpretations. Direct evidence on the records may be lacking and in this case he will usually compensate any mis-ties by drawing the suspected fault through areas of poor record quality. His value for the fault throw is often in error. Frequently he may reverse the upthrown and downthrown blocks. This is often a result of attempting to draw a preconceived structural picture, possibly foisted on him by an overzealous subsurface geologist, who assures him that a suspected fault traversing the area can be upthrown only in a specified direction.

When the subsurface geologist sends a seismograph crew into a prospect area, he will often base his opinion on their efficiency by a comparison of their map with his preconstructed subsurface map. If they picture a high, all is well; if not, he may frequently condemn the shooting as unreliable. Before condemning the shooting, the geologist should determine if the seismograph's subsurface picture is reasonable. Though the geologist's interpretation of the structure before shooting, based on all available information, is in his opinion both reasonable and possible, he should still recognize that there are other possible interpretations, probably more pessimistic than his, that will fulfill all of the geological requirements. If the seismograph interpretation is one of these, he should accept it. He most probably might as well. His company has engaged the crew at considerable expense in the belief that the crew can furnish them with a more detailed and reliable survey of the prospective possibilities of the area than can the subsurface geologist. If, on the other hand, the final seismic map of the area is geologically unreasonable, it is the geologist's duty to bring this to the attention of his management. Management should insist that the final seismic interpretation should be reasonable geologically, since in correcting any of these geological inconsistencies, evidence for new prospective possibilities may be uncovered.

Strangely enough, a seismograph map of ancient vintage may be of as much or more actual use to the exploration geologist as a seismic map recently completed. This is because with the older map his management will usually allow him considerably more latitude in its interpretation. Subsequent drilling will generally reveal inconsistencies and errors in the original seismic survey. If his company and its competitors have been active exploration-wise, there will be very few undrilled closed structural anomalies on the old map. These undrilled anomalies will attract the geologist's attention first. If similar anomalies are shown on the map and are already productive, then of course the untested anomalies become prospects of the first order. The geologist will have more confidence in the anomalies shown if some of those proved productive by drilling were proved after the completion of the shooting.

After the more obvious anomalous areas have been examined, the geologist will examine the old map carefully for the more numerous less obvious clues. Is there an area of anomalous dip that can be recontoured more logically and optimistically? Have recent developments indicated that accumulation is occurring without structural closure, but rather by updip pinch-out across a regional nose? In this case, areas of indicated structural nosing, immediately become prime prospecting grounds. The geologist may indicate faulting in areas of abnormally steep dip. He may indicate structural closure in areas of abnormally flat dip. Only his experience and his imagination will limit him in the myriad possibilities of this type of study.

COMBINATION AND STRATIGRAPHIC TRAPS

The essential requisite of any geological map used in exploration is that it must indicate the anomalies that trap the hydrocarbons. A shallow subsurface map on a datum considerably higher than the producing level is useful in exploration for petroleum at the lower level, only if an anomalous condition is revealed at the shallow level. In the case of structural accumulation, this may take the form of actual closure or a definite and recognizable flattening or nosing of the beds at the shallow level that can be definitely related to the deep structure. In the case of petroleum accumulations trapped by a combination of structure and stratigraphy or by stratigraphy alone, the subsurface structural map will not by itself satisfy the essential requirement of indication of the petroleum accumulation.

PERMEABILITY BOUNDARY LINES

The essential requirement of indication of the petroleum accumulation will generally be satisfied if the basic subsurface structure map is modified by the addition of boundary lines, commonly called pinch-out lines, showing the updip extent of permeability in the various potential producing zones. Where the bands of permeability repeat themselves along a dip profile in strike bands, as in the Denver-Julesburg basin, or where production occurs in lenticular sands regardless of the structure, the actual final map will be simplified by showing all of the limits of permeability, both up and down dip as well as lateral limits. The permeability of the individual zones is determined by core analysis, drill-stem tests, log analysis, and core description.

It can be argued correctly that if the producing reservoir is occurring in strictly stratigraphic traps, where the structure is not a trapping factor, that the addition of structure contours would serve only needlessly to clutter the map. Although this is true by definition, in the case of strictly stratigraphic traps, generally in the same area there are also petroleum accumulations where the structural element is definitely a trapping factor. The ideal exploration tool will be useful in exploration for all types of trap in the area studied.

In areas where there are numerous updip pinch-outs of different potential producing zones at various levels, these may be superimposed on a basic structural map contoured at any depth. However, the structural map must reflect the structure adequately for all of the levels at which the pinch-outs occur. The final map may become complicated if a great many of these pinch-outs occur, and the geologist must carefully differentiate the separate zone pinch-outs by coloring or by some other means.

An example of the use of permeability boundaries in exploration is shown in Figure 7. A four-township area is shown in central Kimball County, Nebraska.

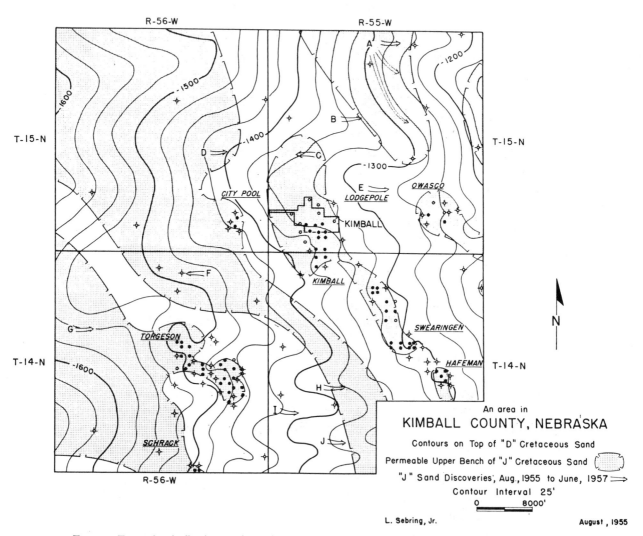

An area in
KIMBALL COUNTY, NEBRASKA

Contours on Top of "D" Cretaceous Sand

Permeable Upper Bench of "J" Cretaceous Sand

"J" Sand Discoveries, Aug., 1955 to June, 1957

Contour Interval 25'

0   8000'

L. Sebring, Jr.   August , 1955

FIG. 7.—Example of effective use in exploratory mapping of permeability boundaries superimposed on subsurface structure map. Depths in feet. Scale in feet.

around the town of Kimball. The map was contoured on a 25-foot interval at the top of the "D" Cretaceous sand and was completed August 11, 1955. There are no closed structural contours on the map and, in fact, the writer mapped only one 25-foot closure in the western three-fourths of the county. At the time of the mapping, he believed that the "D" sand acted as a single permeability unit and that only the uppermost bench of the "J" sand was productive in the area. Therefore, only permeability limits for the upper bench of the "J" sand and for the entire "D" sand were superimposed on the structural map. Since the "J" sand accounts for the bulk of the production in the area, and all of the discoveries since the completion of the map, only permeability limits for what was considered its uppermost bench are shown on Figure 7.

Discoveries made in the mapped area from the date of completion of this map through May, 1957, were taken from the weekly issues of *The Oil and Gas Journal*. Their position is shown on the figure by arrows and letters, A through J.

The multiple discoveries at A, now named Lukassen, Griffith, and Aue fields, were mapped as prospective. The discoveries at C, Morton field, D, Heideman field, and H, Marian field, were also mapped as prospective. At J, although the discovery well was outside the prospective area, this well is the farthest downdip producer in the field. Subsequent development has extended updip across the prospect area and has joined the two discoveries at H, Marian field, and J, Hein field, with the already productive Sloss field at the east. Sloss field is located under the title block in the southeast corner of T. 14 N., R. 55 W., and is not shown. This is by far the most important development in the area mapped, since there are now more than 50 new producing wells completed since the mapping was done and active development is continuing. This entire producing area is now called Sloss field. The discovery at G, Kleinholtz field, could be described as a near-miss, since the area updip from the discovery was mapped as prospective. At I, Evertson field, the discovery might be termed an economic success and a geologic failure. The area was mapped as a "D" sand prospect and it actually produced from the "J" sand. Here the "J" sand possibilities were incorrectly evaluated, because all of the wells surrounding the prospect area showed the "J" sand to be tight by core and drill-stem test. The fact that all of these wells also had shows in the "J" sand should have been given more consideration.

At B, Strain field, and at E and F, which have not been named and may be considered extensions, the map did not show the areas as prospective. At B and E, the writer should have been more optimistic in extending the well defined northwest-southeast Kenton-Houtby-Owasco field trend. (Only the farthest northwest field, Owasco, occurs in the mapped area.) At F, he was almost certainly a victim of incorrect well information. An offset to the discovery well showed the producing sand to be tight by core description and containing no show. The sand appeared permeable on the electric log, and in fact was so interpreted. The failure to report a show condemned the area.

By June 1, 1957, only those new discoveries at D, Heideman field, with eight producing wells, and at H and J, the previously mentioned Sloss area, appear to be commercially significant. At A, there are seven producing wells in scattered, but associated areas, and eventually this area may attain some commercial significance. All of the other new fields had a maximum of two producing wells.

This type of evaluation of a subsurface map has its limitations, because it gives an exaggerated picture of the quality of the mapping. Some dry holes were drilled in these areas that were mapped as prospective. However, if the map had been carefully maintained as the well logs on all of the new wells in the area were obtained, a more accurate delineation of the prospect areas would be possible.

MAP MAINTENANCE

In these days of intense competition, it is essential that the maps used in exploration be maintained currently. As soon as any new information about an area previously mapped becomes available, it should be evaluated immediately and the map changed accordingly. This maintenance should be on the basis of intensity equal to the original study of the area. In other words, if intense field studies were the basis for the original mapping, then these field maps should be revised when additional wells are drilled. On the other hand, if the fields were not worked originally, there would be little point in working the recently completed field wells.

The geologist must use his information as it becomes available to him. If he allows the logs to be filed before he works them, then the chances are that some of these logs will be overlooked and never worked. His map will then suffer in comparison with that of his competition.

For the geologist himself, map maintenance is the best way he can evaluate his own work. If too many of his prospects are being tested unsuccessfully, then he is undoubtedly mapping too optimistically. If, on the other hand, and this is a much more serious error, discovery after discovery is being made in areas where he has mapped no prospects, then his approach is at best too pessimistic, and may be completely in error. The geologist who works a newly received log and finds that it fits his map with only minor modifications of the interpretation, should feel a glow of real pleasure. After all, in a small way, he has justified his place in the industry and he will keep this hard-won position only so long as his successful predictions outnumber his failures by a considerable margin.

Erasure of contours and complete recontouring over large areas because of additional well control will have a sobering effect on any geologist. Indeed it should. There is no better indication that the geologist is mapping incorrectly and is failing at his appointed task.

Basically the best maps will require the least changes and the least time for maintenance. The perfect map would require no changes since all new well datum elevations would fit the map perfectly.

EXPLORATION ATTITUDE

The perfect map is, of course, impossible of attainment. As long as there is drilling, geologists will have to change their maps to compensate for the errors revealed. If the geologist has to err, he should err on the side of optimism. After all, if a geologist condemns every wildcat well to be drilled, he will on the average be correct seven times out of eight. Fortunately, although this pessimist is correct most of the time, he will never add a barrel of oil to the reserves of this country. The geologist whose prospect is drilled and found dry certainly will not feel happy about it. He should feel infinitely worse, however, if a prospect that he has condemned is proved productive.

It has been said that new oil reserves will be found only so long as there are people who believe that additional oil reserves remain to be found. The geologist must be optimistic and positive in his approach. This attitude should apply throughout his exploration technique and even to his basic concepts of the origin, migration, and accumulation of petroleum. Each limitation that the geologist places on these concepts limits the areas where he may search for petroleum. If he believes that oil can originate only in marine beds, he immediately eliminates from his consideration large areas of predominantly non-marine beds. If he believes in long-distance migration of petroleum, he will generally eliminate from exploration consideration areas adjacent to any sizeable accumulation. If he believes that only certain traps can contain hydrocarbons, he will certainly overlook any other type of trap.

The proper attitude for the exploration geologist might be described as reasonable optimism. If he is mapping in an established area of only small oil accumulations, he can reasonably expect to find only small accumulations, as long as he is searching for the same pay zones and using the same basic techniques as his competitors. Perhaps he is searching for oil in a heretofore barren area and by some type of regional mapping discovers a number of structural anomalies. If a large percentage of these anomalies have been drilled unsuccessfully, he can not reasonably hope to be any more successful unless he uses a different technique or a different basic approach.

In the older producing districts it would seem reasonable, therefore, to concentrate the search for large reserves in those districts where they have been, and are being found. Most companies, of course, do this. In virgin areas, which have been explored without success, a new and different basic approach and techniques not previously applied to the area would seem to be more likely of success than a repetition of methods previously applied unsuccessfully.

PREVIOUS MAPPING

Heretofore, this paper has been concerned with subsurface mapping procedures where the geologist started with a blank map and built his map from the ground up. Most major companies, and a great many of the more progressive independents, who have maintained offices in certain areas, ordinarily have basic structural maps of the district. The geologist who has been transferred into the district will generally be required to maintain these maps. This new geologist will have a tendency to interpret these maps in the light of his own experience in other areas. He will often prematurely condemn the maps as useless and his predecessor as an incompetent. The succeeding geologist should not be harsh in his judgment too soon. He should maintain these maps for a reasonable length of time and see how they stand up before condemning them. He is probably in a better position to evaluate the mapping abilities of his predecessor than any other person, but he should reserve judgment until he becomes aware of the peculiarities of his new district. He should judge his predecessor's map in much the same way that he does his own. It should fill all the the basic requirements of any geological map used in exploration. These have already been discussed.

"SELLING" THE PROSPECT

The best geological map ever made is useless unless it is actively used in exploration. The geologist must actively "sell" his map and the prospects it reveals to his management. To a very large degree, the exploration success of any geologist and his company is in direct ratio to the activity of the company. If the geologist is unable to persuade his company to drill or acquire acreage, he will certainly find no oil or gas. The exploration geologist who does not "find production," for whatever reason, has failed at his appointed task.

No honest effort should be spared by the geologist to "sell" his company on his prospects. Probably there are more variations in the manner of presenting and successfully "selling" a prospect among the competing companies than in any other matter. The geologist may be able to acquire acreage or commit his company to a well by picking up the telephone; or he may be required to prepare a leatherette-bound prospect folder complete with brightly colored maps and cross sections and resplendent with fluorescent tape. Whatever the effort required, a successful result justifies it.

Ideally the geologist should be able to sell his prospect on its merits alone, honestly appraised and set forth in a straightforward manner. This will enable his management to appraise it as well, and come to a decision about its merit by a fair comparison with prospects from the other districts.

A written recommendation describing and identifying the prospect, its location, its geology, type and reliability of data, land situation, productive possibilities, expected cost, and recommended action is desirable. Not only will this written recommendation serve to consolidate the thoughts of the geologist, but it will be an adequate record of the original reasons for the action recommended. This recommendation should be accompanied by maps and some other illustrations that will show the geologist's interpretation of the area recommended and its relation to the surrounding area.

It is obvious that the management that requires the least extraneous material, over and above the basic geological requirements of the prospect recommendations, will obtain the most exploration effort from the geologist. A basic company policy that frees the geologist from as many routine non-geological duties as possible, will allow him to devote more of his time to exploration.

Periodic replacement of the original copy, by a revised up-to-date version of the regional subsurface map furnished to the management, should be a very effective long-term selling method. If a reasonable number of the geologist's prospects, which he has been unable to sell his management, produce and the periodic revisions show only minor changes from the preceding version, then the management's resistance to his new recommendations should be worn down by simple attrition.

CONCLUSION

The subsurface structural map, modified by the addition of permeability boundaries, when it is prepared and maintained by a competent geologist with a complete log file and adequate well records, will be an increasingly effective and relatively inexpensive tool for use in exploration for petroleum. Barring the discovery of a direct method of finding oil, the various subsurface methods based on well control will gradually tend to supplant all other methods as more and more wells are drilled in the never-ending search for petroleum.

REFERENCES

Krumbein, W. C., and L. L. Sloss, 1951, *Stratigraphy and Sedimentation.* 497 pp. W. H. Freeman and Company, San Francisco.
Lahee, Frederic H., 1931, *Field Geology.* 789 pp. McGraw-Hill Book Company, Inc., New York.
Leroy, L. W., 1950, *Subsurface Geologic Methods.* 1156 pp. Colorado School of Mines, Golden.
Levorsen, A. I., 1956, *Geology of Petroleum.* 703 pp. W. H. Freeman and Company, San Francisco.
Oil and Gas Journal. All issues: August, 1955, through May, 1957. The Petroleum Publishing Company, Tulsa.